Non Invasive Diagnostic Techniques in Clinical Dermatology

Enzo Berardesca • Howard Maibach
Klaus-Peter Wilhelm
Editors

Non Invasive Diagnostic Techniques in Clinical Dermatology

Springer

Editors
Enzo Berardesca
Clinical Dermatology
San Gallicano Dermatological Institute
Rome, Italy

Klaus-Peter Wilhelm
proDERM Institute for Applied
Dermatological Research
Schenefeld/Hamburg
Germany

Howard Maibach
Dermatology
University of California
San Francisco, CA
USA

ISBN 978-3-642-32108-5 ISBN 978-3-642-32109-2 (eBook)
DOI 10.1007/978-3-642-32109-2
Springer Heidelberg New York Dordrecht London

Library of Congress Control Number: 2013955724

Printed on acid-free paper

Springer is part of Springer Science+Business Media (www.springer.com)

Preface

This project has been undertaken to update the book on Methods and Instrumentation released in the series on "Bioengineering and the Skin" in 1995. However, during the revision process we realized that so much time has passed and many new techniques were developed that it was almost impossible to update the old edition; therefore, on the basis of the previous experience, we devised a table of contents covering not only the "old" and "classic" noninvasive techniques, but focusing on also both new methods and techniques developed recently. The reader will find a new broad section on skin imaging based mainly on techniques developed in recent years, a wide approach to methods for investigating the stratum corneum and the superficial layers of the skin, as well as sections dedicated to particular skin sites of specific interest such as hair and nails. Indeed, the field in this decade has expanded not only from a technical viewpoint in new hardware but also in new applications of existing technology developed to investigate specific areas of "transferred" technology (i.e., from biochemistry or molecular biology) to noninvasively detect and quantify molecules in superficial skin layers to monitor skin reactions. This book remains a small, intuitive, and easy-to-read tool for dermatologists, biologists, pharmacologists, and scientists in general who are willing to approach skin research and noninvasive skin investigation in particular.

Rome, Italy — E. Berardesca
Schenefeld/Hamburg, Germany — K.-P. Wilhelm
San Francisco, CA, USA — H. Maibach

Contents

Part VII Hair and Follicles

Part VIII Nails

Part I

Imaging Techniques

1 Use of Videodermatoscopy in Dermatology

Francesco Lacarrubba, Franco Dinotta, Cecilia Santagati, and Giuseppe Micali

1.1 Introduction

Videodermatoscopy (VD) is a noninvasive technique that allows a rapid and magnified in vivo observation of the skin surface with the visualization of morphologic structures invisible to the naked eye [1–3]. Images obtained by a high-resolution color video camera, equipped with lenses that currently allow magnifications ranging from ×10 to ×1,000, are indirectly visualized on a monitor and stored on a personal computer. This technique allows eventual image processing and comparison of any lesion changes over time, thus significantly improving follow-up assessment. VD represents the evolution of dermatoscopy (also known as dermoscopy), performed with manual devices, which does not require any computer "assistance" but generally reaches magnifications no greater than ×10. VD may be performed directly or, most often, through the technique called epiluminescence microscopy, which involves the application of a liquid transparent medium (oil, gel, alcohol, or water) between the lens and the skin to minimize surface light reflection from the cornified layer. Some systems utilize polarized light, obtaining similar results without the need for liquids. In addition, VD system may be equipped with filtered wavelengths. For instance, a 400 nm light source may be used to obtain imaging enhancement, improving vessel details visualization, for the evaluation of skin microcirculation [4].

VD is widely used in the diagnosis of pigmented skin lesions, as well as in a wide variety of dermatologic conditions (Table 1.1). It may also be useful for prognostic evaluation and monitoring of response to treatment, representing an important and relatively simple aid in daily clinical practice.

In this chapter we will use the term VD, even for those cases in which studies have been performed using manual low-magnification devices.

1.2 Pigmented Skin Lesions

VD allows the identification of different skin structures localized from the epidermis to the superficial/medium dermis usually undetectable by simple clinical observation. Several studies correlated the VD features (retrieved by horizontal observation) and the correspondent histopathologic findings (observed in vertical extension). The introduction of VD has forwarded a great impact in the management of pigmented skin lesions, significantly improving the early detection of melanoma and increasing diagnostic accuracy from 5 to 40 % over clinical visual inspection, depending on lesion type and physician experience [5–9].

F. Lacarrubba (✉) • F. Dinotta • C. Santagati
Dermatology Clinic,
University of Catania, Catania, Italy

G. Micali, MD (✉)
Dermatology Clinic,
University of Catania,
A.O.U. Policlinico – Vittorio Emanuele,
Via Santa Sofia, 78 – 95123 Catania, Italy
e-mail: cldermct@nti.it

E. Berardesca et al. (eds.), *Non Invasive Diagnostic Techniques in Clinical Dermatology*,
DOI 10.1007/978-3-642-32109-2_1, © Springer Berlin Heidelberg 2014

The classic approach to a pigmented lesion by VD is based on a two-step procedure (Fig. 1.1).

The observer should first distinguish melanocytic lesions from non-melanocytic ones. In Tables 1.2 and 1.3, the common VD features of melanocytic and non-melanocytic lesions are described, along with their histopathologic correlation (Tables 1.2 and 1.3). The presence of pigmented network, aggregated globules, streaks, and homogeneous blue pigmentation favors a melanocytic lesion (excluding face, palms, and soles). Moreover, if the lesion does not demonstrate any of the specific structures of non-melanocytic lesion, it should be considered melanocytic and consequently excised to rule out a structureless melanoma [10–13].

The evaluation of vascular structures (Table 1.4) is also useful, especially for the diagnosis of hypo- or nonpigmented lesions [14].

Once a lesion is classified as melanocytic, the second step differentiates benign melanocytic lesions from malignant ones (Table 1.5): in this case, several diagnostic methods may be used. Pattern analysis, first proposed in 1987 and later modified [10, 15, 16], is based on subjective, qualitative, critical, and simultaneous evaluation of several criteria: general appearance of the pigmented skin lesion (uniform or heterogeneous), pattern of pigmentation (type and distribution of color, presence of depigmentation, pigment network, brown globules, and black dots), and lesion margins (regular or irregular for the presence of streaks). Pattern analysis evaluation requires special knowledge of the criteria and specifically trained observers in order to be used with confidence. To make the approach more functional, new diagnostic algorithms have been introduced. They consist of score systems that can be used by less experienced observers, providing a high rate of diagnostic accuracy [6, 11, 17, 18]. They include the ABCD rule, the Menzies method, the seven-point checklist, the CASH (color, architecture, symmetry, and homogeneity) method, and the three-point checklist [19–21]. Pattern analysis, although complex, is the most rapid and complete approach. When the observer achieves good dermoscopic experience, this application becomes intuitive and automatic; it is of great utility, especially in patients with multiple lesions [5].

Table 1.1 VD applications in dermatology

Pigmented skin lesions
Melanocytic (nevi, melanoma)
Non-melanocytic (solar lentigo, seborrheic keratosis, dermatofibroma, basal cell carcinoma)
Nonpigmented skin lesions
Sebaceous hyperplasia
Pyogenic granuloma
Clear cell acanthoma
Xanthomatous neoplasms
Mastocytosis
Sarcoidosis
Median raphe cysts
Eccrine poroma
Keratoacanthoma
Actinic porokeratosis
Nonpigmented facial actinic keratosis
Bowen's disease
Invasive squamous cell carcinoma
Kaposi's sarcoma
Ectoparasitoses
Scabies
Head and pubic lice
Tungiasis
Cutaneous leishmaniasis
Furuncular myiasis
Demodicosis
Cutaneous/mucosal infections
Molluscum contagiosum
Cutaneous warts
Genital warts
Tinea nigra
Lupus vulgaris
Inflammatory disorders
Psoriasis
Lichen planus
Urticaria and urticarial vasculitis
Rosacea
Pityriasis lichenoides et varioliformis acuta
Scalp disorders (see Chap. 39)
Hair loss
Parasitoses
Psoriasis
Hair shaft disorders
Nail disorders
Psoriasis
Onychomycosis
Onychomatricoma
Glomus tumor
Vascular disorders
Port-wine stains
Infantile hemangioma
Pigmented purpuric dermatoses (PPD)

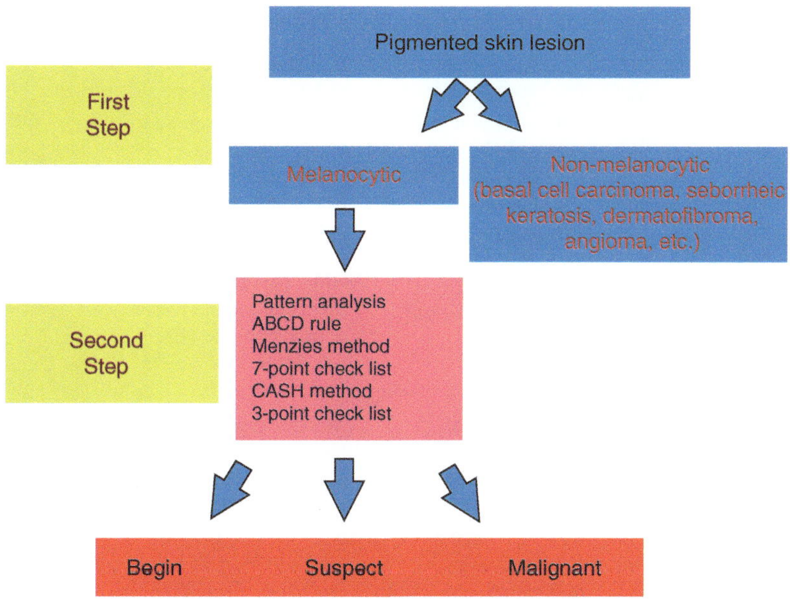

Fig. 1.1 Two-step approach for the diagnosis of pigmented skin lesions

If a non-melanocytic lesion is recognized in the first step, no further evaluation through algorithm methods is needed, as the diagnosis is generally readily made (Table 1.6).

In general, VD examination may confirm or exclude the clinical diagnosis, significantly downsizing unnecessary removal of benign, but clinically equivocal, pigmented skin lesions, such as thrombosed angiomas and seborrheic keratoses. On the other hand, VD may help to increase the index of suspicion in case of melanomas clinically mimicking benign lesions [5].

VD allows close, noninvasive follow-up of pigmented skin lesions in order to detect minimal changes suggesting an early diagnosis of melanoma; monitoring may reduce the number of unnecessary biopsies, representing a helpful tool to decide which lesions should be removed and when. Long-term follow-up allows comparison of atypical nevi over 6–12 months in patients with multiple lesions (i.e., patients with atypical mole syndrome). Short-term follow-up (generally at 3 months) is performed on single suspicious lesions (i.e., anamnesis of changes) that lack features of melanoma. Complete excision should be considered in a lesion that shows significant changes on follow-up [5].

Automated diagnostic systems require no input by the clinician but rather report a likely diagnosis based on computer algorithms. There are many available software approved for medical use; however to date, the true benefits of these systems remain questionable [22].

1.2.1 Special Localizations

1.2.1.1 Face

In this site a specific *pseudo-network* with a broad mesh and holes, due to the numerous follicular and sweat glands openings, is generally seen in both melanocytic and non-melanocytic superficial lesions [23]. In lentigo maligna, in the first stage, hyperpigmented, asymmetric, follicular openings may be detected, with fine streaks, dots, and globules later developing around the follicles and creating the so-called annular-granular pattern along with the formation of rhomboidal structures; in the last stages, hyperpigmentation becomes homogeneous enough to obliterate follicular openings [5].

1.2.1.2 Palms and Soles

In these areas pigmentation is arranged through parallel lines that follow skin grooves [24]. Acral melanocytic nevi may show three different types of VD patterns: the *parallel-furrow pattern*, the *latticelike pattern*, and the *fibrillar pattern*. The *parallel-furrow pattern* is the most frequently observed: it shows linear pigmentation along the

Table 1.2 Common VD features of melanocytic lesions

VD features	Description	Histopathologic correlation	
Pigment network	Grid of brownish lines over a tan background. It can be typical (regular, thin, narrow) or atypical (irregular, thick, wide) in benign and malignant melanocytic lesions, respectively	The lines correspond to melanin pigment contained in keratinocytes or in melanocytes outlining the pattern of the epidermal rete ridges, while the tan areas among them correspond to the dermal papillae tips	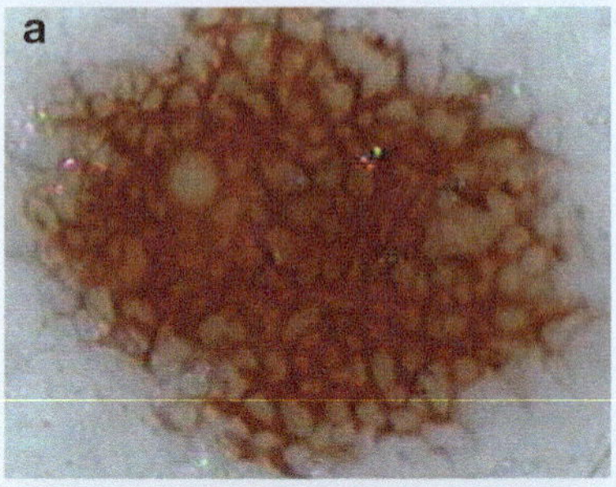
Diffuse pigmentation	According to the localization of the melanin pigment within the skin, different colors can be seen. Even and uneven pigmentation is usually found in benign and malignant lesions	Melanin pigment localized within the epidermal stratum corneum appears black; in the lower epidermal layers light to dark brown; in the papillary dermis gray; pigmentation of the reticular dermis is steel blue	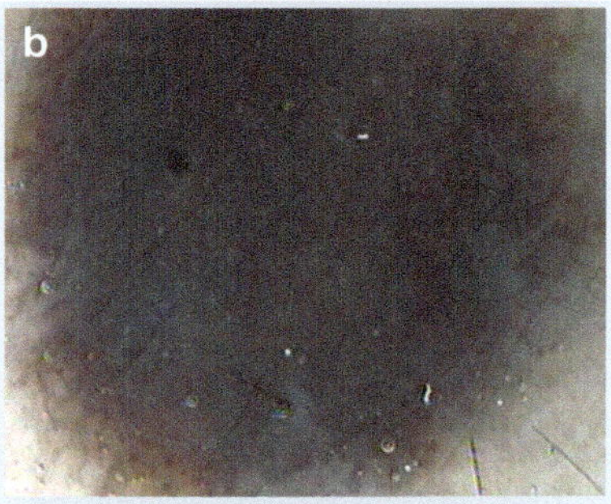
Hypopigmentation	Diffuse or localized areas of decreased pigmentation, commonly observed in benign melanocytic lesions	Decreased melanin pigment	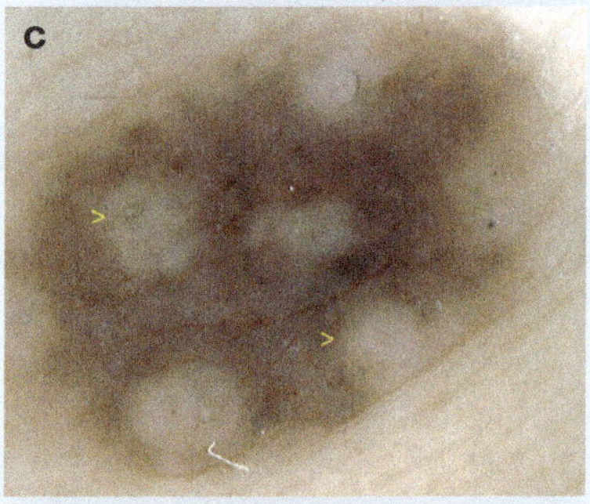
Black dots	Small, round structures that may be regularly or irregularly distributed, respectively, in benign and malignant melanocytic lesions	Focal collections of melanin in the stratum corneum	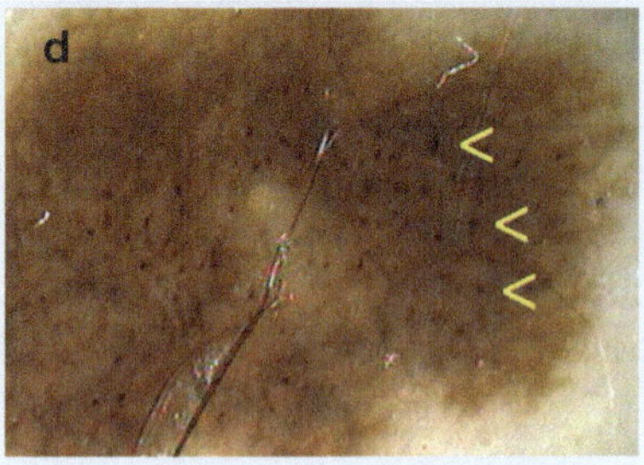

Table 1.2 (continued)

VD features	Description	Histopathologic correlation	
Brown globules	Round to oval, variously sized structures that show a regular or irregular distribution, respectively, in benign and malignant melanocytic lesions	Nests of melanin-containing melanocytes in the lower epidermis	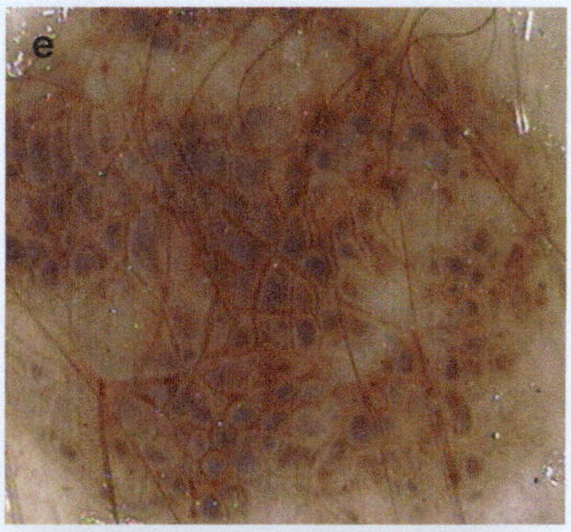
Streaks	This term comprises *radial streaming* and *pseudopods*, which irradiate from the lesion border: the former are narrow, closely arranged, parallel lines, the latter appear as digitiform extensions. Streaks may be regularly or irregularly distributed within both a pigmented Spitz nevus and a melanoma	Peripheral, confluent, and heavily pigmented junctional nests of melanocytes	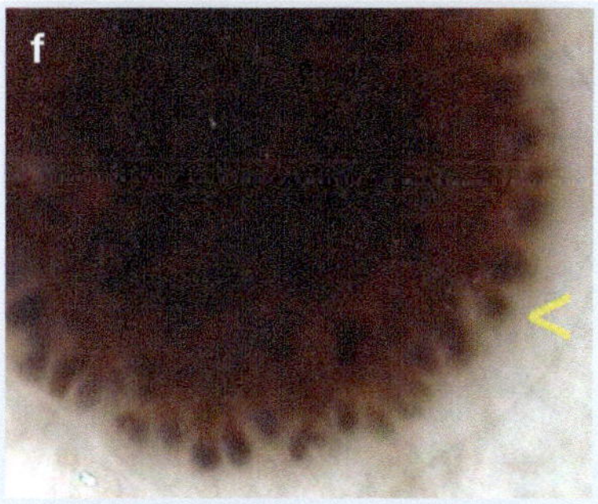
Regression	This term comprises *white scar-like depigmentation*, which corresponds to areas lighter than the normal skin, and the so-called *peppering*, which consists of speckled multiple blue-gray granules within a hypopigmented area. Regression is frequently observed in melanoma	Fibrosis and melanosis	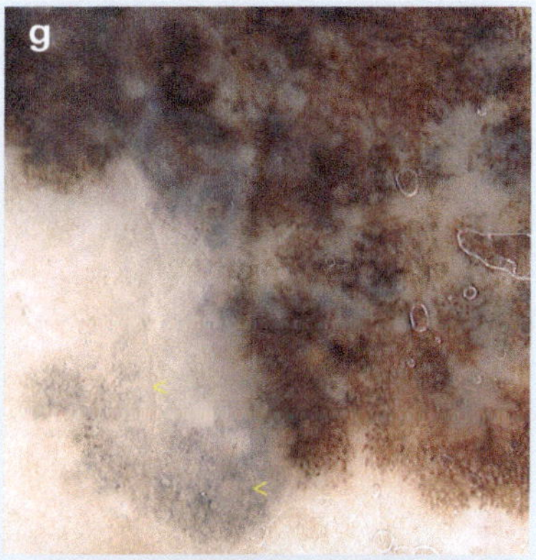
Blue-white veil	Irregular, ill-defined gray-blue to whitish-blue pigmentation dimming the underlying structures It is highly suggestive of melanoma	Acanthotic epidermis with focal hypergranulosis above sheets of melanophages and/or heavily pigmented melanocytes in the superficial dermis	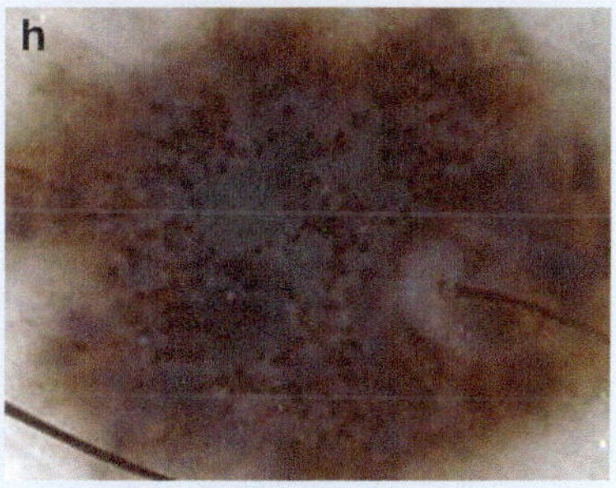

Table 1.3 Common VD features of non-melanocytic lesions

VD features	Description	Histopathologic correlation	
Milia-like cysts (horny pseudocysts)	Round luminescent whitish or yellowish structures. They are mainly observed in seborrheic keratoses	Small intraepithelial cysts filled with keratinized material	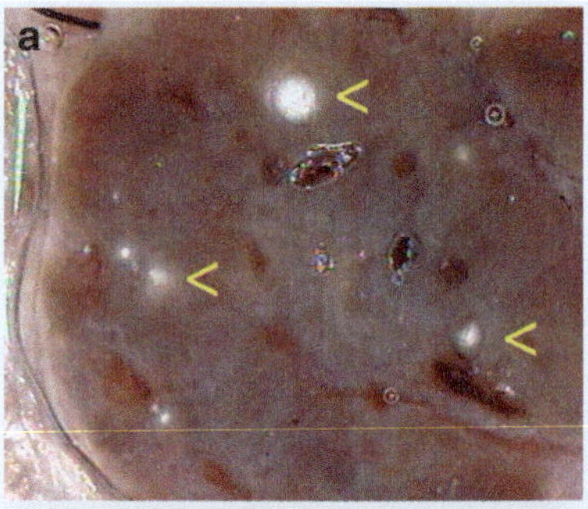
Comedo-like openings	Round or oval-shaped yellow to brown areas. They are mainly observed in seborrhoeic keratoses, occasionally in papillomatous melanocytic nevi, rarely in melanoma	Keratin-filled invaginations of the epidermis	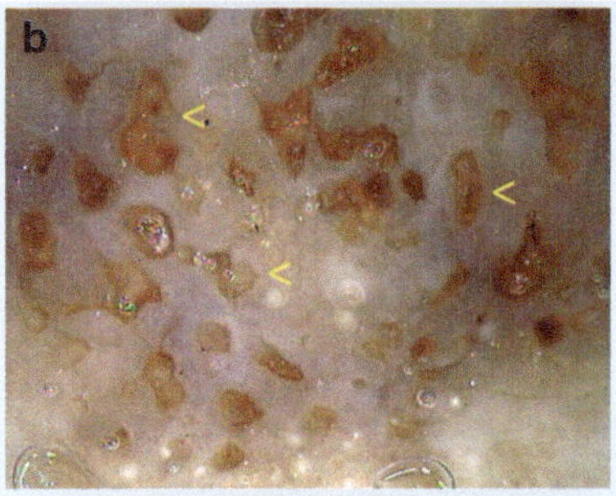
Brain-like appearance	Irregular linear keratin-filled grooves (or sulci) alternating with yellowish to brownish ridges (or gyri). Typical of seborrheic keratoses	Keratin-filled invaginations of the epidermis	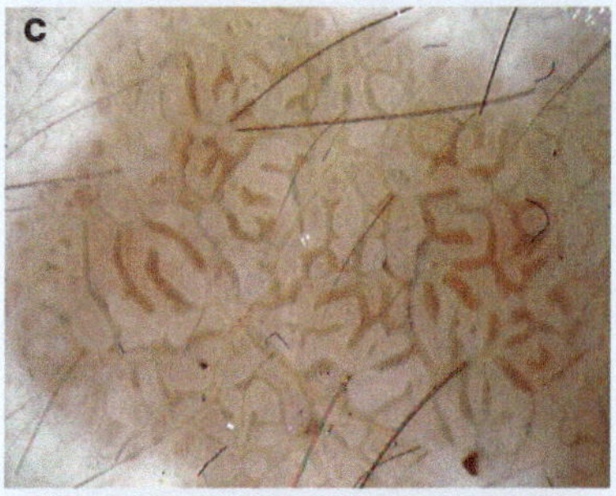
Exophytic papillary structures	Dome-shaped formations. Observed in dermal nevi and in seborrheic keratoses	Pronounced papillomatosis	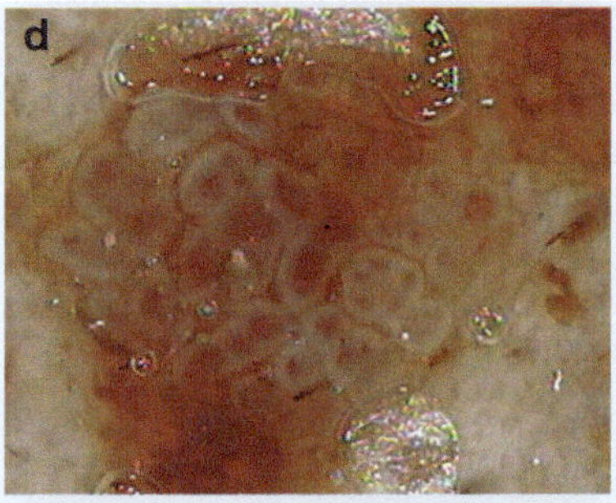
Fingerprint-like structures	Light-brown, delicate, network-like configurations seen at the periphery of a lesion, producing a pattern that resembles fingerprints. They are typical of flat seborrheic keratoses	Thin elongated rete ridges that are heavily pigmented at the basal layer	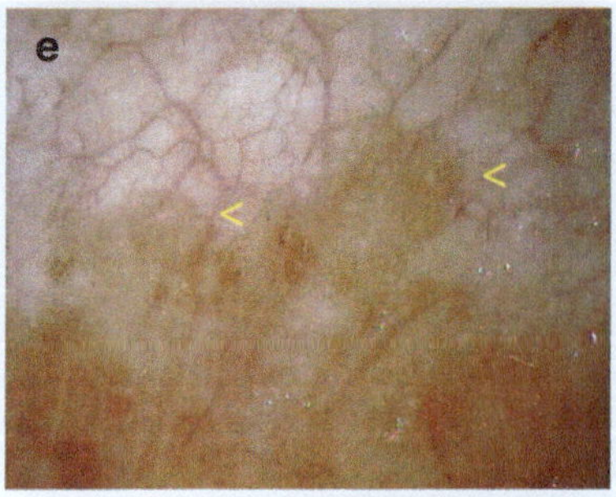

Table 1.3 (continued)

VD features	Description	Histopathologic correlation	
Moth-eaten border	Concave rim which has been compared to a moth-eaten garment. It is typical of flat seborrheic keratoses	Pigment distribution at the basal layer	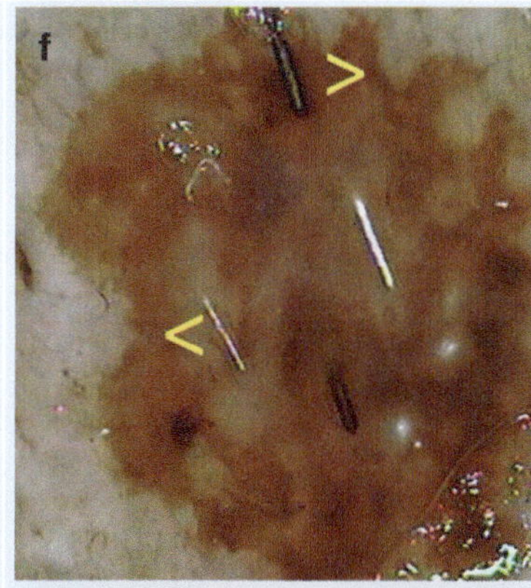
Leaflike areas	Brown-gray or gray-blue regions located at the lesion periphery, forming a leaflike pattern. Typical of basal cell carcinoma. Radial projections meeting at a central axis are defined spoke-wheel areas	Clumps of pigmented basaloid cells	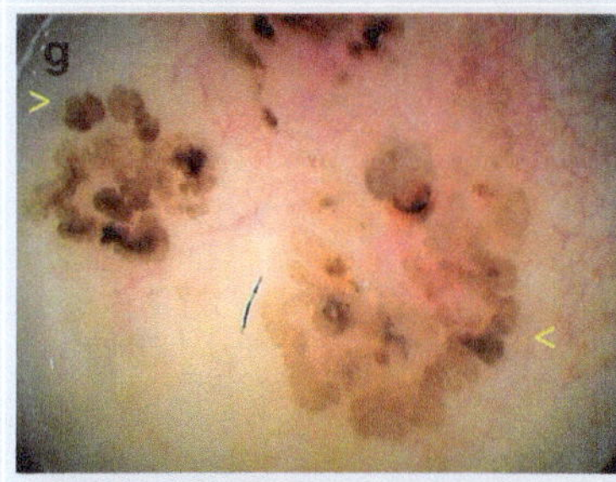
Blue-gray globules and large blue-gray ovoid nests	Well-circumscribed, roundish to oval structures, of different sizes, not intimately connected to a pigmented tumor body. Suggestive of basal cell carcinoma	Pigmented basaloid cells	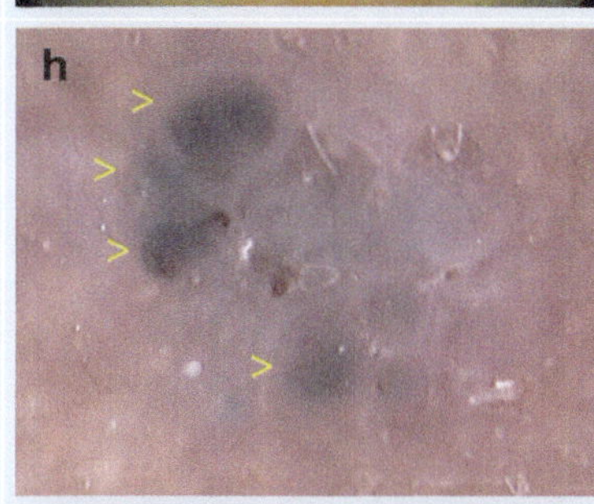
Central white patch	Sharp circumscribed, round to oval, sometimes irregularly outlined, whitish area within the center of a pigmented lesion. Specific for dermatofibroma	Epidermal hyperplasia overlying a variable amount of dermal fibrosis	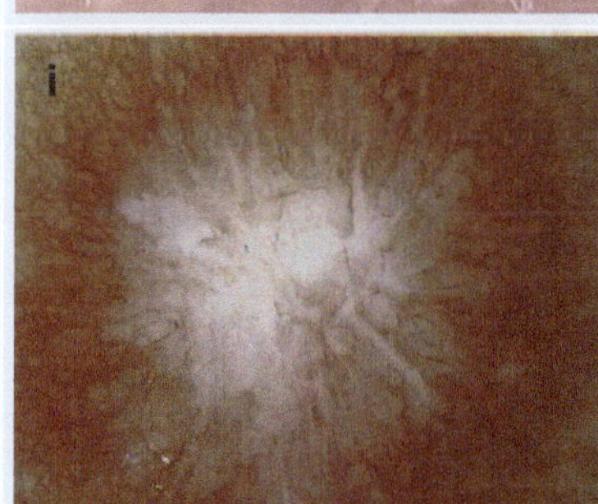
Red-blue areas (red lacunas or red lagoons)	Roundish or oval structures with a reddish or red-bluish coloration. Typical of angiomas. They may acquire a deep blue to black color after thrombosis	Widened vascular lacunae located in the superficial dermis	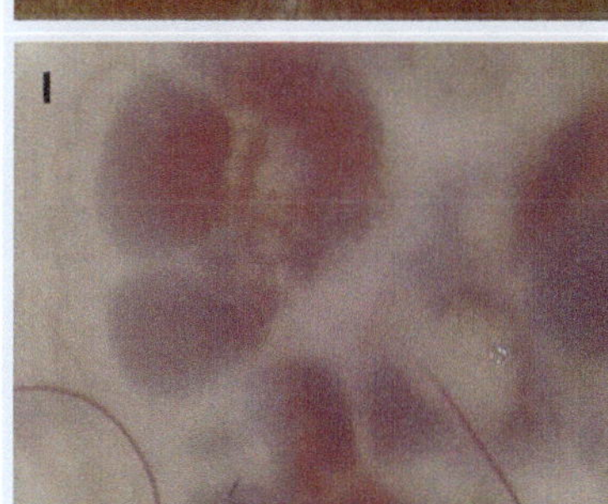

Table 1.4 Common vascular structures in skin lesions

Vascular structure	Description	
Comma-like	Short, strongly curved blood vessels predominantly seen in dermal nevi	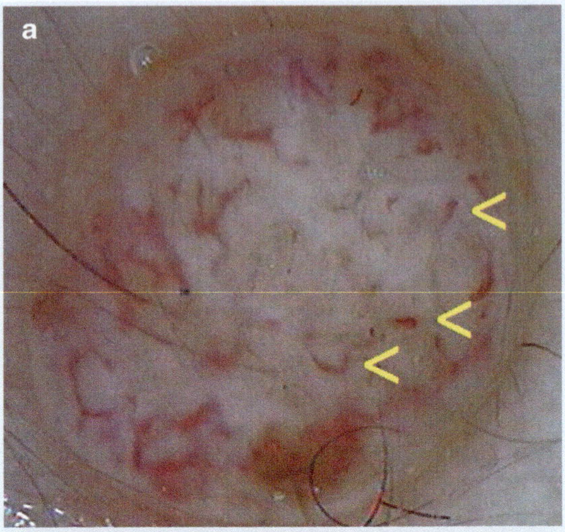
Hairpin-like	Long capillary loops which may be seen in melanoma, keratoacanthoma and seborrheic keratoses	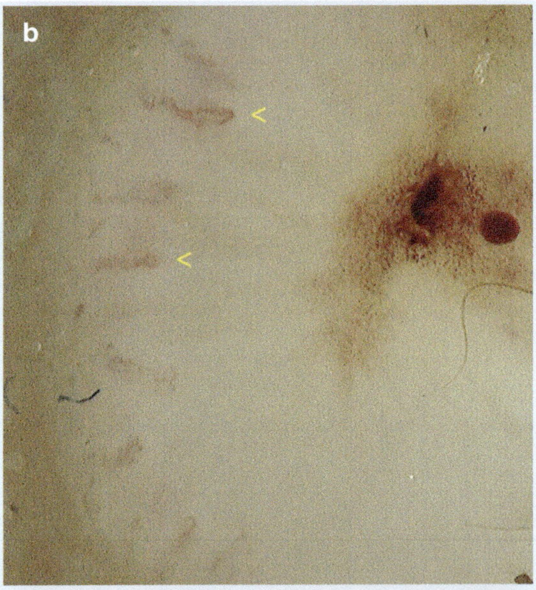
Dotted	Small pinpoint vessels corresponding to short capillary loops, which are commonly seen in all types of tumors, including melanoma and Bowen's disease	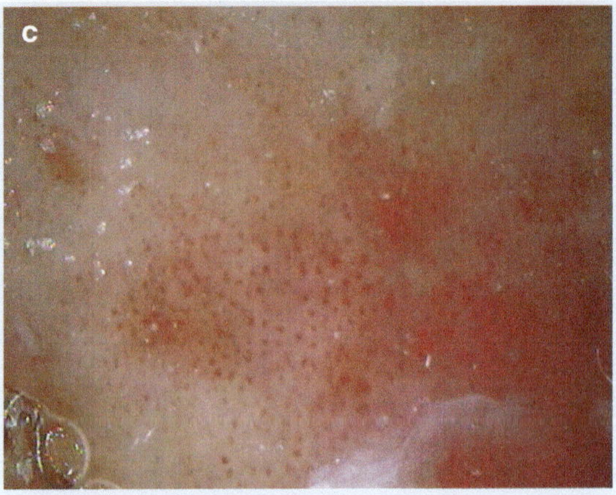

Table 1.4 (continued)

Vascular structure	Description	
Treelike	Thick and arborized structures commonly observed in basal cell carcinoma	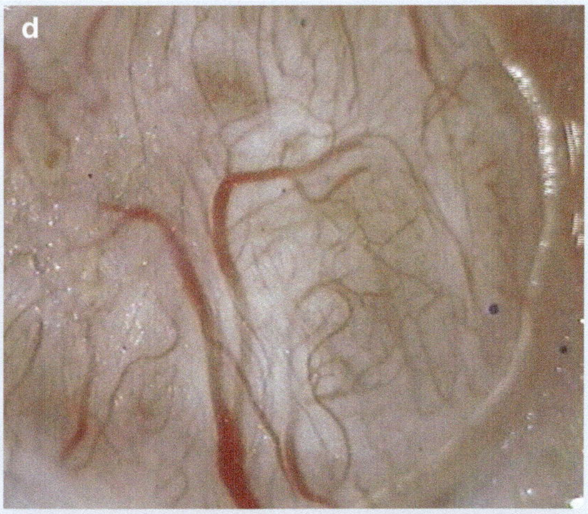
Linear irregular	Predominantly seen in melanoma	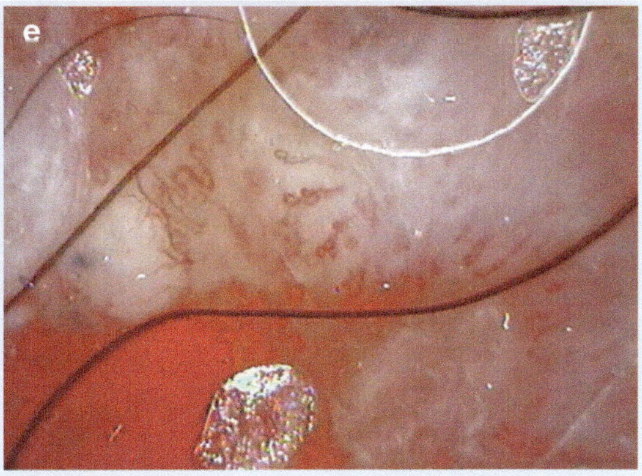

grooves between the skin markings and two linear lines along both sides of each sulcus; single or double dotted lines along the sulci may be present as variants. The *latticelike pattern* shows a linear pigmentation which follows and crosses the surface sulci. The *fibrillar pattern* is characterized by fine fibrillar pigmentation running in a slanting direction to the skin markings. In situ acral melanoma shows the so-called *parallel ridge pattern*, a diffuse and fine reticular, irregularly shaped pigmentation that follows the papillary tips [5]; however, some benign acral lesions may show parallel ridge pattern on VD [25].

1.2.1.3 Other Sites

The use of VD for pigmented lesions of the nails or mucous membranes (both oral and genital) has been significantly less investigated [26–28]. There is a need to intensify research, which would result in creating diagnostic algorithms, useful for early detection of melanoma.

1.3 Nonpigmented Skin Lesions

VD may be useful for the diagnosis of several nonpigmented skin proliferations, showing in some cases specific features. In those cases characterized by indicative but not specific features, VD may help to rule out clinically similar disorders that do not show that pattern.

1.3.1 Sebaceous Hyperplasia

VD features: Central aggregation of white-yellowish globules (*cumulus sign*) surrounded by a crown of vessels (Fig. 1.2). Occasionally, the

Table 1.5 Melanocytic lesions and VD correlations

Melanocytic lesion	VD features	
Junctional melanocytic nevus	Regular, delicate pigment network that gradually fades at its periphery. In its central portion, multiple black dots or a uniform dark pigmentation may be present	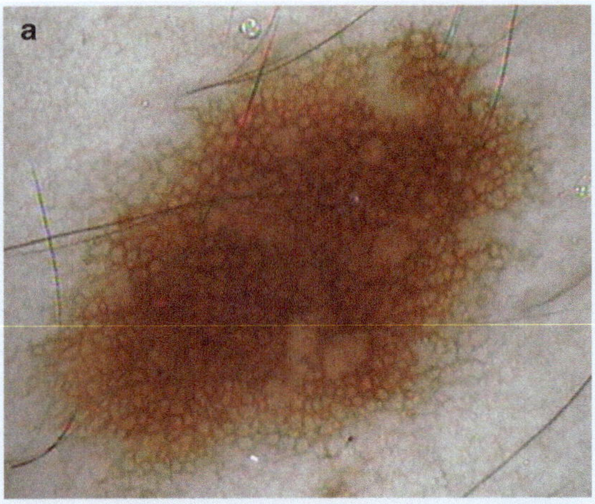
Compound melanocytic nevus	Pigment network and/or diffuse brown pigmentation. It also exhibits regularly distributed brown globules	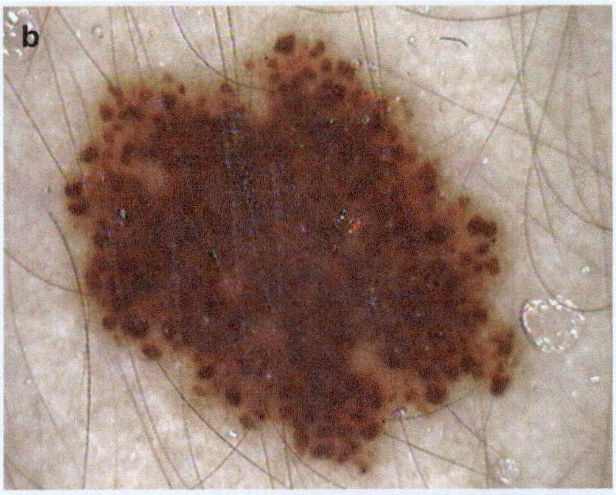
Dermal melanocytic nevus	Brown globules varying in size and color, sometimes arranged to create a "cobblestone" effect. Comma-like vessels are commonly observed	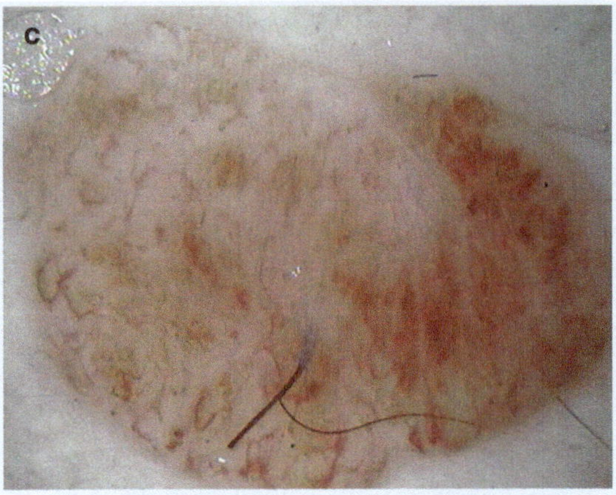

Table 1.5 (continued)

Melanocytic lesion	VD features	
Pigmented Spitz nevus	Characteristic starburst pattern formed by a prominent gray-blue to black diffuse pigmentation and by streaks radially and regularly located along the periphery. In its center, a reticular black-whitish to blue-whitish veil, called reticular depigmentation, may be present. The starburst pattern may rarely be seen also in melanoma; thus, in adult patients, a biopsy should be performed when such spitzoid features are detected by VD	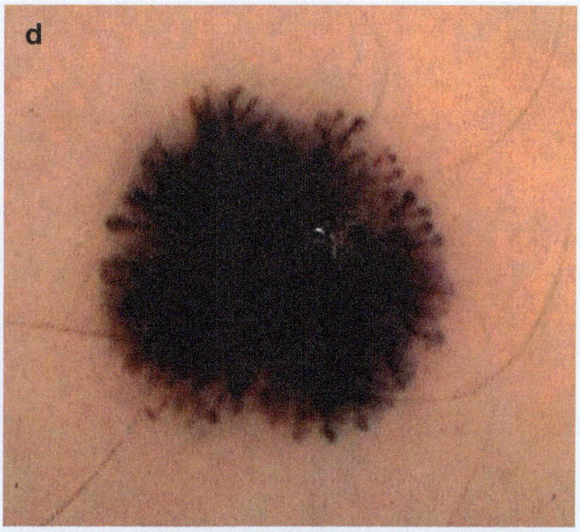
Blue nevus	Homogeneous blue pigmentation with complete absence of other findings. Multiple areas of hypopigmentation corresponding to fibrosis may be observed	
Melanoma in situ and early invasive melanomas	Diffuse pigmentation irregular in color and/or distribution, atypical pigment network with wide and irregular meshes and thick lines ending abruptly at the periphery, dots or globules variously sized and haphazardly distributed, irregularly dispersed streaks	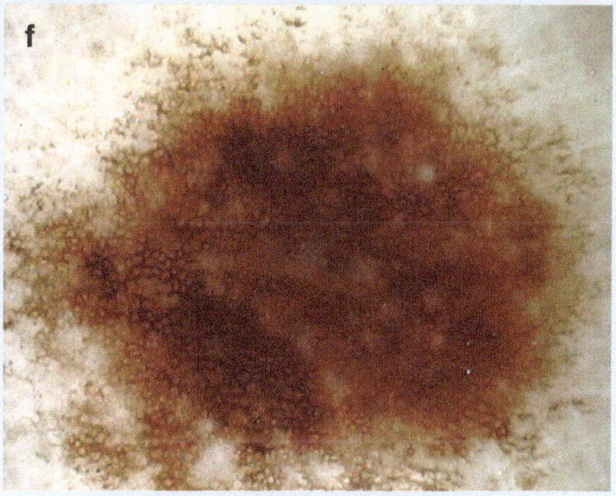

(continued)

Table 1.5 (continued)

Melanocytic lesion	VD features	
Intermediate and thick melanomas	The previous findings are associated with a blue-whitish veil and, frequently, with dotted, linear irregular, and/or hairpin vessels. Regression may also be present and, if extensive, should always prompt biopsy, despite the presence of other VD criteria, to avoid missing a melanoma	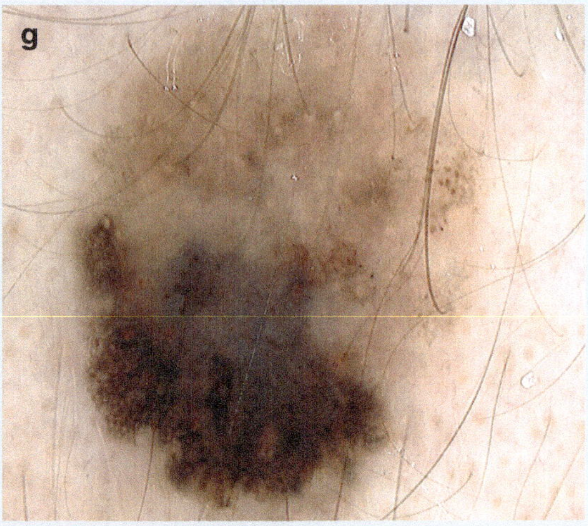
Amelanotic melanoma	Presence of polymorphous vessels, which may be evident as dotted, hairpin, and/or linear irregular. Moreover, milky-red globules or milky-red areas which appear as localized or diffuse areas of reddish-white color may be seen; these findings are thought to represent highly vascularized amelanotic tumor cell complexes	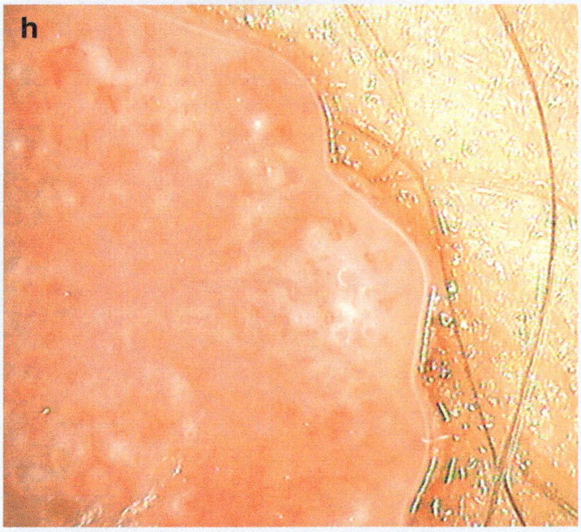

ostium of the gland is visible as a small crater or umbilication in the middle of these yellowish structures (*bonbon toffee sign*) [29, 30].

1.3.2 Pyogenic Granuloma

VD features: Typical pattern characterized by a red to dark pink homogeneous area, corresponding to proliferating vessels, surrounded by a white collarette corresponding to the hyperplastic epithelium. Additional findings include white lines intersecting the lesion (*white rail*) (Fig. 1.3), histologically corresponding to fibrous septa that surround the capillary tufts or lobules, and ulceration [31].

1.3.3 Clear Cell Acanthoma

VD features: At low magnification (×20 to ×50), homogeneous, symmetrical dotted vessels throughout the entire lesion arranged either in a netlike pattern or as pearls on a line. At higher

Table 1.6 Non-melanocytic lesions and VD correlations

Non-melanocytic lesion	VD features	
Basal cell carcinoma	Leaflike areas, spoke-wheel areas, blue-gray globules and/or large blue-gray ovoid nests. The presence of treelike vessels and/or ulcerations is also typical	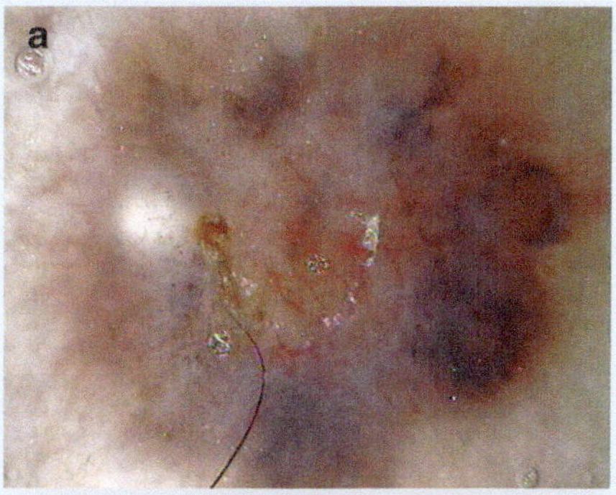
Seborrheic keratosis	Milia-like cysts and comedo-like openings on a background varying from opaque light brown to dark brown or black. A pronounced black pigmentation camouflaging the pathognomonic features may be seen (melanoacanthoma). Other morphological findings that sometimes may be observed include a brain-like appearance and exophytic papillary structures. Some flat seborrhoeic keratoses (also known as solar lentigines) may show either a fingerprint-like pattern or a moth-eaten border	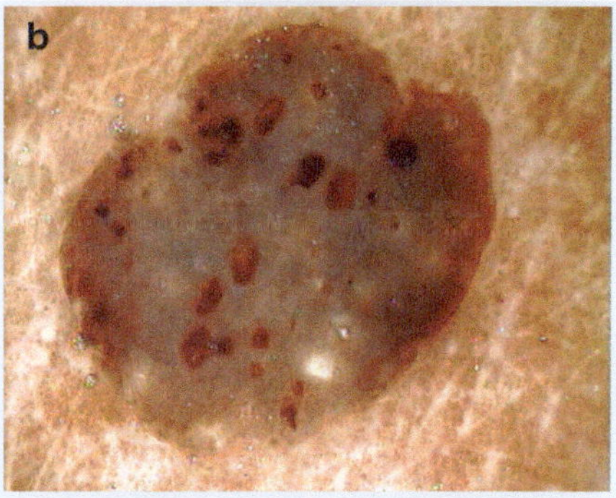
Lichen planus-like keratosis	Brown-gray and blue-gray granular pattern	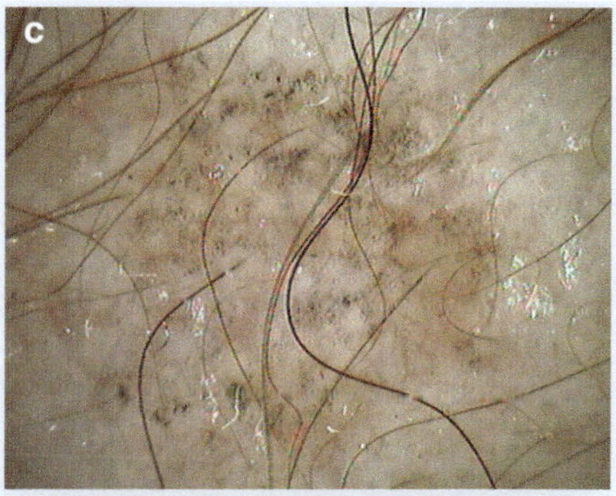
Dermatofibroma	Central white patch surrounded by a delicate, regular, usually light-brown pigment network. Sometimes within the central patch several small round to oval globules of light-brown coloration may be found	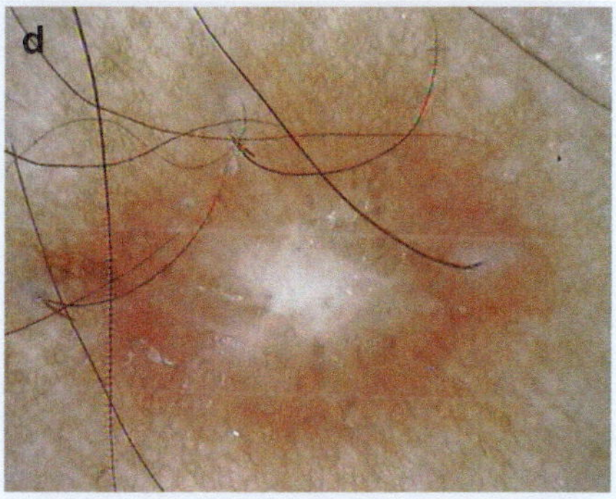

(continued)

Table 1.6 (continued)

Non-melanocytic lesion	VD features	
Senile angiomas	Red-blue areas. Darker red-blue or red-black colors are signs of thrombi within the vascular spaces	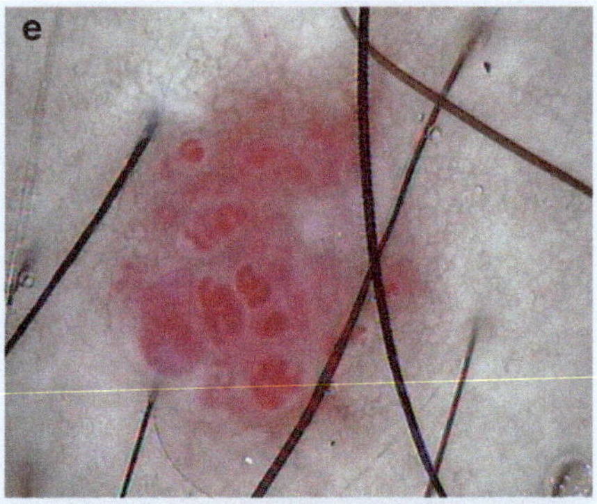
Angiokeratomas	Red-blue or red-black areas associated with whitish keratotic areas or a blue-whitish veil	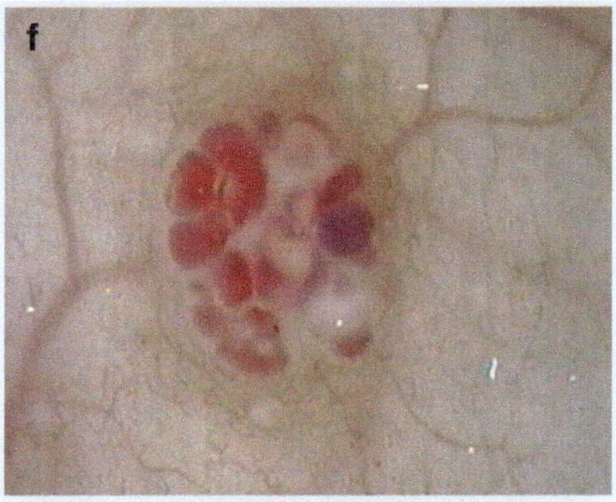
Subcorneal hematomas	Homogeneous dark-brown pattern or dark-red globules	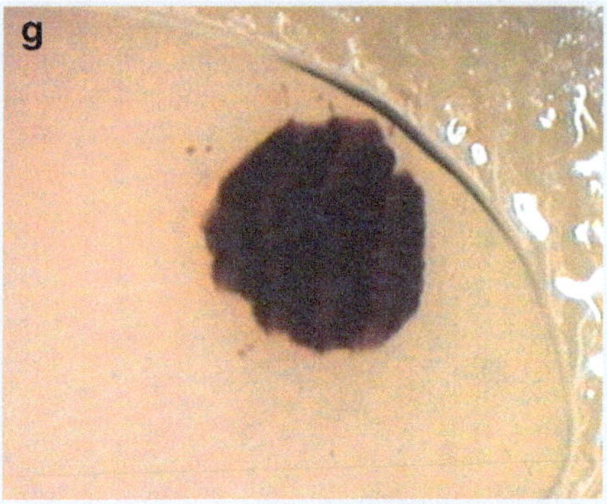

magnification (×200), the dotted vessels display a *bush-like* aspect [32–34].

1.3.4 Xanthomatous Neoplasms

VD features: Xanthoma shows a uniform yellowish pigmentation with subtle erythematous border, compared with a *setting sun* [35, 36]. Moreover, some linear and/or branched vessels may be present. Specific features include brownish globules and dots of dermal hemosiderin in reticulohistiocytoma, peripheral pigmented network with basal hyperpigmentation in xanthomized dermofibroma, isolated dotted vessels, and "cloud" of yellow areas in adult xanthogranuloma [37].

1.3.5 Mastocytosis

VD features: In urticaria pigmentosa and in the papular variant of cutaneous mastocytosis, the

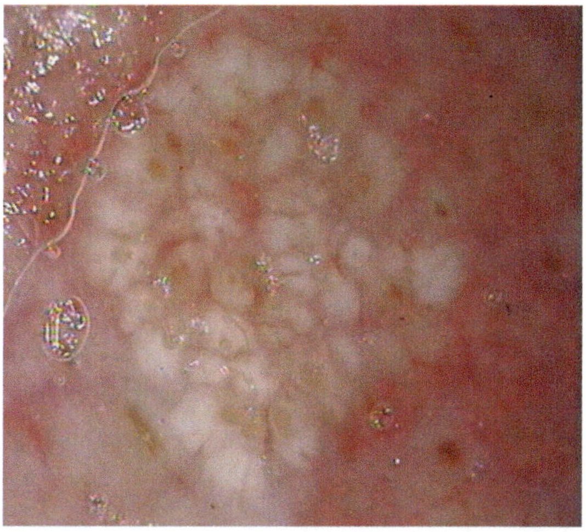

Fig. 1.2 VD of sebaceous hyperplasia: central aggregation of *white-yellowish* globules surrounded by a *crown of vessels* (×20)

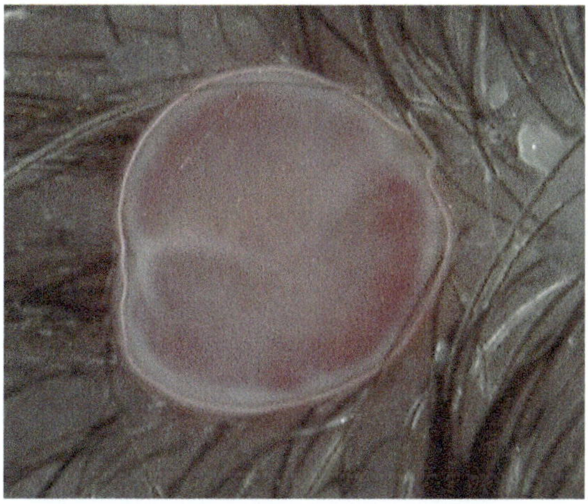

Fig. 1.3 VD of pyogenic granuloma: *reddish* homogeneous area intersected by *white lines* (×20)

most common structures are brown reticular lines. In telangiectasia macularis eruptiva perstans, there are telangiectatic vessels arranged in a reticular pattern [38].

1.3.6 Sarcoidosis

VD features: Translucent yellow to orange globular-like or structureless areas associated with linear vessels should raise the suspicion of a granulomatous skin disease [39].

1.3.7 Median Raphe Cysts

VD may be helpful in the differentiation between cysts and canals of the median raphe of the penis, thus helping to establish the most appropriate therapeutic approach.

VD features: In one case, examination at ×30 revealed that the whitish, cystic lesions were connected and encompassed by a translucent tract, not visible to the naked eye, allowing for a definitive diagnosis of median raphe canal as more appropriate [40].

1.3.8 Eccrine Poroma

VD features: Red lacunae and a polymorphic vascular pattern (irregular linear vessels, glomerular vessels, and hairpin vessels), mimicking those detectable in amelanotic melanoma and basal cell carcinoma [41, 42].

1.3.9 Keratoacanthoma

VD features: Central yellowish to brownish structureless mass of keratin usually surrounded by elongated and sometimes thick hairpin vessels. The latter typically occur within a white halo [42].

1.3.10 Actinic Porokeratosis

VD features: Well-defined whitish-yellow peripheral annular structure, with a brownish pigmentation in the inner side. This structure, which has been described as *the outlines of a volcanic crater as observed from a high point* or as *white track*, surrounds a central whitish or red-whitish, scar-like area, in which scales and dotted or linear vessels may be observed. In some cases, a double *white track* may be observed [43–45].

1.3.11 Nonpigmented Facial Actinic Keratosis

VD features: Peculiar *strawberry* appearance, produced by four features: a pink-to-red pseudo-

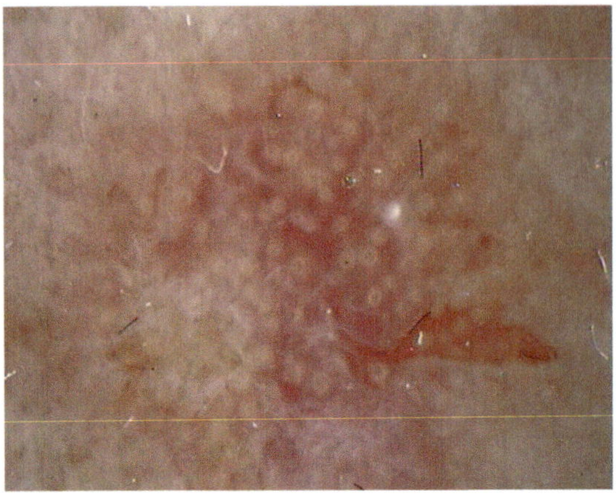

Fig. 1.4 VD of facial nonpigmented actinic keratosis: typical *strawberry* appearance (×20)

network, a white-to-yellow surface scale, fine, linear-wavy vessels, and hair follicle openings filled with yellowish keratotic plugs and/or surrounded by a white halo (Fig. 1.4) [46, 47].

1.3.12 Bowen's Disease

VD features: Dotted/glomerular vessels and a scaly surface occur in up to 90 % of cases. In pigmented Bowen's disease, small brown globules in a patchy distribution, a grey-brown diffuse pigmentation, and a pigmented pseudonetwork represent the most remarkable findings. VD has been considered of help in preoperatory evaluation, in follow-up and in monitoring the nonsurgical outcome of the disease, where the disappearance of vascular structures may indicate adequate treatment [48–51].

1.3.13 Invasive Squamous Cell Carcinoma

VD features: Hairpin, dotted, and/or linear irregular vessels (polymorphous vascular pattern) over a whitish background, with centrally located scales or keratin crusts (structureless yellow to light brown amorphous areas). Ulcerations are often present [42, 47].

1.3.14 Kaposi's Sarcoma

VD features: Homogeneous bluish-reddish pigmentation, which seems to be related to the presence of vascular structures in the deep dermis. Other findings include a scaly surface and small brown globules. The *rainbow pattern*, consisting of multicolored areas showing the colors of a rainbow's spectrum [52], is not specific, as it has been encountered in other conditions (melanoma, stasis dermatitis, lichen planus, and hemosiderotic dermatofibroma) [53].

1.4 Ectoparasitoses

1.4.1 Scabies

VD – especially at high magnification – is a sensible and highly specific technique for the diagnosis of scabies, allowing a rapid inspection of the entire skin surface.

VD features: At low magnification (up to ×40), presence of a small dark brown triangular structure, corresponding to the head of the mite, located at the end of the burrow (*jet with contrail*). At higher magnification (×100 to ×600), VD allows a more detailed inspection and identification of the mite morphology and of its products: the oval translucent body, the legs, the rostrum, the eggs, and the feces (Fig. 1.5). In most cases, it is possible to detect the mite movements inside the burrow.

The exam is not painful, and is thus better accepted, compared to scraping (the traditional diagnostic technique that requires the use of a scalpel blade), especially by children and highly emotional patients. Finally, it may also be used both for the screening of family members and post-therapeutic follow-up [54–58].

1.4.2 Head and Pubic Lice

VD ensures a detailed identification of both mites and nits, allowing a rapid differentiation with pseudo-nits (hair casts, debris of hair spray or gel, or seborrheic scales) [58, 59]. VD may also be useful in therapeutic monitoring and to

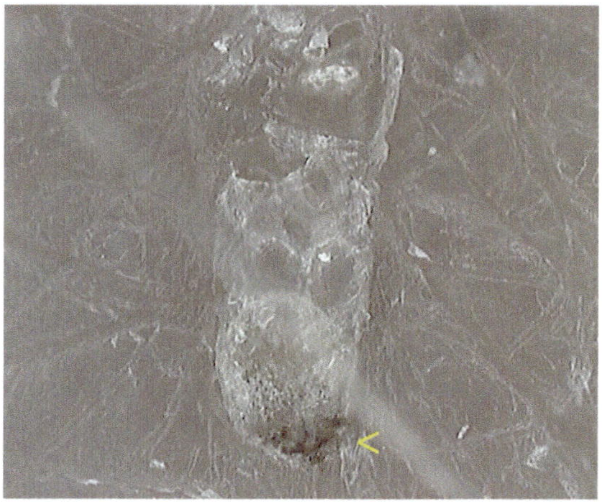

Fig. 1.5 VD of scabies: presence of *Sarcoptes scabiei* at the end of the burrow (bottom). The mite's head appears pigmented (arrow) (×200)

evaluate the pediculicidal activity of different topical products [60].

VD features: Full, viable nits appear as opaque structures, with a rounded free ending, fixed to the hair shaft; empty nits appear as translucent structures with a flat and fissured free ending [58, 59].

1.4.3 Tungiasis

VD features: Brown to black ring with a central pore, corresponding to pigmented chitin surrounding the posterior portion of the parasite's exoskeleton (*Tunga penetrans*). Additional features include a *grey-blue blotch* (representing either developing eggs within the abdomen or hematin in the gastrointestinal tract of the parasite) and *whitish chains* (corresponding to parasite's eggs arranged in a row). After sequential and careful shaving of the epidermis and gently compressing the edges of the wound, a jellylike bag full of ovoid eggs may be seen [61–64].

1.4.4 Cutaneous Leishmaniasis

VD features: Diffuse erythema and vascular structures including comma-shaped vessels, linear or atypical vessels, and arborizing telangiectasia [65]. Also, early lesions may show yellow oval or teardrop-shaped structures (*yellow tears*) composed of follicular plugs produced by the compression of the follicular openings. Advanced stages display hyperkeratosis, vascular abnormalities (hairpin and dotted vessels), and the so-called *white starburst-like* pattern adjacent to the peripheral hyperkeratosis surrounding the central erosion.

1.4.5 Furuncular Myiasis

VD features: View of the posterior segment of the larva (*Dermatobia hominis*) with its breathing spiracles looking like bird's feet. Those structures are seen in the center of a creamy-white body that is surrounded by black dots shaped as a thorn crown and representing small spines in a circular row [66].

1.4.6 Demodicosis

VD features: Identification of Demodex mites, with visualization of Demodex *tails* and Demodex follicular openings. In patients with the inflammatory variant of demodicosis, reticular horizontal dilated blood vessels may be also visualized [67].

1.5 Cutaneous/Mucosal Infections

1.5.1 Molluscum Contagiosum

VD features: Characteristic pattern consisting of a central polylobular white-to-yellow amorphous structure with a peripheral crown of reddish, linear, or branched vessels, which do not usually cross the center of the lesion (*red corona*) (Fig. 1.6) [68, 69].

1.5.2 Cutaneous Warts

VD features: Multiple densely packed papillae, each containing a central red dot surrounded by a whitish halo; irregularly distributed black dots, corresponding to thrombosed vessels, are also

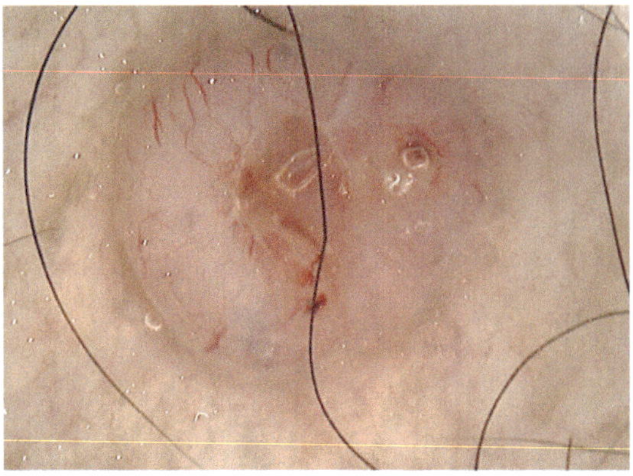

Fig. 1.6 VD of molluscum contagiosum: central yellow amorphous structure with a peripheral crown of branched vessels (×30)

visible [58, 70, 71]. Such hemorrhages are prominent in the plantar wart within a well-defined, yellowish papilliform surface in which skin lines are interrupted. Flat warts are characterized by regularly distributed, tiny, red dots on a light brown to yellow background.

1.5.3 Genital Warts

VD features: Reflect the lesion stage. In papular lesions, VD shows a whitish network circumscribing areas centered by dilated glomerular vessels (*mosaic pattern*). In cauliflower-like lesions, VD shows multiple, irregular whitish projections arising from a common base and comprising elongated and dilated vessels (*fingerlike pattern*) [58, 72–74]. The use of VD may help to differentiate genital warts from vestibular papillae and pearly penile papules [73, 75, 76]. At VD, vestibular papillae present as multiple transparent and cylindrical projections, containing irregular vascular structures, whose bases, however, remain separate; pearly penile papules appear as regular whitish pink cobblestone or grape-like structures in a few rows with central dotted or comma-like vessels in each papule.

1.5.4 Tinea Nigra

VD features: Homogeneous non-melanocytic pigmented pattern with spicules that do not follow the dermatoglyphic lines in the irregular macule [77].

1.5.5 Lupus Vulgaris

VD features: Linear focused telangiectasia on a typical yellow to golden background [58, 78]. Moreover, some milia-like cysts and whitish reticular streaks have also been detected. Nevertheless, none of the observed features is sufficiently specific alone, while their combination may result in increased sensitivity.

1.6 Inflammatory Disorders

In inflammatory diseases, VD evaluation of vascular pattern may be of great importance; in this case, the technique may be compared to capillaroscopy, and for such reasons the use of high magnification (≥×100) is indicated.

1.6.1 Psoriasis

VD features: At low magnification (×10 to ×50), a dotted pattern is evident. Higher magnifications (×100 to ×400) are able to visualize dilated, elongated, and convoluted capillaries showing a typical *glomerular* or *bushy* pattern (Fig. 1.7) [79]. The vessel caliber in psoriasis has been observed to be larger (12–13 μm) than in normal skin (5–6 μm) [79]. In perilesional skin, capillary loops have an elongated "hairpin" disposition, parallel to the cutaneous surface, with a lengthened edge directed towards the board of the lesion.

Identification of *bushy* capillaries by VD may be helpful in addressing the correct diagnosis in unusual presentations, such as in palmar and/or plantar psoriasis [80], psoriatic balanitis [81], and scalp psoriasis [82], particularly in those cases when no other body sites are involved. Finally, the recognition of this pattern is almost important in in vivo therapy

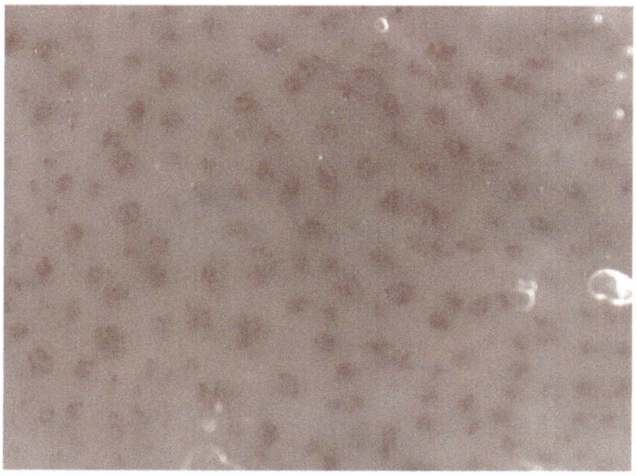

Fig. 1.7 VD of psoriasis plaque: typical *bushy* pattern (×200)

monitoring, disclosing morphological modifications, and loop changes after local and systemic treatments [79].

1.6.2 Lichen Planus

VD features: Easy and rapid recognition of Wickham striae, appearing as pearly whitish structures which secondary develop thin spikes (*comblike* projections) or arboriform ramifications departing from the periphery. Long-standing lesions display a decrease in Wickham striae, gradually surrounded by pigmented structures. Other structures include gray-blue dots, cysts, or vascular structures with a deeper location and thus generally not visible to the naked eye [83, 84].

1.6.3 Urticaria and Urticarial Vasculitis

VD features: Common urticaria reveals a red, reticular network of linear vessels, occasionally associated with dotted vessels. Also, nonvascular areas within vascular structures, as a consequence of massive edema, may be detected [85]. Urticarial vasculitis displays numerous purpuric dots or globules on an orange-brown pigmentation, whose recognition is relevant to their differentiation.

1.6.4 Rosacea

VD features: Dilated vessels, prominent telangiectasia, and large polygonal vascular net. In contrast, no alterations were found in the nail-fold region, suggesting that rosacea specifically affects the facial microvasculature [86].

1.6.5 Pityriasis Lichenoides et Varioliformis Acuta

VD features: Examination at ×20 reveals papules with a central whitish patch and crusted lesions with an amorphous brownish structure; both types of lesion are surrounded by a well-defined ring of pinpoint-like and/or linear vascular structures configuring a targetoid appearance. At higher magnification (×200 to ×300), the ring vascular structures appear dilated and convoluted, showing a glomerular pattern or linear arrangement; moreover, nonblanchable reddish globules, corresponding to micro-hemorrhages in papillary dermis, may be observed [87].

1.7 Scalp Disorders

See Chap. 39.

1.8 Nail Disorders

The use of VD in nail disorders is still in progress. Potentially, VD allows nail examination of all anatomic parts of nail unit, such as nail plate, hyponychium, distal edge, proximal nail fold, nail bed, and matrix [88, 89].

1.8.1 Psoriasis

VD features: Nail matrix psoriasis frequently produces nail plate surface abnormalities, such as pitting, nail crumbling, onycholysis, salmon patches, splinter hemorrhages, and nail bed hyperkeratosis, that may be better defined with the use of VD [89]. Moreover, VD observation

(×40 to ×70) of the hyponychium displays dilated, tortuous, elongated, and irregularly distributed capillaries whose detection can be used to assess the severity of the disease [90].

1.8.2 Onychomycosis

VD features: Nail infections caused by fungi producing melanin show homogenous brownish pigmentation devoid of melanin granules and longitudinal streaks [88, 91, 92].

1.8.3 Onychomatricoma

VD features: Nail plate shows longitudinal white lines, indicating channels containing the tumor projections [93].

1.8.4 Glomus Tumor

VD features: Nail plate VD aids in tumor localization and in the visualization of the vascular pattern of the lesion, consisting of branched telangiectasia. It is also useful to detect residual tumor remnants in the operation field after surgical excision [94].

1.9 Vascular Disorders

1.9.1 Port-Wine Stains

VD features: Superficial lesions located in the papillary dermis (type 1) show roundish and red structures, while deeper lesions (type 2) show prominent, red linear structures arranged in irregular networks. A gray-whitish veil is also consistent with the deep dermal distribution of the lesion. The last feature, along with the presence of white streaks on a variably whitish to blue background and of a pale halo surrounding a comedo-like-shaped center, seems to be predictive of poor laser therapy response [95–98].

1.9.2 Infantile Hemangioma

VD features: Superficial types show a polymorphous vascular structure (globular, circulated, comma-like, and wavy vessels), while deeper types show the same polymorphous vascular structure plus linear and dilated vessels [99].

1.9.3 Pigmented Purpuric Dermatoses

VD features: All pigmented purpuric dermatoses (lichen aureus, Schamberg's disease, Majocchi's disease, Gougerot and Blum syndrome, and eczematid-like purpura of Doucas and Kapetanakis) share irregular, round to oval red dots, globules, and patches with a red-brownish or red-coppery diffuse homogeneous background pigmentation (hemosiderin deposition within the papillary dermis originating from erythrocytes extravased from dilated capillaries) [100].

References

1. Micali G, Lacarrubba F (2003) Possible applications of videodermatoscopy beyond pigmented lesions. Int J Dermatol 42:430–433
2. Micali G, Lacarrubba F (eds) (2010) Dermatoscopy in clinical practice: beyond pigmented lesions. Informa Healthcare Ltd, London
3. Micali G, Lacarrubba F, Massimino D, Schwartz RA (2011) Dermatoscopy: alternative uses in daily clinical practice. J Am Acad Dermatol 64: 1135–1146
4. Lacarrubba F, Potenza MC, Micali G (2012) Enhanced videodermoscopic visualization of skin superficial vascular patterns using a 390–410 nm light. Arch Dermatol 148:276
5. Micali G, Lacarrubba F, Nardone B, Schwartz RA (2008) Dermoscopy for skin cancer. In: Schwartz RA (ed) Skin cancer: recognition and management. Blackwell Publishing Ltd, Oxford
6. Braun RP, Rabinovitz HS, Oliviero M, Kopf AW, Saurat JH (2005) Dermoscopy of pigmented skin lesions. J Am Acad Dermatol 52:109–121
7. Bafounta ML, Beauchet A, Aegerter P, Saiag P (2001) Is dermoscopy (epiluminescence microscopy) useful for the diagnosis of melanoma? Results of a meta-analysis using techniques adapted to the evaluation of diagnostic tests. Arch Dermatol 137:1343–1350

8. Vestergaard ME, Macaskill P, Holt PE, Menzies SW (2008) Dermoscopy compared with naked eye examination for the diagnosis of primary melanoma: a meta-analysis of studies performed in a clinical setting. Br J Dermatol 159:669–676
9. Rosendahl C, Tschandl P, Cameron A, Kittler H (2011) Diagnostic accuracy of dermatoscopy for melanocytic and nonmelanocytic pigmented lesions. J Am Acad Dermatol 64:1068–1073
10. Argenziano G, Soyer HP, Chimenti S, Talamini R, Corona R, Sera F, Binder M, Cerroni L, De Rosa G, Ferrara G, Hofmann-Wellenhof R, Landthaler M, Menzies SW, Pehamberger H, Piccolo D, Rabinovitz HS, Schiffner R, Staibano S, Stolz W, Bartenjev I, Blum A, Braun R, Cabo H, Carli P, De Giorgi V, Fleming MG, Grichnik JM, Grin CM, Halpern AC, Johr R, Katz B, Kenet RO, Kittler H, Kreusch J, Malvehy J, Mazzocchetti G, Oliviero M, Ozdemir F, Peris K, Perotti R, Perusquia A, Pizzichetta MA, Puig S, Rao B, Rubegni P, Saida T, Scalvenzi M, Seidenari S, Stanganelli I, Tanaka M, Westerhoff K, Wolf IH, Braun-Falco O, Kerl H, Nishikawa T, Wolff K, Kopf AW (2003) Dermoscopy of pigmented skin lesions: results of a consensus meeting via the Internet. J Am Acad Dermatol 48:679–693
11. Scope A, Benvenuto-Andrade C, Agero AL, Marghoob AA (2006) Nonmelanocytic lesions defying the two-step dermoscopy algorithm. Dermatol Surg 32:1398–1406
12. Hirokawa D, Lee JB (2011) Dermatoscopy: an overview of subsurface morphology. Clin Dermatol 29:557–565
13. Goodson AG, Grossman D (2009) Strategies for early melanoma detection: approaches to the patient with nevi. J Am Acad Dermatol 60:719–735
14. Zalaudek I, Kreusch J, Giacomel J, Ferrara G, Catricalà C, Argenziano G (2010) How to diagnose nonpigmented skin tumors: a review of vascular structures seen with dermoscopy – part I. Melanocytic skin tumors. J Am Acad Dermatol 63:361–374
15. Pehamberger H, Steiner A, Wolff K (1987) In vivo epiluminescence microscopy of pigmented skin lesions. I. Pattern analysis of pigmented skin lesions. J Am Acad Dermatol 17:571–583
16. Soyer HP, Argenziano G, Chimenti S (2001) Dermoscopy of pigmented skin lesions. An atlas based on the consensus net meeting on dermoscopy 2000. Edra Medica Publishing and New Media, Milan
17. Carli P, De Giorgi V, Soyer HP, Stante M, Mannone F, Giannotti B (2000) Dermatoscopy in the diagnosis of pigmented skin lesions: a new semiology for the dermatologist. J Eur Acad Dermatol Venereol 14:353–369
18. Bowling J, Argenziano G, Azenha A, Bandic J, Bergman R, Blum A, Cabo H, Di Stephani A, Grichnik J, Halpern A, Hofman-Wellenhof R, Johr R, Kittler H, Kopf A, Kreusch J, Langford D, Malvehy J, Marghoob A, Menzies S, Ozdemir F, Peris K, Piccolo D, Pizzichetta MA, Polsky D, Puig S, Rabinovitz H, Rubegni P, Saida T, Scalvenzi M, Seidenari S, Soyer HP, Tanaka M, Zalaudek I, Braun RP (2007) Dermoscopy key points: recommendations from the international dermoscopy society. Dermatology 214:3–5
19. Johr RH (2002) Dermoscopy: alternative melanocytic algorithms-the ABCD rule of dermatoscopy, Menzies scoring method, and 7-point checklist. Clin Dermatol 20:240–247
20. Henning JS, Dusza SW, Wang SQ, Marghoob AA, Rabinovitz HS, Polsky D, Kopf AW (2007) The CASH (color, architecture, symmetry, and homogeneity) algorithm for dermoscopy. J Am Acad Dermatol 56:45–52
21. Zalaudek I, Argenziano G, Soyer HP, Corona R, Sera F, Blum A, Braun RP, Cabo H, Ferrara G, Kopf AW, Langford D, Menzies SW, Pellacani G, Peris K, Seidenari S (2006) Three-point checklist of dermoscopy: an open internet study. Br J Dermatol 154:431–437
22. Stolz W, Schiffner R, Burgdorf WH (2002) Dermatoscopy for facial pigmented skin lesions. Clin Dermatol 20:276–278
23. Wang SQ, Hashemi P (2010) Noninvasive imaging technologies in the diagnosis of melanoma. Semin Cutan Med Surg 29:174–184
24. Saida T, Oguchi S, Miyazaki A (2002) Dermoscopy for acral pigmented skin lesions. Clin Dermatol 20:279–285
25. Tanioka M (2011) Benign acral lesions showing parallel ridge pattern on dermoscopy. J Dermatol 380:41–44
26. Koga H, Saida T, Uhara H (2011) Key point in dermoscopic differentiation between early nail apparatus melanoma and benign longitudinal melanonychia. J Dermatol 38:45–52
27. Olszewska M, Banka A, Gorska R, Warszawik O (2008) Dermoscopy of pigmented oral lesions. J Dermatol Case Rep 2:43–48
28. Ronger-Savle S, Julien V, Duru G, Raudrant D, Dalle S, Thomas L (2011) Features of pigmented vulval lesions on dermoscopy. Br J Dermatol 164:54–61
29. Zaballos P (2010) Sebaceous hyperplasia. In: Micali G, Lacarrubba F (eds) Dermatoscopy in clinical practice: beyond pigmented lesions. Informa Healthcare Ltd, London
30. Oztas P, Polat M, Oztas M, Alli N, Ustun H (2008) Bonbon toffee sign: a new dermatoscopic feature for sebaceous hyperplasia. J Eur Acad Dermatol Venereol 22:1200–1202
31. Zaballos P (2010) Pyogenic granuloma. In: Micali G, Lacarrubba F (eds) Dermatoscopy in clinical practice: beyond pigmented lesions. Informa Healthcare Ltd, London
32. Lacarrubba F, de Pasquale R, Micali G (2003) Videodermatoscopy improves the clinical diagnostic accuracy of multiple clear cell acanthoma. Eur J Dermatol 13:596–598

33. Zalaudek I, Hofmann-Wellenhof R, Argenziano G (2003) Dermoscopy of clear-cell acanthoma differs from dermoscopy of psoriasis. Dermatology 207:428
34. Lacarrubba F, D'Agata O, Dall'Oglio F, Micali G (2010) Clear cell acanthoma. In: Micali G, Lacarrubba F (eds) Dermatoscopy in clinical practice: beyond pigmented lesions. Informa Healthcare Ltd, London
35. Cavicchini S, Tourlaki A, Tanzi C, Alessi E (2008) Dermoscopy of solitary yellow lesions in adults. Arch Dermatol 144:1412
36. Hussain SH, Kozic H, Lee JB (2008) The utility of dermatoscopy in the evaluation of xanthogranulomas. Pediatr Dermatol 25:505–506
37. Mandato F, Biagioli M, Rubegni P (2010) Xanthomatous lesions. In: Micali G, Lacarrubba F (eds) Dermatoscopy in clinical practice: beyond pigmented lesions. Informa Healthcare Ltd, London
38. Vano-Galvan S, Alvarez-Twose I, Heras Ede L, Morgado JM, Matito A, Sánchez-Muñoz L, Plana MN, Jaén P, Orfao A, Escribano L (2011) Dermoscopic features of skin lesions in patients with mastocytosis. Arch Dermatol 147:932–940
39. Pellicano R, Tiodorovic-Zivkovic D, Gourhant JY, Catricalà C, Ferrara G, Caldarola G, Argenziano G, Zalaudek I (2010) Dermoscopy of cutaneous sarcoidosis. Dermatology 221:51–54
40. Lacarrubba F, Tedeschi A, Francesconi L, Micali G (2010) Canal vs cysts of the penile median raphe: advancing diagnostic methods using videodermatoscopy. Pediatr Dermatol 27:667–669
41. Avilés-Izquierdo JA, Velázquez-Tarjuelo D, Lecona-Echevarría M, Lázaro-Ochaita P (2009) Dermoscopic features of eccrine poroma. Actas Dermosifiliogr 100:133–136
42. Zalaudek I, Kreusch J, Giacomel J, Ferrara G, Catricalà C, Argenziano G (2010) How to diagnose nonpigmented skin tumors: a review of vascular structures seen with dermoscopy – Part II. Nonmelanocytic skin tumors. J Am Acad Dermatol 63:377–386
43. Delfino M, Argenziano G, Nino M (2004) Dermoscopy for the diagnosis of porokeratosis. J Eur Acad Dermatol Venereol 18:194–195
44. Panasiti V, Rossi M, Curzio M, Bruni F, Calvieri S (2008) Disseminated superficial actinic porokeratosis diagnosed by dermoscopy. Int J Dermatol 47:308–310
45. Zaballos P (2010) Actinic porokeratosis. In: Micali G, Lacarrubba F (eds) Dermatoscopy in clinical practice: beyond pigmented lesions. Informa Healthcare Ltd, London
46. Zalaudek I, Giacomel J, Argenziano G, Hofmann-Wellenhof R, Micantonio T, Di Stefani A, Oliviero M, Rabinovitz H, Soyer HP, Peris K (2006) Dermoscopy of facial nonpigmented actinic keratosis. Br J Dermatol 155:951–956
47. Zalaudek I, Giacomel J, Schmid K, Bondino S, Rosendahl C, Cavicchini S, Tourlaki A, Gasparini S, Bourne P, Keir J, Kittler H, Eibenschutz L, Catricalà C, Argenziano G (2012) Dermatoscopy of facial actinic keratosis, intraepidermal carcinoma, and invasive squamous cell carcinoma: a progression model. J Am Acad Dermatol 66:589–597
48. Hu SC, Chiu HH, Chen GS, Ke CL, Cheng ST (2008) Dermoscopy as a diagnostic and follow-up tool for pigmented Bowen's disease on acral region. Dermatol Surg 34:1248–1253
49. Bugatti L, Filosa G, De Angelis R (2007) The specific dermoscopical criteria of Bowen's disease. J Eur Acad Dermatol Venereol 21:700–701
50. Zalaudek I, Di Stefani A, Argenziano G (2006) The specific dermoscopic criteria of Bowen's disease. J Eur Acad Dermatol Venereol 20:361–362
51. Bugatti L, Filosa G, Filosa A (2010) Bowen's disease. In: Micali G, Lacarrubba F (eds) Dermatoscopy in clinical practice: beyond pigmented lesions. Informa Healthcare Ltd, London
52. Hu SC, Ke CL, Lee CH, Wu CS, Chen GS, Cheng ST (2009) Dermoscopy of Kaposi's sarcoma: areas exhibiting the multicoloured 'rainbow pattern'. J Eur Acad Dermatol Venereol 23:1128–1132
53. Vázquez-López F, García-García B, Rajadhyaksha M, Marghoob AA (2009) Dermoscopic rainbow pattern in non-Kaposi sarcoma lesions. Br J Dermatol 161:474–475
54. Micali G, Lacarrubba F, Lo GG (2000) Scraping versus videodermatoscopy for the diagnosis of scabies: a comparative study (Letter). Acta Derm Venereol 79:396
55. Lacarrubba F, Musumeci ML, Caltabiano R, Impallomeni R, West DP, Micali G (2001) High-magnification videodermatoscopy: a new noninvasive diagnostic tool for scabies in children. Pediatr Dermatol 18:439–441
56. Micali G, Lacarrubba F, Tedeschi A (2004) Videodermatoscopy enhances the ability to monitor efficacy of scabies treatment and allows optimal timing of drug application. J Eur Acad Dermatol Venereol 18:153–154
57. Zalaudek I, Argenziano G, Di Stefani A, Ferrara G, Marghoob AA, Hofmann-Wellenhof R, Soyer HP, Braun R, Kerl H (2006) Dermoscopy in general dermatology. Dermatology 212:7–18
58. Zalaudek I, Giacomel J, Cabo H, Di Stefani A, Ferrara G, Hofmann-Wellenhof R, Malvehy J, Puig S, Stolz W, Argenziano G (2008) Entodermoscopy: a new tool for diagnosing skin infections and infestations. Dermatology 216:14–23
59. Di Stefani A, Hofmann-Wellenhof R, Zalaudek I (2006) Dermoscopy for diagnosis and treatment monitoring of pediculosis capitis. J Am Acad Dermatol 54:909–911
60. Lacarrubba F, Nardone B, Milani M, Botta G, Micali G (2006) Head lice: ex vivo videodermatoscopy evaluation of the pediculocidal activity of two different topical products. G Ital Dermatol Venereol 141:233–235
61. Bauer J, Forschner A, Garbe C, Röcken M (2005) Variability of dermoscopic features of tungiasis. Arch Dermatol 141:643–644

62. Di Stefani A, Rudolph CM, Hofmann-Wellenhof R, Müllegger RR (2005) An additional dermoscopic feature of tungiasis. Arch Dermatol 141:1045–1046
63. Cabrera R, Daza F (2009) Dermoscopy in the diagnosis of tungiasis. Br J Dermatol 160:1136–1137
64. Moscarella E, Bakos R, Argenziano G (2010) Tungiasis. In: Micali G, Lacarrubba F (eds) Dermatoscopy in clinical practice: beyond pigmented lesions. Informa Healthcare Ltd, London
65. Llambrich A, Zaballos P, Terrasa F, Torne I, Puig S, Malvehy J (2009) Dermoscopy of cutaneous leishmaniasis. Br J Dermatol 160:756–761
66. Abraham LS, Azulay-Abulafia L, Aguiar Dde P, Torres F, Argenziano G (2011) Dermoscopy features for the diagnosis of furuncular myiasis. An Bras Dermatol 86:160–162
67. Segal R, Mimouni D, Feuerman H, Pagovitz O, David M (2010) Dermoscopy as a diagnostic tool in demodicidosis. Int J Dermatol 49:1018–1023
68. Morales A, Puig S, Malvehy J, Zaballos P (2005) Dermoscopy of molluscum contagiosum. Arch Dermatol 141:1644
69. Zaballos P (2010) Molluscum contagiosum. In: Micali G, Lacarrubba F (eds) Dermatoscopy in clinical practice: beyond pigmented lesions. Informa Healthcare Ltd, London
70. Bae JM, Kang H, Kim HO, Park YM (2009) Differential diagnosis of plantar wart from corn, callus and healed wart with the aid of dermoscopy. Br J Dermatol 160:220–222
71. Lee DY, Park JH, Lee JH, Yang JM, Lee ES (2009) The use of dermoscopy for the diagnosis of plantar wart. J Eur Acad Dermatol Venereol 23:726–727
72. Dong H, Shu D, Campbell TM, Frühauf J, Soyer P, Hofmann-Wellenhof R (2011) Dermatoscopy of genital warts. J Am Acad Dermatol 64:859–864
73. Micali G, Lacarrubba F (2011) Augmented diagnostic capability using videodermatoscopy on selected infectious and non-infectious penile growths. Int J Dermatol 50:1501–1505
74. Lacarrubba F, Dinotta F, Nasca MR, Micali G (2012) Enhanced diagnosis of genital warts with videodermatoscopy: histopatologic correlation. G Ital Dermatol Venereol 147:215–216
75. Kim SH, Seo SH, Ko HC, Kwon KS, Kim MB (2009) The use of dermatoscopy to differentiate vestibular papillae, a normal variant of the female external genitalia, from condyloma acuminata. J Am Acad Dermatol 60:353–355
76. Ozeki M, Saito R, Tanaka M (2008) Dermoscopic features of pearly penile papules. Dermatology 217:21–22
77. Xavier MH, Ribeiro LH, Duarte H, Saraça G, Souza AC (2008) Dermatoscopy in the diagnosis of tinea nigra. Dermatol Online J 15:14–15
78. Brasiello M, Zalaudek I, Ferrara G, Gourhant JY, Capoluongo P, Roma P, Argenziano G (2009) Lupus vulgaris: a new look at an old symptom-the lupoma observed with dermoscopy. Dermatology 218: 172–174
79. Micali G, Lacarrubba F, Musumeci ML, Massimino D, Nasca MR (2010) Cutaneous vascular patterns in psoriasis. Int J Dermatol 49:249–256
80. Micali G, Nardone B, Scuderi A, Lacarrubba F (2008) Videodermatoscopy enhances the diagnostic capability of palmar and/or plantar psoriasis. Am J Clin Dermatol 9:119–122
81. Lacarrubba F, Nasca MR, Micali G (2009) Videodermatoscopy enhances diagnostic capability in psoriatic balanitis. J Am Acad Dermatol 61:1084–1086
82. Rosina P, Zamperetti MR, Giovannini A, Girolomoni G (2007) Videocapillaroscopy in the differential diagnosis between psoriasis and seborrheic dermatitis of the scalp. Dermatology 214:21–24
83. Zalaudek I, Argenziano G (2006) Dermoscopy subpatterns of inflammatory skin disorders. Arch Dermatol 142:808
84. Vázquez-López F (2010) Lichen ruber planus. In: Micali G, Lacarrubba F (eds) Dermatoscopy in clinical practice: beyond pigmented lesions. Informa Healthcare Ltd, London
85. Vázquez-López F (2010) Urticaria and urticarial vasculitis. In: Micali G, Lacarrubba F (eds) Dermatoscopy in clinical practice: beyond pigmented lesions. Informa Healthcare Ltd, London
86. Rosina P (2010) Rosacea. In: Micali G, Lacarrubba F (eds) Dermatoscopy in clinical practice: beyond pigmented lesions. Informa Healthcare Ltd, London
87. Lacarrubba F, Micali G (2010) Dermoscopy of pityriasis lichenoides et varioliformis acuta. Arch Dermatol 146:1322
88. Braun RP, Baran R, Le Gal FA, Dalle S, Ronger S, Pandolfi R, Gaide O, French LE, Laugier P, Saurat JH, Marghoob AA, Thomas L (2007) Diagnosis and management of nail pigmentations. J Am Acad Dermatol 56:835–847
89. Tosti A, Piraccini BM, Cadore de Farias D (2010) Nail diseases. In: Micali G, Lacarrubba F (eds) Dermatoscopy in clinical practice: beyond pigmented lesions. Informa Healthcare Ltd, London
90. Iorizzo M, Dahdah M, Vincenti C, Tosti A (2008) Videodermoscopy of the hyponychium in nail bed psoriasis. J Am Acad Dermatol 58:714–715
91. Hirata SH, Yamada S, Almeida FA, Enokihara MY, Rosa IP, Enokihara MM, Michalany NS (2006) Dermoscopic examination of the nail bed and matrix. Int J Dermatol 45:28–30
92. Tosti A, Argenziano G (2002) Dermoscopy allows better management of nail pigmentation. Arch Dermatol 138:1327–1333
93. Piraccini BM, Antonucci A, Rech G, Starace M, Misciali C, Tosti A (2007) Onychomatricoma: first description in a child. Pediatr Dermatol 24:46–48
94. Maehara Lde S, Ohe EM, Enokihara MY, Michalany NS, Yamada S, Hirata SH (2010) Diagnosis of glomus tumor by nail bed and matrix dermoscopy. An Bras Dermatol 85:236–238

95. Vázquez-López F, Manjón-Haces JA, Vázquez-López AC, Pérez-Oliva N (2003) The hand-held dermatoscope improves the clinical evaluation of port-wine stains. J Am Acad Dermatol 48:984–985
96. Eubanks LE, McBurney EI (2001) Videomicroscopy of port-wine stains: correlation of location and depth of lesion. J Am Acad Dermatol 44:948–951
97. Procaccini EM, Argenziano G, Staibano S, Ferrara G, Monfrecola G (2001) Epiluminescence microscopy for port-wine stains: pretreatment evaluation. Dermatology 203:329–332
98. Sevila A, Nagore E, Botella-Estrada R, Sanmartin O, Requena C, Serra-Guillen C, Guillen C (2004) Videomicroscopy of venular malformations (port-wine stain type): prediction of response to pulsed dye laser. Pediatr Dermatol 21:589–596
99. Oiso N, Kimura M, Kawara S, Kawada A (2011) Clinical, dermoscopic, and histopathologic features in a case of infantile hemangioma without proliferation. Pediatr Dermatol 28:66–68
100. Zaballos P, Puig S, Malvehy J (2004) Dermoscopy of pigmented purpuric dermatoses (lichen aureus): a useful tool for clinical diagnosis. Arch Dermatol 140:1290–1291

Visioscan-Driven ULEV Method

2

Gérald E. Piérard, Claudine Piérard-Franchimont, and Sébastien Piérard

2.1 Introduction

Melanocytes and their melanins govern the phototype-related color of the skin. Indeed, the color palette of the skin largely depends on the molecular nature and amount of melanins (eumelanin and pheomelanin) and on the size, shape, and distribution of melanosomes produced by melanocytes and transferred into keratinocytes. Such combinations define what could be called the individual melanotype.

The epidermal melanin unit refers to a microscopic functional entity composed of one single melanocyte and its adjacent keratinocytes into which the melanosomes are transferred. Chronic ultraviolet (UV) light exposures represent positive stimulatory signals to the epidermal melanin units. In such instance, both the active melanocytes are increased in number, and each individual melanocyte is stressed to produce more melanins. In addition, melanosome transfer from melanocytes to adjacent keratinocytes is boosted through the intervention of the protease-activated receptor 2 [1].

G.E. Piérard, MD, PhD (✉)
C. Piérard-Franchimont
Laboratory of Skin Bioengineering and Imaging, Department of Dermatopathology, University Hospital of Liège, CHU Sart Tilman, Liège B-4000, Belgium
e-mail: gerald.pierard@ulg.ac.be

S. Piérard
Telecommunications and Imaging Laboratory, INTELSIG, Montefiore Institute, University of Liège, Liège, Belgium

According to the native individual melanotype, age, and cumulative UV exposures, the skin commonly develops a discrete to severe mottled appearance. Freckles in youths and solar lentigines in older individuals are typical clinical expressions of such events. The clinical aspects are due to an increase in the keratinocyte melanin content (melanotic hypermelanosis) associated or not with melanocytic hyperplasia. The resulting mottled pigmentation is an early key feature of photoaging.

2.2 Clinical Recordings

Photography under UVA light, which is largely absorbed by melanin, was a convenient way to highlight any discrete regional change in skin pigmentation [2–7]. Provided that the lightning was kept constant and the camera calibrated beforehand, this technique revealed with confidence pigmentation changes. Using a UVA source of lighting (Wood's light), any skin blemish was thus conveniently assessed by a regular photography system [2–7]. As such, UV photography was used as a diagnostic tool, but it was rarely employed for measuring the intensity of pigmentation. Indeed, the latter application was not satisfactory because casual equipments generated shadows focally superposed to the skin pigmentation. In order to thwart to this drawback, a CCD camera equipped with an internal UV-emitting unit (Visioscan® VC98, C+K electronic, Cologne, Germany) was designed.

E. Berardesca et al. (eds.), *Non Invasive Diagnostic Techniques in Clinical Dermatology*,
DOI 10.1007/978-3-642-32109-2_2, © Springer Berlin Heidelberg 2014

The Visioscan® VC98 sensor chip is closely applied to the skin surface in order to avoid shadows. The uniform illumination of the skin brings out a sharp video picture of a 6×8 mm area of the skin surface. The high resolution of the video capture allows close assessments of the skin. The computer connection of the Visioscan® VC98 is made through an image digitalization unit configuring a 256 gray level picture, where 0 corresponds to black and 256 to white. The camera-based image analysis leads to the so-called UV light-enhanced visualization (ULEV) method [8, 9].

The ULEV picture results from the combined reflectance and absorption of the incident UV light by the skin structures [10]. Specular reflectance is mainly modulated by the stratum corneum (SC) roughness. The SC surface of scaly dermatoses appears whitish. Once the UV light has crossed the epidermis, the residual UV wavelengths reach the fibrous structures of the dermal extracellular matrix which in turn emit fluorescence. Following its way back to the skin surface, the reflected and/or emitted light is captured by the recording CCD camera. During both the way in and way out through the skin, the UV light is possibly absorbed by two main chromophores, namely, melanins and hemoglobin. Melanin absorption exerts a prominent effect by reducing the light received by the CCD camera. Hemoglobin exerts a similar effect when present in abundance inside superficial hemangiomas. By contrast, the regular amount of blood in superficial blood vessels has no significant effect because fluorescent connective tissue fibers are interposed between the vessels and the skin surface.

The Visioface® Quick (C+K electronic) is another tool providing pictures similar to ULEV. The light booth contains 200 white diodes (LED) uniformly illuminating the exposed skin area. The camera (Canon Powershot A640 with a 10.2 megapixel resolution) and light sources are computer-controlled. After applying adequate software filtering of the cyan hue, the image increases the contrast created by melanin [11].

The Visioscan® VC98 contains a camera-based image analysis system allowing the evaluation of the skin surface. The surface evaluation of living skin (SELS) software generates two relevant parameters in the evaluation of skin surface changes. The smoothness parameter (SESM) is proportional to the depth and width of furrows, and the wrinkling parameter (SEW) is proportional to the number and width of furrows.

2.3 Subclinical Melanoderma

The mottled subclinical melanoderma (MSM) is a mosaic pattern of skin pigmentation typically revealed under ULEV examination [8, 9]. This feature is particularly prominent in photosensitive individuals with a light melanotype and pheomelanin-enriched phenotype [8]. Using the Visioscan-driven ULEV method, the increased contrast between the faint almost invisible spotty melanosis and the surrounding skin is the combined result of a greater fluorescence emission than visible light by collagen and the greater UVA and fluorescence absorption by melanin inside the epidermis. The boundary between MSM macules and the skin in close vicinity is sharply delimited.

Since UV light penetrates less deeply into the skin than visible light, it is considered that both the regular UV photography and the ULEV method mainly catch the intraepidermal pigmentation. Thus, melanin located in the dermis is not detected using these methods. Therefore, a clear distinction is made using ULEV between enhanced melanotic hypermelanosis and any dermal melanoderma. Such distinction is useful because the former type is accessible to depigmenting treatments, while the latter type is unresponsive and persists as a melanin tattoo. These two conditions coexist for instance in facial melasma.

Several MSM patterns were identified [8–13]. They are listed in Table 2.1. Spotty perifollicular dots are commonly seen on the scalp and face (Fig. 2.1), occasionally extending to other seborrheic regions [8, 11–14]. This physiologic perifollicular pattern is recognized as early as during adolescence, and it does not seem altered by aging. By contrast, the interfollicular area exhibits a combination of pinpoint lesions, small macules, and globular macules [8]. Such a pattern appears during adult life and is subject to variations with cumulative photoexposures and aging.

Table 2.1 Patterns of subclinical skin mottling

Pinpoint: minute irregularly distributed darker spots
Follicular dots: speckled perifollicular darker rings
Small macules: small interfollicular darker areas
Globular macules: accretive and circinate confluence of smaller macules
Streaky macules: elongated darker areas along wrinkles
Confluent macules: massive darker areas

Fig. 2.1 Spotty perifollicular MSM on the scalp

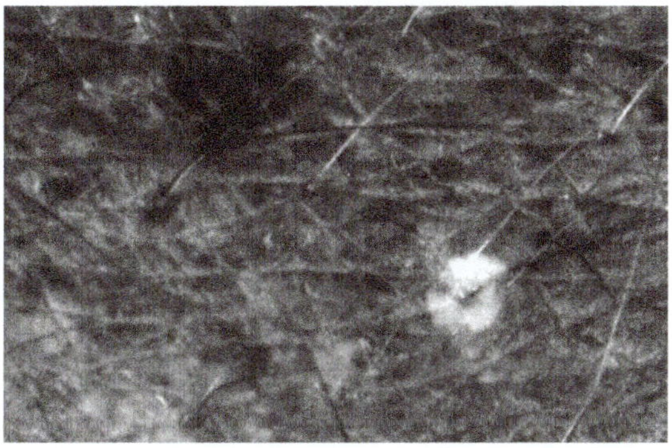

Fig. 2.2 Hypopigmented spot in an area showing darker MSM spots

These discrete hypermelanotic blemishes progressively coalesce, and they appear more prominent on chronically sun-exposed skin. This aspect is regarded as an early sign of photoaging [9, 11, 15, 16]. Still another MSM aspect corresponds to the streaky pattern elongated along wrinkles [8]. This aspect is typically present on the sunny-side slope of facial frown lines [8].

Hypochromic spots associated with MSM (Fig. 2.2) are possibly present in genetic disorders of pigmentation [17] and in ethnic darker skin [18]. They also develop on chronically UV-exposed skin such as in worshipers of recreational sunbeds [19].

2.4 Assessment of Whitening Agents

The ULEV method is employed for assessing the efficacy of cosmetic whitening products [8, 20–24]. Computerized image analysis of the pictures offers objective quantification of the depigmenting effect on the epidermal melanin units. The method was applied for assessing "brown spots" corresponding to solar lentigines or incipient pigmented seborrheic keratoses [20–22]. Such evaluations are more easily handled than the bleaching effect on melasma [22]. Using the ULEV method, the observed effects of whitening agents cannot be distinguished according to the putative biologic effects on tyrosinase, protease-activated receptor-2, or any other step of melanization.

Beyond conventional cosmetic whitening agents, some drugs alter the activity of the epidermal melanin units. Both topical corticosteroids and vitamin D analogues decrease the MSM severity as assessed by the ULEV method [16].

2.5 Skin Surface Microrelief and Scaliness

Any ULEV quantitative assessment of the skin surface microrelief is of interest in evaluating some therapeutic and cosmetic interventions, as well as for the determination of the severity in irritation damages to the skin. Indeed, any measurement where skin is optically monitored using an image digitalization process without any sampling or replica collection represents a technological advance [25–27]. The regular observation of ULEV pictures reveals any hyperkeratotic and scaly aspects of the skin surface (Fig. 2.3). The lesions look whitish contrasting with the gray aspect of the normal-looking skin [25–27].

UVA emitted by the Visioscan® VC98 clearly highlights desquamation in vivo. A semiquantitative assessment is possible, but the data interpretation is

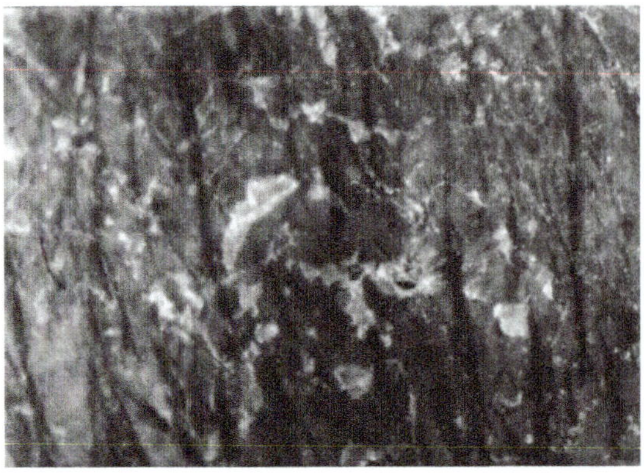

Fig. 2.3 Scaly stratum corneum showing bright white reflectance

not always straightforward. For improving this assessment using the Visioscan® VC98, sampling the SC can be performed using transparent sticky Corneofix® F20 tapes (C+K electronic). After slightly pressing the sampling tape onto the skin surface, the superficial corneocytes adhere to the tape and are harvested for the assessment of their numbers and sizes. As the light is absorbed differently due to the different thickness of corneocyte clumps, they appear as darker pixels in the image. A quantitative assessment is performed looking at the gray level distribution in the histogram. The regular Visioscan® software distinguishes five scaliness levels. The drawback of this automatic cutting is that the shutter does not properly recognize the background in some images. In these instances, the observer can set the shutter manually by comparing the original and the segmented images with a 0.1 % accuracy. Under ULEV examination, dandruff appears as small white objects dispersed along hair shafts [28–30]. Due to the high contrast with hair, a quantitative assessment is made possible without any specific sampling procedure.

The SELS method represents another quantitative assessment of the skin roughness [31].

2.6 Comedogenesis

At the hair follicle openings at the skin surface, microcomedones and keratin-filled funnel-like acroinfundibula are easily identified using ULEV examination. The method proved to be useful in the assessment of comedogenic and comedolytic compounds. In addition, acne physiopathology is highlighted by this method [32].

Cancer patients under targeted chemotherapy to the epidermal growth factor receptor (EGFR) frequently suffer from unusual skin adverse events. The Visioscan® camera revealed specular light reflectance at the site of follicular plugging [33]. The interfollicular stratum corneum showed occasional focal hyperkeratosis. These features increased in severity with the duration of the EGFR inhibitor treatment, indicating follicular involvement as an early adverse event of the therapy. EGFR inhibitor-induced kerosis (follicular hyperkeratosis) is likely responsible for acneiform reactions.

2.7 Sebum Excretion

Sebum flow dynamics can be assessed using lipid-sensitive films [34]. The assessment benefits from image analysis of the sebum-enriched spots. The Visioscan® VC98 camera is conveniently used for that purpose [32]. The camera is covered by the opaque microporous lipid-sensitive Sebufix® F16 foil (C+K electronic) before application onto the skin. After about 30 s, the mean instant sebum follicular output (SFO) corresponding onto the area of the transparent spots of lipid droplets is assessed using computerized image analysis. The face of the Sebufix® F 16 in contact with the skin is glue-free. Thus, sebum fills in the micropores of the sebum-sensitive foil without any restriction (Fig. 2.4). This leads to a short measuring time after which the foil is ready to be evaluated. This requirement is important in order to avoid any artifactual occlusion effect. Indeed, occlusion leads to the stratum corneum swelling and to the increased skin temperature influencing the sebum flow.

The more oily the skin, the shorter collecting time is necessary. Assessments performed any time after cleansing the skin is thus possible. The lipid-sensitive film is lightly present but constantly on the skin leading to reproducible measurements. The special foil typically measures the sebum output from the follicular reservoir.

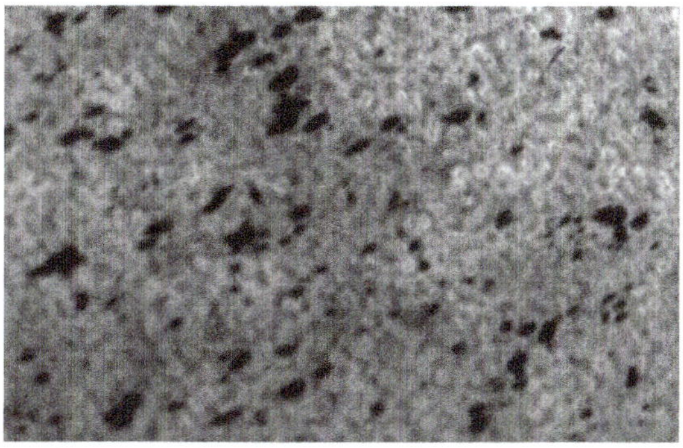

Fig. 2.4 Sebum dots on a lipid-sensitive foil

With this procedure, the sebum output at the skin surface can be monitored live in real time.

2.8 Hair Graying

The loss of melanin content in the hair shaft is a natural manifestation of aging leading to hair graying. Such phenomenon can be observed as early as 20 years of age in Caucasians and about 30 years in Africans; it has been reported that on average, in a cohort of Caucasians, 50 % of people had at least 50 % gray hair when reaching 50 years old [35]. The biologic processes underlying hair graying remain unclear [36]. In addition, the process of hair graying is difficult to assess and quantify in vivo. The ULEV method highlights this physiologic phenomenon by enhancing the contrast between graying hair appearing bright white and the other hair shafts.

2.9 Trichobacteriosis

Some bacterial species tend to clump along hair shafts, particularly in the axillae. They are embedded in a biofilm. This condition corresponds to trichobacteriosis formerly called trichomycosis. The ULEV method is a convenient way to observe these structures appearing as bright white sheaths encasing the base of the hair shafts. This aspect is probably related to the fluorescence emitted by trichobacteriosis under Wood's light. A similar phenomenon can occur at the site of other fluorescent lesions such as erythrasma and pityriasis (tinea) versicolor.

2.10 Risk Assessment of Skin Cancers

A correlation was found between MSM severity and the risk for developing actinic keratoses and basal cell carcinomas on facial skin and the scalp [37–42]. A peculiar aspect corresponds to focal depigmentation suggesting the destruction of some epidermal melanin units. Such finding could help identifying in early adult life subjects at risk of skin cancers.

2.11 Xenobiotic Deposits onto the Skin

Some exogenous compounds adhering to the stratum corneum are conveniently observed using the Visioscan® VC98. For instance, the antiperspirant aluminum hydroxychloride shows peculiar patterns of deposits at the skin surface [43].

Conclusion

UVA photography and its more recent development using a CCD camera equipped with an internal UVA-emitting unit are useful by different aspects. The MSM revealed by the ULEV method is possibly the earliest clinical manifestation of photoaging. In addition, the same method highlights scaliness and desquamation as well as a series of other specific conditions at the skin surface.

Acknowledgments This work was supported by a grant from the "Fonds d'Investissement de la Recherche Scientifique" of the University Hospital of Liège. No other sources of funding were used to assist in the preparation of this manuscript. The authors have no conflicts of interest that are directly relevant to the content of this review. The authors appreciate the excellent secretarial assistance of Mrs. Ida Leclercq and Marie Pugliese.

References

1. Seiberg M, Paine C, Sharlow E et al (2000) The protease-activated receptor 2 regulates pigmentation via keratinocytes-melanocyte interactions. Exp Cell Res 254:25–32
2. Gilchrest BA, Fitzpatrick TB, Anderson RR, Parrish JA (1977) Localization of melanin pigmentation in the skin with Wood's lamp. Br J Dermatol 96:245–248

3. Kikuchi I, Idermori M, Uchimura H, Inoue S (1983) Reflection ultraviolet photography as surface photography of the skin. J Dermatol 10:551–555
4. Masuda Y, Takei K, Mizugaki M (1996) Quantification of the face brown spots and freckles with image analysis method. J Soc Cosmet Chem Jpn 28:147
5. Arai S (1997) Analysis of pigmentation of human skin (UV-light images). In: Wilhem KP, Elsner P, Berardesca Maibach HI (eds) Bioengineering of the skin. Skin surface imaging and analysis. CRC Press, Boca Raton, pp 85–94
6. Fulton JE (1997) Utilizing the ultraviolet (UV detect) camera to enhance the appearance of photodamage and other skin conditions. Dermatol Surg 23:163–169
7. Kollias N, Gillies R, Cohen-Goihman C, Phillips SB, Muccini JA, Stiller M, Drake LA (1997) Fluorescence photography in the evaluation of hyperpigmentation in photodamaged skin. J Am Acad Dermatol 36:226–230
8. Hermanns JF, Petit L, Piérard-Franchimont C et al (2000) Unravelling the patterns of subclinical phaeomelanin-enriched facial hyperpigmentation. Effect of depigmenting agents. Dermatology 201:118–122
9. Petit L, Fougouang L, Uhoda I et al (2003) Regional variability in mottled photo-induced melanoderma in the elderly. Exp Gerontol 38:327–331
10. Szepetiuk G, Piérard S, Piérard-Franchimont C, Caucanas M, Quatresooz P, Piérard GE (2011) Recent trends in specular light reflectance beyond clinical fluorescence diagnosis. Eur J Dermatol 21:157–161
11. Piérard GE, Piérard-Franchimont C, Quatresooz P (2012) Field melanin mapping of the hairless scalp. Skin Res Technol 18(4):431–435
12. Petit L, Piérard-Franchimont C, Saint Léger D et al (2002) Subclinical speckled perifollicular melanosis of the scalp. Eur J Dermatol 12:565–568
13. Xhauflaire-Uhoda E, Hermanns JF, Piérard-Franchimont C et al (2006) Highlighting the rim of perifollicular epidermal unit. Eur J Dermatol 16:225–229
14. Uhoda E, Piérard-Franchimont C, Petit L et al (2005) The conundrum of skin pores in dermoscosmetology. Dermatology 210:3–7
15. Barel AO, Alewaeters K, Clarys P (1999) Optical imaging using UV light for the determination of photoageing. Skin Res Technol 5:126
16. Piérard-Franchimont C, Paquet P, Quatresooz P et al (2007) Smoothing the mosaic subclinical melanoderma by calcipotriol. J Eur Acad Dermatol Venereol 21:657–661
17. Devillers C, Quatresooz P, Hermanns-Lê T et al (2011) Hypomelanosis of Ito: pigmentary mosaicism with immature melanosomes in keratinocytes. Int J Dermatol 50:1234–1239
18. Hermanns JF, Hermanns-Lê T, Piérard GE (2007) Faint innate hypomelanotic spotting in black skin. Eur J Dermatol 17:353–354
19. Piérard-Franchimont C, Piérard GE, Quatresooz P (2012) The skin ivory spot. Predictive sign for skin-field carcinogenesis. Int J Environ Res Public Health 9:362–369
20. Hermanns JF, Petit L, Piérard-Franchimont C et al (2002) Assessment of topical hypopigmenting agents on solar lentigines of Asian women. Dermatology 204:281–286
21. Petit L, Piérard GE (2003) Analytic quantification of solar lentigines lightening by a 2% hydroquinone-cyclodextrin formulation. J Eur Acad Dermatol Venereol 17:546–549
22. Thirion L, Piérard-Franchimont C, Piérard GE (2006) Whitening effect of a dermocosmetic formulation. A randomized double-blind controlled study on melasma. Int J Cosmet Sci 28:263–267
23. Piérard-Franchimont C, Henry F, Quatresooz P, Vroome V, Piérard GE (2008) Analytic quantification of the bleaching effect of a 4-hydroxyanisole-tretinoin combination on actinic lentigines. J Drugs Dermatol 7:873–878
24. Piérard GE, Seite S, Rougier A, Quatresooz P (2011) Analytic assessment under ultraviolet light of actinic lentigines under bleaching treatment. J Cosmet Dermatol 10:102–107
25. Pena Ferreira R, Costa P, Bahia F (2003) Visioscan VC 98 application : a comparison study between coarse and smooth surface. Skin Res Technol 9:204–205
26. Uhoda E, Piérard-Franchimont C, Petit L et al (2003) Skin weathering and ashiness in black Africans. Eur J Dermatol 13:574–578
27. Uhoda E, Petit L, Piérard-Franchimont C et al (2004) Ultraviolet light-enhanced visualization of cutaneous signs of carotene and vitamin A dietary deficiency. Acta Clin Belg 59:97–101
28. Piérard-Franchimont C, Uhoda E, Piérard GE (2005) Quantification of dandruff adherence to hair. Int J Cosmet Sci 27:279–282
29. Piérard-Franchimont C, Xhauflaire-Uhoda E, Piérard GE (2006) Revisiting dandruff. Int J Cosmet Sci 28:311–318
30. Piérard-Franchimont C, Quatresooz P, Piérard GE (2011) Specular light reflectance of flakes in seborrheic dermatitis of the scalp. A pilot study. Clin Exp Dermatol 36:793–796
31. Tronnier H, Wiebuch M, Heinrich U et al (1999) Surface evaluation of living skin. Adv Exp Med Biol 455:507–516
32. Piérard-Franchimont C, Piérard GE (2002) Postmenopausal aging of the sebaceous follicle: a comparison between women receiving hormone replacement therapy or not. Dermatology 204:17–22
33. Piérard GE, Piérard-Franchimont C, Humbert P (2012) Bioimpact of anti-EGFR treatments on the pilosebaceous follicles. Eur J Dermatol 22:54–57
34. Piérard-Franchimont C, Martalo O, Richard A et al (1999) Sebum rheology evaluated by two methods in vivo. Split-face study of the effect of a cosmetic formulation. Eur J Dermatol 9:455–457
35. Keogh EV, Walsh RJ (1965) Rate of graying of human hair. Nature 207:877–878
36. Tobin DJ, Paus R (2001) Graying: gerontology of the hair follicle pigmentary unit. Exp Dermatol 36: 29–54
37. Piérard-Franchimont C, Piérard GE (1998) Heterochromic sun-damaged skin and the risk for skin cancer. Rev Med Liege 53:355–356

38. Hermanns JF, Henry F, Piérard-Franchimont C et al (2001) Analytic quantification of aging of the melanocytic system. Implication in the objective détermination of skin cancer risk. Ann Gerontol 15:233–239
39. Quatresooz P, Petit L, Uhoda I et al (2004) Mosaic subclinical melanoderma : an Achilles heel for UV-related epidermal carcinogenesis. Int J Oncol 25: 1763–1767
40. Quatresooz P, Piérard-Franchimont C, Henry F et al (2006) Subclinical melanoderma and photoinduced cancers. In: Uhoda E, Paye M, Piérard GE (eds) Actualités en ingénierie cutanée, vol 4. Publ. Eska, Paris, pp 67–73
41. Xhauflaire-Uhoda E, Piérard-Franchimont C, Piérard GE, Quatresooz P (2010) Weathering of the hairless scalp: a study using skin capacitance imaging and ultraviolet light-enhanced visualization. Clin Exp Dermatol 35:83–85
42. Quatresooz P, Xhauflaire-Uhoda E, Piérard-Franchimont C, Piérard GE (2009) Epidermal field carcinogenesis in bald-headed. An attempt at finetuning early noninvasive detection. Oncol Rep 21: 1313–1316
43. Mayeux G, Xhauflaire-Uhoda E, Piérard GE (2012) Patterns of aluminum hydroxychloride deposition onto the skin. Skin Res Technol 18:64–68

3 Optical Coherence Tomography

Julia Welzel

3.1 Technique

OCT is an interferometric imaging method. Infrared light is coupled into optical fibers and split into a reference and a probe beam. The probe beam is focused onto the skin, backscattered, and compared with the reference beam. Interference occurs only if both beams match within the coherence length of the light. The coherence length therefore determines the axial resolution, whereas the lateral resolution depends on the focusing objective [13].

For OCT, broadband light sources are used to provide a high axial resolution, because the spectral bandwidth is indirectly proportional to the depth resolution. The wavelength influences the depth penetration of the light and is dependent on the scattering and absorption properties of the sample. A higher wavelength in the range of 1,300 nm provides a deeper detection depth compared to 930 nm, but a lower axial resolution [6].

The sensitivity of an OCT system describes the signal-to-noise ratio, i.e., the weakest signal backscattered that can be distinguished from noise. Sensitivities above 100 dB can be achieved. Analogous to ultrasound, a single-depth profile is called A-scan. By lateral and depth scanning, two-dimensional B-scans and three-dimensional blocks of the tissue can be imaged.

J. Welzel
Department of Dermatology, General Hospital Augsburg, Sauerbruchstrasse 6, D-86179, Augsburg, Germany

The interference can be detected by a time-domain technique, using a scanning reference mirror to detect the reflectivity profile along depth. Another technique to measure interference is the frequency- or spectral-domain OCT, in which the entire spectrum is analyzed with respect to a static reference mirror, resulting in increased sensitivity and image acquisition speed. An inverse Fourier transformation is applied to reconstruct the sample depth scattering profile. In so-called swept-source OCT systems, a narrow linewidth swept laser is used for scanning, with the width of the sweep analogous to the bandwidth of the light source.

In time-domain OCT, the image acquisition time with respect to the scan rate depends on the variation speed of the reference mirror, whereas in frequency-domain OCT, it is a function of the camera speed in the detection spectrometer or of the rate of sweep of the laser. The field of view of the OCT systems depends on the design of scanning and focusing. Lateral image dimensions up to 10 mm have been reached. The depth depends on the light source and detector system and is about 1–2 mm in skin. Three-dimensional images can be reconstructed within some seconds [1, 5, 16].

With these specifications, OCT allows detecting architectural details of the skin but lacks the resolution of single cells.

There are several different systems available for skin imaging.

The OCT system Callisto (Thorlabs AG, Luebeck, Germany) is a frequency-domain OCT system. Working at 930 nm, the axial resolution

E. Berardesca et al. (eds.), *Non Invasive Diagnostic Techniques in Clinical Dermatology*, DOI 10.1007/978-3-642-32109-2_3, © Springer Berlin Heidelberg 2014

is about 5 μm in skin, the depth is about 1.6 mm, and the field of view 10 mm × 10 mm.

Michelson Diagnostics Ltd. (Orpington, UK) provides a swept-source frequency-domain OCT system VivoSight, working at 1,305 nm. Multiple beams focusing at different depth within the probe enhance the lateral resolution to smaller than 7.5 μm. The axial resolution is better than 10 μm. The field of view is 6 × 6 mm, and the depth is about 1.5–2 mm.

AGFA HealthCare (Belgium) constructed a time-domain OCT system SKINTELL, which provides three-dimensional images of 1.8 mm × 1.5 mm and a penetration depth of less than 1 mm. Cross-sectional as well as en face images are taken simultaneously. Continuous focus tracking provides a high lateral resolution of about 3 μm, meaning that the coherence plane and the focal plane are always in the same depth position by moving the optics synchronously with the reference arm. For this system, application of a gel for index matching is recommended.

3.2 Healthy Skin

The layered structure of healthy skin is clearly visible in OCT. At the palms and soles, the thick stratum corneum is the first wavy, signal poor layer (Fig. 3.1). At the other anatomical sites, the

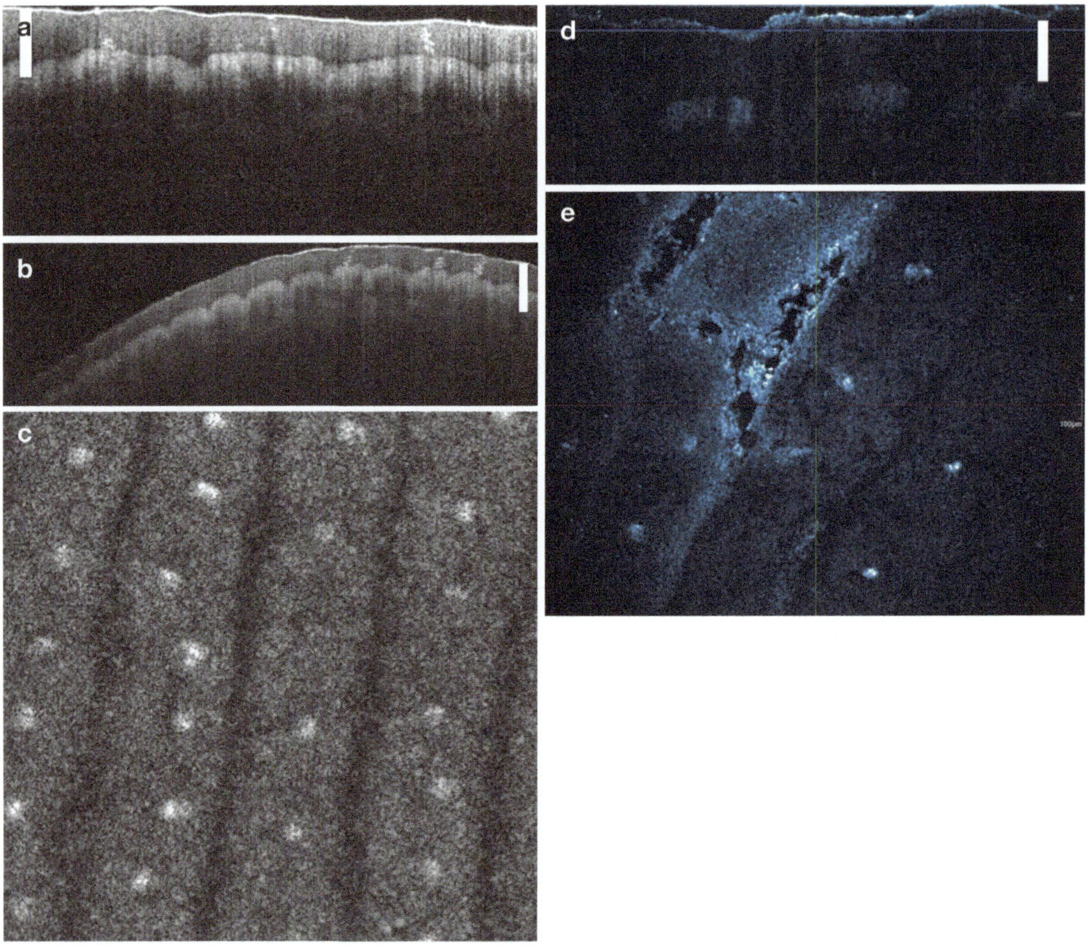

Fig. 3.1 Healthy skin at the fingertip, comparison of the OCT systems: (**a**) Callisto (4 mm × 1.9 mm, bar 0.5 mm), (**b**) VivoSight (6 mm × 2 mm, bar 0.5 mm), (**c**) VivoSight (2 mm × 2 mm horizontal plane), (**d**) SKINTELL (1.8 mm × 0.61 mm cross-sectional image, bar 0.5 mm), and (**e**) SKINTELL (1.8 mm × 1.5 mm horizontal plane). The first wavy layer is the thick stratum corneum with spiral sweat gland ducts inside

stratum corneum is much thinner and therefore only detectable as a splitting of the entrance signal into two lines. The epidermis is slightly more signal intense. The border to the dermis, the so-called dermoepidermal junction, and the stratum papillare of the dermis are signal poor, whereas the reticular dermis is more signal intense. Blood as well as lymphatic vessels are signal-free round or longish structures within the dermis and are compressible. Hair follicles are signal poor, too, and go into the skin at an angle, whereas hairs are signal rich and often cause a signal shadow below (Fig. 3.2) [8, 12, 24, 26, 29–31].

Nail plates appear as well demarcated structures, sometimes with signal-rich parallel layers, in other cases with a granular pattern (Fig. 3.3) [28]. Leukonychia as well as onychomycosis leads to signal-rich artifacts [27].

Due to these differences in scattering, thicknesses of layers and scattering coefficients can be measured by OCT, allowing a quantification of acanthosis, atrophy, or edema [2, 4, 9, 18, 33].

There are several studies on treatment effects like UV irradiation [7, 22], wound healing [3], steroid atrophy, and nail hydration where OCT demonstrates its value for monitoring and quantification of changes.

3.3 Skin Tumors

Epithelial skin tumors are in most cases clearly detectable in OCT. Actinic keratoses show a thickening and stronger scattering of the stratum corneum due to parakeratosis. The epidermis is thickened, but the border to the dermis is still detectable [14] (Fig. 3.4). In contrast, squamous cell carcinoma shows an infiltration into the dermis, leading to a disappearing of the dark line

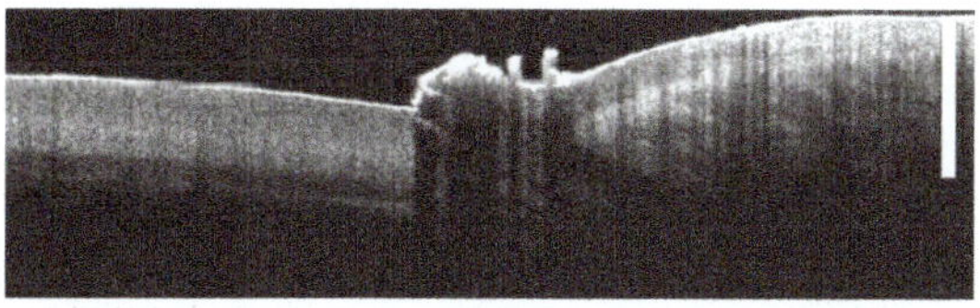

Fig. 3.3 Nail plate and matrix region (Callisto, 6 mm × 1.9 mm, bar 1 mm)

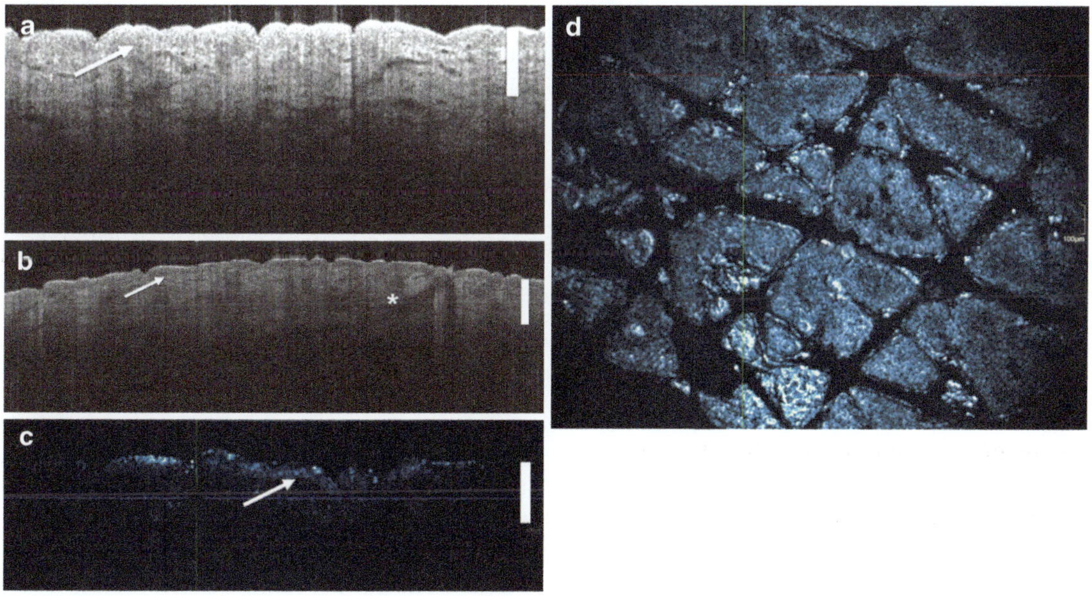

Fig. 3.2 Healthy skin at the forearm, comparison of the OCT systems: (**a**) Callisto (4 mm × 1.9 mm, bar 0.5 mm), (**b**) VivoSight (6 mm × 2 mm, bar 0.5 mm), (**c**) SKINTELL (1.8 mm × 0.6 mm cross-sectional image, bar 0.5 mm), and (**d**) SKINTELL (1.8 mm × 1.5 mm horizontal plane). The first layer is the epidermis. An *arrow* points to the border to the dermis. A *star* marks a hair follicle

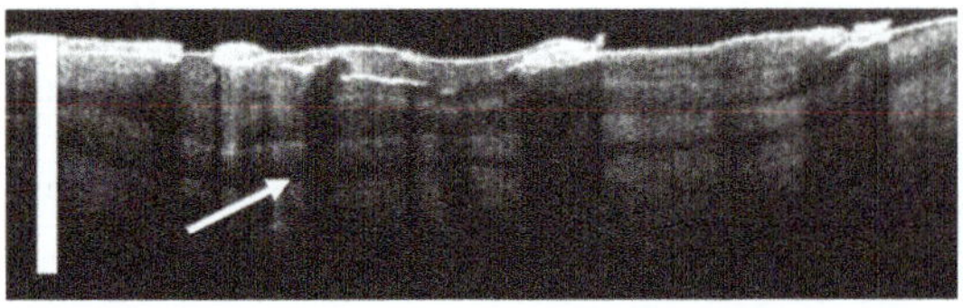

Fig. 3.4 Actinic keratosis at the head (Callisto, 4 mm × 1.9 mm, bar 1 mm). The stratum corneum is thickened with bright parallel lines. The border between epidermis and dermis is still visible (*arrow*)

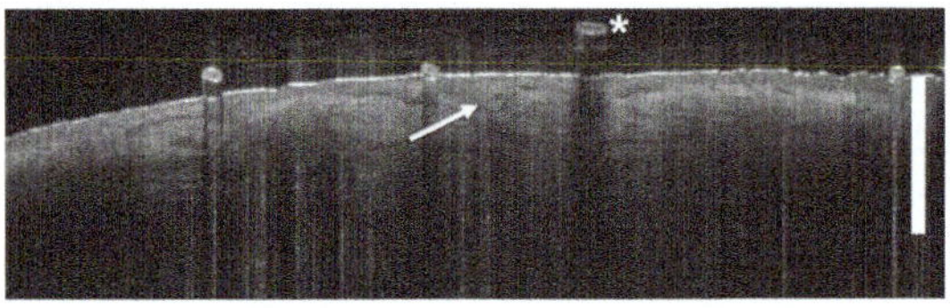

Fig. 3.5 Superficial basal cell carcinoma at the chest (VivoSight, 6 mm × 2 mm, bar 1 mm). An *arrow* marks the tumor proliferates deriving from the epidermis. Hairs (*star*) cause signal shadows

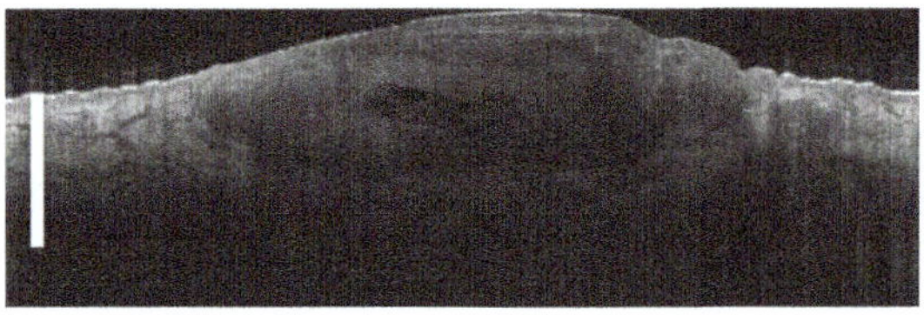

Fig. 3.6 Nodular-cystic basal cell carcinoma at the back (VivoSight, 6 mm × 2 mm, bar 1 mm)

representing the dermoepidermal junction. In cases of severe hyperkeratosis, a signal shadow hides the tumor below. Basal cell carcinomas are characterized by tumor conglomerates with a similar signal intensity compared to the epidermis. Superficial tumor nests derive from the epidermis (Fig. 3.5), whereas deeper parts of the tumor are sharply demarcated from the dermis by a dark line representing the surrounding fibrous stroma (Fig. 3.6). The tumor is accompanied by enlarged blood vessels [11, 19, 20, 23].

Pigmented lesions result in an irregular scattering due to melanin. The dermoepidermal junction often disappears in melanomas because of the infiltrative growth (Fig. 3.7). The junction is still visible in junctional nevi, together with an acanto-papillomatosis, but disappears in compound nevi,

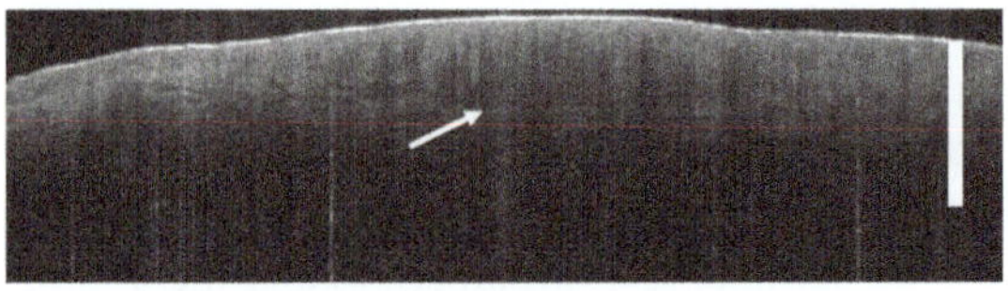

Fig. 3.7 Malignant melanoma at the back (VivoSight, 6 mm × 2 mm, bar 1 mm). An *arrow* points to the lower border of the lesion

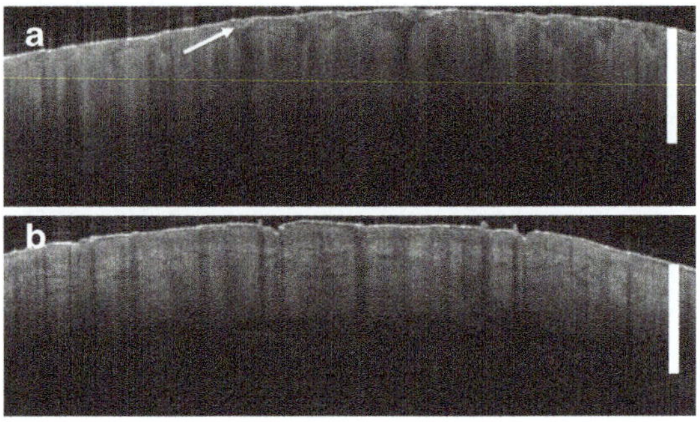

Fig. 3.8 Psoriasis at the dorsum of the middle finger (**a**) compared to healthy skin at the opposite middle finger (**b**). The stratum corneum is thickened and shows bright spots (*arrow*) corresponding to parakeratosis (VivoSight, 6 mm × 2 mm, bar 1 mm)

too. There are no currently understood characteristic or specific features in OCT to distinguish between benign nevi and malignant melanomas [10]. Therefore, OCT is not currently suitable for diagnosing pigmented lesions. In thin melanomas, OCT allows a determination of the tumor thickness, because the scattering inside the tumor differs from the surrounding collagen. Therefore, the lower border is detectable.

Systematic studies on other skin tumors are still lacking.

3.4 Inflammatory Skin Diseases

Psoriasis (Fig. 3.8) as well as eczema shows a thickening of the stratum corneum and the epidermis. Inflammatory infiltration and edema in the dermis lead to a lower scattering. The blood vessels are dilated [21, 32]. In Darier's disease, the acantholytic papules are visible using OCT. In bullous autoimmune diseases like bullous pemphigoid or pemphigus, OCT enables a location of the depth of blistering – intra- or subepidermal [17]. The changes in inflammatory skin diseases can be quantified using OCT by measuring the thickness of layers and the signal attenuation coefficient.

3.5 Other Skin Diseases

Scabies mites are visible in OCT as signal strong structures below the stratum corneum with a burrow behind. In porokeratosis, the cornoid lamellae are typical features in OCT images. In onychomycosis, the nail plate exhibits several white lines and longish structures. These changes are very sensitive for fungal infection, but the specificity is too low for using this method as a native diagnostic tool in onychomycosis [27].

Conclusion

The diagnosis of basal cell carcinoma is the domain of OCT. The technique enables a fast and reliable detection of the tumor together with an assessment of the infiltration depth and the lateral dimensions. The features of superficial basal cell carcinomas in OCT are very typical, allowing differentiation of this entity from clinically similar lesions like Bowen's disease or psoriasis plaques. Especially when treating basal cell carcinomas with nonsurgical procedures like imiquimod or photodynamic therapy, OCT is helpful to avoid repetitive biopsies and to control the efficacy.

In actinic keratoses and squamous cell carcinoma, OCT allows an estimation of infiltrative growth as well as enabling monitoring after treatment.

In inflammatory skin diseases, OCT is a suitable tool for quantification and monitoring over time and under therapy.

OCT has the potential to become widespread as a complimentary tool to histology, dermoscopy, and confocal laser microscopy in general clinical practice. It will become increasingly important, as dermatology is increasingly reliant on nonsurgical techniques. OCT can be combined with other optical techniques like multiphoton tomography or Raman spectroscopy to get both overview and cellular resolution or functional information together with depth resolution [15, 25].

References

1. Alex A, Považay B, Hofer B et al (2010) Multispectral in vivo three-dimensional optical coherence tomography of human skin. J Biomed Opt 15:026025–026015
2. Bechara FG, Gambichler T, Stücker M et al (2004) Histomorphologic correlation with routine histology and optical coherence tomography. Skin Res Technol 10:169–173
3. Cobb MJ, Chen Y, Underwood RA, Usui ML, Olerud J, Li X (2010) Non-invasive assessment of cutaneous wound healing using ultrahigh-resolution optical coherence tomography. J Biomed Opt 11:064002
4. Coßmann M, Welzel J (2006) Evaluation of the atrophogenic potential of different glucocorticoids using optical coherence tomography, 20-MHz ultrasound and profilometry; a double-blind, placebo-controlled trial. Br J Dermatol 155:700–706
5. De Boer JF, Cense B, Park BH, Pierce MC, Tearney GJ, Bouma BE (2003) Improved signal-to-noise ratio in spectral-domain compared to time-domain optical coherence tomography. Opt Lett 28:2067–2069
6. Drexler W (2004) Ultrahigh-resolution optical coherence tomography. J Biomed Opt 9:47–74
7. Gambichler T, Boms S, Stücker M, Moussa G, Kreuter A, Sand M, Sand D, Altmeyer P, Hoffmann K (2005) Acute skin alterations following ultraviolet radiation investigated by optical coherence tomography and histology. Arch Dermatol Res 297:218–225
8. Gambichler T, Moussa G, Sand M, Sand D, Altmeyer P, Hoffmann K (2005) Applications of optical coherence tomography in dermatology. J Dermatol Sci 40:85–94
9. Gambichler T, Matip R, Moussa G, Altmeyer P, Hoffmann K (2006) In vivo data of epidermal thickness evaluated by optical coherence tomography: effects of age, gender, skin type, and anatomic site. J Dermatol Sci 44:145–152
10. Gambichler T, Regeniter P, Bechara FG, Orlikov A, Vasa R, Moussa G, Stücker M, Altmeyer P, Hoffmann K (2007) Characterization of benign and malignant melanocytic skin lesions using optical coherence tomography in vivo. J Am Acad Dermatol 57:629–637
11. Gambichler T, Orlikov A, Vasa R et al (2007) In vivo optical coherence tomography of basal cell carcinoma. J Dermatol Sci 45:167–173
12. Gambichler T, Jaedicke V, Terras S (2011) Optical coherence tomography in dermatology: technical and clinical aspects. Arch Dermatol Res 303:457–473
13. Huang D, Swanson EA, Lin CP, Shuman JS, Stinson WG, Chang W et al (1991) Optical coherence tomography. Science 254:1178–1181

14. Korde VR, Garret TB, Wei X et al (2007) Using optical coherence tomography to evaluate skin sun damage and precancer. Lasers Surg Med 39:687–695
15. König K, Speicher M, Bückle R, Reckfort J, McKenzie G, Welzel J, Köhler MJ, Elsner P, Kaatz M (2009) Clinical optical coherence tomography combined with multiphoton tomography of patients with skin diseases. J Biophotonics 2:389–397
16. Leitgeb R, Hitzenberger C, Fercher A (2003) Performance of fourier domain vs. time domain optical coherence tomography. Opt Express 11:889–894
17. Mogensen M, Morsy HA, Nurnberg BM, Jemec GB (2008) Optical coherence tomography imaging of bullous diseases. J Eur Acad Dermatol Venereol 22: 1458–1464
18. Mogensen M, Morsy HA, Thrane L, Jemec GB (2008) Morphology and epidermal thickness of normal skin imaged by optical coherence tomography. Dermatology 217:14–20
19. Mogensen M, Joergensen TM, Nürnberg BM, Morsy HA, Thomsen JB, Thrane L, Jemec GB (2009) Assessment of optical coherence tomography imaging in the diagnosis of non-melanoma skin cancer and benign lesions versus normal skin: observer-blinded evaluation by dermatologists and pathologists. Dermatol Surg 35:965–972
20. Mogensen M, Nürnberg BM, Forman JL, Thomsen JB, Thrane L, Jemec GBE (2009) In vivo thickness measurement of basal cell carcinoma and actinic keratosis with optical coherence tomography and 20-MHz ultrasound. Br J Dermatol 160:1026–1033
21. Morsy H, Kamp S, Thrane L, Behrendt N, Zayan H, Elmagid EA, Jemec GBE (2010) Optical coherence tomography imaging of psoriasis vulgaris: correlation with histology and disease severity. Arch Dermatol Res 302(2):105–111. doi:10.1007/s00403-009-1000-4
22. Neerken S, Lucassen GW, Bisschop MA et al (2004) Characterization of age-related effects in human skin: a comparative study that applies confocal laser scanning microscopy and optical coherence tomography. J Biomed Opt 9:274–281
23. Olmedo JM, Karen EW, Joseph MS, David LS (2007) Correlation of thickness 758 of basal cell carcinoma by optical coherence tomography in vivo and routine histologic findings: a pilot study. Dermatol Surg 33:421–426
24. Pagnoni A, Knuettel A, Welker P et al (1999) Optical coherence tomography in dermatology. Skin Res Technol 5:83–87
25. Patil CA, Bosschaart N, Keller MD, Van Leeuwen TG, Mahadevan-Jansen A (2008) Combined Raman spectroscopy and optical coherence tomography device for tissue characterization. Opt Lett 33:1135–1137
26. Pierce MC, Strasswimmer J, Hyle Park B, Cense B, de Boer JF (2004) Advances in optical coherence tomography imaging for dermatology. J Invest Dermatol 123:458–463
27. Rothmund G, Sattler E, Kästle R, Fischer C, Haas CJ, Starz H, Welzel J (2013) Confocal laser scanning microscopy as a new valuable tool in the diagnosis of onychomycosis – comparison of six diagnostic methods. Mycoses 56(1):47–55
28. Sattler E, Kaestle R, Rothmund G, Welzel J (2012) Confocal laser scanning microscopy, optical coherence tomography and transonychial water loss for in vivo investigation of nails. Br J Dermatol 166(4):740–746. doi:10.1111/j.1365-2133.2011.10730.x
29. Schmitt AM (2008) Principles and application of optical coherent tomography in dermatology. Dermatology 217:12–13
30. Welzel J, Lankenau E, Birngruber R, Engelhardt R (1997) Optical coherence tomography of the human skin. J Am Acad Dermatol 37:958–963
31. Welzel J (2001) Optical coherence tomography in dermatology : a review. Skin Res Technol 7:1–9
32. Welzel J, Bruhns M, Wolff HH (2003) Optical coherence tomography in contact dermatitis and psoriasis. Arch Dermatol Res 295:50–55
33. Welzel J, Reinhardt C, Lakenau E et al (2004) Changes in function and morphology of normal human skin: evaluation using optical coherence tomography. Br J Dermatol 150:220–225

4 PRIMOS 3D Digital Frame Projection

Sören Jaspers and Christian Benderoth

4.1 Introduction

The introduction of three-dimensional surface scanning by active image triangulation allows for fast, non-invasive in vivo measurements of the skin's surface. This offers significant advantages against classical methods, typically using silicone replica for off-line evaluation. The quantitative determination of skin surface topology, roughness as well as macrostructures like wrinkles or cellulite, is among the most important and probably most-frequently performed investigations in the field of cosmetics, and increasingly in dermatological research as well. A key technology in these applications is phase-measuring fringe projection, as it combines precision with the fast acquisition necessary for in vivo measurement. A wide range of new possibilities results from this, for example, in the field of cosmetic and dermatological research. In the following, we describe the technology, application aspects, results from dermatological research, and latest developments involving portable as well as large-scale scanning devices.

S. Jaspers (✉)
Research and Development, Biophysics Laboratory, Beiersdorf AG, Hamburg, Germany

C. Benderoth
Department of Life Science, GFMesstechnik GmbH, Teltow, Berlin, Germany
e-mail: christian.benderoth@gfm3d.com

4.2 Technical Background

Currently, the PRIMOS system [1] (GFMesstechnik GmbH, Teltow, Germany) is in all probability one of the most widespread in vivo systems for determining skin topology. An important attribute of the PRIMOS system is a sophisticated digital projection device based on DMD technology. This chapter will concentrate on the technical background and the possibilities offered by the various PRIMOS devices currently available. Other instruments based on similar principles include, for example, the FOITS system [2] (Breuckmann GmbH, Meersburg, Germany) or the SkinBio technologies skin profile solution (SkinBio Technologies, Cologne).

4.2.1 Active Image Triangulation

The principle of active triangulation by fringe projection is illustrated in Fig. 4.1. A fringe projector (in the actual device, a DMD™ projector based on a digital micro mirror display from Texas Instruments), casts patterns of parallel stripes onto a shaped object (lenses are omitted in this schematic figure for simplicity). The geometric parameters of the assembly (distance and triangulation angle) cause a certain stripe s to appear at a camera pixel at a coordinate v, directly depending on the distance of the object pixel reflecting the same stripe.

The lateral resolution of the system is simply given by the camera resolution projected on the

DOI 10.1007/978-3-642-32109-2_4, © Springer Berlin Heidelberg 2014

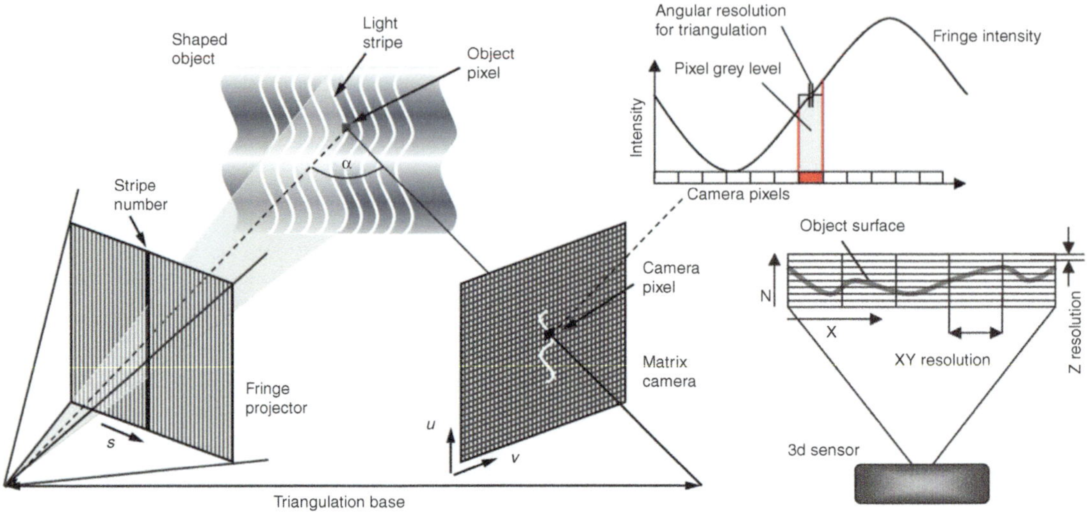

Fig. 4.1 Functional principle of fringe (*stripe*) projection. Fringe intensity and resulting pixel gray levels for fringe projection (*left*), derivation of exact phase values by gray levels (*upper right*), resulting XY and *Z* resolutions within the measuring volume (*lower right*) (Illustration: GFM)

object surface. The distance resolution Δ_Z, as long as classical triangulation system is regarded, is given by a function of the projected pixel sizes of projector and camera and the triangulation angle. Other than the symmetric setup shown in Fig. 4.1, the PRIMOS system has a camera looking perpendicular onto the surface (Fig. 4.2), and the projection is done at an angle β relative to the perpendicular. For this configuration, *Z* resolution can be calculated as $\Delta_Z \approx \Delta_X / \tan(\beta)$ (with a lateral camera resolution Δ_X). For common angles β of 20°…30°, this would return Δ_Z about 2–3 times larger than Δ_X. Very large improvements are achieved, however, using a projected pattern containing well-defined, continuous intensity gradients. In this case, each camera pixel delivers an intensity integral over its projected area on the object, allowing to determine the exact lateral position of the pattern 1–2 orders of magnitude more accurate. *Z* resolutions resulting can surpass lateral resolutions by a factor 10 for a single camera pixel or even up to 50 when integrated over a surface area (Fig. 4.1).

A practical implementation of this principle uses stripe patterns with sine-shaped intensity gradients (often this is referred to as a $\cos^2$ pattern, as this function characterizes a sine wave elevated by 1, with only positive values). As a single pattern of this kind does not result in exploitable gradients in all parts of the surface, several patterns "phase-shifted" by small angles are used, at least four of them in practice.

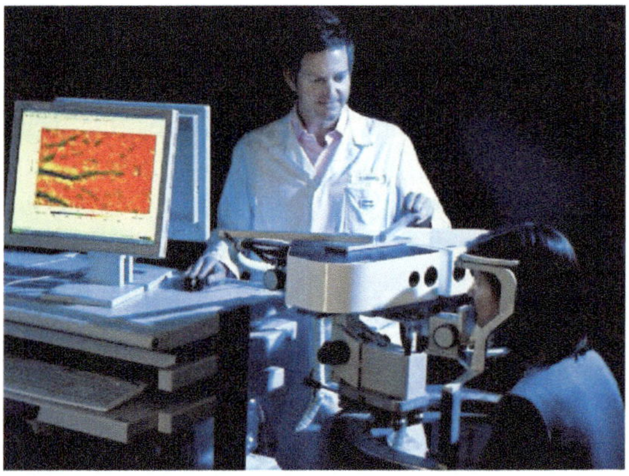

Fig. 4.2 PRIMOS measuring system

Stripe shifts in the camera plane (*v* coordinate) in this case are referred to as "phase shifts," the stripe sequence with its sine wave gradient interpreted as an oscillation and the stripe distance as an angle of 2π. The phase values over the entire surface are referred to as a "phase map" and this essentially contains the height information.

The transformation from the phase map to the desired height map (incurring corrections resulting from the calibration process) is in principle straightforward and is specified by the system's triangulation relations. For instance, transformations as known from two-beam interferometry [3] can directly be applied. Distance (*Z*) values derived from a phase map are, however, relative, as phase values may complete several 2π cycles within the surface area. As long the surface is known not to contain gaps or steps, these parts can easily be assembled for a continuous – relative – height profile. For many life-science applications, e.g., skin roughness measurement or face or body shape recording, this can usually be considered as given, and there is no necessity for absolute distance measurement as well.

Otherwise, sequences of specially encoded binary patterns can be projected in addition to the phase patterns. This allows for a complete registration of any single camera pixel against according projector pixels, hence, an unambiguous acquisition of even entirely fragmented surfaces, at the cost of a longer acquisition time. This option is available with all fringe projection systems but hardly necessary to use in biomedical applications.

In practice, four projected sine patterns are sufficient for such applications. With a projector and camera using the most common video frame rate of 60 Hz, this allows to take a full surface profile in 1/15 s, fast enough for in vivo 3D imaging without extensive precautions. Exploiting the maximum projection frame rate of 180/s. specified, e.g., for the Alligator board by GFM, would even enable a 3D acquisition time of only 1/50 s.

In actual practice, tolerances in the physical parameters of the triangulation system, as well as unavoidable distortions resulting from the lenses, have to be considered in the transformation.

For this reason, calibration algorithms have been developed, using an idealized mathematical model of the system including optics, and deviation parameters corresponding to that model are retrieved by several recordings of test bodies (typically, planes, step, and grid standards). The parameters are calculated for minimum least square errors of both lateral and *Z* linearity.

4.2.2 Digital Micromirror Device (DMD™)

The accuracy of the phase-measuring triangulation approach depends on the correctness of the projected intensity pattern. Digital mirror displays (DMD) allow for a very accurate projection and deliver high light intensities important for low noise figures and tolerance against ambient light. A typical DMD consists of an array of 1,024 × 768 quadratic 16 × 16 μm^2 mirrors that can be individually tilted to plus or minus 10° from the horizontal. The entire mirror array is manufactured from a perfectly pure, monocrystalline silicon wafer by a complex semiconductor manufacturing process of up to 43 process steps. Both the mirror micro system structures including hinge and suspension parts, as well a CMOS driving circuitry in the underlying wafer surface are produced using nothing but masking, etching, and doping techniques. The resulting mirror array is very even and free of irregularities. The silicon surface is turned into a highly reflective mirror by vapor deployed aluminum. Tilting of the individual mirrors is achieved by electrostatic force according to electric surface voltage provided by the CMOS circuitry, which works similar to a static RAM (random access memory) and provides fast writing capabilities for entire display patterns. Basically, the mirrors only allow to switch between full intensity (on) and zero (off). Gray levels are achieved digitally by pulse width modulation (PWM). The typical switching time of 5 μs, compared to the typical duration of a projected image frame of 17 ms (60 Hz frame rate), results in an excellent linearity of the produced gray levels. Altogether, the purely mechanical image formation combined with digital modulation results in a high stability of all parameters, comparable to classical mechanical and optical components. DMD also are among the most durable – if not the most durable – of any display types.

4.2.3 In Vivo Topometry

The optical system allows us to measure an entire area at once, and without any contact to the skin. This constitutes a particular advantage over other methods. Measuring skin or body areas on living persons, however, necessitates a recording process as fast as possible, as the persons will inevitably move and any attempts to avoid this – by mechanical supports, rests, fixations, etc. – normally result in an occlusions of parts of the areas to be recorded. The different PRIMOS setups discussed in the next section are addressing this issue in several ways.

4.2.4 PRIMOS Systems

The PRIMOS systems by GFM implement the aforementioned 3D recording techniques for life-science applications. Various systems with different specifications are available for a wide range of requirements. Typical wrinkle measurement systems have fields of view ranging from 18 × 13 to 40 × 30 mm, with typical vertical resolutions between 2 and 5 μm. Using 4–8 phase shifts according to applications, results in acquisition times of 68–136 ms.

4.3 Illumination

While earlier types of instruments relied on halogen cold-light sources, currently all types have LED (light emitting diodes), contained in the measuring heads themselves. This saves weight, thermal dissipation and power, and eliminates the necessity for occasional lamp replacements.

LEDs are available in different – relatively pure – colors or white tones. For biomedical applications, white light is usually preferred. The recording cameras are black-and-white, as this best suit the requirements of fringe projection measurement. Studies involving skin surface, however, will often profit from a documentation of skin color, in addition to profile. PRIMOS sensors can provide this: they record three additional images with red, green, and blue illumination in addition to the 3D scanning sequences and these are combined to a true color image. This is possible because the projectors provide full color, either by means of a color filter wheel or with the new pico DMD – using three separate LED sources of red, green, and blue, combining into white for the fringe measurements. Color images retained with the separate color method are of high quality. They can be added to the measured 3D data as a surface texture, enabling a realistic on-screen examination.

4.4 Practical Experiences and Quality Aspects of PRIMOS Measurements

Fringe projection technology has meanwhile become a common method for precise quantification of skin surface in the cosmetics industry. A topic of particular importance is the measurement of eye wrinkles. An especially designed support tripod enables a reproducible positioning of the test person. With the recording of PRIMOS measurements in the eye wrinkle area, a number of aspects have to be considered. The quality of measurement achieved is decisive for a correct analysis and quantification of the 3D data. In the following, we will discuss four crucial aspects to be considered during measurements:

1. Achieving a smallest possible height range in the primary images
2. Identification of motion artifacts
3. Identification of mimics differences between images
4. Avoidance of hair artifacts

4.4.1 Achieving a Smallest Possible Height Range in the Primary Image

Anatomic curvature in the eye wrinkle area can lead to large height differences in the unfiltered primary image. This should be adjusted for by an appropriate positioning of the measurement system, which offers the degrees of freedom required for a positioning resulting in a smallest possible height range in the image area (Fig. 4.3). At repetition measurements, the system should be adjusted to approximately the same positioning, as otherwise, deviations in area roughness as well as volume parameters of up to 10 % could occur. Figures 4.4 and 4.5 show a comparison of good and bad primary images. Figures 4.6 and 4.7 show the same images after a flattening of the large area curvature, revealing lots more detail in small-scale height differences that cannot be seen in the primary images (which are spanning a much larger height range by their gray values).

4.4.2 Identification of Motion Artifacts

Deviations similar to the above mentioned can occur during a measurement if the test person moves the head or twinkles with the eyes. Such problems are best identified by parallel fringe patterns they produce in the high-pass filtered height images. Figure 4.8 shows a comparison of images with and without motion artifacts. The fringe patterns resulting are most obvious in the color-coded height images (most relevant are parallel vertical line patterns observed here) (Fig. 4.9).

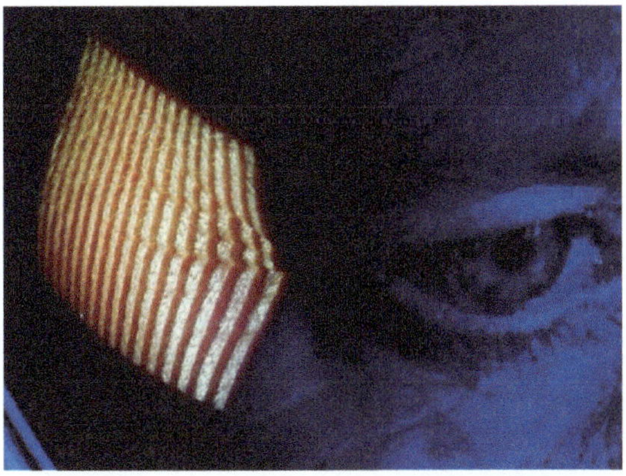

Fig. 4.3 Fringe projection in the eye wrinkle area

4.4.3 Identification of Mimics Differences

Another important aspect for a decent quality measurement is the control of face mimics. Figure 4.10 shows the difference in the distances between wrinkles in the two scans (red arrows).

This difference inevitably causes different formations of the wrinkles and, hence, changed roughness parameters and volumina. This cannot be compensated even by a subsequent matching of the images.

4.4.4 Avoiding Hair Artifacts

For precise measurements, the skin area should be free of hairs. An established best practice is the removal of hairs using an eyebrow trimmer, preceding each measurement. The PRIMOS measuring method also offers the possibility of digital hair removal after the scan. This, however, inevitably also removes a few data (spots covered by individual hairs) and results in a lower sensitivity of the measurement. Figure 4.11 shows the effect of shaving, while Fig. 4.12 shows the effect of digital hair removal.

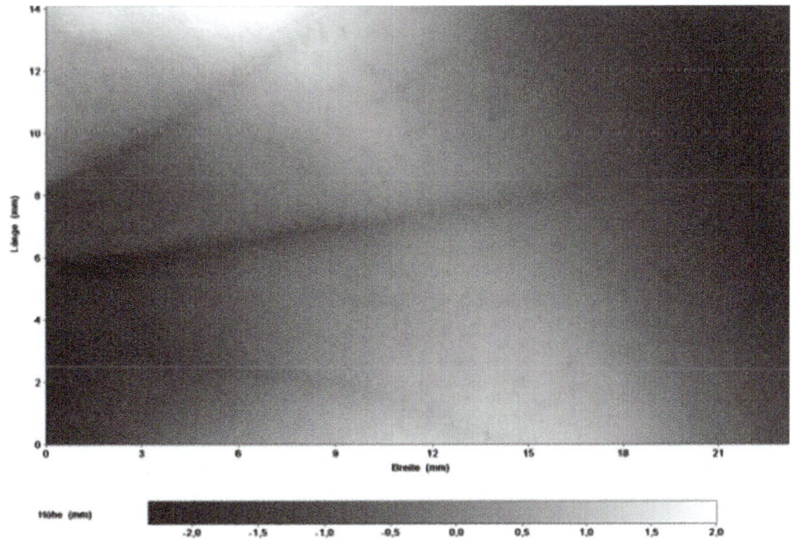

Fig. 4.4 Good primary image (*gray-coded height values*)

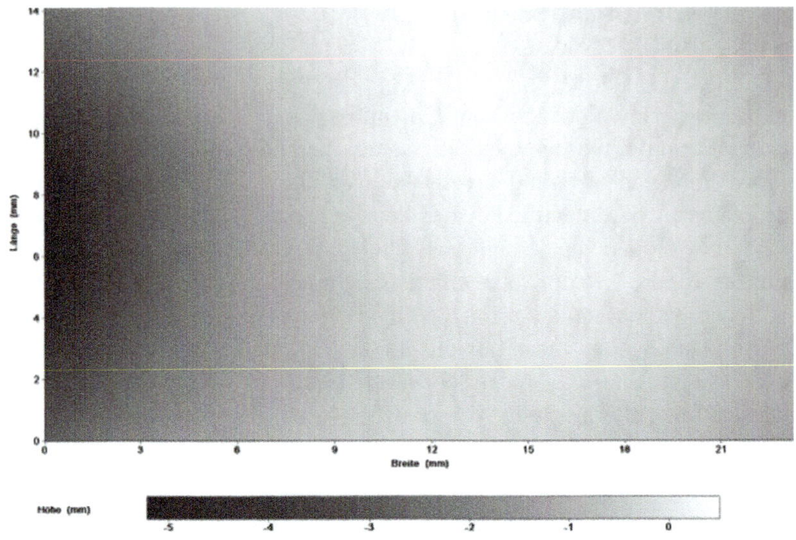

Fig. 4.5 Bad primary image

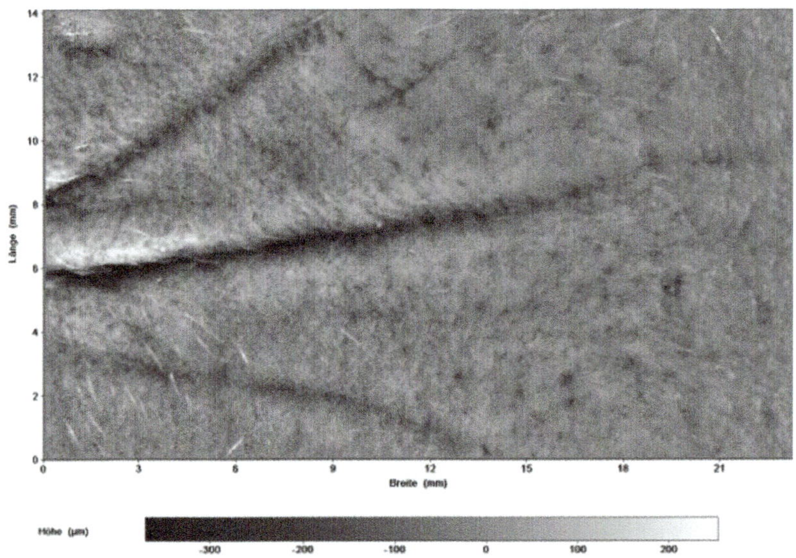

Fig. 4.6 High-pass filtered good primary image

4.5 Handling Aspects

As with any sophisticated measuring device, working with the different PRIMOS systems necessitates a certain amount of accurateness and training. One aspect concerns in choosing a sensible test arrangement. Forcing test persons to unnatural postures inevitably results in muscular tremor – very adverse with measurement of structures in the micron range – and deformations of measured areas. With a relaxed posture, body motion and distortions are greatly reduced. Evidently, a solid stand with a highly reproducible positioning construction and sufficient degrees of freedom is necessary for this. Figure 4.2 shows a typical arrangement for measuring wrinkles in the corner of the eye.

Carrying out series of experiments, involving exact comparisons of even small differences for achieving reliable and reproducible result, necessitates an exact re-identification and readjustment of measured areas, even after longer periods of time.

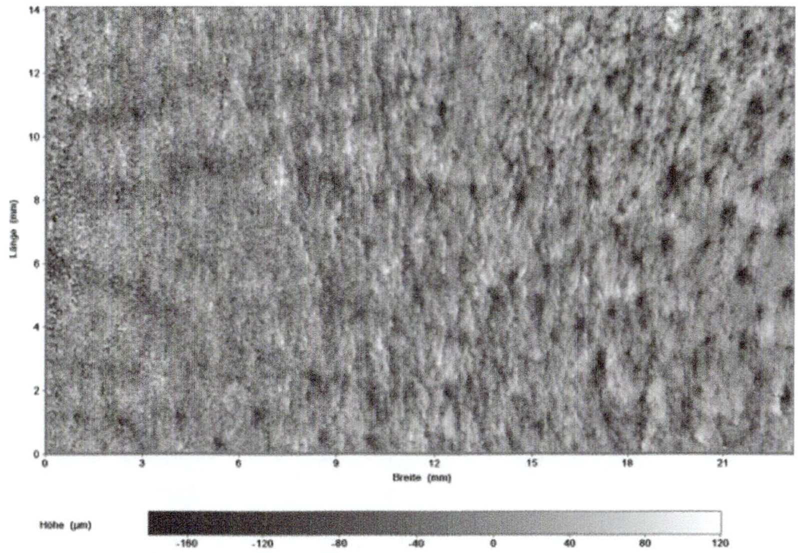

Fig. 4.7 High-pass filtered bad primary image

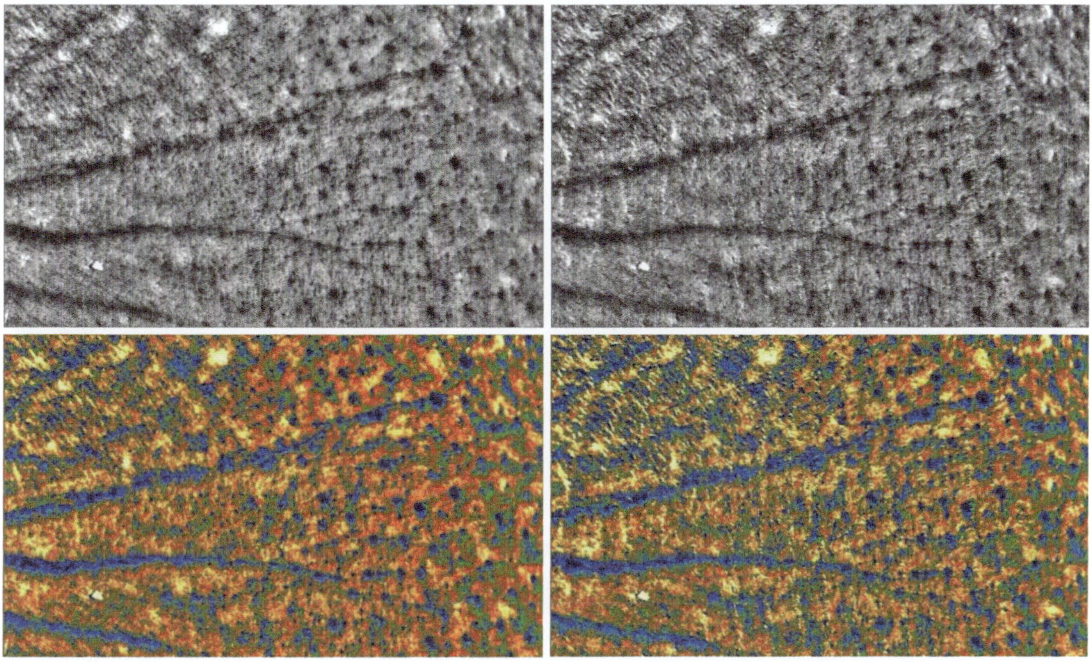

Fig. 4.8 Height images without (*left*) and with motion artifacts (*right*), height values as gray (*top*) and color coded

For this purpose, the PRIMOS system offers a video overlay function that proved to be quite helpful. It works by overlaying the actual camera with a previous scan image and allows for a very accurate matching once the operator has acquired sufficient practice.

An important point worth mentioning is the immediate checking for data quality. There are more parameters to watch with a 3D recording system than there are with a conventional camera. The device allows for an immediate checking for motion artifacts, defocus, under-/overexposure,

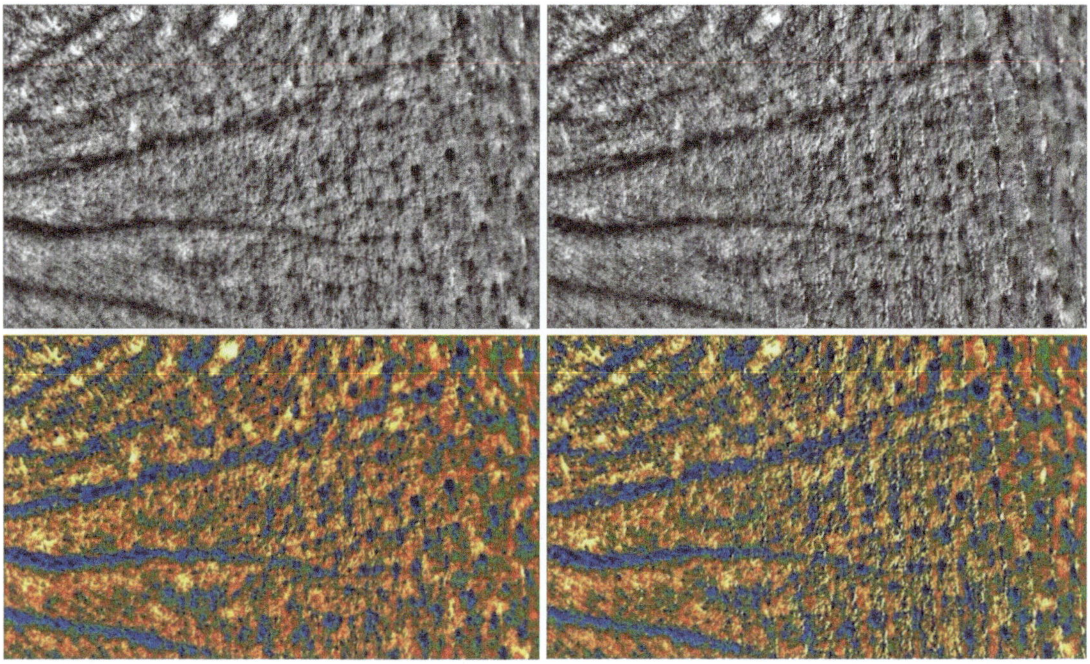

Fig. 4.9 More height images without (*left*) and with motion artifacts (*right*), height values as gray and color coded

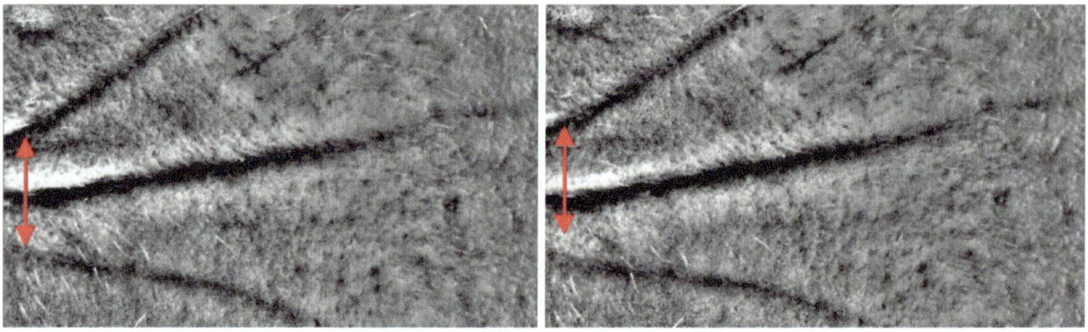

Fig. 4.10 Wrinkle distances changed by mimics

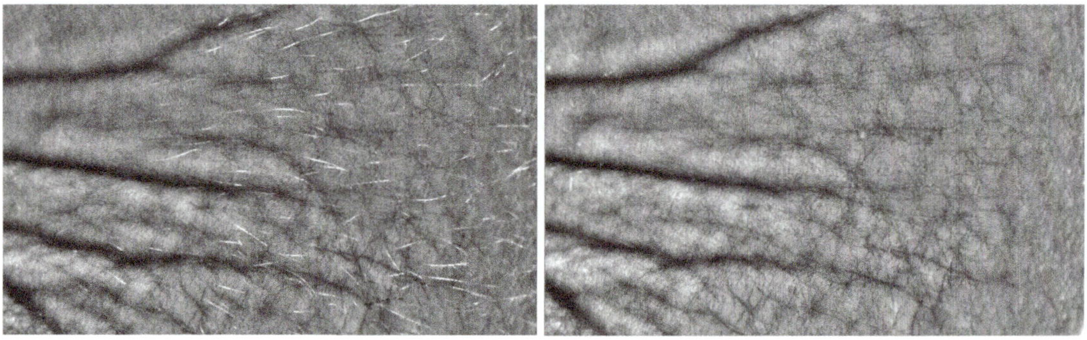

Fig. 4.11 Filtered height image with hairs (*left*) and filtered height image with hairs shaved (*right*)

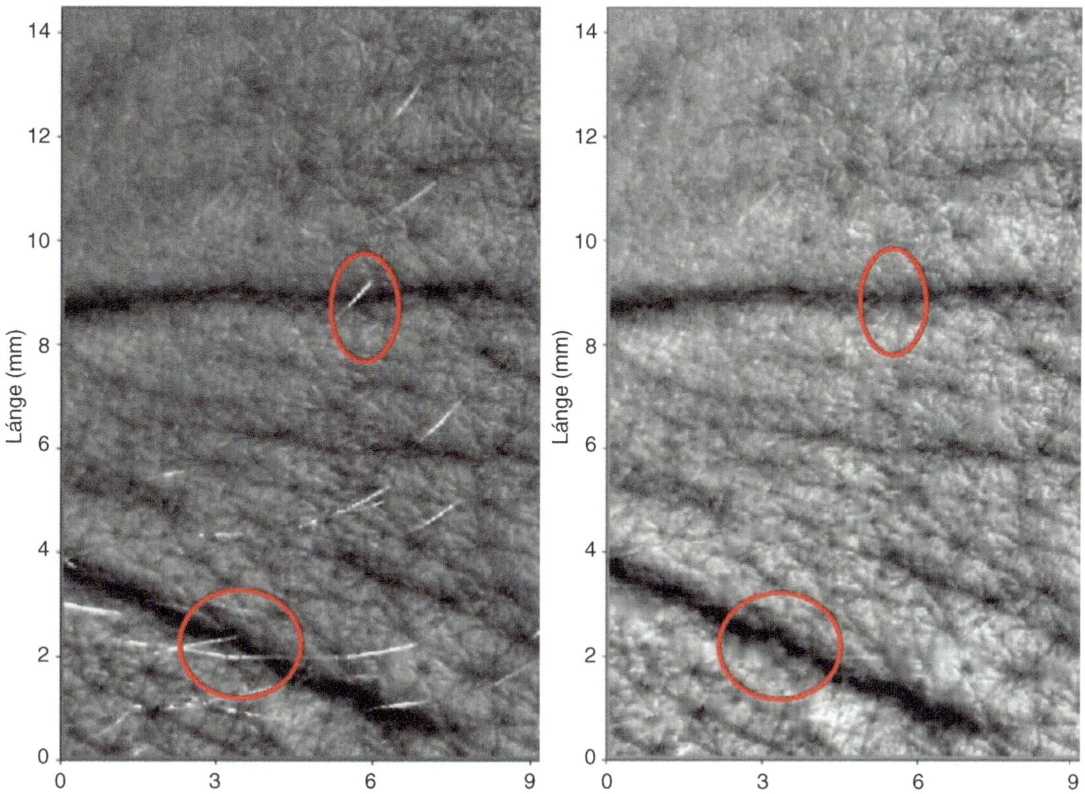

Fig. 4.12 Filtered height image with hairs (*left*) and filtered height image with hairs digitally removed (*right*) (GFM)

and other possible problems. Given the short scanning times of a few seconds including data processing, the measurement can then simply be repeated.

With skin roughness measurement in particular, focus is of major importance. The triangulation-based system allows for a basic distance check by centering a projected crosshair. This, however, does not cover the entire recording area. Moreover, camera contrast with skin measurement may be relatively low and judging crispness by surface detail is very difficult. We therefore conceived a new focusing aid that was implemented by the manufacturer. It adds to the crosshair four small projected marks in the corners of the image. Using this technique, significant improvements were achieved in exploiting the depth-of-focus range.

If this is still not sufficient, patterns of subsequently narrower, static stripe patterns can be projected. This is a bit less quick but allows to judge all points on the area at once.

4.6 Data Analysis

4.6.1 ISO Standard Parameters

Classical surface parameters are defined in ISO 4287. These parameters were originally developed for mechanical (tactile) measuring systems and are defined for a *section line* on a surface. This limitation to line dimension was caused by mechanical constraints (scanning an entire surface with a tactile device takes a long time and the wear on the expensive test pins is considerable). Inevitably, some of the rules and procedures involved are not entirely reasonable, or even applicable, for modern optical topometry systems, partly because of the entirely different physical principle of the scanning process and also – most important – because of the limitation to one dimension and, hence, very small fractions of a surface. We will give a short overview of the use of the classical parameters and then carry on regarding such parameters as are more adequate with optical systems.

Principally, line parameters are classified as form, waviness, and surface roughness. Filter algorithms separating low- and high-surface wave "frequencies," such as "low-pass," "high-pass," and a number of others can be used to separate the different structure classes. A high-pass filter, for example, removing the (low frequency) waviness, serves for evaluating roughness alone. A low-pass filter retains the waviness profile. There is no fixed intersection at which roughness becomes waviness or vice versa, as this depends on the size and nature of the application.

For skin roughness, the most basic approach is to use the same technical roughness parameters as are defined in ISO standards. The current basis for ISO parameters in profilometry is ISO 4287 (1997). Regarded are point deviations y_i from a *reference line* in the surface that is defined by flattening the surface (elimination of low frequencies) and then taking the average of all y_i in the line.

$$R_a = \frac{1}{n}\sum_{i=1}^{n}|y_i| \quad R_q = \sqrt{\frac{1}{n}\sum_{i=1}^{n}y_i^2} \quad R_z = \frac{1}{s}\sum_{i=1}^{s}R_{ti} \qquad (4.1)$$

R_a is the average point deviation, or, average roughness, while R_q is the root mean squared (giving more weight to extreme values). R_z is calculated by dividing the reference line into s sections and calculating the average for the differences between the lowest and highest y in each section. Most commonly, five sections are used.

4.6.2 Selecting Roughness Profile Parameters

We will give a short overview of papers involving the selection of conventional parameters set forth in the DIN specifications. Several appropriate for evaluation of surface features and their changes due to treatment or healing have been selected in [4]. Monti et al. [5] report a statistically significant difference between uninvolved and involved psoriatic skin for R_a but not for R_z.

Other authors report an increase in amplitude parameters and a decrease in the number of peaks in dry atopic skin [6]. Accordingly, spacing parameter S_m should increase. Our own experiments show significant decreases in amplitude parameters in cases of hydration and increases associated with skin ageing [1, 4, 7, 8].

Murahata et al. [9] attempted to correlate subjective evaluations of the skin surface with surface parameters for roughness measured over a length of 7.5 mm on positive replicas. "Plumpness" was used as a parameter for hydration, and scales and cracks for damage to the stratum corneum during visual assessment. On the other hand, R_z and R_a were used as roughness parameters during assessments conducted with measuring instruments. Both roughness parameters were found to correlate significantly with plumpness, but not with scaliness and number of cracks.

4.6.3 Advanced Surface Parameters

Optical systems can deliver the detailed topology of an entire surface area within a short time and with reasonable effort, and this called for new parameters utilizing area features. A workgroup initiated by the EU took up this task [10]. This resulted in a handbook, "Development of Methods for the Characterization of Roughness in Three Dimensions," covering a range of issues related to 3D microtopography with an emphasis on standardization and interpretation. Many meaningful parameters such as S_a, S_q, S_c, and S_z are defined and discussed. The most common parameters are S_a and S_q.

$$S_a = \frac{1}{NM}\sum_{x=0}^{N-1}\sum_{y=0}^{M-1}|z_{x,y}| \quad S_q = \frac{1}{NM}\sum_{x=0}^{N-1}\sum_{y=0}^{M-1}z_{x,y}^2 \qquad (4.2)$$

These 3D parameters, obviously, are extrapolations of the 2D parameters discussed above. They are defined as the sum of z deviations in N rows and M lines. This way, all pixels of a measured surface (with optical systems, several hundred thousand up to several millions) are taken into account. Therefore, these parameters carry a maximum amount of information and, depending on the application, can be a lot more meaningful than their conventional counterparts.

4.6.4 Filtering Aspects

The separation between roughness and waviness is not the only filtering issue to be regarded. At the upper end of the frequency range, system noise may contribute considerable amplitudes, especially with the measurement of very small structures. This applies to mechanical and optical systems as well. One large contribution to noise originates from the considerable amount of signal amplification involved (thermal noise), but there are many others.

An appropriate (usually very moderate) low-pass filtering can be used to keep noise below the surface detail measured. Skin roughness, especially when regarding very small structures, is not characterized by very steep structures. Regarding fringe projection, one may therefore apply a lateral filtering with a wavelength of several pixels and still retain all information about roughness parameters (i.e., Z values) way below this wavelength. The height resolution of the system, being much higher than its lateral resolution, also supports this. Hence, the characteristics of fringe projection systems positively coincide with the requirements of skin roughness measurement.

Another aspect to be considered is "dirt," i.e., little particles on a surface resulting in peaks in the measured profile. Tactile systems are less sensitive to this, as the probe simply pushes such particles aside. With optical measurement, using a special "peak removal" filter usually is a good means against such artifacts.

At the other end of the scale, one may want to retrieve waviness in a certain wavelength range but eliminate contributions from larger scale body shape. The PRIMOS system offers two principal ways achieving this: long wave high-pass filters (of various kinds) or, more sophisticated, a polynomial filter. The polynomial filter is an approximation algorithm yielding a best-fit (two-dimensional) polynomial for the surface area:

$$Z(x,y)=\sum_{i=0}^{n}\sum_{j=0}^{n}a_{i,j}\left(x-\frac{x_{\max}}{2}\right)^{i}\left(y-\frac{y_{\max}}{2}\right)^{j} \quad (4.3)$$

The polynomial is then subtracted from the profile. This algorithm is very computing intense, so it is usually limited to an 8×8 polynomial, roughly analog to a conventional filter with a wavelength of ¼ the area extension.

4.6.5 Volume Parameters

The parameters mentioned so far cannot cover all aspects of practical analyses, and also the simple extension of one-dimensional parameters cannot unleash all possibilities given by a truly area-based scanning system. Analyzing crows-feet, for example, calls for a calculation of wrinkle volume. This, however, is not a standard parameter in current data analysis software. An algorithm for this application could be using edge detection procedures in combination with suitable filter algorithms for separating roughness and wrinkle structures. We have developed software algorithms with these features which have proven to be very meaningful. Interactive determination of the volume parameter is still somewhat time-consuming but the results are very sensitive and at the same time very descriptive.

4.7 Studies

Optical topometry has proven its worth for studies in the field of cosmetic and dermatological research in recent years. The standard systems are commonly used for substantiating smoothing claims and anti-wrinkle claims. Friedman et al. demonstrated the capability of documenting and quantifying the wrinkle structure following multiple treatment sessions with a 1,064 nm QS Nd:YAG laser [11]. They found correlations with clinical and subjective assessments and state that optical topometry provides objective verification and technical understanding of nonablative laser technology. Roques et al. used PRIMOS to measure and evaluate scar surfaces with good definition and reproducibility [12].

Additional types of studies have become possible as a result of modified systems using larger fields of view. In our former studies, objective and quantitative assessments of cellulite and body wrinkles – for example, at the décolleté – have turned out to be very valuable and reliable.

Especially in this area, new developments are promising a much easier in vivo acquisition especially of larger boy areas. We will conclude our report with a short introduction to the latest technology available.

4.8 Recent Developments

The introduction of pico DLP projection technology as well as the integration of computing power into the sensors themselves allows for a number of new, unprecedented applications. Pico projection chips have smaller mirrors (8×8 μm) and a smaller resolution (HVGA), resulting in a chip size only a fraction of the classical DMD. Nevertheless, the principal advantages of the technology are retained, and the smaller projection resolution does not proportionally translate into a smaller *Z* resolution, as the fringe projection is not done crisp anyway. Combined with small "intelligent" cameras, light, handheld measuring devices have been developed (Fig. 4.13). Some of these devices deliver readily rendered 3D data via an Ethernet connection, a functionality resembling conventional image processing cameras, but in three dimensions. On the other hand, pico projectors have been made powerful enough to allow for large measuring areas with only small requirements for environment light dimming, enabling compact, light, and affordable face and body scanners, giving thought to unconventional applications. For example, point-of-sales applications concerning cosmetic products and body shape measurement in gym shops (we will, however, refrain from extending these speculations here).

Yet in scientific applications as well, affordability and network connection open up new possibilities. A combination of multiple scanners for fast, complete, in vivo head, or body scans has become possible. This is of particular advantage, as with in vivo topometry, combining several body scans taken by a single sensor from different directions (the usual procedure with industrial measurements), is normally not possible as the test person will move during the repositioning times required, and this includes changes in the body shape itself, something unheard of with industrial applications. Multiple sensors, however, allow to perform multiple scans from different directions within seconds or even a fraction of a second,

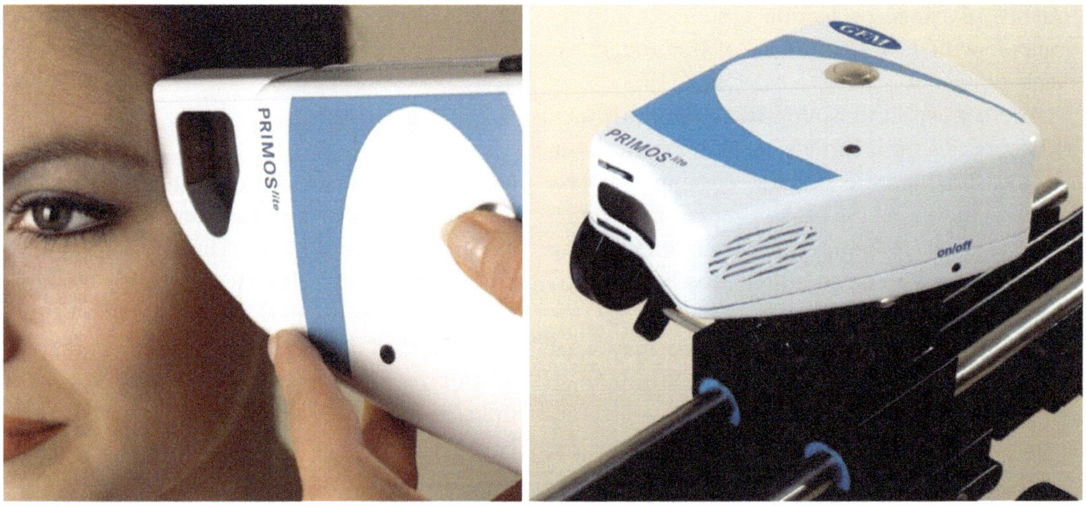

Fig. 4.13 Example of a compact, pico DMD-based sensor in mobile measurement (*left*) and on a stand (*right*) (Photos: GFM)

short enough for most volunteers to stand sufficiently still. Hence, this capability defines body shape measurement entirely new.

Currently, we are carrying out first studies using a four-sensor system by GFM for wide angle face and body measurement (Fig. 4.14). The system has two stacked, motorized lifting columns, and includes individual precision tilting controls for each sensor, fulfilling the aforementioned requirements for positioning with repeated measurements. The system uses new multisensor calibration method allowing to integrate the scanning results on-the-fly [13]. An additional conventional side camera allows for controlling the proper head position (exact height and distance) of the test person. The seat, including an adjustable head rest, is designed for a best possible relaxed position and avoidance of any movements during the (very short) scan. Figure 4.14 shows the complete scanning station. Figure 4.15 shows 2D live images from the sensor cameras as well as the side camera on the control computer screen. Figure 4.16 shows fringes sequentially projected from the four sensors and, to the right, the resulting 3D image, as a 3D simulation including the original color texture of the face (as

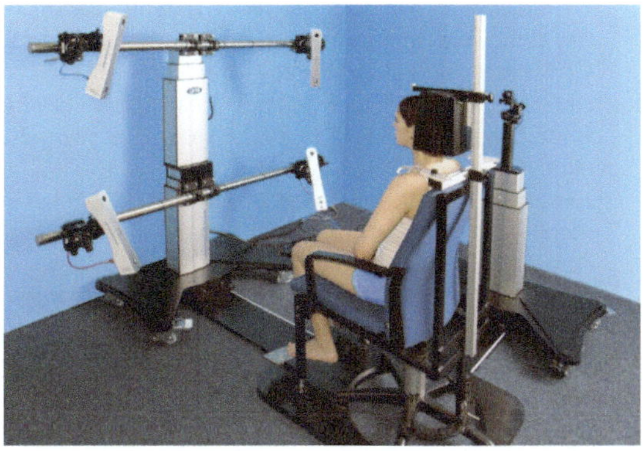

Fig. 4.14 Biomedical body scanning station with four sensors and recallable precision positioning for 180° scanning (Photo: GFM)

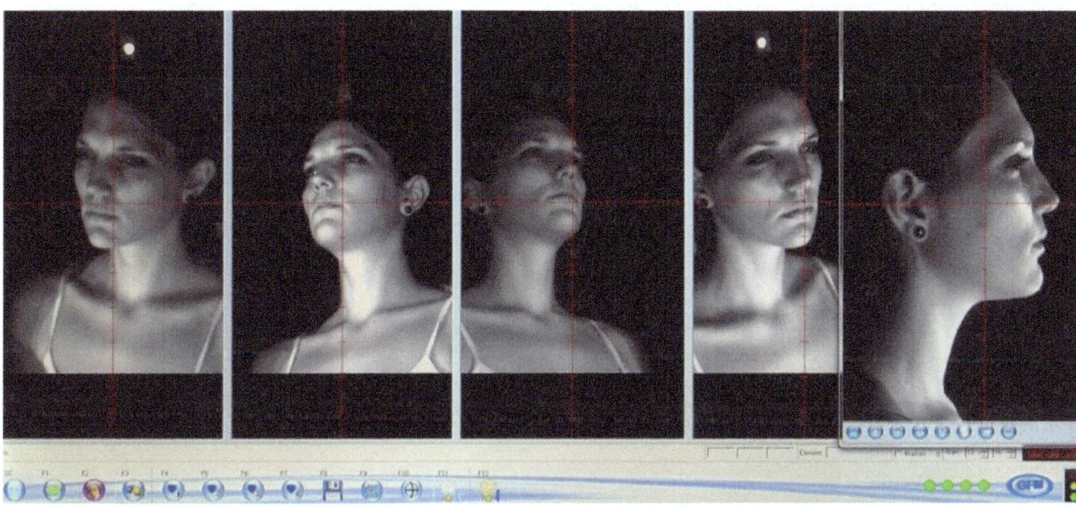

Fig. 4.15 Live camera image for positioning of the four sensor cameras and the side camera (*right*) (Photo: GFM)

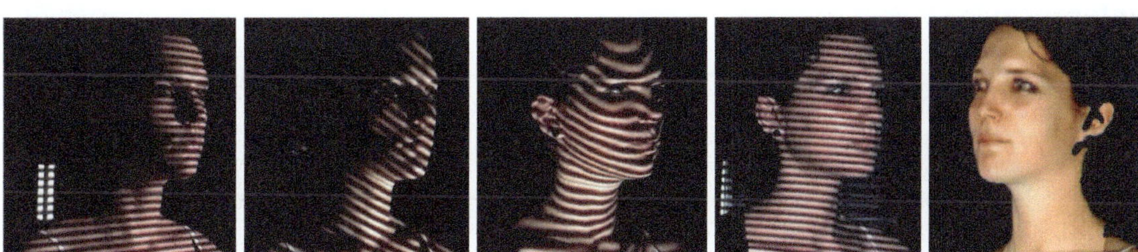

Fig. 4.16 Biomedical face scanning (cf. *bottom*): projected fringes from four different sensors (*left to middle*)

described before, color is recorded using the color capability of the projectors to record three color excerpts in a sequence).

4.9 Conclusions and Outlook

Fringe projection-based topometry opens up a wide range of new approaches offering improved accuracy and repeatability as well as a wider coverage of relevant parameters for cosmetic and dermatological research. A multitude of test designs have been facilitated and new ones have been made possible by the versatility of the systems, offering a multitude of application setups allowing studies of almost any physical body features in small as well as large scales. Recent developments are widening the fields of applications by introducing light and compact measuring heads, suitable for portable as well as complex multisensor devices.

As with any complex biophysical methods, skilled operators as well as the development of standard procedures are important for the quality, reliability, and repeatability of results, especially when carrying out larger studies involving tests on many test persons over a long time.

Essential ingredients are the accuracy and long-term stability of the test systems and the particular methods and procedures developed for carrying out scientific studies. We have shown how this has been approached in a number of recent cases. New possibilities, like multisensor systems and large scale in vivo body shape measurement, will require a continuous adaptation and development of adequate procedures for their application.

References

1. Jaspers S, Hopermann H, Sauermann G, Hoppe U, Lunderstädt R, Ennen J (1999) Rapid *in-vivo* measurement of the topography of human skin by active image triangulation using a digital micromirror device. Skin Res Technol 5:195–207
2. Rohr M, Brandt M, Schrader A (2000) Skin surface – claim support by FOITS. SÖFW J 126(8):2–11
3. Bruning JH, Herriott DR, Gallagher JE (1974) Digital wavefront measuring interferometer for testing optical surfaces and lenses. Appl Opt 13:2693–2703
4. Schreiner V, Sauermann G, Hoppe U (1996) Characterization of the skin surface by ISO parameters for microtopography. In: Wilhelm KP, Elsner P, Berardesca E, Maibach HI (eds) Bioengineering of the skin: skin surface imaging and analysis, vol IV. CRC press, Taylor and Francis. ISBN 0849383757
5. Monti M, Bozzetti M, Motta S, Barbareschi M (1989) Usefulness of surface topography analysis in psoriatic skin. Acta Derm Venereol 69:81
6. Linde YW, Bengtsson A, Loden M (1989) Dry skin in atopic dermatitis II. A surface profilometry study. Acta Derm Venereol 69:315
7. Schreiner V, Hoppe U, Sauermann G, Wenghöfer C (1993) Microtopography of human skin – a validated method of analysis of analogous and Fourier transformed signals. Allergologie 16:165
8. Hoppe U, Sauermann G (1985) Quantitative analysis of skin's surface by means of digital signal processing. J Soc Cosmet Chem 36:105
9. Murahata RI, Crowe DM, Roheim JR (1984) Evaluation of hydration state and surface defects in the stratum corneum: comparison of computer analysis and visual appraisal of positive replicas of human skin. J Soc Cosmet Chem 35:327
10. Stout K, Blunt L (1998) Development of methods for the characterisation of roughness in three dimensions. Penton Press, London. ISBN 1857180232
11. Friedman PM, Skover GR, Payonk G, Kauvar ANB, Geronemus RG (2002) 3D In-vivo optical skin imaging for topographical quantitative assessment of non-ablative laser technology. Dermatol Surg 28(3):199–204
12. Roques C, Teot L, Frasson N, Meaume S (2003) PRIMOS: an optical system that produces three-dimensional measurements of skin surfaces. J Wound Care 12(9):362–364
13. Frankowski G, Hainich R (2012) Network based multi-sensor optical 3D acquisition of complex structures. Proc SPIE Photonics West

5 FOITS (Fast Optical In Vivo Topometry of Human Skin)

Mathias Rohr and Andreas Schrader

5.1 Introduction

First investigations about contactless surface analysis of the skin were initialized in the year 1995. After a successful validation phase, the first publication of the new FOITS technology (Fast Optical In Vivo Topometry of Human Skin) can be found in 1997 [1]. In comparison to the replica-driven technique during the last decade, the touch-free technique of *fringe projection* became state of the art to investigate skin surface [2–5]. Due to many technical advancements (as, e.g., improved camera resolution, the use of blue LED lighting systems, or laser-supported and computer-optimized overlaying procedures), an easy-to-operate system could be realized during the last years. As scientific interest on the mechanisms of wrinkle evaluation has always been pushed, the technical developments led to a tool of high scientific standard [6–9]. From the beginnings in the year 1995, FOITS became worldwide established. Today it is used in various countries (Germany, France, Great Britain, Ireland, Italy, the USA, China, Japan, and South Korea). Consequently, in 2012, the new generation FOITS-2 is now available to realize a one-shot one-face technique with an excellent resolution and time regime.

M. Rohr (✉) • A. Schrader, PhD
Institut Dr. Schrader Hautphysiologie,
Max Planck Str. 6, Holzminden D-37603, Germany
e-mail: mathias.rohr@schrader-institute.de;
http://www.schrader-institute.de

More and more products will find a worldwide distribution on a global marketplace. Demands on product claim support have to fulfill a variety of requests. In particular the analysis of wrinkles is of huge interest. Differences or similarities of varying skin ethnics can be a start point of new product development as well.

Results obtained with the FOITS technique show clear correlation of roughness parameters like Rz or FDD and age of volunteers. Data of more than 2,500 Caucasian women (Germany/Holzminden) and 550 Asian women (China/Beijing) are analyzed in order to present an overview of wrinkles and their history and will be presented in this article as an example of modern wrinkle analysis technique.

5.2 Materials and Methods

5.2.1 FOITS (Fast Optical In Vivo Topometry of Human Skin)

FOITS is a touch-free optical technique with a history of more than a decade to investigate skin surface structures in a direct three-dimensional measurement by fringe projection [10]. The fringe projection technique used is a combination of gray-code and phase-shift technique [6]. In less than a few hundred milliseconds, the absolute space coordinates are measured of all object points in the selected image area with great exactness. The FOITS-measurement system consists of a projection unit and a CCD camera. Both are fixed

E. Berardesca et al. (eds.), *Non Invasive Diagnostic Techniques in Clinical Dermatology*,
DOI 10.1007/978-3-642-32109-2_5, © Springer Berlin Heidelberg 2014

Table 5.1 Synopsis of the technical side parameters of FOITS from the beginning to this day

FOITS	1995	1998	2003	2006
Technique	Gray-code and phase-shift technique			
	Contact-free direct skin measurement in vivo			
	Halogen light		Blue LED technique	
Superimposition	Mechanically aided by online overlay procedure		Laser aided mechanically	Software aided on top of all
Measurement area	Inner side of the forearm	Crow's feet, under the eye, cheek, glabella, lips, nasolabial, décolleté, forearm, leg		
Area of inspection	875 mm^2 (25 × 35 mm)	1,200 mm^2 (30 × 40 mm)		
Area of analysis	20 × 20 mm	20 × 20 mm (or as needed)		
Resolution				
X-direction	~40 μm	~30 μm		
Y-direction	~40 μm	~30 μm		
Z-direction	4 μm	4 μm		
Time to digitize the fine structure	~320 ms	~260 ms		

under the triangulation angle. Concerning the gray-code method, grids with a rectangular brightness distribution by different numbers of lines are projected. The number of lines is doubled at each new projection. This gives a clearly defined hierarchy of lines for each image point. Regarding the phase-shift technique, only one grid with a sinus-like intensity distribution is projected several times with different phase positions. The FOITS technique is able to realize a depth sharpness area of ±10 mm on an inspection area of 30 × 40 mm. The resolution in the vertical *Z*-direction with 0.2 % of the measured area leads to an effective resolution of 4 μm in *Z*-direction. A CCD camera with a horizontal and vertical resolution in *X*- and *Y*-directions of about 30 μm is used. It has to be pointed out that the resolution in *Z*-direction is not limited by 256 gray steps of the CCD camera. The high resolution in the vertical direction is achieved by the analysis of the intensity and phase displacement of the projected grids. The surface structure of the analyzed area causes a deviation of the intensity and phase information of the projected grid structures from the theoretical model structure of a plane surface. With corresponding mathematical algorithms, the absolute 3D coordinates of the inspected area can be calculated of these deviations. A synopsis of the most important experimental side parameters is shown in Table 5.1 starting from the beginning of the first experiments up to these days (Fig. 5.1).

In 1995, the work on this new technique started to validate the method in comparison to existing replica methods. Since 1997 data have been generated using the FOITS technique in routine work. By this optical technique powerful surface information is obtained without any mechanical pre-filtering (influence) as typically has to be considered by taking replica. Starting with analysis of the inner side of the forearm, the crow's feet area became the area of most interest later on. Increasing the power of FOITS technique as described in Table 5.1, more and more areas of analysis could be investigated like the cheek, glabella area, under the eye, nasolabial area, the lips, or all over the body areas like the décolleté and legs. The latest technique used combines the fastest data measurement with the best superimposition technique to guarantee a perfect comparison of baseline and end value data. Superimposition is realized in a combination of laser-aided mechanical alignment of the subject in a first step and a software-driven rotation and shifting procedure of measured data/pictures to find the optimum of superimposition.

5.2.2 Parameter of Analysis

Bringing into focus the periorbital wrinkle area (crow's feet), the morphological structure of

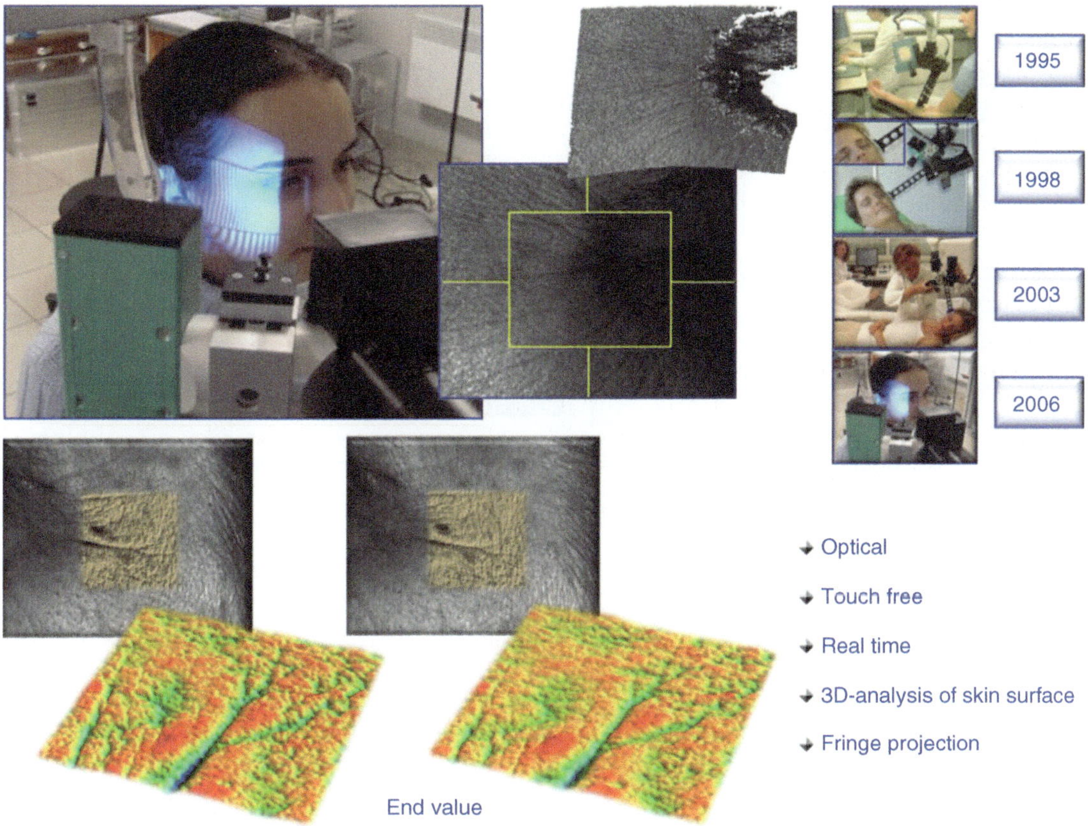

Fig. 5.1 Presentation of various FOITS system from 1995 to present days, example of FOITS-data presentation on an individual subject. 3D data presentation of the crow's feet area before and after 4 weeks of product application

this test area has to be taken into account if wrinkles are investigated. Having this in mind, analysis is carried out perpendicular to the main wrinkle direction based on the Rz parameter (according to DIN 4668 [8]) or the FDD (frequency distribution of depth) analysis. Starting close to the eye, 50 separate lines with a distance of 400 μm are analyzed. The resulting roughness is shown as a function of line number (see Fig. 5.2). Ten successive lines each are averaged, resulting in five areas of evaluation. Separating the area of analysis into these five subareas (Areas 1–5, see Fig. 5.2), the area close to the eye named Area 1 represents the deepest structures, while with Area 5 smaller structures are quantified. An example of this analysis is given in Fig. 5.1. In comparison analysis of the lip area is shown. Due to the smaller test area at all, only four areas are defined with 40 separate lines with a distance of 250 μm. As shown by Fig. 5.2, correlation of line number and Rz results in a more flat link for the lip area in comparison to the crow's feet area.

In order to document the surface structure by a global parameter, the frequency distribution of all depths is used. The frequency distribution of depth (FDD) is calculated in the range from −600 to 600 μm (after polynomial correction) by using interval steps of 5 μm. The defined evaluation area is equivalent to a surface of 2 × 2 cm and represents according to the technical resolution of the camera 640.000 single points. Therefore, a calculated FDD parameter is based on a rearrangement within these 640.000 values of depth.

Working with a distribution function, the zero level has to be kept in mind carefully. Thus the zero level of each volunteer is defined as the first plane

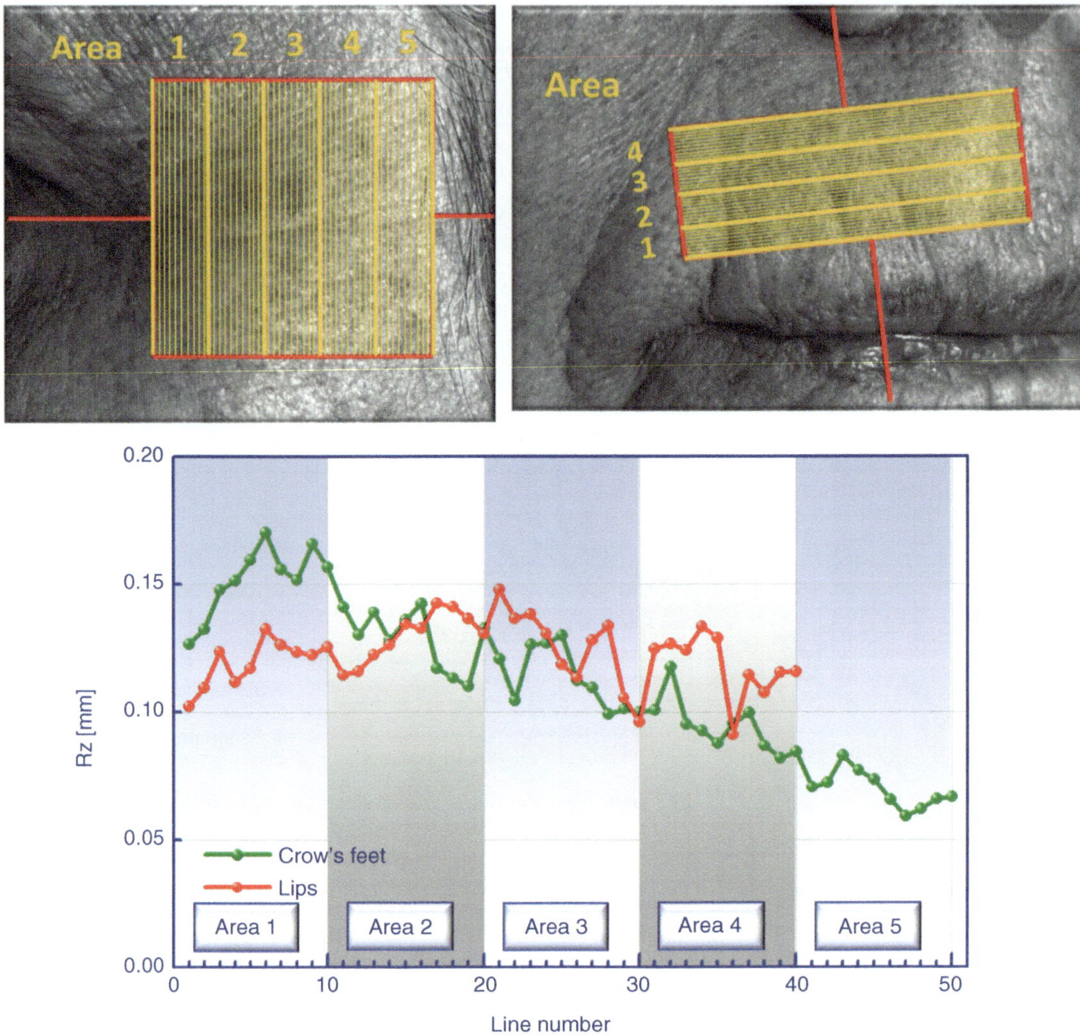

Fig. 5.2 Definition of subarea of analysis. Rz as a function of subarea lines of an individual example in the crow's feet area and lip area

representing a level of about 0.1 % of all single values (about 600 counts). This plane is set as zero, and all further calculations are done with these resulting standardized values. From the surface structure, a frequency distribution of all depths is obtained as shown exemplarily in Fig. 5.3.

According to the selected zero level, a classification of depth is made as follows:

0–50 µm	→ microstructure	(about 5 %)
55–170 µm	→ fine structure	(about 65 %)
<170 µm	→ rough structure	(about 30 %)

The given proportion will give a rough estimation of structure ranges found in the crow's feet area of women with distinct wrinkles and Caucasian skin. Taking into account a product-smoothing effect, the green FDD curve as shown in Fig. 5.4 can be expected. Consequently an improvement of skin structure is defined by a shift of maximum and a change of width of the distribution function. A reduction of rough structures can be expected, while for fine and microstructures, an increase is obtained in case of structural improvements.

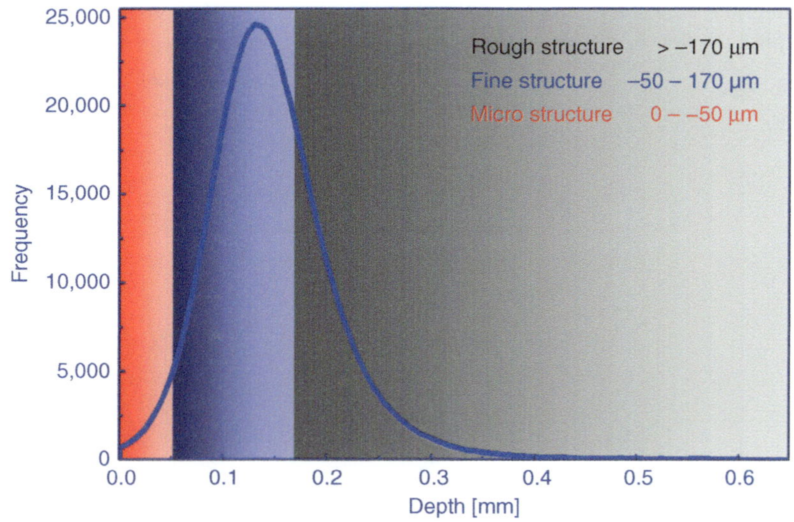

Fig. 5.3 Histogram of depth of a surface profile (periorbital region) and classification of structural regions

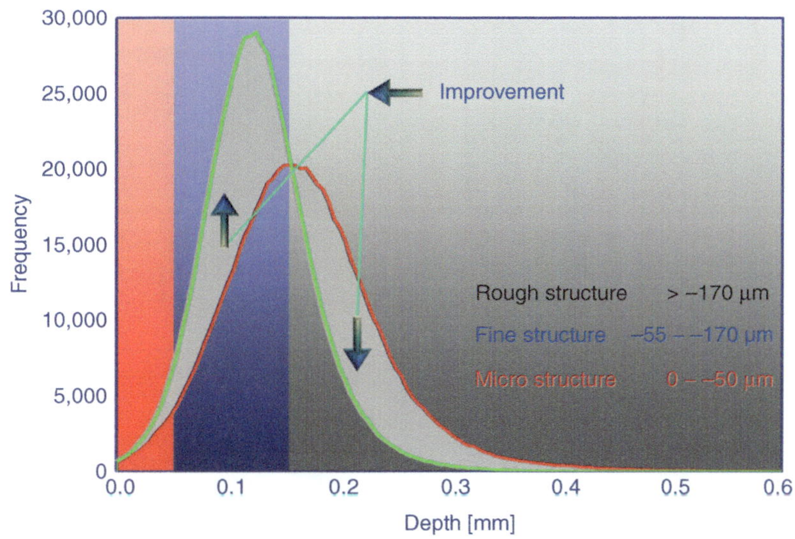

Fig. 5.4 Histogram of depth of a surface profile (crow's feet area), classification of structural regions, as well as visualization of smoothing effect/age effect – baseline, 65-year-old subject; end value, 15-year-old subject

5.2.3 Test Conditions

All data shown in the following are baseline data investigated by FOITS on women with distinct wrinkles. Data were collected after climatizing for 45 min at 50 % rel. hum. and 22 °C. If not, especially remarked data were collected on Caucasian women in Germany/Holzminden. Alternatively data were analyzed taken from women living in China/Beijing in order to open the possibility to compare Asian and Caucasian wrinkle formation [9, 10]. Besides other skin physiological parameters, a particular interest consists in the analysis of age and the appearance of wrinkles.

5.3 Exemplary Results and Discussion

Starting with a general overview, Fig. 5.5 represents start and end point of the age scale. A 15-year-old girl with a smooth skin in the

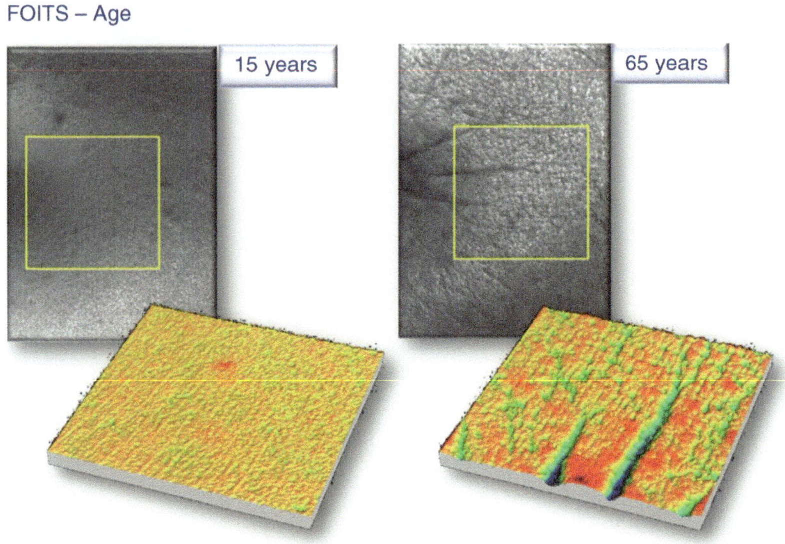

Fig. 5.5 Skin of a 15-year-old girl and a 65-year-old woman and corresponding 3D surface analysis

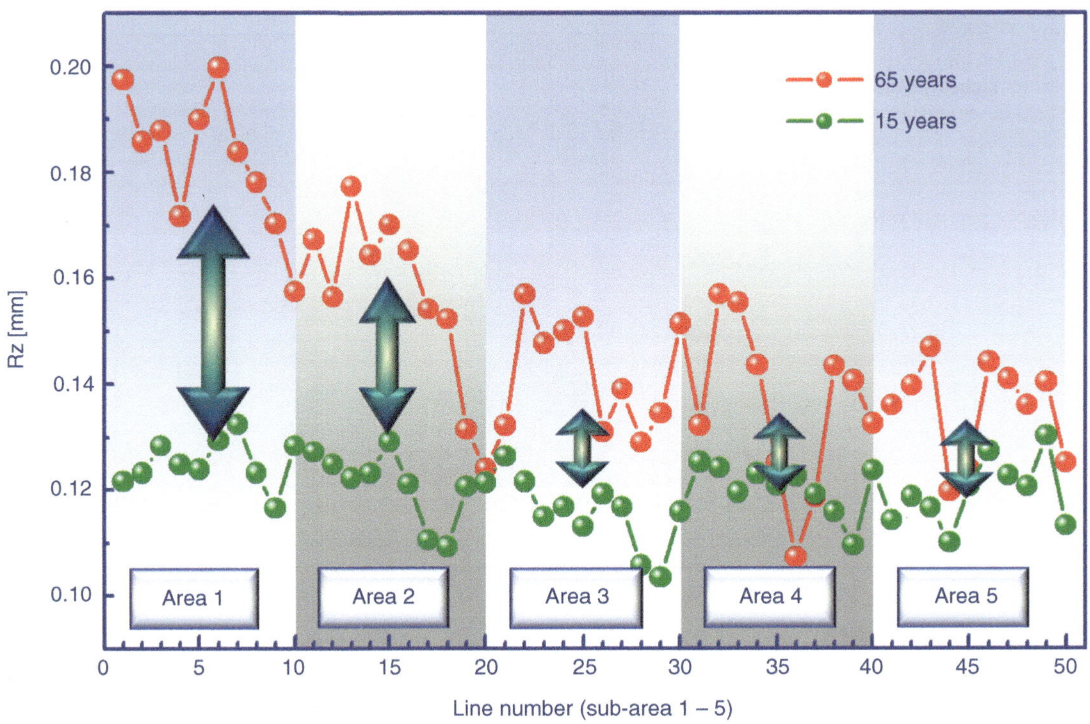

Fig. 5.6 Rz as a function of line number for a 15-year-old subject and a 65-year-old subject

periorbital region is shown in comparison to a 65-year-old woman with distinct wrinkles.

Figures 5.6 and 5.4 present corresponding parameter analysis of analyzed skin areas of these two age groups. Besides the optical impression (Fig. 5.5), a clear difference in Rz and FDD can be identified for different age groups.

For a smooth surface at younger age without any wrinkles, a flat correlation of Rz and line number (subarea) results, while for distinct

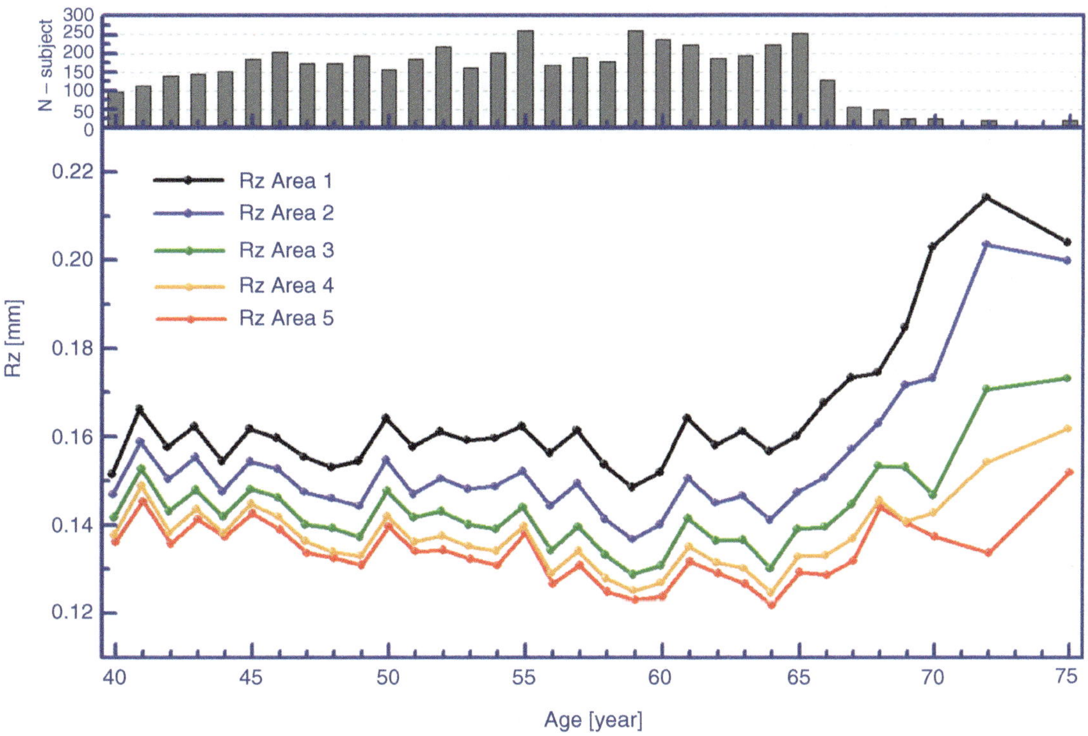

Fig. 5.7 Rz as a function of age on female volunteers calculated for subareas 1–5 presented as a mean value per year from baseline measurements before any product application and after a washout period of 3 days, $n = 2{,}695$ overall

wrinkles an exponential decay curve is obtained (Fig. 5.6). While for Areas 5–3, a more or less constant level of difference can be obtained, the major differences occur for Area 2 and Area 1, respectively. Even more impressive differences occur in the FDD analysis (Fig. 5.4). Distribution of depth gets much broader with age, and the maximum clearly shifts to deeper structures.

Going ahead with the main question of the relationship between wrinkles and age or product efficacy, respectively, Fig. 5.7 shows results of Rz analysis on the basis of 2,695 subjects (5,390 eyes). Data were analyzed as described above. Rz of subareas 1–5 is calculated as a mean value of 10 Rz values. Data were averaged per subarea and per age and finally presented as a correlation of age and Rz individually per subarea in order to open the possibility to get a detailed view of wrinkle formation.

As expected a link of increasing Rz values and age is obtained. This link might be described by an exponential fit at first glance. At closer inspection a more detailed interpretation might occur. From these calculations deeper structures seem to increase at younger age, while fine structures represented by subareas 4 and 5 seem to be on a constant level over a longer period of age. In order to present this detailed analysis, only subarea 1 and subarea 5 are shown in Fig. 5.8.

While for subarea 1 a linear correlation up to 55–60 years is calculated with a slope of zero, for subarea 5 a linear regression ends up with a negative slope of the link of age and Rz. An exponential link could only be found for elder subjects. An explanation of this separated age behavior might be attributed to the fact that for antiageing studies volunteers with distinct wrinkles have to be selected. Consequently a database is analyzed which does not represent a mean "population wrinkle" at younger age groups. Thus for deeper structures (subarea 1), an expected correlation based on a mono-exponential fit

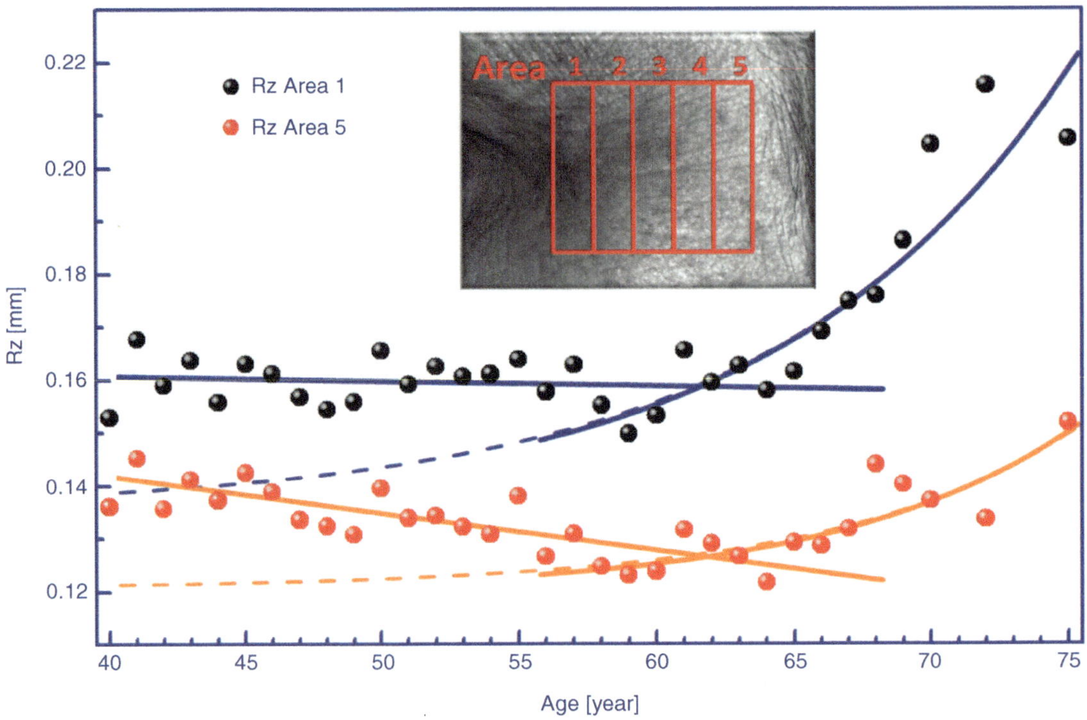

Fig. 5.8 Rz as a function of age on female volunteers calculated for subareas 1 and 5, including linear regression and mono-exponential fit

(dashed blue line) could only be detected on volunteers elder than 57. In order to describe the whole age curve, a combination of a linear and a mono-exponential correlation has to be used (drawn through blue lines, Fig. 5.8). This combination of linear and mono-exponential fit is found for subarea 5 as well (drawn through red lines, Fig. 5.8). Nevertheless interesting differences of subareas 1 and 5 can be identified. While deeper structures seem to be constant in these special panels until an age of approx. 55–60, younger women with distinct wrinkles (in order to be included in antiageing studies) seem to have a rougher skin structure even beside the wrinkles. This means a structure opening the possibility to take part in an antiageing investigation at younger age is linked to a rougher structure overall, while for elder volunteers deeper wrinkles might appear on a smoother skin (lower Rz on subarea 1) as well. This consequence could be linked to the negative slope (decreasing Rz values) of Area 5 up to an age of about 60 years. From an age of approx. 57 years, structures in the crow's feet area increase exponentially. But even at these age groups, differences in subarea 1 and subarea 5 appear. Deeper wrinkles definitely increase stronger and are more pronounced than fine structures represented by subarea 5. Consequently a claim like "looks x-years younger" should be calculated on the basis of the exponential fit information (Fig. 5.8) which mainly is determined by elder subjects. Taking into account that subarea 1 varies most with age, the mentioned claim might be obtained from a decrease in Rz calculated of subarea 1 before and after product application.

Can similar results be expected if the ethnic background is changing? In order to light up this question, a comparison of Asian skin and Caucasian skin is carried out. Measurements of 550 Chinese women from Beijing were performed under identical conditions as described for the 2,695 subjects measured in Germany. A summary of these Asian data is shown in Fig. 5.9.

Besides the fact that less volunteers in comparison to the first analysis of the Caucasian

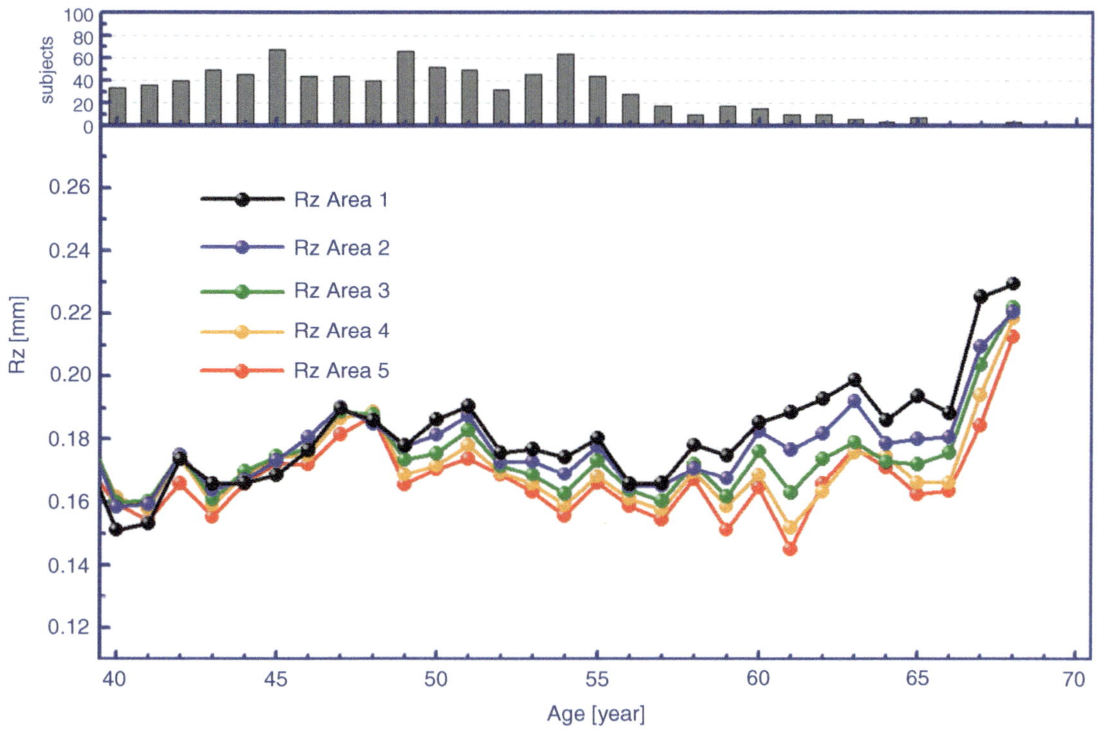

Fig. 5.9 Rz as a function of age on female volunteers calculated for subareas 1–5 presented as a mean value per year from baseline measurements before any product application and after a washout period of 3 days, $n=550$ overall. Place of measurement: Beijing

women leads to more variation, by a first view subareas 1 and 5 are much closer for Asian skin than measured for Caucasian skin. Differences from Area 1 to Area 5 are much less pronounced on Asian women in comparison to Caucasian crow's feet data. A separation may start around the age of 58, while for the obtained data on Caucasian skin, a separation of subareas is documented for all analyzed data for all age groups. On the other hand the described strong increase of Rz values seems to start a few years earlier in Asian skin.

This analysis is a first step to describe differences of skin on different continents. More data are needed to gain more detailed information from this. Nevertheless differences in the link of age and wrinkles are obtained by this first attempt. A more homogenous skin profile seems to be present in the crow's feet area in Asian skin for the first 55 years of life, while the increase of deeper wrinkles seems to be stronger if they occur at all.

Conclusion

Taking into account FDD analysis and analysis of different subareas, a detailed profile of product effectiveness can be calculated on a statistical relevant background from FOITS data. The history of wrinkle appearance is changing with wrinkle deepness. Deeper structures seem to increase earlier and stronger than finer structures. If younger women have deeper wrinkles, they have a rougher skin structure at all.

Ethnic differences in skin surface parameters can be linked to age and characteristics of wrinkles at first glance. Data analysis as a function of subarea and of age will open the possibility to support "look x-years younger" as a new product claim if the panel is analyzed age by age and if an individual age-dependent effectiveness profile is calculated on the basis of a mono-exponential fit based on an elder panel.

Due to detailed possibilities of surface analysis of the FOITS method, new parameters could be innovated for efficacy evaluation during a more than 15-year lasting practice phase. As an example the FDD parameter (frequency distribution of depth) can be mentioned, which allows an efficacy analysis separated in micro-, fine, and rough skin structures on the basis of a statistical relevant method. In the area of analysis which is affected by the FOITS technique, different subareas and degrees of wrinkles are examined. In this way new conclusions about history of wrinkle formation but as well detailed and product-related efficacy analysis can be obtained.

Thus, FOITS has become a worldwide used and accepted classical test for efficacy testing. Many statements concerning product-induced wrinkle reduction are based on results obtained by the fringe projection FOITS technique.

References

1. Rohr M, Schrader K (1997) Fast optical in vivo topometry of human skin (FOITS) – a comparison to LASER-profilometry. In: 5th congress of the international society for skin imaging, Vienna
2. Rohr M, Schrader K (1998) Fast Optical In vivo Topometry of Human Skin (FOITS). SÖFW J 124(2): 52–59
3. Lagarde JM, Rouvrais C, Black D, Dirdollou S, Gall Y (2001) Skin topography measurement by interference fringe projection: a technical validation. Skin Res Technol 7:112–121
4. Jaspers S, Hoperman H, Sauermann G et al (1999) Rapid in vivo measurement of the topography of human skin by active image triangulation using a digital micro-mirror device. Skin Res Technol 5:196–207
5. Ferraq Y et al (2007) Use of 3-D imaging technique for non-invasive monitoring of the depth of experimentally induced wounds. Skin Res Technol 13:399–405
6. Rohr M, Brandt M, Schrader A (2000) Skin surface – claim support by FOITS. SÖFW J 126(8):2–11
7. Breuckmann B (1993) Bildverarbeitung und optische Messtechnik in der industriellen Praxis. Franzis-Verlag GmbH, München
8. DIN 4768 Ermittlung der Rauheitskenngrößen Ra, Rz, Rmax mit elektrischen Tastschnittgeräten, Begriffe, Messbedingungen
9. Nouveau-Richard S et al (2005) Skin ageing: a comparison between Chinese and European populations – a pilot study. J Dermatol Sci 40(3):187–193
10. Rohr M, Yan Qi, Schrader A (2006) Anti-wrinkle performance of cosmetic products differentiated by FOITS, a statement of effectiveness, age and ethnic background. In: IFSCC Congress, Osaka

In Vivo Confocal Microscopy in Skin Oncology

6

Caterina Longo, Barbara De Pace, Simonetta Piana, and Giovanni Pellacani

Abbreviations

AK	Actinic keratosis
BCC	Basal cell carcinoma
LPLK	Lichen planus-like keratosis
RCM	In vivo reflectance confocal microscopy
SCC	Squamous cell carcinoma

6.1 Introduction

In vivo reflectance confocal microscopy (RCM) is an imaging technique that has been recently introduced in clinical practice and research community.

The instrument consists of a head with a laser source that employs the "confocal" concept of light scattering. It is a completely safe and noninvasive method based on the different refractive index of skin structures such as melanin that provides the major source of contrast, keratin, organelles, and collagen. In grayscale confocal images, structures that appear bright (white) have components with high refractive index compared with their surroundings and are similar in size to the wavelength of light [1, 2]. The horizontal scanning of the microscope produces black and white high-resolution images that are consecutively acquired to build up mosaic with an area that is up to 8×8 mm. Several mosaics at different skin depth can be imaged to obtain a complete overview of the lesion that is necessary for assessing the architecture. The instrument is coupled with a handheld dermatoscope that allows a perfect correlation for the confocal findings and guides the clinician during the acquisition. The resolution of RCM is quite similar to the conventional histopathology, and this capability renders the instrument a suitable candidate for skin tumor diagnosis.

RCM has found application in several clinical situations, and it was extensively applied in skin oncology to improve the diagnostic accuracy when combined with dermatoscopy.

In this chapter we will describe the major confocal criteria [3] in melanocytic and nonmelanocytic lesions highlighting the corresponding histopathologic correlates.

C. Longo, MD, PhD
Skin Cancer Unit, Arcispedale Santa Maria Nuova IRCCS, Via Risorgimento 80, Reggio Emilia 42100, Italy
e-mail: longo.caterina@gmail.com

B. De Pace, MD • G. Pellacani, MD (✉)
Department of Dermatology, University of Modena and Reggio Emilia, Via del Pozzo 71, Modena 41124, Italy
e-mail: barbar.depace@gmail.com; pellacani.govanni@gmail.com

S. Piana, MD
Pathology Unit, Arcispedale Santa Maria Nuova IRCCS, Via Risorgimento 80, Reggio Emilia 42100, Italy
e-mail: simonetta.piana@asmn.re.it

E. Berardesca et al. (eds.), *Non Invasive Diagnostic Techniques in Clinical Dermatology*,
DOI 10.1007/978-3-642-32109-2_6, © Springer Berlin Heidelberg 2014

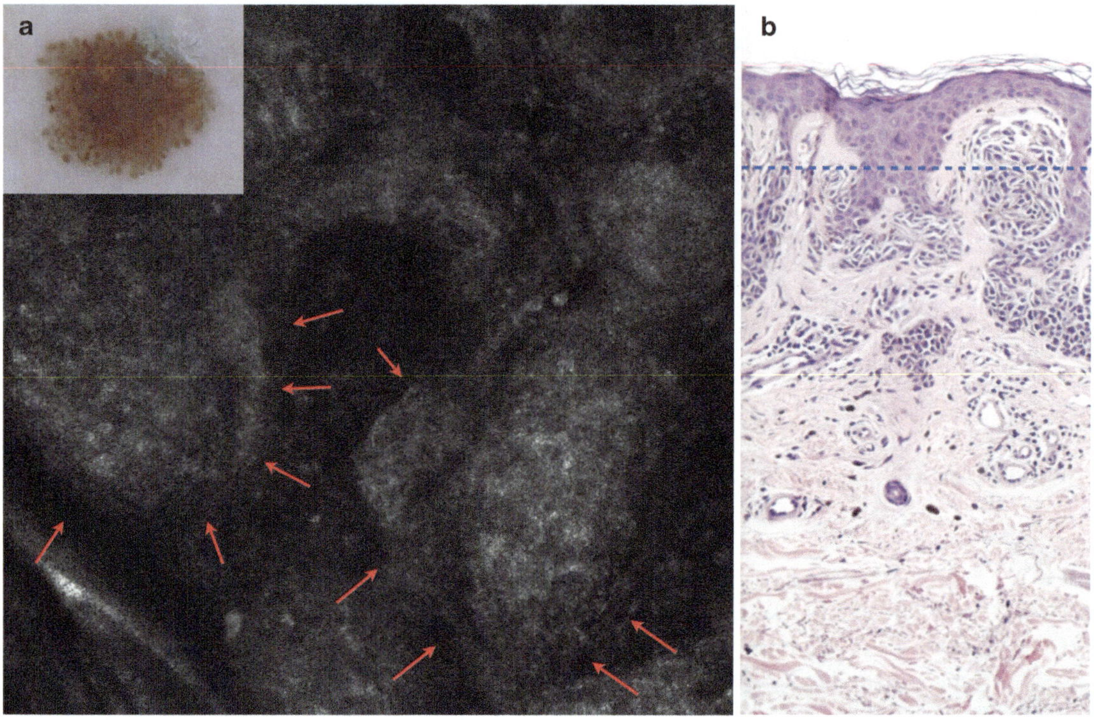

Fig. 6.1 (**a**) RCM image showing regular junctional nests (*arrows*) of a compound nevus that correlate well with the histopathologic ones (**b**)

6.2 Melanocytic Lesions

6.2.1 Common Acquired Nevi

The realm of common acquired nevi includes a variety of clinically different nevi. Traditionally, they are classified based on histopathologic criteria in junctional, compound, and intradermal nevi. Although this classification does not encompass the clinical variability of these lesions, it offers a good understanding of the confocal findings according to different skin depth [4, 5]. Junctional nevi usually do not represent a clinical challenge since dermatoscopy shows a regular pigmented network presenting a light brown to dark brown color or orange hue in fair skin phototype. Confocal microscopy shows the presence of a regular honeycombed pattern and ringed junctional architecture due to the alternation of black/hyporeflective dermal papillae surrounded by bright polygonal cells. Few junctional nests can be present at time and they enlarge the interpapillary space. In compound nevi presenting a mixed dermatoscopic pattern, it is frequent to observe the presence of several junctional nests that give rise to a "meshwork pattern" (Fig. 6.1). When a blue color is present upon dermatoscopy, RCM images highlight the presence of plump bright cells corresponding to melanophages, usually seen into the dermal papilla close to a tiny canalicular vessel [6]. The intradermal nevi with a predominant dermal component usually show a regular epidermis and the presence of large aggregates of roundish monomorphous cells ("clod" pattern) when the nested population is superficially located. In case of a deep dermal component, the limited laser depth penetration hampers the visualization of the melanocytic proliferation. Besides the stereotypical dermatoscopic and confocal aspects of these lesions, some nevi can be challenging from either a clinical point of view or histopathologic ones. Several terms have been coined to refer to these lesions such as "dysplastic" or "atypical" nevi, although a precise correlation between the clinical "atypical" aspects and the "atypical" histologic appearance has not been found yet. Recently a study of our group defined the exact correlations between the histopathologic findings in the

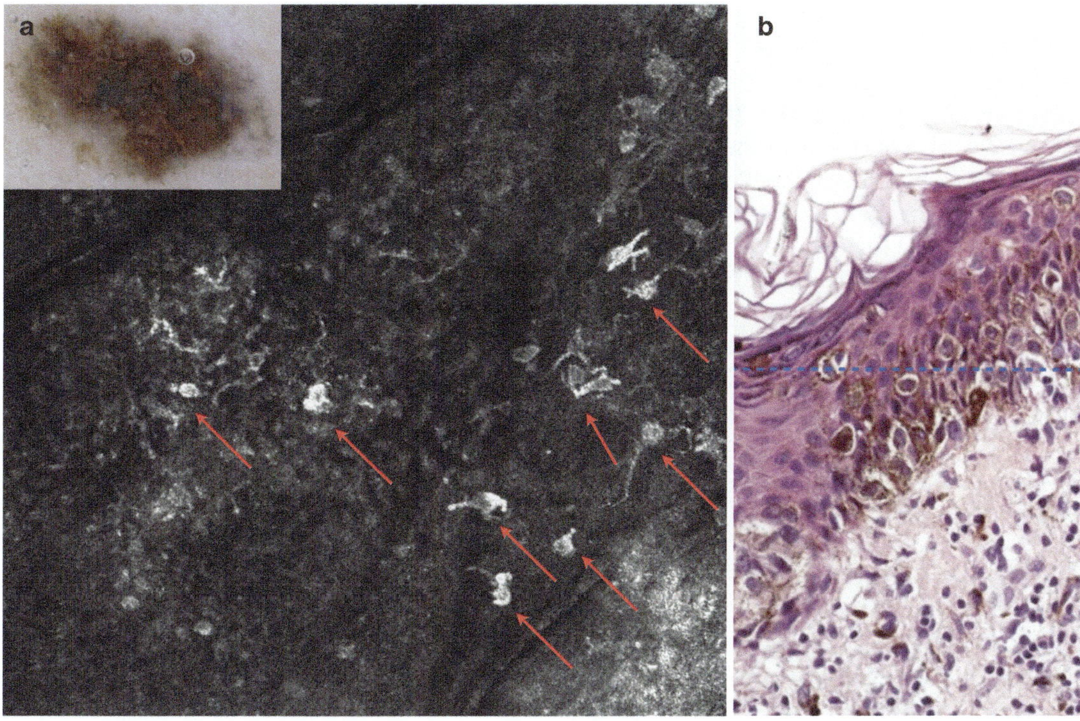

Fig. 6.2 (**a**) Melanoma presenting large bright melanocytes in the epidermis (*arrows*) corresponding to the pagetoid spread (**b**)

so-called dysplastic nevi and the confocal findings such as the bridging of the nests and the inflammatory infiltrate [7]. However, a clear-cut distinction from severe dysplastic nevi to incipient melanoma is not feasible, and it is strongly operator-dependent for either pathologists or clinicians. The morphologic universe of the Spitz tumors represents another difficult scenario. A good correlation between histopathology and confocal microscopy is reached for the pagetoid spread that can be readily detected by the RCM. Unfortunately, a deep assessment of cytologic and architectural changes is not possible by means of RCM, and the management of Spitz tumors is mainly based on clinical aspects and the age of the patient [8].

6.2.2 Melanoma

In the twenty-first century, melanoma diagnosis is based on the integration of clinical and dermatoscopic aspects. With the introduction of dermatoscopy in the clinical setting, the diagnostic accuracy has been greatly improved and the number needed to treat is extremely low in referral centers compared to the past. RCM offers the possibility to improve the specificity compared to dermatoscopy while obtaining the same sensitivity [9, 10]. Dermatoscopy has to be though as the first level for patient's screening since it is a quick, cheap, and easy-to-use tool. In challenging cases, RCM provides more detailed information on the architecture of the lesion and the presence of cytologic atypia. Melanoma represents a family of distinct tumors that differ for their biological behavior and morphologic aspects. One of the histopathologic criteria useful in histopathology is represented by the pagetoid infiltration that represents the suprabasal epidermal location of atypical melanocytes (Fig. 6.2). Pagetoid cells can be readily visualized upon RCM as large and bright cells within the epidermis with a round to oval shape, sometimes presenting dendritic structures [11]. At the dermoepidermal junction, the profile can be altered by a variable nested proliferation of atypical melanocytes. In early melanomas a single-cell proliferation is usually found within the rings that surround the dermal papillae [12–15]. Nests can show up as compact dense nests with atypical melanocytes within the nests or loosely aggregates of nests (sheetlike structures). In deep dermal tumors, the nest tends to be larger outlined by bright collagen septa; this aggregates are called "cerebriform" nests since they resemble the brain cortex. In "pure" nodular melanoma, the dermal component is usually showing aggregates of malignant cells, whereas the pagetoid cells are rarely found with the exception of lesions exhibiting the consumption of the epidermis. Melanoma located on sun-exposed areas such as the face is characterized by a prominent pagetoid spread of dendritic-shaped melanocytes that is preferentially located around the hair follicles that appear as dark roundish structures.

6.3 Benign Epithelial Lesions

6.3.1 Solar Lentigo

The most striking finding in solar lentigo by RCM is observed at the dermoepidermal junction that presents the so-called polycyclic papillary contour or bulbous projections due to the anastomosing elongated structures, separated by dark areas [16]. This bizarre-shaped junction corresponds to the presence of hyperpigmented and elongated rete ridge. The typical RCM aspect of solar lentigo can be found in association with lentigo maligna as a sign of the sun-damaged skin in which lentigo maligna usually arises. For this reason, a careful examination of a wide area is warranted to not miss the atypical melanocytic proliferation of the lentigo maligna.

6.3.2 Seborrheic Keratosis

In clinical practice, the diagnosis of seborrheic keratosis is usually simple and reliably made by a quick dermatoscopic exam. However, a quote of them can be difficult to diagnose. RCM has a limited depth penetration that allows the imaging up to the papillary dermis, but flat or slightly palpable lesions can be confidently diagnosed. Seborrheic keratosis displays a typical architecture at the dermoepidermal junction that shows a ringed pattern with variably spaced papillae due to the presence of acanthosis and papillomatosis. Moreover, elongated cords and bulbous projections represented by round to oval structures contiguous or adjacent to the elongated cords represent common findings. The epidermis shows a cobblestone pattern made by bright polygonal keratinocytes, several cystic inclusions, and milia-like cysts that appear as bright laminar onion-like structures. The top surface is characterized by the presence of holes and fissures (crypts) that appear as round to linear structures, darker than the surrounding epidermal surface.

6.3.3 Lichen Planus-Like Keratosis

Lichen planus-like keratosis (LPLK), also named lichenoid keratosis, represents a keratosis undergoing regression. From a clinical and dermatoscopic point of view, these lesions can be often challenging and in differential diagnosis with lentigo maligna or melanoma with regressive features. At dermatoscopy, LPLK is characterized by the presence of multiple bluish-grayish granules associated or not to the remnant of a keratosis. When the regression is more conspicuous, the granules cover the entire surface and the previous keratosis is no longer visible. This situation complicates the picture even more, and a biopsy is usually performed to rule out the diagnosis of melanoma. The use of RCM in this clinical context is useful since it displays the stereotypical aspect of LPLK seen in histopathology [17, 18]. At the epidermal level, a regular honeycombed pattern is observed along with few bright spots corresponding to inflammatory cells. Typically, at dermoepidermal junction RCM highlights the presence of polycyclic papillary contours that correspond to bulbous epidermal tips with tendency to anastomose. The dermis is occupied by numerous plump bright cells, often clustered and in proximity to vessels. These cells correspond to melanin-laden melanophages.

6.4 Epithelial Tumors

6.4.1 Basal Cell Carcinoma

Basal cell carcinoma (BCC) is the commonest skin tumor in Caucasian and its diagnosis is usually made by clinical inspection and dermatoscopic

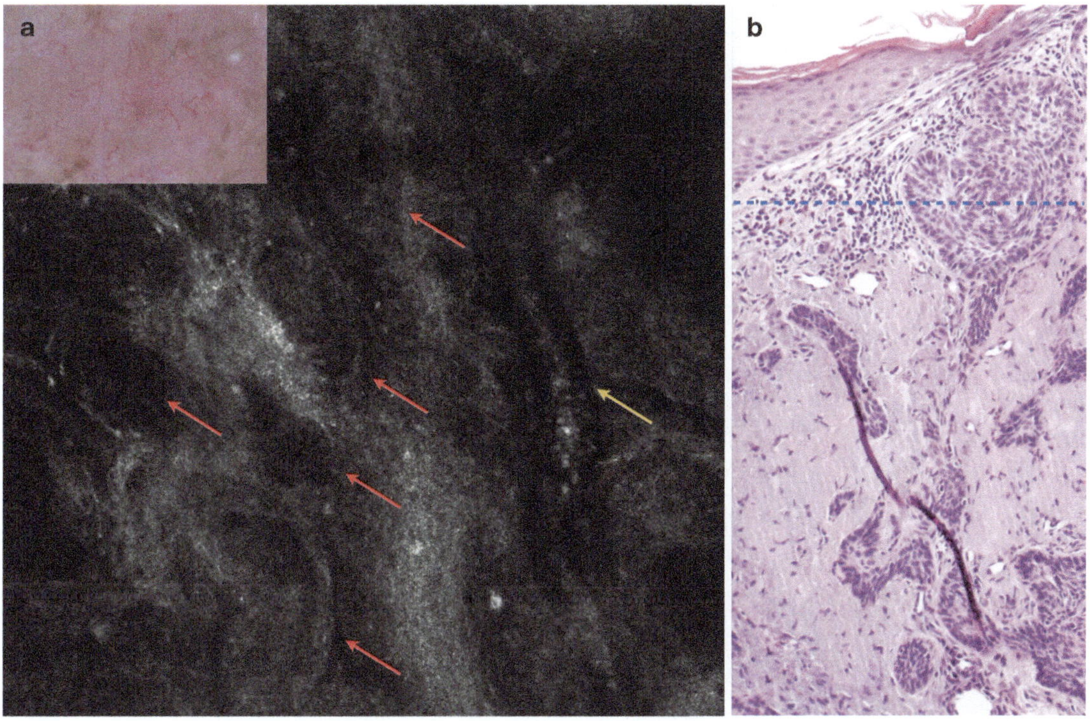

Fig. 6.3 (**a**) Hypopigmented basal cell carcinoma presenting tightly packed aggregates of basaloid cells (*red arrows*) in proximity to a large caliber vessel (*yellow arrow*). (**b**) Histopathologic section showing the typical basaloid island

analysis. A quick and confident diagnosis of BCC is essential to treat these tumors with non-surgical approaches that permit better cosmetic outcomes in case of "cancerization field" or elderly with several comorbidities. RCM provides a rapid diagnosis with high diagnostic confidence since it clearly detects the presence of basaloid islands [19–21]. Upon RCM the basaloid islands appear as tightly packed aggregates with peripheral palisading and lobulated shape. These aggregates are outlined by a dark space (corresponding to mucin) [21] and often surrounded by a prominent vascularity that can be readily detected during live imaging. Interestingly, RCM highlights very well the presence of dendritic melanocytes entrapped within the basaloid islands and observable in a huge number in heavily pigmented BCC. In hypopigmented BCCs the basaloid islands appear as "dark silhouettes" that are hyporeflective area outlined by bright collagen bundles (Fig. 6.3). The shape of the dark silhouettes is variably lobulated and helps to differentiate these aggregates from the surroundings. A florid inflammatory infiltrate is usually seen in association with basaloid islands. The inflammatory cells are plump bright cells that correspond to melanophages.

6.4.2 Actinic Keratosis and Squamous Cell Carcinoma

Actinic keratosis (AK) and squamous cell carcinoma (SCC) usually represent different stages of progression since it has been demonstrated that AK can progress into SCC. Clinically AKs and in situ SCC appear as multiple scaly, reddish macules or papules on sun-exposed sites, whereas invasive SCC is usually a fast-growing papule or nodule with keratotic surface and ulceration.

Upon RCM, AKs show a superficial disruption with single-detached keratinocytes at stratum corneum associated to parakeratosis constituted by nucleated cells with dark center and sharp demarcation [22, 23]. At stratum granulosum, an atypical honeycomb pattern and architectural disarray

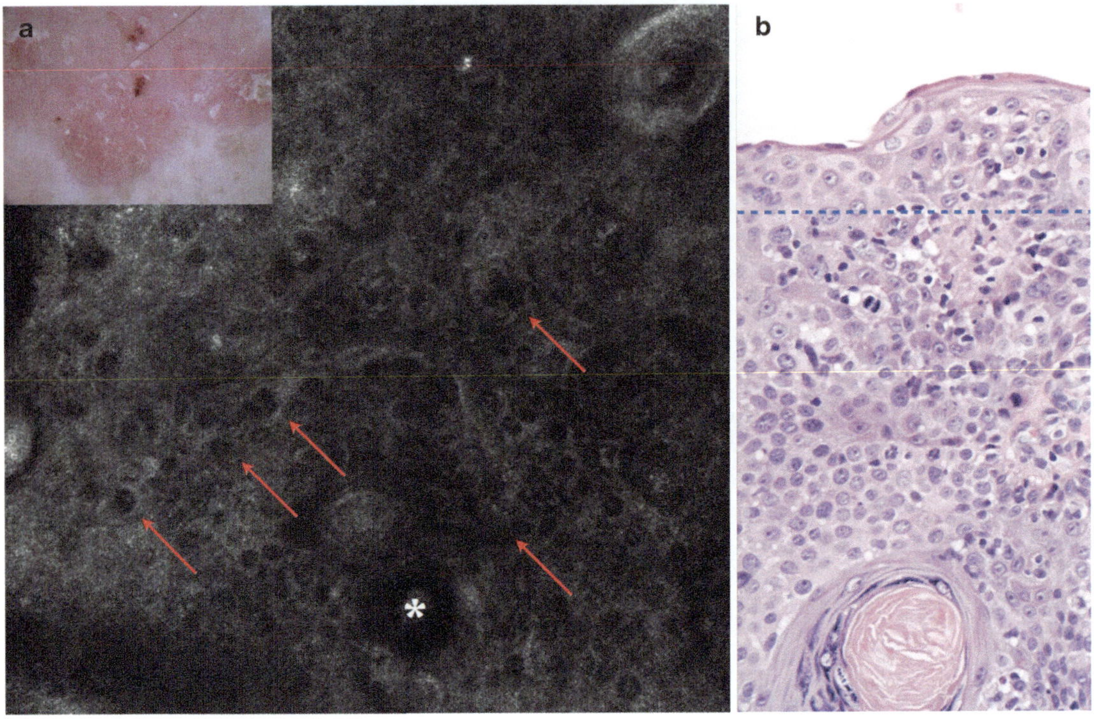

Fig. 6.4 (**a**) Bowen's disease (in situ squamous cell carcinoma) showing an atypical honeycombed pattern with pleomorphic keratinocytes (*arrows*). (**b**) Histopathologic section depicting the keratinocytic atypia at the same depth level (*dashed line*) as seen upon RCM

of variable degree can be correlated to a different degree of keratinocyte dysplasia on histopathologic exam. The atypical honeycombed pattern is formed by variably shaped keratinocytes. In SCC an atypical honeycomb or disarranged pattern of the spinous-granular layer is found along with round nucleated cells corresponding to pleomorphic keratinocytes (Fig. 6.4). Round blood vessels traversing through the dermal papillae perpendicular to the skin surface represent a common finding in SCC and high-grade AK. Few cases of SCC have been reported so far and the specificity of these findings remains to be tested. Moreover, in case of marked hyperkeratosis or abundant scales covering the entire lesion, in-depth imaging is limited by the keratin scattering and this is a limitation when analyzing invasive SCC.

Conclusions

RCM represents a modern imaging tool that permits a good in vivo histopathologic analysis of skin tumors. The great advantage of RCM is not limited to diagnostic purposes, but it is extended to the treatment monitoring since many efforts have been directed to the development of new topical treatment especially for nonmelanoma skin cancer. Moreover, RCM offers the possibility to follow up lesions (i.e., nevi) over time to obtain information on their dynamic evolution. In conclusion, RCM is an emerging tool in the hands of clinicians that holds the capability to explore the histopathologic findings at patient's bedside.

References

1. Rajadhyaksha M, Grossman M, Esterowitz D, Webb RH, Anderson RR (1995) In vivo confocal scanning laser microscopy of human skin: melanin provides strong contrast. J Invest Dermatol 104:946–952
2. Rajadhyaksha M, Gonzalez S, Zavislan JM, Anderson RR, Web RH (1999) In vivo confocal laser microscopy of human skin II: advances in instrumentation and comparison with histology. J Invest Dermatol 113:293–303

3. Scope A, Benvenuto-Andrade C, Agero AL, Malvehy J, Puig S, Rajadhyaksha M, Busam KJ, Marra DE, Torres A, Propperova I, Langley RG, Marghoob AA, Pellacani G, Seidenari S, Halpern AC, Gonzalez S (2007) In vivo reflectance confocal microscopy imaging of melanocytic skin lesions: consensus terminology glossary and illustrative images. J Am Acad Dermatol 57:644–658
4. Ahlgrimm-Siess V, Massone C, Koller S, Fink-Puches R, Richtig E, Wolf I, Gerger A, Hofmann-Wellenhof R (2008) In vivo confocal scanning laser microscopy of common naevi with globular, homogeneous and reticular pattern in dermoscopy. Br J Dermatol 158(5): 1000–1007
5. Pellacani G, Scope A, Ferrari B, Pupelli G, Bassoli S, Longo C, Cesinaro AM, Argenziano G, Hofmann-Wellenhof R, Malvehy J, Marghoob AA, Puig S, Seidenari S, Soyer HP, Zalaudek I (2009) New insights into nevogenesis: in vivo characterization and follow-up of melanocytic nevi by reflectance confocal microscopy. J Am Acad Dermatol 61(6):1001–1013
6. Pellacani G, Longo C, Malvehy J, Puig S, Carrera C, Segura S, Bassoli S, Seidenari S (2008) In vivo confocal microscopic and histopathologic correlations of dermoscopic features in 202 melanocytic lesions. Arch Dermatol 144(12):1597–1608
7. Pellacani G, Farnetani F, Gonzalez S, Longo C, Cesinaro AM, Casari A, Beretti F, Seidenari S, Gill M (2012) In vivo confocal microscopy for detection and grading of dysplastic nevi: a pilot study. J Am Acad Dermatol 66(3):e109–e121
8. Pellacani G, Longo C, Ferrara G, Cesinaro AM, Bassoli S, Guitera P, Menzies SW, Seidenari S (2009) Spitz nevi: in vivo confocal microscopic features, dermatoscopic aspects, histopathologic correlates, and diagnostic significance. J Am Acad Dermatol 60(2): 236–247
9. Gerger A, Koller S, Weger W, Richtig E, Kerl H, Samonigg H, Krippl P, Smolle J (2006) Sensitivity and specificity of confocal laser-scanning microscopy for in vivo diagnosis of malignant skin tumors. Cancer 107:193–200
10. Pellacani G, Guitera P, Longo C, Avramidis M, Seidenari S, Menzies S (2007) The impact of in vivo reflectance confocal microscopy for the diagnostic accuracy of melanoma and equivocal melanocytic lesions. J Invest Dermatol 127:2759–2765
11. Pellacani G, Cesinaro AM, Seidenari S (2005) Reflectance-mode confocal microscopy for the in vivo characterization of pagetoid melanocytosis in melanomas and nevi. J Invest Dermatol 125:532–537
12. Guitera P, Pellacani G, Crotty KA, Scolyer RA, Li LX, Bassoli S, Vinceti M, Rabinovitz H, Longo C, Menzies SW (2010) The impact of in vivo reflectance confocal microscopy on the diagnostic accuracy of lentigo maligna and equivocal pigmented and nonpigmented macules of the face. J Invest Dermatol 130(8):2080–2091
13. Longo C, Rito C, Beretti F, Cesinaro AM, Piñeiro-Maceira J, Seidenari S, Pellacani G (2011) De novo melanoma and melanoma arising from pre-existing nevus: in vivo morphologic differences as evaluated by confocal microscopy. J Am Acad Dermatol 65(3): 604–614
14. Segura S, Pellacani G, Puig S, Longo C, Bassoli S, Guitera P, Palou J, Menzies S, Seidenari S, Malvehy J (2008) In vivo microscopic features of nodular melanomas: dermoscopy, confocal microscopy, and histopathologic correlates. Arch Dermatol 144(10):1311–1320
15. Pellacani G, Cesinaro AM, Seidenari S (2005) In vivo confocal reflectance microscopy for the characterization of melanocytic nests and correlation with dermoscopy and histology. Br J Dermatol 152:384–386
16. Langley RG, Burton E, Walsh N, Propperova I, Murray SJ (2006) In vivo confocal scanning laser microscopy of benign lentigines: comparison to conventional histology and in vivo characteristics of lentigo maligna. J Am Acad Dermatol 55:88–97
17. Bassoli S, Rabinovitz HS, Pellacani G, Porges L, Oliviero MC, Braun RP, Marghoob AA, Seidenari S, Scope A (2012) Reflectance confocal microscopy criteria of lichen planus-like keratosis. J Eur Acad Dermatol Venereol 26(5):578–590
18. Moscarella E, Zalaudek I, Pellacani G, Eibenschutz L, Catricalà C, Amantea A, Panetta C, Argenziano G (2011) Lichenoid keratosis-like melanomas. J Am Acad Dermatol 65(3):e85–e87
19. Nori S, Rius-Díaz F, Cuevas J, Goldgeier M, Jaen P, Torres A, González S (2004) Sensitivity and specificity of reflectance-mode confocal microscopy for in vivo diagnosis of basal cell carcinoma: a multicenter study. J Am Acad Dermatol 51:923–930
20. Agero AL, Busam KJ, Benvenuto-Andrade C, Scope A, Gill M, Marghoob AA, González S, Halpern AC (2006) Reflectance confocal microscopy of pigmented basal cell carcinoma. J Am Acad Dermatol 54:638–643
21. Ulrich M, Roewert-Huber J, González S, Rius-Diaz F, Stockfleth E, Kanitakis J (2011) Peritumoral clefting in basal cell carcinoma: correlation of in vivo reflectance confocal microscopy and routine histology. J Cutan Pathol 38(2):190–195
22. Ulrich M, Krueger-Corcoran D, Roewert-Huber J, Sterry W, Stockfleth E, Astner S (2010) Reflectance confocal microscopy for noninvasive monitoring of therapy and detection of subclinical actinic keratoses. Dermatology 220(1):15–24
23. Rishpon A, Kim N, Scope A, Porges L, Oliviero MC, Braun RP, Marghoob AA, Fox CA, Rabinovitz HS (2009) Reflectance confocal microscopy criteria for squamous cell carcinomas and actinic keratoses. Arch Dermatol 145(7):766–772

In Vivo Reflectance Confocal Microscopy for Inflammatory Skin Diseases' Assessment

7

Marco Ardigò, Marina Agozzino, and Leonardo Abraham

7.1 Introduction

In vivo reflectance confocal microscopy (RCM) is a relatively new technique for real time, en face, non-invasive microscopical imaging of the superficial layers of the skin down to the superficial dermis, with cellular-level resolution close to conventional histopathology. The technology works on the bases of light reflection according to the different reflectance indexes of the different skin structures [1]. RCM gives to clinicians the possibility of a real time and non-invasive "virtual" punch biopsy ranging from 2 to 8 mm in horizontal dimension and 250–300 μm in vertical dimension and providing collection of microscopical features and consequential, immediate "clinical-microscopical" correlation. In specific, RCM has been already successfully tested for the evaluation of several inflammatory, neoplastic skin conditions and has been demonstrated to constitute, in selected cases, an excellent alternative to invasive biopsy. In specific, RCM has been used in several inflammatory skin conditions, such as acute contact dermatitis, discoid lupus erythematosus and psoriasis, and has been correlated with conventional histology in several instances [2–4]. Also pigmentary disorders and more recently hair diseases have been evaluated using confocal microscopy [5, 6].

M. Ardigò, MD (✉) • M. Agozzino
Department of Clinical Dermatology, San Gallicano Dermatological Institute, Via Chianesi, 53, Rome 00144, Italy
e-mail: ardigo@ifo.it; marinaagozzino@libero.it

L. Abraham
Department of Dermatology, Instituto de Dermatologia Prof. Rubem David Azulay, Rio de Janeiro, Brazil
e-mail: dr.lsabraham@gmail.com

7.2 The RCM Technology

The name "confocal" derives from optically "conjugate focal" planes. The device is constituted by a laser point source of light, a condenser, objective lenses and a point detector. A point illumination is achieved by focusing a small source of light into the tissue. The detection of the point is obtained by the introduction of a pinhole in front of the detector that collects light only from the focus blocking light from elsewhere. Scanning the tissue areas in the focal plane non-invasive imaging of a thin (3.5 μm) section can be obtained as grey-scale picture in which bright structures are composed by components with high refractive index compared with their surroundings (backscattered areas). Backscattering is primarily defined by the structures' refractive index (n) compared to the surrounding medium. Highly reflective skin structures include melanin/melanosome ($n = 1.72$), dermal fibres ($n = 1.43$) and keratin ($n = 1.51$) that appear bright when surrounded by epidermis ($n = 1.34$) and dermis ($n = 1.41$). In this way, RCM allows superficial skin non-invasive, real time, high-resolution microscopical evaluation close to histology (lateral resolution of 0.5–1 μm) [7, 8].

E. Berardesca et al. (eds.), *Non Invasive Diagnostic Techniques in Clinical Dermatology*,
DOI 10.1007/978-3-642-32109-2_7,

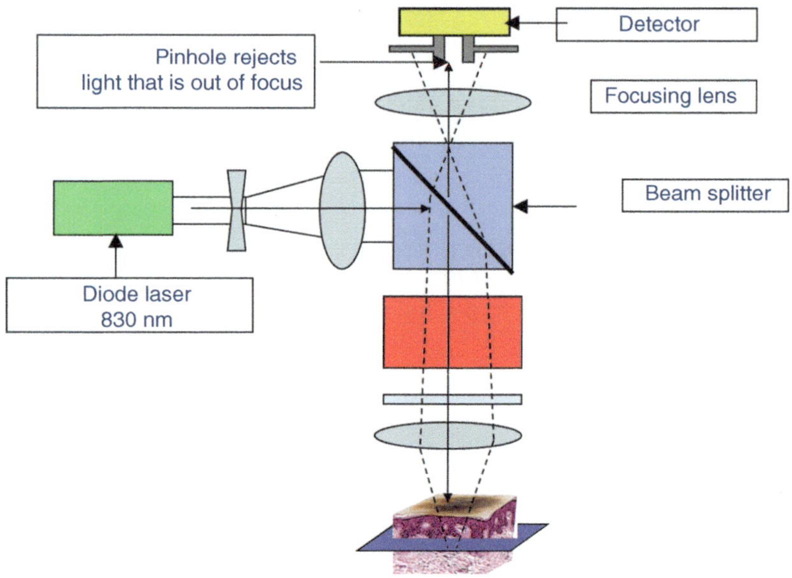

Fig. 7.1 Schematic of in vivo reflectance confocal microscope

A drop of immersion oil or water has to be applied to the skin lesion in order to have a refractive index sufficiently close to that of the stratum corneum. The immersion lens requires to be used with water/water gels placed between the window and objective lens with a refractive index close to that of the epidermis (Fig. 7.1).

RCM requires the use of a metal ring with a plastic window that have to be attached to the skin in order to "connect" the optic to the skin areas to be evaluated. Also, handled devices with similar technical characteristics and that do not require to be "connected" to the skin exist [7].

7.3 RCM for Inflammatory Skin Diseases

In literature, RCM has been already demonstrated to be able to identify both main and secondary microscopical features characterizing specific inflammatory process with high grade of correspondence and correlation to optical histology [3, 4]. This lets the reader, according to the identification of the mainly represented RCM features in the lesion, first to collocate the skin disorder in one of the three main categories of inflammatory skin diseases identifiable with RCM: i.e. presence of:

1. Thick corneum and epidermis → hyperkeratotic and acanthotic dermatitis
2. Spongiosis → spongiotic dermatitis
3. DEJ obscuration due to inflammatory cells → interface dermatitis

Second, RCM let to add to the main criterion the collection of secondary RCM features (i.e. inflammatory cells in the epidermis, in the adnexal epithelium or perivascular in the upper dermis, papillomatosis, dilated vessels) necessary for a better definition of the inflammatory process and consequently better support the clinical diagnosis.

7.3.1 Inflammatory Diseases

As previously mentioned, collecting RCM features and using the pattern method, specific inflammatory processes can be identified. In specific, lupus erythematosus, lichen planus, psoriasis and spongiotic dermatitis have been studied with RCM disclosing a high grade of correspondence between confocal criteria and histological criteria with promising clinical application in diagnosis and follow-up.

Confocal microscopy of psoriasis is mainly characterized by the presence of increased thickness of the stratum corneum (>40 μm) and epidermis (>90 μm) according to the different skin sites as the main criterion [4]. Moreover, increased density and diameter of dermal papillae already visible at the level of the upper layers of

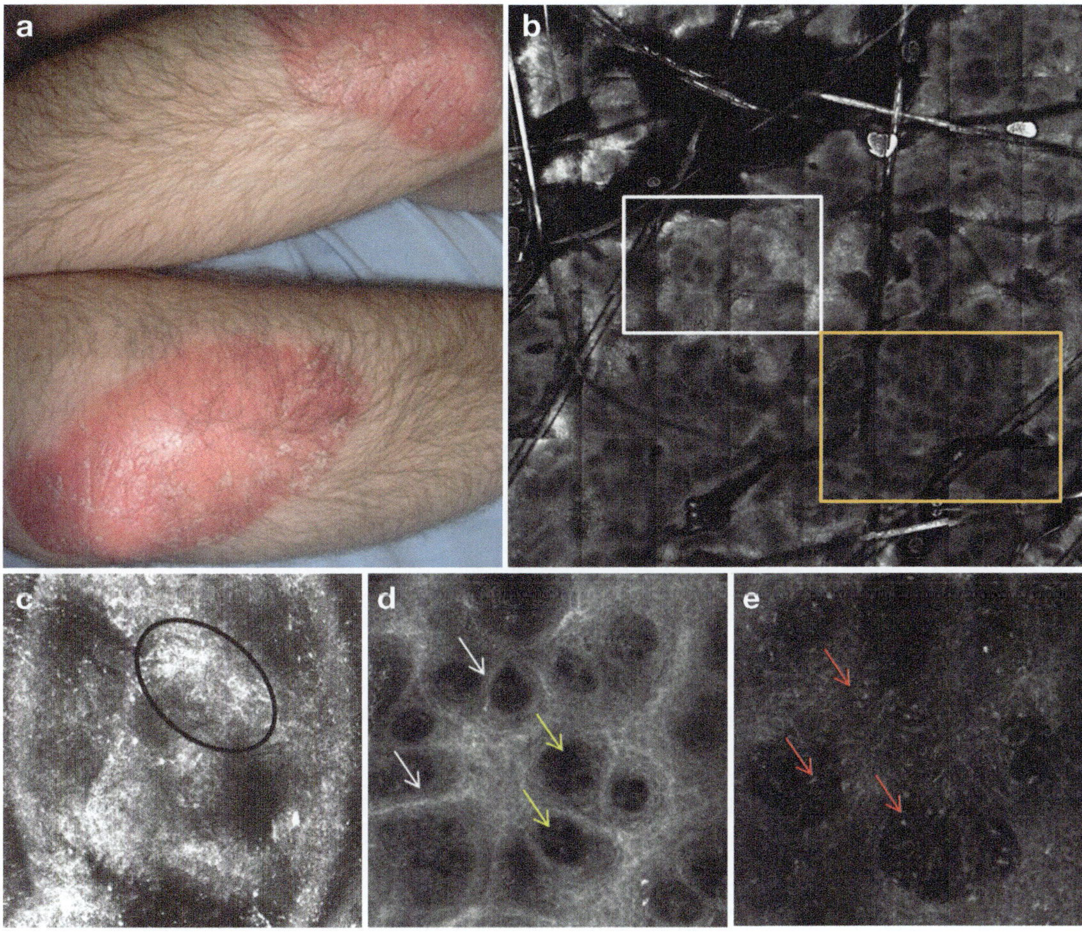

Fig. 7.2 Plaque psoriasis (**a**); 4-by-4 mm RCM mosaic shows hyperkeratotic areas (*white square*), up located, enlarged dermal papillae with no rimming (*yellow square*) (**b**); 0.5-by-0.5 mm single image taken at the level of the stratum corneum with parakeratosis (*black circle*) (**c**); image from the DEJ shows thin inter-papillary septa (*white arrows*) associated with enlarged DP filled by dilated blood vessels (*yellow arrows*) (**d**); image taken at the level of the upper dermis evidences the presence of inflammatory cells (*red arrows*); (**e**) presence of inflammatory cells in the papillary dermis around blood vessels

the stratum spinosum, as sign of papillomatosis, are also typical for psoriasiform and other hyperkeratotic diseases. Dermal papillae are filled by dilated vessels that can be easily visualized using RCM [4] (Fig. 7.2).

The main confocal feature characterizing interface dermatitis (i.e. lupus erythematosus and lichen planus) is the obscuration of the DEJ by inflammatory cells distributed focally or in sheet at the interface between the dermis and epidermis. Inflammatory cells, in those cases, can be also seen at the level of the epidermis as well as in the upper dermis around vessels as secondary criteria. Distinctive RCM criteria are dermal fibres that are generally more thickened on lupus than in lichen and necrotic keratinocytes that are usually more visible and represented in lichen than in lupus; adnexa infundibular hyperkeratosis is more commonly seen in lupus [3].

Spongiotic dermatitis under confocal microscopy is characterized by the presence of inflammatory cells up-migrated into the epidermis as single or in cluster to the vesicle formation. Spongiosis is seen under RCM as the presence of darker areas in comparison with the surrounding epithelium, with broad band intercellular spaces and associated with round to oval bright cellular structures between keratinocytes spaces [9, 10].

Dilated vessels and dermal inflammation can be also associated as secondary criteria.

7.3.2 Hair Diseases

In literature, only few papers on RCM microscopical anatomy of scalp and pathophysiology of related diseases have been published [11–14], but preliminary data about RCM application in trichology are extremely promising.

As previously mentioned, the intrinsic horizontal, in vivo RCM approach to the tissue supports the evaluation of the skin layer by layer from the top to the upper dermis similarly to as required in optical histology of scalp diseases. In specific, the scalp tissue can be visualized, starting from the stratum corneum down to the upper dermis (around 250 um), letting to evaluate all the thickness of the epidermis, the upper dermis, the hair shafts and the upper part of the adnexal infundibular epithelium (focusing on the opening) and the surrounding stroma. In trichology, RCM can be considered as an intermediate step between trichoscopy and horizontal histology.

Using RCM, information about adnexal structures distribution and density, hair shaft integrity and dimension, microscopical distribution and amount of inflammatory cells in different epidermal layers, superficial adnexal structures and upper dermis can be easily obtained. Moreover, dermal scarring, inflammation and vessels can be also examined.

In detail, during the last 10 years, RCM data about the normal skin anatomy have been collected comprehensive of adnexal structures. RCM features of hair shaft as well as glands and their infundibular epithelium have been studied demonstrating that RCM is able to visualize the hair shaft structures showing the medulla (visible as a strongly bright structure), the shaft and the cuticle. Also the infundibular epithelium and the openings can be evaluated with RCM.

Starting from the experience on normal, RCM features of abnormal hair shaft have been also anecdotally reported (i.e. uncombable hair syndrome, pili torti, trichothiodystrophy) describing the alteration in hair and/or medulla thickness and uniformity of the cuticula [13, 14].

Also alopecias have been considered for RCM. In cicatricial and non-cicatricial alopecia, dermoscopical/confocal correlation has been done starting from the comparison between dermoscopy and RCM mosaics. The advantage deriving from the use of RCM stays in the possibility of an immediate and non-invasive close-up to the tissue adding to dermoscopy, microscopical information concerning the histological correspondence of dermoscopical structures (yellow dots) [15] and white dots [16] (vellus hair, etc.). Contemporary, location and amount of inflammatory cell involving adnexa, epidermis and upper dermis can be also examined (Fig. 7.2).

Moreover, in the case of cicatricial alopecia, presence of a scarred upper dermis can be easily evaluated as well as the remnant of thick collagen bundles in the site of the previously present hair follicle (Fig. 7.3).

7.3.3 Pigmentary Diseases

Acquired hyperpigmentations are characterized microscopically by pigment deposition in the epidermis (brown macules) or in the dermis (grey/blue macules) or in both the skin layers (brown to blue macules). Identification of its prevalent localization influences significantly the treatment selection [17]. The two most common hyperpigmentary conditions are melasma/chloasma and postinflammatory pigmentation. On the other side, the most common skin disorder characterized by the development of white skin macules, due to the reduction or disappearing of melanocytes from the dermo-epidermal junction, is vitiligo. The effective possibility to evaluate pigmentary skin diseases using RCM has been already demonstrated with success in literature. In specific cases, skin biopsy can be useful, but avoided by the patient and not applicable routinely in the therapeutical follow-up. RCM have been also demonstrated to be powerful in the follow-up and patient's management giving the possibility of a fine modulation of the treatment and reducing side effects and monitoring the response to treatment [5, 6].

In specific, RCM used on vitiligo lesional and nonlesional skin provides non-invasively, microscopical information useful for disease

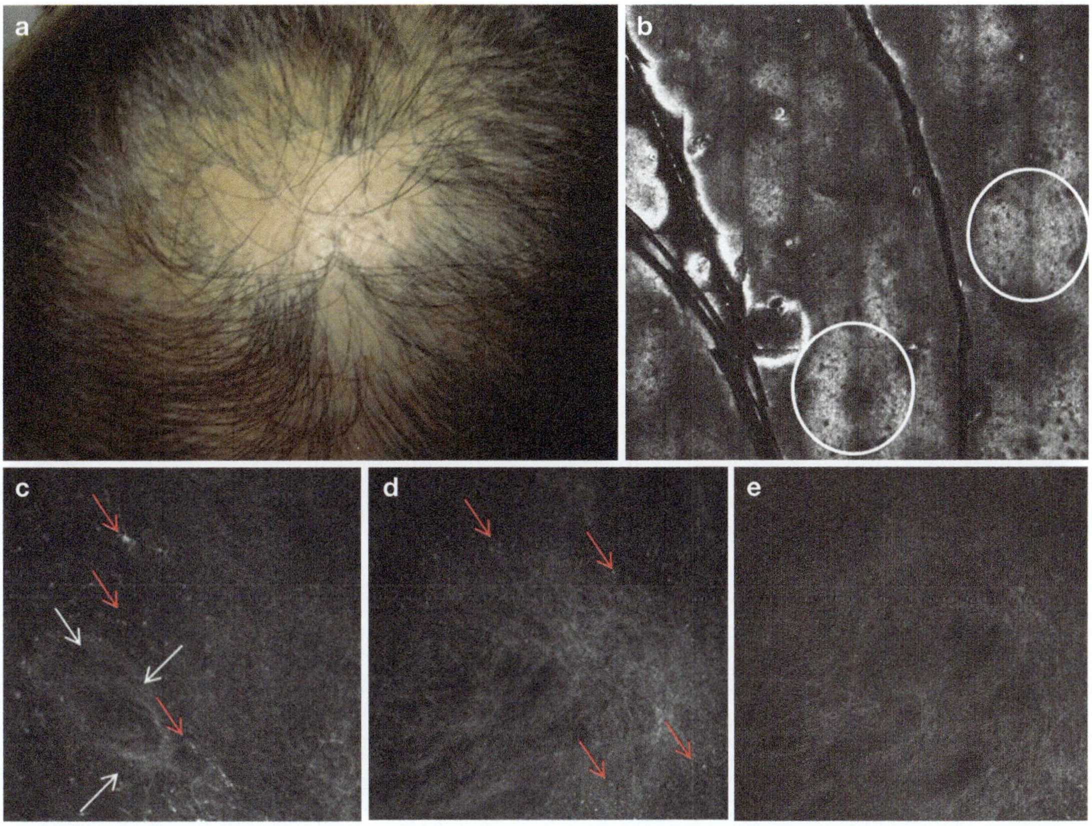

Fig. 7.3 Cicatricial alopecia, lupus erythematosus type (**a**); 4-by-4 mm RCM mosaic taken at the level of the upper dermis shows thicker and bright dermal fibres as sign of scar (*white circles*) (**b**); 0.5-by-0.5 mm single image taken at the level of the stratum corneum with inflammatory cells (*red arrows*) in the epidermis and obscuring the DEJ (*white arrows*) (**c**); image from the upper dermis showing inflammatory cells between dermal fibres (*red arrows*) (**d**); image of the upper dermis showing detail of the thick and disarrayed dermal fibres, a sign of scarring alopecia (**e**)

management. Vitiligo lesions show disappearance of the normal brightness at dermo-epidermal junction level normally visible in normal skin and due to the presence of pigmented keratinocytes and melanocytes at the DEJ. At RCM only remnant of a "shadow" of the pre-existing papillary ring can be seen. Moreover, bright keratinocytes, seen in normal skin above the DEJ in higher phototypes, are generally absent in vitiligo lesions. The disappearance of brightness (i.e. pigment) at the dermo-epidermal junction level or above fits perfectly with the progressive loss of melanocytes and the reduction of epidermal pigmentation previously demonstrated with histopathology and histochemistry [5].

After treatment (UVB-narrow band), repigmented areas show a variable number of activated melanocytes located at the dermo-epidermal junction. Activated melanocytic cells can be seen as bipolar or stellate dendritic structures usually located around adnexal structures [5].

Differently, melasma/chloasma is characterized by an increased deposition of pigment at the different skin layers. Generally the pigment deposition involves prevalently the epidermis and the dermo-epidermal junction, but the presence of pigment in the dermis is constantly seen: recently, presence of bright pinpoint elements that can correspond to "free" melanosomes in the upper dermis that have been lost from junctional defective or injured melanocytes [6] (Fig. 7.4). Postinflammatory pigmentations may occur at any site of previous existing acute or chronic inflammatory processes. RCM of postinflammatory pigmentation is characterized by a typical distribution of pigment prevalently at the dermo-epidermal junction with the evidence

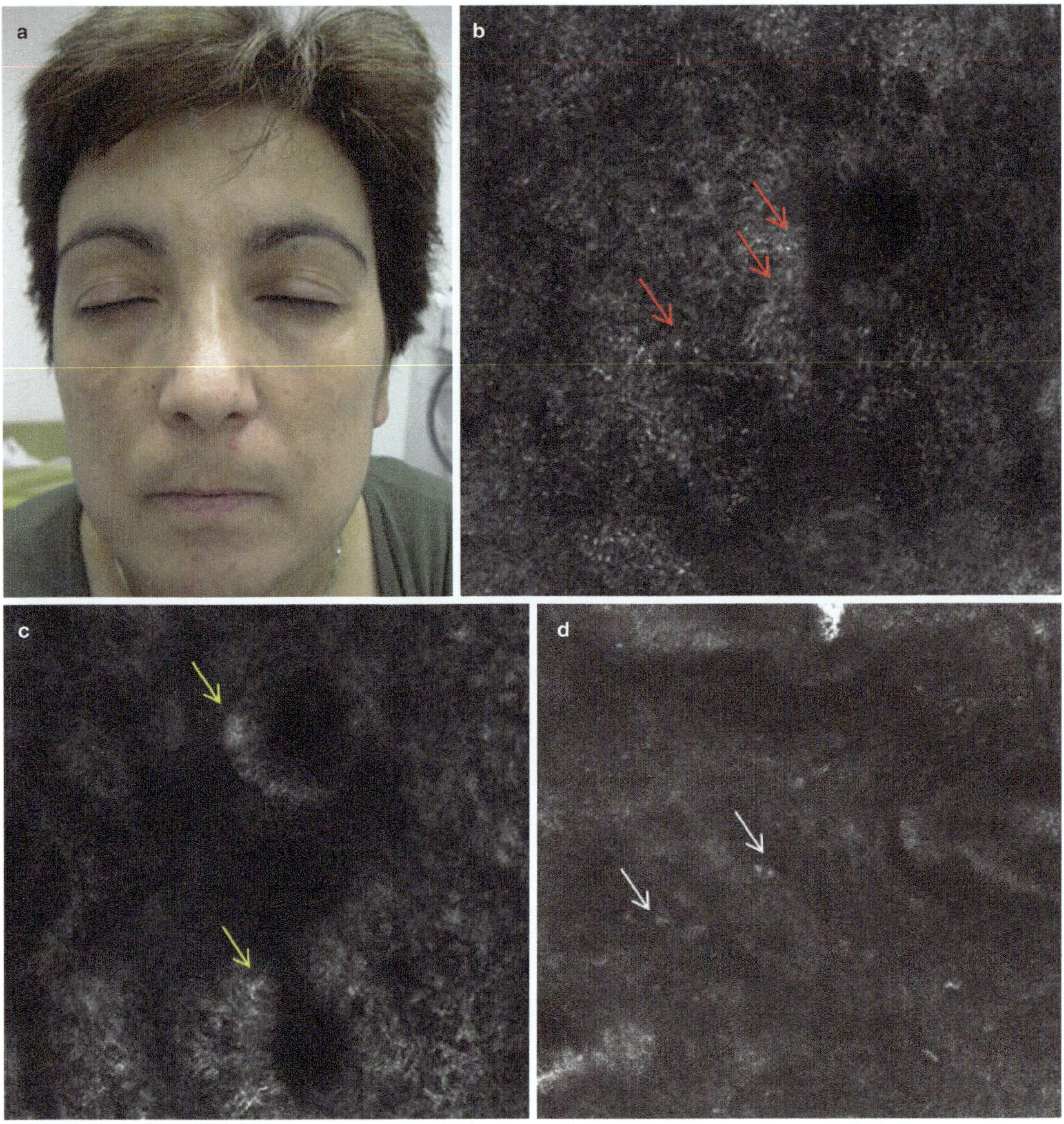

Fig. 7.4 Melasma (**a**); 0.5-by-0.5 mm single image taken at the level of the spinous layer shows strongly (pigmented) keratinocytes (*red arrows*) (**b**); adnexal structures epithelium, corresponding to the DEJ on the face, is strongly bright because of the increased pigmentation (*yellow arrows*) (**c**); melanophages are visible at the level of the upper dermis (*white arrows*) (**d**)

of strongly bright rims associated with a variable involvement of the epidermis that is characterized by evident increased brightness of keratinocytes of the spinosum in early stage of the pigmentation but by a less commonly seen bright cobblestone pattern than in melasma. Typically, no or closely absent bright melanophages in the upper dermis are seen in the postinflammatory pigmentation pattern of distribution of pigment. The last does not fit with the histopathological description of the postinflammatory pigmentation in which melanophages have been described involving the deeper part of the papillary dermis and/or in reticular dermis [6]. The explanation of it can be related to the limit of penetration of the RCM in the skin tissue, but let the definition of the two patterns of distribution of pigment in melasma vs postinflammatory inflammation.

7.4 Limits of RCM in Inflammatory Skin Diseases' Evaluation

Both literature and clinical routine application have been demonstrated how RCM represents a valid device for inflammatory cases management. On the other side, some limits exist affecting the method. First of all, the limit of penetration to the papillary dermis (250 μm) does not permit the visualization of the inflammatory process in the dermis when deep (i.e. lupus tumidus, sarcoidosis). Moreover, in severely acanthotic epidermis, the dermis cannot be visualized with RCM, and limited information to the epidermis can be obtained. Also in trichological application, the major limit of RCM for adnexal structure evaluation is represented by the tissue depth that can be visualized excluding the possibility to get information about the hair follicle and glands and the impossibility of subtype of leucocyte discrimination. Discrimination between the different subtype of leukocytes and the different dermal fibres is affected by the obvious absence of a specific staining (as haematoxylin and eosin for optical histology), limiting the interpretation of the significance of the inflammatory cells infiltrate (i.e. eosinophils in allergic processes). Some software are under evaluation and test for virtual staining of the tissue.

Conclusion

RCM opened and is still opening new frontier for dermatologist involved in the diagnosis and management of inflammatory skin diseases giving the possibility of better sustained clinical and therapeutical decision and letting treatment follow-up with the advantage of non-invasive, infinitively repeatable exams for microscopical information.

References

1. Rajadhyaksha M, Anderson RR, Webb RH (1999) Video-rate confocal scanning laser microscope for imaging human tissues in vivo. Appl Optics 38:2105–2115
2. Gonzalez S, Gonzalez E, White WM, Rajadhyaksha M, Anderson RR (1999) Allergic contact dermatitis: correlation of in vivo confocal imaging to routine histology. J Am Acad Dermatol 40(5 Pt 1):708–713
3. Ardigo M, Maliszewski I, Cota C, Scope A, Sacerdoti G, Gonzalez S et al (2007) Preliminary evaluation of in vivo reflectance confocal microscopy features of discoid lupus erythematosus. Br J Dermatol 156(6):1196–1203
4. Ardigo M, Cota C, Berardesca E, González S (2009) Concordance between in vivo reflectance confocal microscopy and histology in the evaluation of plaque psoriasis. J Eur Acad Dermatol Venereol 23:660–667
5. Ardigo M, Malizewsky I, Dell'Anna ML, Berardesca E, Picardo M (2007) Preliminary evaluation of vitiligo using reflectance confocal microscopy. J Eur Acad Dermatol Venereol 21:1344–1350
6. Ardigo M, Cameli N, Berardesca E, Gonzalez S (2010) Characterization and evaluation of pigment distribution and response to therapy in melasma using in vivo reflectance confocal microscopy: a preliminary study. J Eur Acad Dermatol Venereol 24(11):1296–1303
7. Rajadhyaksha M, González S, Zavislan JM et al (1999) In vivo confocal scanning laser microscopy of human skin II: advances in instrumentation and comparison to histology. J Invest Dermatol 113:293–303
8. Rajadhyaksha M, Grossman M, Esterowitz D et al (1995) Video-rate confocal scanning laser microscopy for human skin: melanin provides strong contrast. J Invest Dermatol 104:946–952
9. Swindells K, Burnett N, Rius-Diaz F, Gonzalez E, Mihm MC, Gonzalez S (2004) Reflectance confocal microscopy may differentiate acute allergic and irritant contact dermatitis in vivo. J Am Acad Dermatol 50(2):220–228
10. Astner S, Gonzalez S, Gonzalez E (2006) Noninvasive evaluation of allergic and irritant contact dermatitis by in vivo reflectance confocal microscopy. Dermatitis 17(4):182–191
11. Ardigò M, Torres F, Abraham LS, Piñeiro-Maceira J, Cameli N, Berardesca E, Tosti A (2011) Reflectance confocal microscopy can differentiate dermoscopic white dots of the scalp between sweat gland ducts or follicular infundibulum. Br J Dermatol 164(5):1122–1124
12. Ardigò M, Tosti A, Cameli N, Vincenzi C, Misciali C, Berardesca E (2011) Reflectance confocal microscopy of the yellow dot pattern in alopecia areata. Arch Dermatol 147(1):61–64
13. Rudnicka L, Olszewska M, Rakowska A (2008) In vivo reflectance confocal microscopy: usefulness for diagnosing hair diseases. J Dermatol Case Rep 2(4):55–59
14. Agozzino M, Tosti A, Barbieri L, Moscarella E, Cota C, Berardesca E, Ardigò M (2011) Confocal microscopic features of scarring alopecia: preliminary report. Br J Dermatol 165(3):534–540
15. Tosti A, Whiting D, Iorizzo M et al (2008) The role of scalp dermoscopy in the diagnosis of alopecia areata incognita. J Am Acad Dermatol 59:64–67
16. Kossard S, Zagarella S (1993) Spotted cicatricial alopecia in dark-skin. A dermoscopic clue to fibrous tract. Australas J Dermatol 34:49–51
17. Taieb A, Picardo M (2007) The definition and assessment of vitiligo: a consensus report of the Vitiligo European. Pigment Cell Res 20:27–35

8 In Vivo Reflectance Confocal Microscopy for Oral Mucosa Assessment

Maria Contaldo, Marina Agozzino, and Marco Ardigò

8.1 Introduction

On the basis of its proven effectiveness in healthy and pathological skin imaging [1–6], in vivo reflectance confocal microscopy (RCM) may also be advantageously used to evaluate healthy and pathological oral mucosa as previously reported in literature [7, 8]. The possibility of a non-invasive imaging of oral mucosa represents an interesting solution for screening, diagnosis and follow-up of that small mucosal sites where biopsy—representing at the moment the gold standard for diagnosis in stomatology—is accompanied by defects as time consuming (10 days or more) for histopathological analysis, morbidity and postoperative complications due to the surgical wound, low patient's compliance and need for multiple biopsies for extended or plurifocal lesions. Thus, in order to enhance the diagnostic management of oral mucosa diseases, RCM is a valid support to orient the clinicians toward a non-invasive diagnosis. Potentially, each oral disease could benefit by the use of a non-invasive device that allows to reduce the diagnosis timing, to be less invasive and more comfortable and to evaluate the follow-up and the drug response.

Furthermore, RCM applied to oral mucosa showed better resolution and contrast, and deeper penetration through the epithelial and subepithelial layer, than in skin analysis, thanks to the absence of cornified superficial layer (with no consequent light backscattering). However, on the other hand, because of the peculiar anatomical and topographical oral mucosa features (limited accessibility, presence of concavity and convexity and mucosal sites not everywhere supported by bones and rigid structures), a dedicated handheld confocal microscope is needed for oral mucosa imaging to allow to overcome limitations due to the motions of patient and operator, its not-always-easy manageability, its shortness that does not allow to reach such sites too far (retromolar trigone, soft palate, tonsils) and its rigidity that arrests its tips to curved surfaces, such as hard palate and gingiva (Fig. 8.1).

8.2 RCM Description of Normal Oral Mucosa

RCM has already been described in literature to be useful for oral mucosa microscopical imaging [7, 8]. Both distinctive features and similarities between normal oral mucosa and normal skin can be observed with RCM. As already assumed, oral mucosa is made up of stratified squamous epithelium, underlined by submucosa and, according

M. Contaldo
Department of Odontostomatological, Orthodontic and Surgical Sciences, Second University of Naples, Naples, Italy
e-mail: maria_contaldo@hotmail.it

M. Agozzino • M. Ardigò, MD (✉)
Department of Clinical Dermatology, San Gallicano Dermatological Institute, Via Chianesi, 53, Rome 00144, Italy
e-mail: marinaagozzino@libero.it; ardigo@ifo.it

E. Berardesca et al. (eds.), *Non Invasive Diagnostic Techniques in Clinical Dermatology*, DOI 10.1007/978-3-642-32109-2_8, © Springer Berlin Heidelberg 2014

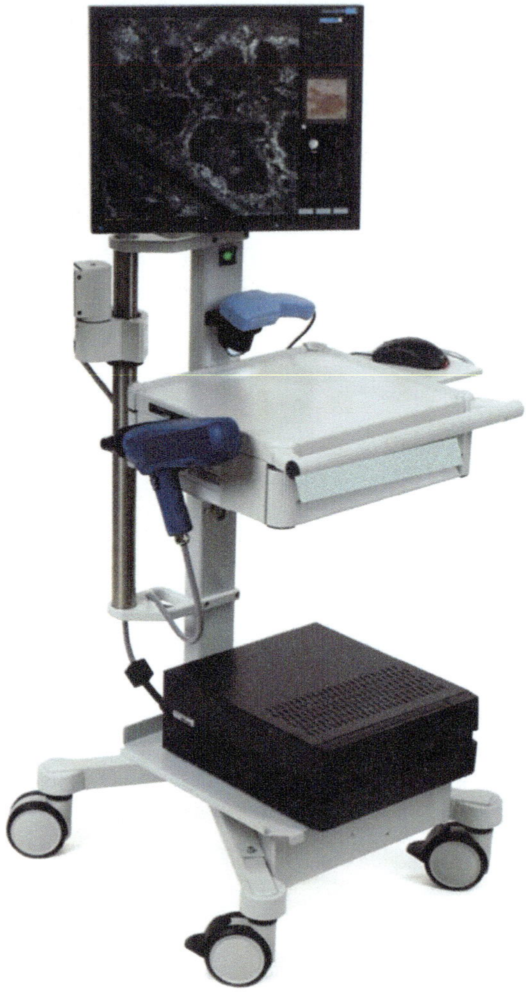

Fig. 8.1 Handheld reflectance confocal microscope that allows to image majority of the oral mucosa sites, thanks to its design

to its histological features and functions, it is traditionally divided into lining mucosa, masticatory mucosa and specialised mucosa. The differences between the different mucosa subtypes can be easily visualised by RCM, showing different descriptors.

8.2.1 Lining Mucosa

Lining mucosa, which covers structures such as lips, cheeks, floor of the mouth, ventral tongue and soft palate (Fig. 8.2a), is made up by a non-cornified stratified squamous epithelium. On RCM, its superficial layer can be visualised as predominantly constituted by large cells recognisable by well-defined and hyper-reflecting borders, centred by a bright roundish structure that accordingly to cellular proportions can be interpreted as the nucleolus. The nucleolus is surrounded by a dark, roundish halo corresponding to the nucleus as well known in RCM analysis of the skin. The cytoplasm is large, dark to mildly refractile and filled by bright grainy structures usually not visible in the normal skin (Fig. 8.2b). Going toward the lower layers, cells decreased in diameter until at the chorion-epithelial junction (CEJ), where cells appear more ellipsoidal and smaller, with polygonal shape and connected one to the other by a thin strongly bright outline, endowed by cytoplasm slightly darker and neither nucleoli nor intracellular bright grainy structures. This architecture describes a frosted glass-like pattern (Fig. 8.2c) in contrast with the honeycombed pattern usually visible in the skin. Dermal papillae appear as roundish to oval or elongated dark areas delimited by basal keratinocytes and filled by clearly visible blood vessels. Dermal papillae are not rimmed by a bright contour because of the absence of pigment. Vessels appear generally horizontally oriented (parallel to the en face section) (Fig. 8.2d). Upper chorion shows connective fibres, recognisable as a bright reticulate structure underneath the epithelium, with no possibility to distinguish individual mesenchymal cells (Fig. 8.2e).

8.2.2 Masticatory Mucosa

Masticatory mucosa is present on gingiva and hard palate, and it shows similarity closer to the skin than lining mucosa, as the dense dermal papillae, but also differences, as lack of reamed papillae, pigment and annexes. In detail, due to its protective role from mechanical stresses, masticatory mucosa is more rigid than covering mucosa, and it is tough and tightly bound to underlying bones by dense connective tissue. At confocal imaging, its superficial layers, made up by cornified stratified squamous epithelium, show a thin *stratum corneum* as uniform, strongly

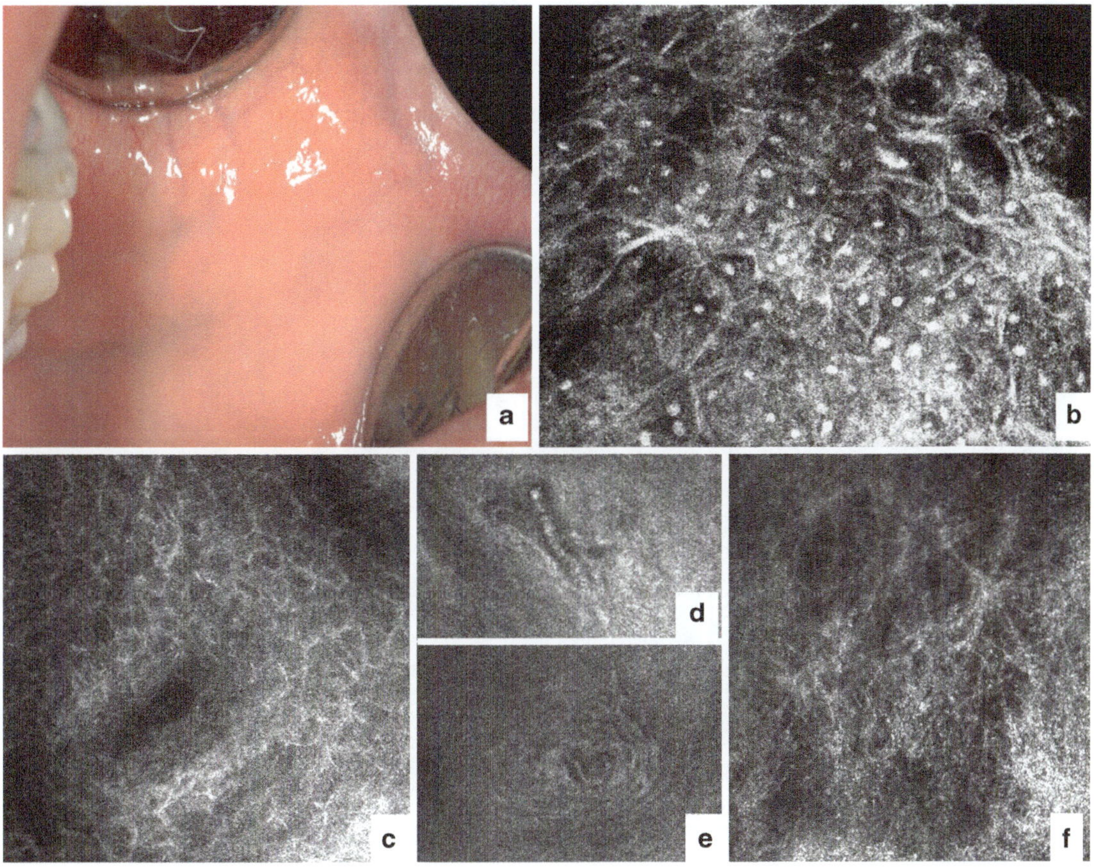

Fig. 8.2 Covering oral mucosa. (**a**) Clinical examination. (**b**) Superficial layer. A carpet of large cells characterised by bright outlines and small roundish highly refractile nucleolus surrounded by nuclear dark halo. Grainy cytoplasm inside. (**c**) Going deeper toward lower epithelial layers, cells appear smaller and organised in a frosted glass-like pattern. (**d**, **e**) Detail of a dermal papilla with horizontal blood vessels inside. (**f**) Submucosal layers shows connectival fibres recognisable as bright reticuli

bright with no visible cells, except in some roundish and less refractile areas, where islands of small keratinocytes of the lower layers could be visualised as honeycomb-like pattern. Neither bright nucleoli nor dark halos corresponding to nuclei can be observed. At the *stratum spinosum*, keratinocytes become identifiable as small polygonal cells with strong bright borders and strongly grainy grey cytoplasm, regularly organised. At this level highly densely expressed dermal papillae are just visible because of the high epithelial connectival interdigitation of this site to allow the absorption of mechanical stress and to offer most nourishment to the epithelium. Also at the CEJ, masticatory mucosa keratinocyte arrangement resembles the honeycombed pattern, as seen in the skin, due to the similar maturation pattern of these two anatomical sites. According to its function and its close attachment to the underlying bones, gingival upper chorion shows regular dense bright connectival fibres.

8.2.3 Specialised Mucosa

Specialised mucosa is located on the dorsal and lateral surface of the tongue and contains lingual papillae and taste buds (Fig. 8.3a). Four kinds of lingual papillae exist, according to their shape and function: *filiform papillae*, *circumvallate papillae*, *fungiform papillae* and *foliate papillae*. *Filiform papillae* are visible on the dorsal surface as flexible

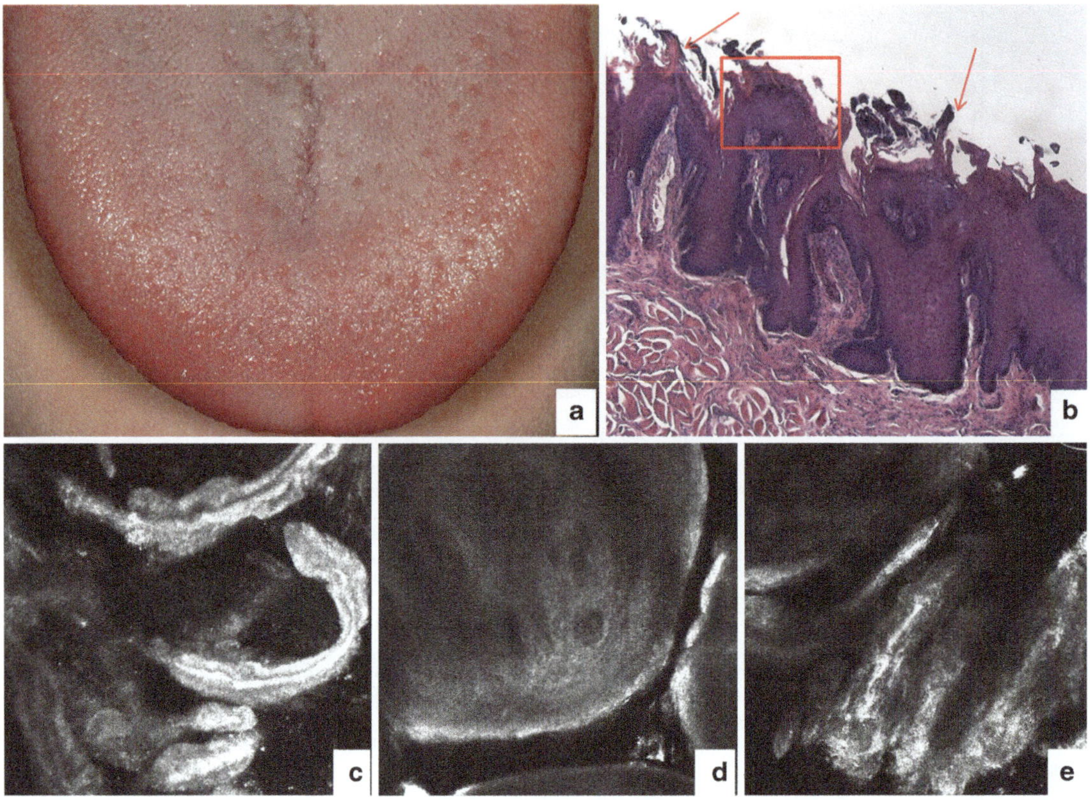

Fig. 8.3 Specialised oral mucosa of the tongue. (**a**) Clinical picture. (**b**) Filiform papillae. (**c**) Fungiform papillae. (**d**) Foliate papillae

and elongated structures (Fig. 8.3b). *Fungiform papillae* appear as well-defined roundish islands separated by dark clefts (Fig. 8.3c). Pressing them with the RCM optic, taste buds can be visualised. Similarly to conventional histology, taste buds appear at RCM as flask-like structures made up by spindle-shaped cells, converging into a hyper-refractile "pore" and full of granules with large spherical bright nucleoli, surrounded by the dark nuclear halos. *Foliate papillae* appear as densely packed, elongated structures larger than the filiform papillae and can be visible on the tongue margins (Fig. 8.3d).

8.3 Preliminary RCM Description of Oral Diseases

As in skin, inflammatory and neoplastic diseases, thoroughly documented in literature [1–6], RCM may also be useful to orient diagnosis and to suggest therapeutic guidance of lesions and pathologies affecting oral mucosa.

In clinically suspicious "white lesions" affecting the oral mucosa, RCM can help to distinguish between non-neoplastic alterations (as leukoplakia, oral lichen planus, frictional hyperkeratosis) and early cancer on the basis of identification criteria that are proven to appear only in cancerised mucosa, in our preliminary observations.

Leukoplakia is a chronic white mucosal patch which is not due to any identifiable disease [9]. The term, purely clinical, needs histological examination in order to exclude malignancies and other benignant diseases as frictional keratosis. Since leukoplakia is considered a precancerous lesion and can express early signs of transformation as dysplasia, RCM imaging allows to study in a non-invasive way its histological features, orienting the stomatologist to the correct treatment. RCM leukoplakia images show normal, thick stratum corneum, typical of these lesions, and

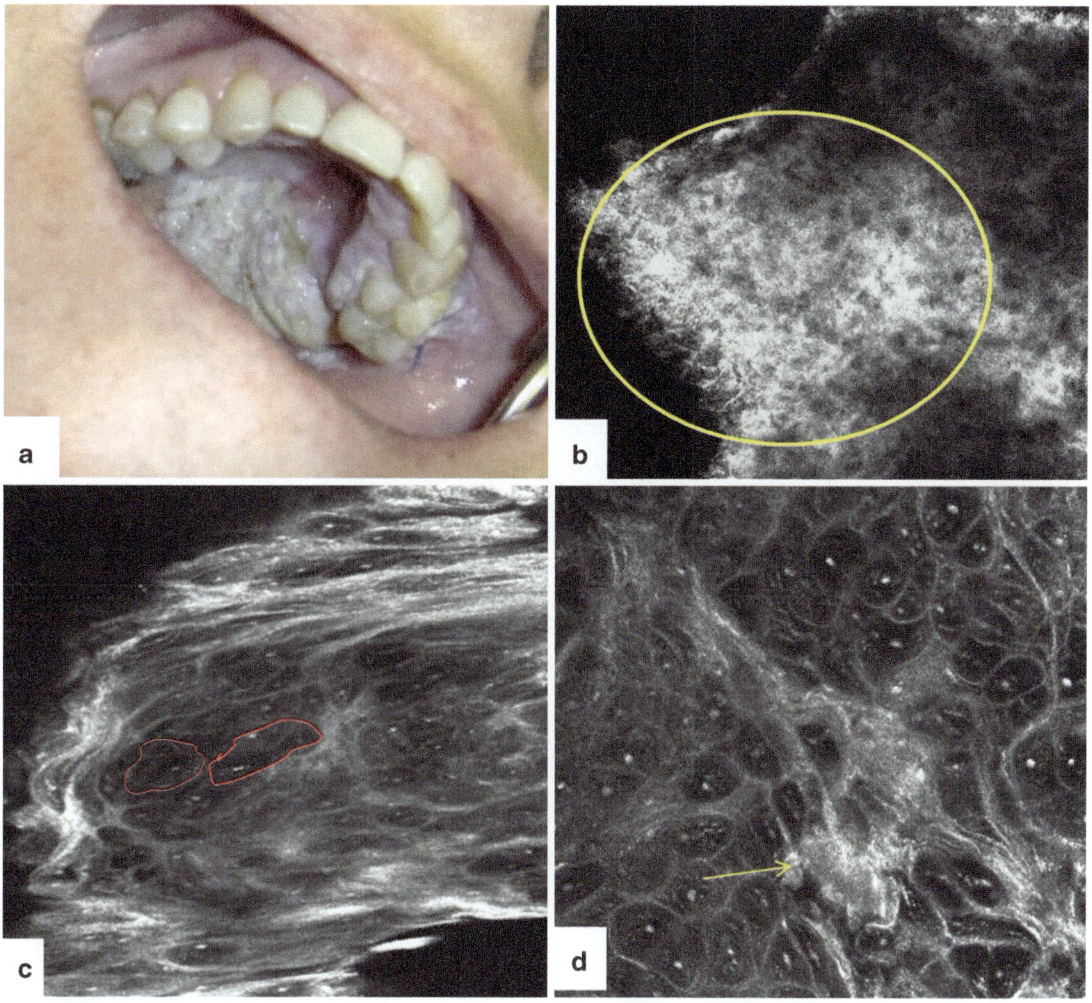

Fig. 8.4 Oral squamous cell carcinoma (OSCC). (**a**) Clinical picture showing …. (**b**) At RCM multiple signs of cancerisation can be seen: Pleomorphism is revealed by the presence of different-in-size monstrous cells, ranging from small to very large and elongated ones, disarranged and not orderly organised. (**c**) Detail showing polynucleolated cells and a great amount of keratinisation

epithelial thickening with presence, sometimes, of inflammatory cells at the CEJ. When malignancy does not occur, RCM is not associated to cellular morphology atypia in terms of diameter, pleomorphism and structure (multiple nucleoli) as expression of epithelial cancer.

On the opposite, when suspicious lesions (Fig. 8.4a) are observed using RCM (in which the subsequent histopathology reveals an *oral squamous cell carcinoma* (OSCC)), in addition to hyperkeratinisation (Fig. 8.4b) present also in leukoplakia, cellular pleomorphism, polynucleolated cells, neo-angiogenesis and irregular maturation of the layers are also found (Fig. 8.4c, d). Pleomorphism is revealed by the presence of different-in-size monstrous cells, ranging from small to very large and elongated ones, disarranged and not orderly organised (Fig. 8.4b). In some of them, more than one nucleolus is present (Fig. 8.4c); this is a sign never found in normal mucosa or leukoplakia without dysplasia. In the submucosa, when visible, lots of vessels are identified as not always horizontally oriented as in healthy mucosa and showing various diameters and irregular arrangement. In addition, during the real-time scan of this layer, it is possible to note an improved flow rate, more dynamic than in healthy mucosa. These features are an expression of the neo-angiogenesis occurring in OSCC. As regards cell maturation, no regular stratification of the layer is present since large and small cells can equally be visualised both at the upper layer and at the lower ones (Fig. 8.4d).

Evaluating the potential role of RCM in inflammatory processes affecting the oral mucosa, *oral lichen planus* (OLP), as example, suggests analogies with skin lichen planus, since OLP shows at RCM some of the features observed in interface dermatitis, as inflammatory cell carpet appearing in the intraepithelial and subepithelial layers, plus epithelial thickening. Moreover, in preliminary evaluations of vesicular-bullous diseases as *pemphigus vulgaris* oral, RCM allows the visualisation of acantholytic cells and the absence of a prevalent inflammatory process in order to distinguish from other erosive inflammatory lesions as oral lichen planus. RCM allows the interpretation of the erosion in order to exclude signs of malignancy Fig. 8.5.

Conclusions

The encouraging preliminary results on RCM applied to oral mucosa open the possibilities of routine application of this technology to the clinical practice. Before that, evaluation of a large number of cases is needed in order to better define first the differences between benign and malignant lesions, second, identify specific descriptors for inflammatory as well as neoplastic oral diseases. The necessity of histological examination of equivocal lesions still exists, but RCM in this phase could be used as a valid method for biopsy site selection in order to reduce the number of needed biopsies. Moreover, the non-invasive and real-time RCM approach can orient the surgeon to excise the lesion with free disease margins more accurately, and first of all, it allows the clinician to accelerate time of diagnosis.

These data could be useful also to evaluate the responsiveness to drug and the illness versus healing. Therefore, RCM seems to be able to distinguish precancerous and tumoral lesion, an important step toward early non-invasive diagnosis.

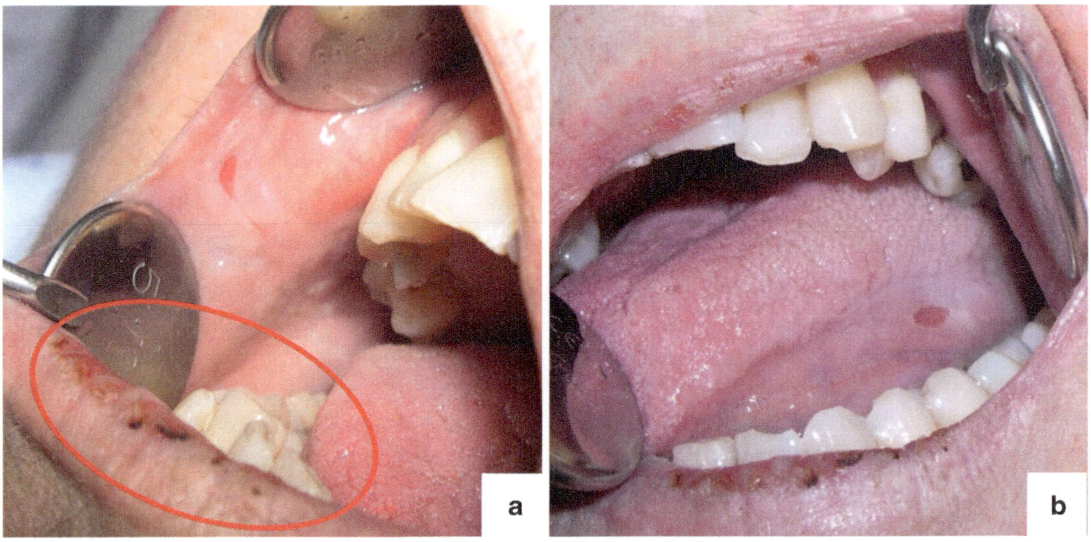

Fig. 8.5 Pemphigus vulgaris oralis (PVO). (**a**, **b**) Clinical picture showing erosion and disruption of an oral blister in PVO. (**c**) presence of multiple acantholitic keratinocytes (*white arrows*) at the level of the stratum spinosum. (**d**) Dense carpet of inflammatory cells fulfilling the epithelial lower layers (*white arrows*)

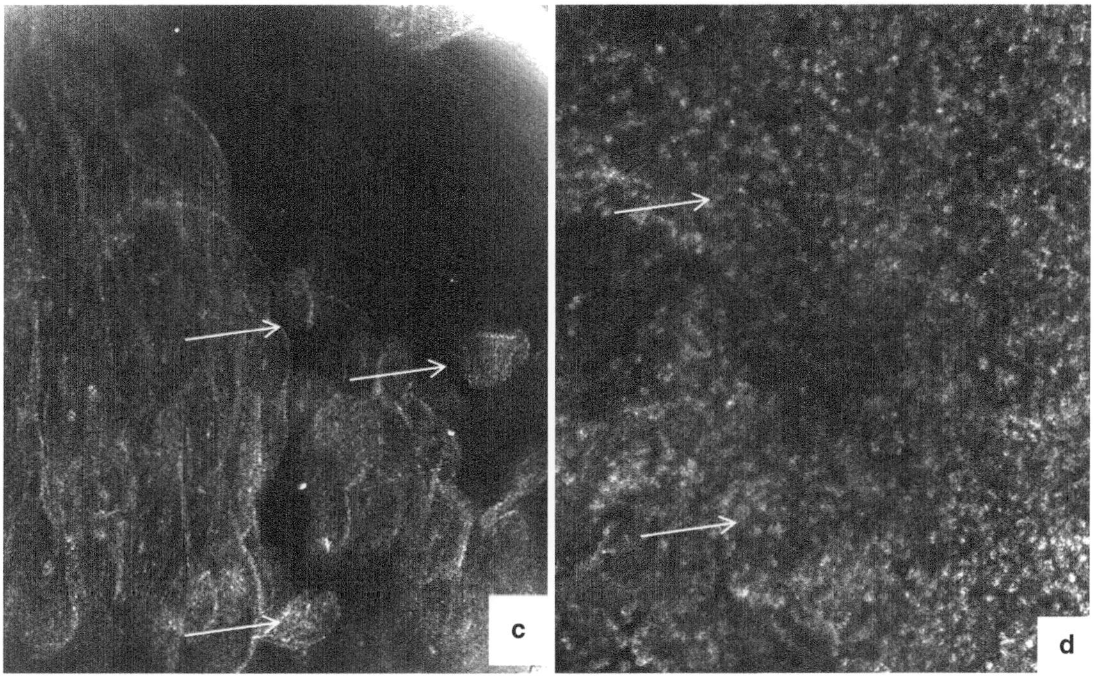

Fig. 8.5 (continued)

References

1. Ardigò M, Cota C, Berardesca E, González S (2009) Concordance between in vivo reflectance confocal microscopy and histology in the evaluation of plaque psoriasis. J Eur Acad Dermatol Venereol 23(6): 660–667
2. Pellacani G, Vinceti M, Bassoli S, Braun R, Gonzalez S, Guitera P et al (2009) Reflectance confocal microscopy and features of melanocytic lesions: an internet-based study of the reproducibility of terminology. Arch Dermatol 145(10):1137–1143
3. Ulrich M, Roewert-Huber J, González S, Rius-Diaz F, Stockfleth E, Kanitakis J (2011) Peritumoralclefting in basal cell carcinoma: correlation of in vivo reflectance confocal microscopy and routine histology. J Cutan Pathol 38(2):190–195
4. Ulrich M, Kanitakis J, González S, Lange-Asschenfeldt S, Stockfleth E, Roewert-Huber J (2012) Evaluation of Bowen disease by in vivo reflectance confocal microscopy. Br J Dermatol 166(2): 451–453
5. Moscarella E, González S, Agozzino M, Sánchez-Mateos JL, Panetta C, Contaldo M et al (2012) Pilot study on reflectance confocal microscopy imaging of lichen planus: a real-time, non-invasive aid for clinical diagnosis. J Eur Acad Dermatol Venereol 26(10):1258–1265.
6. Ardigò M, Maliszewski I, Cota C, Scope A, Sacerdoti G, Gonzalez S et al (2007) Preliminary evaluation of in vivo reflectance confocal microscopy features of discoid lupus erythematosus. Br J Dermatol 156(6): 1196–1203
7. White WM, Rajadhyaksha M, González S, Fabian RL, Anderson RR (1999) Noninvasive imaging of human oral mucosa in vivo by confocal reflectance microscopy. Laryngoscope 109:1709–1717
8. Just T, Stave J, Pau HW, Guthoff R (2005) In vivo observation of papillae of the human tongue using confocal laser scanning microscopy. ORL J Otorhinolaryngol Relat Spec 67:207–212
9. Kramer IR, Lucas RB, Pindborg JJ, Sobin LH (1978) WHO collaborating center for oral precancerous lesions: definition of leukoplakia and related lesions: an aid to studies on oral precancer. Oral Surg Oral Med Oral Pathol 46:518

Multiphoton Laser Microscopy and Fluorescence Lifetime Imaging for the Assessment of the Skin

9

Stefania Seidenari

Abbreviations

BCC	Basal cell carcinoma
FLIM	Fluorescence lifetime imaging
MM	Melanoma
MPT	Multiphoton tomography
NADH	Reduced nicotinamide adenine dinucleotide
NADPH	Reduced nicotinamide adenine dinucleotide phosphate
NIR	Near infrared
SHG	Second harmonic generation
UV	Ultraviolet

Multiphoton laser microscopy or tomography (MPT) associated to fluorescence lifetime imaging (FLIM) is one among the very few noninvasive imaging methods enabling an in vivo optical biopsy. Thanks to its spatial resolution similar to that of histopathology at high power magnification (<1 μm lateral, <2 μm axial) [1] and to the possible use of colour coding of the fluorescence decay time [2, 3], this technique enables the morphology of the tissue to be immediately recognizable.

MPT is an optical imaging system that excites fluorescence from the tissue through the simultaneous absorption of two or more photons of infrared light [4]. Whereas for conventional confocal fluorescence microscopy, fluorophores are excited by absorption of individual photons in the visible or ultraviolet spectrum, MPT excitation entails the simultaneous absorption of two or more photons of longer wavelength. This process requires a high intensity of excitation light and is confined to the tightly focused excitation spot. The longer wavelength infrared radiation undergoes less scattering than visible light and can thus facilitate high resolution imaging deeper into biological tissue [1, 4–13].

MPT can exploit autofluorescence of intrinsic tissue fluorophores, i.e. naturally occurring molecules that can be imaged using MPT without the need for exogenous contrast agents. Fluorophores are integral components of the molecules to which they confer the characteristic autofluorescence. They include NADH (reduced nicotinamide adenine dinucleotide), NADPH (reduced nicotinamide adenine dinucleotide phosphate), flavines, keratin, melanin, elastin, collagen, porphyrin, tryptophan, cholecalciferol and lipofuscin [1, 4–13]. After energy absorption, fluorophores can then emit energy in turn, generating a visible signal at defined and characteristic wavelengths, different from those of absorption. The quantity and the wavelength of the emitted energy depend on the chemical characteristics of the fluorophore, on its environment and particularly on the type of the surrounding molecules [14–16].

Efficient MPT excitation usually requires ultrashort femtosecond laser pulses, which are also able to produce the nonlinear effect of second harmonic generation (SHG). The SHG signal comes

S. Seidenari
Department of Dermatology,
University of Modena and Reggio Emilia,
via del Pozzo 71, Modena 41100, Italy
e-mail: stefania.seidenari@skincenter.it,
stefania.seidenari@unimore.it

E. Berardesca et al. (eds.), *Non Invasive Diagnostic Techniques in Clinical Dermatology*,
DOI 10.1007/978-3-642-32109-2_9, © Springer Berlin Heidelberg 2014

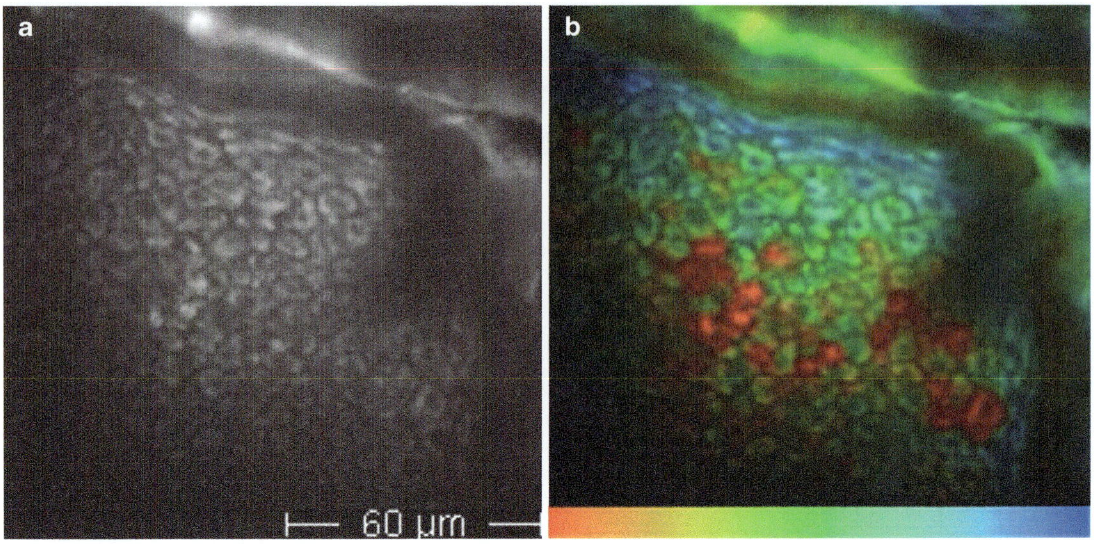

Fig. 9.1 On the *left* (**a**), multiphoton tomography intensity image of healthy epidermis. In (**b**) pseudo-colour-coded fluorescence lifetime image of the same epidermal area, where *red-coloured* melanocytes, characterized by a short lifetime decay time, are well distinguishable from *green-coloured* keratinocytes with a medium lifetime value. Pseudo-colour scale: 0–2,000 ps

from periodic structures such as collagen and is characterized by an emission wavelength corresponding to half of that of the incident photon; this particular signal allows the visualization of dermal collagen bundles and their distinction from cellular components and elastin fibres [1, 5, 6]. With MPT bidimensional images are acquired which correspond to optical sectioning parallel to the tissue surface (reported to a defined xy-plane). Pictures obtained at various depths, called z-stacks, can be acquired by sequentially modifying the depth of the focal plane in the tissue, reaching levels of 200 µm measured from the departure point at the skin surface [8, 10, 13]. Grey scale images are generated, reproducing the fluorescence intensity in different tissue components (Fig. 9.1a).

9.1 Excitation Wavelength

By modulating the excitation wavelength, different skin structures can be selectively excited, obtaining an enhancement of their morphology. When exploring the skin, a wavelength of 760 nm is first chosen for a proper imaging of epidermal structures [8, 10, 13]. When reaching the dermo-epidermal junction, the excitation wavelength has to be increased up to 800–820 nm, for selective melanin imaging. By this wavelength, most keratinocytes progressively become invisible. On the contrary, since melanin has an absorption spectrum that decreases from the UV region to NIR, with a selective excitation wavelength of 800 nm [17–19], melanin and melanin-containing cells (melanocytes and melanin-containing keratinocytes) will appear as single cells in the basal layer showing intense fluorescence. This characteristic can be employed to recognize melanocytes and melanin granules in cell cultures (Fig. 9.2).

A wavelength of 800 nm is generally employed to adequately visualize the extracellular matrix of the dermis. At this wavelength, collagen fibres that generate the SHG signal are selectively excited, whereas at 760 nm, dermal autofluorescent components such as elastin are enhanced in the image.

9.2 Fluorescence Lifetime Imaging

Fluorescence lifetime imaging (FLIM) is based on the measurement of the decay rate of the fluorescence signal following a short pulse of

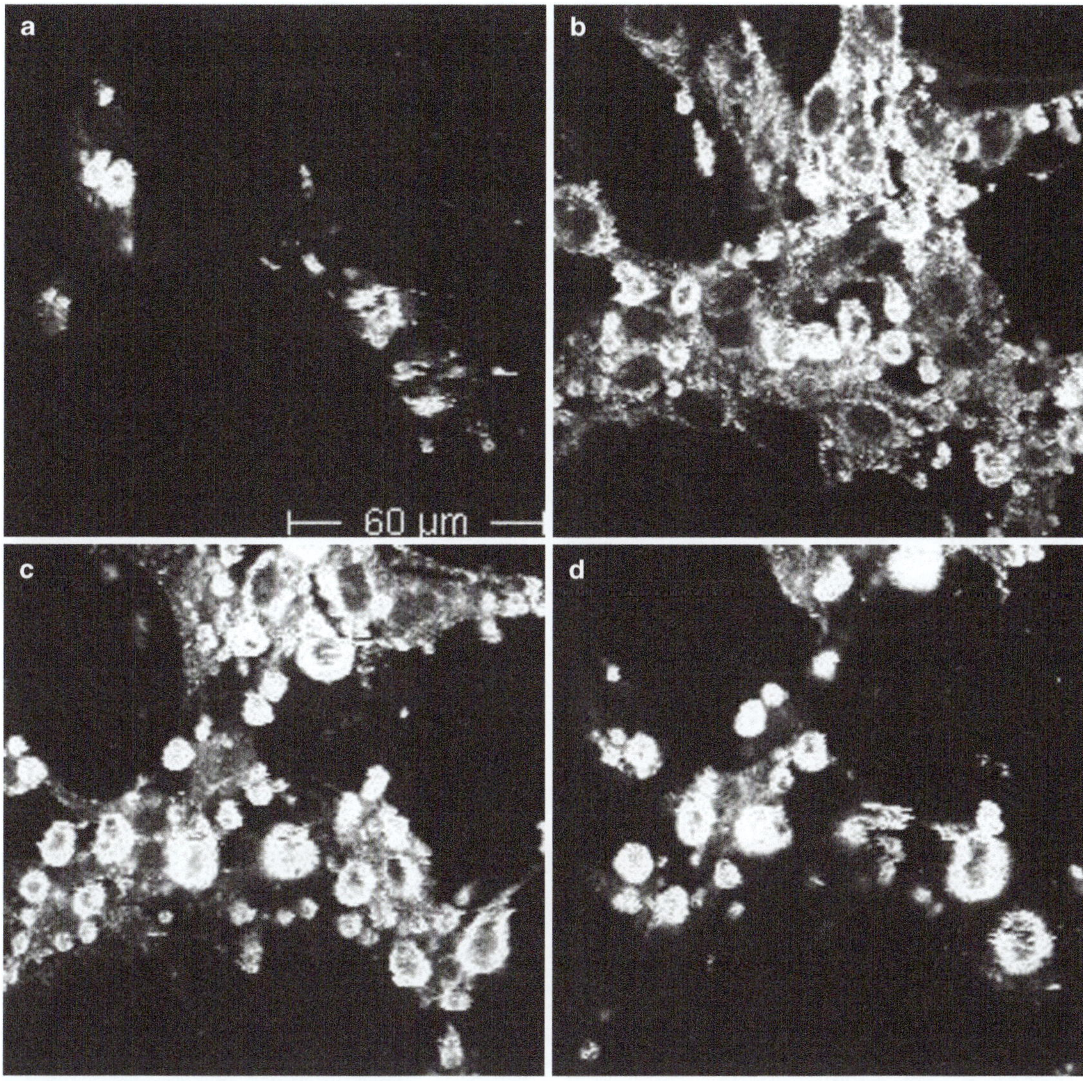

Fig. 9.2 Multiphoton tomography intensity images. Melanin imaging and excitation wavelength. In a melanocyte cell culture, by a 720-nm excitation wavelength (**a**), cells and melanin granules are not recognizable; (**b**) at 760 nm, both melanocytes and melanin granules are clearly visible; (**c**, **d**) by increasing the excitation wavelength to 780 and 800 nm, melanin granules are enhanced, but melanocyte bodies fade (**c**) and are not more visible (**d**)

excitation light [11, 15, 19]. This represents an additional technique for use in combination with MPT (MPT/FLIM), further improving the understanding of skin morphology in detail. Further discrimination between fluorophores can be gained using multispectral FLIM, which is based on the analysis of multiple emission spectral channels [20–22].

Whereas MPT conventionally relies on the assessment of the intensity of endogenous fluorescence emitted by fluorophores in the skin, FLIM provides additional information based on the contrast generated by differences in the decay rate of the fluorescence intensity, not only providing morphological information but also a quantitative assessment of metabolic changes in living tissue. Since FLIM is immune to intensity artefacts, it enables a more robust numerical description of the images than intensity imaging [2, 20–22]. Images presented in this chapter are obtained by a FLIM system, developed jointly by the Photonics Group of the Imperial College of London and

JenLab GmbH (Jena, Germany), which has been incorporated into the commercially available *DermaInspect*® (JenLab GmbH, Jena, Germany). The distribution of fluorescence lifetimes within an image is visualized through a histogram that plots the fluorescence lifetime (x-axis) against the number of corresponding pixels occurring at that lifetime (y-axis). Pseudo-colour images are generated, where each image pixel contains information about its fluorescence decay time corresponding to a specific colour. Thus, four dimensional data sets are generated, where the tissue is not only studied according to its structure in the x-, y- or z-axis but also according to the fluorescence dynamics of its components corresponding to different states or metabolic characteristics of the tissue [2, 20–22]. Besides essential information on fluorescence decay rates, false colour coding, enhancing image contrast and providing a more user-friendly visualization method permits the immediate identification of cellular, subcellular or extracellular structures in the image. As an example, when employing a 0–2,000 scale, keratinocytes, exhibiting fluorescence decay time values around 1,000 ps, are represented in the blue-green range, whereas melanocytes, with medium to short fluorescence decay time, are coded in the yellow-red range (Fig. 9.1b) [23]. Fluorescence lifetime scale intervals may be varied according to the need for enhancing particular structures or cells belonging to a certain typology or to identify subcellular particles. Figure 9.3 shows upper melanoma layers represented by two different scales. On the left, employing a 0–2,000 colour scale, melanoma cells are coloured red, corresponding to short-lifetime values, and this permits the immediate categorization into melanin-containing cells; on the right, short-lifetime values are expanded employing a 0–400 scale, enabling the enhancement of intracellular structures.

9.3 Application Fields

Cell Cultures. The MPT/FLIM technique has numerous applications in dermatology. Morphologic and metabolic characteristics of different cell types can be studied employing cell cultures,

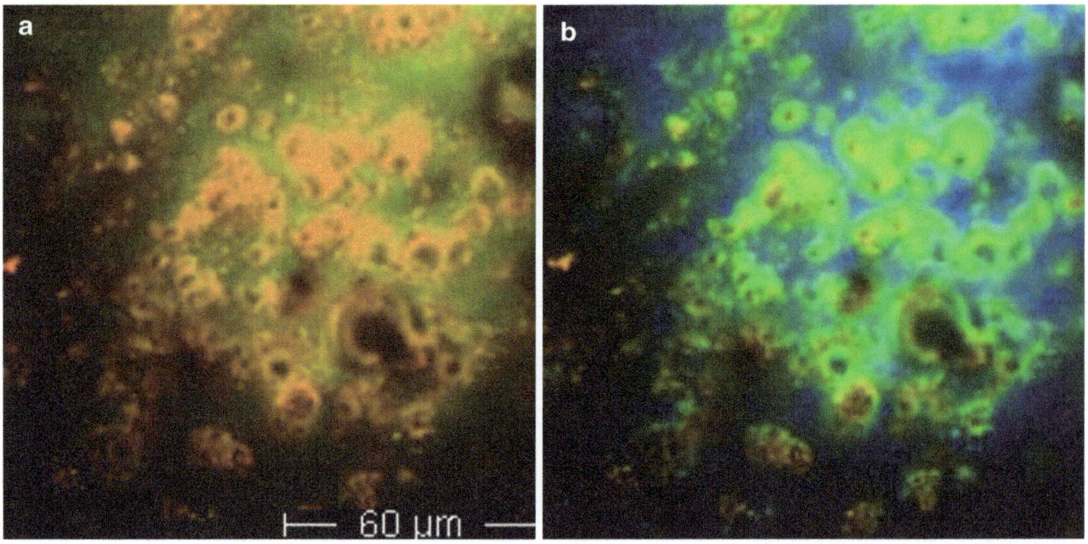

Fig. 9.3 On the *left* (**a**), *upper* melanoma layers represented by a 0–2,000 ps pseudo-colour scale. Melanoma cells, displaying short lifetime values, are shown in *orange* and are immediately recognizable. On the *right* (**b**), a 0–400 ps scale is employed to increase the contrast inside cell components and to recognize the characteristics of the cytoplasm. Images presented in this chapter were obtained by a FLIM system, developed jointly by the Photonics Group of the Imperial College of London and JenLab GmbH (Jena, Germany), which has been incorporated into the commercially available *DermaInspect*® (JenLab GmbH, Jena, Germany) and were calculated using the software SPCImage (Becker & Hickl GmbH)

where isolated cells can be examined out of the tissue environment. Fibroblast cultures are widely used as an experimental model to study the expression of specific genes or the effect of drugs with potential chemotherapeutic activity and to check the mutagenicity and carcinogenicity of different substances. Using MPT/FLIM, a precise and rapid assessment of the morphologic and metabolic changes fibroblasts undergo after exposure to various environmental factors can be achieved without the need of cell processing and staining [24] (Fig. 9.4).

Healthy Skin and Skin Ageing. Healthy epidermis shows a homogenous distribution of cells which are divided by thin nonfluorescent intercellular spaces [7–13, 23, 25]. Since nuclei are deficient in autofluorescence signals, they appear as dark areas inside the fluorescent cytoplasm (Figs. 9.1a and 9.5). The stratum corneum shows strong keratin-based fluorescent hexagonal-shaped large flat cells. At a depth of about 20 μ from the skin surface, we find the stratum granulosum, where keratinocytes appear as large, oval cells with autofluorescence in the cytoplasm and dark nuclei. Melanin blotches inside the keratinocytes may be unique, involving a large part of the cytoplasm and sparing the nucleus, or may be organized in multiple spots, conferring a granular appearance to the keratinocyte. Sometimes the melanin blotch forms a cap over the nucleus, which is not recognizable because of the horizontal optical section of the sample. Stratum spinosum cell diameter decreases and cell density increases going more in depth towards the basal layer. At 50–100 μ from the skin surface, we find the basal layer with brightly fluorescent small polygonal keratinocytes and dark nuclei. At the dermo-epidermal junction, we find dark round/oval-shaped areas, interrupting the basal layer, corresponding to the top of the dermal papillae, where we can appreciate the presence of fibres [23].

The morphology and the metabolic states of healthy skin can be assessed in vivo, according to skin site and age. In fact, cell diameter and density vary according to epidermal cell depth and skin site [23]. In the elderly, epidermal cells show morphologic alterations, presenting irregular shape, size and intercellular distance and a decreased number at the basal layer (Fig. 9.5d). The modifications of the metabolic activities characteristic of the ageing process are reflected by variations in FLIM values, which increase at both the upper and lower layers in elderly subjects [23]. Further efforts are needed to increase knowledge on variations the epidermis undergoes according to environmental influences, for use as

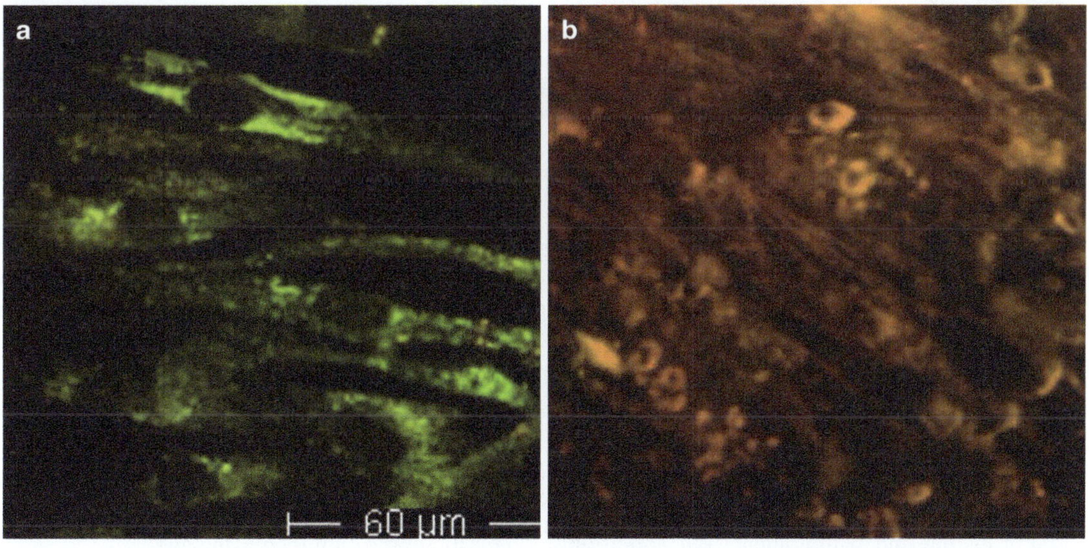

Fig. 9.4 On the *left* (**a**), a 3-day old fibroblast culture; on the *right* (**b**), a 1 month old culture stimulated with ascorbic acid where the fibres produced by the fibroblast are clearly visible as *red* (short-lifetime) filaments

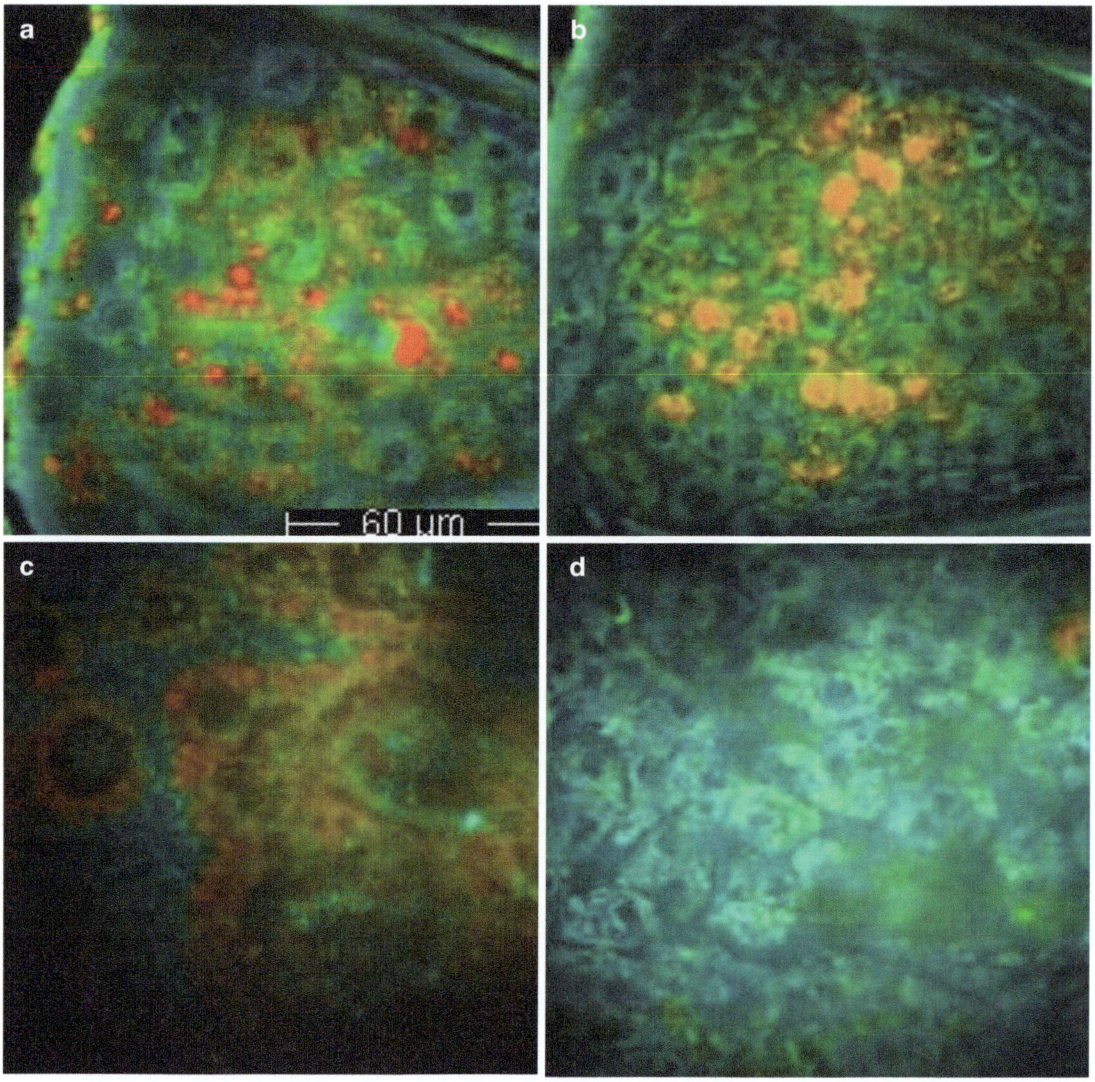

Fig. 9.5 (**a**–**c**) Pseudo-colour FLIM stack of healthy epidermis. (**a**) *Upper* epidermal layers of a young subject, (**b**) *lower* epidermal layers, (**c**) dermo-epidermal junction and (**d**) *upper* epidermal layers in an elderly subject. In (**a**), keratinocytes show regular contours and intercellular distance, whereas in (**d**) cells present morphologic alterations and a shift of the fluorescence decay time towards higher values (*blue-coloured keratinocytes*)

the SHG component of the signal when exciting at up to 800–820 nm. When employing the filter eliminating the SHG signal, collagen fibres only contribute to the signal with their modest autofluorescence component, thus fading and enabling the recognition of fibres containing elastin. The latter appear as fine-curled fibres with a nonhomogeneous calibre (Fig. 9.6).

Skin ageing and pathological skin conditions of the dermis bring about morphological changes in the dermal collagen and elastin fibre network. papillary dermis. Conversely, in the elderly, SHG signals can only be detected in the uppermost part of the dermis, whereas large amounts of fluorescent elastic fibres, corresponding to solar elastosis, are found in the dermis [28]. The trend of decreasing SHG signals and increasing autofluorescence signals is correlated with the histological findings of the decrease in collagen fibres and the increase in elastic fibres with increasing age.

MPT/FLIM is a procedure holding great promise for increasing diagnostic accuracy of skin tumours.

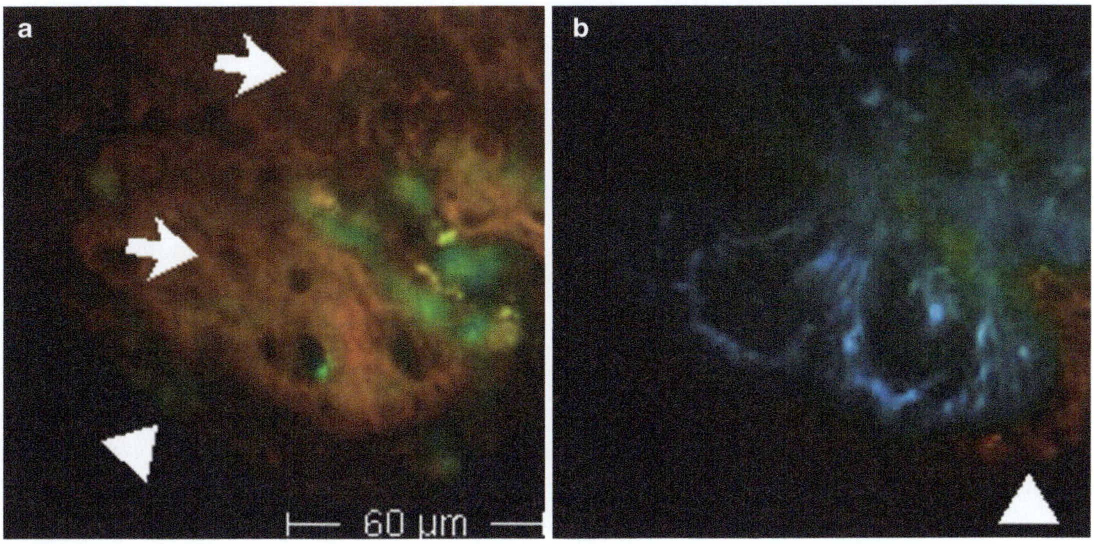

Fig. 9.6 MPT/FLIM: pseudo-colour-coded fluorescence lifetime imaging (pseudo-colour scale 0–2,000 ps). Enhancement of elastin fibres by inserting a filter for SHG signals originating from collagen fibres. A nevus at a depth of –60 **μ**. (**a**) Collagen fibres (*white arrows*) appear as thick *red* filaments inside the dermal papillae; (**b**) after filtering SHG signals, elastin fibres appear as tiny blue (long-lifetime) filaments; melanocytes surround the papillae (*triangle*)

Basal Cell Carcinoma (BCC). BCC represents the most frequent skin tumour. Besides the diagnostic difficulties for superficial lesions which lack sensitive dermoscopic criteria, the identification of tumour boundaries, and consequently of the excision margins, represents practical problems in the daily dermatological routine. The use of MPT alone has already proved effective in identifying some diagnostic aspects of BCC. Investigating BCC specimens by MPT, histopathological features characteristic of BCC, such as elongated cells, nuclei polarization and peripheral palisading, can be recognized in the images [29, 30]. Lin et al. achieved a discrimination of BCC from normal dermal stroma by MPT imaging of formalin-fixed specimens of nine nodular type BCCs [31]. BCCs appeared as clumps of autofluorescent cells with large nuclei and peripheral palisading. By FLIM implementation and valuable information about time-resolved analysis of the fluorescence signal, further criteria are added for diagnostic purposes and tumour margin assessment. Using a multidimensional nonlinear laser imaging approach to visualize *ex vivo* samples of BCC, Cicchi et al. observed a blue-shifted fluorescence emission, a higher fluorescence response at 800 nm excitation wavelength and a slightly longer mean fluorescence lifetime in BCCs [32]. A wide-field (single photon excitation) study by Galletly et al. employing FLIM, imaged unstained excision biopsies of 25 BCCs with FLIM following excitation of autofluorescence with a 355 nm pulsed ultraviolet laser [3]. A significant reduction in mean fluorescence lifetimes between areas of BCC and those of surrounding uninvolved skin was demonstrated. Investigating BCC by MPT, Seidenari et al. identified specific descriptors, which were never observed in healthy skin but were present in all BCCs [30]. Further examination of BCC by MPT/FLIM revealed that some of these features correspond to traditional histopathological diagnostic criteria, such as aligned elongated monomorphous cells, peripheral palisading of tumour cells and typical tumour nests, whereas others, such as ‘blue cells’ and ‘phantom islands’, cannot be recognized in fixed and stained specimens (Fig. 9.7) [33]. ‘Blue cells’ correspond to basaloid cells showing fluorescence lifetime values higher than those of normal keratinocytes. The biological significance of the lifetime modifications in cancer cells is thought to be due to a shift from oxidative metabolism to

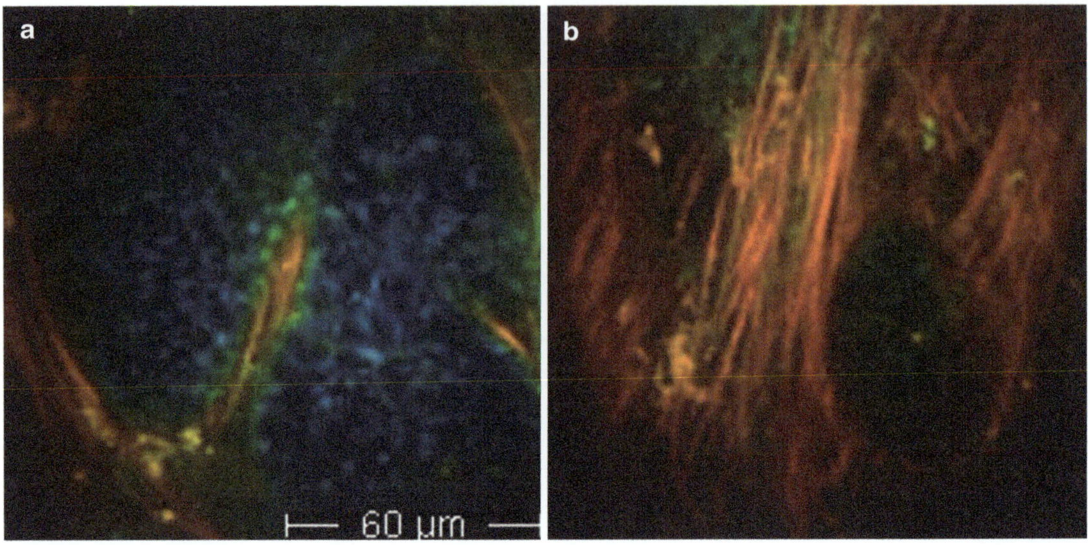

Fig. 9.7 On the *left* (**a**), basal cell carcinoma nests constituted by '*blue cells*'; on the *right* (**b**), 'phantom islands', i.e. basal cell carcinoma nests disappearing at an excitation wavelength of 800 nm; at the same time, collagen fibres are enhanced and appear as *red* filaments circumscribing the nests

glycolysis associated to variations in the amount of free and bound NADH [14]. Thus, FLIM imaging may represent a method to measure the oxidative stress and/or preferential metabolic pathway in living tissues [14]. 'Phantom islands' are generated by shifting the excitation wavelength to 800–820 nm. In this way, keratinocytes and BCC cells almost disappear, whereas the visualization of fibres is enhanced. Short-lifetime red (collagen) fibres tightly embrace the basaloid longer lifetime blue cell nests, corresponding to empty spaces ('phantom islands') (Fig. 9.7). Employing the MPT/FLIM technique, the separation between healthy skin and the tumour mass is achieved in all cases, making this technique a consistent tool for the definition of tumour margins.

Melanocytic Nevi. Common nevi, which are characterized by a proliferation of typical melanocytes, in single cells or in aggregates, are benign lesions which are clinically relevant as their appearance may be very similar to malignant melanocytic lesions. By MPT/FLIM, specific descriptors, referring both to cytology and architecture, enable their diagnosis [34]. Melanin-containing keratinocytes are recognizable in FLIM images as long-lifetime cells containing short-lifetime spots or blotches (red to yellow in the 0–2,000 ps pseudo-colour scale) characterizing melanin, whereas epidermal melanocytes in the upper and the lower layers appear as tiny short-lifetime cells, which are smaller than keratinocytes presenting a well visible small nucleus and homogeneously distributed melanin (Fig. 9.8a). At the dermo-epidermal junction level, the tips of the dermal papillae, appearing as roundish structures surrounded by a regular orange boundary of melanin-containing keratinocytes and melanocytes, are visible. 'Edged papillae' are characterized by a well-demarcated rim of orange cells surrounding a dark space containing fibres or cells, whereas 'non-edged papillae' lack a well-defined boundary (Fig. 9.9). Junctional nevus cell nests appear as aggregates of short-lifetime cells rising from the periphery of the papillae (Fig. 9.9). In compound nevi, short-lifetime 'dermal cell clusters' are also visible (Fig. 9.8b). By MPT/FLIM a regular architecture is generally observable in benign melanocytic lesions.

Melanoma. Melanoma (MM) is a lethal cancer. Despite a high cure rate for thin melanomas, advanced melanomas have a poor prognosis. Thus, early identification of melanoma represents a crucial end point for physicians. Due to its high resolution, MPT/FLIM holds a great potential for improving MM diagnosis [19, 35].

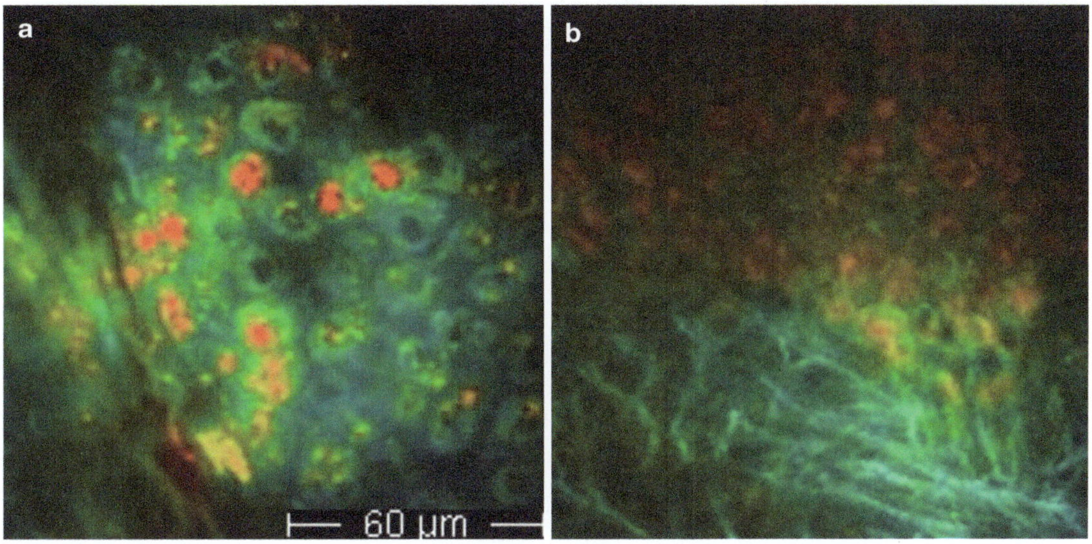

Fig. 9.8 On the *left* (**a**), pseudo-colour-coded fluorescence lifetime image of *upper layers* in a nevus, where *small red-coloured* melanocytes, characterized by a short-lifetime decay time, are well distinguishable from *large green-coloured* keratinocytes with a medium lifetime value. The pseudo-colour scale employed to represent this picture ranges from 500 to 2,000 ps. On the *right* (**b**), a dermal melanocyte cluster in a compound nevus; *blue* elastin fibres are visible in the dermis

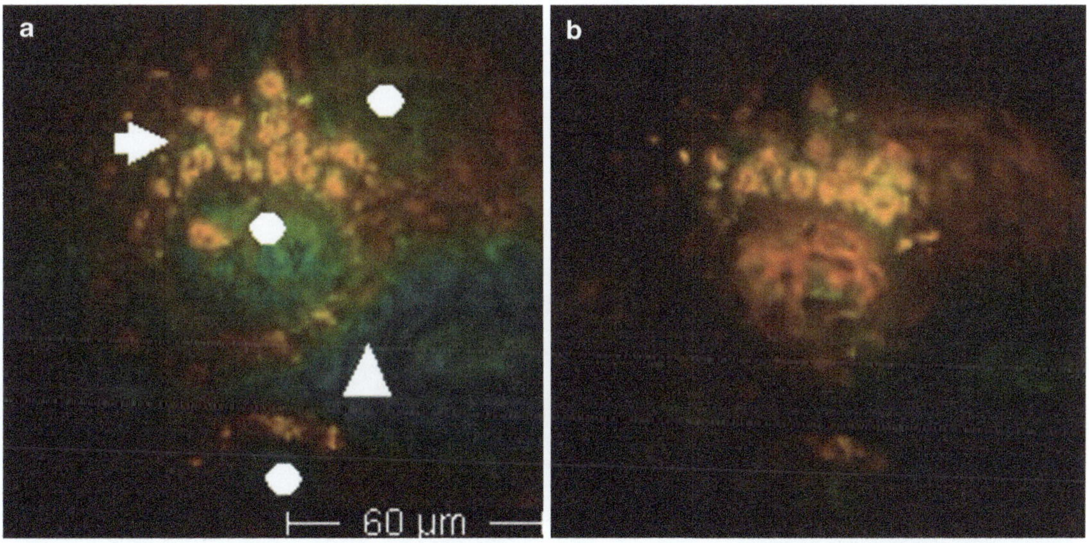

Fig. 9.9 Selective melanin imaging and enhancement of collagen fibres by switching the excitation wavelength from 760 (**a**) to 800 nm (**b**). A nevus at a depth –60 m. Pseudo-colour scale: 0–2,100 ps. (**a**) By a 760-nm excitation wavelength, dermal papillae (*white dots*) are surrounded by *red* short-lifetime cells (melanin-containing cells), forming a junctional nest (*arrow*), whereas keratinocytes (*white triangle*) appear *green*. (**b**) By an excitation wavelength of 800 nm, melanin-containing cells maintain their visibility, whereas keratinocytes disappear. Inside the papillae, collagen fibres appear as *thick red* (short-lifetime) filaments

By MPT/FLIM, keratinocytes may not be recognizable at the superficial layers of a MM. When present, they are infiltrated by atypical melanocytes [36]. When displaying images by a 0–2,000 ps fluorescence lifetime range, MM cells appear orange to red, due to their short fluorescence decay time, and this immediately enables the determination of the melanocytic nature of the lesion (Fig. 9.3a). To increase contrast for the distinction of image details, a 0–400 scale is then applied (Fig. 9.3b). At the superficial level, atypical MM cells appear as large short-lifetime cells with a nucleus with undefined contours and a nonhomogeneous cytoplasm with speckled melanin. A peripheral cytoplasmic halo is often observable due to variable lifetime values of different cell areas (Fig. 9.3). Deeper MM layers show atypical short-lifetime cells smaller in size with respect to those observable on the surface; these are pleomorphic, variable in size and irregular in shape with a nonhomogeneous distribution. At this level, atypical short-lifetime cells may form aggregates or nests and may infiltrate dermal papillae and hair follicles. When dermal papillae are recognizable, these are not usually surrounded by a rim of regularly arranged melanocytes; they are irregularly sized and shaped and are separated by interpapillary spaces of different thickness. An architectural disorder is always visible. In MM metastases, besides atypical short-lifetime cells, long-lifetime (lacking melanin) cell nests are intermingled with long-lifetime (collagen) fibres (Fig. 9.10).

9.4 Advantages, Disadvantages and Future Objectives of MPT/FLIM

The morphology of MPT/FLIM images, acquired with a spatial resolution similar to that of histopathology at high power magnification, is extremely coherent with that of the histopathological analysis, suggesting that this new technique, once fully developed, may replace histology [1, 10].

Intensity images are visualized during the examination process in real time. FLIM images require a further elaboration to allow the operational system to develop a diagram of the fluorescence decay time and to produce a pseudo-colour image [3, 22]. Not only does MPT/FLIM generate high resolution images providing highly informative morphologic details but it also enables the study of the kinetics of the fluorescence decline and the calculation of the mean fluorescence intensity and lifetime values for selected areas of interest such as the cytoplasm of single cells, providing numerical data for an objective assessment of physiologic and pathologic skin conditions [2–4, 22]. The subcellular spatial resolution study of tissue samples in three dimensions is made possible by the acquisition of a sequence of horizontal optical sections (stack).

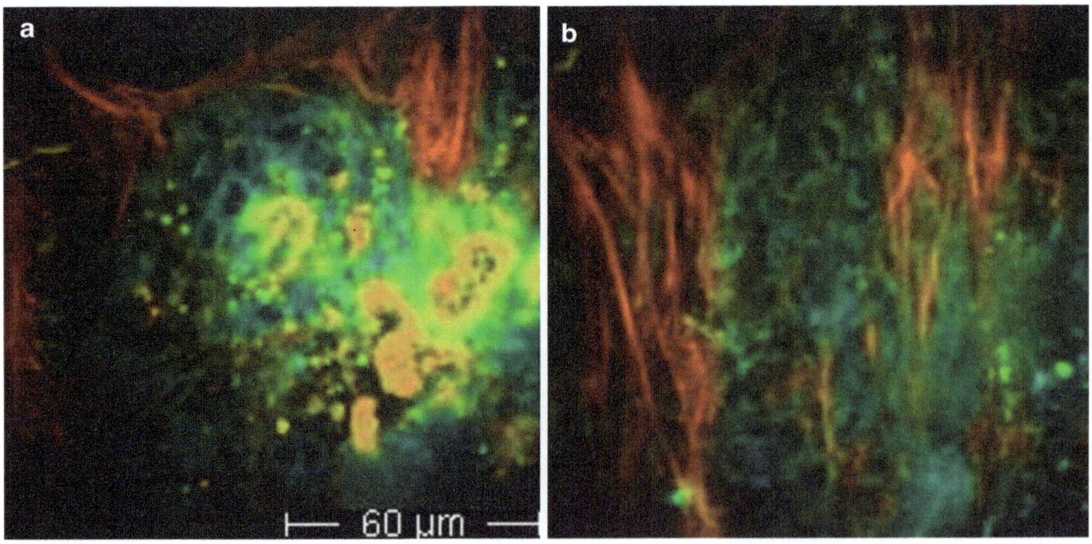

Fig. 9.10 Melanoma metastases. On the *left* (**a**), atypical melanocytes (*orange* short-lifetime cells) and collagen fibres; on the *right* (**b**), nests of atypical long-lifetime cells (atypical melanocytes not producing melanin) intermingled with fibres

Moreover, the nonlinear excitation produced by the NIR radiation allows a higher visualization of the deep dermis. The potential of this technique to provide an improvement in diagnostic accuracy is especially valuable for skin tumours, where an 'optical biopsy' enabling a diagnosis on the basis of both architectural and cellular morphology and numerical parameters is immediately available.

Long-term studies of cutaneous affections can be performed by the repetition of the in vivo examination on the same explored skin site within short or long intervals, considering that this technique has proved noninvasive [9].

On the other hand, the MPT/FLIM technique is at present unsuitable for continuous clinical use, since it presents some drawbacks consisting in a long acquisition/elaboration time of the images and the small field of view. Thus, important goals are both to reduce the acquisition time to avoid imaging deformities caused by voluntary and involuntary movements and to implement a device enabling the imaging of skin areas, large enough to be representative of the entire lesion.

References

1. König K, Riemann I (2003) High-resolution multiphoton tomography of human skin with subcellular spatial resolution and picoseconds time resolution. J Biomed Opt 8(3):432–439
2. Elson D, Requejo-Isidro J, Munro I, Reavell F, Siegel J, Suhling K, Tadrous P, Benninger R, Lanigan P, McGinty J, Talbot C, Treanor B, Webb S, Sandison A, Wallace A, Davis D, Lever J, Neil M, Phillips D, Stamp G, French P (2004) Time-domain fluorescence lifetime imaging applied to biological tissue. Photochem Photobiol Sci 3:795–801
3. Galletly NP, McGinty J, Dunsby C, Teixeira F, Requejo-Isidro J, Munro I, Elson DS, Neil MA, Chu AC, French PM, Stamp GW (2008) Fluorescence lifetime imaging distinguishes basal cell carcinoma from surrounding uninvolved skin. Brit J Dermatol 159:152–161
4. Denk W, Strickler JH, Webb WW (1990) Two-photon laser scanning fluorescence microscopy. Science 248:73–76
5. Kollias N, Zonios G, Stamatas GN (2002) Fluorescence spectroscopy of skin. Vib Spectrosc 28(1):17–23
6. Zipfel WR, Williams RM, Christie R, Nikitin AY, Hyman BT, Webb WW (2003) Live tissue intrinsic emission microscopy using multiphoton-excited native fluorescence and second harmonic generation. Proc Natl Acad Sci U S A 100(12):7075–7080
7. Riemann I, Dimitrow E, Fischer P, Reif A, Kaatz M, Elsner P, König K (2004) High resolution multiphoton tomography of human skin in vivo and in vitro. SPIE-Proc 21–28
8. Schenke-Layland K, Riemann I, Damour O, Stock UA, König K (2006) Two-photon microscopes and in vivo multiphoton tomographs-powerful diagnostic tools for tissue engineering and drug delivery. Adv Drug Deliv Rev 58(7):878–896
9. Lin SJ, Jee SH, Dong CY (2007) Multiphoton microscopy: a new paradigm in dermatological imaging. Eur J Dermatol 17(5):361–366
10. König K (2008) Clinical multiphoton tomography. Review. J Biophotonics 1(1):13–23
11. Cicchi R, Sestini S, De Giorgi V, Massi D, Lotti T, Pavone FS (2008) Nonlinear laser imaging of skin lesions. J Biophotonics 1(1):62–73
12. Tsai TH, Jee SH, Dong CY, Lin SJ (2009) Multiphoton microscopy in dermatological imaging. J Dermatol Sci 56(1):1–8
13. König K, Raphael AP, Lin L, Grice JE, Soyer HP, Breunig HG, Roberts MS, Prow TW (2011) Applications of multiphoton tomographs and femtosecond laser nanoprocessing microscopes in drug delivery research. Adv Drug Deliv Rev 63(4–5):388–404
14. Skala MC, Riching KM, Gendron-Fitzpatrick A, Eickhoff J, Eliceiri KW, White JG, Ramanujam N (2007) In vivo multiphoton microscopy of NADH and FAD redox states, fluorescence lifetimes, and cellular morphology in precancerous epithelia. Proc Natl Acad Sci U S A 104(49):19494–19499
15. Sanchez WY, Prow TW, Sanchez WH, Grice JE, Roberts MS (2010) Analysis of the metabolic deterioration of ex vivo skin from ischemic necrosis through the imaging of intracellular NAD(P)H by multiphoton tomography and fluorescence lifetime imaging microscopy. J Biomed Opt 4:046008
16. Palero JA, Bader AN, de Bruijn HS, Der Ploeg van den Heuvel A, Sterenborg HJ, Gerritsen HC (2011) In vivo monitoring of protein-bound and free NADH during ischemia by nonlinear spectral imaging microscopy. Biomed Opt Express 2(5):1030–1039
17. Teuchner K, Freyer W, Leupold D, Volkmer A, Birch DJ, Altmeyer P, Stücker M, Hoffmann K (1999) Femtosecond two-photon excited fluorescence of melanin. Photochem Photobiol 70(2):146–151
18. Teuchner K, Ehlert J, Freyer W et al (2000) Fluorescence studies of melanin by stepwise two-photon femtosecond laser excitation. J Fluoresc 10:275–281
19. Dimitrow E, Riemann I, Ehlers A, Koehler MJ, Norgauer J, Elsner P, König K, Kaatz M (2009) Spectral fluorescence lifetime detection and selective melanin imaging by multiphoton laser tomography for melanoma diagnosis. Exp Dermatol 18(6): 509–515
20. Becker W, Bergmann A, Biskup C (2007) Multispectral fluorescence lifetime imaging by TCSPC. Microsc Res Tech 70(5):403–409
21. Talbot CB, Patalay R, Munro I, Warren S, Ratto F, Matteini P, Pini R, Breunig HG, König K, Chu AC,

Stamp GW, Neil MA, French PM, Dunsby C (2011) Application of ultrafast gold luminescence to measuring the instrument response function for multispectral multiphoton fluorescence lifetime imaging. Opt Express 19(15):13848–13861
22. Patalay R, Talbot C, Alexandrov Y, Munro I, Neil MA, König K, French PM, Chu A, Stamp GW, Dunsby C (2011) Quantification of cellular autofluorescence of human skin using multiphoton tomography and fluorescence lifetime imaging in two spectral detection channels. Biomed Opt Express 2(12):3295–3308, Epub 2011 Nov 10
23. Benati E, Bellini V, Borsari S, Dunsby C, Ferrari C, French P, Guanti M, Guardoli D, Koenig K, Pellacani G, Ponti G, Schianchi S, Talbot C, Seidenari S (2011) Quantitative evaluation of healthy epidermis by means of multiphoton microscopy and FLIM. Skin Res Technol 17:295–303
24. Rice WL, Kaplan DL, Georgakoudi I (2010) Two-photon microscopy for non-invasive, quantitative monitoring of stem cell differentiation. PLoS One 5(4):e10075
25. Seidenari S, Arginelli F, Bassoli S, Cautela J, French P, Guanti M, Guardoli D, König K, Talbot C, Dunsby C (2012) Multiphoton laser microscopy and fluorescence lifetime imaging for the evaluation of the skin. Dermatol Res Pract 2012:810749
26. Koehler MJ, Preller A, Kindler N, Elsner P, König K, Bückle R, Kaatz M (2003) Intrinsic, solar and sunbed-induced skin aging measured in vivo by multiphoton laser tomography and biophysical methods. Skin Res Technol 15(3):357–363
27. Koehler MJ, Hahn S, Preller A, Elsner P, Ziemer M, Bauer A, König K, Bückle R, Fluhr JW, Kaatz M (2008) Morphological skin ageing criteria by multiphoton laser scanning tomography: non-invasive in vivo scoring of the dermal fibre network. Exp Dermatol 17(6):519–523
28. Lin SJ, Wu R Jr, Tan HY, Lo W, Lin WC, Young TH, Hsu CJ, Chen JS, Jee SH, Dong CY (2005) Evaluating cutaneous photoaging by use of multiphoton fluorescence and second-harmonic generation microscopy. Opt Lett 30(17):2275–2277
29. Paoli J, Smedh M, Wennberg AM, Ericson MB (2008) Multiphoton laser scanning microscopy on non-melanoma skin cancer: morphologic features for future non-invasive diagnostics. J Invest Dermatol 128(5):1248–1255
30. Seidenari S, Arginelli F, Bassoli S, Cautela J, Cesinaro AM, Guanti M, Guardoli D, Magnoni C, Manfredini M, Ponti G, König K (2013) Diagnosis of BCC by multiphoton laser tomography. Skin Res Technol 19(1):e297–e304
31. Lin SJ, Jee SH, Kuo CJ, Wu RJ, Lin WC, Chen JS, Liao YH, Hsu CJ, Tsai TF, Chen YF, Dong CY (2006) Discrimination of basal cell carcinoma from normal dermal stroma by quantitative multiphoton imaging. Opt Lett 31(18):2756–2758
32. Cicchi R, Massi D, Sestini S, Carli P, De Giorgi V, Lotti T, Pavone FS (2007) Multidimensional non-linear laser imaging of basal cell carcinoma. Opt Express 15(16):10135–10148
33. Seidenari S, Arginelli F, French P, König K, Magnoni C, Manfredini M, Ponti G, Talbot C, Dunsby C (2012) Multiphoton laser tomography and fluorescence lifetime imaging of basal cell carcinoma: morphologic features for non-invasive diagnostics. Exp Dermatol 21(11):831–836
34. Arginelli F, Manfredini M, Bassoli M, Dunsby C, French P, König K, Magnoni C, Ponti G, Talbot C, Seidenari S (2013) High resolution diagnosis of common nevi by multiphoton laser tomography and fluorescence lifetime imaging. Skin Res Technol 19(2):194–204
35. Dimitrow E, Ziemer M, Koehler MJ, Norgauer J, König K, Elsner P, Kaatz M (2009) Sensitivity and specificity of multiphoton laser tomography for in vivo and ex vivo diagnosis of malignant melanoma. J Invest Dermatol 129(7):1752–1758
36. Seidenari S, Arginelli F, Bassoli M, Dunsby C, French P, König K, Magnoni C, Ponti G, Talbot C. Multiphoton laser tomography and fluorescence lifetime imaging improve the diagnostic accuracy of melanoma. Submitted (in press)

Application of Laser Scanning Microscopy in Dermatology and Cutaneous Physiology

10

J. Lademann, S. Lange-Asschenfeldt, M. Ulrich, M.E. Darvin, K. König, M.C. Meinke, and B. Lange-Asschenfeldt

10.1 Introduction

The skin is not only the largest organ of the human body but also represents a barrier to the environment, protecting our body from water loss, penetration of environmental pollutants and microorganisms [1–3]. The skin has also a social function, as people frequently associate success in their private and professional lives with young-looking, smooth skin [4]. Therefore, not only medical but also cosmetic aspects are to be considered in skin treatment. While topically applied medical products must penetrate through the skin barrier in order to become effective in the living cells, cosmetic products do not necessarily induce any medical effects [5]. In most cases, these products remain on the skin surface or in the upper corneal layers in order to protect the skin (e.g., sunscreens) or to stabilise the dermal barrier (e.g., skin protection creams or lotions). Sometimes, in particular with respect to anti-aging creams, it is difficult to distinguish medical from cosmetic products.

The development, optimisation and evaluation of topically applied medical and cosmetic products focuses, inter alia, on analysing the kinetics of local penetration and deposition of formulations in the skin as well as changes in skin physiological parameters during therapy [6–8]. For this purpose, a number of experimental ex vivo models have been used for their investigation. With the introduction of laser scanning microscopy, a number of these clinical investigations may now be performed non-invasively in human skin [9]. Moreover, based on its high resolution laser scanning microscopy allows the analysis of cellular structures and thereby detection of histopathological changes in vivo and in real time [10]. Laser scanning microscopes can be operated in fluorescence and in reflectance mode.

In fluorescence mode, exogenous fluorescent dyes are required for visualisation of cellular structures. In this way, the differential penetration and redistribution of topically applied fluorescent dyes or fluorescence-labelled substances can be visualised using a fluorescence confocal laser scanning microscope [11]. Reflectance confocal laser scanning microscopy (RCLSM) on the other hand is based on the detection of the reflected light coming from horizontal planes within the tissue. With RCLSM imaging, no

J. Lademann (✉) • M.E. Darvin • M.C. Meinke
Department of Dermatology,
Center of Experimental and Applied Cutaneous Physiology, Venerology and Allergology,
Charité – Universitätsmedizin Berlin,
Charitéplatz 1, 10117 Berlin, Germany
e-mail: juergen.lademann@charite.de

S. Lange-Asschenfeldt • M. Ulrich
B. Lange-Asschenfeldt
Department of Dermatology, Skin Cancer Center,
Charité – Universitätsmedizin Berlin,
Charitéplatz 1, 10117 Berlin, Germany

K. König
JenLab GmbH, Schillerstr 1, Jena 07745, Germany

Department of Biophotonics and Laser Technology,
Saarland University, Campus A5.1,
66123 Saarbrücken, Germany

E. Berardesca et al. (eds.), *Non Invasive Diagnostic Techniques in Clinical Dermatology*,
DOI 10.1007/978-3-642-32109-2_10, © Springer Berlin Heidelberg 2014

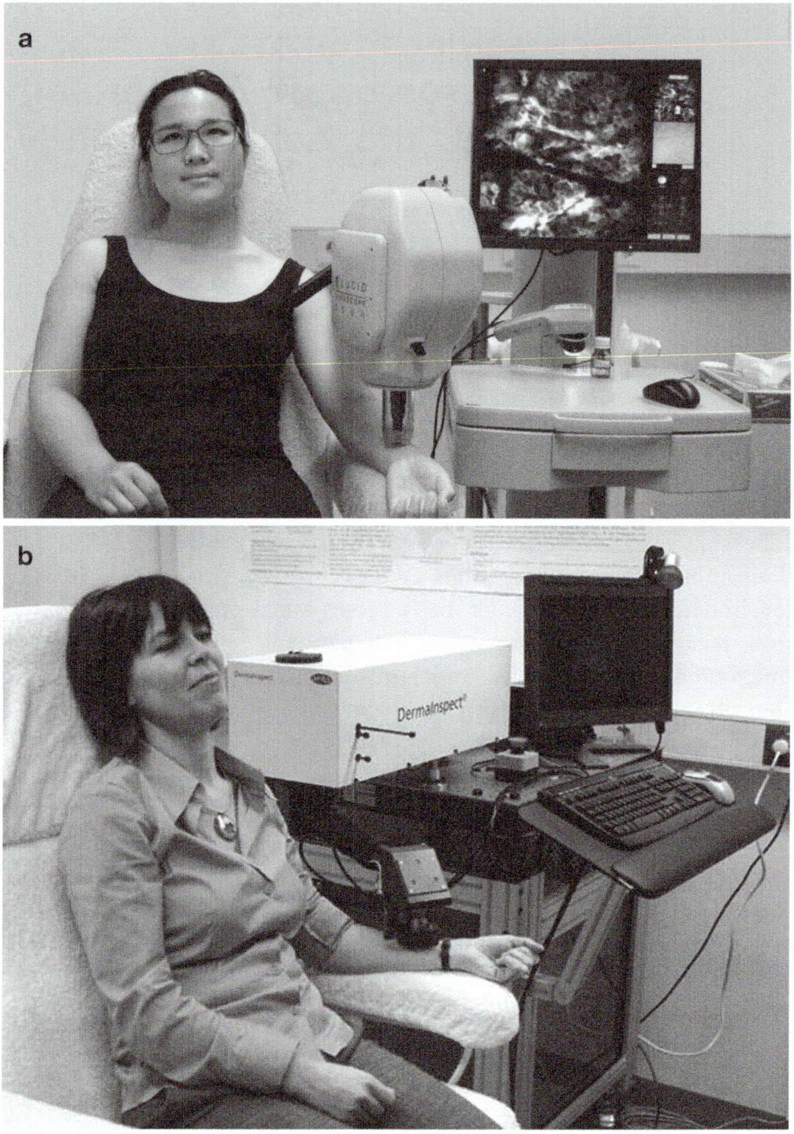

Fig. 10.1 (**a**) – In vivo laser scanning microscope Stratum and volunteer; (**b**) – multiphoton tomograph and volunteer

exogenous contrast is required for image generation. Recent technical advancements in optical research have combined the advantages of both techniques for evaluation of human skin [12].

In this chapter, the use of confocal laser scanning microscopy in fluorescence mode (FCLSM) and in reflectance mode in dermatology and cutaneous physiology is described.

Figure 10.1a depicts the in vivo laser scanning microscope Stratum (Optilas, Melbourne, Australia) [13]. This system consists of a handheld scanning device, which is used for tissue illumination and fluorescence detection. Following the intracutaneous administration of an exogenous fluorophore, a coherent argon-ion laser light source (λ=488 nm) is used to illuminate the fluorescent molecules. Emitted tissue fluorescence is propagated through the optical fibres and detected by the optical system. By using a fluorescence barrier filter, backscattered or reflected light will not be propagated, enabling the high resolution of FCLSM images. The flexible design of the hand piece allows the examination of any anatomical body site in patients and volunteers.

Figure 10.1b shows an application of the multiphoton tomograph (Jenlab GmbH, Jena, Germany) [14]. The optical principle is the simultaneous absorption of two or more photons by a fluorophore and the simultaneous emission of one photon with a shorter wavelength. The system employs a titanium sapphire laser (Mai Tai XF, Spectra-Physics, USA) for tissue illumination with a spectral wavelength ranging from 710 to 920 nm, generating femtosecond pulses. For in vivo examination of human skin, ultrashort pulses with low average energy are used in order to avoid tissue damage during the imaging process. The presence of endogenous fluorescent biomolecules/structures with fluorescent properties is used for image generation.

10.2 Representation of Dermal Structures by Histology and Laser Scanning Microscopy

Figure 10.2 shows a histological section of a skin sample derived from a healthy volunteer's forearm. Routine histology allows the visualisation of all cell layers in vertical sections thereby clearly distinguishing, i.e., the stratum corneum (SC), the epidermis (EP) and the dermis (DE). In contrast to routine histology, images obtained by fluorescence confocal laser scanning microscopy are oriented within the horizontal plane, parallel to the skin surface, yielding en face optical sections.

Figure 10.3 represents the various cell layers imaged by multiphoton laser scanning microscopy, namely, the stratum corneum (a), the stratum spongiosum (b) and stratum basale (c).

Figure 10.4 shows the stratum corneum (a, c) and the basal layer (b, d) including the papillary structure, comparing histological sections obtained in parallel to the skin surface (a, b) to images taken with the fluorescence laser scanning microscope Stratum (c, d).

The invasive character of biopsies does not allow an assessment over time, and the tissue removal and histological processing may result in artefacts, further limiting its clinical applicability.

Moreover, for ethical reasons taking large number of biopsies from a study patient or volunteer might be discussed critically based on the consecutive scar formation or possible cosmetic disfigurement.

In contrast, laser scanning microscopy is performed non-invasively and in vivo, providing images at a resolution comparable to routine histology. Because of this non-invasive character of CLSM, this technique allows the serial examination of the same skin area over time without tissue damage or processing artefacts.

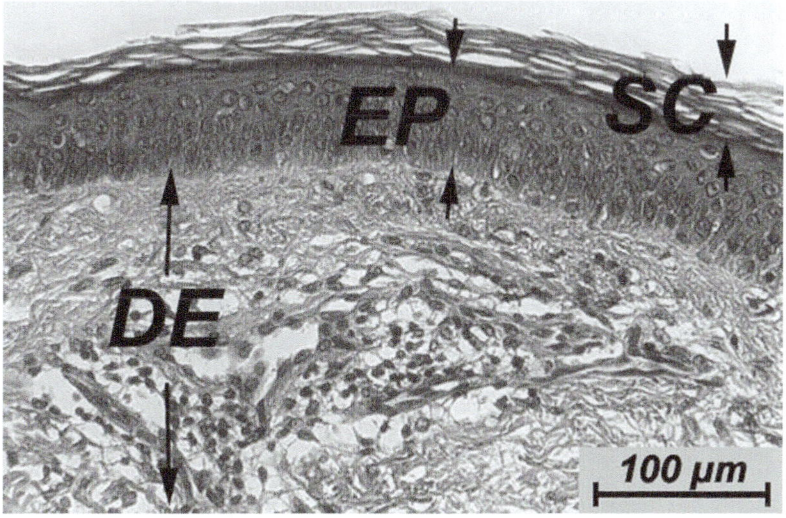

Fig. 10.2 Histological section showing the stratum corneum (*SC*), the epidermis (*EP*) and the dermis (*DE*)

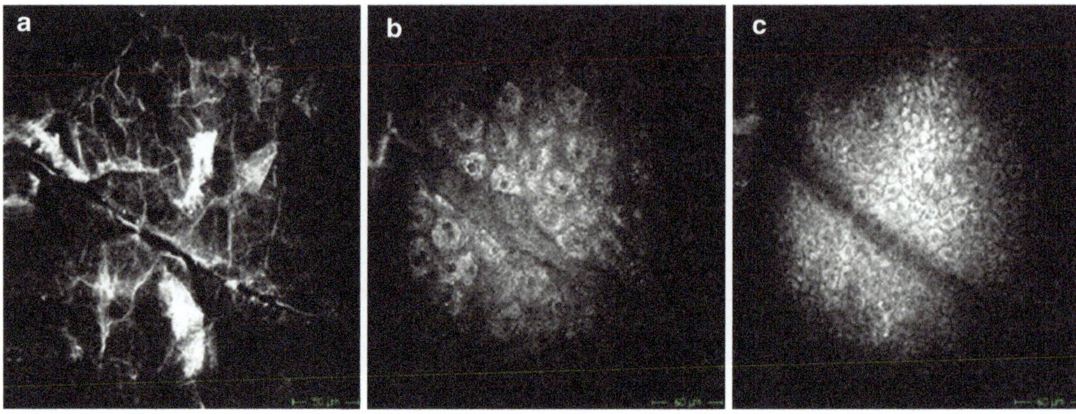

Fig. 10.3 Series of multiphoton/images of the cellular structures; (**a**) – stratum corneum, (**b**) – stratum spongiosum, (**c**) – stratum basale

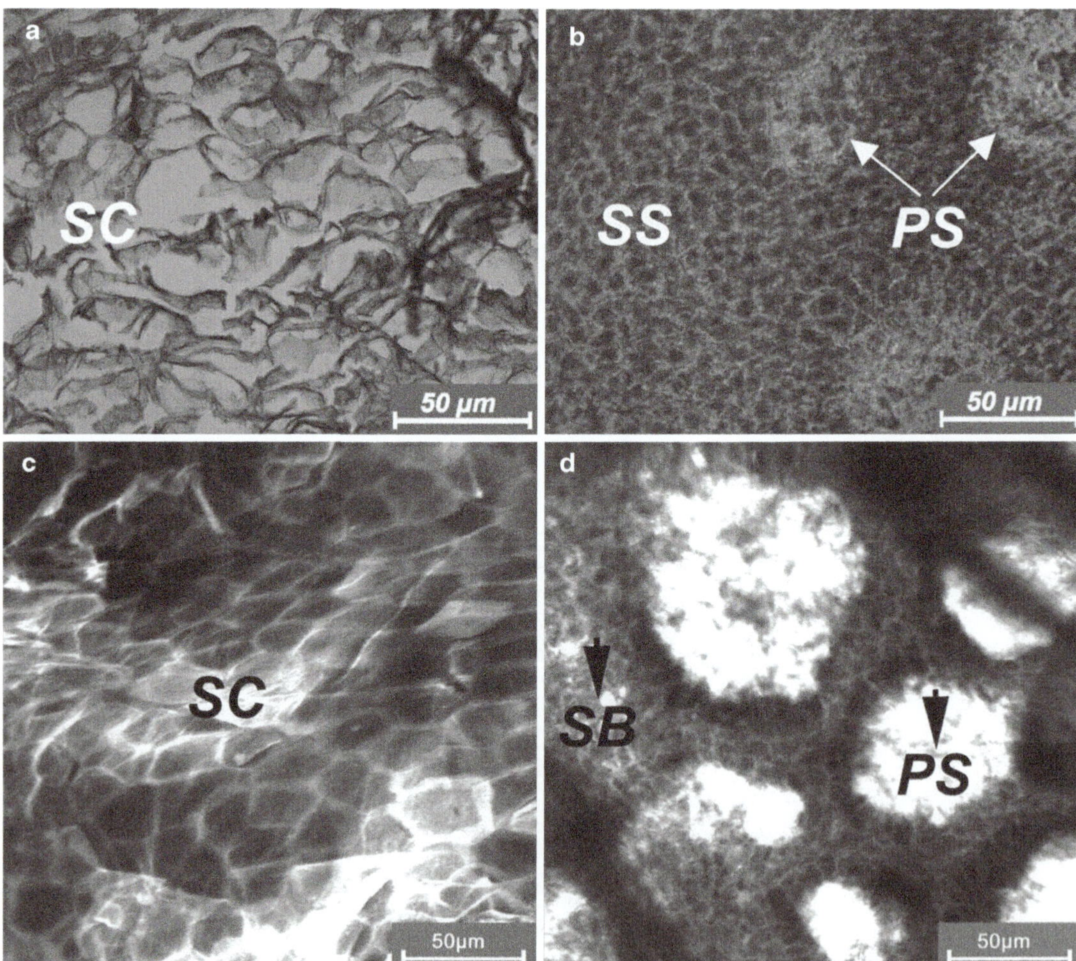

Fig. 10.4 Comparison between a horizontal histological section (**a**, **b**) and FCLSM images (**c**, **d**); stratum corneum (*SC*), stratum spinosum (*SS*), stratum basale (*SB*), papillary structure (*PS*)

10.3 Characterisation of the Skin Barrier

An intact skin barrier is of vital importance for the protection of the human body. The reasons for disturbances of the skin barrier can be manifold, ranging from genetic predisposition, exogenous factors or the presence of acute to chronic inflammatory skin conditions [15]. A common way to analyse the skin barrier is the measurement of the transepidermal water loss (TEWL) [16, 17]. These TEWL measurements are based on the concept that even the skin barrier of healthy skin is penetrated by small amounts of water vapour. This water vapour can be measured using a sensor containing a highly sensitive capacitor, changing its capacity depending on the water evaporating from the skin. TEWL values rise in correlation with increasing barrier defects. The advantage of this method is that the measurements yield numerical values, which are well suited to evaluate kinetic processes, e.g., in wound healing.

However, TEWL measurements are extremely susceptible to failure [18]. For these measurements, standardised conditions in terms of temperature and air humidity are indispensable and the volunteers must not sweat. Topically applied water-containing formulations can strongly influence the TEWL measurements since the water included within the formulations can evaporate and interfere with measurements. Moreover, skin care products can form protective films on the skin surface, thereby retaining the moisture in the skin barrier. These protective films can be easily destroyed when the sensor is positioned on the measuring site during TEWL measurements [19]. Damage to the film can give rise to an increase in the TEWL values, suggesting that a barrier defect has occurred as a result of the topical treatment. Topically applied products show a verifiable improvement in the barrier properties but can lead to misinterpretation of the TEWL values. This situation is exemplarily represented in Fig. 10.5. After application of a care cream for 2 weeks, a protective film developed on the surface of the stratum corneum (Fig. 10.5a). The TEWL measurements show increased values after the treatment: This, in turn, indicates that the barrier damage is increasing, too. In contrast to these results, all volunteers with extremely dry skin prior to treatment reported that their skin properties had distinctly improved during the cream application. This subjective clinical self-assessment was confirmed by objective laser scanning microscopy analyses. Figure 10.5 depicts the status of a volunteer's skin prior to (b) and after treatment (c) with the care product. The dry skin (Fig. 10.5b) shows an irregular, mountain-like arrangement of the corneocytes, while the skin structure after the treatment

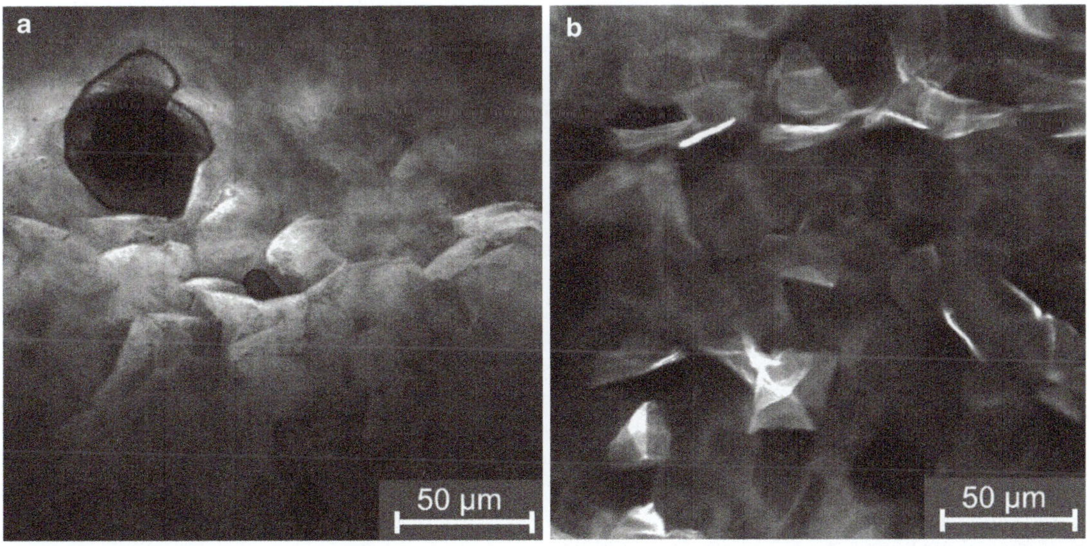

Fig. 10.5 FCLSM images: (**a**) – protective film of a formulation on the skin; (**b**) – dry skin; (**c**) – normal skin

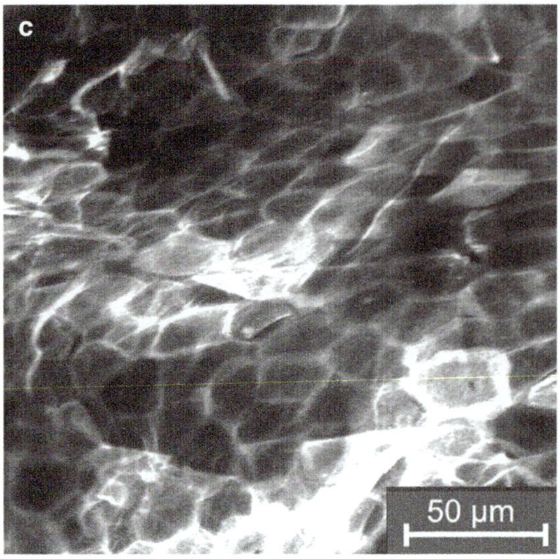

Fig. 10.5 (continued)

(Fig. 10.5c) reveals a regular honeycomb-like structure of the corneocytes as typical for healthy skin.

The skin barrier can also be characterised by the thickness of the corneal layers. Particularly in epidermal wound healing, where the stratum corneum is regenerating, determination of the thickness may give valuable information with respect to the evolution of therapy and optimisation. In Fig. 10.2 the stratum corneum is clearly visible in the histological section. However, the thickness of the stratum corneum of this histological section does not necessarily correlate with the situation in vivo, since the sample is subjected to hydration when the tissue is processed for routine histology and the sections are stained. Apart from the optical coherent tomography, laser scanning and multiphoton microscopy can be used to determine the layer thickness of the stratum corneum [20]. At the beginning of the epidermal imaging, the laser is focused on the surface of the stratum corneum (Fig. 10.6a). Subsequently the laser focus is moved deeply into the stratum corneum until the first cell layer of the stratum granulosum (SG) becomes visible as shown in Fig. 10.6b. In the bottom left-hand corner, the corneocytes of the stratum corneum are well recognisable, while moving deeper within the epidermis, the stratum granulosum comes into sight in the top right-hand corner. The distance of the laser focus from the surface of the stratum corneum to the SC/SG boundary layer corresponds to the real in vivo layer thickness of the skin barrier.

Furthermore, artificial interruptions of the skin barrier as induced when using microneedles can be visualised and analysed by the application of a fluorescent dye [21].

10.4 Penetration of Topically Applied Substances

There are two pathways for topically applied drugs to penetrate through the skin barrier. Drugs can penetrate the skin barrier by either the intercellular pathway within the lipid layers around the corneocytes or by entering the hair follicles [22].

The penetration into the hair follicles, however, does not necessarily mean that the drug passes the barrier of the hair follicles reaching the living cells [23]. The barrier structure of the upper part of the hair follicle in the upper part is comparable to that of the stratum corneum. In deeper parts, tight junctions make up the barrier [24]. The kinetics of the drug penetration is dependent on the composition of the skin barrier as well as the type of the topically applied substance. Once the drug passes the barrier, it will enter the circulatory system, either via the lymphatic system or by cells that have taken up the drug. The process of intercellular penetration can be easily analysed using FCLSM since the process takes place within the technically limited imaging depths of the system. While the stratum corneum is roughly 15 μm thick, the skin at the palms of the hands and soles of the feet is significantly thicker measuring up to 150 μm. The human vellus hair follicles reach as deep as 250 μm and the terminal hair follicles even up to 1 mm into the skin [25]. Consequently, the drug penetration into the hair follicles can only be visualised and analysed in the upper part of the follicles because of the limited penetration depth of the laser light (approximately 150 μm). For this reason, in vivo FCLSM is of limited use for quantitative penetration studies of drugs into the hair follicles. Therefore, other non-invasive methods like differential stripping were developed to analyse also the follicular penetration [26].

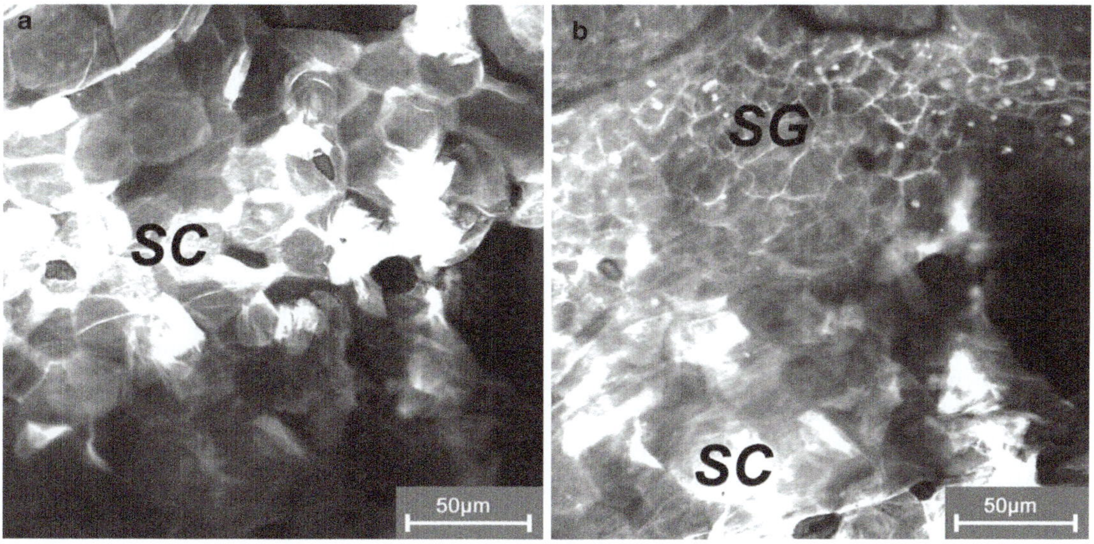

Fig. 10.6 FCLSM images: separating skin layers (stratum corneum (*SC*) (**a**), stratum corneum (*SC*)/stratum granulosum (*SG*) (**b**)) for determination of the thickness of the skin barrier

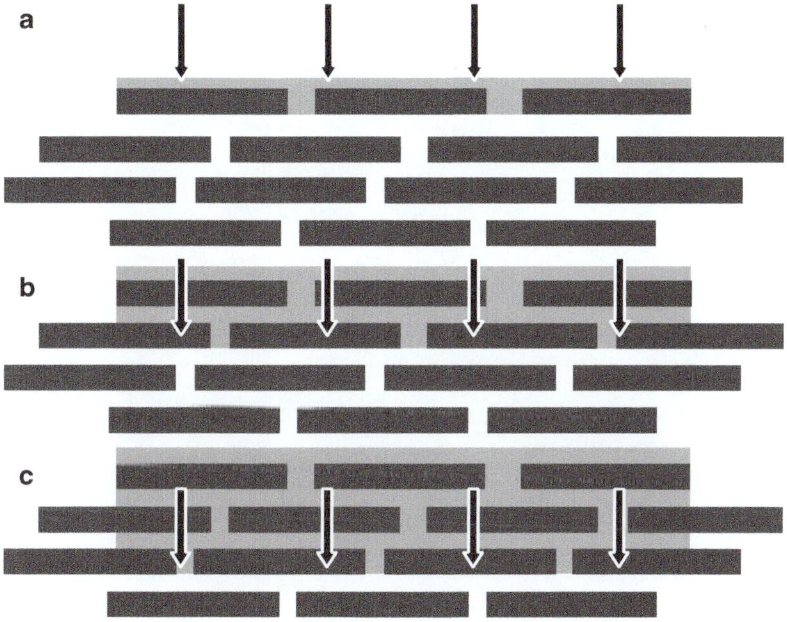

Fig. 10.7 Bricks-and-mortar model according to Peter Elias; the substance is topically applied (**a**) and penetrates deeper into the stratum corneum (**b**, **c**)

In vivo FCLSM is a promising tool for the investigation of drug penetration into and past the stratum corneum. Since the flat corneocytes are permeable to laser and fluorescence light within the wavelength spectrum of 480–600 nm, the distribution of fluorescent or fluorescence-labelled substances can be easily detected.

The structure of the stratum corneum can be schematically described as a brick-and-mortar model according to Elias et al. [27]. In this model, the corneocytes represent the bricks and the lipid layers the mortar as illustrated in Fig. 10.7a. Topically applied substances (Fig. 10.7b) penetrate into deeper layers of the stratum corneum as shown in Figs. 10.7b, c.

In Fig. 10.8, the penetration process of substances is demonstrated by in vivo laser scanning microscopy. Immediately after application, the localisation of the formulation is restricted to the skin surface including the first cell layers of the corneocytes. After a penetration time of 20 min, additional layers of corneocytes are visible. As the structure of these cell layers is offset, they can be easily counted. Figure 10.8 shows a formulation detectable as deep as in the sixth cell layer of corneocytes [9]. Differences in the penetration kinetics of different formulations can be analysed by this method. However, the penetrating substance cannot be analysed quantified within the epidermis since the fluorescent signals cannot be allocated to single layers. However, when the penetration kinetics is studied by in vivo laser scanning microscopy, the amount of topically applied substances in the different depths of the stratum corneum at different times can be quantified by the tape stripping procedure. The quantification of drugs using the tape stripping procedure is a noninvasive method that has been successfully used for the quantification of various drugs [28–31].

In case of an intact barrier, only 0.1–1 % of the topically applied drug passes through the skin barrier dependent on the chemical properties of the tested drug [32]. The detection of these small amounts of drugs in the viable dermis is almost impossible by fluorescence laser scanning micros-

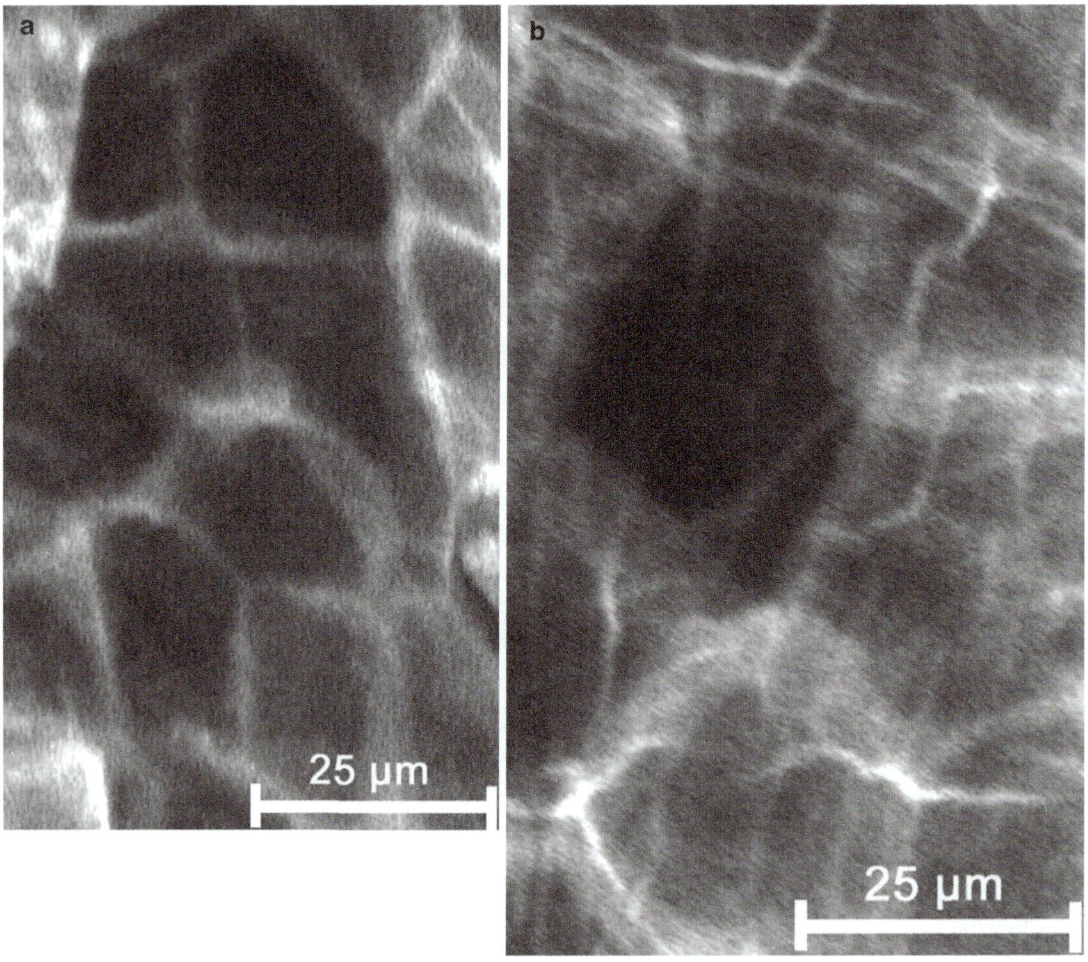

Fig. 10.8 FCLSM images: penetration of formulations into the SC; (**a**) – immediately after application; (**b**) – 20 min. after application

copy. The rich blood perfusion of the dermis and the drainage of the tissue by the lymphatic system as well as the phagocytosis of immune cells diminish the concentration of the drugs rapidly, thus preventing their large-scale accumulation.

The penetration of zinc oxide (ZnO) nanoparticles, which are commonly used in sunscreens, is discussed controversially in the literature [33, 34]. Non-linear properties of ZnO nanoparticles such as a second harmonic generation (SHG) and hyper-Rayleigh scattering were used for their visualisation [35, 36]. In this regard, a specifically developed channel was embedded into the multiphoton microscope. The detection limit of 0.08 fg/μm^3 was estimated for ZnO nanoparticles (30 nm in diameter). Taking advantage of the high sensitivity obtained, it was clearly shown in in vivo measurements that ZnO nanoparticles penetrate only in the outermost layers of the SC, furrows, wrinkles and orifices of the hair follicles, but do not get into contact with the viable epidermis [37].

The kinetics of the drug distribution within the stratum corneum provides clues about the penetration properties of the drug through the skin barrier. In damaged skin, significant amounts of topically applied drugs could reach the living cells. Recently it was reported that cold electrical plasma, which can be used for skin disinfection, strongly stimulates the penetration of topically applied substances [38]. Figure 10.9 represents

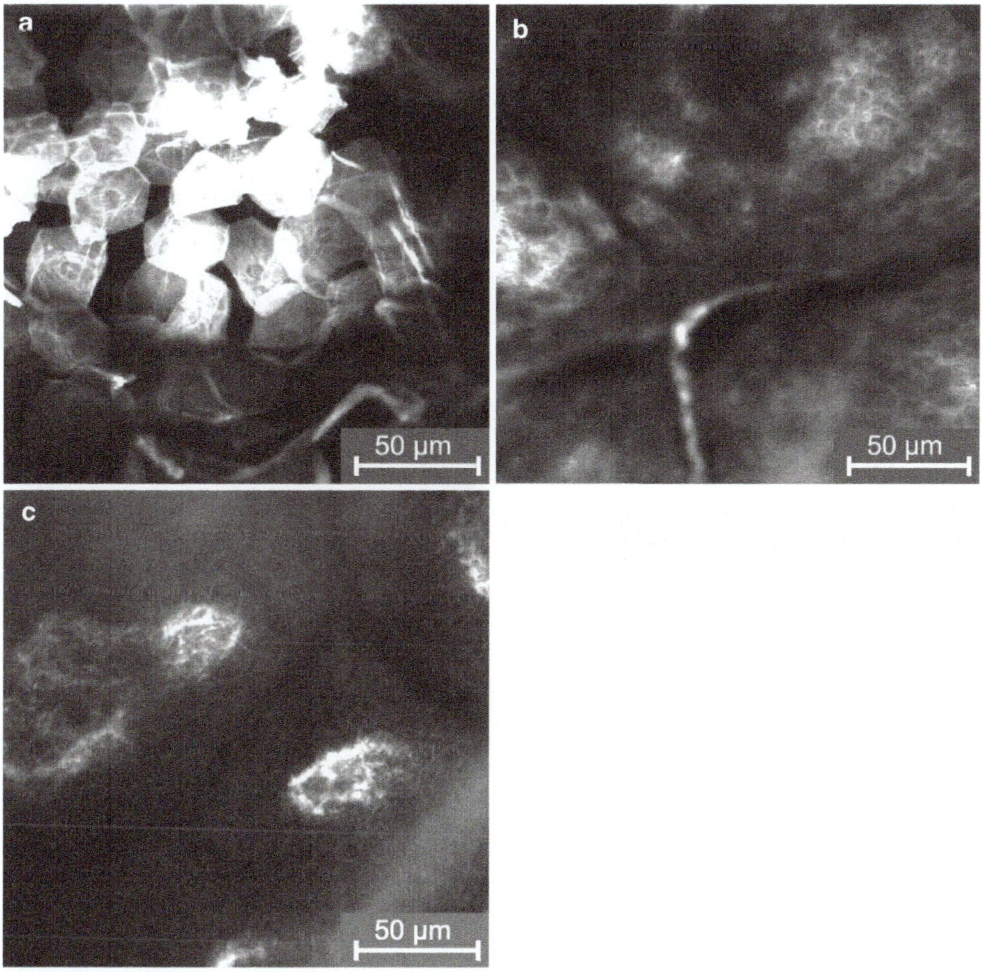

Fig. 10.9 FCLSM images: before plasma treatment fluorescent dye could be detected only on the skin surface (**a**); after plasma treatment it could be detected also in the stratum basale (**b**) and around the papillary structure (**c**)

the distribution of a formulation-containing fluorescent dye on the surface of porcine ear skin, which is an appropriate in vitro model for human skin. Without plasma treatment, the fluorescence signal can be detected only on the skin surface 60 min after application, yet.

When the skin had been treated with the plasma jet after topical application of the formulation, the fluorescent dye was detected also in the basal layer and in the papillary dermis at high concentrations, as evidenced by the strong fluorescence signal.

10.5 Distinction Between Intact and Inflammatory Skin by In Vivo Multiphoton Microscopy

Figure 10.3 illustrates the different cell layers of healthy skin. In the case of inflammatory processes such as atopic dermatitis or psoriasis, the structure of the skin barrier – the stratum corneum – is strongly disturbed, impairing the visualisation of a regular stratum corneum. Underneath within the viable epidermis, inflammatory cells are detected as demonstrated in Fig. 10.10. In addition, enlarged intercellular spaces may be observed, likely corresponding to spongiosis. Figure 10.11 compares the basal layer of a psoriatic lesion to the basal layer of corresponding healthy skin, whereby the enlarged intercellular spaces can be clearly recognised.

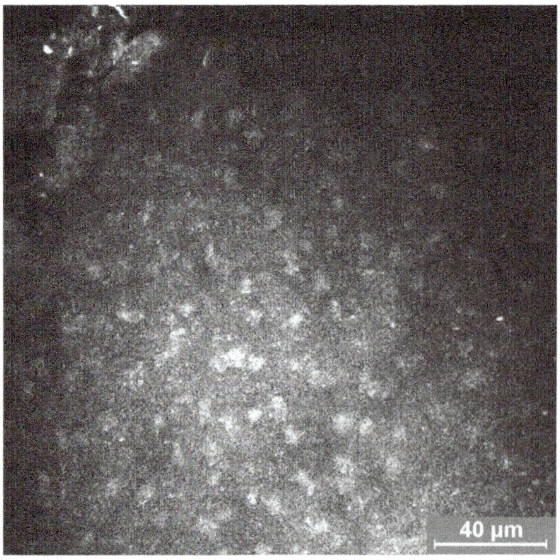

Fig 10.10 Inflammatory cells detected by MPT

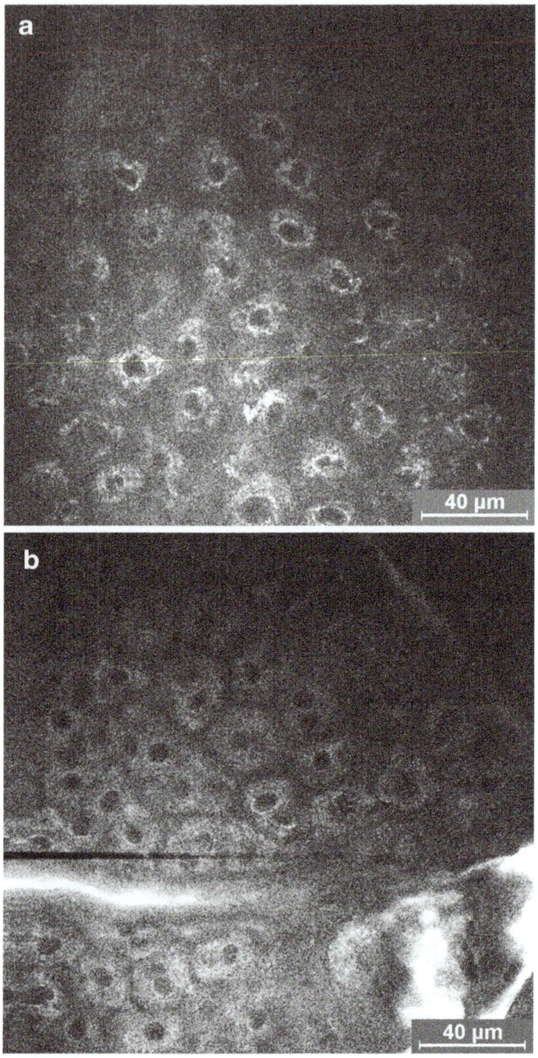

Fig. 10.11 Enlarged intracellular space of psoriatic lesion (**a**) compared to healthy skin (**b**) imaged by multiphoton tomography

10.6 Distinction Between Different Types of Skin Cancer

Actinic keratosis and basal cell carcinoma represent the most common subtypes of non-melanoma skin cancer and may be difficult to distinguish on clinical examination alone. However, since

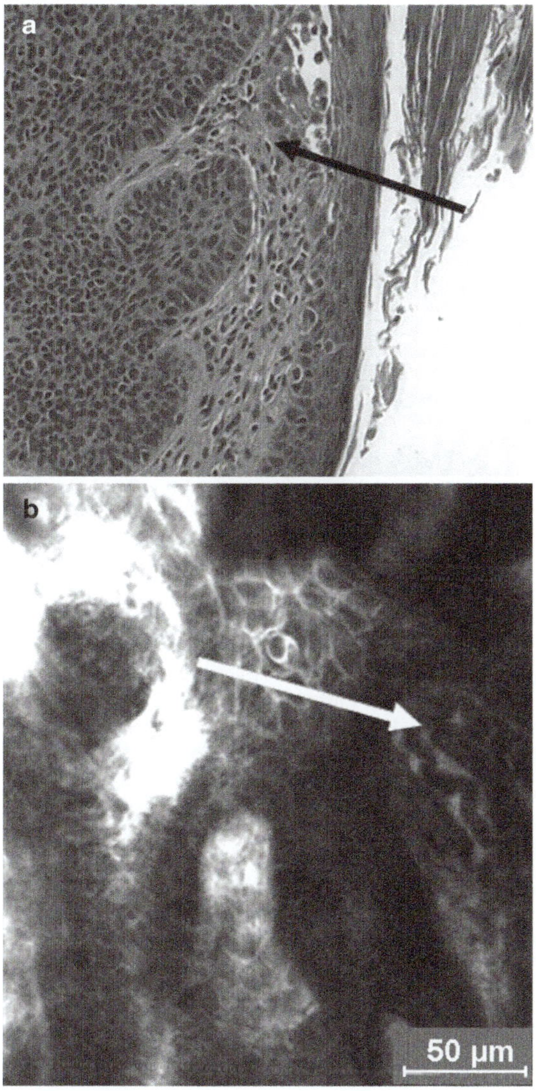

Fig. 10.12 Basal cell carcinoma: histological section (**a**) compared to in vivo RCLSM measurement (increased tortuosity of blood vessels) (**b**)

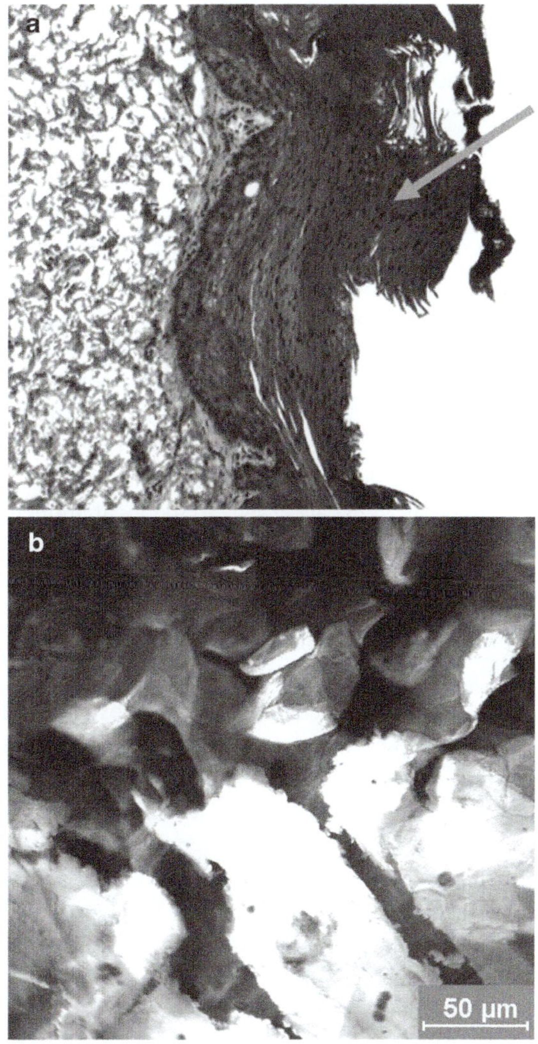

Fig. 10.13 Histological section showing actinic keratosis (**a**); RCLSM image of actinic keratosis with visible hyperkeratoses (**b**)

respective treatment regimen may vary considerably, the correct diagnosis is essential for appropriate therapeutic management. While actinic keratosis has been recognised as an early-stage, in situ squamous cell carcinoma, basal cell carcinoma has been defined as a semimalignant type of skin cancer, with a variety of clinical presentations and histological variants [13]. Histologic hallmarks of actinic keratoses are frequently hyperkeratosis, parakeratosis and the presence of keratinocyte atypia, while basal cell carcinoma shows lobular arrangements of neoplastic cells derived from the basal layer of the skin and the follicular epithelium. While hyperkeratosis is more characteristic for actinic keratosis, histological evaluations have shown an increased tortuosity of blood vessels as more typical for basal cell carcinoma. Figure 10.12a shows a histological section through a basal cell carcinoma. The increased tortuosity of the blood vessels is marked by an arrow. Corresponding findings may also be detected by in vivo laser scanning microscopy as shown in Fig. 10.12b. Figure 10.13a showing a histological section of an actinic keratosis is presented, while the corresponding in vivo laser scanning microscopic image is shown in Fig. 10.13b. Both figures illustrate that both cellular and structural features of these two types of cancer may be documented by in vivo laser scanning microscopy. Considering these findings, in vivo laser scanning microscopy may be used as an adjunct diagnostic tool, thereby avoiding unnecessary biopsy procedures. Considering a rapid, in vivo diagnosis, CLSM may minimise delays before the initiation of appropriate therapy. In addition CLSM may be useful for monitoring the treatment outcome of non-invasive therapies, ultimately avoiding excisions altogether. The latter may be of particular importance with respect to the increased development of topical therapies for treatment of skin cancer, where the local application of drugs permits therapeutic methods which render the excision of the lesions unnecessary.

Conclusions

In vivo laser scanning microscopy is well suited to characterise the structural integrity of the skin barrier. It permits to evaluate the influence of topically applied drugs and cosmetic products on barrier repair. The kinetics of the penetration of fluorescent or fluorescence-labelled substances can be investigated by this method non-invasively. In addition, in vivo laser scanning microscopy can be used as an adjunct to clinical diagnosis and for monitoring therapeutic outcome based on the non-invasive morphological analysis. Current literature suggests that confocal laser scanning microscopy may serve as a substitute to invasive histological investigations. In comparison to histological analysis, in vivo laser scanning microscopy offers a painless, non-invasive and rapid diagnostic tool investigating human skin in its native state and in vivo. Limitations of confocal laser scanning microscopy include the limited optical penetration depending on the wavelength used by the respective system. The fluorescence mode using an argon laser irradiating at 488 nm is limited to 150–200 μm penetration depth, whereas the reflectance mode using a wavelength above 700 nm can penetrate deeper.

The disadvantage of in vivo laser scanning microscopy, however, is that it is still very expensive. In the light of the rapid development and progressive miniaturisation of the measuring systems, it can be expected that these systems will get smaller in size, ultimately decreasing respective costs for acquisition, such that they can be widely applied in dermatology and cosmetology.

References

1. Schittek B (2011) The antimicrobial skin barrier in patients with atopic dermatitis. Curr Probl Dermatol 41:54–67
2. Seyfarth F, Schliemann S, Antonov D, Elsner P (2011) Dry skin, barrier function, and irritant contact dermatitis in the elderly. Clin Dermatol 29(1):31–36
3. Lademann J, Richter H, Schanzer S, Knorr F, Meinke M, Sterry W, Patzelt A (2011) Penetration and storage of particles in human skin: perspectives and safety aspects. Eur J Pharm Biopharm 77(3):465–468
4. Martin KI, Glaser DA (2011) Cosmeceuticals: the new medicine of beauty. Mo Med 108(1):60–63
5. Situm M, Bulat V, Buljan M, Puljiz Z, Situm V, Bolanca Z (2010) Senile lentigo – cosmetic or medical issue of the elderly population. Coll Antropol 34(Suppl 2):85–88
6. Lademann J, Jacobi U, Surber C, Weigmann HJ, Fluhr JW (2009) The tape stripping procedure–evaluation of some critical parameters. Eur J Pharm Biopharm 72(2):317–323
7. Wei H, Chen Y, Xu L, Zheng J (2007) Percutaneous penetration kinetics of lidocaine and prilocaine in two local anesthetic formulations assessed by in vivo microdialysis in pigs. Biol Pharm Bull 30(4): 830–834
8. Lademann J, Richter H, Schaefer UF, Blume-Peytavi U, Teichmann A, Otberg N, Sterry W (2006) Hair follicles – a long-term reservoir for drug delivery. Skin Pharmacol Physiol 19(4):232–236
9. Lademann J, Otberg N, Richter H, Meyer L, Audring H, Teichmann A, Thomas S, Knuttel A, Sterry W (2007) Application of optical non-invasive methods in skin physiology: a comparison of laser scanning microscopy and optical coherent tomography with histological analysis. Skin Res Technol 13(2):119–132
10. Roberts MS, Dancik Y, Prow TW, Thorling CA, Lin LL, Grice JE, Robertson TA, Konig K, Becker W (2011) Non-invasive imaging of skin physiology and percutaneous penetration using fluorescence spectral and lifetime imaging with multiphoton and confocal microscopy. Eur J Pharm Biopharm 77(3):469–488
11. Meyer LE, Otberg N, Sterry W, Lademann J (2006) In vivo confocal scanning laser microscopy: comparison

of the reflectance and fluorescence mode by imaging human skin. J Biomed Opt 11(4):044012
12. Sattler E, Kastle R, Arens-Corell M, Welzel J (2012) How long does protection last? – In vivo fluorescence confocal laser scanning imaging for the evaluation of the kinetics of a topically applied lotion in an everyday setting. Skin Res Technol 18(3):370–377
13. Astner S, Dietterle S, Otberg N, Rowert-Huber HJ, Stockfleth E, Lademann J (2008) Clinical applicability of in vivo fluorescence confocal microscopy for noninvasive diagnosis and therapeutic monitoring of nonmelanoma skin cancer. J Biomed Opt 13(1):014003
14. Konig K (2008) Clinical multiphoton tomography. J Biophotonics 1(1):13–23
15. Larsen TH, Gregersen P, Jemec GB (2001) Skin irritation and exposure to diisocyanates in orthopedic nurses working with soft casts. Am J Contact Dermat 12(4):211–214
16. Laudanska H, Reduta T, Szmitkowska D (2003) Evaluation of skin barrier function in allergic contact dermatitis and atopic dermatitis using method of the continuous TEWL measurement. Rocz Akad Med Bialymst 48:123–127
17. Pinnagoda J, Tupker RA, Agner T, Serup J (1990) Guidelines for transepidermal water loss (TEWL) measurement. A report from the Standardization Group of the European Society of Contact Dermatitis. Contact Dermatitis 22(3):164–178
18. Vergou T, Schanzer S, Richter H, Pels R, Thiede G, Patzelt A, Meinke MC, Sterry W, Fluhr JW, Lademann J (2012) Comparison between TEWL and laser scanning microscopy measurements for the in vivo characterization of the human epidermal barrier. J Biophotonics 5(2):152–158
19. Teichmann A, Sadey Pour Saleh H, Schanzer S, Richter H, Lademann J (2006) Investigations of bedan face cream. Kosmetische Medizin 4:180–182
20. Weigmann HJ, Schanzer S, Teichmann A, Durat F, Antoniou C, Schaefer H, Sterry W, Lademann J (2007) Ex-vivo spectroscopic quantification of sunscreen efficacy: proposal of a universal sun protection factor. J Biomed Opt 12(4):044013
21. Bal SM, Kruithof AC, Zwier R, Dietz E, Bouwstra JA, Lademann J, Meinke MC (2010) Influence of microneedle shape on the transport of a fluorescent dye into human skin in vivo. J Control Release 147(2):218–224
22. Blume-Peytavi U, Massoudy L, Patzelt A, Lademann J, Dietz E, Rasulev U, Garcia Bartels N (2010) Follicular and percutaneous penetration pathways of topically applied minoxidil foam. Eur J Pharm Biopharm 76(3):450–453
23. Knorr F, Lademann J, Patzelt A, Sterry W, Blume-Peytavi U, Vogt A (2009) Follicular transport route–research progress and future perspectives. Eur J Pharm Biopharm 71(2):173–180
24. Leclerc EA, Huchenq A, Mattiuzzo NR, Metzger D, Chambon P, Ghyselinck NB, Serre G, Jonca N, Guerrin M (2009) Corneodesmosin gene ablation induces lethal skin-barrier disruption and hair-follicle degeneration related to desmosome dysfunction. J Cell Sci 122(Pt 15):2699–2709
25. Patzelt A, Knorr F, Blume-Peytavi U, Sterry W, Lademann J (2008) Hair follicles, their disorders and their opportunities. Drugs Discov Today 5(2):e173–e181
26. Patzelt A, Richter H, Buettemeyer R, Huber HJ, Blume-Peytavi U, Sterry W, Lademann J (2008) Differential stripping demonstrates a significant reduction of the hair follicle reservoir in vitro compared to in vivo. Eur J Pharm Biopharm 70(1):234–238
27. Elias PM, Steinhoff M (2008) "Outside-to-inside" (and now back to "outside") pathogenic mechanisms in atopic dermatitis. J Invest Dermatol 128(5):1067–1070
28. Jacobi U, Meykadeh N, Sterry W, Lademann J (2003) Effect of the vehicle on the amount of stratum corneum removed by tape stripping. J Dtsch Dermatol Ges 1(11):884–889
29. Lademann J, Ilgevicius A, Zurbau O, Liess HD, Schanzer S, Weigmann HJ, Antoniou C, Pelchrzim RV, Sterry W (2006) Penetration studies of topically applied substances: optical determination of the amount of stratum corneum removed by tape stripping. J Biomed Opt 11(5):054026
30. Meinke M, Abdollahnia M, Gahr F, Platzek T, Sterry W, Lademann J (2009) Migration and penetration of a fluorescent textile dye into the skin–in vivo versus in vitro methods. Exp Dermatol 18(9):789–792
31. Weigmann H, Lademann J, Meffert H, Schaefer H, Sterry W (1999) Determination of the horny layer profile by tape stripping in combination with optical spectroscopy in the visible range as a prerequisite to quantify percutaneous absorption. Skin Pharmacol Appl Skin Physiol 12(1–2):34–45
32. Feldmann RJ, Maibach HI (1969) Percutaneous penetration of steroids in man. J Invest Dermatol 52(1):89–94
33. Baroli B (2010) Penetration of nanoparticles and nanomaterials in the skin: fiction or reality? J Pharm Sci 99(1):21–50
34. Burnett ME, Wang SQ (2011) Current sunscreen controversies: a critical review. Photodermatol Photoimmunol Photomed 27(2):58–67
35. Johnson JC, Yan H, Schaller RD, Petersen PB, Yang PD, Saykally RJ (2002) Near-field imaging of nonlinear optical mixing in single zinc oxide nanowires. Nano Lett 2(4):279–283
36. Segets D, Tomalino LM, Gradl J, Peukert W (2009) Real-time monitoring of the nucleation and growth of ZnO nanoparticles using an optical hyper-rayleigh scattering method. J Phys Chem C 113(28):11995–12001
37. Darvin ME, Konig K, Kellner-Hofer M, Breunig HG, Werncke W, Meinke MC (2012) Safety assessment by multiphoton fluorescence/SHG/HRS tomography of ZnO nanoparticles used in cosmetic products. Skin Pharmacol Physiol 25(4):219–226
38. Lademann O, Richter H, Kramer A, Patzelt A, Meinke MC, Graf C, Gao Q, Korotianskiy E, Rühl E, Weltmann KD, Lademann J, Koch S (2011) Stimulation of the penetration of particles into the skin by plasma tissue interaction. Laser Phys Lett 8(10):758–764

11 Telethermography

Aldo Di Carlo

Telethermography (TT) is a technique in which an infrared camera captures infrared or caloric energy from a distant object and transforms it into video images, without any contact (Fig. 11.1) [1]. Each of the light points that compose these images has a correspondent thermal (infrared) point of the source. TT allows to overcome the limits of traditional thermometric methods (e.g., thermometers, thermocouples, and thermistors) that required, first of all, to plot the temperature point by point and, in a second time, the recomposition of them in a "thermal map." In fact, TT gives the possibility to visualize at a glance all the thermal gradients on the skin surfaces under observation and, moreover, to check their modifications either during physiopathological processes or during experimental studies (e.g., vasoactive drugs).

11.1 Physical Basis

TT is based on the principle that above the zero point of the absolute temperature (−273 °C), anybody emits a quantity of infrared radiation that is proportional to its absolute temperature according to the Stefan-Boltzmann law ($R = \varepsilon\sigma T^4$, where σ is a constant, T is the absolute temperature, and ε is the emissivity of the body; ε varies from 0 to 1 and for skin is equal to 1).

A. Di Carlo
San Gallicano Dermatological Institute IRCCS, Rome, Italy
e-mail: dicarlo@ifo.it

11.2 Instruments

TT employs a thermal camera that recalls a video camera in which special lens made of silicon or germanium captures the energetic infrared band of the electromagnetic spectrum. This band is rapidly decomposed by means of two rotating orthogonal prisms into packets of infrared energy that are then transmitted to special sensors made of crystals of indium antimonide or cadmium telluride. These crystals convert the electromagnetic incident energy into electrical signals, which are transmitted to a cathode ray tube and transformed into video images. Inside, special insulating devices, such as liquid nitrogen (−196 °C) or thermocouples (Peltier effect), eliminate the interference of ambient temperature and obtain maximum sensitivity. The system has a spatial resolution of approximately 0.001 mm

Fig. 11.1 A FLIR® thermographic camera provided by a PC for elaboration and storage of the IR images

E. Berardesca et al. (eds.), *Non Invasive Diagnostic Techniques in Clinical Dermatology*, DOI 10.1007/978-3-642-32109-2_11, © Springer Berlin Heidelberg 2014

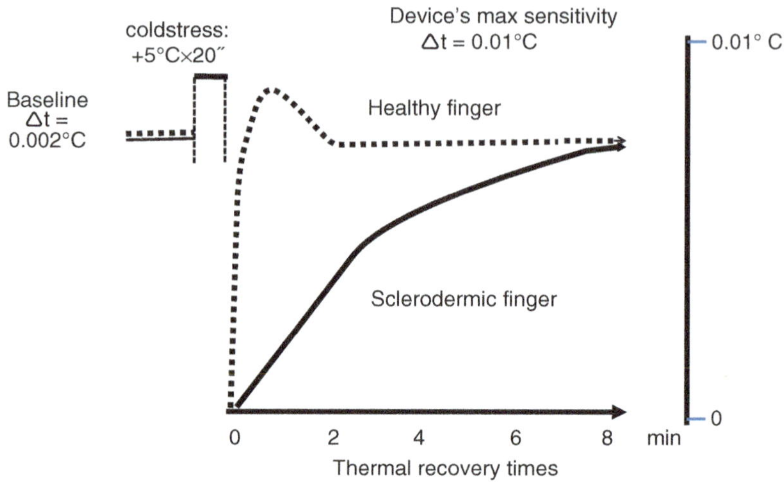

Fig. 11.2 Principle of thermal stress

per pixel and a time resolution of a few milliseconds.

Thermal images in gray appear on the monitor as a continuous varying of tones, from gray black (for the coldest points) to gray white (for the hottest points). The video images can also be displayed in bands of colors, generally on a scale of eight colors or more, the yellow color conventionally indicating the hottest band and blue color indicating the coolest. The differences of temperature between two different tonalities of gray or two bands of colors are defined as "thermal gradients," and are expressed as °C or fractions, and express the sensitivity of the instruments. The present sensitivity ranges from a minimum of three tenth of °C to maximum values, other than dozen of °C, according the circumstances. To let possible to repeat and compare the results obtained, the instruments are continuously automatically adjusted to the absolute temperature (margin radiometric error of adjustment, ±1 °C).

To have a good image, the operator adjusts the visual thermal field to the cutaneous area under observation, so that the image could contain all the proper thermal points, from the hottest points to the coolest ones. These instruments are connected to a PC, and specific software enables the images to be acquired, processed, and registered. TT is a safe and repeatable method, giving us the possibility for many applications in the field of dermatology and cosmetology.

11.3 Thermography with Thermal Stress

The application of TT in the field of clinical and experimental dermatology and in cosmetology is limited by the fact that the thermal gradients of the skin are often much less than 0,02 °C, that are the limits of the current thermographic devices. This fact limits the possibility to study anatomical variations but also physiological and pathological processes in dermatology.

To overcome this problem, some methods have been introduced by many authors, similar to what has happened in other fields in which TT is also applied (industry, antiquities, veterinary, etc.). These methods are each time called as dynamic, or active, or pulsed TT [2–7], in contrast with the "direct TT." Among these, the thermal stimulation method (TS) has been previously described by us [8–10] (Fig. 11.2). TS permits to perform a thermal stress under controlled conditions, respectively, at +5 °C ×20″, for a "cold stress," or at +40 °C ×20″ in case of "hot stress." By practice, after performed thermal stress, it is possible, starting from the time 0 (= the end of stress), to follow on the monitor the times of recovery of thermal gradients, second by second (Figs. 11.3a–c). These thermal recovery times (TRT) will be very short in case of high basal thermal gradients and will be longer in case of low basal thermal gradients. So by this method, thermal gradients are evaluated through minutes

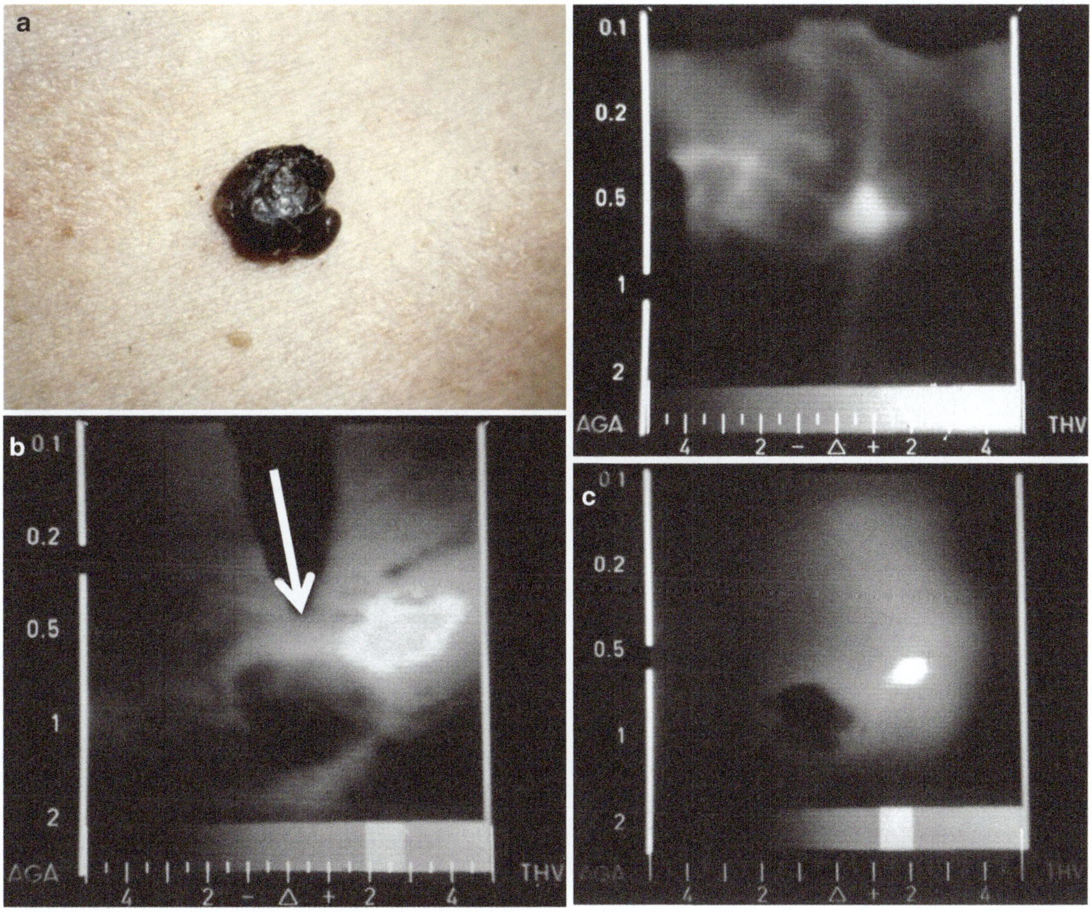

Fig. 11.3 (**a**) Melanoma of the back. A bright hyperthermic stria is direct toward the nearer nodal station. Note also the hyperthermic perilesional halo. (**b**) Thermographic pattern of melanoma. Note the hyperthermic gradients of the tumor surface and the hyperthermic gradients of the halo (*arrow*) (*comet sign*). (**c**) Another thermographic aspect of melanoma. "*Candle sign*"

or seconds, instead of degrees centigrade (°C), permitting to reach a great improvement as sensitivity and accuracy, as compared to the "direct" TT. Using a TS at 5 °C ×20″, the healthy skin shows generally a TRT ≥2 min. A reduction of TRT of thermal gradients (<2 min) means that there is an increased thermal conductivity in this area of the skin. This fact could depend on different causes: vascular, metabolic, tumoral, and inflammatory.

In case of vasoconstrictive conditions, e.g., the smoking effect, a "hot" stress is more convenient (+40 °C ×20″).

Another advantage is the fact that using TS method is no more necessary to provide to the climatization of the subjects in the room, as it normally is requested for the direct TT.

11.4 Clinical Applications

11.4.1 Melanoma

The clinical use of telethermography in the study of skin cancers was introduced by Brasfield, the reported malignant melanoma as having a very high hyperthermic gradients [11]. Then, other researchers confirmed this report and also described a specific thermographic pattern of the melanoma [12, 13]. Afterwards, TT was criticized from other authors, because of its reduced sensitivity and specificity and for the high frequency of false-positive and false-negative. This was observed not only in the case of melanoma but also in other tumors of external organs [14, 15]. The application of TT assisted by TS is very useful in these cases.

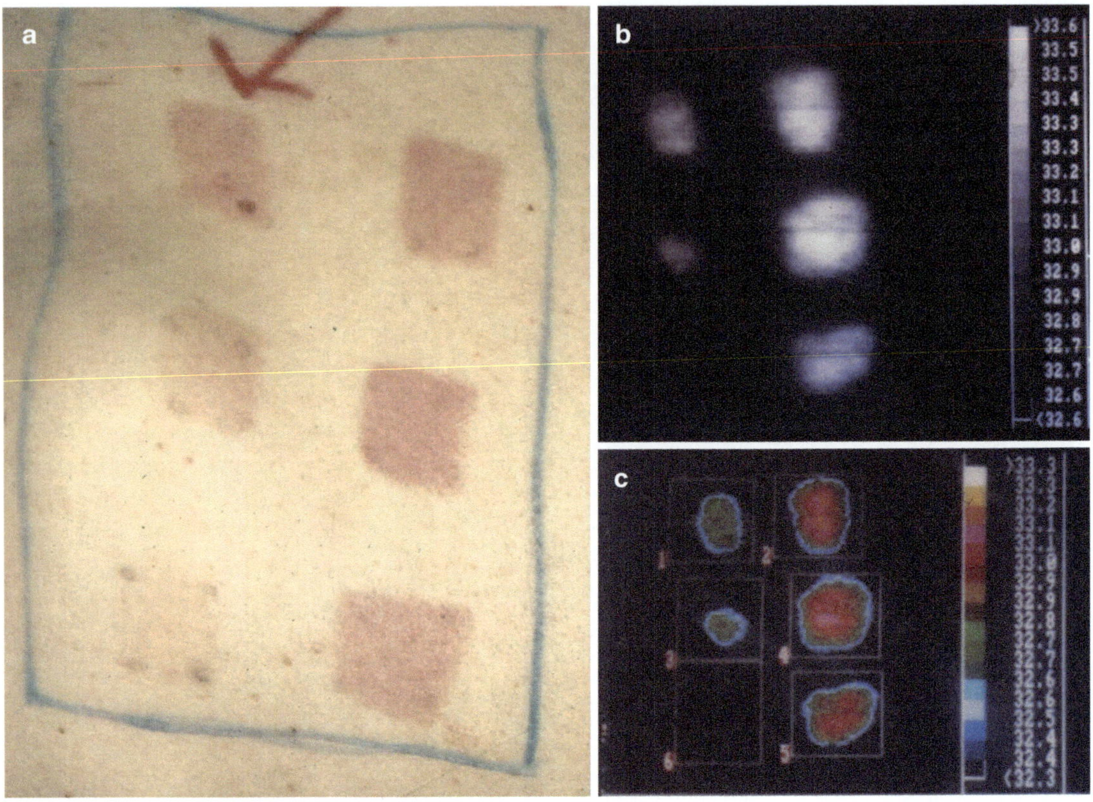

Fig. 11.4 (**a**) Clinical evaluation of the MED (*arrow*). (**b**) Thermographic evaluation of MED in the same subject. (**c**) The same thermogram of (**b**) that shows semiquantitative data of MED

In a more extensive study, a number of 402 melanomas at different thickness (Breslow's criteria) were evaluated. As results, a rapid TRT was seen in all the melanomas, except in cases of "in situ" melanoma, or at a Breslow's thickness <0.76 mm [16]. This fact means that a reduced critical mass of the tumor does not permit at direct TT to identify the tumor. Hence, TT assisted by TS method allows a precise definition of the thermographic pattern of the malignant melanoma, characterized by a hyperthermic halo and its extension toward the nearer nodal stations, named by the authors as *thermal flame or comet sign* (Fig. 11.3a–c).

11.4.2 Patch Tests and MED

TT with TS is useful for objectively evaluating patch tests in allergic contact dermatitis, particularly when clinical evaluation is difficult to perform, as happens in case of atopic or brown-skinned individuals [17–20]. In the field of photodermatology, TT with TS allows to quantify the minimal erythema dose (MED), until now evaluated on subjective criteria and only on fair skin (Fig. 11.4a–c).

11.4.3 Microangiopathies

Under conditions of thermal equilibrium of the subject in the room (T=+21 °C, RH=50 %), skin temperature is determined by the blood flow in the papillary dermis [21]. This is the nutritional flow of the skin that is much lower than that of other organs (e.g., heart muscle), because of the reduced metabolic request of the skin. For this reason, minimal changes in the thermal gradients are barely detectable by direct TT. Nevertheless, this subtle part of the skin, also called "epidermal-dermal unit," is the site of numerous dermatological microangiopathies, as are observed in the aging and diabetic skin, connective tissue disease (e.g., scleroderma) [22], as well as in smokers [23, 24]. In a previous study on diabetic subjects, we found, by means of TT assisted by TS, a significantly prolonged TRT in the group of individuals with insulin-dependent diabetes (ID), while in non-ID diabetic subjects the TRT had the same values of healthy individuals [25]. In the aging skin, it has been observed a different pattern between photoaging and chronoaging, in relation to differences in vascular damage [26] (Fig. 11.5a–c).

In the so-called cellulite, a characteristic "leopard-like" image has been described, consisting of small, medium, and large areas, corresponding to different grades of this anti-esthetic condition. This provides the new possibility to have an objective evaluation of the real effects of the remedies that are continuously proposed in the market [27] (Fig. 11.6a–d).

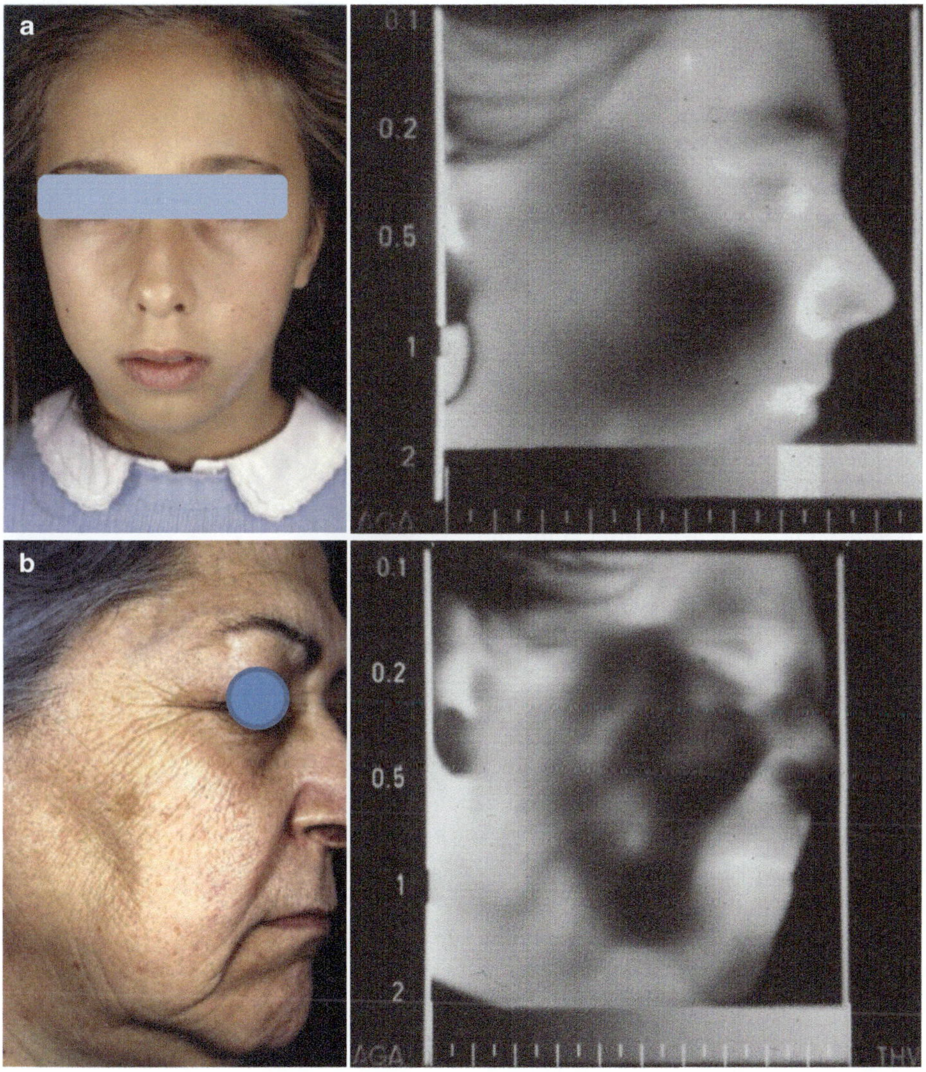

Fig. 11.5 (**a**) Thermographic pattern of the face of a young girl. Note the regular decalage of thermal gradients. (**b**) Thermographic aspect of the face of an old clerk (chronoaging). Note the altered thermographic pattern. (**c**) Thermographic pattern of the face of an old farmer (photoaging) Note the severe alterations of the vascular pattern

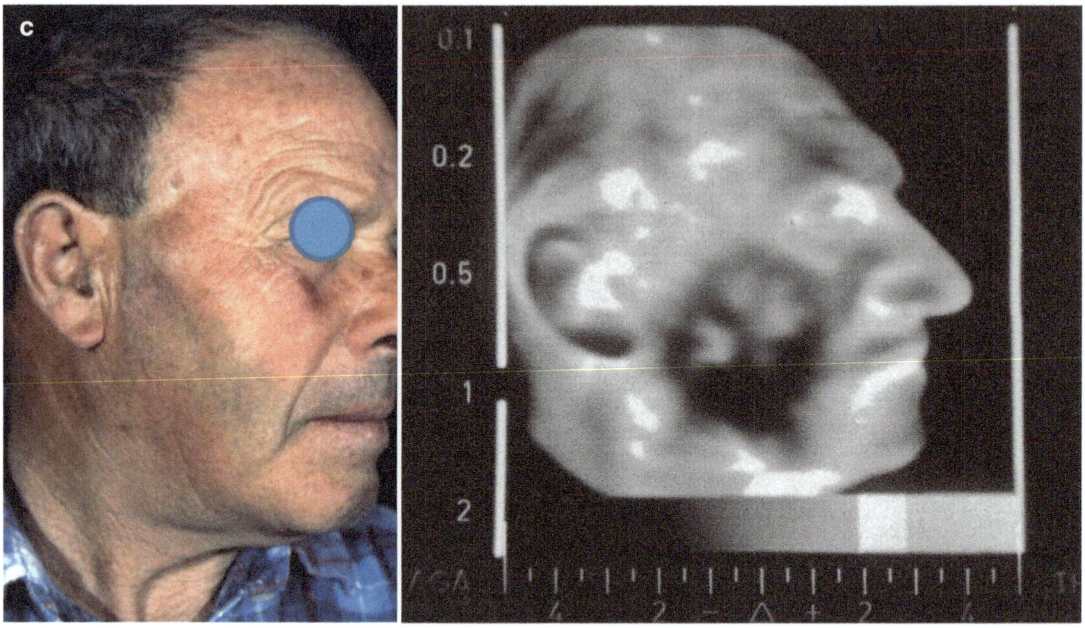

Fig. 11.5 (continued)

11.4.4 Blanching Effect

In the evaluation of the "potency" of topical corticosteroids, it is possible to better study the vasoconstriction effect, or "blanching" effect according to McKenzie's criteria, using a "hot" stimulus (+40 °C ×20″). This method allows for the evaluation not only of the "potency" of different steroid's classes in terms of TRT but also of the "reservoir effect" at skin level, a particular pharmacological aspect until now evaluated through complex experimental investigations or based on theoretical studies [28, 29] (Fig. 11.7a, b). Other skin diseases that have a vascular pattern such as psoriasis could be well followed after a specific therapy [30].

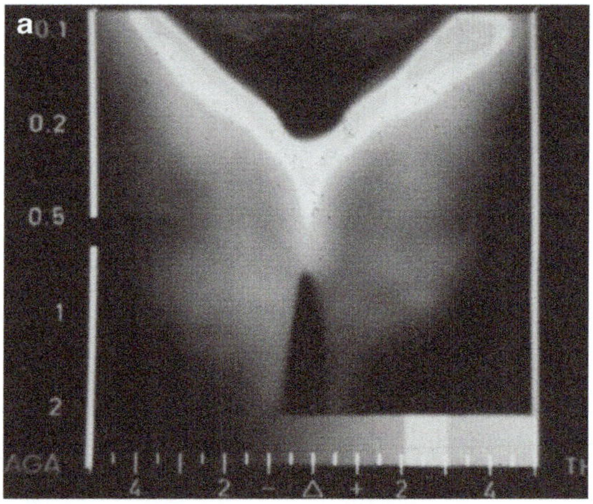

Fig. 11.6 (**a**) Normal subject in the "cellulite" sites. (**b**) "Cellulite" of a young girl. Regular hot spots showing a low thermal gradients. *Moucheté pattern*. (**c**) "Cellulite" of an obese girl. Note the large hyperthermic areas. (**d**) Pattern of cellulite of an obese menopausal woman. Note the deeply disorganized thermographic pattern

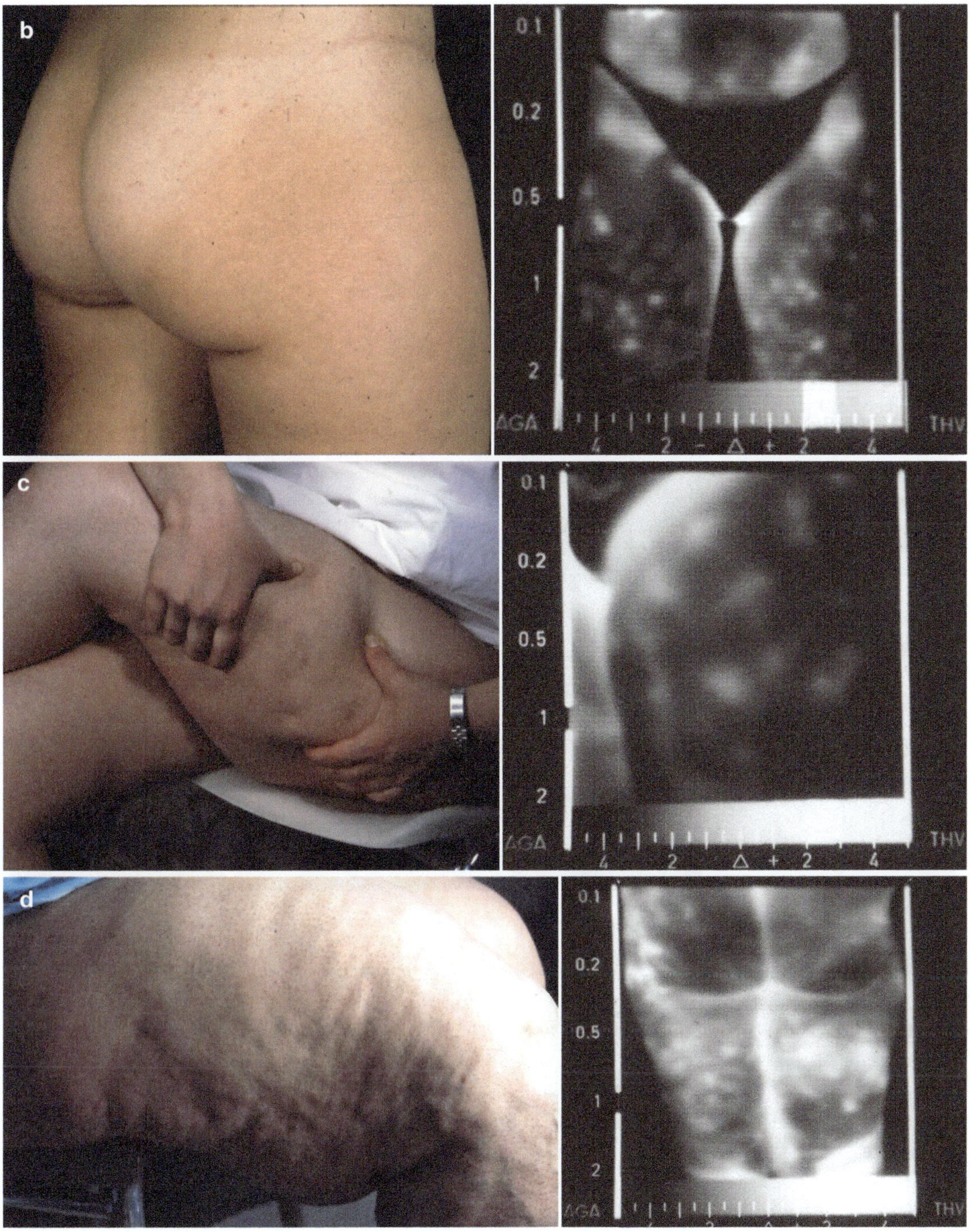

Fig. 11.6 (continued)

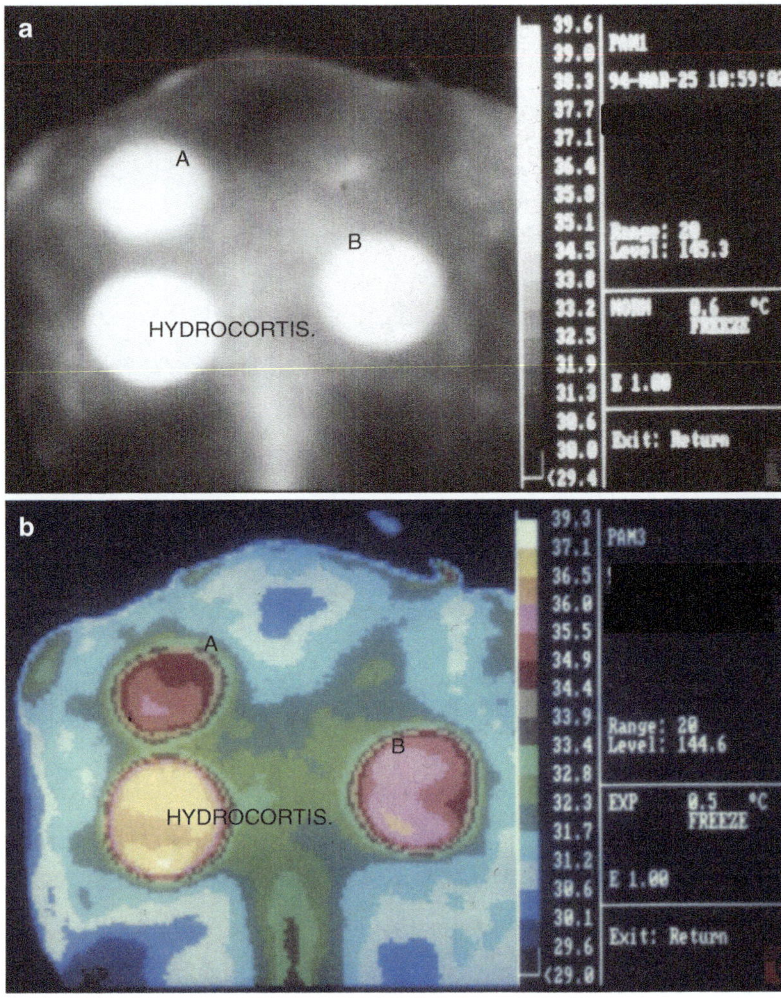

Fig. 11.7 (**a**) Evaluation of "potency" of two steroids vs. hydrocortisone. Thermal stress performed at +40 °C ×20″. (**b**) The same case of the (**a**) Color thermogram. Note the possibility to "measure" the potency of three different steroids

11.4.5 Dermocosmetology

By means of TT assisted by TS, we have studied the percutaneous absorption of cosmetic preparations. Following the papers of Engel and other authors [31, 32], we compared the effect of a simple cosmetic cream to a cream containing a vasoactive substance (buflomedil chlorhydrate) in ten subjects and, parallelly, in other ten subjects the same cosmetic cream vs. a polyethylene sheet 20×20 cm applied to the back skin for 2 h. Then we followed by on time the modification of the thermal gradients. The results showed that cosmetic cream had a hyperthermic effect on the skin; yet, the hyperthermic gradient was surprisingly higher than vasoactive cream and similar to the plastic sheet. These data could suggest that the hyperthermic effect is attributable mainly to the chemical-physical activity of these preparations, more than pharmacological activity of their components. This phenomenon probably occurs through a blocking of the *perspiratio insensibilis*, with a consequent increase in the temperature of the underlying tissue ("greenhouse effect") [33]. This is related to a vasodilatation of arterioles and capillaries of the papillary dermis that causes a rise of the flow of the water and also electrolytes in the tissue (Fig. 11.8a–d). Clinically, this effect translates into a more moisturized, velvety, and soft skin. On these bases, it is possible to

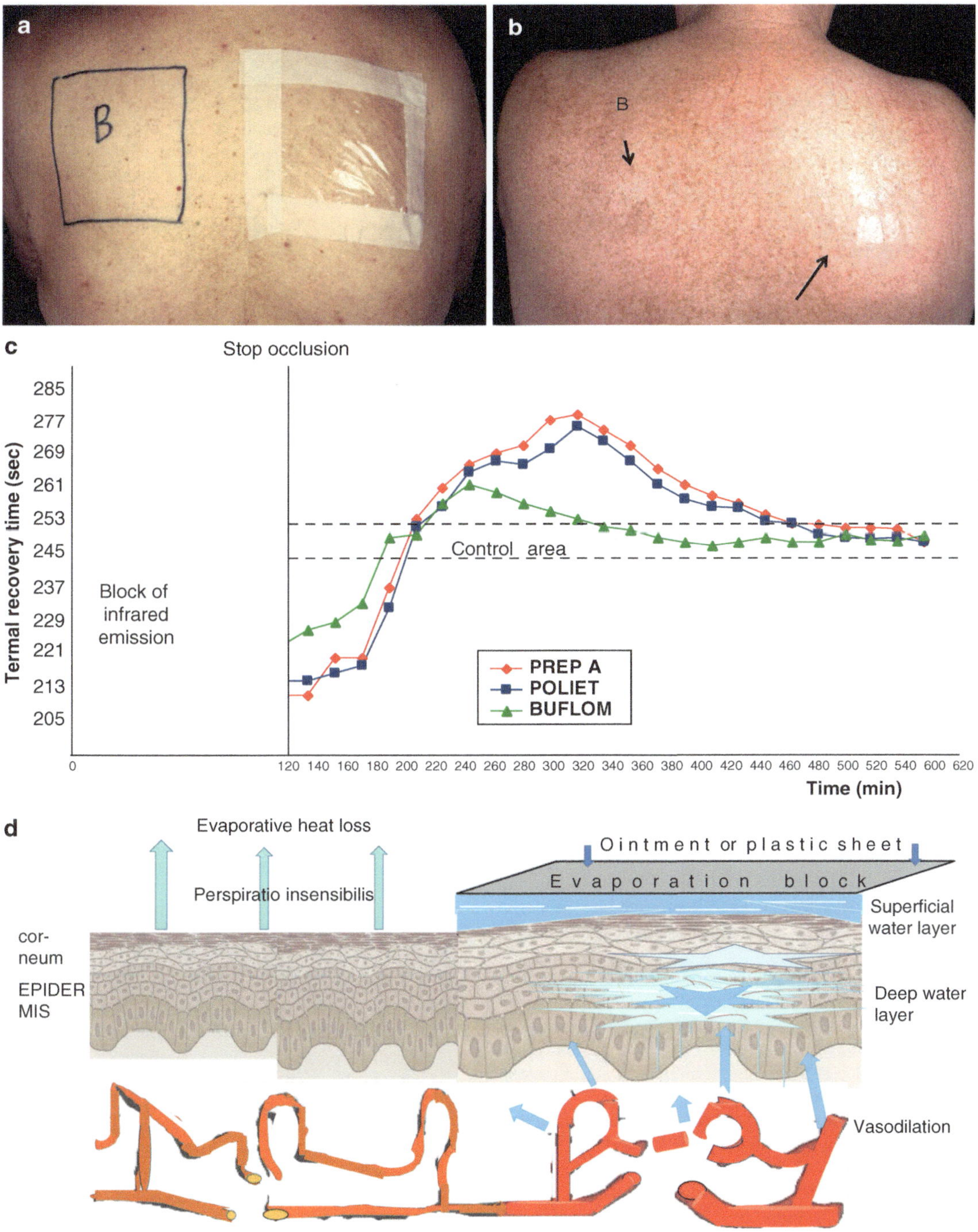

Fig. 11.8 (**a**) Vasoactive cream (*B* buflomedil) vs. polyethylene sheet. (**b**) Observe the soft, velvet skin, soon after having removed the plastic sheet, maintained in occlusion for 2 h. (**c**) The diagram shows the different times of thermal recovery after thermal stimulation at +40 °C ×20″. (**d**) A schematic drawn to explain a possible local thermal effect provoked by the application of topical substances (*greenhouse effect*)

hypothesize that the same effect happens after the application of the so-called beauty masks. These data allow to be considered the cosmetic preparations as possible active chemical-physical barriers, suggesting the need for further investigations under this particular view.

In conclusion, TT assisted by TS can be considered a very useful method for studying not only neoplastic and microvascular skin diseases but also for clinical and investigative applications in the field of esthetic medicine.

References

1. Bjork NA (1967) AGA thermovision, a high speed infrared camera with instantaneous picture display. J Radiol Electrol Med Nucl 48:30–33
2. Parker WJ, Jenkins RJ, Butler CP et al (1961) Flash method of determining thermal diffusivity, heat capacity, and thermal conductivity. J Appl Phys 32: 1679–1684
3. Santa Cruz GA, González SJ, Bertotti J et al (2009) First application of dynamic infrared imaging in boron neutron capture therapy for cutaneous malignant melanoma. Med Phys 36:4519–4529
4. Anbar M (2002) Assessment of physiologic and pathologic radiative heat dissipation using dynamic infrared imaging. Ann N Y Acad Sci 972:111–118
5. de Weerd L, Weum S, Mercer JB (2009) The value of dynamic infrared thermography (DIRT) in perforator selection and planning of free DIEP flaps. Ann Plast Surg 63:274–279
6. Van Dyk H, Lemaster RL (2010) An investigation of the use of active infrared thermography to detect localized surface anomalies in lumber. Scanning 32: 219–223
7. Pirtini Çetingül M, Herman C (2011) Quantification of the thermal signature of a melanoma lesion. Int J Therm Sci 50:421–431
8. Ippolito F, Di Carlo A (1980) La metodica della criostimolazione nell'indagine termografica in dermatologia. Boll Ist Dermatol S Gall X:165–183
9. Ippolito F, Di Carlo A (1981) L'indagine termografica nella oncologia dermatologica. Ann It Derm Clin Sper 35:137–141
10. Di Carlo A (1995) Thermography and the possibilities for its applications in clinical and experimental dermatology. Clin Dermatol 13:329–336
11. Brasfield RD, Laughlin JS, Sherman RS (1964) Thermography in the management of cancer. Ann N Y Acad Sci 121:235–247
12. Michel U, Hornstein OP, Schönberger A (1986) Thermographico-histologic study of the lymph drainage areas in malignant melanoma. Hautarzt 37: 12–16
13. Amalric R, Altschuler C, Giraud D et al (1984) Value of infrared thermography in the assessment of malignant melanoma of the skin. In: Ring EFJ, Phillips B (eds) Recent advances in medical thermology. Plenum Press, New York, pp 623–629
14. Cristofolini M, Perani B, Piscioli F et al (1981) Uselessness of thermography for diagnosis and follow-up of cutaneous malignant melanoma. Tumori 67:141–143
15. Moskowitz M et al (1976) Lack of efficacy of thermography as a screening tool for minimal and stage I breast cancer. N Engl J Med 295:249–252
16. Ippolito F, Di Carlo A (1991) La teletermografia nella diagnosi del melanoma cutaneo. G Ital Dermatol Venereol 126:327–335
17. Phipatanakul CS, Slavin RG (1972) Use of thermography in clinical allergy. J Allergy Clin Immunol 50:264–275
18. Agner T, Serup J (1988) Contact thermography for assessment of skin damage due to experimental irritants. Acta Derm Venereol 68:192–195
19. Baillie AJ, Biagioni PA, Forsyth A et al (1990) Thermographic assessment of patch-test responses. Br J Dermatol 122:351–360
20. Laino L, Di Carlo A (2010) Telethermography: an objective method for evaluating patch test reactions. Eur J Dermatol 20:175–180
21. Strandness DE, Sumner DS (1975) Cold injury. In: Hemodynamics for surgeons. Grune and Stratton Eds, New York, pp 582–620
22. Moore TL, Vij S, Murray AK et al (2009) Pilot study of dual-wavelength (532 and 633 nm) laser Doppler imaging and infrared thermography of morphoea. Br J Dermatol 160:864–867
23. Bornmyr S, Svensson H (1991) Thermography and laser-Doppler flowmetry for monitoring changes in finger skin blood flow upon cigarette smoking. Clin Physiol 11:135–141
24. Di Carlo A, Ippolito F (2003) Early effects of cigarette smoking in hypertensive and normotensive subjects. An ambulatory blood pressure and thermographic study. Minerva Cardioangiol 51:387–393
25. Ippolito F, Di Carlo A, Maggio C (1984) La termografia con metodica di criostimolazione nello studio del microcircolo nel soggetto diabetico. Giorn Ital Dermatol Venereol 119:49–55
26. Di Carlo A, Carducci M, Mussi A (1994) Prospettive dermocosmetologiche nell'invecchiamento cutaneo. Giorn It Dermatol Venereol 129:523–528
27. Ippolito F, Di Carlo A (1984) La thermographie. Son utilité comme critère de diagnostic et d'efficacité dans le traitment de la cellulite. J Med Esth XI(42):81–86
28. Ippolito F, Di Carlo A (1988) La teletermografia, nuova metodica per la valutazione dell'attività dei corticosteroidi topici. Giorn It Dermatol Venereol 123:291–295
29. Ippolito F, Di Carlo A, Frascione P (1989) Studio termografico sull'assorbimento percutaneo dei corticosteroidi. Giorn It Dermatol Venereol 124:9–12

30. Ippolito F, Di Carlo A, Carducci M, Leone G, Frascione P (1989) Cyclosporin a and psoriasis: a thermographic study. Acta Derm Venereol Suppl (Stockh) 146:155–158
31. Engel JM (1984) Physical and physiological influence of medical ointments on infrared thermography. In: Ring EFJ, Phillips B (eds) Recent advances in medical thermology. Plenum Press, New York, pp 177–183
32. Lehmuskallio E, Anttonen H (1999) Thermophysical effects of ointments in cold: an experimental study with a skin model. Acta Derm Venereol 79:33–36
33. Ippolito F, Di Carlo A (1989) Valutazione con metodica teletermografica della cinesi di penetrazione percutanea di una soluzione acquosa di liposomi complessi di Buflomedil cloridrato. In: Microcircolazione 1989. Tesi e Monduzzi Eds, Bologna, pp 61–65

12 Capillaroscopy

Marianna Mazzullo, Giuseppa Marina Miceli, Alessio Adamo, and Salvatore Amato

12.1 History

The structure of the microcirculation was discovered in 1662 by a physician from Bologna, Marcello Malpighi, who described the tiny vessels of the frog lung, comparing them to the human hair.

Using his father's microscope, Jan Swammerdam later started to study tadpoles and their blood. Herman Boerhaave (1668–1738) (Boravius) was the first to use conjunctival biomicroscopy to observe intravasal aggregation of erythrocytes in the eyes of feverish patients.

Lazzaro Spallanzani (1729–1799) described white blood cells in the blood of the salamander but mistakenly thought them to be air bubbles.

In 1912, Lombard discovered that morphology of skin capillaries could be studied under the microscope in the periungual site (proximal nail fold), after applying a drop of paraffin oil. Lombard was the father of capillaroscopy test on mankind.

M. Mazzullo, MD • G.M. Miceli, MD • S. Amato (✉)
Medicina – U.O. Dermatologia, ARNAS Civico, Arnas, Piazza Nicola Leotta 4, 90100 Palermo, Italy
e-mail: dermatologia@ospedalecivicopa.org; marinamiceli@live.it

A. Adamo
Via Pantelleria 98, 91100 Trapani, Italy
e-mail: a_adamo@msn.com

In 1949, Knisely and Bloch expressed two concepts of fundamental importance:

1. The overlapping between the status of a district's microcirculation and that of all other districts; the integrity of a district is based on the integrity of its microcirculation.

At the beginning of the twentieth century, Brown and O'Leary used the capillaroscopy test to observe in detail the anomalies that characterized the involvement of the microcirculation during Raynaud's phenomenon in systemic sclerosis.

In 1973, Maricq and collaborators published an article that first described the existence of specific patterns of progressive systemic sclerosis (PSS) and the changes in blood flow during cold exposure in patients with primary and secondary Raynaud's phenomenon.

12.2 Capillaroscopy

The "-scopy" suffix indicates a diagnostic method based on human observation while using optical instruments. This technique evolved coupling the optical system with the electronic system to obtain digital images, visualize them on the screen, and save them on a magnetic support. The goal was to obtain a digital image as similar as possible to the vision of the human eye. This was extremely important to allow performing a timely and reliable follow-up.

E. Berardesca et al. (eds.), *Non Invasive Diagnostic Techniques in Clinical Dermatology*, DOI 10.1007/978-3-642-32109-2_12, © Springer Berlin Heidelberg 2014

12.3 Applications

- Angiology
- Phlebology
- Vascular surgery
- Plastic surgery
- Rheumatology
- Odontostomatology

The following are some of the possible applications:

- Microangiopathies of primary dermatology expression (psoriasis, lichen, oncological dermatology, mouth burning syndrome, oral lichen, mal perforant ulcer, alopecia areata, pyoderma gangrenosum, livedo reticularis , angiodermatitis of the lower limbs)
- Systemic microangiopathies (Raynaud's syndrome and diseases, scleroderma, Sjögren's syndrome, systemic lupus erythematosus, rheumatoid arthritis, vasculitis, polyarteritis nodosa)

12.4 Observation Parameters

12.4.1 Morphological or Static Observation Parameters

- Locoregional microangiotectonics
- Visibility
- Loop morphology
- Loop orientation
- Capillary density
- Loop length
- Loop diameter
- Loop-dermal papilla distance

12.4.2 Functional or Dynamic Observation Parameters

- Presence of erythrocyte aggregation
- Flow velocity reduction
- Stasis
- Erythrocytes "come and go" movement
- Rapid blood flow
- Intermittent blood flow
- Constant blood flow

12.5 Performing the Test

The test must be performed in a controlled temperature setting. The patient must refrain from smoking for at least 1 h before the test and remain inside the test room for at least 15 min in order to balance his/her body temperature.

Multiple skin areas must be examined starting from nail folds of all fingers, fingertips, thenar and hypothenar eminence, and lower lip mucosa.

In vasculopathies, start reviewing the paranecrotic area then move to the foot dorsum, anterior, lateral, and posterior tibial region.

In stasis microangiopathies, start the test from the foot dorsum and then move up the lower limb until reaching the posterolateral regions of the thigh.

Start with small enlargement magnification (30× to 50×) and continue the detailed test using large magnification (150× to 200×).

To assess the morphology aspects of a district microcirculation, use the 150× to 200× lens to explore a field of view (FOV) of approx 1.5/1.8 mm. The probe must be moved in slow uniform motion to correctly visualize all the morphostructural characteristics of the capillaroscopy field.

Standard white light capillaroscopy allows to detect the blood flow, albeit in a slightly unclear fashion as the erythrocytes have an average diameter of 7–8 μm. On the other hand, the erythrocyte aggregation phenomena (sludge) and the capillaries or venules with the slower flow or temporary standstill are well visible.

The new fluorescence method allows to amplify these phenomena and visualize them more precisely, detecting small capillaries of a diameter of approx 10 μm, inside which erythrocytes flow in a line and appear as small floating points with highly variable speed in the capillaries or their segments.

The 400× lens can be considered for research. An optimal view requires good manual skills and solid experience. This lens is particularly appropriate for in-depth review of pathological alterations in individual microvessels and to measure capillary density. Special effort should be made to always acquire a consistent number of images and videos in order to closely assess the dynamics.

12.6 Follow-Up

Follow-up is one of the main aspects of video diagnostics in an effective prevention. It should be performed in an appropriate and constant fashion using dedicated software that allow performing a rapid and rational assessment. This allows to carefully and rapidly review the onset and/or evolution of the pathological processes, if any.

A key factor for appropriate follow-up is test repeatability, i.e., to be able to perform the test in the same area and in similar operational conditions as the previous one. To obtain the best results, it is important to use systems that allow to identify the exact area to test under standard light conditions. These factors exclusively rely on the operator and on the device used for the test.

VALLO UNGUEALE III DITO – MANO SX

Fig. 12.1 Descriptive localization

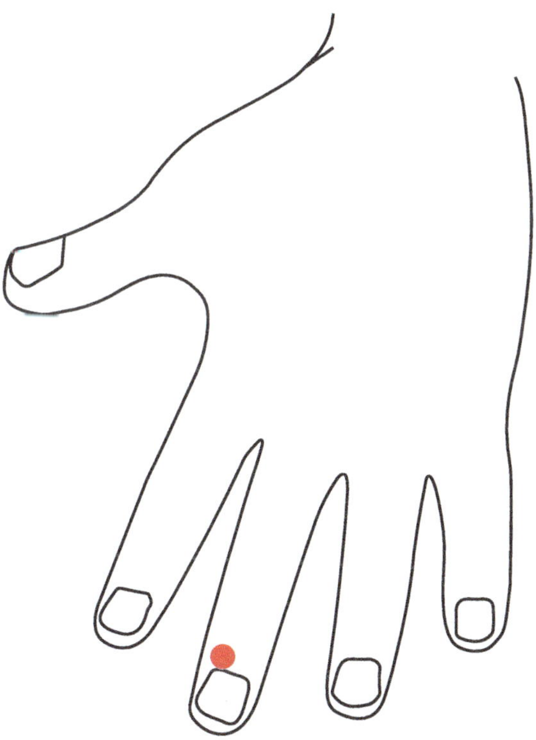

Fig. 12.2 Schematic localization

There are different methods to achieve a precise localization of reference for subsequently acquired microscopic images. *Descriptive localization* (Fig. 12.1) is the most used method but at the same time also the less indicated to precisely identify the area under examination.

Other localization systems, strongly recommended for a reliable and appropriate follow-up, are *schematic localization* (Fig. 12.2) and *clinical localization* (Fig. 12.3).

Schematic localization is performed with dedicated software that can visualize preloaded stylized patterns – body maps of various anatomical areas of the body and timely retrieve them, identifying the area to test and acquiring new microscopic images. It is recommended to make sure that the software allow the insertion of alternative patterns for purpose of customizing the application.

Clinical localization is undoubtedly the best system for two main reasons:

1. It is easier to identify potential superficial landmarks on the patient's clinical image.
2. The patient's clinical image shows a global view of the area under examination and highlights other potentially important aspects for follow-up and diagnosis.

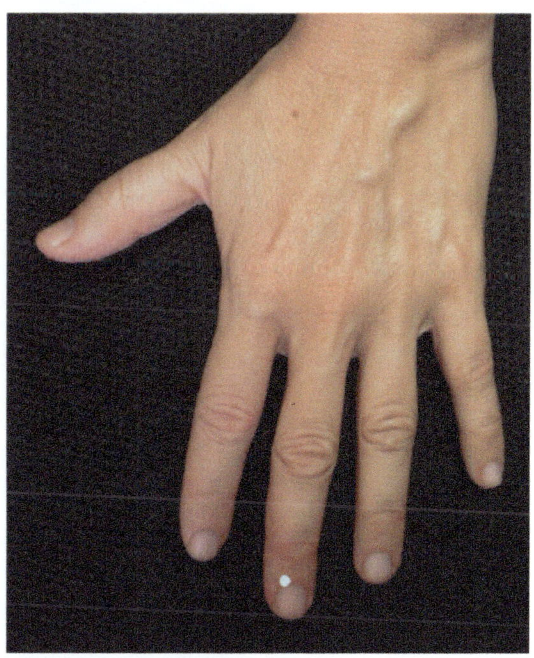

Fig. 12.3 Clinical localization

This procedure is performed using digital cameras or integrated systems that in addition to performing the microscopic test also allow to acquire clinical images. This option is highly recommended because it enables to save the image much more rapidly. Clinical image acquisition for purpose of localization is only performed when starting the patient's follow-up.

After the localization is performed using one of the above mentioned procedures, the capillaroscopy is carried out using the selected magnification, usually between 150× and 200×. The acquired microscope image (Fig. 12.4) is saved and, using the localization procedure, automatically associated to the previously created reference.

The use of specifically designed software for these applications is of extreme importance as it allows saving time during the test and concentrating more on the clinical assessment.

Subsequent consults of the same patient simply involve acquiring microscope images and performing an immediate comparison with all previously saved images for every individual localization. This way any morphologic alteration occurred between the two consults is immediately highlighted (Fig. 12.5).

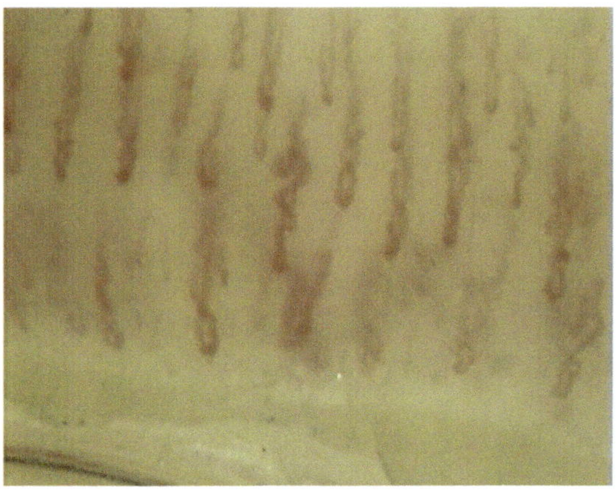

Fig. 12.4 150× microscope image

12.7 Methods to Perform a Videocapillaroscopy

The more widespread and simple methods to perform a capillaroscopy fall under two main categories:

1. Capillaroscopy with optical stereomicroscope
2. Optical capillaroscopy with digital image visualization, aka videocapillaroscopy

The most common devices are:

- Optical stereomicroscope
- Halogen illumination polychromatic videocapillaroscope

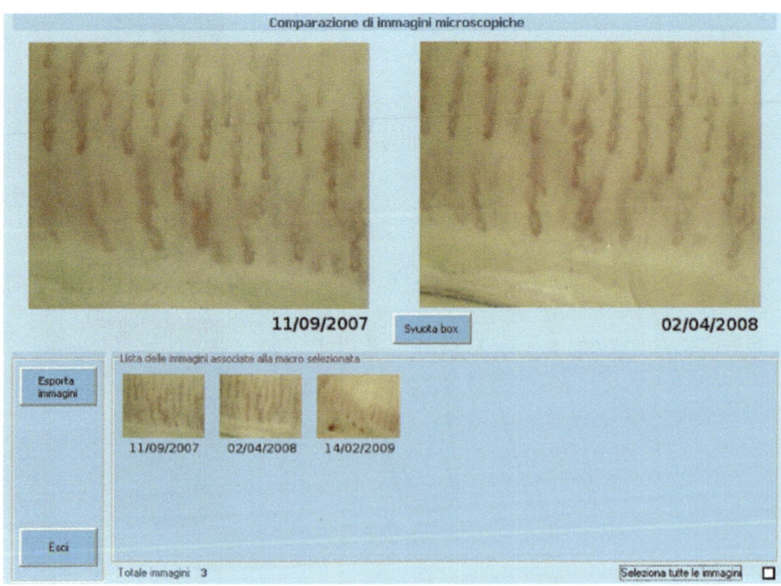

Fig. 12.5 Image comparison

- Diode illumination polychromatic videocapillaroscope
- Monochrome illumination videocapillaroscope

12.7.1 Basic Principles

The basic principle shared by all methods is the illumination of the test area with polarized polychromatic, nonpolarized, and high-intensity monochromatic light, and measuring the reflected light that, by means of a system of lens, is appropriately focused and on which the optical magnification of the test area is based (Fig.12.6). The source of light can be:

- Halogen – a high-intensity filament bulb generates polychromatic light
- Diode – generated by a semiconductor (light-emitting diode, LED)

Both lights can be appropriately filtered to generate polarized or monochromatic light.

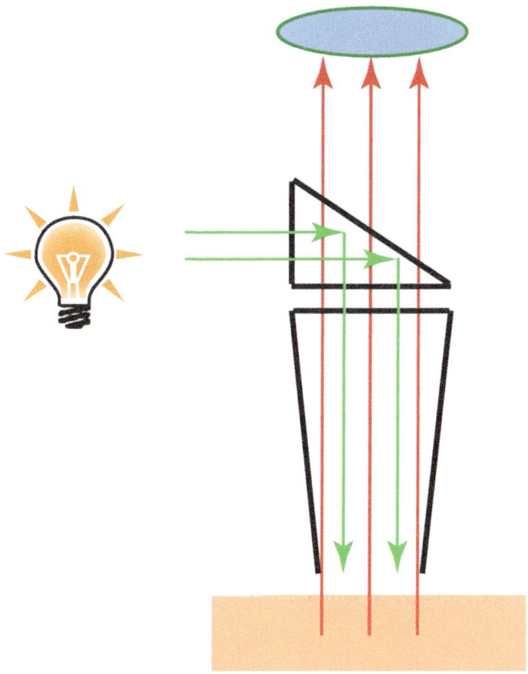

Fig. 12.6 Emission/reflection

12.7.2 Capillaroscopy with Optical Stereomicroscope

The capillaroscopy with an optical stereomicroscope (Figs. 12.7 and 12.8) allows a good visualization of capillaries as the stereoscopic effect generates a 3D view of the image and a good depth of field of view. It is generally utilized on nail folds as its application is very complex in other areas of the body.

The system works with halogen or diode polychromatic and monochromatic light.

The stereomicroscope can be equipped with an optical divider to connect a camera that, with an appropriate software and hardware, is interfaced with a computer for image digitalization and storage. The image is viewed through the eyepiece, with a three-dimensional perception, and at the same of the monitor.

If the lighting system is equipped with a polarizer, the test can be performed straightforward. Otherwise, corneal layer transparency must be improved by placing one or two drops of cedar, paraffin, or glycerol oil on the test site. The oil on

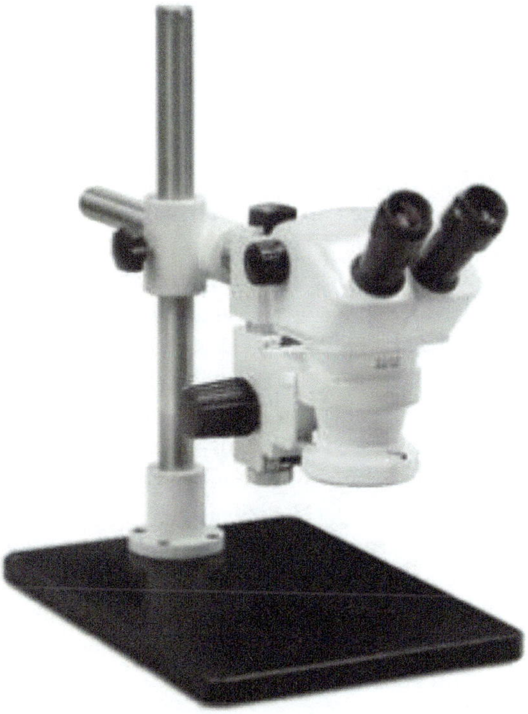

Fig. 12.7 Stereomicroscope

Fig. 12.8 Stereomicroscope

Table 12.1 Refractive index

Refractive index	
Air	1
Water	1.33
Cedar or synthetic oil	1.51
Glass	1.53

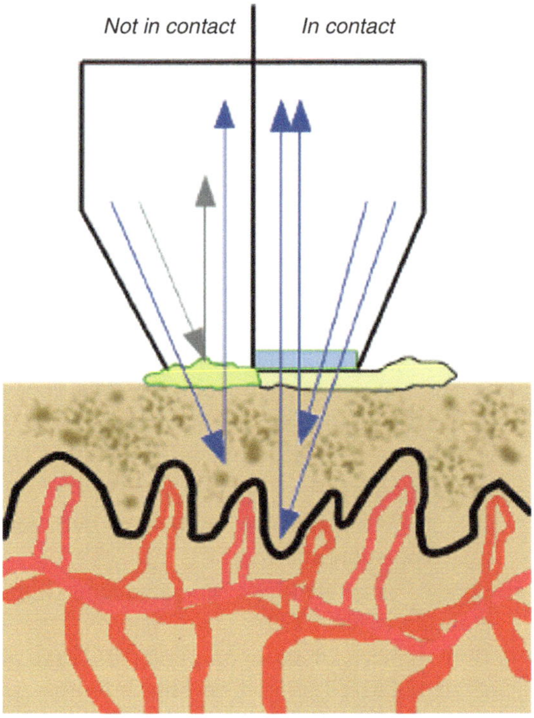

Fig. 12.9 "Not in contact" and "in contact"

the skin should be distributed uniformly and regularly in order to achieve a homogeneous thin layer. The purpose of creating the oil layer on skin is acquiring a skin refractive index (Table 12.1) as close as possible to the optic one, decreasing superficial reflection (scattering) that creates a dazzling effect and generates images with no detail.

12.7.3 White Light Videocapillaroscopy

In videocapillaroscopy, the operator's eye is replaced by an electronic sensor known as CCD (*charge-coupled device*). This device transforms the intensity of the magnetic radiation the light reflected by subcutaneous structures, in electric signals processed and transmitted to the computer that visualizes them as images. The optical part is very similar to the one used by the stereomicroscope, and all procedures performed with the stereomicroscope can be applied. The basic difference is the scanning system of the probe:

- In immersion, *not in contact* and *in contact*
- With polarized light

The *not in contact* videocapillaroscopy in immersion (Fig. 12.9), although using a spacer that maintains a fixed distance, makes sure the superficial layer of the area under examination is completely free. The only advantage of this technique is it places no pressure on the skin surface thus avoiding obstructions of the capillary lumen caused by mechanical compression.

The *in contact* videocapillaroscopy in immersion (Fig. 12.9) is performed using a glass slide secured on the end of the spacer. This system allows an improved quality visualization because it uniforms the oil surface increasingly reducing light reflections that can often be extremely disturbing in the *not in contact* systems. This technique requires higher manual skills and experience from operators, but results are definitely better.

12.7.4 Polarized Light Videocapillaroscopy

The basic principle is the same as the white light system, simply adding two filters (Fig. 12.10) for two different purposes:

- Polarizing filter – placed between the polychromatic source and the object to observe – incident light
- Analyzing filter – placed between the object and the eyepiece – reflected light

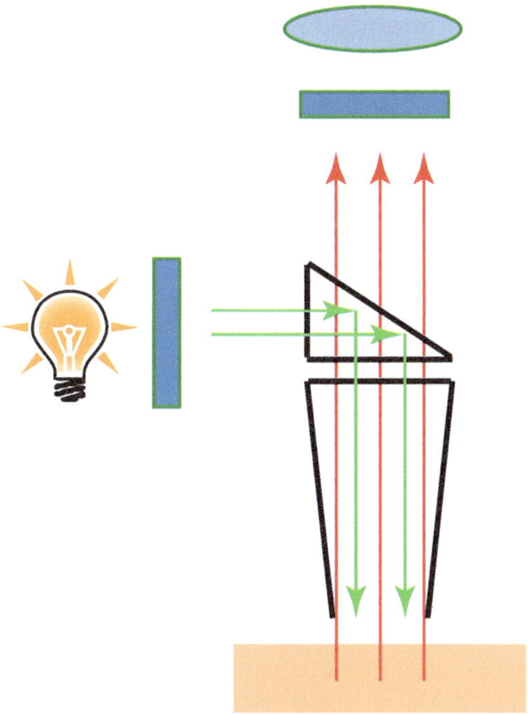

Fig. 12.10 Polarization

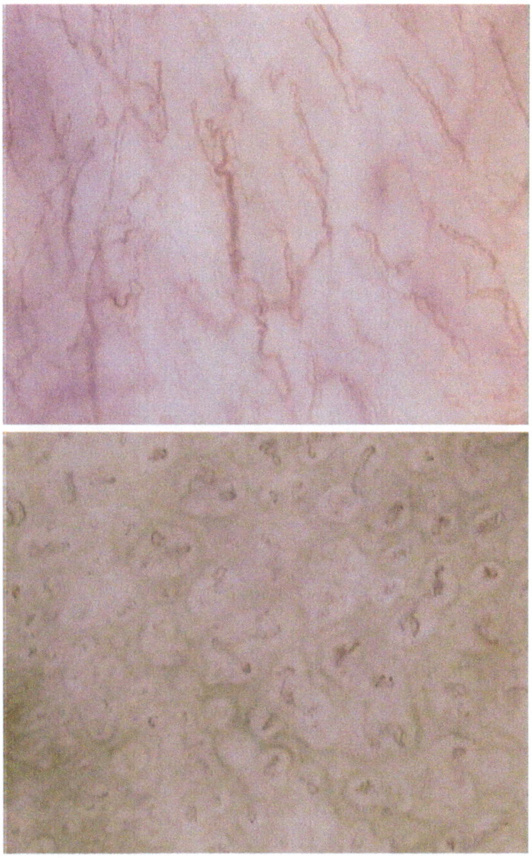

Fig. 12.11 Standard videocapillaroscopy

The two filters are identical, and their principle consists in filtering and making visible only light rays with a specific phase angle. The angle, thus the degree of linear polarization, is the result of the rotation of one of the two filters. This eliminates all interferences caused by reflection phenomena of the corneal layer.

This method makes it considerably easier to perform the test because it does not require the use of oil. It does however present two strong disadvantages:

1. It reduces the luminosity according to Malus' law.
2. The colors tend to blue and green or red and magenta based on well-defined principles of physics. For example, in the presence of a natural pigment (melanin), colors tend to blue/green or red/magenta based on the more or less grainy pigment.

12.7.5 Videocapillaroscopy in Fluorescence

Fluorescence is the characteristic of certain substances – absorbing electromagnetic radiations in the ultraviolet region of the spectrum and emitting them with a longer wavelength (Stokes's law) and therefore in the field of visible light. This technique however requires systems currently very complex and difficult to use in standard diagnostic practice.

The fluorescence technique that is starting to be used in videocapillaroscopy takes advantage of the absorption phenomenon at a specific wavelength. This becomes more effective and simple technology that can be used to replace standard white light videocapillaroscopy.

The test in fluorescence allows to visualize additional morphology and dynamic capillary details compared to classic (standard) videocapillaroscopy (Fig. 12.11), empathizing the whole vascular component of the microcirculation at the level of the dermal-epidermal junction and of the deep surface of the dermis (Fig. 12.12).

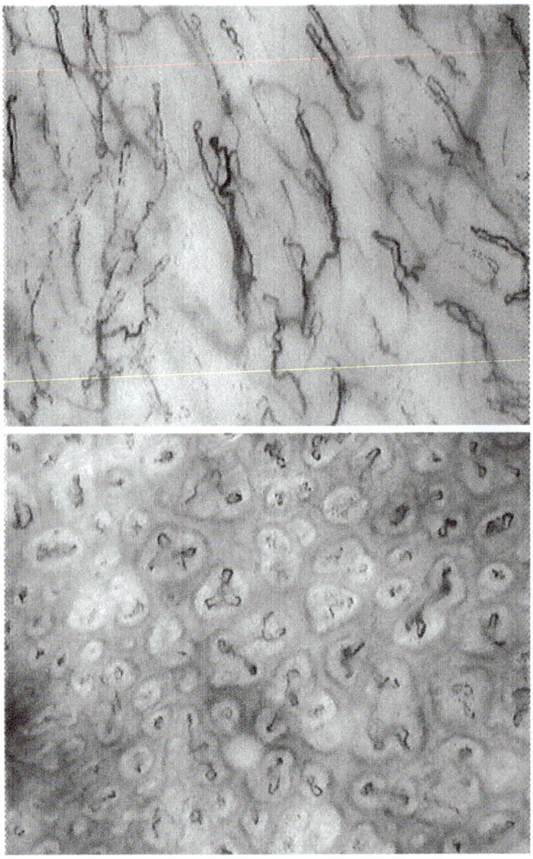

Fig. 12.12 Videocapillaroscopy in fluorescence

Recommended Reading

Bollinger A, Fagrell B (1990) Clinical capillaroscopy, a guide to its use in clinical research and practice. Hsgrefe and Huber Publishers, Toronto

Campanale G (1987) Alterazioni capillaroscopiche nelle malattie diffuse del connettivo. Recenti Progressi Medicina 78(9)

Curri SB (1992) Una nuova tecnica capillaroscopica per lo studio della microcircolazione cutanea in sedi corporee di interesse dermo-cosmetologico: La Videocapillaroscopia a Sonda Ottica. XIV Congr Naz Soc It Med Est 16:7–11

Cutolo M, Sulli A, Briata M, Seriolo B, Accardo S (1996) Videocapillaroscopia periungueale e Sclerosi Sistemica Progressiva: proposta di classificazione tra differenti quadri microangiopatici. Reumatismo 212: 48–53

Grassi W, Core P, Carlino G, Cervini M (1991) La capillaroscopia della mucosa orale nella sclerosi sistemica. Il Reumatologo 8:28–32

Grassi W, Core P, Cervini M (1997) Atlante interattivo di capillaroscopia clinica in reumatologia CDRom, Centro scientifico editore, anno di edizione

Grassi W, Del Medico P (2004) Atlante di capillaroscopia, Edra Medical Publishing, anno di edizione

Grassi W, Core P, Carlino G, Cervini M (1993) Labial capillary microscopy in systemic sclerosis. Ann Rheum Dis 52:564–569

Guzzo G, Senesi M, Giordano M, Mantagnassi M (1996) La capillaroscopia in medicina, Collana medico scientifica, Edra Medical Publishing. tecnica, indicazioni, utilità', limiti, prospettive

Stefania Seidenari (1998) Diagnostica non invasiva in dermatologia. Edra Medical Publishing, Milano

Sonography of the Skin

13

Stephan El Gammal, Claudia El Gammal,
Peter Altmeyer, and Michael Vogt

13.1 Introduction

It is a peculiarity of the dermatological specialty that the whole spectrum of diseases, from slight irritations to malignant transformations, lies directly before the doctor's eyes. Clinical examination by inspection and palpation thus plays a crucial role in dermatological diagnosis.

When examining tumorous or inflammatory diseases, in-depth expansion of the process is a further important clinical parameter. Sonography can deliver this missing information.

Sonography has already conquered many specialties in medicine. Ultrasound imaging systems currently used to study inner organs are – due to their poor resolution – of minor interest for dermatologists who want to study the skin. Until 1975 ultrasound transducers with a center frequency and bandwidth of maximal 7.5 MHz were available. Attempts to evaluate inflammatory and tumorous processes in the skin using 1.5–5 MHz transducers delivered very unsatisfactory results [49]. The pioneers Alexander and Miller (1979) were the first scientists measuring skin thickness using 15 MHz pulsed ultrasound [3].

In the 1980s, first dedicated 15–20 MHz ultrasound imaging systems were developed.

Figure 13.1 visualizes the number of Medline publications focusing on different noninvasive skin imaging methods in the past 30 years. Ultrasound is the oldest method to look at skin layers *beneath* the skin surface and was widely introduced in clinical dermatology in the 1990s. In the past decade, scientific research focused on epiluminescence microscopy because the method delivers additional information to differentiate between malignant and benign pigmented skin tumors. New noninvasive "subsurface" skin imaging techniques have been developed recently, improving evaluation of skin diseases prior to biopsy. Many of these interesting methods are presented elsewhere in this book.

During the past 25 years, 25 MHz sonography of the skin has gained increasing importance as a noninvasive imaging method in dermatology. Clinical applications are the preoperative determination of the extension of skin tumors [14, 15, 16, 20, 24, 26, 27, 30, 33, 54], the monitoring of inflammatory lesions [13, 28, 31, 50, 58, 59] and sclerotic processes [2, 11, 32, 36, 41, 44, 47, 55], and the objective judgement of skin tests, such as patch test reactions [52, 53,

S. El Gammal (✉)
Department of Dermatology, Dermatological Clinic, Diakonie Klinikum Bethesda, Euelsbruchstr. 39, Freudenberg 57258, Germany
e-mail: stephan@elgammal.de

C. El Gammal
Department of Dermatology, Medical Care Center, Diakonie Klinikum Jung-Stilling, Siegen 57072, Germany

P. Altmeyer
Dermatological Clinic of the Ruhr-University, St. Josef Hospital, Bochum 44791, Germany

M. Vogt
Institute for High Frequency Techniques, Ruhr-University, Bochum 44780, Germany

E. Berardesca et al. (eds.), *Non Invasive Diagnostic Techniques in Clinical Dermatology*, DOI 10.1007/978-3-642-32109-2_13,

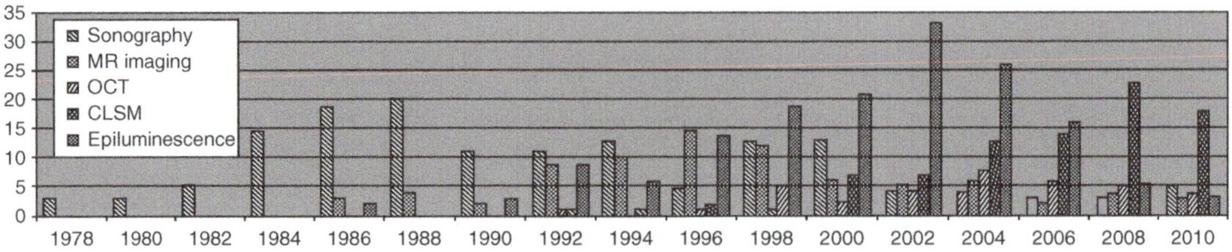

Fig. 13.1 Medline publications on noninvasive skin imaging methods. *MR* imaging magnetic resonance imaging, *OCT* optical coherence tomography, *CLSM* confocal laser scanning microscopy

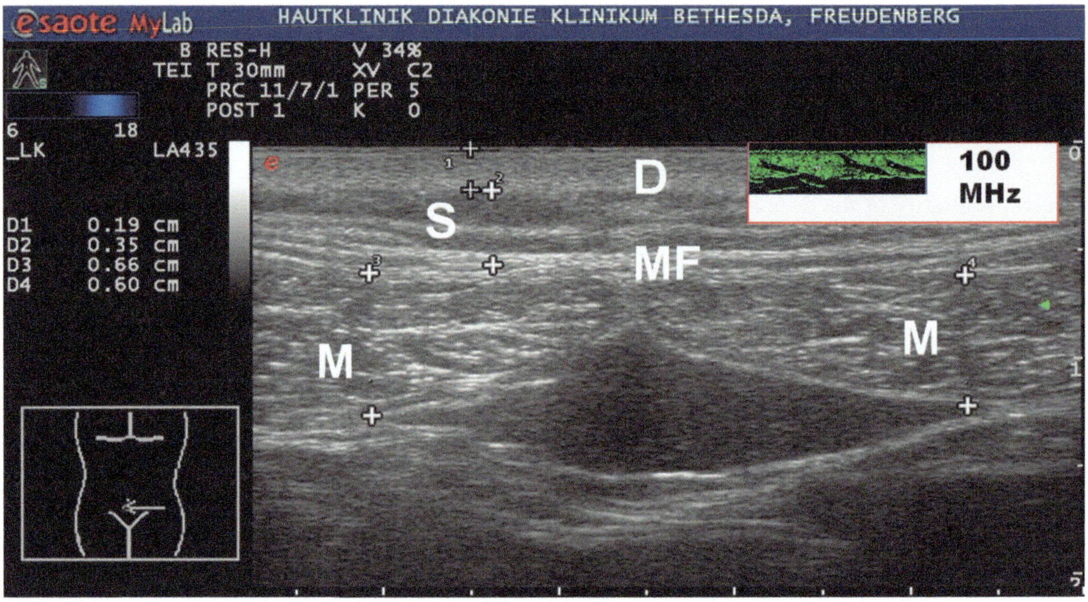

Fig. 13.2 Comparison of 15 and 100 MHz sonography (*inset*). At 15 MHz, the dermis is an ill-defined line at the upper part of sonogram due to insufficient resolution; at 100 MHz, structures within the dermis are visualized (e.g., obliquely oriented hair follicles). *D* dermis, *S* subcutis, *MF* muscle fascia, *M* musculus rectus abdominis. The distance between two neighboring lines (*right border* of 15 MHz picture) is 5 mm

56] and tuberculin test reaction [5], to name but a few. Today, 25 MHz ultrasound equipment has become widely available at reasonable cost.

Sonograms of normal skin show at their upper border a thin, very echorich line, the so-called skin entry echo. Beneath, a broad, echorich band with scattered reflexes is seen, which corresponds to the dermis (Fig. 13.2). The subcutaneous fatty tissue is echolucent and traversed by obliquely oriented echorich connective tissue septa. The epidermis cannot be visualized, and certainly structures within the epidermis cannot be differentiated [14, 16, 24], due to the lack of resolution using commercially available transducers with center frequencies below 20 MHz.

To investigate the epidermis, the resolution must be improved. The axial resolution is mainly influenced by the bandwidth (Fig. 13.3). The lateral resolution is proportional to the center frequency and indirectly proportional to the focal length [14, 16–19]. By raising the center frequency and bandwidth of the ultrasound transducer, resolution increases.

But signal penetration depth into the skin is reduced (Fig. 13.4), a burden we have to accept for more detailed sonograms.

We modified the 100 MHz transducer technology in such a way that skin structures up to 2 mm depth can be visualized [19, 46] at very high resolution (Fig. 13.2 inset).

Considering that dermatology includes *all structures* from the skin surface down to the muscle fascia, lower frequencies are necessary to study deep skin structures (Fig. 13.5).

Today 10–18 MHz sonography is used to evaluate the peripheral lymph nodes of patients with skin tumors [6–8, 10, 12] or pathological states of the testes [29, 40, 51].

Since 1995, sonography of the skin and subcutis (including the peripheral lymph nodes) has been integrated into the educational curriculum for dermatologists in Germany, a testimony that this method has become a routine diagnostic tool in dermatology.

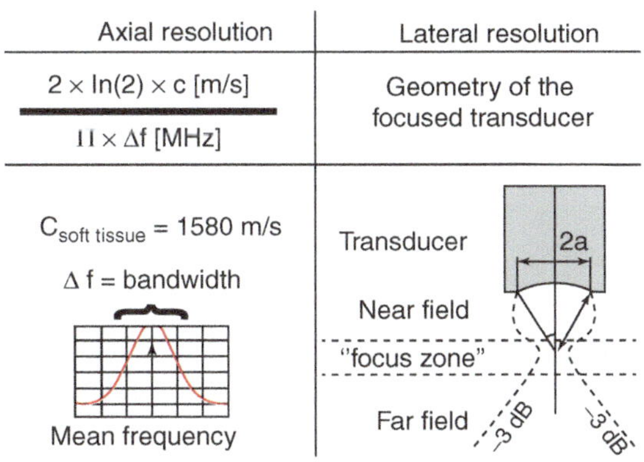

Fig. 13.3 Physical parameters influencing the axial and lateral resolution. The axial resolution is mainly influenced by the bandwidth of the transducer, and the lateral resolution by the geometry of the transducer in the focus zone. The near and far fields of the transducer bear multiple artifacts

13.2 Methods and Patients

To study the subcutis, we used an Esaote MyLab60 ultrasound unit equipped with a 3–9 and a 9–18 MHz linear array applicator. Ultrasound gel was used as coupling medium.

Examination of the dermis, epidermis, and stratum corneum was done using high-resolution sonography. An experimental ultrasound imaging unit was developed which can be operated with different transducers in a frequency range of 20–250 MHz. Technical details have been published elsewhere [17, 19, 46]. To study the epidermis, we used a 100 MHz ceramic transducer. The characteristics of this transducer have been published elsewhere [18].

To obtain an excellent lateral resolution (Fig. 13.6b), we used a highly focused transducer (with a short depth in the focus zone of only 400 μm). In order to obtain sharp sonograms not only from a stripe of 400 μm but from a wider part of the skin, we introduce a mechanical focusing procedure, which we call Brightness/Depth-Scan (B/D-Scan; Fig. 13.6c). The principle of this method is to compose the sonogram of several 400 μm wide image stripes, which are recorded one after the other, each in

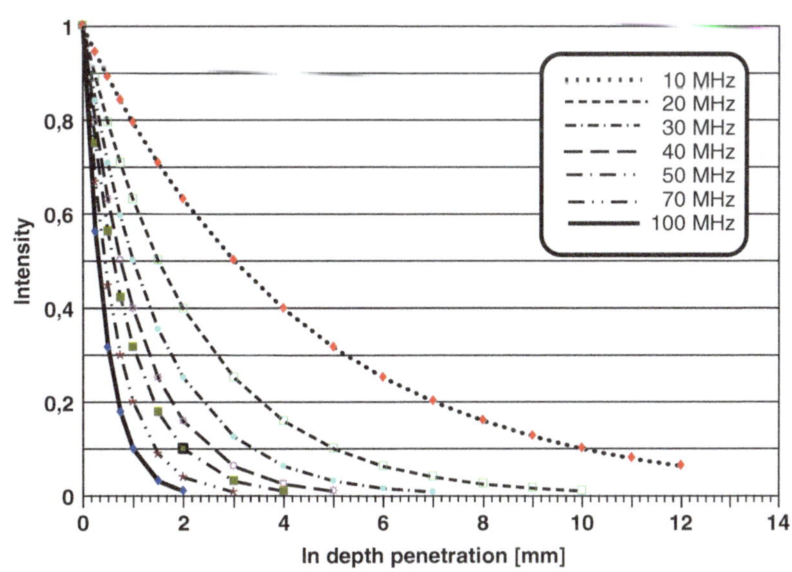

Fig. 13.4 Center frequency and in-depth penetration of the ultrasound beam. The curves have been plotted down to 0.01 (1 %, –40 dB) of the initial energy I_0 (100 %, 1.00). When the center frequency is raised (and thereby the axial resolution is improved), the in-depth penetration of the ultrasound beam decreases significantly

Range	In-depth penetration	Resolution (axial)
5 MHz	> 3 cm	0,4 mm
20 MHz	8 mm	80 µm
50 MHz	4 mm	40 µm
100 MHz	2 mm	11 µm

Range	Examined Structure
5–15 MHz	Subcutaneous lymphnodes large arteries and veins subcutis and fatty tissue, muscles
15–25 MHz	Dermis (overiew and sonometry) subcutis (detailed architecture) arterioles and venoses
25–50 MHz	Epidermis (palms and soles) Skin appendages dermis (detailed architecture)
50–150 MHz	epidermis (detailed architecture) mucosa upper dermis

Fig. 13.5 Sonography in dermatology. The frequency range is a compromise between depth and resolution of the structure of interest

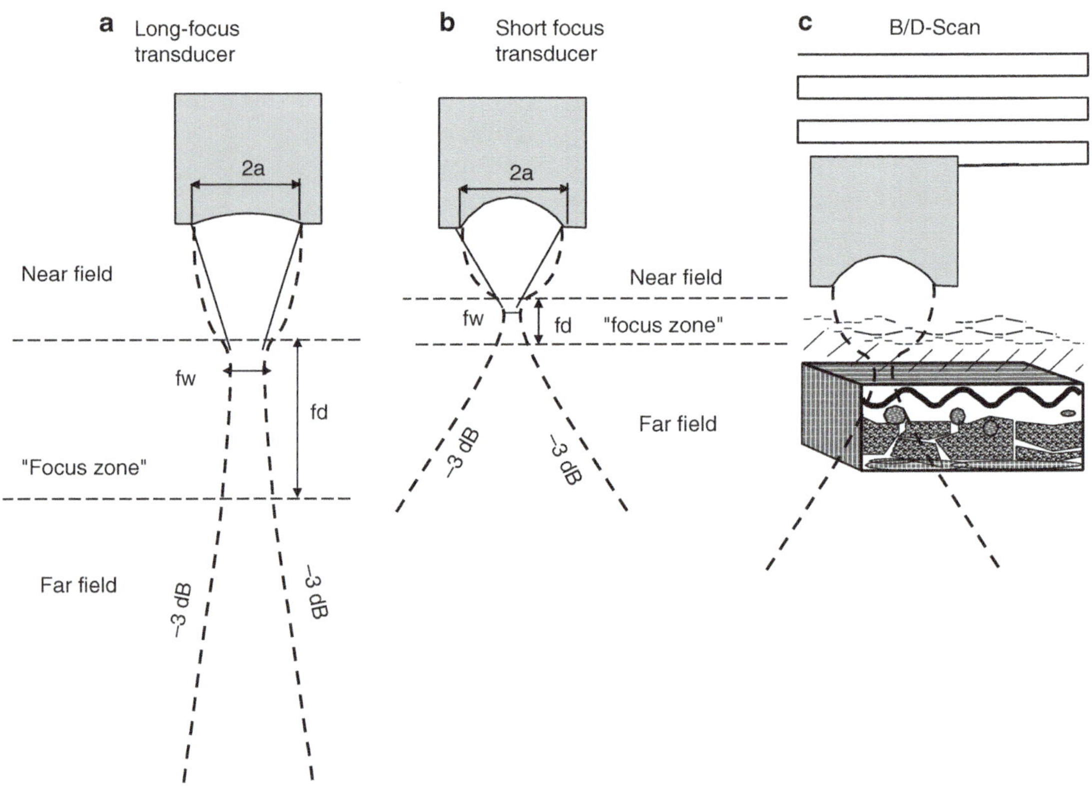

Fig. 13.6 (**a**) Usually long-focused transducer is used in mechanical scanners. The focal zone is defined by the focal width *Fw* and focal depth *Fd*. (**b**) To improve lateral resolution (*Fw*), we used a short-focused transducer. The focal depth *Fd* is reduced to 400 µm. *Fw* = 30 µm. (**c**) B/D-Scan. The final sonogram is composed of several 400 µm wide image stripes, which are recorded one after the other, each in the focus zone of the transducer

the focus zone of the transducer [17]. After the uppermost stripe is recorded by lateral movement of the transducer over the selected area, the transducer is moved vertically 400 μm towards the skin surface before the next image stripe is recorded. To eliminate movement artifacts between adjacent stripes, the overlapping parts (near field and far field) are used for adjustment by the computer program, which puts together the final image.

Image processing involved two steps for every picture. First, the internal echoes of the 100 MHz transducer were eliminated. The oscillation curves of all neighboring A-scans of the image were averaged, and the mean oscillation curve was then subtracted from every single A-scan. Secondly, the A-scans were demodulated. The envelope curve was determined by two complex fast Fourier transformations for every A-scan. This procedure provides optimal results but does not allow to promptly view the recorded data. For this purpose, we additionally implemented a fast method of demodulation consisting of a digital rectification of the high-frequency A-scan combined with a non-recursive digital filter of the order of ten. This linear phase filter has a pass-band cutoff frequency of 150 MHz.

For clinical studies, we also used the DUB50 Profi (taberna pro medicum, Lüneburg, Germany), a commercially available system. To obtain sonograms of good quality, the returning echo signal was digitized using a 1 GHz 8 bit high-speed analog-to-digital converter. Our applicator can be equipped with a 50 MHz or 75 MHz transducer.

13.2.1 Patients and Volunteers

All volunteers and patients gave informed consent for all examinations.

Subcutis: Over a time period of 12 years, sonograms of different pathological conditions were collected in a huge data bank. More than 800 patients with malignant melanoma were followed up on a regular basis using lymph node sonography (inguinal, axilla, neck region). We studied lymph nodes in 500 patients documenting palpation and visualization of lymph nodes (benign and malignant lymphadenopathy) using sonography.

Normal Palmar Skin: Ten right-handed volunteers with healthy skin (five men and five women, age 29–76, mean 56.4) were investigated. The index finger tip of the left hand was occluded for 30 min with an emulsion using a Finn Chamber.

Normal Glabrous Skin: Sonograms were taken from volunteers with healthy skin (age 20–32, mean 24.1) on abdomen about 3 cm lateral the umbilicus ($n=8$), the upper back over the scapula ($n=11$), the dorsal forearm ($n=9$), and the calf ($n=14$). In several persons, we took sonograms on the volar wrist at the transition from palmar to glabrous skin.

Psoriasis Vulgaris and Lichen Planus: 35 untreated, infiltrated, and slightly scaly psoriatic lesions on the extremities of 18 patients with chronic plaque-type psoriasis vulgaris and 10 lichen planus papules of 6 patients were investigated. Sonograms were taken in the center and the margin of the lesions and in the surrounding normal skin.

Skin Tumors: Sonograms were taken from skin tumors (basal cell carcinoma, malignant melanoma, seborrheic keratosis, nevocellular nevi) and normal adjacent or contralateral skin. Then the tumors were excised for histology (see below). Only skin tumors with a poor subtumoral infiltrate in histology were included in our study: a total of 27 superficial basal cell carcinoma, 13 malignant melanoma, 16 nevocellular nevi, and 11 seborrheic keratoses were evaluated using statistical methods.

13.2.2 Image Processing and Statistical Evaluation

For image analysis, we used the program AnalySIS® (Soft Imaging Software GmbH, Münster, Germany). The structure of interest within the sonographic image (e.g., entry echo or echopoor band) was manually delimited by a polygon, using a position cursor. Its mean diameter in y-axis was calculated as the average length of all A-scans in the polygon. The mean gray level of the polygon area was also determined; its value ranges from 0 (black) to 255 (white).

In every sonogram, structures of interest (thickness of the skin entry echo, echodensity of the skin entry echo, echodensity of the echopoor tumor area, echodensity of the subtumoral region) were measured in the skin tumor region and in the adjacent or contralateral normal skin. Values were calculated as an index (% of the normal skin) to correct for differences between different anatomical locations.

Finally, the *U*-test (Mann–Whitney-Wilcoxon) for unpaired observations was used to compare the parameters mentioned above. P-values of <0.05 were considered significant.

13.2.3 Correlation with Histology

In 79 patients (psoriasis plaques, $n=11$; lichen planus papules, $n=6$; basal cell carcinoma, $n=27$; malignant melanoma, $n=13$; nevocellular nevi, $n=16$; seborrheic keratoses, $n=6$), a biopsy was taken after sonography. In order to obtain exactly correlating sonographic and histologic images, a 10 mm long line was drawn on the skin in the plane of the B-scan. After local anesthesia of the area, the skin was cut along this line down to the subcutis. Then a spindle-shaped excision was performed with this cut in the center. The two halves of the tissue spindle were separated and their central cutting planes placed on cardboard. This prevents warping of the tissue during formalin fixation. In the histological sections, the thickness of the epidermis (from the stratum granulosum to the lowest points of the rete pegs), the thickness of the inflammatory infiltrate (from the uppermost parts of the dermal papillae downwards), and the thickness of both the epidermis plus the infiltrated dermis were measured.

By means of the linear regression analysis, we checked whether there is a significant correlation between the thickness measurements in sonographic images and the corresponding histologic sections.

13.3 Results

13.3.1 Subcutis

7.5–15 MHz sonography is very suitable to study structures in the deep fatty tissue. Such structures are large arteries and veins, connective tissue routes, muscle fascia, and lymph nodes (Fig. 13.5). Soft tissue tumors, such as lipoma, are easily identified. They are seen as ovaloid structures with a homogeneous fine echo texture (Fig. 13.7) within the rather echopoor fatty tissue. Solid- or liquid-filled subcutaneous tumors are seen as ovaloid echopoor structures. If they are calcified, small echo reflexes with fine dorsal shadow stripes can be present. Structures with a wall or capsule show typical sonographic artifacts: below the structure, there is an echo enhancement (dorsal echo signal augmentation) which is delimited by a border shadow. The atheroma in Fig. 13.8 exhibits these typical sonographic artifacts.

Connective tissue diseases such as scleroderma, dermatolipo(fascio)sclerosis in venous leg disease, and lipo(lymph)edema or inflammatory diseases such as acute cellulite can be easily quantified, to name just a few.

Acute cellulite is accompanied by a localized swelling and inflammation of the skin. Sonography exhibits multiple "lymph fissures" (Fig. 13.9a) in the upper subcutis. Typically these "lymph fissures" disappear when applying a gentle pressure on the tissue, because the liquid is moved laterally. In chronic lymphedema, multiple lymph fissures are observed at all levels of the subcutaneous fatty tissue. In acute thrombophlebitis, lymph fissures surround the involved vein, and the vein is no longer compressible (Fig. 13.9b).

Lymph node sonography has become a routine method in dermatology used for skin tumor staging and skin tumor follow-up programs. In most body regions, normal lymph nodes are too small to be visualized using 7.5 or 10 MHz sonography.

Activation or malignant proliferation brings about an enlargement of the lymph node. We examined lymph nodes in more than 300 patients (axilla and inguinal region) and found that, in

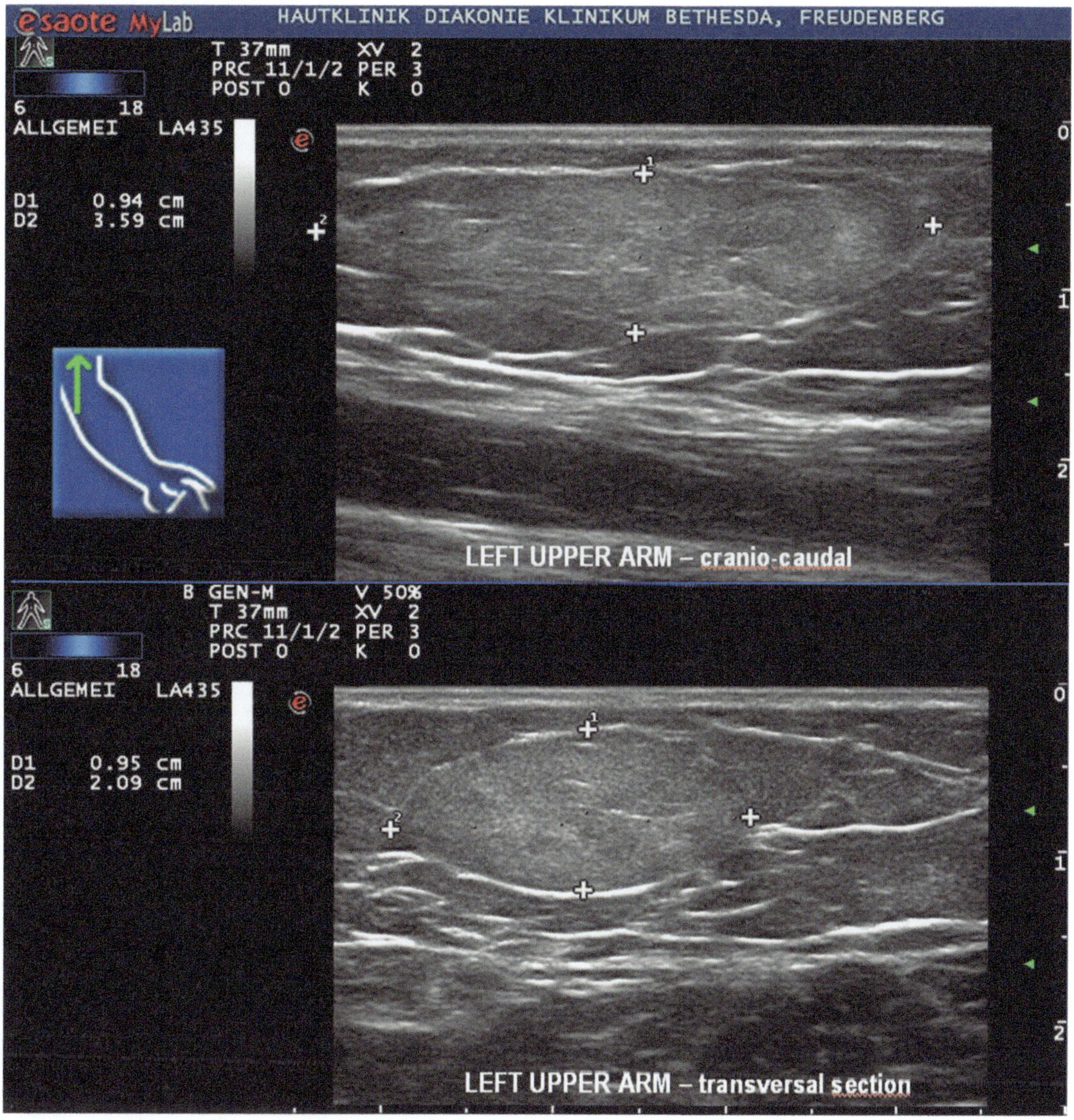

Fig. 13.7 Lipoma on the lower arm of a 67-year-old woman. The two sonograms were registered perpendicularly. *L* lipoma, *S* subcutis, *MF* muscle fascia. Distance between two graduation marks = 5 mm

particular concerning small lymph nodes, sonography is by far superior to palpation (Fig. 13.10). This finding has also been confirmed by other authors. Loose et al. [42] reported concerning detection of regional metastatic lymph nodes in patients with malignant melanoma that in the groin 61 % and in the neck 67 % of the sonographically enlarged lymph nodes were also palpable, while in the axilla only 45 % were palpable.

After the inflammatory process has ended, most lymph nodes become smaller again. Normally this takes 6–8 weeks. Sometimes, however, postinfectious lymph nodes remain enlarged. It has been speculated that new lymph follicles develop in the medulla region, due to a persisting stimulus. These lymph nodes often exhibit an echorich medulla region and are often referred to as "reactive lymph nodes."

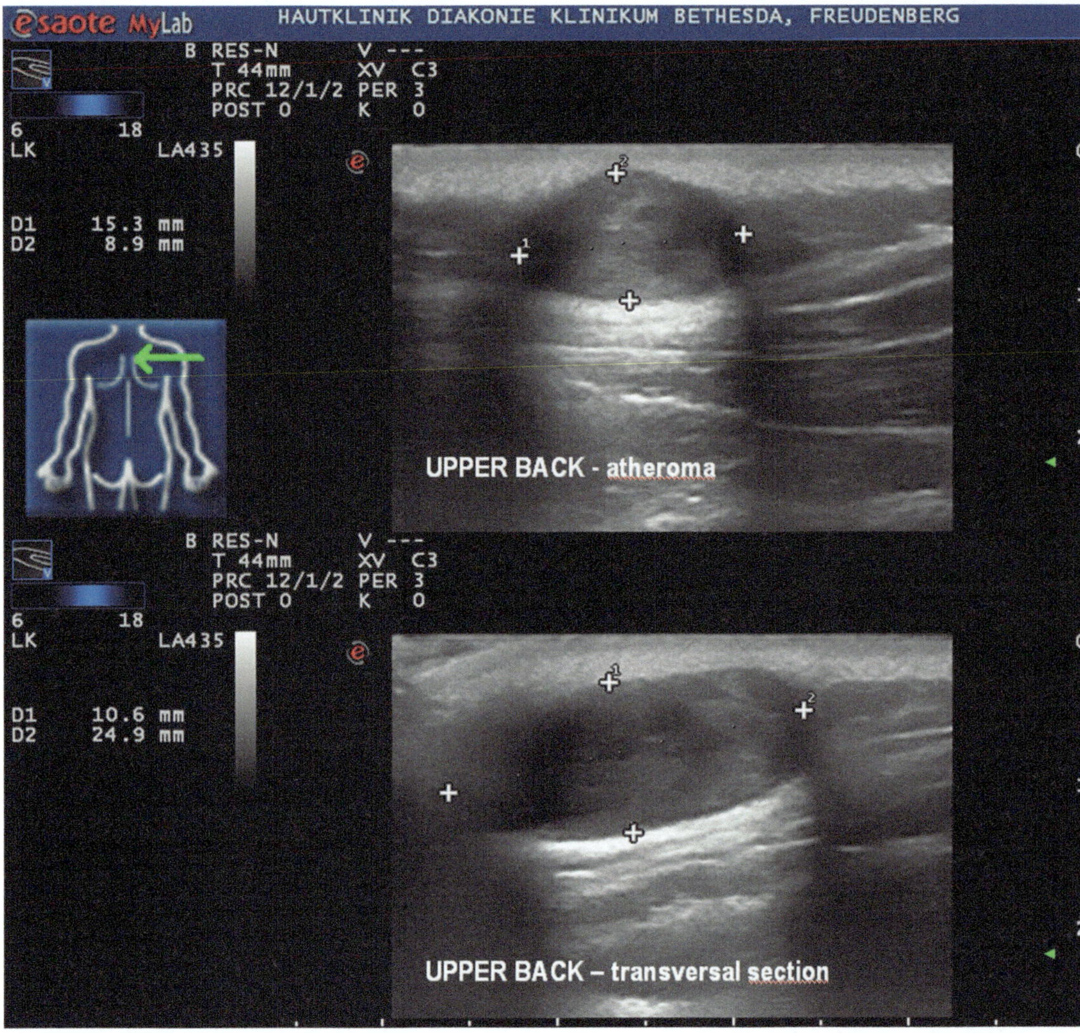

Fig. 13.8 Atheroma on the back of an 85-year-old woman. An echo signal enhancement and a border shadow beneath the structure are typically observed in solid (or liquid) cysts. Distance between two graduation marks = 5 mm

Figure 13.11 shows an enlarged lymph node from the groin region with a tiny medulla and an enlarged cortex region. Changes of the size of the lymph node by keeping the shape constant are typical for benign lymphadenopathies.

A simple way to evaluate changes of the shape of a lymph node is calculation of the Solbiati Index [57]. The suspect lymph node is visualized in two perpendicular axes, the maximal longitudinal diameter (L) and the minimal width (W) of the lymph node are measured, and the L/W quotient is calculated. An L/W quotient smaller than 2 means that the lymph node is more ball-shaped (L/W = 1).

Lymph nodes can also be the battlefield of primary and secondary malignancies. Vassallo et al. [60] studied benign and malignant lymphadenopathies sonographically, excised them, and classified them after histopathological examination. They could show that 85 % of all examined benign lymphadenopathies had an L/W >2. In 86 % of all primary malignant lymphadenopathies (e.g., lymphoma) and 85 % of all secondary malignant lymphadenopathies (lymph node metastases), the L/W was <2. Obviously, tumor masses destruct the inner architecture, such as trabecula of the lymph

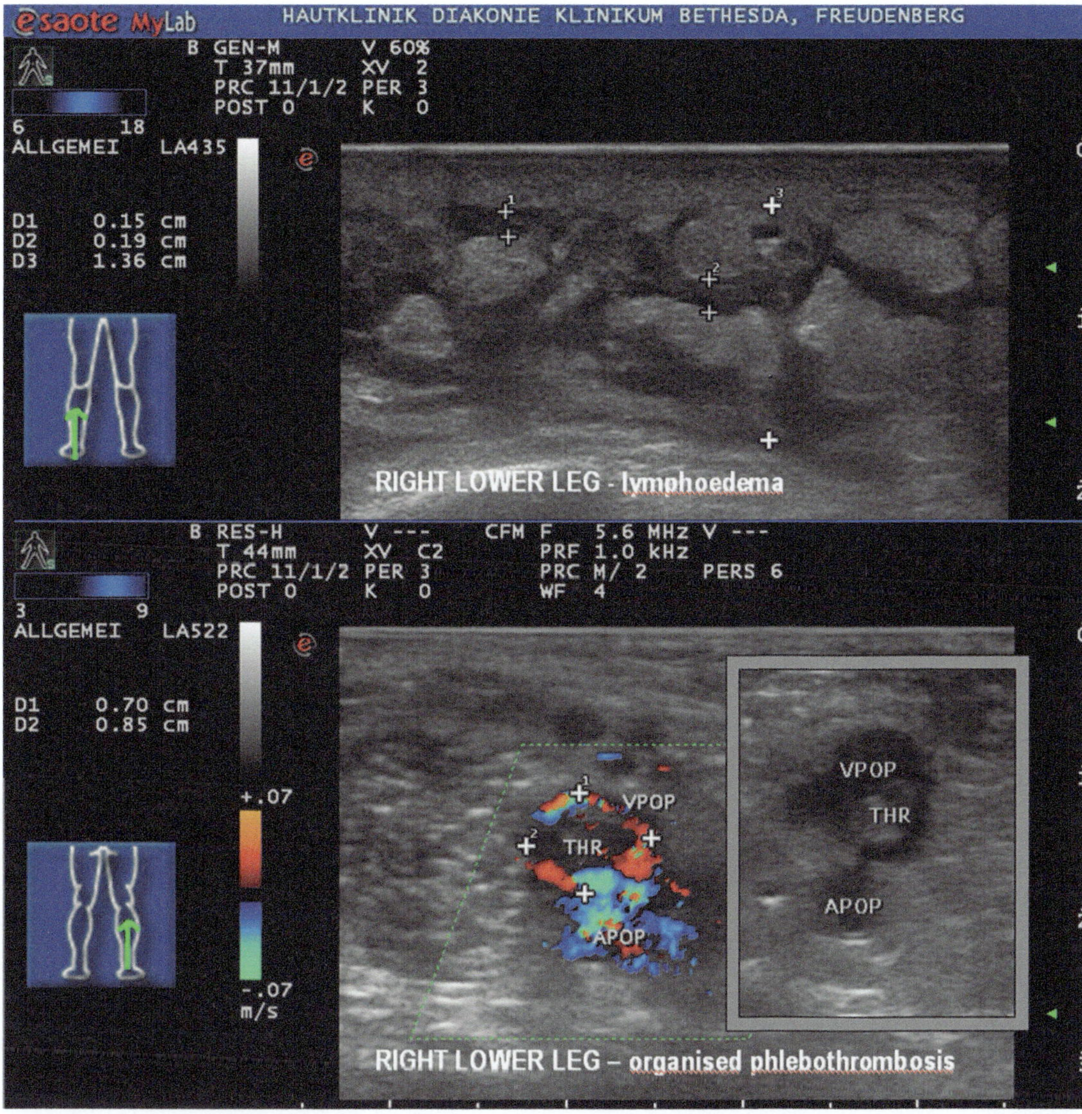

Fig. 13.9 (**a**) Chronic lymphedema. Lymph fissures are seen at all levels of the subcutaneous fatty tissue. They disappear when applying local pressure. *L* lymph fissures. (**b**) Blood vessels, in particular veins, can easily be studied using compression sonography. Vessels with thrombotic material are incompressible. Blood flow is visible around the thrombotic material in duplex mode (*inset*). Distance between two graduation marks = 5 mm

node. Therefore, the shape of the lymph node changes in malignant lymphadenopathies.

In metastatic lymph nodes, the medulla and hilus region are often no longer visible. In primary malignant lymphadenopathy, the medulla region is often tent-shaped and the cortex region eccentrically enlarged. The lymph node has changed its shape considerably.

Figure 13.12 shows that in benign lymphadenopathy, the enlarged lymph nodes keep their shape, while the relation between the medulla and cortex region may vary. In malignant lymphadenopathy, the shape is altered, because the lymph node becomes more ball-shaped and/or because there is an eccentric lymph node enlargement. However, because there is an overlap of

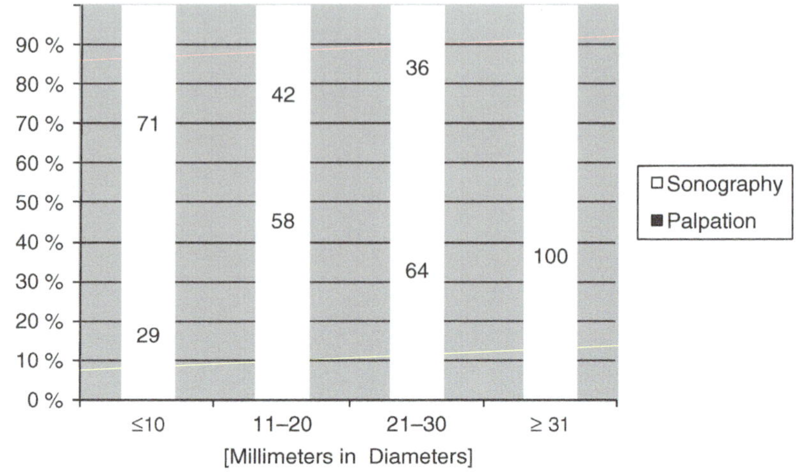

Fig. 13.10 Large lymph nodes are easily palpated and visualized by sonography ($n = 300$). Small lymph nodes may be missed by palpation. Only 25 % of the lymph nodes smaller than 10 mm of diameter which were seen sonographically were also detected by palpation

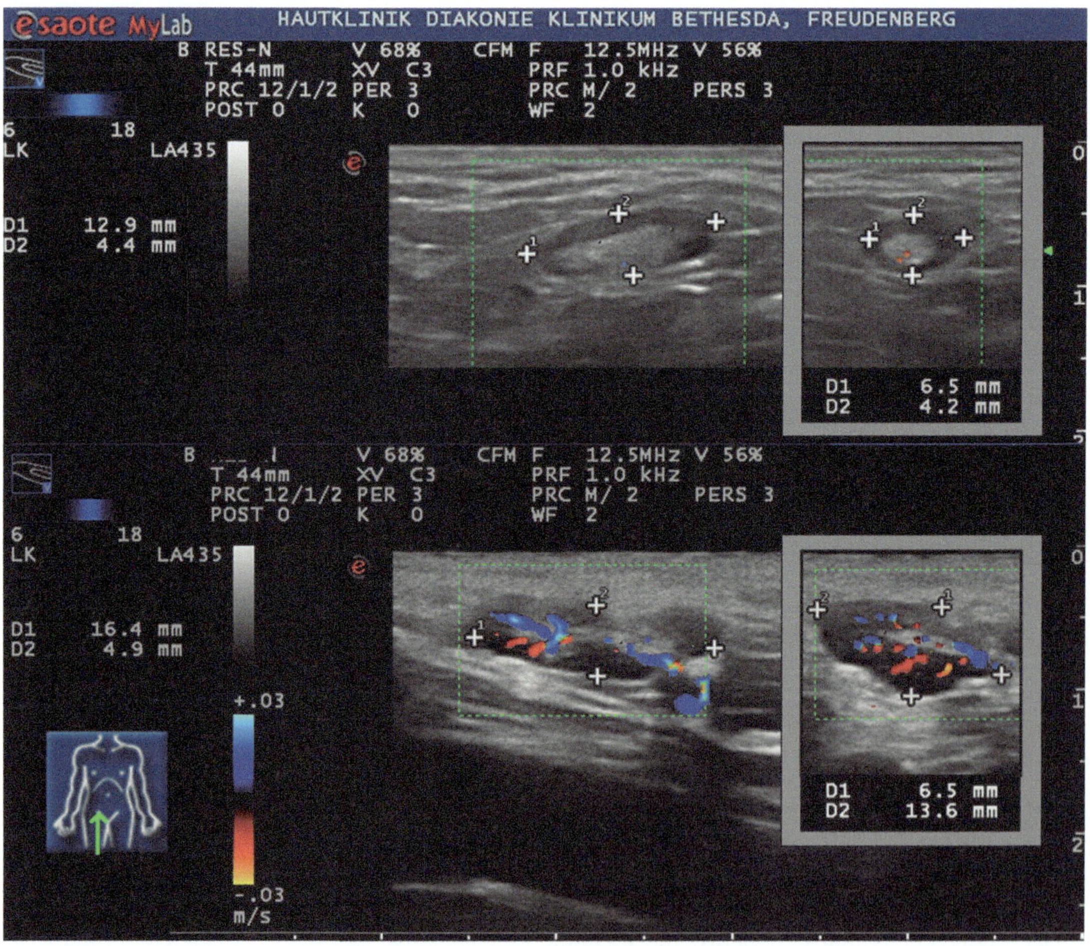

Fig. 13.11 Benign lymphadenopathy. Distance between two graduation marks = 5 mm. *Upper pictures*: Sonograms in 2 perpendicular axes, left groin. Solbiati Index = 2.3. *Lower pictures*: Normal lymph nodes exhibit a central vascularization. Solbiati Index = 3.3

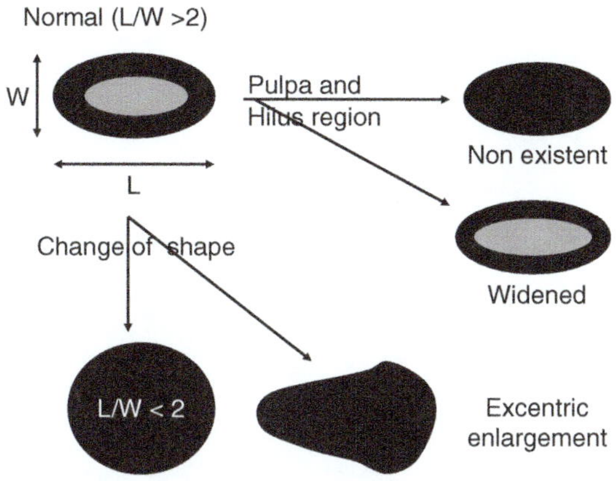

Fig. 13.12 Changes in size and shape of lymph nodes. *L* largest longitudinal diameter, *W* width of the lymph node. The L/W quotient (Solbiati Index) >2 is often found in benign lymphadenopathies and an index <2 in malignant lymphadenopathies (the lymph node becomes ball-shaped)

15 % between both groups of enlarged lymph nodes, it is wise to repeat sonography after 6 weeks. Malignant lymph nodes will continue to enlarge, while benign lymphadenopathies tend to become smaller after inflammation. In case of uncertainty, lymph node biopsy should be taken into consideration.

13.3.2 Sonography of the Dermis and Epidermis

13.3.2.1 Normal Glabrous Skin

Sonograms of normal skin show at their upper border a thin, very echorich line, the so-called skin entry echo. The skin entry echo is due to an impedance jump at the intersection between the water (water is used as coupling medium) and the waterproof stratum corneum. The thin stratum corneum of glabrous skin obviously cannot be differentiated in 100 MHz sonograms. There is a tiny echopoor band (EPB) between the entry echo and the dermal reflexes. Below, a broad, echorich band with scattered reflexes is seen, which corresponds to the dermis (Fig. 13.13). The subcutaneous fatty tissue is echolucent with obliquely oriented echorich connective tissue septa.

The reticular dermis is visible as an echorich zone with densely scattered, confluent echo reflexes. It is sharply demarcated from the very echopoor subcutaneous fat. Within the dermis, hair follicle complexes are visible as homogeneous echopoor structures (Fig. 13.13). To find out which structures of the hair follicle complex can be visualized sonographically, "noninvasive" three-dimensional reconstructions were performed. The principles of three-dimensional reconstruction have extensively been discussed elsewhere [14, 16, 21]. In Figure 13.14a, the sonograms of the left column stand perpendicular to the sonograms of the right column. The upper left sonogram shows a sharply delineated trifoliate echopoor structure at the dermis-subcutis interface. The middle left sonogram shows the hair follicle canal. The lower left sonogram exhibits an ill-defined structure in the dermis corresponding to the lobular structure of the sebaceous gland. The three-dimensional reconstruction exhibits that the hair canal has an angle of 30° to the skin surface. The lobular structure of the sebaceous gland can be visualized. Using three-dimensional reconstructions, anagen follicles could be differentiated from telogen follicles [21].

13.3.2.2 Normal Palmar Skin

In 100 MHz sonograms of palmar skin, an echorich entry echo is seen at the upper border (Figs. 13.15 and 13.16). Where the dermatoglyphics are crossly cut (which is mostly the case), the entry echo is wavy; in parts with longitudinally cut dermatoglyphics, it is a straight line. Below the entry echo, there is an echopoor band, which we will call EPB 1 (echopoor band 1). It is followed by an echorich line, which runs parallel to the entry echo but is less intense (Figs. 13.15 and 13.16). El Gammal et al. [19] could show by removal (tape stripping) and swelling (occlusion with petrolatum) of the horny layer that the EPB 1 truly represents the stratum corneum.

Below the EPB 1, a second echopoor band is seen, which we will call EPB 2 (echopoor band 2). Neither removal nor swelling of the horny layer changes its thickness significantly [19]. The EPB 2 is separated from the EPB 1 by an echorich line. Obviously this line represents

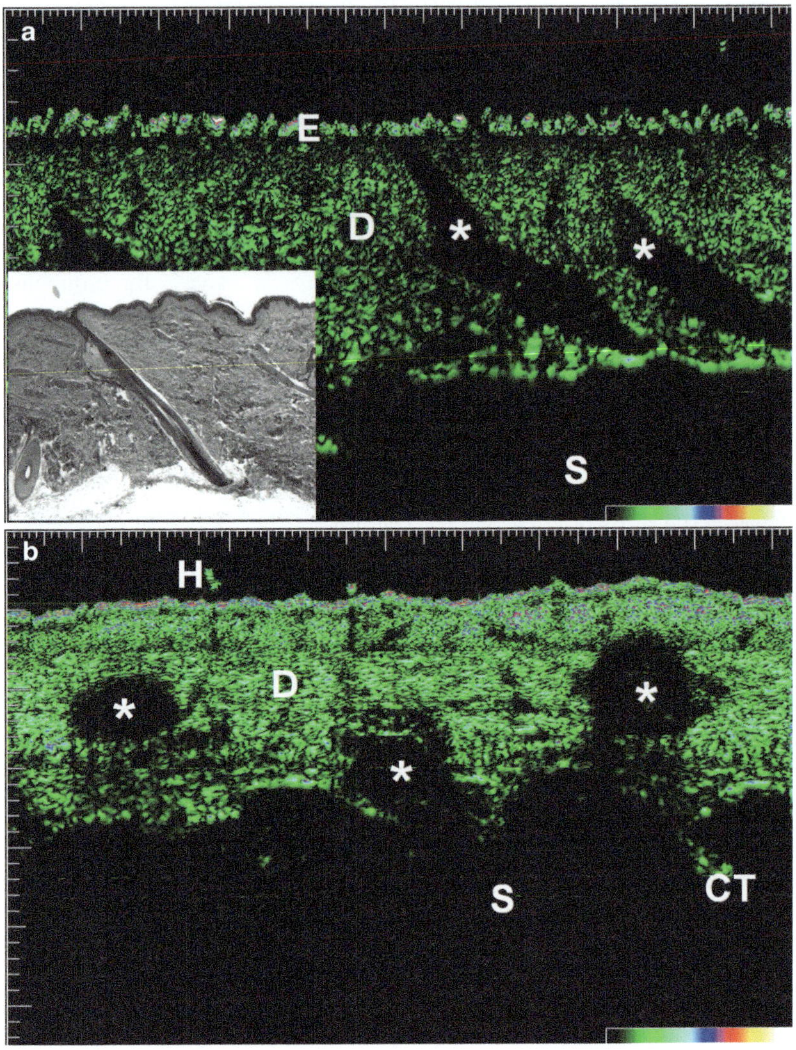

Fig. 13.13 Hair follicles are represented as echopoor structures in the echorich dermis. (**a**) Thigh; * longitudinal section of hair follicles; *Inset*: correlating histology. (**b**) Thigh; * cross sections of hair follicles. *E* Skin entry echo, *D* dermis, *S* subcutaneous fatty tissue, *CT* connective tissue septa in the subcutis, *H* hair (cross-sectioned). Distance between two graduation marks = 100 μm

the interface between the waterproof stratum corneum and the moist, living part of the epidermis. The lower border of the EPB 2 is defined by the scattered reflexes of the dermis (Figs. 13.15 and 13.16). This border is too straight to correspond to the undulating dermo-epidermal junction (Fig. 13.16); it rather represents the interface between the papillary dermis and reticular dermis.

In the EPB1, twisted, echorich, about 100 μm wide structures are seen, which cross the EPB 1 vertically. The distance between two of them is 800–950 μm or a multiple. Each of them ends in a small dip on top of a dermatoglyphic crest (Fig. 13.16). These structures represent eccrine sweat gland ducts. In the EPB 2, they are rarely visible, and in the echorich dermis, they cannot be detected either.

Figure 13.17 shows the transition from palmar to glabrous skin on the wrist: the upper and lower borders of the EPB 1 (skin entry echo and echorich line below) merge into one echorich line in glabrous skin, so that the EPB 1 disappears. The thin stratum corneum of glabrous skin obviously cannot be differentiated in 100 MHz sonograms. The EPB 2 remains as the only echopoor band between the entry echo and the dermal reflexes. While the thickness of the entry echo is about the same as in palmar skin [19], the echopoor band (EPB) of glabrous skin is markedly

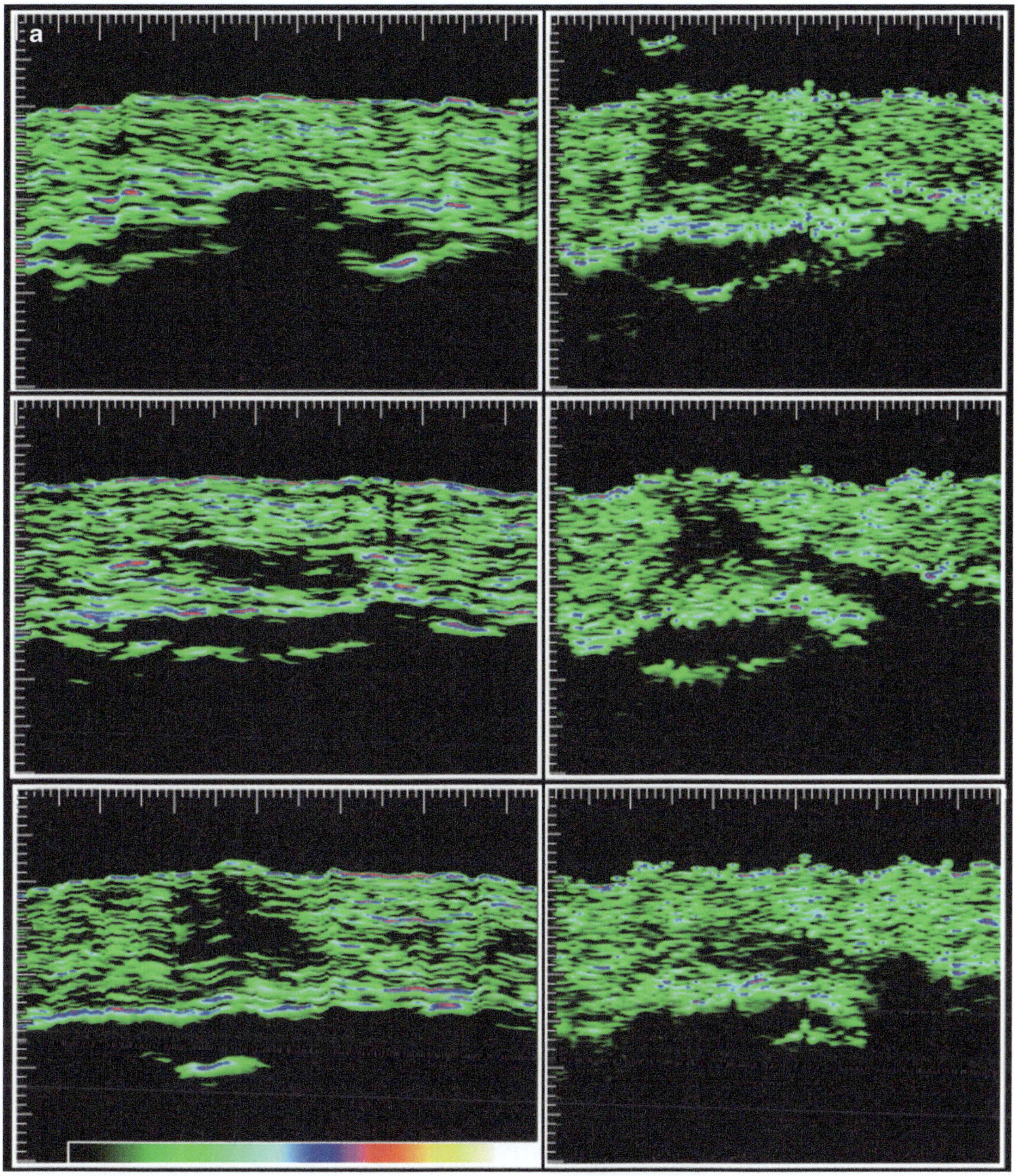

Fig. 13.14 Serial sections (**a**) and three-dimensional reconstruction of the hair follicle complex on the calf of a 23-year-old woman at 50 MHz. Voxel cube of 4 mm (thickness) × 6,5 mm × 6,5. Sections were taken at 50 µm intervals. (**a**) The sonograms of the left column are perpendicular to the sonograms of the right. *Left upper sonogram*: sharply delimited trifoliate echopoor structure at the dermis-subcutis interface. *Left middle sonogram*: hair follicle canal. *Left lower sonogram*: ill-defined lobular structure of the sebaceous gland. *Right sonograms*: longitudinal section of the hair follicle. Note that the hair is slightly bent. (**b**) 3D reconstruction. The hair is oriented 30° to the skin surface. The hair can be seen in the middle of the hair canal (line reconstruction) and is invisible where it is more obliquely oriented (echo reflexes do not return to the transducer). The hair becomes visible again when it is oriented parallel to the skin surface. *Red* = lobular structure of the sebaceous gland. Distance between two graduation marks = 100 µm

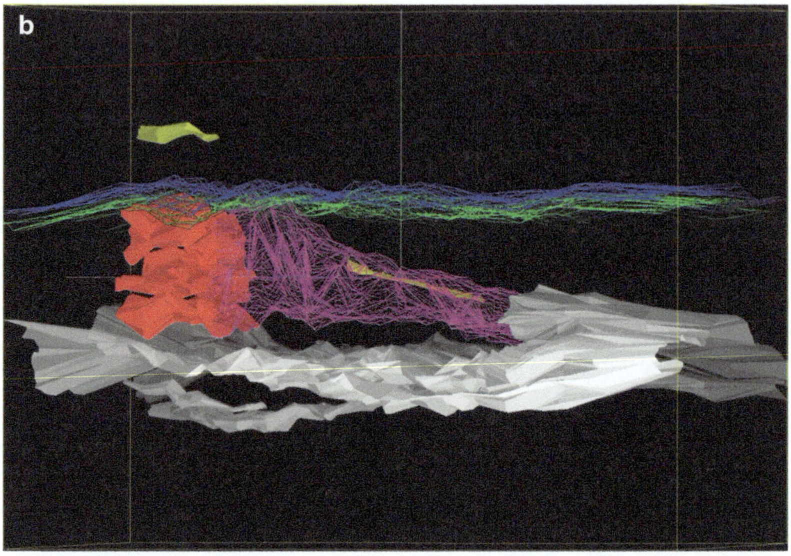

Fig. 13.14 (continued)

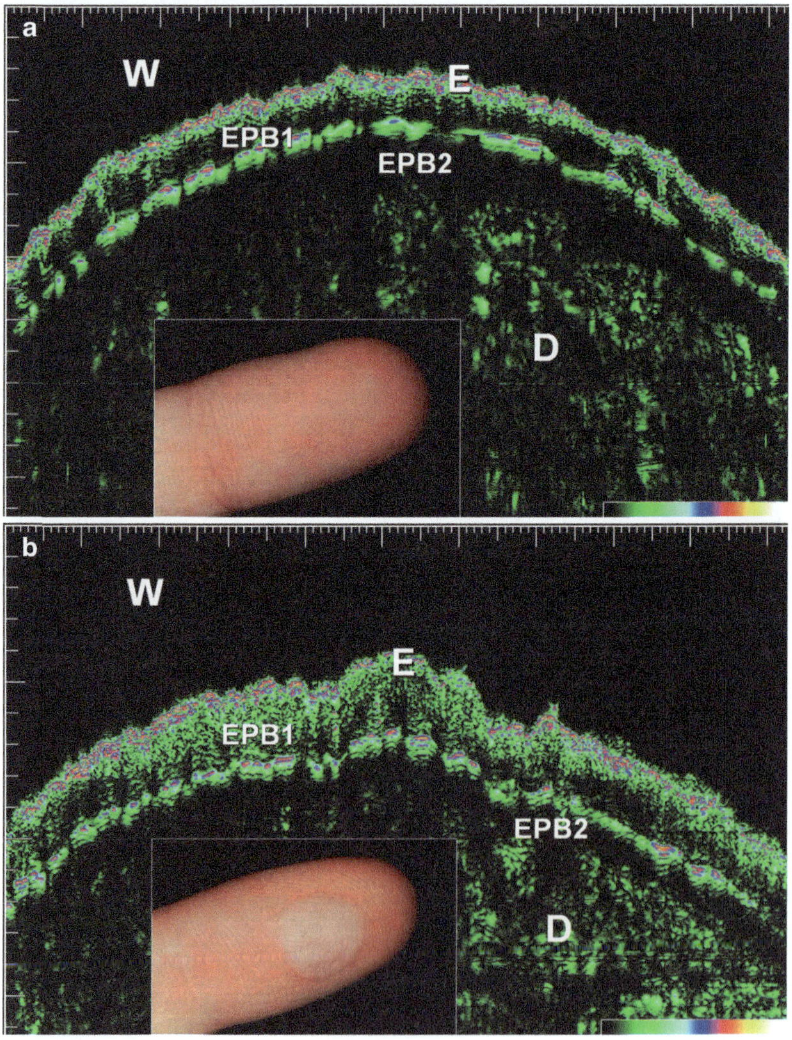

Fig. 13.15 Palmar side of left index finger of a 30-year-old woman (distal phalanx) before (**a**) and after (**b**) 30 min occlusion of an emulsion. *W* water (coupling medium), *E* entry echo, *EPB1* echopoor band 1, *EPB2* echopoor band 2, *D* dermis. Distance between two graduation marks = 100 μm. *Insets*: photos of the finger before and after occlusion

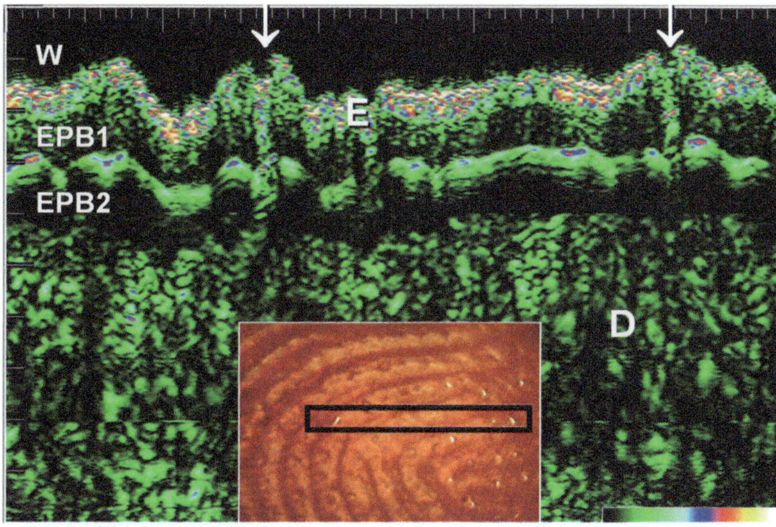

Fig. 13.16 Palmar side of the right index finger of a 39-year-old man at high magnification. *W* water (coupling medium), *E* entry echo, *D* dermis, *arrows* sweat gland duct orifices. Distance between two gradation marks = 100 μm. Inset: Epiluminescence microscopic picture of the left index finger. On the crests of the dermatoglyphics, white points in a row are seen, which represent sweat gland ducts

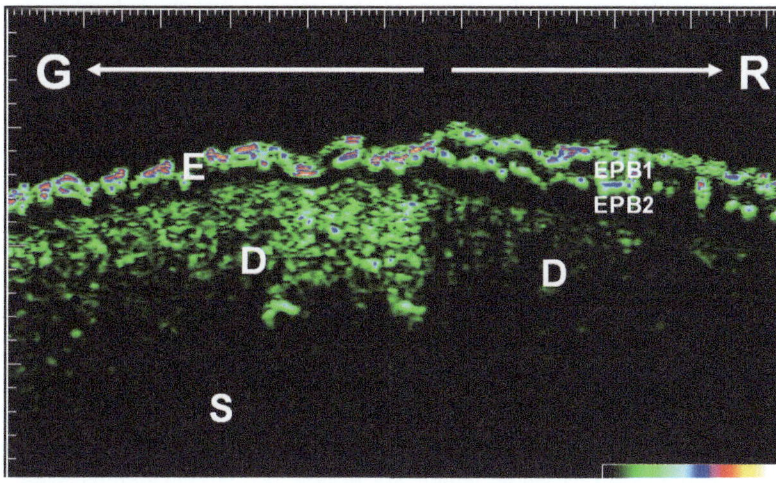

Fig. 13.17 At the transition from glabrous to the palmar skin, a second echorich line separates the EPB1 (upper echopoor band) from the EPB2 (lower echopoor band). Wrist of a 33-year-old woman. *E* skin entry echo, *G* glabrous skin, *R* ridged (palmar) skin, *D* dermis, *S* subcutis. Distance between two graduation marks = 100 μm

thinner than on the palms. On the lower extremities, it is thicker than on the trunk [19].

Concurrent with the sudden appearance of EPB1, when moving from glabrous to ridged skin (palmar and plantar), the echogenicity of the dermis is reduced compared to the glabrous skin. This is due to the strong absorption of the ultrasound energy in the upper layers of the stratum corneum.

13.3.2.3 Inflammatory Skin Diseases

Psoriasis Vulgaris

Compared to normal skin, lesions of psoriasis vulgaris exhibit distinct alterations of the upper dermis.

At the border of a psoriatic lesion, the echopoor band (EPB) of normal skin widens into a broad echopoor band. The thickness of this band correlates very well with the thickness of the acanthotic epidermis plus the dermis with the inflammatory infiltrate in the corresponding histology ($r=0{,}94$). No significant correlation was observed between the thickness of the echopoor band and the epidermis, respectively, and the infiltrated dermis alone [19]. The lower, quite straight border of EPB2 is defined by the scattered reflexes of the dermis.

Furthermore, distinct alterations of the entry echo are observed in psoriasis lesions as compared to normal skin. In untreated scaly plaques, directly below

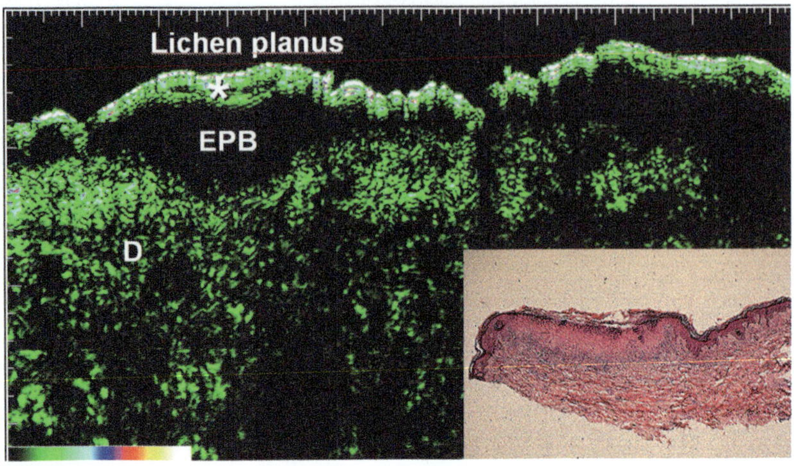

Fig. 13.18 Lichen planus papule on the thigh of a 65-year-old woman. In the middle of the papule, an echopoor line (*) is seen between the skin entry echo and the echorich line beneath. Corresponding histology (inset) exhibits that this region has a significant hyperkeratosis. The histology exhibits, furthermore, that the EPB2 corresponds to the str. Malpighii and the inflammatory infiltrate in the upper dermis. *EPB2* echopoor band 2, *D* dermis. Distance between two graduation marks = 100 μm

the entry echo, several parallel, echorich lines are seen. They melt with the entry echo into an echorich band, which is significantly thicker than the entry echo in normal skin (111 ± 16 μm, $n=35$, $p<0{,}001$) and has a much more irregular surface.

The following observations show that the described band with varying echodensity represents the hyperkeratotic horny layer [18]: after application of petrolatum under occlusion for 60 min on a hyperkeratotic psoriatic plaque, the thickness of the band increases, and its echodensity markedly decreases. In its whole length, it can now be distinguished as an echopoor band with the entry echo as upper and a thin echorich line as lower border. Repeated tape stripping of the scaly surface results in a gradual decrease in the thickness of the echorich band. When the scales are removed entirely, only a single echorich line remains.

In conclusion, the horny layer is echorich in untreated scaly psoriatic plaques; after treatment with petrolatum, its echodensity decreases. Furthermore, the acanthotic epidermis and the dermis with the inflammatory infiltrate are represented as one echopoor band [18, 19].

Lichen Planus

In lichen planus papules, the echopoor band (EPB) of normal glabrous skin focally widens into a spindle-shaped echopoor area. The maximal thickness of this band correlates well with the maximal thickness of the acanthotic epidermis plus the dermis with inflammatory infiltrate in the corresponding histology ($r=0{,}86$). Thick lichen planus papules (Fig. 13.18) often exhibit an echopoor line (EPB1) beneath the skin entry echo [19].

13.3.2.4 Skin Tumors

To study whether the improved resolution at 100 MHz has an impact on visualization of skin tumor details, we evaluated thin basal cell carcinoma using image analysis. Correlation of sonograms and histology reveals that the tumor parenchyma and stroma seen histologically as separate structures (inset, Fig. 13.19) are summed up to a uniform spindle-shaped echopoor area in the upper dermis in sonograms (Fig. 13.19). Tumor parenchyma and stroma are represented as one echopoor area.

To evaluate the potential of 100 MHz sonography, we studied thin pigmented skin tumors (basal cell carcinoma, malignant melanoma, seborrheic keratosis, nevocellular nevi) using sonography and histology. Sonograms were evaluated by comparing the skin tumor region with the adjacent or contralateral normal skin (% change of the normal skin). Most parameters were not significant

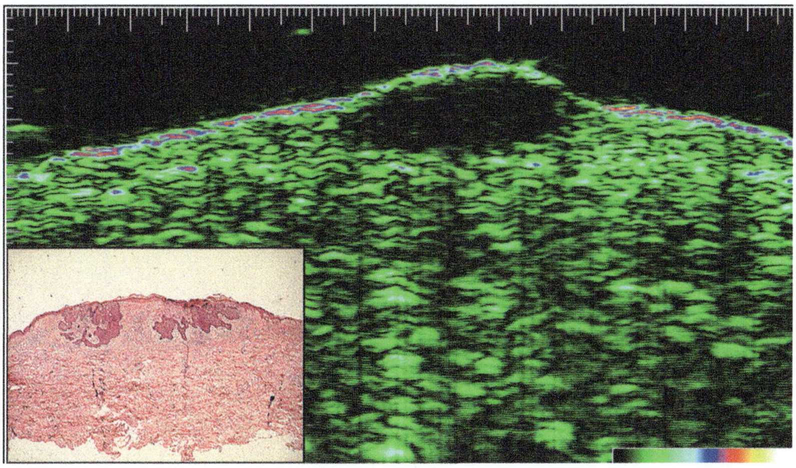

Fig. 13.19 Basal cell carcinoma on the back of a 78-year-old woman. Corresponding histology (inset) reveals that tumor parenchyma and stroma (together 912 μm thick) are summed up to a spindle-shaped echopoor area (maximal thickness 770 μm) in the upper dermis of the sonogram. Distance between two graduation marks = 100 μm

Table 13.1 Statistical analysis of different regions of interest in pigmented skin tumors

		NCN	MM	SebK	Bas
Skin entry echo	Thickness	n.s.	n.s.	$p=0.02$	n.s.
	Mean echo intensity of the neighboring normal epidermis [%]	n.s.	n.s.	n.s.	n.s.
Tumoral echopoor area	Mean echo intensity of the neighboring normal dermis [%]	33 ± 9 %	16 ± 6 %	13 ± 6 %	25 ± 9 %
Subtumoral region	Mean echo intensity of the neighboring normal dermis [%]	71 ± 23 %	50 ± 15 %	25 ± 3 %	86 ± 24 %

n.s. non significant, *NCN* nevocellular nevus, *MM* malignant melanoma, *SebK* seborrheic keratosis, *Bas* basal cell carcinoma

(Table 13.1). In seborrheic keratosis, the thickness of the skin entry echo was significantly increased. Furthermore in seborrheic keratosis, the mean echo intensity of the subtumoral region correlated with the thickness of the skin entry echo ($r=0.92$).

In all tumors, the echopoor tumor region (Fig.13.20a) was not statistically significant from each other, save seborrheic keratosis versus nevocellular nevus ($p=0.05$). The subtumoral area was significantly ($p<0.01$) different for all tumors (Fig. 13.20b), save for malignant melanoma versus nevocellular nevus ($p<0.05$).

We can conclude that differentiation between skin tumors is not possible by studying the echodensity of the tumoral echopoor region. On the other hand, the echodensity of the subtumoral (normal) dermis gives an *indirect hint* about the absorption characteristics of the tumor region: our findings suggest that the absorption of echo reflexes is significantly different for all examined pigmented skin tumors, save malignant melanoma versus nevocellular nevus.

13.4 Discussion

Dermatology deals with all skin structures from the skin surface to the muscle fascia. We therefore need imaging tools to look beneath the skin. Sonography is a particularly interesting method, because it is noninvasive and harmless (no radiation) and can be repeated if necessary.

7.5–15 MHz sonography has become a great tool to study the subcutis and subcutaneous pathologies. Typical examples are connective tissue diseases, cellulite, blood vessel diseases, soft tissue tumors, and lymph node pathology [7–9, 12, 18].

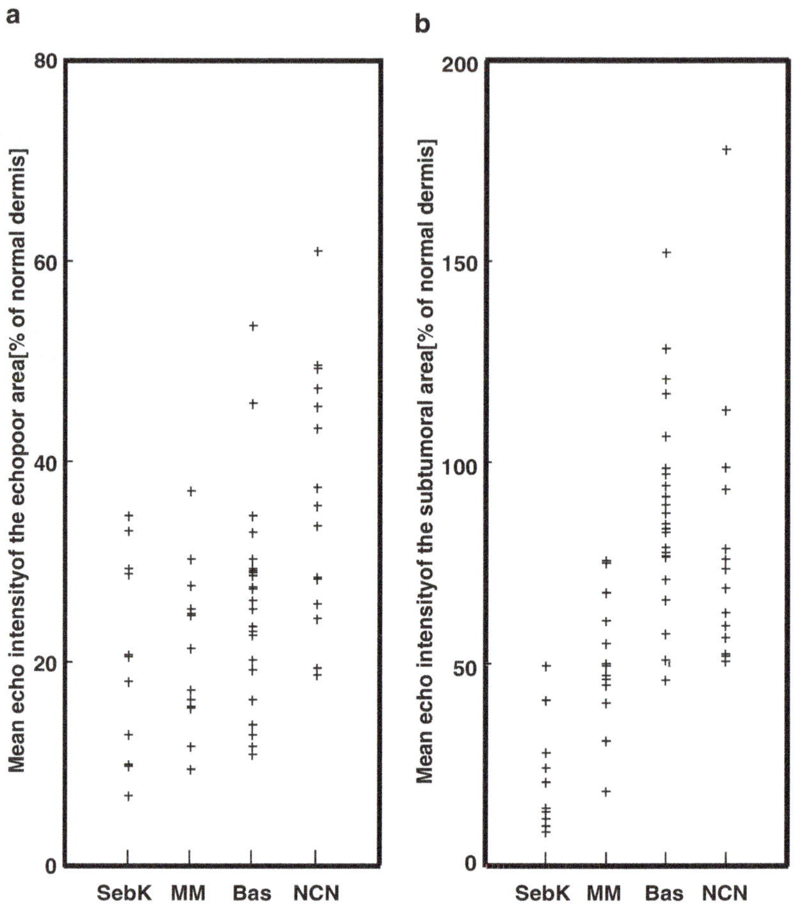

Fig. 13.20 Evaluation of skin tumors, focusing on (**a**) the tumoral echopoor area and (**b**) the subtumoral region. (**a**) In all tumors, the echopoor tumor region is not statistically significant from each other, save SebK versus NCN ($p=0.05$). (**b**) The subtumoral area is significantly ($p<0.01$) different for all tumors, save for MM versus NCN ($p<0.05$). *SebK* seborrheic keratosis, *MM* malignant melanoma, *Bas* basal cell carcinoma, *NCN* nevocellular nevus

20 MHz sonography is useful to study the dermis. Different pathological processes (virtually all skin tumors, inflammatory infiltrates, edema, scar tissue, elastosis) as well as skin appendages and large blood vessels are represented as echopoor areas [4, 14, 16, 24] within the echorich dermis.

Our 100 MHz experimental sonography unit allows a far more detailed visualization of the upper skin layers as compared to 20 MHz sonography. Especially with regard to the in vivo assessment of the horny layer, 100 MHz sonography is a valuable tool. Whereas in normal glabrous skin the stratum corneum is too thin ([39]: about 12–15 μm; [35]: mean thickness of 15 μm for dry stratum corneum and 48 μm after hydration) to be separated from the entry echo, in palmar skin and hyperkeratotic states, it is represented as a distinct band and its thickness can be easily determined. Our results suggest that the echodensity of the horny layer depends on its water content: psoriatic scales, appearing silvery because of the included air, are, for example, much more echorich than the moist stratum corneum of the palms. The significant impedance gap between the stratum corneum and the Malpighian layer – visible as an echorich line – can be explained by the different hydration states of these layers. As demonstrated by various methods, e.g., MR imaging [1, 48] or measurement of the transepidermal water loss [38], there is a sudden increase in water content at the border between the horny layer and the viable epidermis.

Today most authors agree that the skin entry echo is an artifact caused by the change in impedance between the coupling water and the horny layer [25, 34, 53]. This hypothesis is confirmed by our results: the thickness and echodensity of the entry echo remain constant, no matter whether the horny layer is stripped, occluded with topical agents, or entirely removed [19]. In most 20 MHz studies, the echo signal was strongly amplified to reach a high signal depth penetration. This however leads to a significant blurring of the entry echo. This effect as well as the low lateral resolution of only 200 μm entails that the entry echo is represented as a 100–250 μm thick, relatively homogeneous band [34]. During image acquisition of the 100 MHz sonograms, we applied the B/D-Scan technology, which allows to select a specific amplification for each of the 4–8 horizontal stripes, which compose the sonographic image. An overamplification of the entry echo is thus avoided. Its thickness is about 80 μm, and due to the high lateral resolution of 27 μm, it reflects even fine irregularities of the skin surface like the dermatoglyphics or the rough surface of psoriatic lesions.

In 20 MHz sonograms of normal glabrous skin, the dermal reflexes are directly adjacent to the entry echo; the viable epidermis, which is about 80 μm thick, cannot be visualized. At 100 MHz, the resolution is sufficient to show in normal skin a thin echopoor band above the dermal reflexes. Its thickness and its straight lower border suggest that it represents the viable epidermis together with the papillary dermis. At the transition from normal skin to a psoriatic plaque, this band widens into a 400–500 μm thick echolucent band. An echolucent band of comparable thickness can be observed also in 20 MHz sonograms of psoriatic lesions [13, 24, 31, 53, 58, 59]. Conflicting theories have been proposed regarding its nature. While some authors equate it with the sum of acanthosis and the upper dermis with the inflammatory infiltrate [24, 31, 43], others interpret it as correlate of the papillary dermis [13, 58, 59]. Our results favor the first hypothesis: comparison with the corresponding histology revealed an excellent correlation between the thickness of this band and the histometric thickness of the Malpighian layer plus the inflammatory infiltrate. Moreover, in 100 MHz sonograms, this echopoor band always has a fairly straight upper and lower border. If it represented the viable epidermis only, we would expect an undulating lower border; if it was the correlate of the papillary dermis, the upper border would be wavy, especially in psoriatic lesions where there are prominent rete pegs. The lateral resolution of the 100 MHz transducer is high enough to depict structures of this dimension, as the cross sections of the dermatoglyphics in palmar skin demonstrate. We can conclude that both the viable epidermis and the infiltrated dermis are echopoor and cannot be differentiated from each other.

These reflections illustrate that it is not only a question of resolution whether a structure is visualized sonographically. As we could show, the resolution of 100 MHz sonography allows to detect structures as small as a sweat gland duct in the horny layer. On the other hand, the viable epidermis cannot be distinguished from the papillary dermis, and as we learned from the study of skin tumors, the stroma of basal cell carcinoma cannot be distinguished from the tumor cell nests (Fig. 13.19).

How can this be explained? According to Fields and Dunn [22], echoes are only reflected from the border between two tissues, when they have a different acoustic impedance at the applied frequency. Obviously, there is no difference in impedance between the viable epidermis, compact tumor masses, dense lymphocytic infiltrate, and fine fibrillary connective tissue at 100 MHz but only between these structures and the reticular dermis. These acoustic tissue properties are the reason why Gupta et al. [27] and Semple et al. [54] using 40–60 MHz transducers could not improve imaging of skin tumors as compared with 20 MHz.

An entirely exact correlation of histometry and sonometry cannot be expected as various artifacts influence the measurements in both methods. Histological processing leads to tissue shrinkage; the stratum corneum shows the characteristic basket-weave structure which does not correspond to the in vivo anatomy. Sonographic examination requires water as coupling medium which itself may lead to swelling of the horny layer. Sonometry is also influenced by the sound speed which is taken as basis for distance calculations. In dermatological sonography, distance calculations from the echo signal time lapse are usually based on the sound speed of the dermis (1,580 m/s) [3, 5]. In the nail plate, however, Finlay et al. [23] found a sound speed of 2,140 m/s comparing 20 MHz sonography and thickness measurements by a micrometer screw. Jemec and Serup [37] divided the nail into two compartments with different speeds, an upper dry one (3,103 m/s) and a lower humid inner one (2,125 m/s). Obviously, a similar situation must be postulated for the stratum corneum which consists of keratin and has a low content of water thereby having similar properties than the upper nail compartment.

We can conclude that high-resolution sonography enables in vivo examination of fine structural details of the epidermis. In order to obtain images with significant information, understanding of the basic physics of ultrasound principles is indispensible. Only optimal pre-amplification, implementation of a B/D-Scan to increase in-depth penetration at 100 MHz, and knowledge about the influence of resolution on the echo-character enable correct interpretation of the sonograms.

References

1. Ablett S, Burdett NG, Carpenter TA, Hall LD, Salter DC (1996) Short echo time MRI enables visualisation of the natural state of human stratum corneum water in vivo. Magn Reson Imaging 13:357–360
2. Akesson A, Forsberg L, Hederström E, Wollheim E (1986) Ultrasound examination of skin thickness in patients with progressive systemic sclerosis (scleroderma). Acta Radiol Diagn 27:91–94
3. Alexander H, Miller DL (1979) Determining skin thickness with pulsed ultrasound. J Invest Dermatol 72:17–19
4. Altmeyer P, Hoffmann K, Stücker M, Goertz S, El-Gammal S (1992) General phenomena of ultrasound in dermatology. In: Altmeyer P, El-Gammal S, Hoffmann K (eds) Ultrasound in dermatology. Springer, Berlin/Heidelberg/New York, pp 55–79
5. Beck JS, Speace VA, Lowe JG, Gibbs JH (1986) Measurement of skin swelling in the tuberculin test by ultrasonography. J Immunol Methods 86:125–130
6. Beyer D, Peters PE, Friedmann G (1982) Leistungsbreite der Real-time-Sonographie bei Lymphknotenerkrankungen. Rontgenpraxis 35:393–402
7. Blum A, Dill-Müller D (1998) Sonographie der Lymphknoten und der Subkutis in der Dermatologie; Teil 1. Hautarzt 49:942–949
8. Blum A, Dill-Müller D (1999) Sonographie der Lymphknoten und der Subkutis in der Dermatologie; Teil 2. Hautarzt 50:62–73
9. Blum A, Schlagenhauff B, Stroebel W, Breuninger H, Rassner G, Garbe C (2000) Ultrasound examination of regional lymph nodes significantly improves early detection of locoregional metastases during the follow-up of patients with cutaneous melanoma. Cancer 88:2534–2539
10. Brockmann WP, Maas R, Voigt H, Thoma G, Schweer S (1985) Veränderungen peripherer Lymphknoten im Ultraschall. Ultraschall 6:164–169
11. Cole CW, Handler SJ, Burnett K (1981) The ultrasonic evaluation of skin thickness in sklerederma. J Clin Ultrasound (NY) 9:501–503
12. Dill-Müller D, Maschke J (2007) Ultraschalldiagnositk in der Dermatologie. JDDG 5:689–708
13. Di Nardo A, Seidenari S, Giannetti A (1992) B-scanning evaluation with image analysis of psoriatic skin. Exp Dermatol 1:121–125
14. El Gammal S, Auer T, Hoffmann K, Altmeyer P, Paßmann C, Ermert H (1993) Grundlagen, Anwendungsgebiete und Grenzen des hochfrequenten (20–50 MHz) Ultraschalls in der Dermatologie. Zbl Haut 162:817–838
15. El Gammal S, Auer T, Hoffmann K, Matthes U, Altmeyer P (1992) Möglichkeiten und Grenzen der hochauflösenden (20 und 50 MHz) Sonographie in der Dermatologie. Akt Dermatol 18:197–208
16. El Gammal S, Auer T, Hoffmann K, Matthes U, Hammentgen R, Altmeyer P, Ermert H (1993) High-frequency ultrasound: a non-invasive method for use in dermatology. In: Frosch P, Kligman AM (eds) Non-invasive methods in dermatology. Springer, Heidelberg/New York, pp 104–129
17. El Gammal S, Auer T, Hoffmann K, Paßmann C, Ermert H (1995) High resolution ultrasound of the human epidermis. In: Serup J, Jemec GBE (eds) In vivo examination of the skin: a handbook of non-invasive methods. CRC Press, Ann Arbor/London/Tokyo, p 125
18. El Gammal S, El Gammal C, Altmeyer P, Vogt M, Ermert H (2007) Sonography of the skin in health and disease. In: Wilhelm KP, Elsner P, Berardesca E, Maibach HI (eds) Bioengineering of the skin – skin imaging and analysis. Informa Healthcare, New York, pp 353–375
19. El Gammal S, El Gammal C, Kaspar K, Pieck C, Altmeyer P, Vogt M, Ermert H (1999) Sonography of the skin at 100 MHz enables in-vivo-visualization of stratum corneum and viable epidermis in palmar skin and psoriatic plaques. J Invest Dermatol 113: 821–829
20. El Gammal S, Hoffmann K, Auer T, Korten M, Altmeyer P, Höss A, Ermert H (1992b) A 50 MHz high-resolution imaging system for dermatology. In:

Altmeyer P, el-Gammal S, Hoffmann K (eds) Ultrasound in dermatology. Springer, Heidelberg/New York, pp 297–322

21. El Gammal S, Hoffmann K, Kenkmann J, Altmeyer P, Höss A, Ermert H (1992c) Principles of three-dimensional reconstructions from high resolution ultrasound in dermatology. In: Altmeyer P, el-Gammal S, Hoffmann K (eds) Ultrasound in dermatology. Springer, Heidelberg/New York, pp 355–387
22. Fields S, Dunn F (1993) Correlation of echographic visualizability of tissue with biological composition and physiological state. J Acoust Soc Am 54: 809–812
23. Finlay AY, Moseley H, Duggan TC (1987) Ultrasound transmission time: an in vivo guide to nail thickness. Br J Dermatol 117:765–770
24. Fornage BD, McGavran MH, Duvic M, Waldron CA (1993) Imaging of the skin with 20-MHz US. Radiology 189:69–76
25. Gniadecka M, Gniadecki R, Serup J, Søndergaard J (1994) Ultrasound structure and digital image analysis of the subepidermal low echogenic band in aged human skin: diurnal changes and interindividual variability. J Invest Dermatol 102:362–365
26. Gropper CA, Stiller MJ, Shupack JL, Driller J, Rorke M, Lizzi F (1993) Diagnostic high-resolution ultrasound in dermatology. Int J Dermatol 32:243–250
27. Gupta AK, Turnbull DH, Foster FS, Harasiewicz KA, Shum DT, Prussick R, Watteel GN, Horst LN, Sauder DN (1996) High-frequency 40-MHz ultrasound. A possible noninvasive method for the assessment of boundary of basal cell carcinoma. Dermatol Surg 22: 131–136
28. Gupta AK, Turnbull DH, Harasiewicz KA, Shum DT, Watteel GN, Fister FS, Sauder DN (1996) The use of high-frequency ultrasound as a method of assessing the severity of a plaque of psoriasis. Arch Dermatol 132:658–662
29. Haller J, Gritzmann N, Czembirek H, Schmidbauer CH, Leitner H, Sommer G, Tscholakoff D (1987) Der okkulte und der klinisch verdächtige Hodentumor. Abklärung mittels Realtime-Sonographie. Radiologie 27:113–117
30. Harland CC, Bamber JC, Gusterson BA, Mortimer PS (1993) High frequency, high resolution B-scan ultrasound in the assessment of skin tumours. Br J Dermatol 128:525–532
31. Hoffmann K, Dirschka T, Schwarze H, el-Gammal S, Matthes U, Hoffmann A, Altmeyer P (1995) 20 MHz sonography, colorimetry and image analysis in the evaluation of psoriasis vulgaris. J Dermatol Sci 9:103–110
32. Hoffmann K, el-Gammal S, Gerbaulet U, Schatz H, Altmeyer P (1992a) Examination of circumscribed scleroderma using 20 MHz B-scan ultrasound. In: Altmeyer P, el-Gammal S, Hoffmann K (eds) Ultrasound in dermatology. Springer, Heidelberg/New York, pp 231–243
33. Hoffmann K, el-Gammal S, Winkler K, Jung J, Pistorius K, Altmeyer P (1992b) Skin tumours in high-frequency ultrasound. In: Altmeyer P, el-Gammal S, Hoffmann K (eds) Ultrasound in dermatology. Springer, Heidelberg/New York, pp 181–201
34. Hoffmann K, Stücker M, Dirschka T, Görtz S, el-Gammal S, Dirting K, Hoffmann A, Altmeyer P (1994) Twenty MHz B-scan sonography for visualization and skin thickness measurement of human skin. J Eur Acad Dermatol 3:302–313
35. Idson B (1978) Hydration and percutaneous absorption. Curr Probl Dermatol 7:132–134
36. Ihn H, Shimozuma M, Fujimoto M, Sato S, Kikuchi K, Igarashi A, Soma Y, Tamaki K, Takehara K (1995) Ultrasound measurement of skin thickness in systemic sclerosis. Br J Rheumatol 24:535–538
37. Jemec GB, Serup J (1989) Ultrasound structure of the human nail plate. Arch Dermatol 125:643–646
38. Kalia YN, Pirot F, Guy RH (1996) Homogeneous transport in a heterogeneous membrane: water diffusion across human stratum corneum in vivo. Biophys J 71:2692–2700
39. Kligman AM (1964) The biology of the stratum corneum. In: Montagna W, Lobitz WC (eds) The epidermis. Academic, New York, p 387
40. Leopold GR, Woo VL, Scheible W, Nachtsheim D, Gosink B (1979) High-resolution ultrasonography of scrotal pathology. Radiology 131:719–722
41. Lévy J, Gassmüller J, Audring H, Brenke A, Albrecht-Nebe H (1993) Darstellung der subkutanen Atrophie bei der zirkumskripten Sklerodermie im 20 MHz-B-scan Ultraschall. Hautarzt 44:446–451
42. Loose R, Weiss J, Kühn W, Simon R, Teubner J, Georgi M (1992) Comparison of ultrasound with clinical findings in the early detection of regional metastatic lymph nodes in patients with malignant melanoma. In: Altmeyer P, El Gammal S, Hoffmann K (eds) Ultrasound in dermatology. Springer, Heidelberg/New York, pp 93–99
43. Murakami S, Miki Y (1989) Human skin histology using high-resolution echography. J Clin Ultrasound 17:77–82
44. Myers SL, Cohen JS, Sheets PW, Bies JR (1986) B-mode ultrasound evaluation of skin thickness in progressive systemic sclerosis. J Rheumatol 13:577–580
45. Paßmann C, Ermert H, Auer T, Kaspar K, el-Gammal S, Altmeyer P (1989) In vivo ultrasound biomicroscopy. Proc IEEE Ultrasonics Symposium 1015–8
46. Paßmann C, Ermert H, Auer T, Kaspar K, el-Gammal S, Altmeyer P (1993) In vivo ultrasound biomicroscopy. Proc IEEE Ultrasonics Symposium 1015–8
47. Querleux B, Lévêque JL, de Rigal J (1988) In vivo cross-sectional ultrasonic imaging of the skin. Dermatologica 177:332–337
48. Querleux B, Richard S, Bittoun J, Jolivet O, Idy-Peretti I, Bazin R, Lévêque JL (1994) In vivo hydration profile in skin layers by high-resolution magnetic resonance imaging. Skin Pharmacol 7:210–216
49. Rukinava B, Mohar N (1979) An approach of ultrasound diagnostic techniques of the skin and subcutaneous tissue. Dermatologica 158:81–92

50. Schwartz SR, Murray RA (1991) Assessment of epithelial thickness by ultrasonic imaging. Decubitus 4:29–30
51. Schwerk WB, Schwerk WN (1989) Sonographie des Skrotalinhaltes (III-1.9.4.). In: Braun B, Günther R, Schwerk B (eds) Ultraschalldiagnostik. Lehrbuch und Atlas. Ecomed, München, pp 1–70
52. Seidenari S, Di Nardo A (1992) B scanning evaluation of irritant reactions with binary transformation and image analysis. Acta Derm Venereol Suppl (Stockh) 175:9–13
53. Seidenari S (1995) High-frequency sonography combined with image analysis: a non-invasive objective method for skin evaluation and description. Clin Dermatol 13:349–359
54. Semple JL, Gupta AK, From L, Harasiewicz KA, Sauder DN, Foster FS, Turnbull DH (1995) Does high-frequency (40–60 MHz) ultrasound imaging play a role in the clinical management of cutaneous melanoma? Ann Plast Surg 34:599–605
55. Serup J (1984) Decreased skin thickness of pigmented spots appearing in localized scleroderma (morphoea) - measurement of skin thickness by 15 MHz pulsed ultrasound. Arch Dermatol Res 276:135–137
56. Serup J, Staberg B (1987) Ultrasound for assessment of allergic and irritant patch test reactions. Contact Dermatitis 17:80–84
57. Solbiati L, Rizzatto G, Bellotti E (1988) High resolution sonography of cervical lymph nodes in head and neck cancer: Criteria for differentiation of reactive versus malignant lymph nodes. In: Proceedings of the 74th Meeting of the Radiologic Society of North America, Chicago. Radiology, 169: abstract 113ff
58. Stiller MJ, Gropper CA, Shupack JL, Lizzi F, Driller J, Rorke M (1994) Diagnostic ultrasound in dermatology: current uses and future potential. Cutis 53: 44–48
59. Vaillant L, Berson M, Machet L, Callens A, Pourcelot L, Lorette G (1994) Ultrasound imaging of psoriatic skin: a non-invasive technique to evaluate treatment of psoriasis. Int J Dermatol 33:786–790
60. Vassallo P, Edel G, Roos N, Naguib A, Peters PE (1993) In-vitro high resolution ultrasonography of benign and malignant lymph nodes. A sonographic-pathologic correlation. Invest Radiol 28: 698–705

14 New Ultrasound Techniques in Clinical Dermatology

Francesco Maria Solivetti, Fulvia Elia, and Carlo De Mutiis

Abbreviation

ECT Electrochemotherapy

14.1 Introduction

Although dermatologic ultrasound is a branch of eidology that traditionally had not seen rapid technological progress or great advancements in knowledge, new applications have been developed in recent years, as the result of both technological advancements, including very high-frequency ultrasound, and the renewed interest of clinicians. With particular regard to psoriatic arthropathy, biological drugs have been shown to be an effective form of treatment; however, these drugs are costly and can have important side effects. For this reason, researchers have attempted to develop methods to evaluate their effectiveness at an early stage of treatment using ultrasound, which is extremely sensitive, nonionizing, repeatable, and not very expensive. Similarly, for in-transit metastases of malignant skin melanoma, electrochemotherapy has been shown to be an effective and inexpensive form of palliative treatment, and ultrasound is being used to evaluate the response to treatment. Finally, the most recently developed ultrasound equipment has allowed us to discover new patterns of tumours that are relatively common yet which have not been extensively studied, such as calcifying epithelioma of Malherbe (or "pilomatricoma"), mainly in the area of paediatric dermatology. In this chapter, we address the use of ultrasound in evaluating calcifying epithelioma of Malherbe, in-transit metastases, and psoriasis, which in our opinion represent new and important areas of interest, worthy of more in-depth exploration.

F.M. Solivetti, MD (✉)
Radiology Department,
San Gallicano Dermatologic Institute, IFO – Rome,
Via Chianesi, 53, Rome 00144, Italy

Via Città della Pieve 19, Rome 00191, Italy
e-mail: solivetti@ifo.it

F. Elia, MD • C. De Mutiis
Radiology Department,
San Gallicano Dermatologic Institute, IFO – Rome,
Via Chianesi, 53, Rome 00144, Italy

14.2 Calcifying Epithelioma of Malherbe

Calcifying epithelioma of Malherbe (also known as "pilomatricoma", "pilomatrixoma", or "trichomatricoma") is a neoplastic disease which has been known for some time [1]. It is generally defined as a hamartomatous neoformation of the pilaris matrix.

It is considered to be relatively rare, with an incidence of 1 case per 800–1,000 cases of skin cancer and with at least 20 new cases per year reported in the literature [2, 3], with a higher incidence among females. It mainly affects persons in the first 20 years of life, with 40 % of cases reported among persons younger than 10 years of age [4], although it can appear at any age.

E. Berardesca et al. (eds.), *Non Invasive Diagnostic Techniques in Clinical Dermatology*,
DOI 10.1007/978-3-642-32109-2_14, © Springer Berlin Heidelberg 2014

Calcifying epithelioma of Malherbe usually occurs on parts of the body exposed to sun, mainly on the upper body, in particular, the face (47 % of cases) [5] and the shoulders or upper trunk. Clinically, it manifests as a hard lump, with clearly defined contours which are often lobulated; in some cases it is dark in colour. It is slow growing and remains the same size for many years [6]. Pain is reported relatively rarely (36 % of cases), and in 20 % of cases there are signs of phlogosis [7].

The lesion is usually solitary, although multiple lesions have been reported, as has familial occurrence [8]. Calcifying epithelioma of Malherbe is known to be associated with other conditions, such as myotonic dystrophy (or "Steinert's disease") and Gardner syndrome [9]. The association with Gardner syndrome is possibly related to an anomaly on the long arm of chromosome 19 [4, 10] and is frequently associated with the presence of mutations of the β-catenin gene [CTNNB1] [11].

Based on a review of the literature, it can be deduced that the lesions are generally relatively small (less than 1 cm in diameter), although in an important study, conducted among 26 individuals, the diameter was greater than 2 cm in three individuals and between 11 and 20 mm in other 11, with the lesion beginning on the skin and expanding toward to the subcutaneous fat [12]. Although the tumour is normally benign, it is usually excised surgically, so as to avoid the risk of spreading (though this risk is very low) and more often because of diagnostic doubts upon clinical examination.

In recent years, more has been learned of the potential biological aggressiveness of this neoplasia, in terms of both recurrence at the local level and distant metastases [2, 3, 13]. Although the percentage of tumours that are aggressive is not clear from the literature, it has been reported to reach nearly 2 % [8]. However, some others claim that the aggressiveness of this neoplasm has not been objectively demonstrated [13]. Secondary lesions are more frequently reported in the bone, lungs, and lymph nodes, though in some cases they have been reported to be ubiquitous, rapidly growing [3], and not very responsive to treatment [14]. Some authors, moreover, have hypothesised that there is an association between an insufficient initial excision and local recurrence and the occurrence of distant metastases [15].

Histologically, the lesion appears as a dermal nodule that is well circumscribed and consists of basophilic cellular elements of the pilaris matrix. The so-called shadow or ghost cells (i.e. lacking a nucleus) are typical and are an expression of trichilemmal keratinization [16]. Calcification and inflammatory phenomena are common, at times with aspect of a granulomatous reaction to a foreign body. Regarding the development of the neoplasm, the lesions apparently begin as a cyst whose walls are formed of matrix cells, with a characteristic basaloid aspect, with a tendency to become eosinophilic anucleate cells (i.e. ghost cells).

Because clinical diagnosis can be problematic, in the past, diagnosis was attempted using "soft X-ray" to identify calcifying foci. However, this method does not allow calcifying epithelioma of Malherbe to be distinguished from other pathologies with a calcific component [17]. The use of traditional radiology, computed tomography, and magnetic resonance, as well as needle aspiration biopsy, has not produced satisfactory results over the years [17–21], in that none of these methods is sufficiently sensitive or specific. According to the literature, on ultrasound, the lesion typically appears as a small, hypoechoic nodule, located superficially, between epidermis and derma, with contours that are not necessarily well defined. A calcified area is almost always present, and its aspect varies: it can consist of single or grouped foci of varying shapes, be situated centrally or periphericallly, and be punctiform or affect the entire lesion [22]; in some cases, there is a peripheral hypoechoic halo; also in some cases there are Doppler perilesional flow signals.

In the past, high-frequency ultrasound was rarely used for studying this pathology. In fact, few studies have been published [17, 23–27]. In those studies that have been performed, the maximum frequency used was 12 MHz, which is not sufficient for evaluating small and superficial

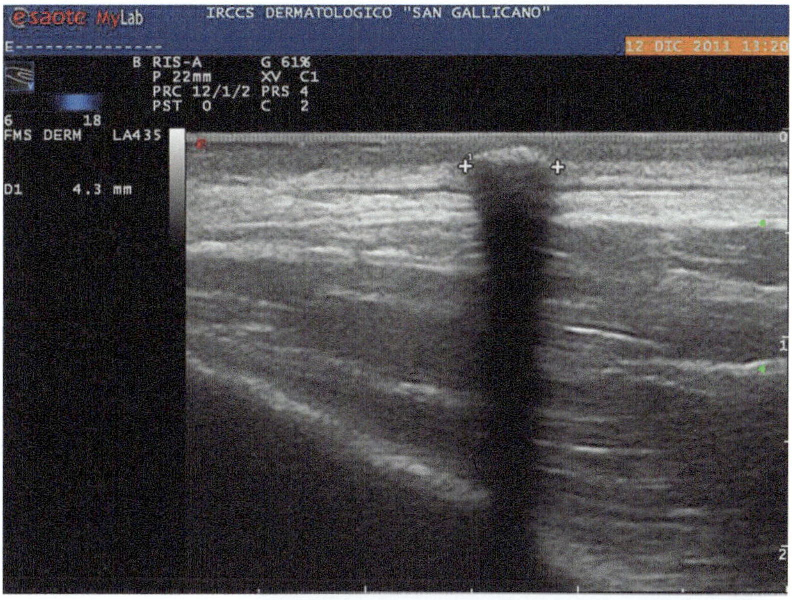

Fig. 14.1 Completely calcified nodule, with an internal structure that cannot be evaluated, because of eco-attenuation, due to a macrocalcification

lesions. In a recently published study [28], we found that the use of ultrasound with higher frequencies (18–22 MHz) allowed this pathology to be more accurately diagnosed. It also revealed, in addition to the two known forms [i.e. completely calcified (Fig. 14.1) and partially calcified (Fig. 14.2)], other ultrasound patterns: a pseudo-cystic form (Fig. 14.3), which lacked posterior wall reinforcement, with relatively thick walls; a complex form (Fig. 14.4); and a pseudo-neoplastic form. The pseudo-cystic form, in accordance with histological results [29], seems to be the expression of more or less vast areas in which shadow cells are decidedly prevalent in the phase preceding the formation of calcifications, whereas the pseudo-neoplastic form seems to be prevalent in elderly persons, in whom pain and inflammation are more common, given the longer clinical history of the pathology.

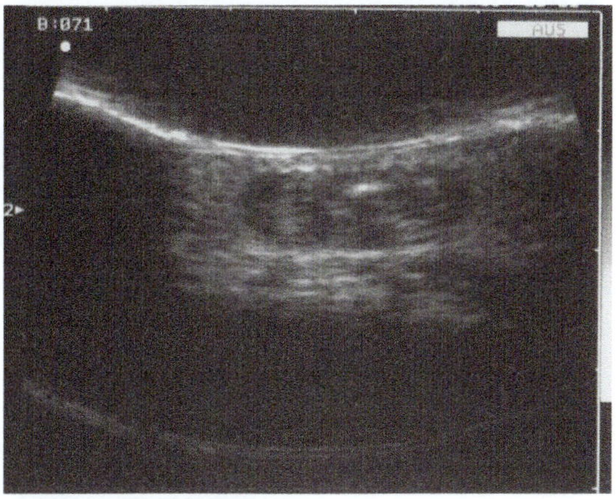

Fig. 14.2 Partially calcified nodule, with an internal structure that can be evaluated only in part; the nodule is solid and hypoechoic, with well-defined contours and few relatively coarse calcific components

14.3 In-Transit Metastases

Malignant cutaneous melanoma is an aggressive pathology which is relatively common and whose occurrence has been increasing; in fact, its incidence has doubled in the past 10–20 years [30, 31]. It is also the third most common neoplasm with skin metastases, which are found in 24 % of cases [32]. Melanoma colonies in soft tissues are known as "satellite metastases" if they are no farther than 2 cm from the primitive neoplastic lesion, whereas " in-transit metastases " are cutaneous and subcutaneous lesions that occur along the lymphatic pathway between the primitive lesion and the draining lymph nodes and are more than 2 cm from the primitive lesion [33].

Ultrasound has been used in this sector for some time. In particular, Fornage was the first to propose using ultrasound for detecting these lesions (in-transit or satellite) [34], followed by others [35]. High-frequency ultrasound has facilitated the early diagnosis of in-transit metastases. In particular, frequencies between 10 and 20 MHz have in some cases made it possible to differentiate diagnostically between recurrence/metastases of malignant cutaneous melanoma and skin metastases of other malignant pathologies and above all between metastases and other diagnoses (i.e. fibromas, lipomas, and aberrant scars) [36]; it can also be used to detect lesions that are only a few millimetres in size.

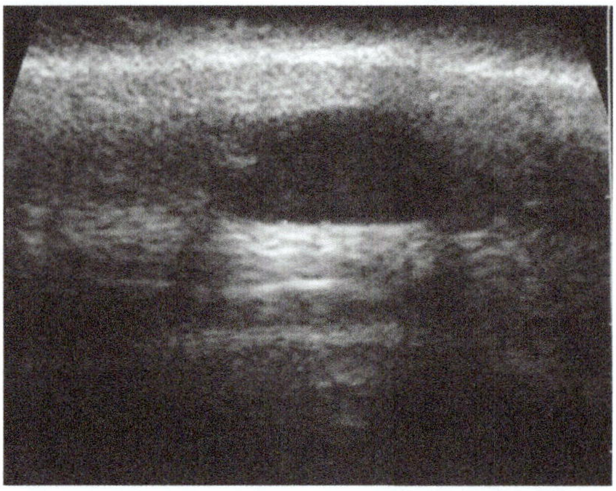

Fig. 14.3 Nodule with fluidic area, with a pseudo-cystic form; the nodule extends to the derma and has well-defined contours and scarce posterior wall reinforcement

In-transit metastases typically appear as a semi-superficially located nodule, which is solid, hypoechoic, more or less round, and usually less than 1 cm in diameter; they have clearly defined contours, which are at times polycyclic; they are not necessarily solitary and are often grouped or confluent (Figs. 14.5 and 14.6). The occasional presence of internal fluidic areas seems to be related to necrotic phenomena and is thus more common in larger nodules, in which blood intake is insufficient (Figs. 14.7 and 14.8).

These lesions normally show good transmission of ultrasound and reinforced posterior wall echo. Evaluation with colour and power Doppler can reveal internal vascular signals, although in small lesions vascular signals may be totally undetectable because of technological limitations of the instrument (Fig. 14.8). Moreover, high-frequency ultrasound allows accurate cytological sampling to be performed in cases of dubious diagnoses [37], although some authors have reported otherwise [38].

Few studies have been performed on the use of ultrasound contrast medium for the diagnoses of these lesions. The most well known of these studies, conducted by Lassau in 2007 [39], showed that perfusion studies, combined with

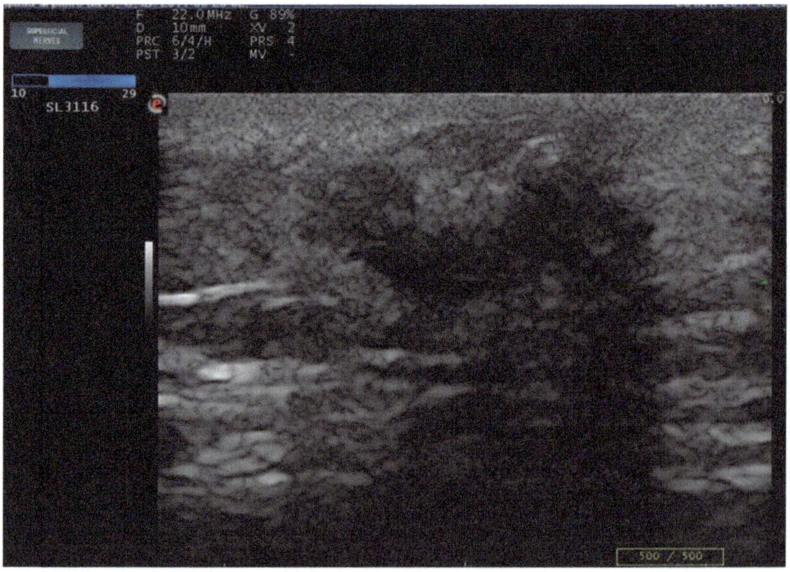

Fig. 14.4 Complex nodule, with calcified area, poorly defined contours, and with small cystic areas

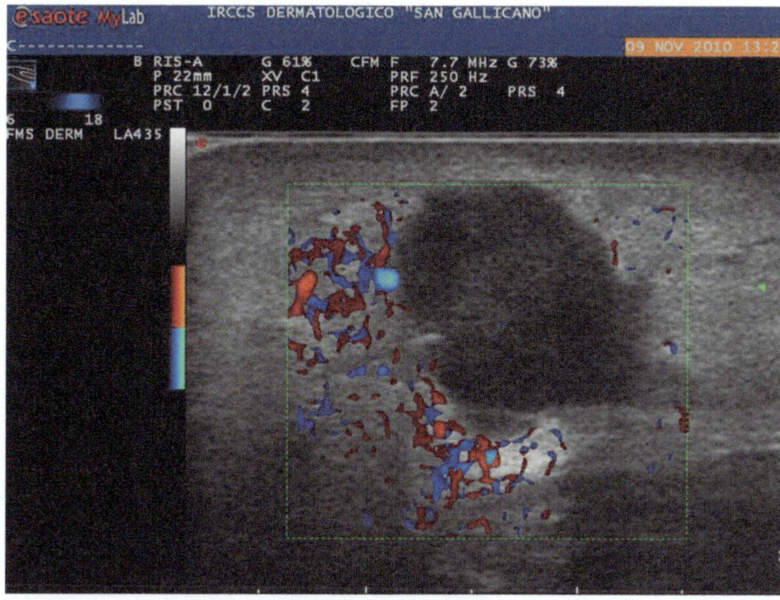

Fig. 14.5 Single in-transit metastasis; the lesion is large and clinically evident; it is hypoechoic, with no colour signal; it is located subcutaneously, solid, and with slightly irregular contours

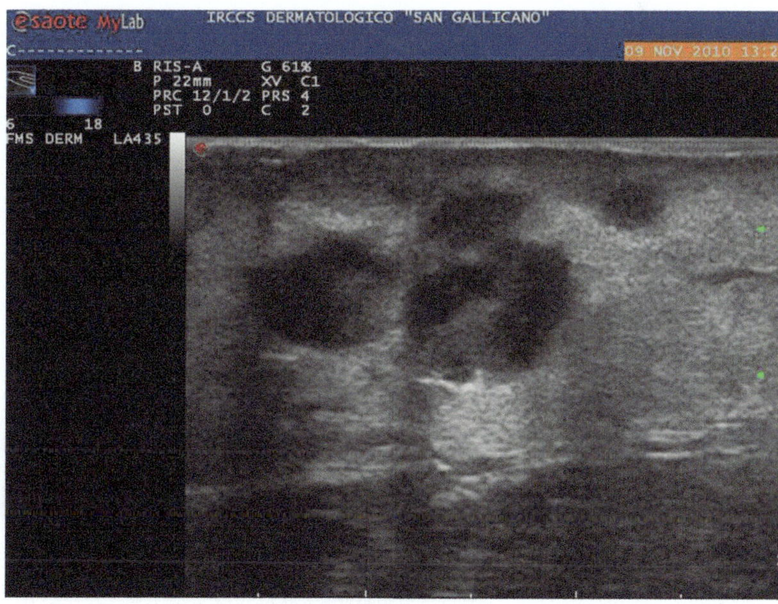

Fig. 14.6 Grouped in-transit metastases; the single lesion shows a classic pattern

normal morphological studies, provide better diagnostic definition in dubious cases and a more accurate evaluation of the residual tumour mass following treatment.

Since the introduction of efficient, relatively non-invasive, well-tolerated, and inexpensive palliative treatment, such as electrochemotherapy (ECT), the detection of clinically nonevident in-transit metastases has increased in importance. In particular, ultrasound can be used to both detect and locate these lesions, as well as to measure their size and the distance from the epidermis, given that it is possible to choose the most appropriate needle length for ECT. For lesions that have been treated properly with ECT, ultrasound can show (even shortly after treatment) that the lesion has substantially disappeared, with evidence of hyperechoic dishomogeneous tissue at

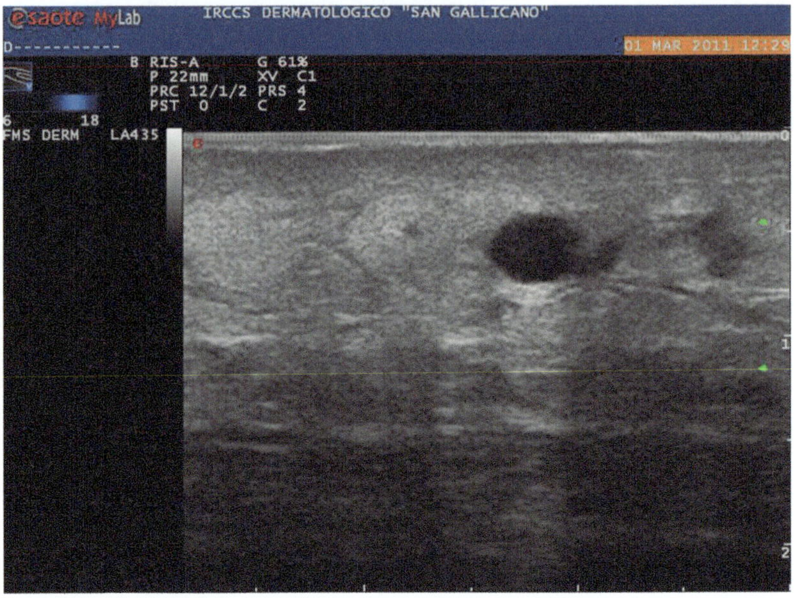

Fig. 14.7 Very small impalpable in-transit metastasis, with a pseudo-cystic form and posterior wall reinforcement

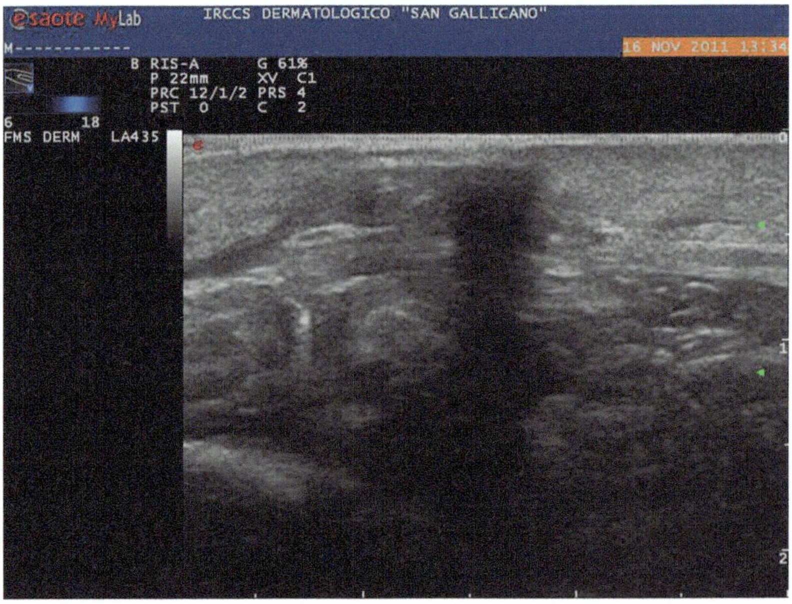

Fig. 14.8 Lesion treated with ECT; the nodule is no longer visible; residue of a calcified form; disorganised adipose tissue is present

the tumour site, at times with high absorption of the ultrasonic rod, with no colour signals (Fig. 14.9). Also in our experience, in-transit metastases that have responded to treatment do not show, after contrast medium administration, any significant enhancement, whereas for lesions that do not completely respond to treatment, a more or less significant enhancement persists after contrast medium is administered, in addition to some residue of the nodular lesion in B-mode.

14.4 Psoriasis

Psoriasis is a disease with a genetic component, mainly affects the skin, with clinical and morphological aspects that can vary. Not infrequently it involves the joints, in the form of arthritis, as well as the nails [40]. Of particular interest is the use of ultrasound in the study of nail psoriasis [41, 42]. Even this specific form can have varied

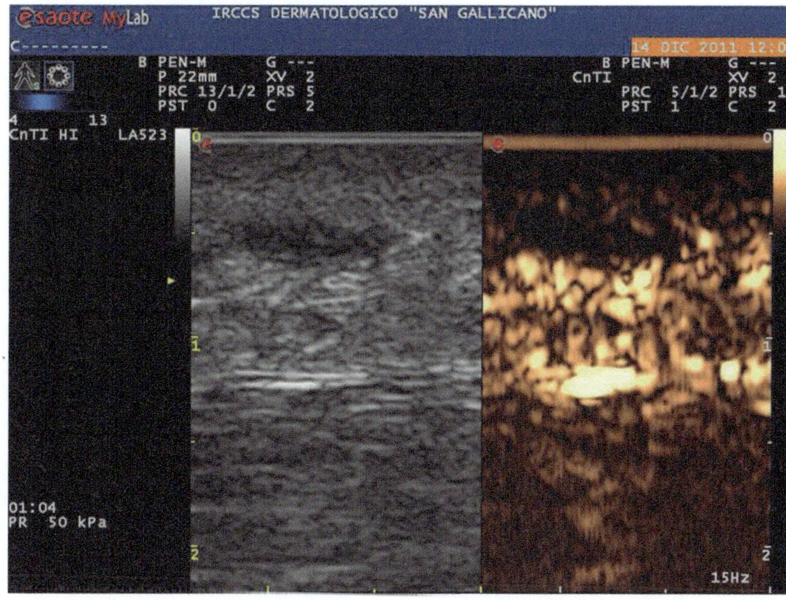

Fig. 14.9 Nodular lesion after treatment with ECT for a few days; a faded hypoechoic area can be seen; following the administration of contrast medium, no enhancement is evident

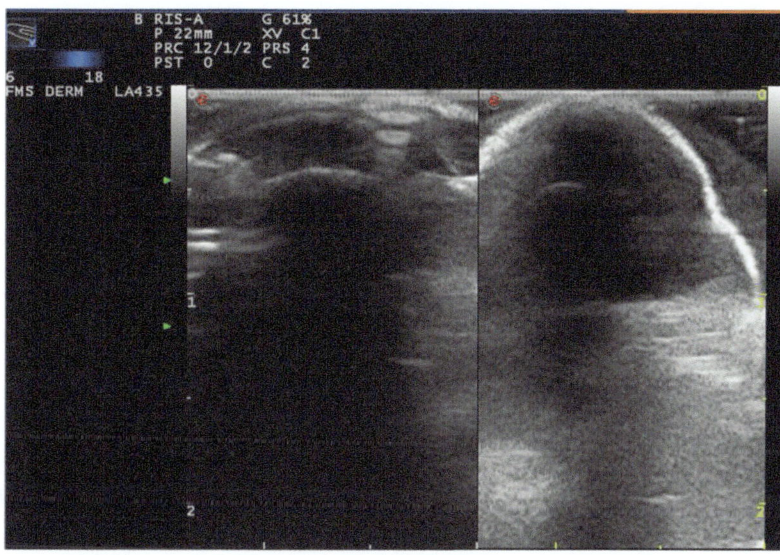

Fig. 14.10 Transverse and longitudinal section of nail; thickening of the nail affected by psoriasis is evident, with a dishomogeneous aspect and a moderate increase in the attenuation coefficient of the ultrasounds

clinical aspects with respect to classic pitting, and morphology and morphological changes can be accurately evaluated with ultrasound, generally at a frequency of around 13–15 MHz, so as to have adequate penetration.

The nail consists of the root, the body, and the free edge; the body is divided into the dorsal and ventral plates, whereas the root consists of the nail (or "germinal") matrix, which is proximal to and under the lunula, and, more distally, the so-called nail bed. In cases of nail psoriasis, ultrasound shows a diffuse thickening of the nail, with an increased distance between the external edge and the anterior surface of the distal phalanx; this is associated with hyperechogenicity and irregularity of the ventral plate, with normality of the dorsal plate (the nail itself has a trilaminar appearance, with a hypoechoic central band). Afterwards, there is a loss of definition of the ventral plate; still later, both plates have an undulated appearance, and finally there is a loss of definition of both plates (Fig. 14.10).

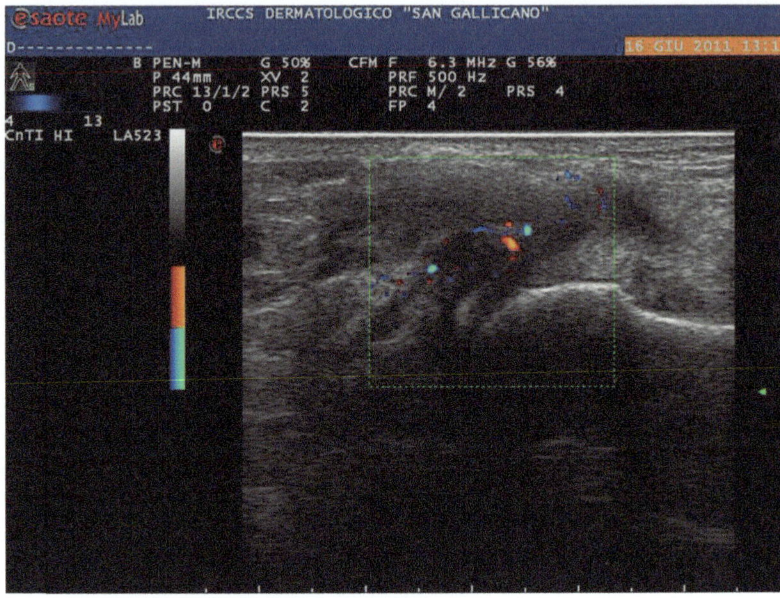

Fig. 14.11 Psoriatic arthritis of a distal interphalangeal joint; a moderate endoarticular fluid layer is present, with a moderate synovial reaction; internal colour signals are present

Psoriatic arthritis is a chronic inflammatory arthropathy, seronegative for rheumatoid arthritis and which affects periarticular structures, such as the tendons and bursa. It can result in periosteal neoformation and bone involvement, which, if not diagnosed and treated early, can develop not infrequently into a disability, often permanent and at times severe [43]. Although radiological findings of arthritic damage continue to be the gold standard for the diagnosis of psoriatic arthritis from a medical-legal standpoint, the study of potentially affected joints using ultrasound is becoming increasingly important. This is because ultrasound, in addition to its wide-scale use and low cost, can reveal structural alterations at their onset, even in the absence of symptoms and when clinical examination cannot yet reveal the early signs of joint involvement, in other words, that which is considered as "early arthritis" [44]. In particular, ultrasound can detect, in addition to late alterations of the bone outline, early alterations in the soft tissues, such as joint effusion (Fig. 14.11), synovial hypertrophy, and alterations in the enthesis. Ultrasound can also provide functional information, based on the evaluation of vascularity, of the dynamic behaviour of the tendons, and of overall articulation [45].

For psoriatic arthritis, a fundamental application is ultrasound after contrast medium administration, not only for diagnosis, for which recent studies have reported results that are comparable to those obtained using magnetic resonance, but most of all for follow-up during therapy with biological drugs. Through this evaluation, in which an enhancement score is attributed (Figs. 14.12 and 14.13), and the study of wash-in and wash out curves, it is possible to establish the extent of phlogosis and disease activity. In this way the response to therapy can be evaluated, and the results are similar to those of nuclear magnetic resonance, yet the cost is definitely lower and there is greater accessibility to this method; the only true limit is that the operator has to be an expert [46].

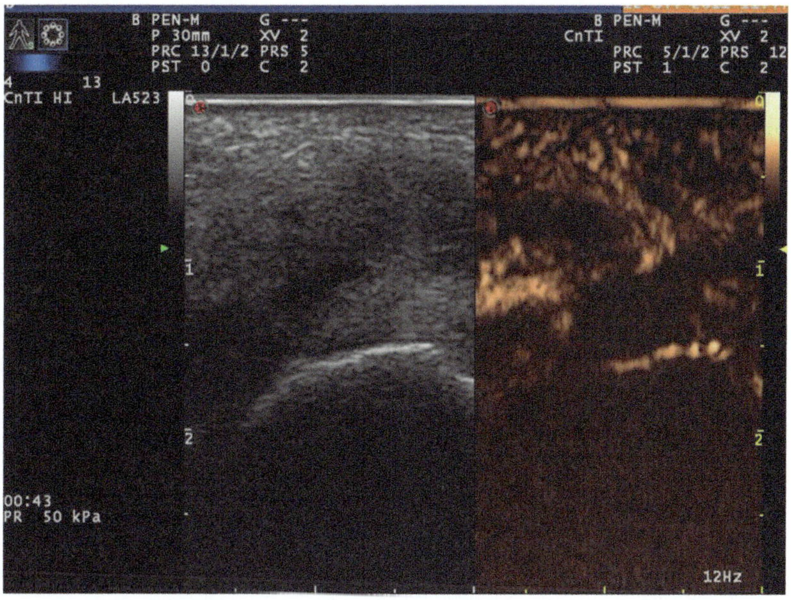

Fig. 14.12 Psoriatic arthritis of interphalangeal joint: a joint fluid layer can be seen, with synovial reaction; marked enhancement after the administration of second generation contrast medium

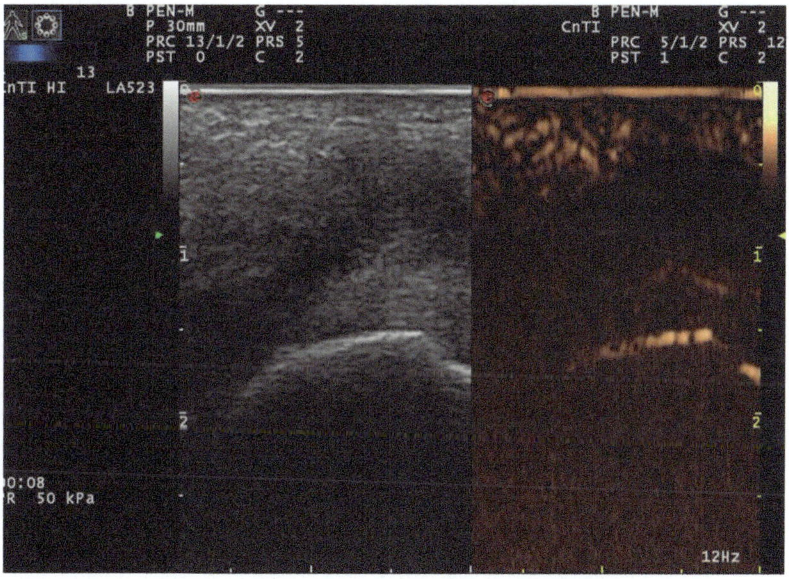

Fig. 14.13 Small joint before and after contrast medium administration. Following the treatment with biological drugs, in this patient there is a permanent joint fluid layer with synovial hypertrophy, yet the clinical symptoms have disappeared, and enhancement, after contrast medium administration, has disappeared

References

1. Malherbe A, Chenantais J (1880) Note sur l'epithèlioma calcifiè des glandes sèbacée. Prog Med 8:826
2. Harbon S, Choisnard S, Carbillet JP, Agache P, Laurent R, Ricbourg B (1990) Epithélioma calcifié de Malherbe. Revue de quatre-vingts cas. Ann Chir Plast Esthet 35(4):277–282
3. Niedermeyer HP, Peris K, Hofler H (1996) Pilomatrix carcinoma with multiple visceral metastases. Cancer 77(7):1311–1314
4. Berberian BJ, Colonna TM, Battaglia M, Sulica VI (1997) Multiple pilomatricomas in association with myotonic dystrophy. J Am Acad Dermatol 37: 268–269
5. Taaffe A, Wyatt EH, Bury HP (1988) Pilomatricoma (Malherbe). A clinical and histopathologic survey of 78 cases. Int J Dermatol 27:477–480
6. Nield DV, Saad MN, Ali MH (1986) Aggressive pilomatrixoma in a child: a case report. Br J Plast Surg 39:139–141
7. Pirouzmanesh A, Reinish JF, Gonzalez-Gomez I, Smith EM, Meara JG (2003) Pilomatrixoma: a review of 346 cases. Plast Reconstr Surg 112(7):1784–1789
8. Forbis R, Helwig EB (1961) Pilomatrixoma. Arch Dermatol 83:606–618
9. Pujol RM, Casanova JM, Egido R, Pujol J, de Moragas JM (1995) Multiple familial pilomatricomas: a cutaneous marker for Gardner syndrome? Pediatr Dermatol 12(4):331–335
10. Harper PS (1972) Calcifying epithelioma of Malherbe. Association with myotonic muscular dystrophy. Arch Dermatol 106:41–44
11. Sherrod QJ, Chiu MW, Gutierrez M (2008) Multiple pilomatricomas: cutaneous marker for myotonic dystrophy. Dermatol Online J 14(7):22
12. Darwish AH, Al-Jalahema EK, Dhiman AK, Al-Khalifa KA (2001) Clinicopathological study of pilomatricoma. Saudi Med J 22(3):268–271
13. Detlefs RL (1984) Pathology quiz case 2. Arch Dermatol 120:782–783
14. Mir R, Cortes E, Papantoniou PA, Heller K, Muehlhausen V, Kahn LB (1986) Metastatic trichomatricial carcinoma. Arch Pathol Lab Med 110(7): 660–663
15. Vico P, Rahier I, Ghanem G, Nagypal P, Deraemaecker R (1997) Pilomatrix carcinoma. Eur J Surg Oncol 23(4):370–371
16. Hashimoto T, Inamoto N, Nakamura K, Harada R (1987) Involucrin expression in the skin appendage tumours. Br J Dermatol 117(3):325–332
17. Rossi E, Carbone M, Iurassich S, Amodio F, Gatta G, Vallone G (1998) Epitelioma calcifico di Malherbe: correlazione tra segni clinici, reperti istologici e immagini ecografiche in 4 casi. Radiol Med 96(4): 410–411
18. Lim HW, Im SA, Lim GY, Park HJ, Lee H, Sung MS, Kang BJ, Kim JY (2007) Pilomatricomas in children: imaging characteristics with pathologic correlation. Pediatr Radiol 37(6):549–555
19. Martino G, Braccioni A, Cariati S, Calvitti M, Veneroso S, Tombesi T, Vergine M (2000) Il pilomatricoma o epitelioma calcifico di Malherbe: Descrizione di un caso e revisione della letteratura. G Chir 21(3):104–109
20. Hoffman V, Roeren T, Moller P, Heuschen G (1998) MR imaging of a pilomatrixoma. Pediatr Radiol 28:272
21. Layfield LJ, Glasgow BJ (1993) Aspiration biopsy cytology of primary cutaneous tumors. Acta Cytol 37(5):679–688
22. Cammarota T (1998) Ecografia in Dermatologia. Poletto Editore, Milano
23. Hughes J, Lam A, Rogers M (1999) Use of ultrasonography in the diagnosis of childhood pilomatrixoma. Pediatr Dermatol 16:341–344
24. Ulrich J, Wesarg I (2001) High-frequency ultrasound in the diagnosis of pilomatrixoma. Pediatr Dermatol 18(2):163
25. Hwang JY, Lee SW, Lee SM (2005) The common ultrasonographic features of pilomatricoma. J Ultrasound Med 24(10):1397
26. Whittle C, Martinez W, Baldassare G, Smoje G, Bolte K, Busel D, Gonzalez S (2003) Pilomatrixoma: ultrasound diagnosis. Rev Med Chil 131(7):735–740
27. Buchwald HJ, Spraul CW, Kampmeier J, Lang GK (2002) Ultrasound biomicroscopy in eyelid lesions – a clinical study in 30 patients. Klin Monbl Augenheilkd 219:95–100
28. Solivetti FM, Elia F, Drusco A, Panetta C, Amantea A, Di Carlo A (2010) Epithelioma of Malherbe: new ultrasound patterns. J Exp Clin Cancer Res 29(42):47
29. Ackerman AB, De Viragh PA, Chongchitnant N (1993) Neoplasm with follicular differentiation. Lea & Febiger, Philadelphia, p 477, cap 21
30. Seigler HF (1986) Investigative studies in the diagnosis and treatment of melanoma. Invest Radiol 21: 596–600
31. Adam YG, Efron G (1983) Cutaneous malignant melanoma: current views on pathogenesis, diagnosis and surgical management. Surgery 93:481–494
32. Clemente GC, Mihm MC Jr, Cainelli T, Cristofolini M, Cascinelli N (1997) Melanoma e Nevi. Effetti s.r.l. Editore. Milano, Italy
33. Das Gupta T, Brasfield R (1964) Metastatic melanoma: a clinicopathological study. Cancer 17:1323
34. Fornage B, Lorigan JG (1989) Sonographic detection and fine-needle aspiration biopsy of nonpalpable recurrent or metastatic melanoma in subcutaneous tissues. J Ultrasound Med 8:421–424
35. Veronesi U, Testori A (2000) Il Melanoma. Diagnosi e trattamento specialistico. Masson Editore, Milano
36. Borgstein PJ, Meijer S, van Diest P (1999) Are locoregional cutaneous metastases in melanoma predictable? Ann Surg Oncol 6(3):315–321
37. Solivetti FM, Di Luca SA, Pirozzi G, Coscarella G, Brigida R, Eibenshutz L (2006) Sonographic

evaluation of clinically occult in-transit and satellite metastases from cutaneous malignant melanoma. Radiol Med 111:702–708

38. Voit C, Schoengen A, Schwurzer M, Weber L, Mayer T, Proebstle TM (1999) Detection of regional melanoma metastases by ultrasound b-scan, cytology or tyrosinase RT-PCR of fine-needle aspirates. Br J Cancer 80(10):1672–1677
39. Lassau N, Chami L, Peronneau P (2007) Imaging of melanoma: accuracy of ultrasonography before and after contrast injection for diagnostic and early evaluation of treatments. Bull Cancer 94(1):93–98
40. McGonagle D, Gibbon W, O'Connor P, Green M, Pease C, Emery P (1998) Characteristic magnetic reson ance imaging entheseal changes of knee synovitis in spondyloarthropathy. Arthritis Rheum 41:694–700
41. Tan AL, McGonagle D (2010) Psoriatic arthritis: correlation between imaging and pathology. Joint Bone Spine 77(3):206–211
42. Gutierrez M, Filippucci E, De Angelis R, Bertolazzi C, Becciolini A, Ariani A, Grassi W (2009) The ultrasound assessment of the psoriatic arthritis: from joint to skin. Reumatismo 61(4):309–315
43. Gladman DD, Brockbank J (2000) Psoriatic arthritis. Expert Opin Investig Drugs 9:1511–1522
44. Bonifati C, Elia F, Francesconi F, Ceralli F, Izzi S, Solivetti FM, De Mutiis C (2011) The diagnosis of early psoriatic arthritis in an outpatient dermatological centre for psoriasis. J Eur Acad Dermatol Venereol. doi:10.1111/j.1468-3083.2011.04138.x
45. Scarpa R, Cuocolo A, Peluso R, Atteno M, Gisonni P, Iervolino S et al (2008) Early psoriatic arthritis: the clinical spectrum. J Rheumatol 35(1): 137–141
46. Solivetti FM, Elia F, Teoli M, De Mutiis C, Chimenti S, Berardesca E et al (2010) Role of contrast-enhanced ultrasound in early diagnosis of psoriatic arthritis. Dermatology 220(1):25–31

15 Multimodal Magnetic Resonance Imaging of the Human Skin

Bernard Querleux

Abbreviations

3D	3 dimensional
FOV	Field of view
MR	Magnetic resonance
RF	Radio frequency
SAT	Subcutaneous adipose tissue
SD	Standard deviation

15.1 Introduction

Cross-sectional imaging of the skin in vivo with noninvasive methods started about 25 years ago. Since that date, a lot of imaging methods have been proposed, most of them based on ultrasound, optics, or magnetic resonance (MR) imaging [1]. Nowadays, skin imaging is reaching a certain degree of maturity which allows us to consider that we are moving from imaging of skin structures to imaging of skin functions.

In vivo imaging methods can be defined according to three main characteristics: the spatial resolution (size of details), the chemical resolution (differentiation of chemical components), and the temporal resolution (dynamic follow-up).

According to these criteria, MR imaging has a specific place. MR imaging can be considered as a very versatile tool capable to explore skin and underlying tissues' anatomy at different scales. Also, MR imaging offers many possibilities to characterize molecular interactions through numerous MR parameters such as relaxation times, free/bound water differentiation, and spectroscopic data. Finally, first attempt to measure in vivo dynamics of water in skin through diffusion-weighted MR images illustrates how MR imaging is approaching a dynamic physiological process.

In this chapter, we will address these 3 aspects with each time, in a first part, a short review, and in a second part, a focus on one detailed example.

B. Querleux
Department of Research and Innovation, L'OREAL, Aulnay-sous-bois, France
e-mail: bquerleux@rd.loreal.com

15.2 Spatial Resolution/Field of View: Versatility of MR Imaging

15.2.1 Review

MR imaging of the skin is still a great challenge when submillimeter pixel size is required for imaging in detail the different skin layers. In the first studies, authors reported on developments of add-on devices based on a high-strength surface gradient coil [2, 3] or a dedicated surface radiofrequency (RF) detector applied for imaging normal [4–18] and diseased skin [19–28].

At such a submillimeter pixel size in a plane perpendicular to the skin surface, epidermis, dermis, and hypodermis are clearly delineated as

E. Berardesca et al. (eds.), *Non Invasive Diagnostic Techniques in Clinical Dermatology*, DOI 10.1007/978-3-642-32109-2_15, © Springer Berlin Heidelberg 2014

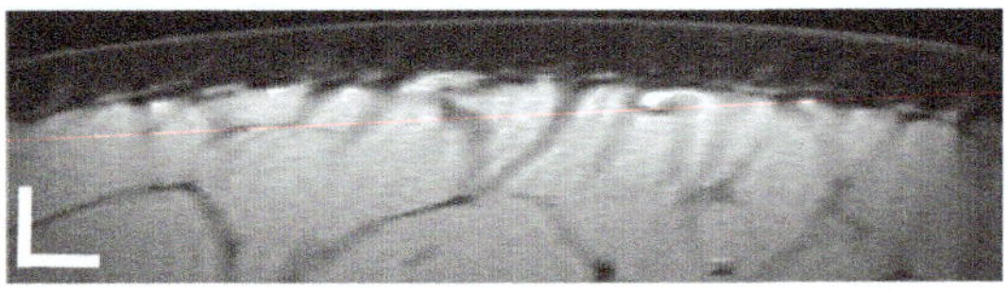

Fig. 15.1 MR image of the skin acquired in 3 min with an add-on device comprising a specific high-strength surface gradient coil and a 3 cm in diameter surface RF coil. The in-plane resolution is $78 \times 390\ \mu m^2$, and the slice thickness is 3 mm. Bars = 2 mm

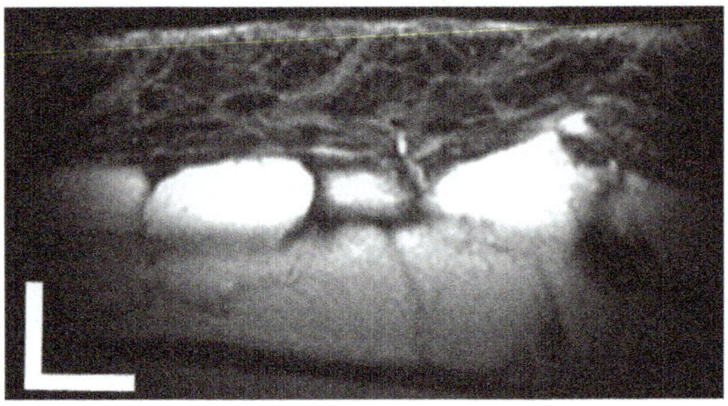

Fig. 15.2 MR image of the skin acquired in 10 min with an ultrasensitive RF detector. The in-plane resolution is $39 \times 78\ \mu m^2$, and the slice thickness is 900 μm. Bars = 1 mm

well as cutaneous appendages within the dermis. Epidermis appears as a bright thin layer and dermis as a hypointense structure. At high spatial resolution, hypodermis can be visualized as a bright thick layer where fibrous septa are visualized as thin hypointense structures (Fig. 15.1).

Nowadays, standard whole-body MR scanners operating at 1.5 and 3 T have gradient coils which intensity is sufficient for skin studies. Thus MR imaging is becoming much easier and only requires the use of a surface coil with a diameter ranging between 2 and 8 cm in diameter. Research focuses now on the design of ultrahigh sensitive radio-frequency detectors made of high-temperature superconducting coils [29, 30]. With such an improved image quality, new details of the skin architecture are visualized whose analysis is still in progress (Fig. 15.2).

In dermatological practice, MR imaging as most of noninvasive cross-sectional imaging methods has not yet fully demonstrated its usefulness for diagnosis [31]. However MR imaging can help accurately define the location and limits of glomus tumors before excision [22] or help in the quantitative follow-up of treatments.

On another aspect, MR imaging is a very powerful method to explore not only the skin but also the underlying structures as it is well known that skin properties and functions are depending on hypodermis and muscle architecture [32].

The potential of MR imaging to study skin and hypodermis tissues at different field of view (FOV) by acquiring two-dimensional (Fig. 15.3) or three-dimensional (Fig. 15.4) images at FOVs varying from $20 \times 50 \times 20\ mm^3$ to $240 \times 240 \times 300\ mm^3$ has been demonstrated. Information on the specific invaginations of fat tissue within the dermis is nowadays well documented as well as some results on the 3D architecture of fibrous septae.

15.2.2 Focus on Age-Related Skin Sagging of the Face

We present more in detail results on skin sagging of the face according to the age analyzed by quantifying the subcutaneous adipose tissue (SAT) volume, as the literature is particularly poor [33]. After informed consent, we acquired frontal MR images of the face (Fig. 15.5) on more than 75 volunteers with age ranging from 18 to 70 years. Image processing was used to extract SAT volume, and 3D quantification was performed on reconstructed 3D images of SAT volume (Fig. 15.6).

Mean SAT volume is of about 300 cm^3, and no difference according to the age was observed in this population whose main inclusion criteria concerned a restricted range of body mass index of 23 ± 1. We thus analyzed evolution of SAT volume in the cheeks area compared to the chin area. Our results demonstrate that, according to the age, there is a decrease of the SAT volume on the cheeks and an increase of the SAT volume on the chin. We thus defined the ptosis index according to Eq. (15.1):

$$I_{\text{Ptosis}} = \frac{\text{Volume}_{\text{Chin}}}{\text{Volume}_{\text{Cheeks}}} \tag{15.1}$$

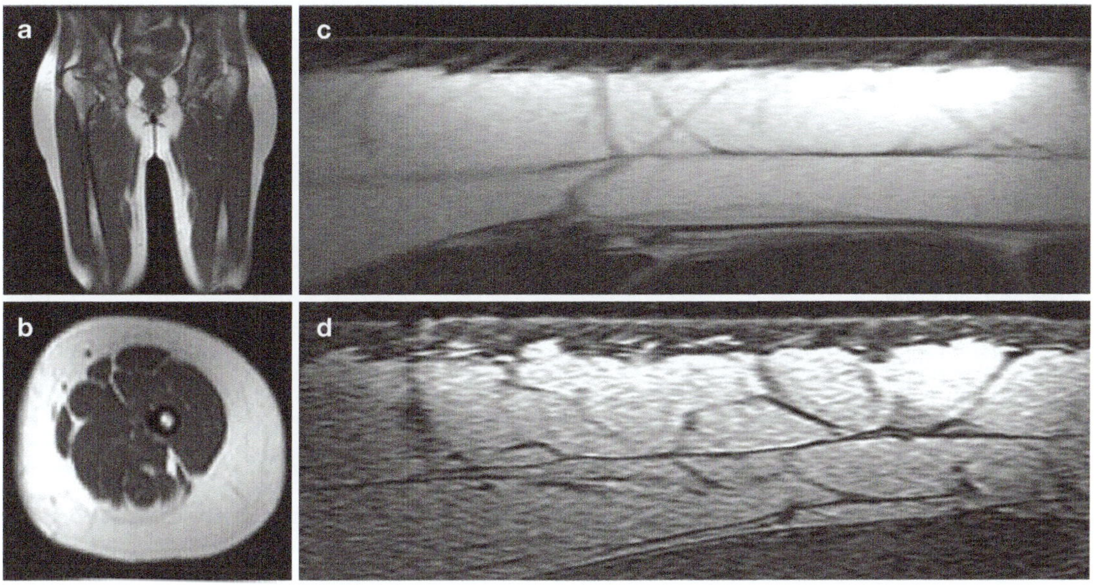

Fig. 15.3 2D MR images of the skin and underlying soft tissues at different field of views. Fat tissue appears as a homogeneous tissue at low spatial resolution. Fat lobules are well differentiated at high spatial resolution. (**a**) Sagittal view, (**b**) axial view of the thigh, (**c**) high-resolution image (in-plane resolution 78 × 390 μm²; slice thickness 3 mm), (**d**) high-resolution image (in-plane resolution 78 × 300 μm²; slice thickness 500 μm)

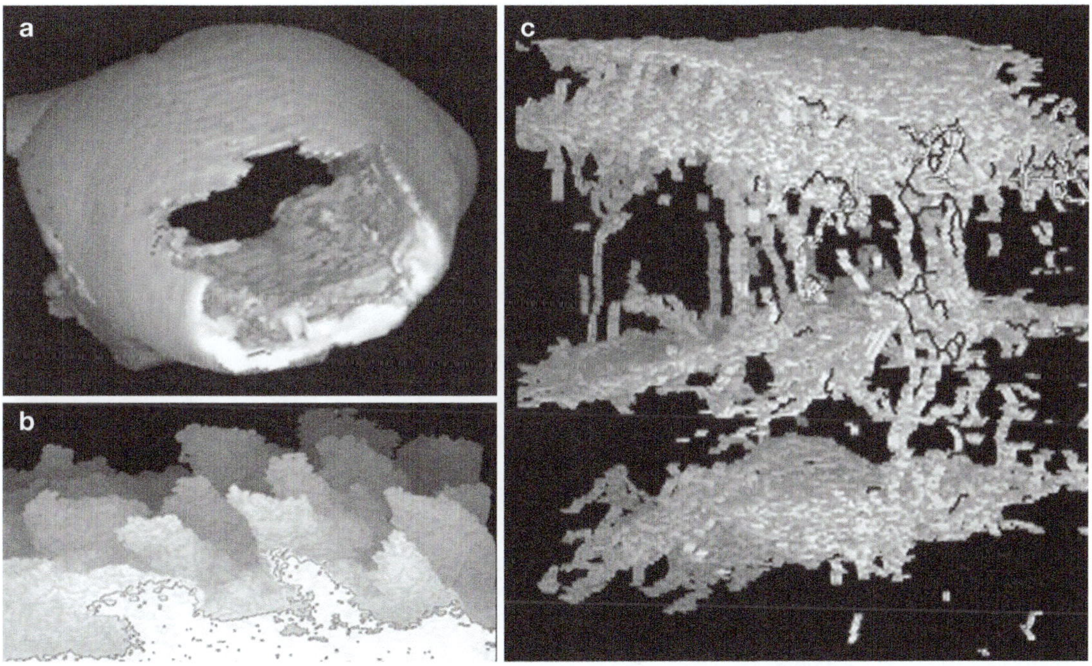

Fig. 15.4 3D visualization of hypodermis at different fields of view. (**a**) Fat volume on a whole thigh, (**b**) invaginations of hypodermis within the dermis, (**c**) 3D architecture of fibrous septae within the hypodermis

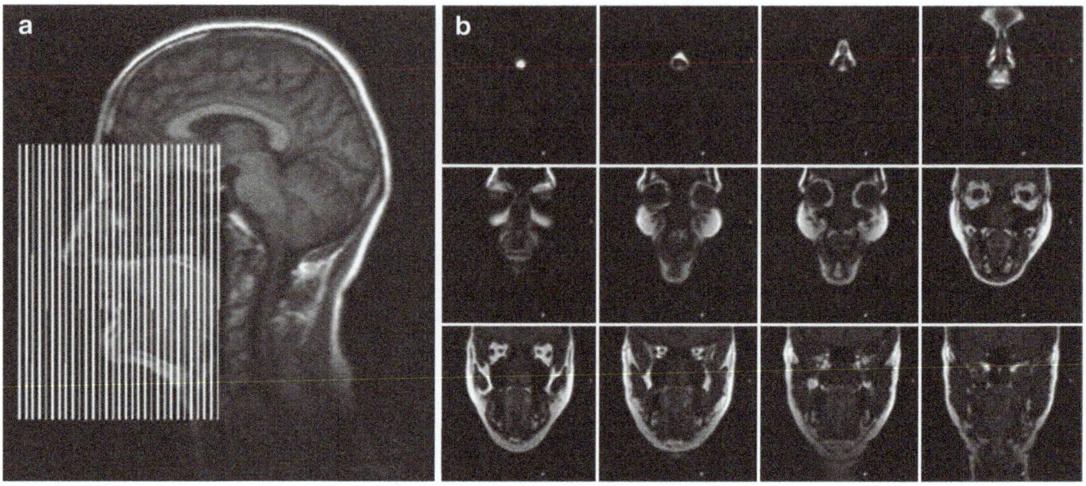

Fig. 15.5 MR imaging of the face. (**a**) Sagittal view. Volume of interest is graphically selected by the operator; (**b**) frontal MR images of the face. 1 over 8 contiguous slices is presented

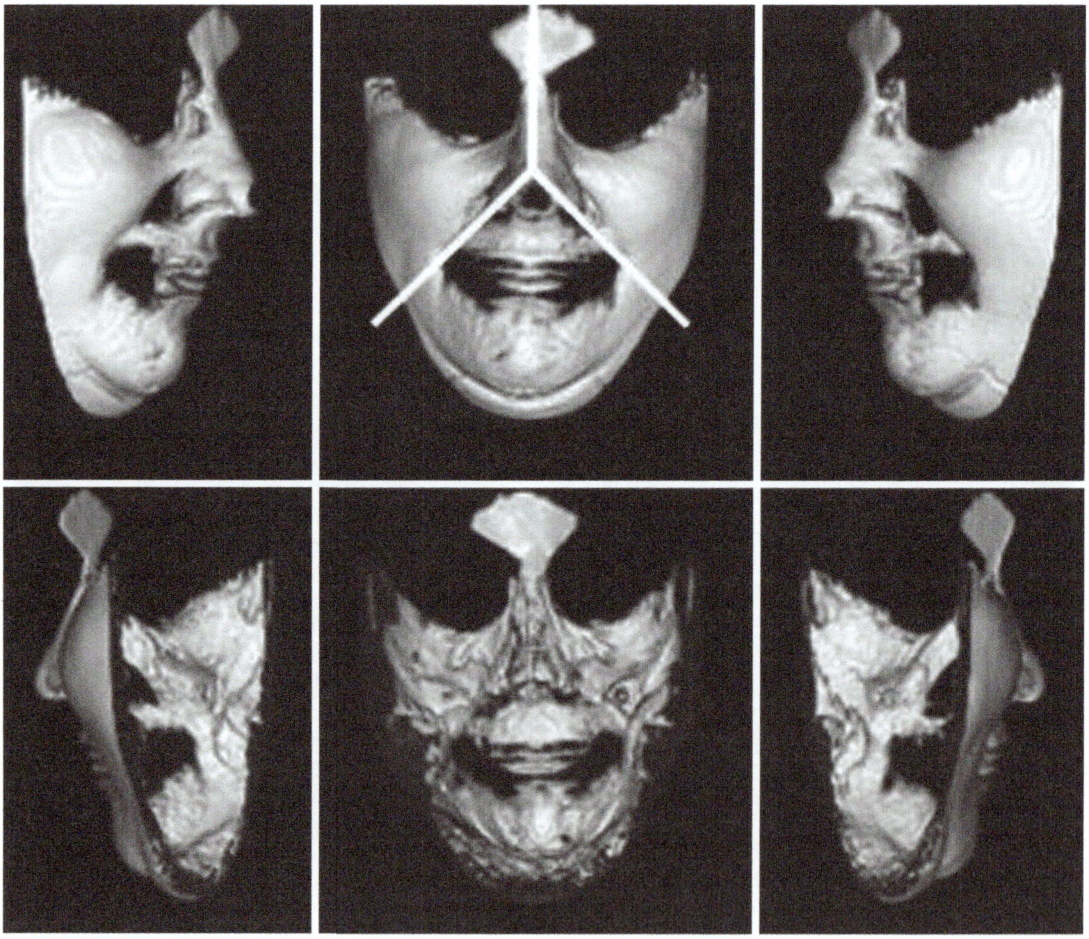

Fig. 15.6 3D visualization of the adipose tissue of the face. Cheek sites and chin site were differentiated through specific anatomical marks

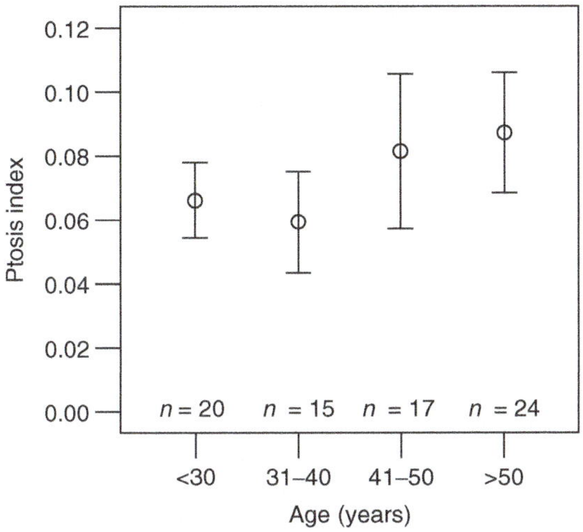

Fig. 15.7 Ptosis index versus age. A significant increase ($p<0.01$) versus age is established, with a clear increase in the early 40s

With this index, our results indicate that the displacement of SAT from the cheeks toward the chin starts at the beginning of the fourth decade (Fig. 15.7) [34].

In conclusion, MR imaging has confirmed to be the reference method for quantifying underlying skin structures at an optimized spatial resolution. Such results obtained on a large number of volunteers will help us to go a step further in the knowledge of age-related changes on the face.

15.3 Molecular Characterization of the Skin

15.3.1 Review

Whatever the MR pixel size is, physics of the MR signal is related to molecular interactions of some nuclei which can be detected. In living tissues, due to sensitivity requirements, most MR imaging studies concern proton interactions.

Detectable protons in skin and hypodermis are located in water-rich structures such as living epidermis and dermal matrix and lipid-rich structures such as pilosebaceous units and fat invaginations within the dermis [8, 30].

As skin is composed of about 70 % of water protons, most MR studies focused on quantifying water content and water state in the different skin layers. We previously reported on relaxation times and free water quantification in the different skin layers [35] and have shown that an increase of the free water content in the outermost past of aged dermis could be related to a decrease of macromolecular-water sites following degradation of the collagen with age [36], in good agreement with other studies [37].

Magnetization transfer was proposed for differentiating the free water pool from the bound water pool [38], while more recently 1D profile at high [39] and ultrahigh spatial resolution in the direction perpendicular to the skin surface with the Garfield technique has been described for an accurate dynamic follow-up of water ingress in skin [40–42].

15.3.2 Focus on In Vivo Water Mobility Quantification

Mobility of water in the skin, which is certainly related to its viscoelastic properties, can be assessed by diffusion-weighted MR imaging through the measurement of the apparent diffusion coefficient. Thus, the aim of the study was to quantify diffusion coefficients in the different skin layers according to the age.

After informed consent, 20 healthy volunteers were enrolled in this study, 10 young subjects (25 ± 3 years) and 10 aged subjects (65 ± 3 years). The study was approved by the hospital's ethics committee.

We acquired anatomic and diffusion images (Fig. 15.8) with an anisotropic pixel size; the highest resolution was set at 35 μm in the direction in depth of the skin.

Results are summarized in Table 15.1. Water mobility in epidermis is higher than water mobility in the outer dermis, while water mobility in the inner dermis is much lower.

According to the age, there is a global trend of an increase of water mobility. These results are statistically significant in living epidermis and papillary dermis. To the best of our knowledge, it is the first time that an MR parameter differentiates young living epidermis from aged living epidermis.

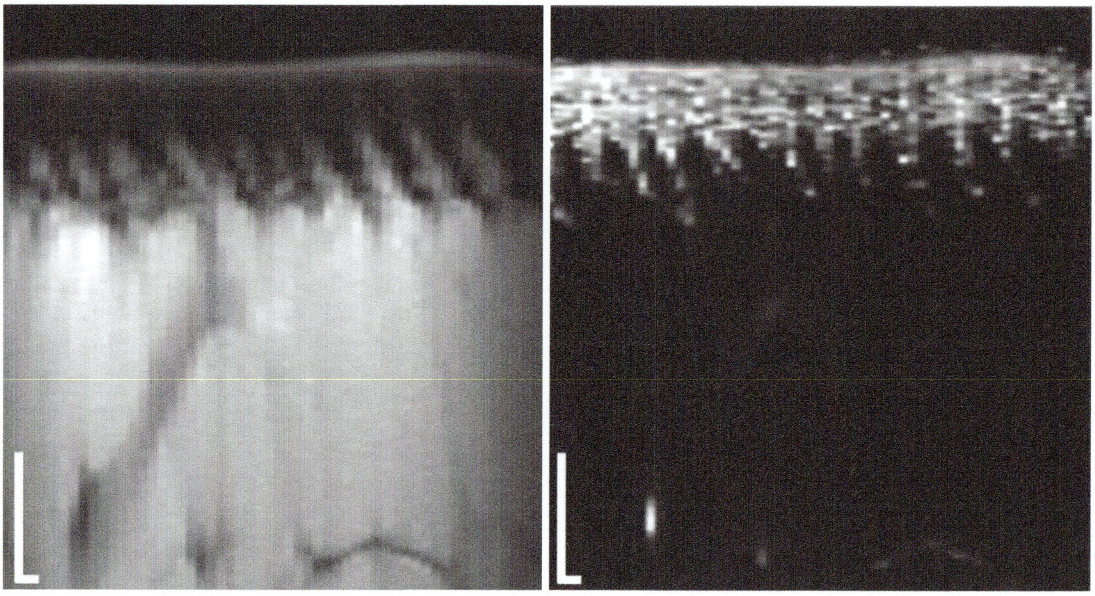

Fig. 15.8 Morphological (*left*) and diffusion-weighted (*right*) images of normal skin. The in-plane resolution is $35 \times 625 \mu m^2$, and the slice thickness is 6 mm. Bars = 2 mm

Table 15.1 Apparent diffusion coefficient (mean ± SD) of the different skin layers according to age

	Epidermis		Outer dermis[a]		Inner dermis	
	<30	>60	<30	>60	<30	>60
D (10^{-5} m²/s)	2.81 ± 0.25	3.17 ± 0.26	2.33 ± 0.29	2.85 ± 0.34	0.90 ± 0.65	1.40 ± 0.60
	$p=0.02$		$p=0.006$		Nonsignificant	

[a]Outer dermis corresponds to a subepidermal layer about 200 μm thick

Such results should contribute to a better understanding of the aged-related mechanisms of transepidermal water loss and of skin biomechanical alterations in chrono- and photoaging.

Conclusion

Characteristics provided by new MR whole-body scanners are now adapted to the acquisition of MR images at a submillimeter scale as requested by skin anatomy. MR protocols are becoming simpler as they only necessitate standard head coil or surface coil detectors according to the skin site.

In this overview, we pointed out the multimodal potential of MR imaging. As a consequence, there are no longer limitations for a more widespread use of MR imaging in dermatology and cosmetology.

Acknowledgment The author thanks all the members of IR4M UMR8081 CNRS, the University Paris Sud, for their outstanding contribution to these works. Most of them were conducted during collaborative projects.

References

1. Wilhelm KP, Elsner P, Berardesca E, Maibach HI (eds) (2007) Bioengineering of the skin: skin imaging and analysis, 2nd edn. Informa Healthcare, New York
2. Bittoun J, Saint-Jalmes H, Querleux B, Darrasse L, Jolivet O, Idy-Peretti I et al (1990) In vivo high-resolution MR imaging of the skin in a whole-body system at 1.5 T. Radiology 176(2):457–460
3. Querleux B (2004) Magnetic resonance imaging and spectroscopy of skin and subcutis. J Cosmet Dermatol 3(3):156–161
4. Hyde JS, Jesmanowicz A, Kneeland JB (1987) Surface coil for MR imaging of the skin. Magn Reson Med 5(5):456–461

5. Hawnaur JM, Dobson MJ, Zhu XP, Watson Y (1996) Skin: MR imaging findings at middle field strength. Radiology 201(3):868–872
6. Song HK, Wehrli FW, Ma J (1997) In vivo MR microscopy of the human skin. Magn Reson Med 37(2):185–191
7. Stock W, Mäurer J, Felix R (1997) Characteristics of normal cross-sectional anatomy of the skin in high resolution magnetic resonance tomography. Hautarzt 48(1):26–30
8. Weis J, Ericsson A, Hemmingsson A (1999) Chemical shift artifact-free microscopy:spectroscopic microimaging of the human skin. Magn Reson Med 41(5):904–908
9. Weis J, Ericsson A, Aström G, Szomolanyi P, Hemmingsson A (2001) High-resolution spectroscopic imaging of the human skin. Magn Reson Imaging 19(2):275–278
10. Liffers A, Vogt M, Ermert H (2003) In vivo biomicroscopy of the skin with high-resolution magnetic resonance imaging and high frequency ultrasound. Biomed Tech (Berl) 48(5):130–134
11. Mirrashed F, Sharp JC (2004) In vivo morphological characterisation of skin by MRI micro-imaging methods. Skin Res Technol 10(3):149–160
12. Denis A, Loustau O, Chiavassa-Gandois H, Vial J, Lalande Champetier de Ribes C, Railhac JJ et al (2008) High resolution MR imaging of the skin: normal imaging features. J Radiol 89(7–8 Pt 1):873–879
13. Aubry S, Casile C, Humbert P, Jehl J, Vidal C, Kastler B (2009) Feasibility study of 3-T MR imaging of the skin. Eur Radiol 19(7):1595–1603
14. Stefanowska J, Zakowiecki D, Cal K (2010) Magnetic resonance imaging of the skin. J Eur Acad Dermatol Venereol 24(8):875–880
15. Barral JK, Bangerter NK, Hu BS, Nishimura DG (2010) In vivo high-resolution magnetic resonance skin imaging at 1.5 T and 3 T. Magn Reson Med 63(3):790–796
16. Sharma R (2010) Gadolinium toxicity: epidermis thickness measurement by magnetic resonance imaging at 500 MHz. Skin Res Technol 16(3):339–353
17. Sans N, Faruch M, Chiavassa-Gandois H, de Ribes CL, Paul C, Railhac JJ (2011) High-resolution magnetic resonance imaging in study of the skin: normal patterns. Eur J Radiol 80(2):e176–e181
18. Laistler E, Loewe R, Moser E (2011) Magnetic resonance microimaging of human skin vasculature in vivo at 3 Tesla. Magn Reson Med 65(6):1718–1723
19. Zemtsov A, Reed J, Dixon L (1992) Magnetic resonance imaging evaluation helps to delineate a recurrent skin cancer present under the skin flap. J Dermatol Surg Oncol 18(6):508–511
20. Ono I, Kaneko F (1995) Magnetic resonance imaging for diagnosing skin tumors. Clin Dermatol 13(4): 393–399
21. Mäurer J, Knollmann FD, Schlums D, Garbe C, Vogl TJ, Bier J, Felix R (1995) Role of high-resolution magnetic resonance imaging for differentiating melanin-containing skin tumors. Invest Radiol 30(11): 638–643
22. Drapé JL, Idy-Peretti I, Goettmann S, Guérin-Surville H, Bittoun J (1996) Standard and high resolution magnetic resonance imaging of glomus tumors of toes and fingertips. J Am Acad Dermatol 35(4): 550–555
23. Idy-Peretti I, Bittoun J, Alliot FA, Richard SB, Querleux BG, Cluzan RV (1998) Lymphedematous skin and subcutis: in vivo high resolution magnetic resonance imaging evaluation. J Invest Dermatol 110(5):782–787
24. Mäurer J, Hoffmann KT, Lissau G, Schlums D, Felix R (1999) High-resolution magnetic resonance imaging for determination of thickness and depth of invasion of skin tumours. Acta Derm Venereol 79(6): 478–479
25. Mäurer J, Strauss A, Ebert W, Bauer H, Schroeder RJ, Felix R (1999) Mn-TPPS4 in the diagnosis of malignant skin tumors. In vivo studies with high resolution magnetic resonance tomography in melanotic melanoma. Radiologe 39(5):422–427
26. Hong H, Sun J, Cai W (2008) Anatomical and molecular imaging of skin cancer. Clin Cosmet Investig Dermatol 1:1–17
27. Aubry S, Leclerc O, Tremblay L, Rizcallah E, Croteau F, Orfali C, Lepage M (2012) 7-Tesla MR imaging of non-melanoma skin cancer samples: correlation with histopathology. Skin Res Technol 18:413–420
28. Tawfik AM, Kreft A, Wagner W, Vogl TJ (2011) MRI of a microcystic adnexal carcinoma of the skin mimicking a fibrous tumour: case report and literature review. Br J Radiol 84(1002):e114–e117
29. Ginefri JC, Darrasse L, Crozat P (2001) High-temperature superconducting surface coil for in vivo microimaging of the human skin. Magn Reson Med 45(3):376–382
30. Bittoun J, Querleux B, Darrasse L (2006) Advances in MR imaging of the skin. NMR Biomed 19(7): 723–730
31. Rallan D, Harland CC (2004) Skin imaging: is it clinically useful? Clin Exp Dermatol 29(5):453–459
32. Querleux B, Cornillon C, Jolivet O, Bittoun J (2002) Anatomy and physiology of subcutaneous adipose tissue by in vivo magnetic resonance imaging and spectroscopy: relationships with sex and presence of cellulite. Skin Res Technol 8(2):118–124
33. Gosain AK, Klein MH, Sudhakar PV, Prost RW (2005) A volumetric analysis of soft-tissue changes in the aging midface using high-resolution MRI: implications for facial rejuvenation. Plast Reconstr Surg 115(4):1143–1152; discussion 1153–5
34. Querleux B, Bazin R, Dubuisson RM, Durand E, Glutron D, Bittoun J, Herment A, Baldeweck T, Fonolla-Moreno A, Mauvais-Jarvis E, Marion C (2007) Age-related effects on subcutaneous adipose tissue volume and localizazion on the face: an MRI study. 21st World Congress of Dermatology, Buenos Aires, October 1–5, 2007.

35. Richard S, Querleux B, Bittoun J, Idy-Peretti I, Jolivet O, Cermakova E, Lévêque JL (1991) In vivo proton relaxation times analysis of the skin layers by magnetic resonance imaging. J Invest Dermatol 97(1):120–125
36. Richard S, Querleux B, Bittoun J, Jolivet O, Idy-Peretti I, de Lacharriere O, Leveque JL (1993) Characterization of the skin in vivo by high resolution magnetic resonance imaging: water behavior and age-related effects. J Invest Dermatol 100(5):705–709
37. Gniadecka M, Quistorff B (1996) Assessment of dermal water by high-frequency ultrasound: comparative studies with nuclear magnetic resonance. Br J Dermatol 135(2):218–224
38. Mirrashed F, Sharp JC (2004) In vivo quantitative analysis of the effect of hydration (immersion and Vaseline treatment) in skin layers using high-resolution MRI and magnetisation transfer contrast. Skin Res Technol 10(1):14–22
39. Querleux B, Richard S, Bittoun J, Jolivet O, Idy-Peretti I, Bazin R, Lévêque JL (1994) In vivo hydration profile in skin layers by high-resolution magnetic resonance imaging. Skin Pharmacol 7(4):210–216
40. Backhouse L, Dias M, Gorce JP, Hadgraft J, McDonald PJ, Wiechers JW (2004) GARField magnetic resonance profiling of the ingress of model skin-care product ingredients into human skin in vitro. J Pharm Sci 93(9):2274–2283
41. McDonald PJ, Akhmerov A, Backhouse LJ, Pitts S (2005) Magnetic resonance profiling of human skin in vivo using GARField magnets. J Pharm Sci 94(8): 1850–1860
42. Ciampi E, van Ginkel M, McDonald PJ, Pitts S, Bonnist EY, Singleton S, Williamson AM (2011) Dynamic in vivo mapping of model moisturiser ingress into human skin by GARfield MRI. NMR Biomed 24(2):135–144

16 Skin Capacitance Imaging

Jean Luc Lévêque

16.1 Introduction

Before the 1980s, skin was the only organ of the human body impossible to explore and study quantitatively, in vivo, in its three dimensions. Since then, two types of methods were developed and are now routinely used. The first ones are dedicated to the visualisation and characterisation of the different structural layers of the skin (ultrasounds, magnetic resonant imaging, confocal microscopy, optical coherent tomography, multiphoton imaging). These are based on the recording of either vertical or horizontal virtual sections of the different skin layers. Other types of methods are dedicated to skin surface imaging. Skin colour, temperature, micro-relief and blood flow can now be recorded, pixel by pixel, with a high 2D resolution.

The present method, capacitance imaging (CI), belongs to the latter category. CI supplies high-resolution images of the skin capacitance [1, 2]. These images hold information simultaneously on two characteristics of the skin surface: the capacitance or hydration map of the area under examination and its micro-topography. This chapter deals with the presentation of various examples of application of CI in skin research. The functioning principle of CI will be first be exposed.

J.L. Lévêque
29 Avenue de Wagram, 75017 Paris, France
e-mail: jll-skindata@wanadoo.fr

16.1.1 Functioning Principle

CI is based on the use of a 1.8×1.28 cm^2 silicon chip onto which 92,160 micro-capacitors are distributed every 50 μm [2, 3]. The silicon plates are covered by a thin layer of silicon oxide for protection. Applied onto the skin, each micro-capacitor measures skin capacitance at a given point, forming all together a capacitance image of the skin surface with a 2D spatial resolution of 50 μm. Each sensor cell contains an active capacitance feedback circuit whose effective feedback is modulated by the capacitance of the skin in contact with the measuring window. This electronic circuit, for each pixel, is integrated on the same board. The signal is coded in 256 grey levels with the darker pixels representing the highest capacitance (or hydration) and vice versa, white ones the lowest. The device for skin recording is called "SkinChip®" (L'Oréal, Paris). It can be plugged to the USB port of any computer equipped with a dedicated software, Windows®/Microsoft compatible. When the measuring window is applied onto the skin surface for 5 s, the corresponding capacitance (or hydration) map of the skin is displayed as image in less than 1 s and is ready to be stored. In another mice click, different parameters related to the investigated skin area (see later) are computed and displayed through an Excel sheet.

E. Berardesca et al. (eds.), *Non Invasive Diagnostic Techniques in Clinical Dermatology*,
DOI 10.1007/978-3-642-32109-2_16, © Springer Berlin Heidelberg 2014

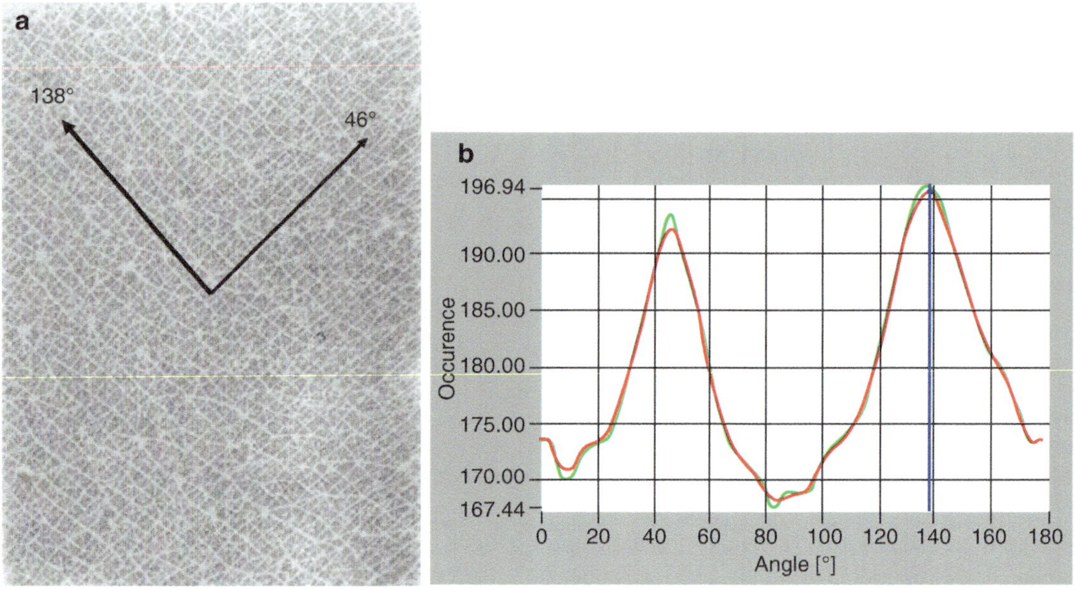

Fig. 16.1 Main directions of the primary lines of the ventral forearm, as they automatically appear according to the software dedicated to the skin micro-relief. (**a**) Main directions, (**b**) automatic detection of the angles relative to the transversal axis of the forearm axis

16.1.2 Image Analysis Software

Beside the basic software corresponding to acquisition of images, different software were developed for getting some objective characteristic parameters of the skin surface.

Skin surface hydration relates to the "mean grey level" (MGL) corresponding to the mean of the capacitance signals delivered by each pixel. MGL is automatically displayed as well as the standard deviation of the grey levels histogram which can also be used as objective information of the grey-level unevenness of the image.

For better assessing evenness of an image (in terms of grey levels), the latter is divided in adjoining overlapping 9 × 9 or 30 × 30 pixel areas. Then, variance of MGL of these sub-images is displayed. This variance represents another type of unevenness of the global skin image.

Main orientations of the primary lines were also determined through a three-step process: There is first a preconditioning of the image where the background in homogeneity of the images is corrected. Then, the 256-grey-level image was reduced to a 5-grey-level image. Finally co-occurrence matrices were calculated at different angles, and the maxima corresponding to the main orientations of the lines are displayed (see Fig. 16.1).

Primary line density is calculated from the preconditioned image. Following a thresholding of the image, a thinning process is applied and the "corners", representing the crossings of the primary lines, are detected and displayed. Their density is proportional to the density of the lines, an important characteristic of the skin surface. Another skin parameter, the furrow edge length (FEL), may also be obtained. It correlates with the corner density (CD).

Finally, four types of information can be automatically extracted from the images: density of the lines (CD), main orientations in degrees, mean hydration (MGL) and hydration evenness. The questioning of the parameter of choice for measuring skin hydration will be rediscussed later (see next section).

16.1.3 Hydration Measurement

The excellent correlation obtained between MGL given by SkinChip and the corresponding

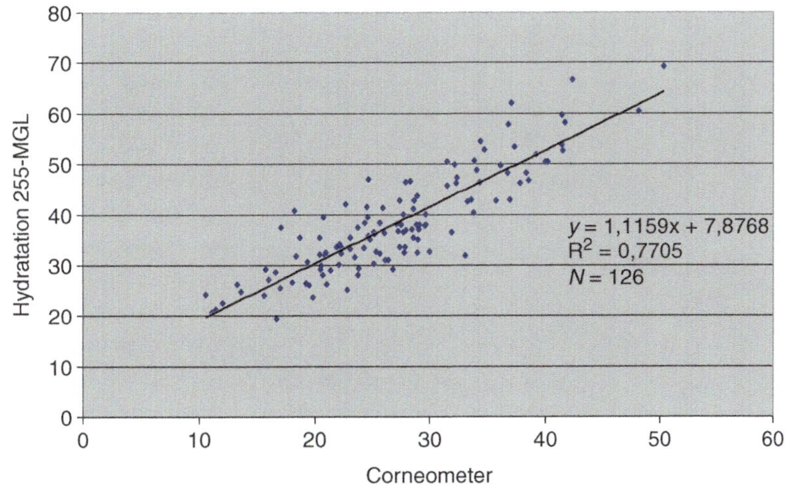

Fig. 16.2 Linear regression between the Corneometer values and the hydration or capacitance parameter (255-MTH20) given by SknChip

"Corneometer" (C+K electronic, Cologne, Germany) values has been published since long [3]. This linear correlation has a negative slope due to the coding scale (0 corresponding to black-hydrated pixels and 255 to the white-dry ones). For obtaining a direct correlation, 255-MTH20 must be plotted. This parameter then directly represents skin capacitance or hydration. Figure 16.2 illustrates the direct correlation recently obtained on 120 measurements [4]. This excellent correlation is not surprising because both techniques and its measure in fact the same parameter corresponding to two different skin areas: 2.3 cm^2 for CI versus 0.5 cm^2 for Corneometer.

Because of the skin relief (lines, pores) and presence of hair, there is a non-negligible part of the skin under measurement which has a very poor contact with the measuring window. These areas therefore tend to alter the actual skin capacitance measurement. To overcome this difficulty, it was proposed to systematically suppress 20 % of the clearer pixels, those corresponding to the noncontact areas of the measuring window of SkinChip [3]. This new parameter, called MTH20, for mean thresholded histogram at 20 %, is also well correlated with Corneometer but has a higher range of variation, meaning a better sensitivity in the hydration measurements.

CI is much more than an alternative or an equivalent method to the basic capacitance or conductance methods since affording the measurement of the actual capacitance of the skin surface. Another important difference with the other capacitance or conductance methods lies in the possibility of extracting the grey-level histogram which illustrates the evenness of skin hydration (see section "ageing skin") or to specifically measure sweating. Moreover, as it will be shown later, CI gives immediate information about skin micro-relief.

16.2 Skin Surface Investigation: Hydration

16.2.1 Sweating

At the onset of sweating process, which is normally imperceptible, black droplets appear on the capacitance images (Fig. 16.3). Progressively, these droplets become larger and larger and finally merge in a continuous black area (Grove G, 2009) [5]. Contribution of sweating to the skin grey-level histogram is easy to observe because it corresponds to the lowest values of the histogram (areas of very high capacitance). Sweat amount can therefore be easily obtained by isolating and then measuring this part of the histogram by thresholding (Grove G, unpublished data) [5].

By juxtaposing sequentially different SkinChip images, recorded as a function of time after the onset of sweating, Grove produced a picture very informative and useful in the study of a pathological disease [6].

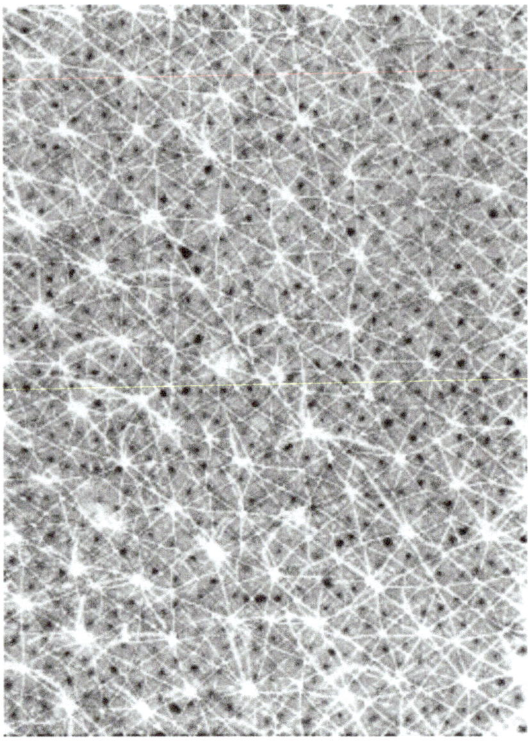

Fig. 16.3 Onset of the sweating process on ventral forearm

16.2.2 Effect of Cosmetics

By using CI, increase of skin surface hydration is quite easy to "see" and record following treatment with a topical product. As mentioned above, quantitative results appear on line with those obtained with Corneometer or other capacitance or conductance devices. However, as compared to these basic techniques, CI offers a very important information with regard the repartition of hydration onto the skin surface.

Figures 16.4a, b give an example of a negative result obtained on the forearm which was treated for two weeks by a cosmetic product. In this capacitance image, the hydrated areas were coded in green for well distinguishing the locations of dryness (white areas). Following treatment, despite a decrease in the total dry area, large areas of dryness still remained. An efficient treatment would have suppressed the large white areas corresponding to a given level of dryness.

Examination of the skin hydration map is also very advantageous, mostly for studying both evenness of skin hydration in aged people and

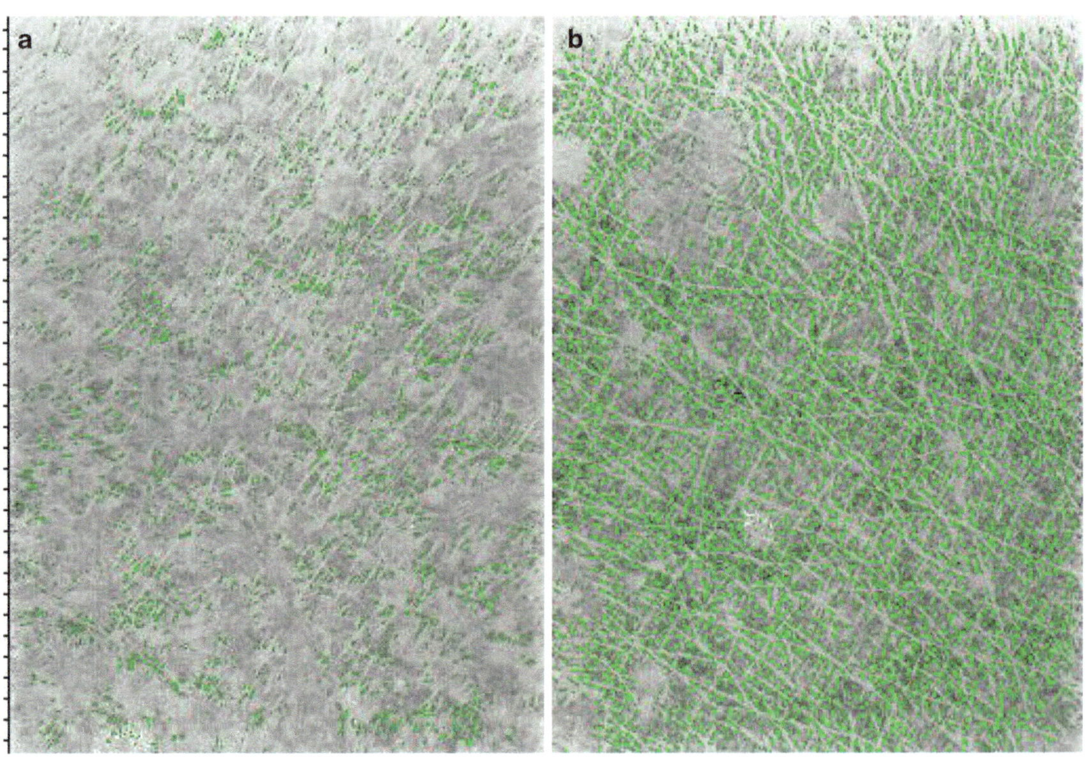

Fig. 16.4 Example of a bad result obtained after a 21-day treatment of the forearm skin by a moisturiser. (**a**) Capacitance image of a dry forearm skin at T0. (**b**) Result obtained after treatment at T21. *Green areas* correspond to the more hydrated areas

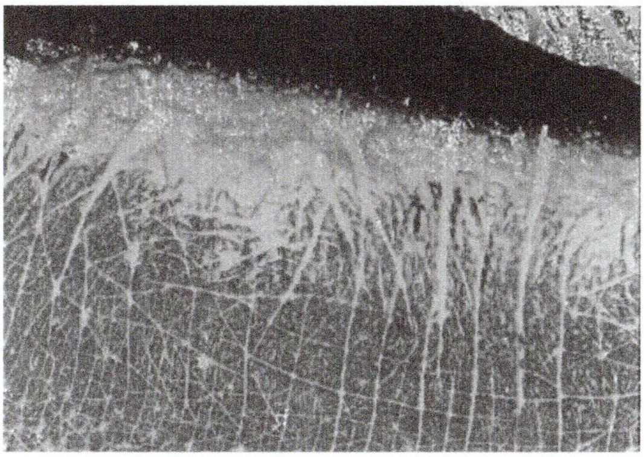

Fig. 16.5 Normal hydration state of the lower lip. The most external part of the lip is darker (more hydrated) than the internal part (See Caisey et al. [8])

depicting the hydration state at the surface of various skin lesions (see later).

16.2.3 Hydration of Lips

CI is particularly adapted to the visualisation and study of the surface of the lip. Numerous studies on lips were carried out by using CI such as their micro-reliefs and their hydration, impacts of age, menopause, treatment by hormonal replacement therapy and moisturising effect of some lipstick… It was shown that the upper lip is more hydrated than the lower one and that MGL, which measure lip capacitance, and clinical score of lip dryness [6] appeared not correlated. In this study, the surface pattern was classified into three main groups, irrespective of age. In another study [7], ageing showed no influence on lips capacitance, whereas the difference in surface capacitance between upper and lower lip was confirmed.

A surprising result concerns the unexpected distribution of stratum corneum hydration between the internal (drier) and external part (more hydrated) of the lip surface (see Fig. 16.5). This result is explained by the structural differences of the living epidermis between these two close areas [8].

16.2.4 Skin Photo-Ageing

The effect of chronic daylight exposures to the skin surface of the chest was studied on 64 women [9]. The severity of photo-ageing was clinically scored on different delineated skin areas, and photographs and capacitance images were recorded. Same were obtained from adjacent light-protected areas, as controls. From the capacitance images, skin hydration parameters, evenness of hydration and density of lines were recorded.

Results show that the decrease in the density of micro-relief lines is related to age (see later) and not to photo-ageing. Mean hydration of the skin is independent of both age and light exposure. On the opposite, evenness of skin surface hydration appears markedly linked to the severity of photo-ageing, independently of age. Uneven hydration is due to juxtaposition of both hydrated and dry tiny areas. Hydrated areas often correspond to pigmented or inflamed areas and the driest to hyperkeratotic and/or desquamative ones. It is worth noting that hydration of pigmented areas is sometime very low, as in lentigines.

Unevenness of pigmentation is one of the hallmarks of photo-aged skin. This work demonstrates that the unevenness of the hydration of the skin surface is another very relevant feature of photo-ageing. Interestingly, these two characteristics are not linked.

16.2.5 Skin Lesions

The capacitance of various skin lesions was investigated during clinical examination [9]. In overall, psoriatic lesions usually show a lowered capacitance, admixed with foci of moderately higher capacitance [10]. Some other lesions present higher capacitance corresponding to inflamed areas. It is worth noting that sweating is often markedly impaired within the lesions. As said above, lentigines always displayed a lowered capacitance. Nevi are different, appearing with either a dark or white appearance, according to their inflammatory status. Pityriasis versicolor lesions are clearly visible on capacitance image (although often invisible by naked eye for the smaller ones) because their anhydrotic pattern makes their surface drier than the surrounding skin [11]. Characteristics of other lesions (viral warts, pigmented tumours) were investigated and published [12]. The inflammatory and noninflammatory types of melanocytic nevi and seborrheic keratosis are clearly distinguished [13].

An acne papule is revealed by its target-like black area with a white central small circle representing the dry comedone. Follicular hyperkeratosis is represented by white areas of different sizes. Acne is a typical condition where skin capacitance imaging can highlight the heterogenous patchwork of the electrical properties of skin [14].

Systematic study of the different types of lesion (as it has been done for psoriatic lesions) is still to be undertaken to more deeply understand and interpret these features [14].

16.2.6 Surfactants

Effect of mild surfactants on SC was studied by Uhoda et al., taking advantage of the great sensitivity of CI to detect presence (or not) of water on a tiny area [15]. Indeed, some other methods give averaged information on a large surface, blurring focal and minute phenomena. This author demonstrated that irritation by surfactants is a two-phase process: at the onset of the process, darker areas corresponding to over-hydrated and swollen corneocytes that are present. This phase is followed by a rapid drying out process, and 2 days later, the area becomes of a white or very clear image, corresponding to dryness.

CI appears a method of choice for studying early signs of irritation in vivo on tiny areas.

16.2.7 Exposure to Cold and Heat

Exposing skin to high cold or heat produces an inflammation which confers to stratum corneum (SC) a dry condition in the following days or weeks. Figure 16.6 represents the consequence of a weak burn on the top of hand 1 month later. The burned area in white is clearly visible although neither scar nor desquamation can be distinguished by eye. Consequence of such a weak burn is a totally dry SC. This result is surprising since 1 month would normally ensure a complete renewal of SC. It was previously shown that the same phenomenon occurs, following a treatment of the skin surface by a cold liquid gas [3].

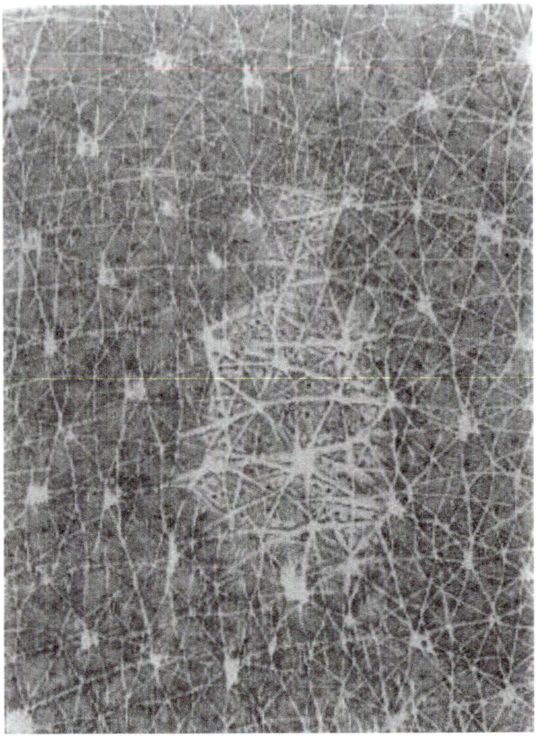

Fig. 16.6 Capacitance image of the skin surface of the top of hand 1 month after a weak burn. Skin surface exhibits a well-defined dry stratum corneum. This scar is quite invisible by eye but appears a little bit rough at palpation

The systematic follow-up of these phenomena versus time would be useful for understanding how these thermal aggressive treatments modify the epidermal turn over.

16.3 Skin Surface Investigation: Micro-relief

16.3.1 Cosmetics

Numerous cosmetics are dedicated to aged skin. According to claims usually made, these products are supposed conferring skin to a beautiful aspect, making it more radiant, firmer, smoother, etc. Most of these skin attributes depend, at least partly, on the quality of skin micro-relief which is hidden when the skin gets very dry and disorganised in aged people. Measurement of skin micro-relief becomes therefore of great importance for assessing the efficacy of a product. Techniques, like replicas or fringe projection, are adequate for quantifying micro-relief.

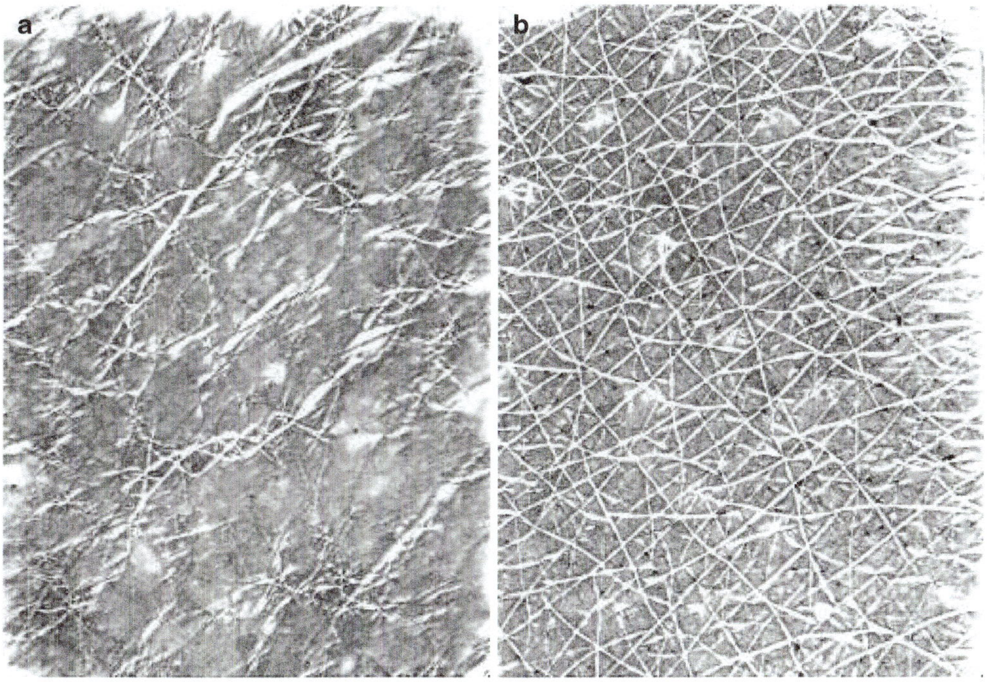

Fig. 16.7 Example of a good result obtained after a two-week treatment of the top hand skin by a cosmetic. (**a**) At T0 skin micro-relief is disorganised, hardly visible. (**b**) Two weeks later, skin micro-relief appears much more better, quite normal

They are however cumbersome, time-consuming and subjected to artefacts. Moreover, they are hardly usable in routine studies, when carried out on numerous volunteers. These drawbacks are absent with CI. In a few seconds, CI allows evaluating if micro-relief has (or not) changed after a treatment, when chronically applied. Figure 16.7b is an illustration of an efficient treatment.

16.3.2 Skin Ageing

Influence of age on the skin micro-relief was quantitatively described since long by Corcuff [16]. In a few words, density of the primary lines decreases and their mean depth increases. More importantly, there are two main orientations at 90° for the lines, but the percentage of people showing these two axes decreases versus age. Some anisotropy progressively takes place in the orientations of lines in elderly people.

Conclusions relative to the density of the lines and their relative orientations at 90° were confirmed by Berardesca et al. on two groups of women of different ages [17]. The same work shows that the “corner density”, or CD, which represents the density of lines is higher on the ventral forearm than on the dorsal site. This result was confirmed by Diridollou et al. on different ethnies [18]. CD is an intrinsic skin parameter varying on the ventral forearm, values ranging 400 (young) to 250 (aged) per cm^2 [5].

16.3.3 Skin Deformation

Under deformation, skin micro-relief parameters (density, mean depth, orientation of the primary lines) are modified. This was described by Corcuff et al. in the case of the forearm extension for three groups of people of different ages [19]. It was shown that great differences occur between the groups of young and very aged people. For elderly women, the single main axis progressively rotates to parallel the forearm axis.

The same type of study was carried out by means of CI in the case of the forearm rotation, compression and extension (Lévêque JL, 2011). As in the

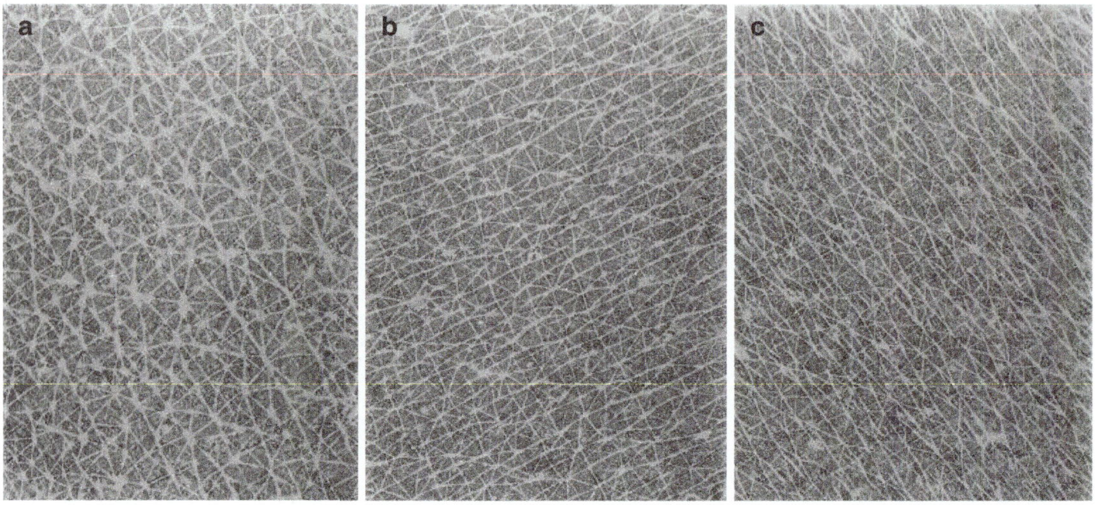

Fig. 16.8 Effect of a rotation of the forearm of an aged person on the micro-relief of the ventral skin. (**a**) Normal, (**b**) rotation on the right, (**c**) rotation on the left). In the normal position, the two main orientations of the skin micro-relief are equivalent (31° and 122°). In the b situation, orientation at 122° is preponderant. In the c situation, orientation at 31° is preponderant

Corcuff's study, the skin of elderly people exhibits great changes versus adult ones, particularly the orientation of lines (see Fig. 16.8)

16.3.4 Lips

As said above, lip's micro-relief was investigated and classified by different teams [12, 13]. CI was used to revisit the patterns of lips: Our own observations confirmed that the major patterns correspond to three main types of network where lines cross each other, either on the whole lip or on their extremities [6].

Lip's lines have a function of preservation of the epidermal structure when lips are stretched. The "opening" of the lines after extension is easy to observe by CI (Fig. 16.9)

16.3.5 Scars

CI also appears an easy parameter for routine observation of skin scars linked to previous wounds or surgery. Striae distensae also exhibit the same type of structural abnormality. In these skin tension areas, the isotropic orientation of the collagen bundles is altered, and a preferential direction of collagen may be viewed. Figure 16.10 shows the transversal organisation of collagen bundles along the scar of the calf skin, after wound.

Fig. 16.9 Opening of the lip lines after transversal extension (**a**) normal, (**b**) stretched)

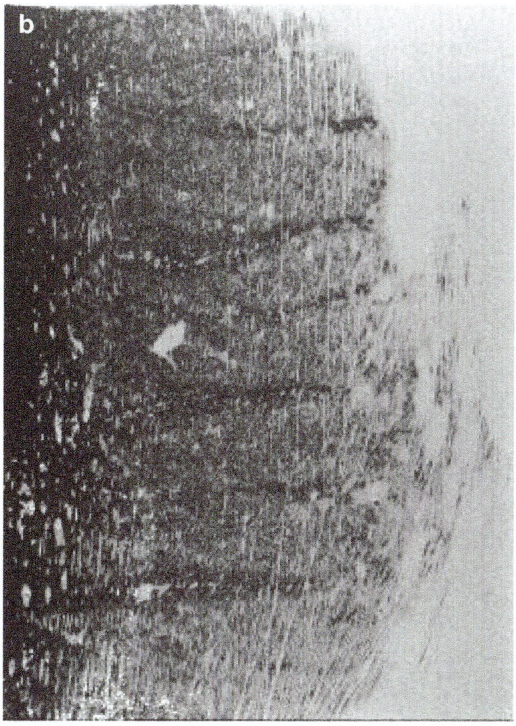

Fig. 16.9 (continued)

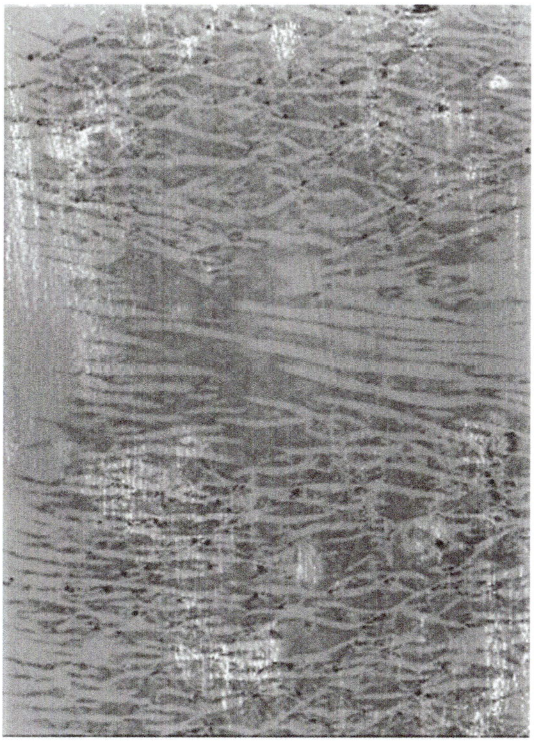

Fig. 16.10 Capacitance image of an ancient scar of the calf skin. On this image, the horizontal orientation of the collagen bundles is clearly visible

Conclusion

A first great asset of CI is its great simplicity of use: CI can be used routinely in experiments including large cohorts of volunteers. Obtaining, storing an image takes less than second. Moreover, this image is immediately scored through objective parameters like density of lines, orientations and hydration. Other software can also supply information on evenness of hydration, a characteristic of photo-aged skin.

In dermatology, CI brings insights into physiopathological disorders, revealing some unexpected features. Moreover, some details that are invisible by the naked eye may be recorded.

Cosmetology has been, up today, the main field of CI use. The greatest advantage of this technology, for studying skin surface hydration, as compared to other capacitance and conductance methods, lies in its capacity to afford in real time an image which, in turn, yields quantitative data on both hydration of the skin andits micro-relief.

References

1. Piérard GE, Lévêque JL (2004) What is SkinChip? From silicon image sensor technology to SkinChip. Dermatology 208(4):291–292
2. Lévêque JL, Querleux B (2003) SkinChip, a new tool for investigating the skin surface in vivo. Skin Res Technol 9(4):343–347
3. Batisse D, Giron F, Lévêque JL (2006) Capacitance imaging of the skin surface. Skin Res Technol 12(2):99–104
4. Lévêque JL, Laquieze S, Piquet A, Bazin R (2010) Complémentarité de deux methods d'imagerie de surface pour l'étude de la peau sèche et de ses traitements. Congrès SF2IC, Bordeaux
5. Lévêque JL (2006) Capacitance imaging of the skin surface. In: Wilhelm KP, Berardesca E, Elsner P, Maibach HI (eds) Bioengineering and the skin. Skin imaging and analysis. Informa Healthcare, New York, pp 331–337
6. Leveque JL, Gubanova E (2004) Influence of age on lips and perioral skin. Dermatology 208:307–313
7. Caisey L, Gubanova E, Camus C, Lapatina N, Smetnik V, Lévêque JL (2008) Influence of age and hormonal therapy on the functional properties of lips. Skin Res Technol 14(2):220–225
8. Caisey L, Gubanova E, Baras D, Lévêque JL (2008) Unexpected distribution of surface hydration level of the lip. J Eur Acad Dermatol Venereol 22(10): 1159–1162

9. Bazin R, Laquieze S, Rosillo A, Lévêque JL (2010) Photoaging of the chest analyzed by capacitance imaging. Skin Res Technol 16(1):23–29
10. Xhauflaire-Uhoda E, Pierard-Franchimont C, Pierard GE (2006) Skin capacitance mapping of psoriasis. J Eur Acad Dermatol Venereol 20:1261–1265
11. Uhoda E, Pierard-Franchimont C, Pierard GE (2005) Pityriasis versicolor anhydrotique. Dermatol Actual 89:16–17
12. Lévêque JL, Xhauflaire-Uhoda E, Piérard GE (2006) Skin capacitance imaging, a new technique for investigating the skin surface. Eur J Dermatol 16(5):500–506
13. Xhauflaire-Uhoda E, Piérard GE (2006) Contrasted skin capacitance imaging of seborrheic keratoses and melanocytic nevi. Dermatology 212(4):394–397
14. Xhauflaire-Uhoda E, Pierard GE (2007) Skin capacitance imaging of acne lesions. Skin Res Technol 13:9–12
15. Uhoda E, Lévêque JL, Pierard GE (2005) Silicon image sensor technology for in vivo detection of surfactant-induced corneocyte swelling and drying. Dermatology 210(3):184–188
16. Corcuff P, de Rigal J, Makki S, Lévêque JL, Agache P (1983) Skin relief and aging. J Soc Cosmet Chem 34: 177–190
17. Bérardesca E, Primavera G, Zahouani H, Lévêque JL (2005) Capacitance imaging of the skin: new parameters for characterizing the skin surface texture. Bioengineering and the Skin, Philadelphia
18. Diridollou S, de Rigal J, Querleux B, Leroy F, Holloway Barbosa V (2007) Comparative study of the hydration of the stratum corneum between four ethnic groups: influence of age. Int J Dermatol 46(Suppl 1): 11–14
19. Corcuff P, de Lacharriere O, Leveque JL (1991) Extension-induced changes in the microrelief of the human volar forearm with age. J Gerontol 46(6): M223–M227

17 Tissue Viability Imaging for Assessment of Skin Erythema and Blanching

Gert Nilsson

17.1 Introduction

Tissue Viability Imaging (TiVi) is an emerging *polarization spectroscopy camera* technology that takes subjectivity out of skin testing by making it possible to quantify what is generally observed by the naked eye. In addition to mapping skin *erythema* and *blanching*, optional toolboxes provide for in vivo objective and quantitative assessment of wrinkle appearance, pigmentation, surface smoothness, and other skin parameters of interest. TiVi was designed with ease of use in mind and productive in gaining investigator-independent data from large panels of test subjects. By reducing data from a sequence of images to curves and indexes, TiVi is a versatile tool in skin testing including evaluation of skin care products, assessment of pharmaceuticals, and in grading the performance of sensitive skin to various challenges. When investigating the skin microcirculation, TiVi measures – in contrast to laser Doppler technology – only the concentration of red blood cells (RBCs) in tissue and not their velocity, thereby making the images closer related to what is observed by the naked eye. Furthermore, TiVi is not sensitive to movement artifacts and the distance to the object, but the amount of skin pigmentation modulates the output signal. This disadvantage is eliminated when recording only changes in skin RBC concentration as response to, e.g., topical application of a vasoactive agent, since these changes generally appear on a timescale much shorter than that at which the melanin content of the skin is altered.

17.2 Operating Principle

Since advanced digital cameras were introduced in the 1990s, these devices have developed quickly and are increasingly used as part of advanced imaging systems in a variety of medical and other applications. To reduce the adverse effects of specular reflections of the illuminating light in the surface of the object, a cross-polarization filter technique is frequently used [1].

In TiVi, a high-performance digital camera is attached to a table-mounted stand (Fig. 17.1). The Illuminator comprising 96 white light-emitting diodes (LEDs) is attached to the camera lens, thereby providing a uniform and stable illumination of the skin surface under investigation. Separate polarizing filters in front of the LEDs and the camera lens, respectively, form integral parts of the Illuminator. The system can be set to operate in cross- or co-polarized mode by turning the part of the Illuminator that hosts the LEDs. By setting the system to cross-polarized mode, surface specular reflections are effectively suppressed and depth sensitivity enhanced, while setting the system to co-polarized mode enhances surface structures in the photo.

G. Nilsson, PhD
Department of Research and Development,
WheelsBridge AB, Lövsbergsvägen 13,
Linköping 58937, Sweden
e-mail: gerni@glocalnet.net,
http://www.wheelsbridge.se

E. Berardesca et al. (eds.), *Non Invasive Diagnostic Techniques in Clinical Dermatology*,
DOI 10.1007/978-3-642-32109-2_17, © Springer Berlin Heidelberg 2014

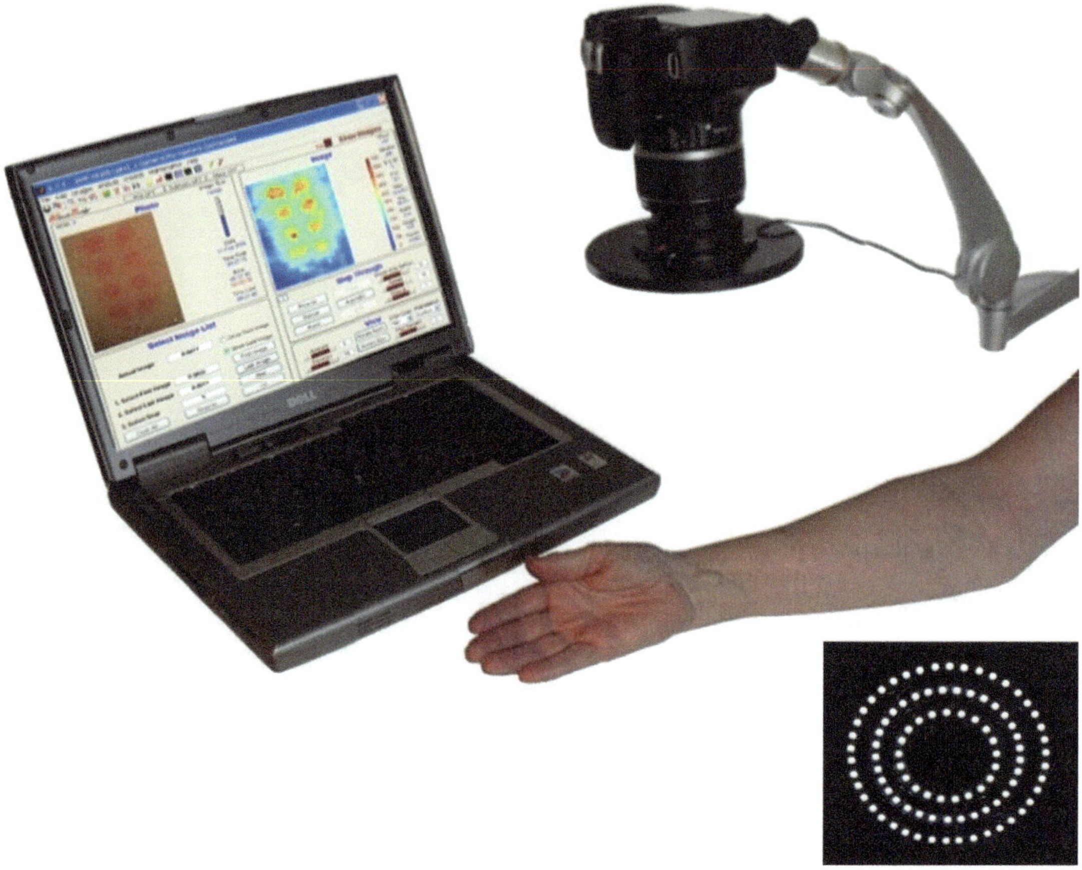

Fig. 17.1 The Tissue Viability Imager. *Lower right*: Bottom view of Illuminator

When reaching the target, the white illuminating light is partly reflected by the upper layer of the skin and partly diffusely scattered in the deeper dermal layers where the microvascular network is located (Fig. 17.2).

Most of the directly reflected light preserves its state of linear polarization, while the light diffusely scattered in the tissue successively becomes randomly polarized. The backscattered linearly polarized light directed towards the detector is effectively blocked by the filter in front of the camera if the polarization direction of this filter is perpendicular to that of the linearly polarized illuminating light (cross-polarized mode). A portion of the randomly polarized backscattered light passes through this filter and reaches the detector. This arrangement gives the impression that the camera can see through the top layer of the skin and probe the microcirculation in the deeper dermal layers. The green component of the light reaching the detector is attenuated due to a high absorption in RBCs, while the red component is virtually unaltered because of its low absorption in the red blood cells. Surrounding tissue absorbs green and red light to approximately the same amount (Fig. 17.3).

The TiVi system takes advantage of this wavelength dependence in red blood cell absorption. By first separating the color matrixes and then applying an algorithm in which the value of each picture element in the green color matrix is subtracted from the corresponding value in the red color matrix, an output matrix representing the local RBC concentration is generated. After color-coding the elements of this output matrix, the Tissue Viability Image is generated.

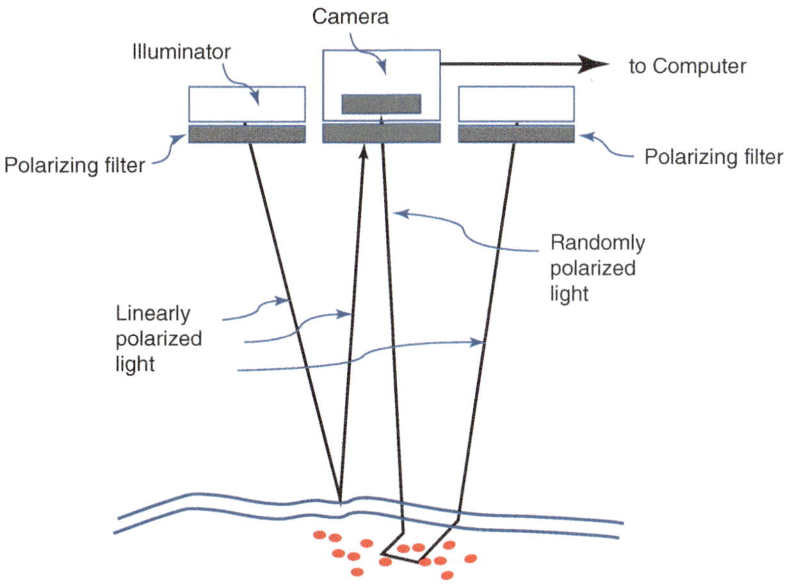

Fig. 17.2 Operating principle of TiVi

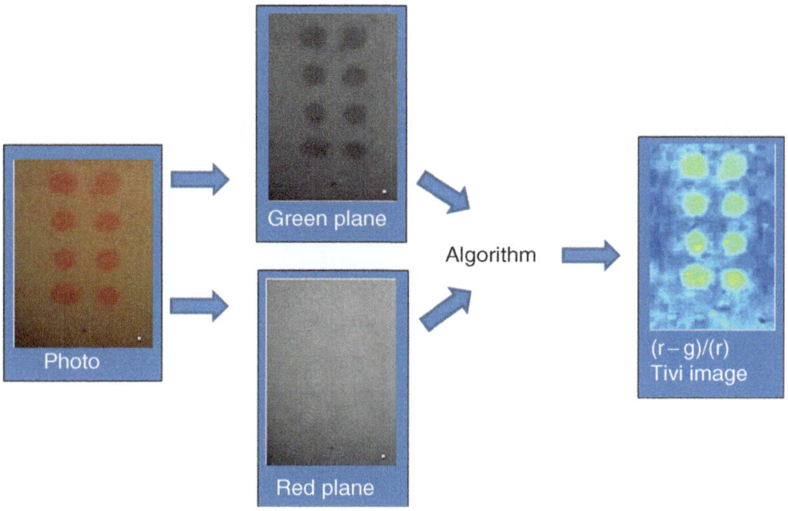

Fig. 17.3 The true color photo is split up in its *red* and *green plane*. The *erythema* in the photo produced by topical application of methyl nicotinate results in a dark area in the *green plane* due to high absorption of *green light* in the RBCs, while the *red plane* is virtually unaffected

17.3 Theory

A two-layer model is used to theoretically analyze the spectral content of the backscattered light at different skin RBC concentrations. The thin model top layer comprises the epidermis (approximately 50–100 μm on forearm skin) which primarily acts as an absorption filter at the actual wavelengths. The model bottom layer is infinitely deep and is composed of a uniform mixture of tissue and RBCs at different concentrations. The amount of diffusely backscattered light from the bottom layer can be described by Kubelka-Munk theory [2].

The wavelength-dependent intensity of backscattered light from the skin with a polarization perpendicular to that of the incident-polarized light is described by equation [1]:

$$I_{\text{per}}(\Delta\lambda) \propto k_0(\Delta\lambda) I_0 T_{\text{epid}}(\Delta\lambda) R_d(\Delta\lambda) \qquad (17.1)$$

where $k_0(\Delta\lambda)$ represents the fraction of the total light intensity I_0 within the wavelength interval $\Delta\lambda$, incident on the skin, $T_{\text{epid}}(\Delta\lambda)$ represents the transmission properties of the epidermal layer (model top layer), and $R_d(\Delta\lambda)$ represents the fraction of diffusely reflected light within the wavelength interval $\Delta\lambda_x$ from dermis (model bottom layer). The *Tissue Viability index* ($\mathrm{TiVi}_{\mathrm{index}}$) representing the local RBC concentration is defined as:

$$TiVi_{index} = k_{gain}\left(\frac{I_{per}(\Delta\lambda_r) - k_1 I_{per}(\Delta\lambda_g)}{I_{per}(\Delta\lambda_r)}\right) \quad (17.2)$$

where $I_{\mathrm{per}}(\Delta\lambda_r)$ and $I_{\mathrm{per}}(\Delta\lambda_g)$ represent the wavelength-dependent intensity of backscattered light in the red and green wavelength regions, respectively, with a polarization perpendicular to that of the incident-polarized light, k_1 is a constant that can be fitted for best algorithm performance, and k_{gain} is the gain factor. For the digital camera employed, $\Delta\lambda_r$ and $\Delta\lambda_g$ are typically in the order of 100 nm. Inserting Eq. 17.1 into Eq. 17.2 cancelling the I_0 factor and setting $k = k_1 k_0(\Delta\lambda_g)/k_0(\Delta\lambda_r)$, the algorithm expands to:

$$TiVi_{index} = k_{gain}\left(\frac{T_{epid}(\Delta\lambda_r)R_d(\Delta\lambda_r) - kT_{epid}(\Delta\lambda_g)R_d(\Delta\lambda_g)}{T_{epid}(\Delta\lambda_r)R_d(\Delta\lambda_r)}\right) \quad (17.3)$$

By this arrangement $\mathrm{TiVi}_{\mathrm{index}}$ becomes independent of the total light intensity I_0 and relates only to the diffusely backscattered light from the bottom layer, the amount of which to some degree is modulated by the absorption in the model top layer. Since the numerator approaches zero for bloodless tissue, $\mathrm{TiVi}_{\mathrm{index}}$ becomes close to zero for low RBC concentrations. The exact tuning is balanced by fitting the constant k. First considering a thin epidermal layer with low melanin content, the value of T_{epid} can be set to 1 and Eq. 17.3 is reduced to:

$$TiVi_{index} = k_{gain}\left(\frac{R_d(\Delta\lambda_r) - kR_d(\Delta\lambda_g)}{R_d(\Delta\lambda_r)}\right) \quad (17.4)$$

In Eq. 17.4, $\mathrm{TiVi}_{\mathrm{index}}$ depends only of the fraction of backscattered light $R_d(\Delta\lambda_r)$ and $R_d(\Delta\lambda_g)$ within the red and green wavelength intervals. These fractions can be linked to the RBC concentration by use of Kubelka-Munk [2] theory and its modification [3] for an infinite uniform tissue layer (model bottom layer):

$$R_d(\Delta\lambda) = 1 + \frac{K(\Delta\lambda)}{S(\Delta\lambda)} - \left(\frac{K(\Delta\lambda)^2}{S(\Delta\lambda)^2} + 2\frac{K(\Delta\lambda)}{S(\Delta\lambda)}\right)^{\frac{1}{2}} \quad (17.5)$$

where K and S are Kubelka-Munk absorption and scattering coefficients. Using a derivation from diffusion theory [3, 4], and assuming isotropic scattering, it can be shown that:

$$\frac{\mu_a(\Delta\lambda)}{\mu_s(\Delta\lambda)} = \frac{3K(\Delta\lambda)}{8S(\Delta\lambda)} \quad (17.6)$$

where $\mu_a(\Delta\lambda)$ is the absorption coefficient and $\mu_s(\Delta\lambda)$ is the scattering coefficient. This implies that the diffusely reflected light can be expressed as:

$$R_d(\Delta\lambda) = 1 + \frac{8\mu_a(\Delta\lambda)}{3\mu_s(\Delta\lambda)} - \left(\frac{(8\mu_a(\Delta\lambda))^2}{(3\mu_s(\Delta\lambda))^2} + 2\frac{8\mu_a(\Delta\lambda)}{3\mu_s(\Delta\lambda)}\right)^{\frac{1}{2}} \quad (17.7)$$

The total absorption and scattering coefficients can each be considered to be composed of two parts – one that relates to the RBC ($\mu_{a\mathrm{RBC}}(\Delta\lambda)$ and $\mu_{s\mathrm{RBC}}(\Delta\lambda)$) and one that relates to the remaining tissue ($\mu_{a\mathrm{TISSUE}}(\Delta\lambda)$ and $\mu_{s\mathrm{TISSUE}}(\Delta\lambda)$). The total tissue absorption $\mu_a(\lambda)$ and scattering coefficient $\mu_s(\lambda)$ can be regarded as a linear combination of these two parts [5]:

$$\mu_a(\lambda) = RBC_f\,\mu_{aRBC}(\lambda) + (1 - RBC_f)\,\mu_{aTISSUE}(\lambda) \quad (17.8)$$

$$\mu_s(\lambda) = RBC_f\,\mu_{sRBC}(\lambda) + (1 - RBC_f)\,\mu_{sTISSUE}(\lambda) \quad (17.9)$$

where RBC_f represents the fraction of RBCs occupying the tissue volume. Equations 17.4, 17.5, 17.6, 17.7, 17.8, and 17.9 thus relate the $TiVi_{index}$ and fractions of backscattered light ($R_d(\Delta\lambda_r)$ and $R_d(\Delta\lambda_g)$) to the RBC concentration in dermal tissue. With data from the literature [5] inserted in the equations, the theoretical relationship between tissue relative fraction of RBC and $TiVi_{index}$ is plotted for 0, 70, and 100 % oxygen saturation, respectively, in Fig. 17.4.

The influence of a thin melanin layer in the epidermis can be modeled as indicated in Eq. 17.1, by way of an absorption filter with the transmission function:

$$T_{epid}(\Delta\lambda) = e^{-2\mu_{aEPID}(\Delta\lambda)x} \tag{17.10}$$

where μ_{aEPID} is the wavelength-dependent absorption coefficient of the epidermal layer and x is the layer thickness. Inserting Eq. 17.10 into Eq. 17.3 yields:

$$TiVi_{index} = k_{gain}\frac{R_d(\Delta\lambda_r) - ke^{-2(\mu_{aEPID}(\Delta\lambda_g) - \mu_{aEPID}(\Delta\lambda_r))x}R_d(\Delta\lambda_g)}{R_d(\Delta\lambda_r)} \tag{17.11}$$

In forearm skin and in the skin of the back where most skin testing is performed, the epidermal layer thickness amounts to 50–100 μm, resulting in a numerical value in $e^{-2(\mu_{aEPID}(\Delta\lambda_g) - \mu_{aEPID}(\Delta\lambda_r))x}$ of about 0.90 to 0.95 (Caucasian skin) based on values inserted from the literature [5]. If the melanin content in the epidermal layer is higher or if melanin is present also in the subepidermal layers, a greater influence of melanin on the $TiVi_{index}$ must, however, be expected. The influence of epidermal layer thickness on the $TiVi_{index}$ is calculated from Eq. 17.11 and displayed for values of x ranging from 50 to 100 μm (Caucasian skin, oxygenated blood) in Fig. 17.5.

It can be concluded from Fig. 17.5 that a change in epidermal layer thickness (or alternatively an increase in melanin concentration) offsets the $TiVi_{index}$, while the gain factor is affected to a minor extent only. Consequently changes in RBC concentration are affected only to a minor degree by alterations in skin pigmentation for low and moderate melanin levels.

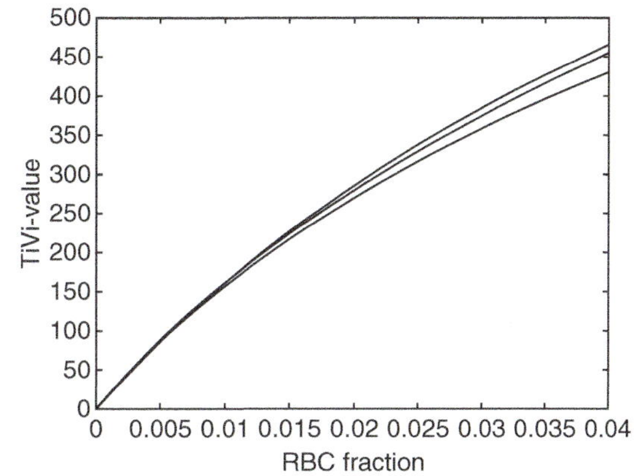

Fig. 17.4 Theoretical relationship between the RBC fraction and the $TiVi_{index}$ (0 and 100 % oxygen saturation correspond to *lower* and *upper curve*, respectively. Absorption in the epidermal layer is neglected)

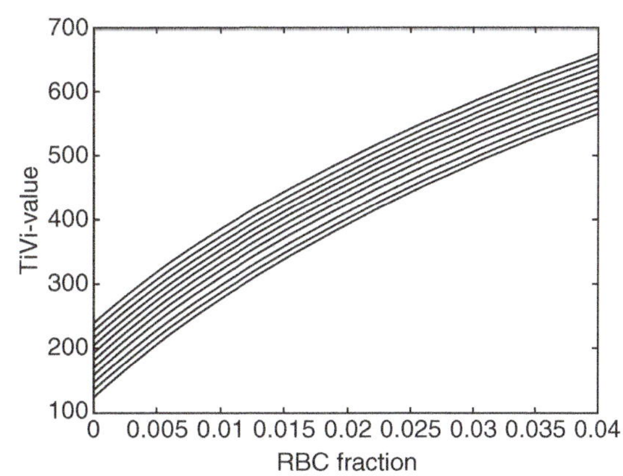

Fig. 17.5 Theoretically calculated influence of epidermal thickness (50 μm *lower curve*, 100 μm *upper curve*, Caucasian skin)

The nonlinearity in the TiVi value–RBC fraction relationship can in practice be compensated for by multiplying Eq. 17.11 by a factor raised to a power of the calculated TiVi value.

17.4 Instrumentation

Based on the theory for detection of diffusely backscattered light in tissue, a system for visualization of RBC concentration in dermal tissue was developed

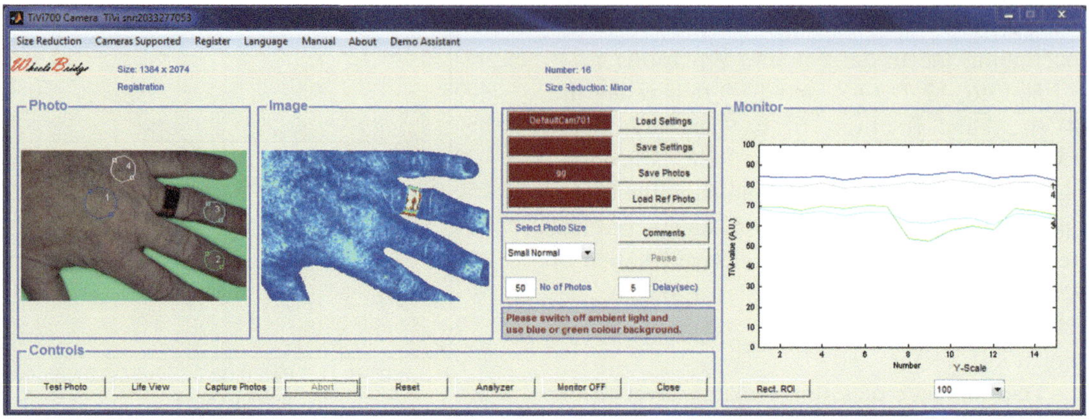

Fig. 17.6 The TiVi701 Camera software user interface

by way of polarization light spectroscopy imaging (TiVi700, WheelsBridge AB, Linköping, Sweden [6]). The system comprises an 18 megapixel digital camera (Canon EOS 550D, Canon Inc., Japan) equipped with polarization filters and controlled remotely by a USB-connected interface with a portable PC (Dell Latitude (64 bit), E5510, Dell Inc., Round Rock, TX, USA). Camera parameter settings, image acquisition and processing, and generation of result diagrams are all accomplished by the dedicated TiVi700 system software based on MATLAB® (MATLAB, MathWorks Inc., Natick, MA, USA). When connected to the computer, photo capturing is controlled entirely from the computer keyboard through the TiVi701 camera software (Fig. 17.6). The Illuminator-equipped camera may be operated on battery power, allowing photos to be stored temporarily in its internal memory while the camera is disconnected from the computer.

Parameter settings for a photo-capturing sequence include photo size, number of photos, and time interval, as well as camera lens zoom position and polarization. During a photo capture session, the TiVi701 software displays each successive photo along with its corresponding TiVi image on the computer screen. If regions of interest (ROI) are drawn in the photo, the average TiVi values within these ROI can be successively displayed in the monitor panel so that dynamic skin reactions can be observed in real time. In many applications, a series of photos may be taken in a timed sequence, allowing evaluation of dynamic skin reactions to topical or systemic administrations of vasoactive agents that induce either vasodilatation or vasoconstriction of skin microvasculature. In the TiVi700 Analyzer software (Fig. 17.7), integrated software wizards and statistical measures may be used subsequently to visualize and quantify *erythema* and *blanching* reactions. Assessment of these physiologic processes relies on the conversion of TiVi image numeric data to curves and indexes. In comparison to the conventional process of assessing a specific skin reaction by unaided visual inspection, an objective and precise TiVi analysis can be done independent of the skill of the individual investigator. To use the integrated wizards effectively, the object in a sequence of photos must appear in the same position in all photos. When photos are recorded at different points in time, the first photo in a sequence can be used as the *reference photo*. When successive photos are to be captured, preliminary test photos are displayed as superimposed images on the *reference photo*, allowing the subject to be repositioned continuously until an acceptable overlap of the object in the *reference photo* occurs. Further alignment of the object in a photo sequence can be accomplished through the use of an integrated software preprocessor prior to execution of the final image processing. The resulting aligned photo sequence may then be analyzed by the TiVi700 Analyzer software.

The use of the integrated wizard can be demonstrated by the analysis of a photo sequence showing progressively increasing *erythema* induced by topical application of methyl nicotinate

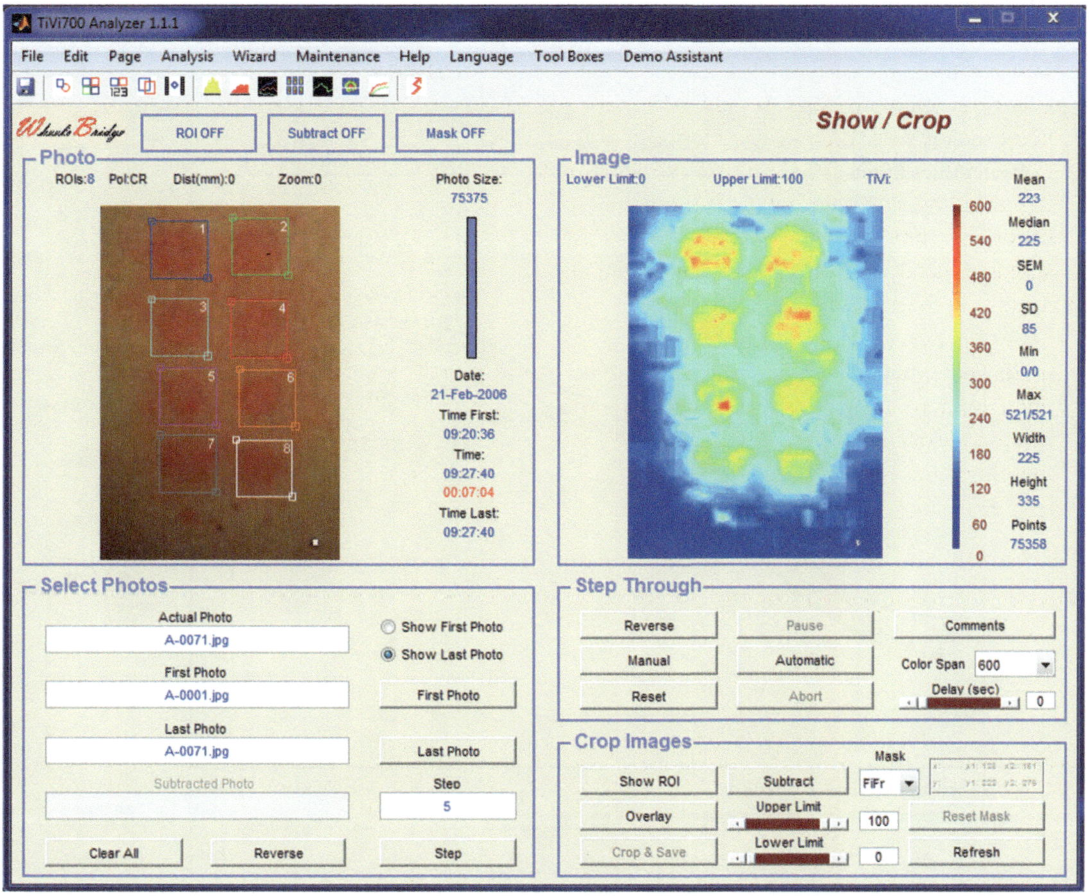

Fig. 17.7 The TiVi700 Analyzer user interface displaying photo and TiVi image in a patch testing procedure using the Finn Chamber

(MN). Using the last photo in the series, ROI are drawn around individual patch areas showing pronounced *erythema* reactions (Fig. 17.7). The wizard then evaluates the selected patch areas, automatically calculating the average TiVi values proportional to local RBC concentrations and qualified by a user-defined threshold value. The results are displayed as time traces in the TiVi Chart (Fig. 17.8).

Average TiVi values found in each of the first image's ROI are subtracted from those of the corresponding ROI in the remaining images. The resulting values are then automatically plotted to produce the curves illustrated in Fig. 17.8 above. These curves represent the progression of increasing *erythema* independent of skin pigmentation as postulated by the TiVi theoretical model. Any temporal and spatial variations in skin microvascular activity which may not be the direct result of the administered vasoactive material can be taken into account by selecting a reference ROI in an unexposed or untreated area of the skin. The average TiVi values derived from this site may then be used to produce a reference trace in the chart so that differences between this trace and all other traces can be calculated and displayed.

As an alternative to selecting ROI for analysis, the *Cross-Section Visualizer* function may be applied to the same set of photos to display the progression of *erythema* as a function of elapsed time along a user-selected cross section of the image (Fig. 17.9).

From the cross-section map, it can be concluded that during the initial 2- to 5-min time period, the *erythema* effect occurs mainly within the individual patches, while in the later stages of the experiment, the applied substance begins to produce vasodilatation in the skin between the

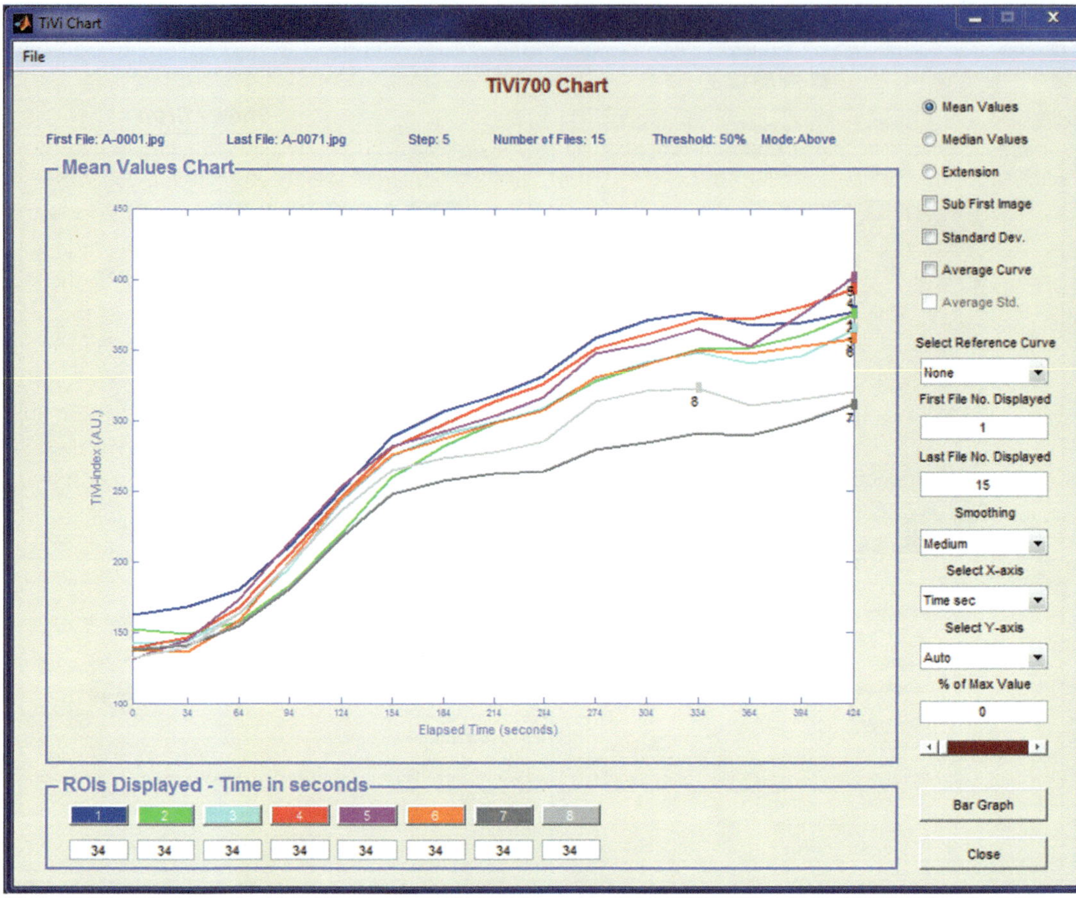

Fig. 17.8 TiVi Chart displaying the increasing TiVi values within the individual patches as a function of time.

individual patches. This method of presenting the results makes it possible to assess the pharmacodynamics of a vasoactive substance that may produce an initial effect (after about 2 min in this particular case) and then a more widespread vasodilation later on.

The dynamics of a progressive *erythema* or *blanching* effect produced by a topically applied substance may also be observed over a prolonged period of time. TiVi images derived from hundreds of sequential photos taken over several hours can be used to produce a time-compressed video clip which makes it possible to analyze slow-developing and long-lasting effects.

Rapid microvascular events, such as capillary refilling, may also be observed and evaluated by first capturing video at a frame rate of 25 frames per second. The video sequence can then be used to generate a set of individual TiVi images which in turn can be processed by any of the TiVi system's analytical functions to evaluate rapid changes in skin microcirculation.

17.5 Validation

The linear relationship between the TiVi-generated value and the actual concentration of RBCs was verified by way of a fluid model composed of tightly wound latex tubing simulating the blood vessels of the microcirculations [5]. The system was perfused by fresh human blood mixed with saline to produce RBC concentrations ranging from 0 to 4 % in steps of 0.2 %, which was considered to cover the physiological

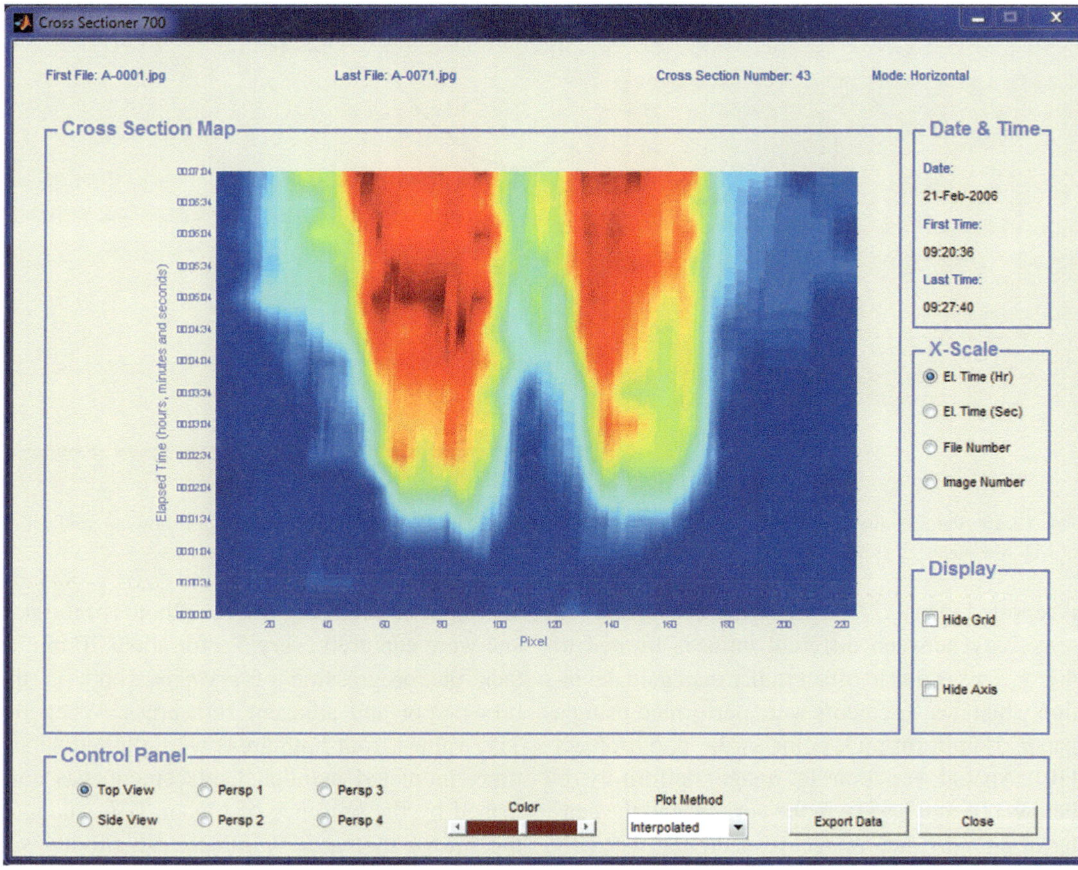

Fig. 17.9 *Erythema* development along a line crossing two patches as a function of elapsed time (from 0 (*lower part of map*) to about 10 min (*upper part of map*)) following removal of occlusion cover

range of skin tissue. Maximal oxygen saturation (approximately 100 %) of the mixture was provided by stirring and exposure to ambient air [7], while minimal oxygen saturation (approximately 0 %) was attained by bubbling N_2 gas through the blood samples. Five photos were captured for each RBC fraction and cropped to a region of interest covering the center of the fluid model. The average $TiVi_{index}$ values were then plotted as a function of the actual RBC fraction from 0 to 4 %. For 0 and 100 % oxygen saturation, the correlation coefficient was calculated to 0.998 ($n=4$) and 0.997 ($n=20$), respectively. The average $TiVi_{index}$ value calculated for the lower oxygen saturation was on the average 91.5 % of that calculated for the higher oxygen saturation, corresponding to a deviation of less than 3.9 % within the physiological range (70–100 % oxygen saturation) well in accordance with theory. Therefore, in practice, TiVi can be considered independent of tissue oxygen saturation.

The average depth of penetration for both red and green photons returning to the skin surface has been investigated through the use of a Monte Carlo simulation model [5, 8] adapted to accept both linear and cross-polarized light. With the use of cross-polarized filters, the average photon penetration depth increased from 345 to 482 μm for red wavelengths and from 316 to 387 μm for green wavelengths. This increase in depth sensitivity has also been confirmed in enhanced visual scoring of skin *erythema* while using a polarized light visualization system [9].

The average systematic drift in TiVi-system sensitivity over a 3-month time period, during which systems were moved between laboratories,

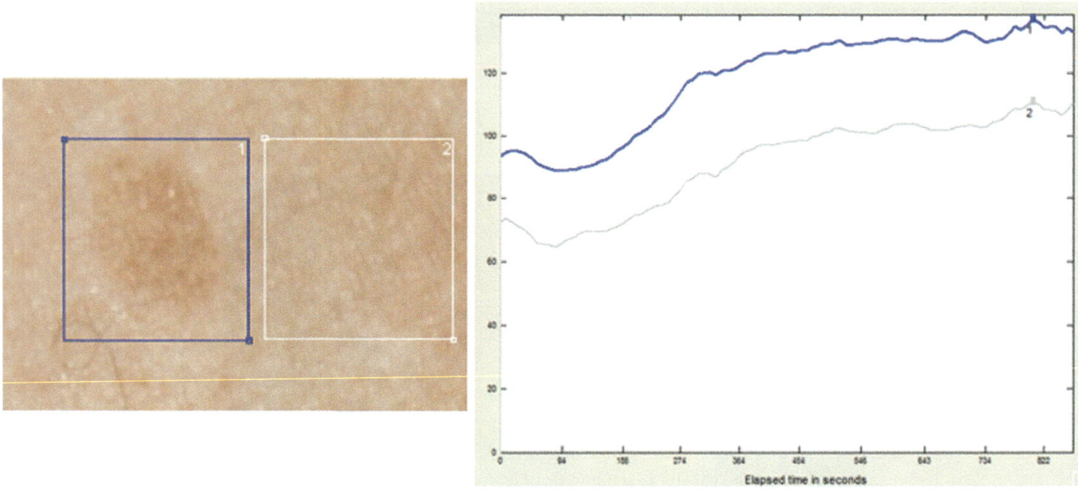

Fig. 17.10 Parallel displacement of TiVi value traces measured in and adjacent to a brown spot following application of MN to Caucasian skin

is reported to be 0.27 %, while the discrepancy in sensitivity between different units is limited to 4.1 %. This is due to offset rather than gain deviation when measurements were performed using a paper of uniform pink color as the photo object [10]. Spatial variation in image uniformity is below 3.08 and 1.93 % in the corners and the center of an individual image, respectively. The distance dependence was assessed by recording images alternatively at distances of 15 and 25 cm and calculating the TiVi values produced by *erythema* inducing MN applied to forearm skin. No significant difference in average TiVi values at the two distances could be demonstrated ($n=25$). In contrast to laser Doppler methodology, the TiVi mechanism is not sensitive to movement artifact because no Doppler components are recorded. Since all TiVi image points are recorded simultaneously, there is no interpretation ambiguity regarding spatial and temporal variations in skin microcirculation, as may be the case with raster scan laser Doppler perfusion imaging devices. On the other hand, since TiVi is based on spectroscopy, it is sensitive to all objects of a specific color.

In order to validate the predicted upward parallel displacement of the TiVi value with epidermal melanin content (Fig. 17.5), MN was applied on Caucasian skin having both a brown spot with elevated pigmentation and an adjacent area of low pigmentation (Fig. 17.10). Photos of the skin site were captured every 5 s for about 10 min to tack the progression of *erythema* both in the brown-spot and adjacent skin areas. When the TiVi values from both areas were plotted versus time, predicted parallel displacement was confirmed by the presence of higher TiVi values in the brown-spot area, while the sensitivity was found to be about the same for both areas. Subtracting the first photo TiVi values from each trace therefore results in an accurate measure of the *erythema* reaction independent of epidermal melanin content for low and moderate pigmentation levels. For higher pigmentation levels, however, the light cannot pass through the epidermis, and no information about the RBC concentration can be obtained.

17.6 Applications

Establishment of healthy skin RBC concentration reference values is important for the design of versatile test procedures for evaluating skin damage precipitated by the use of vibrating tools, exposure to damaging chemicals, the presence of peripheral vascular disorders, and other causes. One of the first applications of emerging TiVi technology was the mapping of spatial and temporal variations of RBC concentrations found in healthy Caucasian skin at the dorsum of the hand while at rest and during post-occlusive hyperemia [11]. A lower skin RBC concentration (179–184 TiVi units (tvu)) was observed at the back of the hand and base of the thumb compared with areas adjacent to the nailfold region of the fingers (190–213 tvu) when recording was made at an ambient temperature of 21–23 °C. Values in the same range were recorded from the volar side of forearm skin (192 ± 19 tvu) and at the skin of the back (178 ± 17 tvu) in the resting state. The short-term variation (within 70 s) was 2 % in all areas of the dorsal side of the hand, while day-to-day variations were in the range 5–7 % in the back of the hand and up to 10 % in areas adjacent to the nailfold region. In the post-occlusive hyperemia phase, up to a 60 % increase in skin RBC concentration was observed in the early part of the reactive hyperemia phase. This increase in skin RBC concentration successively decreased but remained about 18 % above the pre-occlusive level after 30 min. Reperfusion of the skin microvasculature following tissue occlusion measured in forearm skin revealed a strong correlation and high reproducibility between changes in perfusion and RBC in the post-occlusive hyperemia phase [12], which increased with the length of the duration of the occlusion.

To demonstrate the potential of TiVi in the objective and operator-independent assessment of skin *blanching*, TiVi was used for quantification of human skin *blanching* with the Minolta chromameter CR200 (Minolta, Tokyo, Japan) as an independent colorimeter reference method [13]. Desoximetasone gel 0.05 % (Topicort®, TaroPharma, Taro Pharmaceuticals USA Inc., Hawthorn, NY, USA) was applied topically on the volar side of the forearm under occlusion for 6 h in healthy adults. The relative uncertainty in the *blanching* estimate produced by TiVi was about 5 % and similar to that of the chromameter operated by a single user and taking the a* parameter as a measure of *blanching* [14]. In a separate study, the induction of *blanching* in the occlusion phase was mapped using a transparent occlusion cover. The successive induction of skin *blanching* during the occlusion phase could thereby be mapped using TiVi. After an occlusion time of about 6 h, the RBC concentration was reduced to approximately 80 % of the initial value.

To study the barrier function of the skin, MN which is a water-soluble compound that rapidly penetrates the skin may be used [15]. In order to minimize environmental influence, the MN solution is applied to the skin under occlusion using Finn Chambers on Scanpor® application systems (Epicutan, Tuusula, Finland). The average maximum change in recorded TiVi value (128 ± 28 tvu) was found to be about 70 % of the TiVi value recorded from unprovoked skin. The average variability in *erythema* intensity between different patches on the skin of the back of the same individual, following 30 s application time of MN (10 mml) was calculated to 22.1 % (var. coeff.). The average time to reach 20, 80, and 100 % of the end point *erythema* was calculated to be 114 ± 31, 264 ± 47, and 400 ± 65.2 s, respectively. The *erythema* extension generally spread outside the patch areas in which the MN was applied after about 5 min following the removal of occlusion and start of measurement, possibly caused by axon reflexes, histamine release, or lateral diffusion of the compound. The time course of *erythema* intensity induction can be modulated by lowering of fat content in topical formulations. A benzyl nicotinate formulation containing 10 % fat induced *erythema* more rapidly and with higher intensity than the formulation with higher fat content (50–80 %) as demonstrated by TiVi [16]. In the same study, a corticosteroid (betamethasone 17-valerate) in vehicles with different lipid content was used to induce skin *blanching*. A significant reduction in lag time to the appearance of *blanching* as measured with TiVi was demonstrated when the lipid content was reduced from 80 to 10 %.

In UVB phototesting of the skin, TiVi has been used to measure the *erythema* profile induced by a divergent UVB beam, from which the minimal *erythema* dose can be determined [17]. TiVi has further been used to map and quantify the local vasodilatation and vasoconstriction produced by acetylcholine and sodium nitroprusside, respectively, and noradrenaline and phenylephrine when these vasoactive substances were forced to migrate into the skin by way of iontophoresis [18]. Capsaicin-evoked *erythema* as measured with TiVi and visual flare were found to be normal in patients with multiple chemical sensitivity and in eczema patients with airway symptoms from odorous chemicals, but *erythema* intensity generally decreased with age [19].

In a comparative study the skin microcirculation responses were investigated during the reactive hyperemic phase following release of a pressure cuff applied around the upper arm and inflated to 90 and 130 mmHg, respectively, using a laser Doppler line scanner (LDSLS) [20], a laser Speckle perfusion imager (LSPI) [21], and TiVi. The time course for perfusion tracks recorded by LDSL and LSPI was similar with a reduction during the occlusion phase and a distinct reactive hyperemia peak (pressure 130 mmHg) and no reactive hyperemia (pressure 90 mmHg), respectively, which was reduced to pre-occlusion values after about 2 min. The time course of the TiVi track differed from this pattern and demonstrated a continuous increase in *erythema* (pressure 90 mmHg) and a steady RBC concentration level (pressure 130 mmHg), respectively, during the occlusion phase [22]. This difference in track signature is explained by the fact that for the lower pressure, the venous return is blocked while the arteries supplying the skin with RBCs are still open; however, at the higher cuff pressure, the RBC volume is not altered during the occlusion phase. Following topical application of MN, all methods displayed distinct and transient *erythema*. When clobetal propionate solution (Dermovate®, 0.5 mg/ml, GlaxoSmithKline Ltd., London, UK) was topically applied on the skin to produce local vasoconstriction, both LDLS and LSPI proved to be insensitive to the action of the vasoconstrictor, while TiVi showed clear boundaries of the reaction, thus demonstrating its ability to map both *erythema* and *blanching*. Further comparison studies of TiVi and laser Doppler flowmetry involving assessment of pharmacologically induced vasodilatation and vasoconstriction in human skin demonstrated dose-dependent and time-related vasodilator responses to prostaglandin E2 and vasoconstrictor responses to norepinephrine [23], while TiVi measurements showed a higher sensitivity to norepinephrine-induced vasoconstriction.

17.7 Toolboxes

As the TiVi system was being validated for use in different applications, it became apparent that dedicated versions were needed for solving particular problems and that the basic principle of digital camera polarization spectroscopy can potentially be applied for evaluating other skin parameters. These specific embodiments of the TiVi were developed as toolboxes that can be optionally integrated with the basic TiVi system. The general purposes of each of these toolboxes are listed below.

The *Skin Damage Visualizer* is intended primarily for direct assessment of skin damage at the worksite or in occupational health applications. Skin sites identified as having either increased or reduced RBC concentration are evaluated to determine both the extent of skin damage and the degree of change over time through the use of integrated library and trend monitor functions.

The *Skin Color Tracker* allows for monitoring changes of skin color within a user-defined color space. Visualizing these color changes over time through the use of the integrated library and trend monitor functions makes it possible to assess skin color changes that are too subtle to be observed by unaided visual inspection.

The *Spot Analyzer* is a fully automatic and user-independent system for assessing facial spot *erythema* intensity and extension. This toolbox is designed to be used as a complement to the visual scoring procedure.

The *Surface Analyzer* is designed for analysis of skin surface texture, employing information derived from photos captured in both cross- and co-polarization modes.

The *Skin Pigmentation Analyzer* creates two dimensional maps of local skin pigmentation in which the contribution from RBCs are effectively suppressed.

The *Wrinkle Analyzer* calculates the depth, volume, and asymmetry of individual wrinkles and furrows based on local intensity variations in the photo color planes.

The *Microstructure Analyzer* evaluates microscopic structures such as individual hair shafts and stubble with a resolution of about 3–4 μm per pixel with the TiVi microscope adapter attached to the camera lens system.

The *TiVi Video* Analyzer produces video clips captured at a frame rate of 25 frames per second and is intended for investigation of rapid microvascular events such as capillary refilling.

17.8 Summary and Conclusion

The emerging technology of polarization spectroscopy imaging has been integrated into a portable device (Tissue Viability Imager) that allows for investigation of skin microcirculation and other skin parameters in the laboratory as well as at the worksite. Early applications demonstrate how this emerging technology can be used in the assessment of skin *erythema* and *blanching* in skin product development and testing.

The ease of use of the TiVi technology makes it well suited for productive and user-independent investigations of data from large panels of test subjects.

References

1. Jasques SL, Ramella-Roman JC, Lee K (2002) Imaging skin pathology with polarized light. J Biomed Opt 7:329–340
2. Ishimaru A (1978) Wave propagation and scattering in random media, vol, vol 1. Academic Press, London, Single scattering and transport theory
3. Brinkworth BJ (1964) A diffusion model of the transport of radiation from a point source in the lower atmosphere. Br J Appl Phys 15:733–741
4. Fabbri F, Francesschini MA, Fantini S (2003) Characterization of spatial and temporal variations in the optical properties of tissue-like media with diffuse reflectance imaging. Appl Optics 42:3063–3072
5. O'Doherty J, Henricson J, Anderson C, Leahy MJ, Nilsson GE, Sjöberg F (2007) Sub-epidermal imaging using polarized light spectroscopy for assessment of skin microcirculation. Skin Res Technol 13:472–484
6. http://www.wheelsbridge.se (instrument supplier web-site)
7. Nilsson GE, Tenland T, Öberg PÅ (1980) Evaluation of a laser Doppler flowmeter for measurement of tissue blood flow. IEEE Trans Biomed Eng BME-27(10):597–604
8. Ramella Roman JC, Prahl SA, Jacques SL (2005) Three Monte Carlo programs of polarized light transport into scattering media: part 1. Opt Express 13:4420–4438
9. Farage M (2008) Enhancement of visual scoring of skin irritant reactions using cross-polarized light and parallel-polarized light. Contact Dermatitis 58:147–155
10. Nilsson GE, Zhai H, Chan HP, Farahmand S, Maibach HI (2009) Cutaneous bioengineering instrumentation standardization: the tissue viability imager. Skin Res Technol 15:6–13
11. Zhai H, Chan HP, Farahmand S, Nilsson GE, Maibach HI (2009) Tissue viability imaging: mapping skin erythema. Skin Res Technol 15:14–19
12. Farnebo S, Thorfinn J, Henricson J, Tesselaar E (2010) Hyperaemic changes in forearm skin perfusion and RBC concentration after increasing occlusion times. Microvasc Res 80(3):412–416
13. Zhai H, Chan HP, Farahmand S, Nilsson GE, Maibach HI (2009) Comparison of tissue viability imaging and colorimetry: skin blanching. Skin Res Technol 15:20–23
14. Westerhof W (2006) Colorometry. In: Serup J, Jemec GBE, Grove GL (eds) Handbook of non-invasive methods and the skin, 2nd edn. CRC Press, USA, Boca Raton, FL, pp 635–647
15. Magnusson BM, Nilsson GE, Anderson C (2006) The polarization spectroscopy camera – a promising tool for assessment of erythematous reactions to topically applied agents. Perspectives in percutaneous perfusion, La Grande Motte, 18–22 Apr 2006
16. Wirén K, Frithiof H, Sjöqvist C, Lodén M (2009) Enhancement of bioavailability by lowering of fat content in topical formulations. Br J Dermatol 160(3):552–556
17. O'Doherty, J., Henricson, J., Enfield, J., Nilsson G.E., Leahy, M and Anderson C. Use of Tissue Viability Imaging (TiVi) in the assessment of divergent beam UVB phototesting, Submitted to Acta-Dermato-Venereologica, March 2009.
18. Henricson J, Nilsson A, Tesselar E, Nilsson GA, Sjöberg F (2009) Tissue viability imaging: microvascular response to vasoactive drugs induced by microdialysis. Microvasc Res 78(2):199–205
19. Holst H, Arendt-Nielsen L, Mosbech H, Serup J, Elberling J (2011) Capsaicin-induced neurogenic inflammation in the skin in patients with symptoms induced by odorous chemicals. Skin Res Technol 17(1):82–90
20. http://www.moor.co.uk (instrument supplier web-site)
21. Briers JD (2001) Laser Doppler, speckle and related techniques for blood perfusion mapping and imaging. Physiol Meas 22:R35–R66
22. O'Doherty J, Clancy NT, Enfield JG, McNamara P, Leahy M (2009) Comparison of instruments for investigation of microcirculatory blood flow and red blood cell concentration. J Biomed Opt 14(3): 034025
23. Petersen LJ, Zacho HD, Lyngholm AM, Arendt-Nielsen L (2010) Tissue viability imaging for assessment of pharmacologically induced vasodilatation and vasoconstriction in human skin. Microvasc Res 80(3):499–504

Part II

Skin Analysis

18 Interaction Between Free Radicals and Antioxidants in Human Skin

J. Lademann, M.E. Darvin, J.W. Fluhr, and M.C. Meinke

18.1 Introduction

Free radicals are produced in the human body as a result of the metabolism [1, 2]. They are important for signalling processes in the organism [1, 3] as well as disinfector against viruses and bacteria [4]. If their concentration exceeds a critical value, these highly reactive molecules can destroy cells and cell compartments and cause serious damage to the human body [5, 6]. An enhanced radical formation is due to inflammatory processes in the human body or to environmental influences [7, 8]. In this context, high doses of solar radiation are dangerous [9]. Enhanced UV doses do not only evoke sunburns and skin aging but induce even skin cancer if significant amounts of free radicals are generated in the skin [10]. With the antioxidant protection system, the human organism has developed a defense mechanism against the destructive effect of the free radicals. The most important antioxidants in this defense system are the vitamins, the carotenoids, and some enzymes [11, 12]. The antioxidants are capable of neutralizing the free radicals before the same start damaging the body. Most of these antioxidants cannot be generated by the human body automatically, but must be taken in with food rich in fruit and vegetables. If the free radicals exceed a critical level, the antioxidants are destroyed, too. Thus, the detection of antioxidants in tissue permits at the same time conclusions to be drawn about the formation of radicals [13].

So far, free radicals have exclusively been detected in vitro by means of electron paramagnetic resonance (EPR) spectrometry. In order to analyze the antioxidants in tissue intricate and expensive methods, e.g., high-pressure liquid chromatography (HPLC) or mass spectroscopy were necessary. For both of these methods, it is indispensable to take tissue samples, in most cases biopsies. This, however, strongly restricts the kinetic investigations of the interaction between the free radicals and antioxidants, as it is impossible for ethical reasons to take large numbers of biopsies from the same volunteers. Recently, spectroscopic and spectrometric methods have been developed, which permit both the free radicals and the antioxidants to be detected in tissue in vivo noninvasively. In this chapter, the application of the in vivo EPR technology and the spectroscopic detection of antioxidants are described.

18.2 Detection of Free Radicals and Antioxidants in the Skin

18.2.1 Resonance Raman Spectroscopic Measurements

In 1998, Bernstein et al. had described the resonance Raman spectroscopic detection of carotenoids in the eye, in which the carotenoid level is approxi-

J. Lademann (✉) • M.E. Darvin • J.W. Fluhr
M.C. Meinke
Department of Dermatology, Venerology and Allergology, Center of Experimental and Applied Cutaneous Physiology, Charité – Universitätsmedizin Berlin, Charitéplatz 1, 10117 Berlin, Germany
e-mail: juergen.lademann@charite.de

E. Berardesca et al. (eds.), *Non Invasive Diagnostic Techniques in Clinical Dermatology*,
DOI 10.1007/978-3-642-32109-2_18, © Springer Berlin Heidelberg 2014

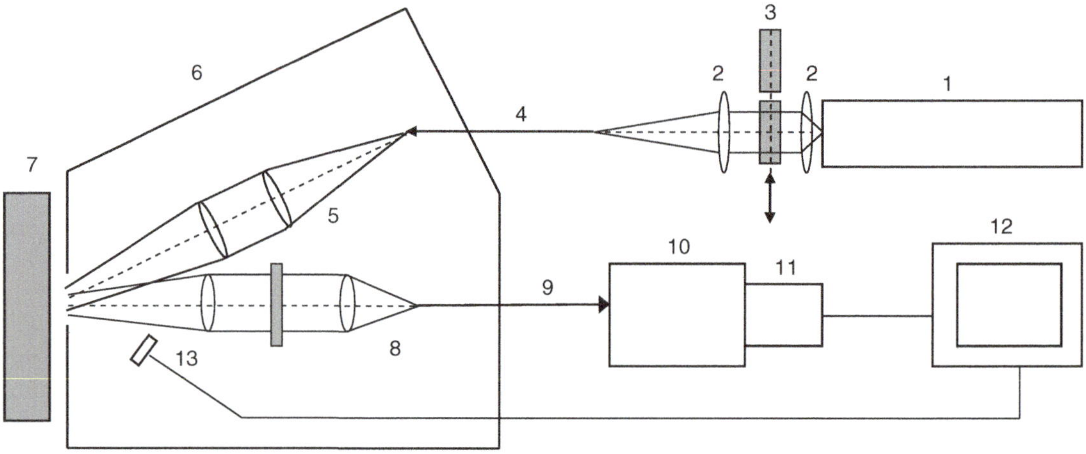

Fig. 18.1 Scheme of the resonance Raman spectrometer setup [14]

mately one order of magnitude higher than in other organs of the human body [15]. In 2000, Hata et al. reported the possibility to measure carotenoids in human skin noninvasively by the use of resonance Raman spectroscopy [16]. In 2004, the possibility to detect dermal carotenoids selectively using this measuring principle was reported [17]. Later, the measurement stability was improved, and subsequently a setup was developed for the noninvasive in vivo detection of carotenoids in human skin [18, 19]. Figure 18.1 shows the scheme of the measuring setup. A photograph of the setup is presented in Fig. 18.2. For this purpose an argon laser system (*1*) was applied emitting at two wavelengths (488 and 514.5 nm), simultaneously (Fig. 18.1). Then, the laser beam was collimated by the lens system (*2*), filtered (*3*), and focused onto an optical fiber (*4*). The 488 nm wavelength corresponds to the absorption maximum of both carotenoids β-carotene and lycopene. The wavelength of 514.5 nm is in the absorption maximum of lycopene and is absorbed by β-carotene to a small extent, only [20]. The absorption spectra of both substances and the corresponding excitation wavelengths are presented in Fig. 18.3.

Based on the two-wavelength measuring principle, it is possible to quantitatively determine the β-carotene and lycopene concentrations.

From the base of the measuring system housing besides the laser system also the detection and control unit, the laser radiation is led through optical fibers (*4* and *9* in Fig. 18.1) into the optical measuring system, which is integrated into the handpiece (*6*). The handpiece contains the excitation and detection channels (*5* and *8*) and the channel for measurement of backscattered light (*13*). To improve the measurement stability, the spot of excitation beam on the sample (*7*) was widened to 6.5 mm in diameter. This optical measuring system has two detection channels (*8* and *13*). Both channels collect the light backscattered from the skin, reflecting it to the base through the optical fibers. There, detection channel no. 1 (*8*) is subjected to a spectral analysis determining the β-carotene and lycopene concentrations. Detection channel no. 2 (*13*) is used to determine the skin type of the volunteer by measuring the laser energy backscattered from the skin. The analysis of the skin type, which is categorized into types I through IV according to Fitzpatrick, is necessary because light skin (type I) absorbs the laser radiation less than very dark skin (type IV). The radiation scattered by the skin, which includes the elastic scattering at the excitation wavelength, the fluorescence, and the Stokes Raman scattering, is filtered and collected by the lens system (*8*) at the entrance of the optical fiber bundle (*9*). The filtered signal is led to a spectrometer (*10*) equipped with a CCD camera (*11*). The obtained spectrum is then analyzed and displayed on the computer (*12*). The excitation radiation at 488 and 514.5 nm penetrates approximately 150 μm deep into the human skin, accessing the epidermis and parts of the dermis, but not the blood vessels.

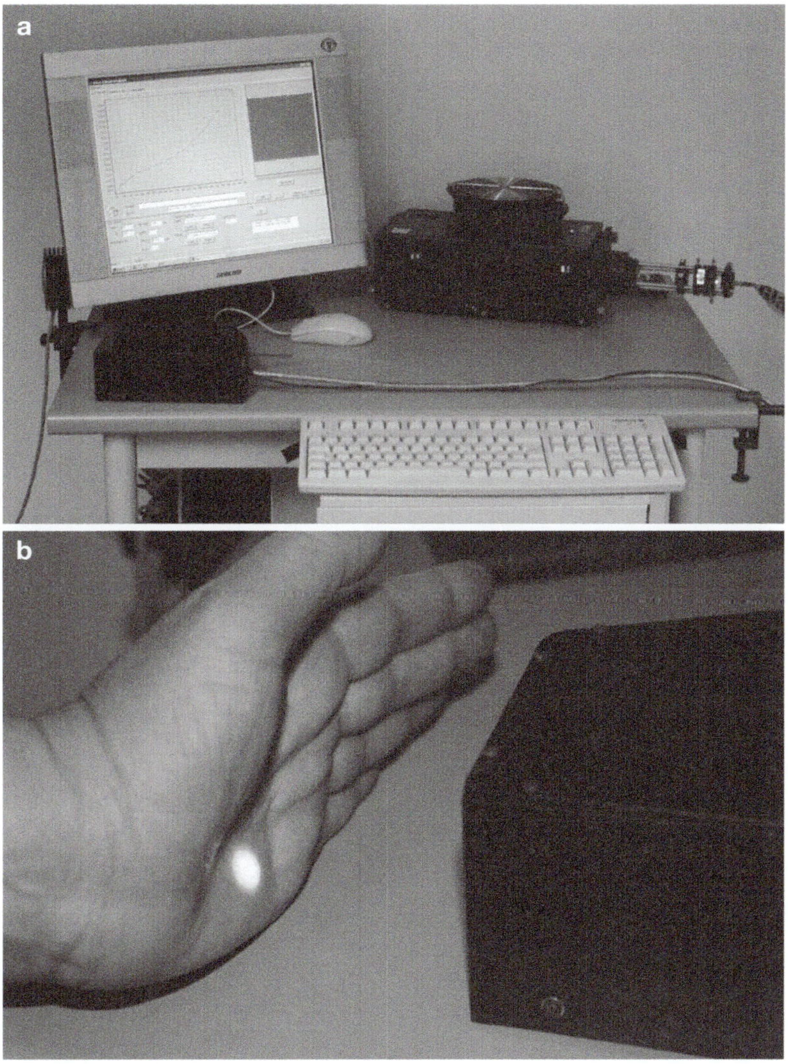

Fig. 18.2 Photos of the resonance Raman spectrometer

This is of particular importance, since the system was specifically developed for analyzing the skin.

A typical resonance Raman spectrum of the carotenoids in the skin of a volunteer (skin type II) is presented in Fig. 18.4. The small carotenoid signals can be recognized on a high-fluorescence background. Deduction of the background and spectral processing yields a clearly analyzable resonance Raman spectrum as shown in Fig. 18.5.

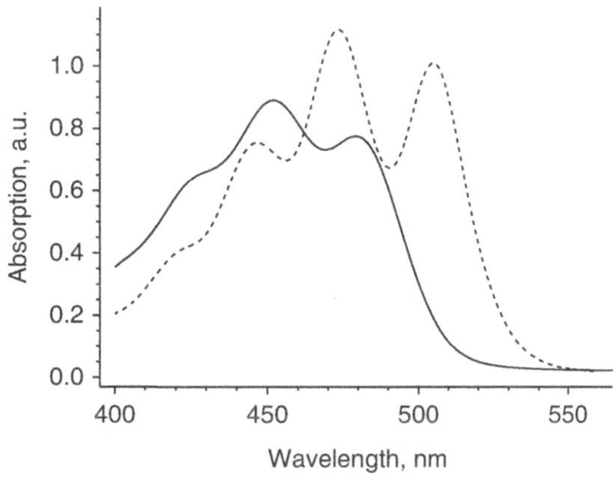

Fig. 18.3 Absorption spectra of β-carotene (*full line*) and lycopene (*dotted line*) [14]

18.2.2 Reflectance Measurements

Recently, a miniaturized spectroscopic skin scanner has been developed, which is capable of

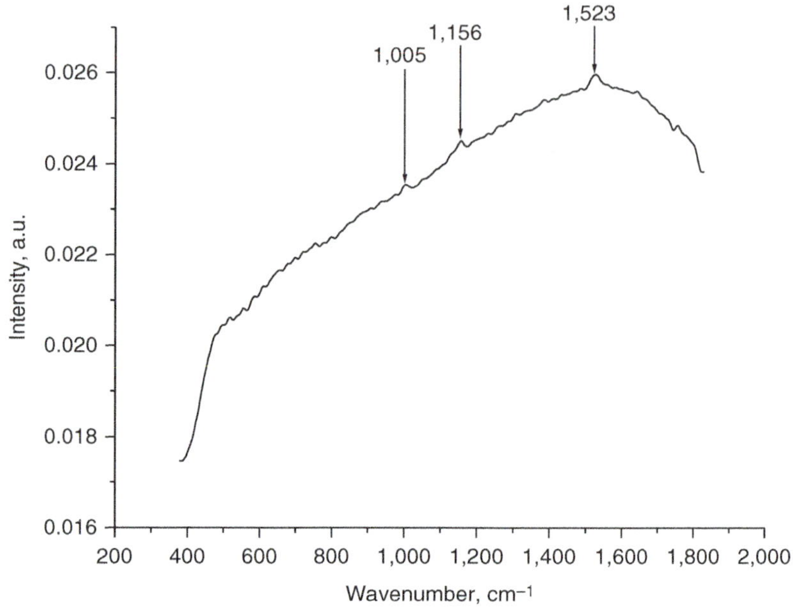

Fig. 18.4 In vivo resonance Raman spectrum of human skin. Excitation 514.5 nm [21]

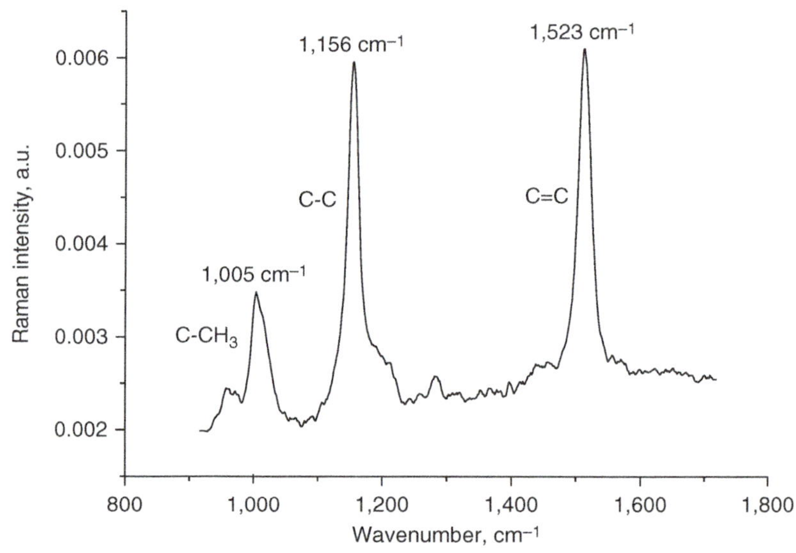

Fig. 18.5 In vivo resonance Raman spectrum of human skin after deduction of the fluorescent background and after signal processing [21]

analyzing by reflectance measurements the carotenoid level in human skin in vivo. This scanner system uses a light-emitting diode (LED) operating in the bright optical range of the spectra. A LED emitting a blue spectrum ranging from 440 to 490 nm (10 % of maximum intensity) overlaps the maximum of the absorption of carotenoids. Like in the case of the resonance Raman spectrometer, the diffuse light backscattered from the skin is detected, and the carotenoid concentration is determined based on the analyses of the profile of the diffuse reflection spectrum. The dip in the diffuse reflection spectrum measured between 458 and 472 nm (maximum of the absorption of carotenoids) was recalculated to the concentration of dermal carotenoids [22]. This measuring system analyzes the carotenoids as a whole not distinguishing between β-carotene and lycopene. A schematic of the skin scanner is shown in Fig. 18.6. A photograph of the scanner is presented in Fig. 18.7.

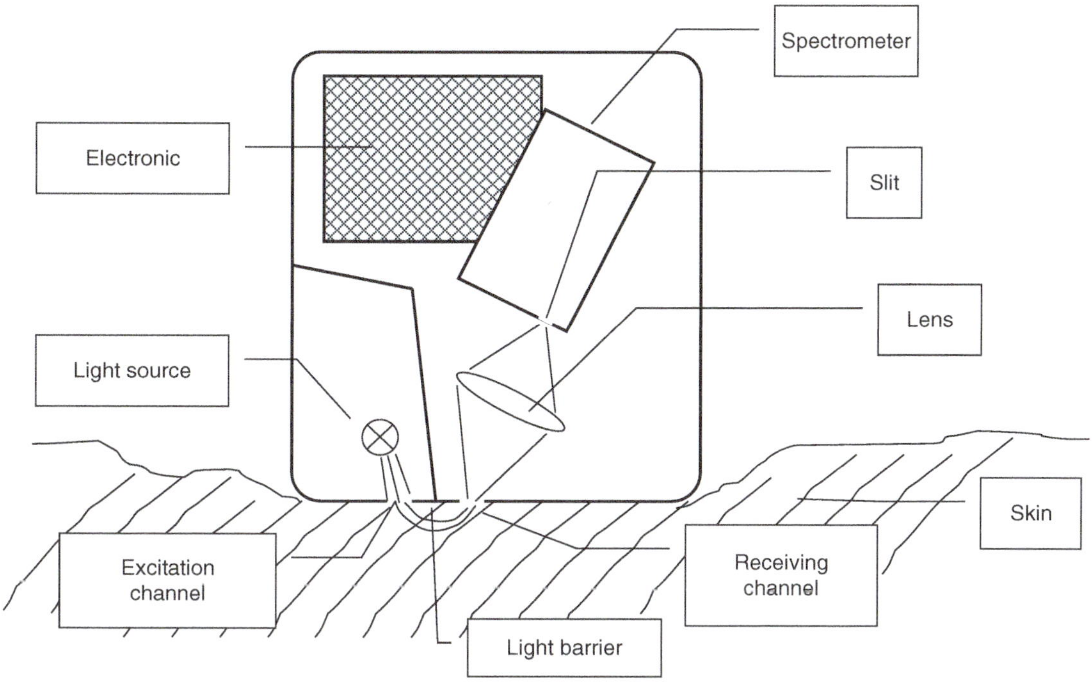

Fig. 18.6 Scheme of skin scanner setup for carotenoid measurements [21]

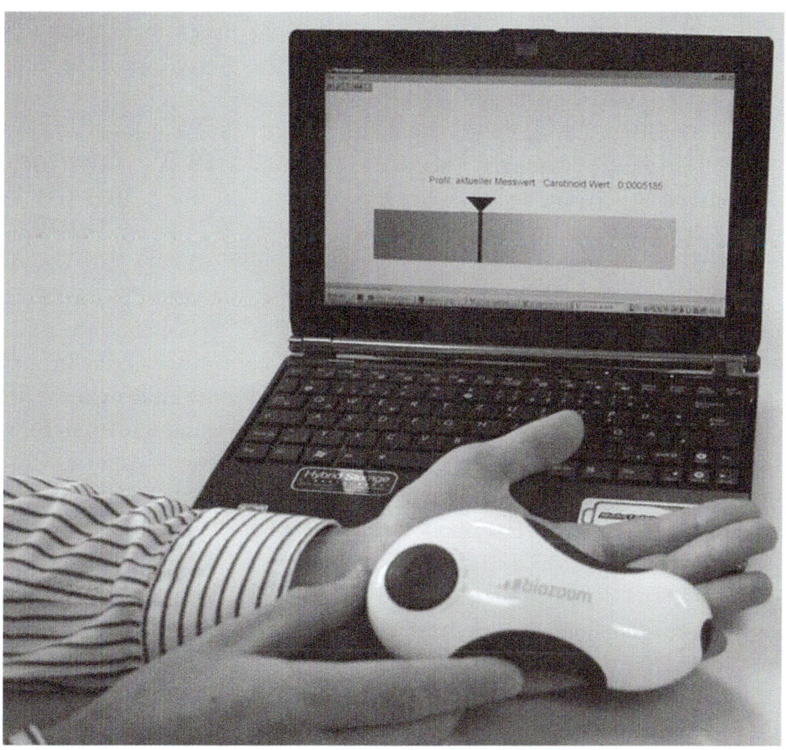

Fig. 18.7 Photograph of skin scanner

When the results obtained by the skin scanner were compared with those obtained by the resonance Raman spectrometer on the same skin area of the volunteers, a very good correlation of the determined carotenoid levels ($R = 0.88$) was established. The limitations of this method are discussed in detail by our group [22].

18.2.3 Electron Paramagnetic Resonance (EPR) Spectroscopic Measurements

Free radicals are short-lived, instable molecules, which quickly react with other compounds. Therefore, the EPR technique utilizes spin probes for detecting the free radicals [23]. These spin probes are specific radicals with a considerable longer lifetime ranging between some minutes and some hours. Reacting with the short-lived free radicals, these spin probes are degraded. Consequently, the decline of the spin trap concentration is a measure for the free radicals formed. On the other hand, a slow decrease in the concentration of many spin probes can be observed, even if the samples are not affected by external stress factors such as UV irradiation. This signal reduction is due to the antioxidants in the tissue sample reacting with the spin probe. Thus, the kinetics of the signal reduction is a measure for the concentration of the antioxidants in the tissue sample. This means that EPR measurements can be applied for the detection of both free radicals and antioxidants.

There are different types of spin traps, many of which are nitroxides and suitable for in vitro application, only. 2,2,6,6-tetramethylpiperidine-1- oxyl (TEMPO) and 3-carboxy-2,2,5,5-tetramethylpyrrolidine-1-oxyl (PCA) are two of the few substances, which can be used for in vivo investigations, too [24]. For the EPR measurements, the spin probe must be introduced into the tissue to be investigated. In vitro investigations can be done by submerging the tissue samples, which are in most cases biopsies, directly into the spin probe solution. In vivo investigations require the spin probe to be topically applied. The penetration of the spin probe can be stimulated by selecting a suitable formulation or nanotransporter, in which the spin probe is applied on the skin [25]. PCA only slowly reacts with the skin antioxidants and is therefore used for the measurements of radical formation [26, 27], whereas the TEMPO is used to measure antioxidants in the skin [26].

In almost any case, an X-band EPR spectrometer is used for in vitro EPR measurements on tissue samples. Operated in the frequency range of approximately 9.4 GHz, this system is very sensitive and provided with a very small sample compartment, only, so that it is unsuitable for in vivo investigations. For such in vivo investigations an L-band EPR spectrometer (frequency range around 1.3 GHz) must be applied. The size of the sample compartment of this spectrometer system can be dimensioned to well accommodate the arms and legs of volunteers (Fig. 18.8). However, due to the considerably larger distance between the magnets, the L-band spectrometers provide a far lower measuring sensitivity. Figure 18.9 shows an EPR spectrum of the skin.

The decrease in intensity of TEMPO in the skin follows an simple exponential function $I(t) = I_0 \exp(-kt)$ where k is the rate constant [28, 29]. The higher the rate constant, the higher is the antioxidative status of the skin.

Since this chapter refers to dermal in vivo measuring methods, only the L-band technology is discussed hereinafter.

18.3 Carotenoids as Markers for the Whole Antioxidative Status of Human Skin

Both, the resonance Raman spectroscopy and reflectance spectroscopic measurements permit only carotenoids to be detected in tissue, whereas the EPR method measures the antioxidative status although even EPR does not guarantee that all antioxidants are covered. Comparison of both resonance Raman measurements on cutaneous carotenoid concentration and EPR spectroscopic determined antioxidative status on the same volunteers, and skin areas showed a significant correlation if the lifestyle of the volunteers is stable and the nutrition is well balanced [29]. This indicates that the carotenoids can be used as marker substances for the entire antioxidative status of human skin. This is not surprising as the different antioxi-

Fig. 18.8 In vivo EPR measurement on a volunteer

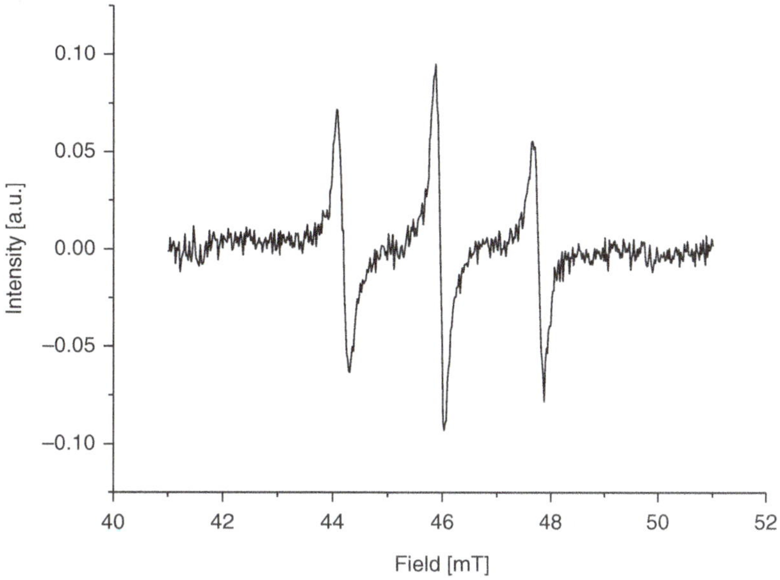

Fig. 18.9 EPR spectrum of TEMPO measured in the skin of the forearm of a volunteer (L-band system)

dants form protective chains in the tissue thus protecting each other against the harmful effect of the free radicals [30–32]. If carotenoids are applied via supplementation, the rate constant and therefore the antioxidative status increase as well [33].

18.4 The Human Antioxidative Status

The carotenoid concentration in the same volunteer is different depending on the body site [19]. The concentrations at the balls of the thumbs and the feet as well as at the forehead exceed those at the forearm or the cheek by about 20–30 %. As a matter of principle, the carotenoid concentration is higher in the body sites with a high density of sweat glands. This can be explained by the fact that the carotenoids are supplied onto the skin surface mainly with the sweat [34, 35]. In long-term investigations it is important, therefore, to analyze always the same skin area. Preferably this should be the palm, which is easily accessible and clearly demarcated from other skin areas and has a high density of sweat glands and, consequently, a high carotenoid level. Moreover, the

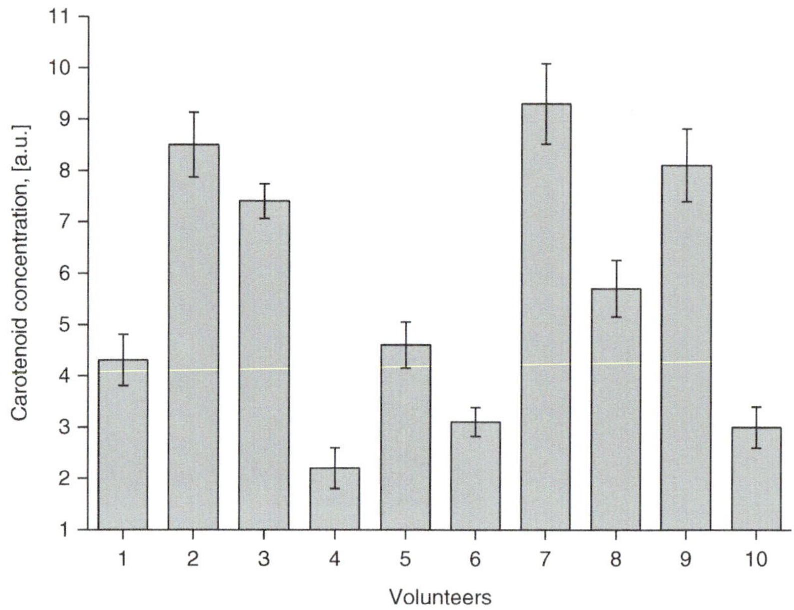

Fig. 18.10 Mean carotenoid concentrations measured on 10 volunteers during a period of 5 days

color of the palm is always light, so that the measuring results are not or only insignificantly influenced by the skin type. The carotenoid concentrations in the individual volunteers differ depending on their specific lifestyles and dietary habits [36–38]. Investigations on 150 volunteers have shown that smokers have a significantly lower level of 21 % than nonsmokers [37]. Interestingly, the carotenoid values are influenced by gender. Women exhibited significantly higher values than men ($p<0.05$) of about 13 %. This correlates with investigation of blood plasma where the differences in carotenoids for men and women are correlated with a different uptake of carotenoids from fruit and vegetables [39, 40]. Furthermore, the study has shown that the age of volunteers has no influence on the total carotenoid concentration in the skin; however, for lycopene, a low, but significant, correlation with age was found.

In Fig. 18.10 the mean carotenoid concentrations of 10 volunteers measured on their palm and examined on 5 consecutive days are presented. The carotenoid concentration of the volunteer with the highest values exceeds that of the volunteer with the lowest value by approximately 400 %. In parallel to the spectroscopic measurements, the volunteers were asked to complete questionnaires on their dietary habits and lifestyles. As a result, it turned out that the carotenoid values in smokers and individuals living on unhealthy food had been low, whereas individuals on a healthy diet exhibited the highest carotenoid concentrations [36].

The observation according to which dietary and stress factors have an immediate effect on the carotenoid concentration in the volunteers' skin was confirmed in a 1-year study performed on the staff of the Center of Experimental and Applied Cutaneous Physiology of the Charité [36]. The palms of the staff were measured every working day at a specific time. In that study, too, questionnaires were distributed to gather information about the dietary behavior and stress situations. The results of these investigations furnish clear evidence that the antioxidative status of the skin is influenced by two opposing processes. While the intake of healthy food rich, for instance, in fruit and vegetables causes the carotenoids in human skin to accumulate, any kind of stress leads to a degradation of the carotenoids in the skin. Essential stress factors are, e.g., illness, alcohol consumption, smoking, and stress on the job and in private life. Thus, healthy food alone does not suffice to keep the body healthy; any external stress factors must be reduced to the largest possible extent. While everybody can influence his own dietary habits rather easily, stress situations are more difficult to cope with.

18.5 Degradation of Antioxidants in the Skin by Stress Factors

In the aforementioned studies the carotenoid concentrations in the skin of volunteers were measured without influencing the volunteers' dietary habits and lifestyles. These data were merely documented in questionnaires to obtain a correlation between the carotenoid values and the lifestyle and stress conditions. In further studies, the volunteers were subjected to defined stress situations. The results of these investigations shall be briefly discussed hereinafter.

As described earlier in this chapter, a high level of free radicals in human skin is generated particularly due to the UV irradiation of the sun. Therefore, volunteers of different age were irradiated with the minimal erythema dose (MED), i.e., with exactly that UV dose, which is sufficient to induce mild sunburn in the respective volunteer. The β-carotene and lycopene concentrations prior to and at various times after irradiation were determined using the resonance Raman spectrometer. As a result, it turned out that the lycopene concentration declined immediately upon irradiation, while the β-carotene concentration declined not before approximately 30 min [41]. An obvious explanation of this fact is that lycopene is considerably less stable than β-carotene as described in the literature [42, 43], too. While the degradation of the carotenoids occurs relatively quickly, within a period of 1 or 2 h, it takes up to 3 days until the carotenoid concentration has reached its initial level, again. A similar situation was observed for the consumption of alcohol. After the intake of approx. 1 mL ethanol/1 kg weight, the carotenoid concentration in the skin declined drastically within 10 min. In this case, too, it took up to 3 days until the initial carotenoid concentrations had been restored in the respective volunteers. In addition to measuring the carotenoid concentrations, the MED prior to and after alcohol consumption was determined, too. As a result, it became apparent that the UV dose required to induce sunburn was distinctly reduced. Consequently, the volunteers, who had consumed alcohol, were affected quicker by sunburn.

Unexpectedly it turned out that even infrared radiation can reduce the antioxidative status of the skin [44–46]. This is the more surprising as contrary to the UV radiation, the photon energy in the infrared spectral range does not suffice to generate free radicals directly. Obviously, there are mechanisms in the human body, which accumulate the energy of the infrared radiation and prompt a radical formation process. The mitochondria, for example, are known for triggering such processes. These results are in agreement with the findings of Zastrow et al., who determined the free radicals' action spectrum for the whole solar spectrum including the infrared range [9].

With the small handheld skin scanner now available, the investigations had not to be restricted to the laboratory, any more, but could be performed under daily routine conditions. It could be demonstrated with this scanner system that the carotenoid concentration in women is slowly declining during pregnancy and even drastically shortly prenatal. In the skin of the newborns, however, extraordinarily high carotenoid concentrations were detected, which drastically declined in the first days of their life because of the extremely high stress. Only with the breast milk the newborns are fed new antioxidants, which increase the antioxidant concentration in the skin, including carotenoids.

18.6 The Accumulation of Antioxidants in the Skin by Topical and Systemic Application

The intake of fruit and vegetables as well as of food supplements is an important way to accumulate antioxidants in the skin. This process is of particular importance as the skin represents the barrier to the environment. At this dermal barrier, the stress initiated by solar UV radiation-induced free radicals and other environmental toxins is highest. Another method to increase the antioxidant level in the skin is to apply them topically. Currently a wide range of cosmetic creams and lotions that contain antioxidants is commercially available. The importance of these products is steadily increasing, specifically in terms of antiaging strategies. The measuring methods described in this chapter lend themselves excellently to analyze the accumulation of antioxidants by topically applied substances. As in the case of the systemic application, resonance Raman and reflectance spectroscopic measurements can be used for such investigations only if the applied substances contain also carotenoids. If not so, the measurements must be done by electron paramagnetic resonance spectroscopy. The topical application offers the advantage that the antioxidants are completely accumulated immediately after application of the product. If administered systemically, however, the antioxidant concentration in the skin gradually increases with time. On the other hand, the obvious advantage of topical application is neutralized by the fact that the antioxidants remain in the skin only for a limited period of time. Topically applied products are initially on the skin surface or penetrate into the upper cell layers of the stratum corneum. It is in the rarest of cases that they penetrate through the skin barrier (stratum corneum) and if so at relatively low concentrations, only. But every washing process or textile contact removes large parts of these products from the skin. Contrary to that, Raman spectroscopic investigations of the sustainability of the systemic administration of food supplements have shown that enhanced carotenoid concentrations were still detectable after the end of the systemic application for maximally 6 weeks [30, 32]. This is obviously due to the fact that carotenoids well accumulate in fatty tissue, where a reservoir for these antioxidants is formed from which they are slowly released after the end of the systemic application and penetrate out onto the skin together with the sweat and sebum. However, this is not to suggest that the intake of food supplements is the method suited best to accumulate antioxidants in human skin. If applied in enhanced concentrations, antioxidants, too, can lead to the formation of free radicals. The risk of radical formation by antioxidants is higher with single components at higher concentrations than with mixtures of substances in the physiological range. As a general rule, a food supplement is the more effective and safe, the more its composition and concentration corresponds to the natural intake of antioxidants with daily food. For the consumer, it is rather difficult to recognize whether or not a product meets these requirements. Consequently, the introduction of quality standards for food supplements is indispensable.

None of these problems arise by eating fruit and vegetables. With the satiety the human organism has created its own natural limit preventing the excessive intake of antioxidants at concentrations likely to be critical. Moreover, fruit and vegetables contain always a mixture of substances instead of a single antioxidant component.

18.7 Antioxidants and Skin Aging

The intake of healthy food, but also of food supplements, and the administration of antioxidant-containing creams are always intended to strengthen the antioxidant protective system of the human organism, thus protecting the body against illness and aging processes. As a result of Raman spectroscopic investigations carried out on volunteers at the Charité for several years, it was found that individuals with high carotenoid values in their skin looked younger for their age [47]. Based on this result it was tried to substantiate this subjective assessment by objective measurements. It had been intended to recruit for the study healthy volunteers, aged 50 years, which had not changed their lifestyle for years. This meant, for instance, that a former smoker who later desisted from smoking could not be included in the study. As it proved very difficult to find 20 volunteers who fulfilled these criteria, the age segment was extended to a range between 40 and 50 years. The experiments were carried out on the forehead, being a dermal area exposed to light. Using an optical skin surface profilometer Primos 4.0 (GFMesstechnik GmbH, Teltow, Germany), skin aging was described by means of the dermal roughness. This roughness is a measure for the depth and density of furrows and wrinkles. The same skin area, which had been analyzed for its dermal roughness, was also subjected to measurements of the carotenoid concentration. In a first step the values determined for the dermal roughness were compared to the age of the volunteers. If the group of volunteers had covered an age segment between 18 and 80 years, a high correlation between the dermal roughness and age should have been expected. But as the age segment between 40 and 50 years was rather narrow, no correlation between the dermal roughness and age of the volunteers' could be detected in the study. However, a comparison of the concentration of the carotenoid lycopene in the skin with the dermal roughness clearly showed that volunteers with high lycopene concentrations in their skin were distinctly less affected by skin aging than volunteers with low carotenoid levels [47]. This objective assessment confirms the subjective observation that the skin aging process can be essentially reduced by an efficient antioxidative protection system. This is understandable as the solar UV radiation is the main cause of premature skin aging. The free radicals formed in the skin by the solar radiation destroy, inter alia, the collagen and elastin fibers or impede their regeneration. A defense system with a high antioxidant level can neutralize these free radicals before they become harmful thus protecting the collagen and elastin fibers of the skin, which makes the skin appear much younger compared to the actual age. Thus, a healthy diet rich, for instance, in fruit and vegetables proves to be an excellent prevention strategy against premature skin aging. It is important to impart this knowledge particularly to children enabling them to influence their appearance over the years. Although a change of lifestyle at the age of 50 or 70 after having eaten unhealthy food for decades will have a positive effect in the years to come, it is impossible to "eat yourself young."

Conclusions

In summary, a variety of methods is currently available to analyze the interaction between antioxidants and free radicals in the skin. The research results obtained noninvasively, presented in this chapter, clearly demonstrate that a healthy diet is a prerequisite for creating an antioxidative protection system in the human body, with the influence of stress factors having duly to be considered. While it should be no problem for anybody to live on healthy food, the stress factors we are subjected to from day to day are often beyond our control.

In any case, the accumulation of antioxidants in human skin, whether through a healthy diet or the additional intake of food supplements and the topical application of antioxidant substances, represents an essential antiaging strategy.

References

1. Droge W (2002) Free radicals in the physiological control of cell function. Physiol Rev 82:47–95
2. Iannone A, Marconi A, Zambruno G, Giannetti A, Vannini V, Tomasi A (1993) Free radical production during metabolism of organic hydroperoxides by normal human keratinocytes. J Invest Dermatol 101: 59–63
3. Jackson MJ (1999) Free radicals in skin and muscle: damaging agents or signals for adaptation? Proc Nutr Soc 58:673–676
4. Akaike T (2001) Role of free radicals in viral pathogenesis and mutation. Rev Med Virol 11:87–101
5. Halliwell B (2007) Biochemistry of oxidative stress. Biochem Soc Trans 35:1147–1150
6. Valko M, Izakovic M, Mazur M, Rhodes CJ, Telser J (2004) Role of oxygen radicals in DNA damage and cancer incidence. Mol Cell Biochem 266:37–56
7. Maeda H, Akaike T (1998) Nitric oxide and oxygen radicals in infection, inflammation, and cancer. Biochemistry (Mosc) 63:854–865
8. Stone V, Johnston H, Clift MJ (2007) Air pollution, ultrafine and nanoparticle toxicology: cellular and molecular interactions. IEEE Trans Nanobioscience 6:331–340
9. Zastrow L, Groth N, Klein F, Kockott D, Lademann J, Renneberg R, Ferrero L (2009) The missing link–light-induced (280–1,600 nm) free radical formation in human skin. Skin Pharmacol Physiol 22:31–44
10. Sander CS, Hamm F, Elsner P, Thiele JJ (2003) Oxidative stress in malignant melanoma and non-melanoma skin cancer. Br J Dermatol 148:913–922
11. Thiele JJ, Schroeter C, Hsieh SN, Podda M, Packer L (2001) The antioxidant network of the stratum corneum. Curr Probl Dermatol 29:26–42

12. Darvin M, Zastrow L, Sterry W, Lademann J (2006) Effect of supplemented and topically applied antioxidant substances on human tissue. Skin Pharmacol Physiol 19:238–247
13. Darvin ME, Haag SF, Meinke MC, Sterry W, Lademann J (2011) Determination of the influence of ir radiation on the antioxidative network of the human skin. J Biophotonics 4:21–29
14. Darvin ME, Zastrov L, Gonchukov SA, Lademann J (2009) Influence of IR radiation on the carotenoid content in human skin. Optics and Spectroscopy 107(6):917–920
15. Bernstein PS, Yoshida MD, Katz NB, McClane RW, Gellermann W (1998) Raman detection of macular carotenoid pigments in intact human retina. Invest Ophthalmol Vis Sci 39:2003–2011
16. Hata TR, Scholz TA, Ermakov IV, McClane RW, Khachik F, Gellermann W, Pershing LK (2000) Non-invasive raman spectroscopic detection of carotenoids in human skin. J Invest Dermatol 115:441–448
17. Ermakov IV, Ermakova MR, Gellermann W, Lademann J (2004) Noninvasive selective detection of lycopene and beta-carotene in human skin using raman spectroscopy. J Biomed Opt 9:332–338
18. Darvin ME, Gersonde I, Albrecht H, Gonchukov SA, Sterry W, Lademann J (2005) Determination of beta carotene and lycopene concentrations in human skin using resonance raman spectroscopy. Laser Phys 15:295–299
19. Darvin ME, Gersonde I, Meinke M, Sterry W, Lademann J (2005) Non-invasive in vivo determination of the carotenoids beta-carotene and lycopene concentrations in the human skin using the raman spectroscopic method. J Phys D: Appl Phys 38: 2696–2700
20. Darvin ME, Gersonde I, Albrecht H, Meinke M, Sterry W, Lademann J (2006) Non-invasive in vivo detection of the carotenoid antioxidant substance lycopene in the human skin using the resonance raman spectroscopy. Laser Phys Lett 3:460–463
21. Darvin ME, Gerzonde I, Ey S, Brandt NN, Albrecht H, Gonchukov SA, Sterry W, Lademann J (2004) Noninvasive detection of beta-carotene and lycopene in human skin using Raman spectroscopy. Laser Phys 14(2):231–233
22. Darvin ME, Sandhagen C, Koecher W, Sterry W, Lademann J, Meinke MC (2012) Comparison of two methods for non-invasive determination of carotenoids in human and animal skin: Raman spectroscopy versus reflection spectroscopy. J Biophotonics. 5(7):550–558
23. Herrling T, Fuchs J, Rehberg J, Groth N (2003) Uv-induced free radicals in the skin detected by esr spectroscopy and imaging using nitroxides. Free Radic Biol Med 35:59–67
24. Fuchs J, Groth N, Herrling T (1998) Cutaneous tolerance to nitroxide free radicals in human skin. Free Radic Biol Med 24:643–648
25. Haag SF, Chen M, Peters D, Keck CM, Taskoparan B, Fahr A, Teutloff C, Bittl R, Lademann J, Schafer-Korting M, Meinke MC (2011) Nanostructured lipid carriers as nitroxide depot system measured by electron paramagnetic resonance spectroscopy. Int J Pharm 421:364–369
26. Haag SF, Bechtel A, Darvin ME, Klein F, Groth N, Schafer-Korting M, Bittl R, Lademann J, Sterry W, Meinke MC (2010) Comparative study of carotenoids, catalase and radical formation in human and animal skin. Skin Pharmacol Physiol 23:306–312
27. Meinke MC, Haag SF, Schanzer S, Groth N, Gersonde I, Lademann J (2011) Radical protection by sunscreens in the infrared spectral range. Photochem Photobiol 87:452–456
28. Fuchs J, Groth N, Herrling T (2002) In vivo measurement of oxidative stress status in human skin. Methods Enzymol 352:333–339
29. Haag SF, Taskoparan B, Darvin ME, Groth N, Lademann J, Sterry W, Meinke MC (2011) Determination of the antioxidative capacity of the skin in vivo using resonance raman and electron paramagnetic resonance spectroscopy. Exp Dermatol 20:483–487
30. Meinke MC, Darvin ME, Vollert H, Lademann J (2010) Bioavailability of natural carotenoids in human skin compared to blood. Eur J Pharm Biopharm 76:269–274
31. Blume-Peytavi U, Rolland A, Darvin ME, Constable A, Pineau I, Voit C, Zappel K, Schafer-Hesterberg G, Meinke M, Clavez RL, Sterry W, Lademann J (2009) Cutaneous lycopene and beta-carotene levels measured by resonance Raman spectroscopy: high reliability and sensitivity to oral lactolycopene deprivation and supplementation. Eur J Pharm Biopharm 73:187–194
32. Darvin ME, Fluhr JW, Schanzer S, Richter H, Patzelt A, Meinke MC, Zastrow L, Golz K, Doucet O, Sterry W, Lademann J (2011) Dermal carotenoid level and kinetics after topical and systemic administration of antioxidants: enrichment strategies in a controlled in vivo study. J Dermatol Sci 64:53–58
33. Meinke MC, Friedrich A, Tscherch K, Haag SF, Darvin ME, Vollert H, Groth N, Lademann J, Rohn S (2013) Influence of dietary carotenoids on radical scavenging capacity of the skin and skin lipids. Eur J Pharm Biopharm 84(2):365–373
34. Lademann J, Caspers PJ, van der Pol A, Richter H, Patzelt A, Zastrow L, Darvin M, Sterry W, Fluhr JW (2009) In vivo raman spectroscopy detects increased epidermal antioxidative potential with topically applied carotenoids. Laser Phys Lett 6:76–79
35. Fluhr JW, Caspers P, van der Pol JA, Richter H, Sterry W, Lademann J, Darvin ME (2011) Kinetics of carotenoid distribution in human skin in vivo after exogenous stress: disinfectant and wira-induced carotenoid depletion recovers from outside to inside. J Biomed Opt 16:035002

36. Darvin ME, Patzelt A, Knorr F, Blume-Peytavi U, Sterry W, Lademann J (2008) One-year study on the variation of carotenoid antioxidant substances in living human skin: influence of dietary supplementation and stress factors. J Biomed Opt 13:044028-(1–9)
37. Meinke MC, Taskoparan B, Gersonde I, Lademann J, Darvin ME (2011) Influence on the carotenoid levels of skin arising from age, gender, body mass index in smoking/non-smoking individuals. Free Radic Antioxid 1:15–20
38. Mayne ST, Cartmel B, Scarmo S, Lin H, Leffell DJ, Welch E, Ermakov I, Bhosale P, Bernstein PS, Gellermann W (2010) Noninvasive assessment of dermal carotenoids as a biomarker of fruit and vegetable intake. Am J Clin Nutr 92:794–800
39. Tucker KL, Chen H, Vogel S, Wilson PW, Schaefer EJ, Lammi-Keefe CJ (1999) Carotenoid intakes, assessed by dietary questionnaire, are associated with plasma carotenoid concentrations in an elderly population. J Nutr 129:438–445
40. Brady WE, Mares-Perlman JA, Bowen P, Stacewicz-Sapuntzakis M (1996) Human serum carotenoid concentrations are related to physiologic and lifestyle factors. J Nutr 126:129–137
41. Darvin ME, Gersonde I, Albrecht H, Sterry W, Lademann J (2006) In vivo raman spectroscopic analysis of the influence of uv radiation on carotenoid antioxidant substance degradation of the human skin. Laser Phys 16:833–837
42. Ribaya-Mercado JD, Garmyn M, Gilchrest BA, Russell RM (1995) Skin lycopene is destroyed preferentially over beta-carotene during ultraviolet irradiation in humans. J Nutr 125:1854–1859
43. Di Mascio P, Kaiser S, Sies H (1989) Lycopene as the most efficient biological carotenoid singlet oxygen quencher. Arch Biochem Biophys 274:532–538
44. Darvin ME, Gersonde I, Albrecht H, Zastrow L, Sterry W, Lademann J (2007) In vivo raman spectroscopic analysis of the influence of ir radiation on the carotenoid antioxidant substances beta-carotene and lycopene in the human skin. Formation of free radicals. Laser Phys Lett 4:318–321
45. Darvin ME, Haag S, Meinke M, Zastrow L, Sterry W, Lademann J (2010) Radical production by infrared a irradiation in human tissue. Skin Pharmacol Physiol 23:40–46
46. Darvin ME, Haag SF, Lademann J, Zastrow L, Sterry W, Meinke MC (2010) Formation of free radicals in human skin during irradiation with infrared light. J Invest Dermatol 130:629–631
47. Darvin M, Patzelt A, Gehse S, Schanzer S, Benderoth C, Sterry W, Lademann J (2008) Cutaneous concentration of lycopene correlates significantly with the roughness of the skin. Eur J Pharm Biopharm 69:943–947

Raman Spectroscopy

19

Anke Sieg

19.1 Introduction

Raman spectroscopy, a vibrational spectroscopy technique, provides detailed information about molecular composition and molecular structure. When light interacts with a sample, part of the energy is absorbed, the other is scattered. From the latter portion, most photons are elastically scattered with the same energy and wavelength as the incident light (Rayleigh scattering). A small fraction, approximately 1 in 10 million photons, is inelastically scattered, with the scattered photons having a frequency different from (and usually lower than) the incident light. This fraction is the Raman scattering.

Thanks to technical progress in recent years, tissue characterization by Raman spectroscopy has received increased attention. In the past decade, the technique has become a routine method in vitro, and in vivo analysis of skin has become a reality [1]. Raman spectroscopy is complementary to IR spectroscopy, which analyses the absorbed fraction of light. Both techniques induce specific changes in vibrational modes of molecular structures: IR-active bonds are those that change their dipole moment upon absorption of electromagnetic radiation; Raman-active bonds are those that change their polarizability. This makes Raman spectroscopy particularly attractive for living tissue, where the strong IR signal of water may easily overwhelm those of other tissue components. Lipid and lipophilic structures are particularly suited for analysis by Raman spectroscopy, so that the Raman technique is able to provide particularly valuable information about the structure and composition of the stratum corneum (SC).

A. Sieg, PhD
Science and Technology, Healthcare Industries,
Dow Corning Europe SA, Parc Industriel – Zone C,
Rue Jules Bordet, Seneffe B-7180, Belgium
e-mail: anke.sieg@dowcorning.com

19.2 Technical Approaches

The most used technical set-ups for skin analysis are Fourier transform (FT)-Raman spectroscopy, resonance Raman spectroscopy and Raman microscopy with or without confocal capability. Common components of all instruments are one or two lasers as radiation source, optics to illuminate the sample and to focus the scattered radiation and a detector. FT-Raman spectrometers use high excitation wavelengths (e.g. 1,064 nm) and employ a Michelson interferometer, which allows detecting photons scattered over the whole wavelength range simultaneously by a single detector. The Raman spectrum of a sample is obtained by inverse Fourier transformation of the signal intensities. Alternatively, dispersive spectrometers, in which the scattered radiation is dispersed into its constituent wavelengths with the aid of a monochromator via diffraction grating, can collect Raman spectra much faster when equipped with multichannel CCD detectors [2]. The latter option is employed in the confocal Raman

E. Berardesca et al. (eds.), *Non Invasive Diagnostic Techniques in Clinical Dermatology*,
DOI 10.1007/978-3-642-32109-2_19, © Springer Berlin Heidelberg 2014

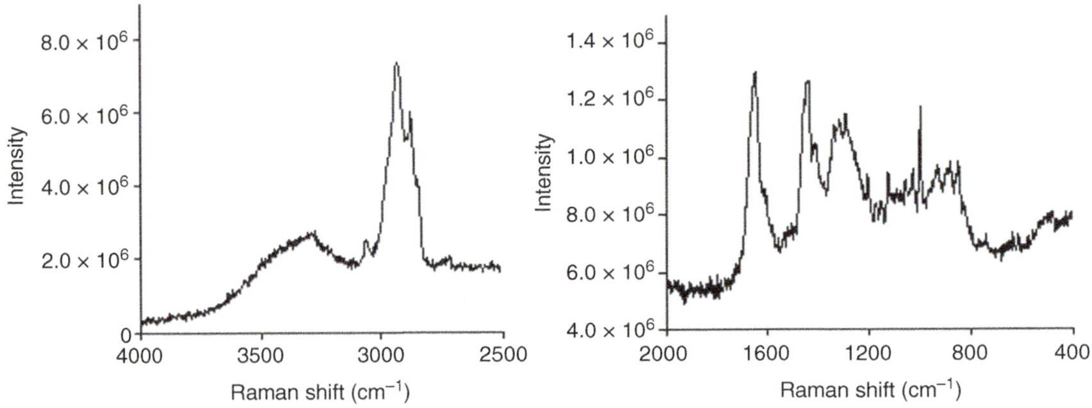

Fig. 19.1 Typical, background-corrected Raman spectra from a single scan of human skin in vivo, collected at ~4 μm depth. *Left*: high-wavenumber region (671-nm laser excitation, exposure 1 s). *Right*: fingerprint region (785-nm laser excitation, exposure 10 s)

instrument described below, in which a Raman spectrometer is combined with a confocal microscope in one instrument. FT-Raman spectroscopy has the advantage of better spectral resolution and lower fluorescence background and that the signal can be collected from larger sample areas. However, dispersive CCD devices with lasers in the visible or NIR range outweigh these advantages by the speed with which spectra can and need to be collected in a clinical environment [2].

In resonance Raman spectroscopy, the intensity of the incoming laser is adjusted so that the scattered light coincides with an electronic transition of the molecule of interest. This requires the use of a tunable laser in the visible or UV range. The main advantage of the resonance Raman approach is the large increase of intensity for the peaks of interest; however, increased fluorescence can also compromise the quality of the analysis. This technique has been mainly used to study the content of carotenoid antioxidant substances in the skin [3, 4].

Much of the earlier work in dermatology employed mainly FT-Raman spectroscopy to investigate skin tissue in vitro and ex vivo [2, 5]. A major breakthrough for skin analysis was the combination of Raman spectroscopy with confocal microscopy, allowing a completely non-invasive, real-time "optical biopsy" in vivo [1, 6, 7]. Since 2004, specialized confocal Raman instrumentation has become available as clinical research tool, and the applications for investigating the skin structure, diagnosing skin pathologies, measuring skin hydration or penetration of actives for product development and cosmetic claim substantiation have steadily increased.

A typical confocal Raman set-up for in vivo use employs one or two lasers in the visible or near-infrared range, such as 671 and 785 nm, respectively [1]. The 671-nm laser allows analysis of the high-wavenumber region (2,500–4,000 cm^{-1}), mainly used for skin hydration measurements, while the 785-nm laser serves for analysis of the fingerprint region (400–2,000 cm^{-1}). Figure 19.1 shows typical spectra of human stratum corneum collected with both lasers. The use of shorter wavelengths allows faster collection of Raman spectra but increases autofluorescence of the skin as background noise. Laser light in the near-infrared range penetrates deeper into the tissue and generates less autofluorescence but requires longer exposure times.

The optics used in the confocal microscope determines the depth resolution [8, 9]. The system described by Caspers reaches a depth resolution of approximately 4 μm [6]. With oversampling in steps of 2 μm, one obtains a representative tissue profile even of body sites where the skin is relatively thin (cheek, forearm) (Fig. 19.2). Other non-invasive techniques such as optical coherence tomography (OCT) do not yet reach this level of resolution [6].

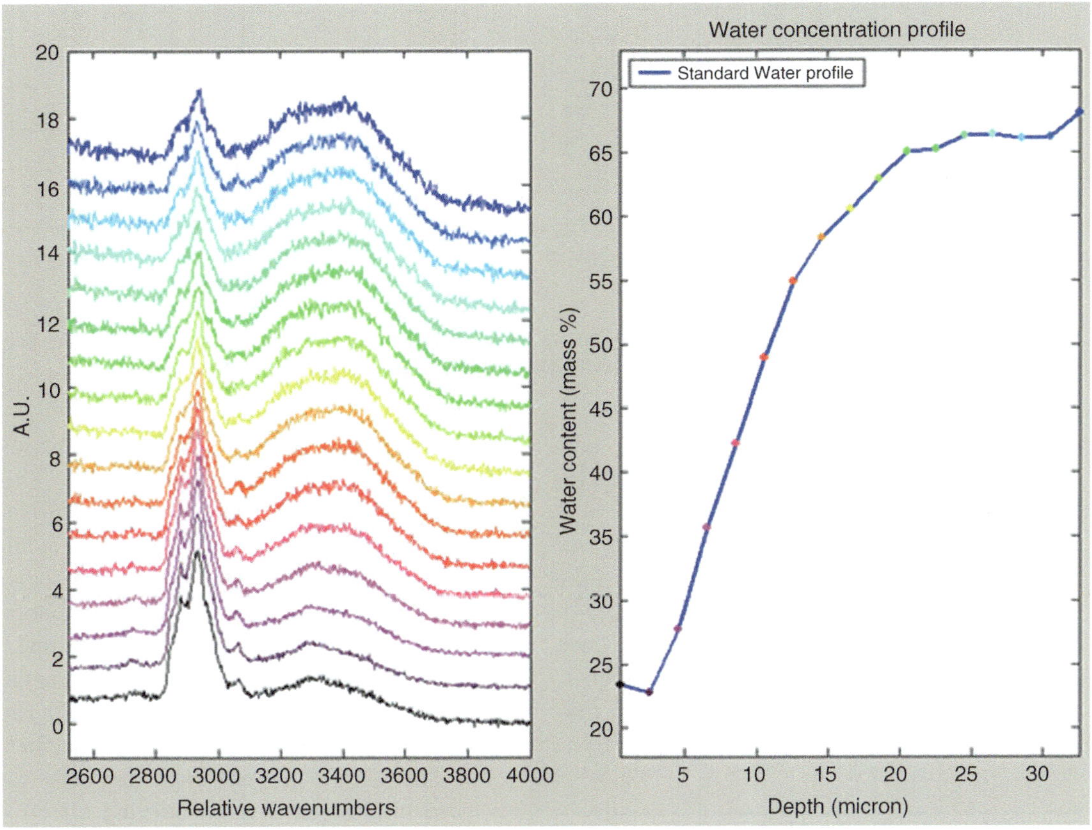

Fig. 19.2 Typical water profile of human stratum corneum on the volar forearm, obtained with confocal Raman spectroscopy

The rest of this chapter provides an overview of investigations using Raman spectroscopy that are relevant to dermatology.

19.3 Structure Analysis of the Skin

Initial work to analyse the structure of stratum corneum, nail and hair with FT-Raman spectroscopy was published in the 1990s [10–12], and spectral bands to the different skin components were assigned by Barry [13]. A lot of effort has been further focused on the analysis of skin lipids and, more specifically, ceramides [14–16]. Complementary to other techniques (such as X-ray, DSC, FT-IR spectroscopy), Raman spectroscopy has provided key insights into the organization of skin lipid bilayers and their role as barrier as well as penetration pathway across the stratum corneum. In this context, the work of Williams and Barry as well as others has been fundamental to understand the effect of penetration enhancers on stratum corneum lipid organization [17, 18]. Finally, FT-Raman spectroscopy was also employed in the analysis of archaeological skin samples of the Iceman [19].

Environmental stress and protective mechanisms of the skin are subject of current research. Antioxidant systems in the skin, and particularly carotenoids, have been well characterized by Raman spectroscopy [3, 4, 20–22]. A separate chapter of this book is dedicated to this topic. Further, UV damage expressed via the isomerization of urocanic acid from the trans to cis configuration has been reported by Japanese researchers [23].

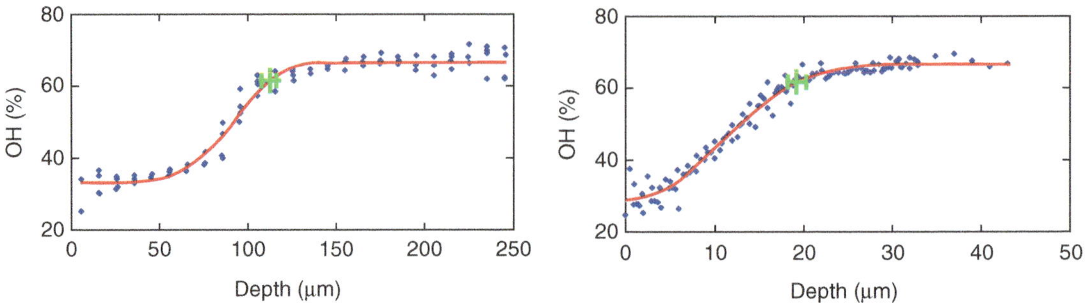

Fig. 19.3 Average hydration profile of the thumb (*left*) and volar forearm (*right*) measured by confocal Raman spectroscopy. The upper point of inflexion is mathematically determined via an automated algorithm [6]

Confocal Raman spectroscopy is able to provide representative hydration profiles of the stratum corneum from different body sites (see next section). Analysing the shape and its points of inflexion (Fig. 19.3) allows determination of the SC thickness at different body sites [1, 6, 24]. Several semiautomated algorithms have been developed and validated against other techniques, in order to render this application useful for routine clinical use [25, 26].

19.4 Skin Hydration

A lot of effort has been focused on obtaining a deeper insight into skin hydration and the effect of topical moisturizers. "Traditional" methods such as electrical measurements of the stratum corneum hydration provide a fast and easy measure, but the value of this information (in particular related to interference with materials applied on the skin and the depth of measurement) has been more and more questioned as the new technologies emerge [6, 27]. Confocal Raman spectroscopy has allowed a major breakthrough in understanding swelling of the stratum corneum upon occlusion as well as in measuring the benefit different moisturizers deliver short or long term in treating dry skin conditions. For example, Stamatas et al. [28] investigated the uptake of lipids into the stratum corneum as well as their occlusive effect on the stratum corneum. Crowther et al. [6] were able to show the long-lasting benefit of a moisturizing cream with niacinamide on skin hydration via an increased SC thickness that leads to improved barrier function and higher total water content in the SC. This effect became significant after 2 weeks of treatment and remained measurable after one additional week of regression, while the other products did not show sustained improvement on skin hydration after the treatment was stopped. Thus, the study confirmed a possible treatment to effectively break the "dry skin cycle" [29]. The findings were not mirrored by skin capacitance measurements, which underlines the need for advanced clinical tools in identifying effective treatments for impaired skin barrier conditions.

The impact of another moisturizing agent, the biomimetic biopolymer pMPC, was investigated with a slightly different confocal Raman set-up [30]. Egawa and Tagami [31] studied SC hydration and NMF content of the cheek and volar forearm on a Japanese population and identified effects of age, anatomical site and season.

Very recent research went beyond the stratum corneum to investigate hydration of deeper skin layers. Using confocal Raman microscopy in vitro and dynamic vapour sorption measurements, Zhang et al. [32] investigated hydration mechanisms of human collagen and dermal pig tissue - with relevance for future applications in wound healing and skin aging.

Nearly all major cosmetic companies have now employed confocal Raman spectroscopy in clinical investigations to understand mechanisms of skin hydration and skin aging. Claims made on these findings as well as further research reports in this field will continue to emerge in the near future.

19.5 Diagnosis of Skin Diseases

While research around skin hydration is mainly driven by the cosmetic industry, clinical dermatologists also seek better tools to non-invasively diagnose skin diseases. The potential of Raman spectroscopy to help in this field was first demonstrated in 1995 [33]. Since then, researchers have focused their attention on major skin conditions. Psoriatic scales [34], psoriatic skin models [35] and clinically uninvolved skin of patients with atopic dermatitis [36] were compared to healthy skin in vitro and ex vivo. Several differences in the conformation and bands of lipids and proteins were found, leading to a less-organized SC structure and, thus, a weaker skin barrier. Very recent efforts focused on the early screening to detect the prevalence to develop atopic dermatitis in children and to adapt the treatment depending on the genotype [37]. This study revealed that for the screening of the 3 filaggrin genotype, a confocal Raman set-up is not necessarily required but that a simpler Raman instrument with simpler resolution may be of advantage due to the presence of small, highly localized tyrosine spots in the skin. This could ease the accessibility of the technique for a routine clinical environment since handheld, less-expensive Raman devices for field applications are already available for non-medical applications.

Other researchers used Raman spectroscopy to localize melanin distribution in the skin and to understand excessive pigmentation in patients with melasma [38]. Azrad et al. [39] detected alterations of the protein content in skin tissues of patients with bullous pemphigoid. Finally, in the attempt to improve early diagnosis of skin cancer, the water content of biopsies from benign and malignant tumours was measured with FT-Raman spectroscopy [40].

19.6 Percutaneous Absorption

Besides profiling of hydration and skin components, confocal Raman spectroscopy also allows tracking the penetration of substances into the skin in vivo. A separate chapter of this book is dedicated to this topic. While this application presents particular interest for cosmetic and drug delivery applications, the confocal Raman technique also provides valuable information for toxicological purposes [41, 42].

Stimulated Raman scattering (SRS) recently emerged as new technology to track skin penetration of actives and excipients. SRS is derived from coherent anti-Stokes Raman scattering (CARS). CARS employs multiple photons to address the molecular vibrations and requires two synchronized laser pulses of different wavelengths. Pump field and Stokes field interact with the sample to generate a signal at anti-Stokes frequency which is orders of magnitude stronger than the spontaneous Raman emission. CARS is particularly suited for high-resolution, 3D imaging with sectioning capability without the need for further labelling; the signal can be easily detected in the presence of fluorescent background [43]. However, CARS spectra are very complex, image interpretation is difficult and the concentration dependence of the compound to be tracked is not linear. SRS microscopy refines this method in that the two synchronized laser pulses are tuned to a difference in wavelength that matches an intrinsic molecular vibrational frequency in the sample. As a result, the image interpretation is simpler, the signal intensity is linear to the concentration in the sample and high sensitivity for the compounds of interest is achieved. A separate chapter in this book is dedicated to this technique.

SRS has been used for ex vivo investigations in mouse skin, where the penetration of ibuprofen and ketoprofen was simultaneously imaged to that of propylene glycol [44]. This study was able to outline the contribution of the follicular and intercellular pathway to the overall penetration of the drugs. The SRS images further showed much faster penetration of propylene glycol across the stratum corneum, leaving the actives crystallized behind in the stratum corneum. Understanding the fate of the different formulation components once applied onto the skin is key to effective formulation design. For formulators, therefore, these technical developments are exciting prospects for improving the efficacy of topical treatment and enhancing patient engagement in long-term dermatological therapy.

19.7 Summary and Conclusions

Over the last decade, Raman spectroscopy has emerged to a non-invasive, established clinical research tool that benefits clinical dermatology as well as cosmetic and pharmaceutical product development. Applications range from skin structure analysis, response to environmental stress, diagnosis of skin disorders and skin hydration to penetration of actives for drug delivery and toxicology. Confocal Raman spectroscopy has provided a major breakthrough in this but represents for in vivo use, a significant investment. To increase success and widespread use beyond research and specialized labs, the development of fibre optics and portable, affordable equipment is required. Algorithms to automatically and reliably analyse large data sets from clinical trials need to be extended beyond hydration.

Recent SRS studies demonstrated concomitant analysis of the penetration of drug and key excipients. This opens the way for more detailed investigation on how key excipients interact with the drug active as well as the skin barrier. This will provide valuable insights to design more effective topical formulations and to achieve improved patient engagement in the long-term treatment of common skin conditions.

References

1. Caspers PJ, Lucassen GW, Carter EA, Bruining HA, Puppels GJ (2001) In vivo confocal Raman microspectroscopy of the skin: noninvasive determination of molecular concentration profiles. J Invest Dermatol 116(3):434–442
2. Wolthuis R, Bakker Schut TC, Caspers Peter J, Buschman HPJ, Römer TJ, Bruining HA et al (1999) Raman spectroscopic methods for in vivo and in vitro tissue characterization. In: Mason WT (ed) Fluorescent and luminent probes for biological activity – a practical guide to technology for quantitative real-time analysis, 2nd edn. Academic, London, pp 433–455
3. Darvin ME, Lademann J (2010) Resonance Raman spectroscopy of human skin for the in vivo detection of carotenoid antioxidant substances. In: Tuchin VV (ed) Handbook of photonics for biomedical science. CRC Press, Boca Raton, pp 229–252
4. Ermakov IV, Sharifzadeh M, Ermakova M, Gellermann W (2005) Resonance Raman detection of carotenoid antioxidants in living human tissue. J Biomed Opt 10(6):064028, 1–18
5. Williams AC, Barry BW (1999) Raman spectroscopy. In: Bronaugh RL, Maibach HI (eds) Percutaneous absorption drugs-cosmetics-mechanisms-methodology, 3rd edn. Marcel Dekker, New York, pp 499–514
6. Crowther JM, Sieg A, Blenkiron P, Marcott C, Matts PJ, Kaczvinsky JR et al (2008) Measuring the effects of topical moisturizers on changes in stratum corneum thickness, water gradients and hydration in vivo. Br J Dermatol 159(3):567–577
7. Caspers PJ, Lucassen GW, Puppels GJ (2003) Combined in vivo confocal Raman spectroscopy and confocal microscopy of human skin. Biophys J 85(1): 572–580
8. Everall N (2004) Depth profiling with confocal Raman microscopy, part II. Spectrosc (Duluth, MN, United States) 19(11):16, 8 ,20–26
9. Everall N (2004) Depth profiling with confocal Raman microscopy, part I. Spectrosc (Duluth, MN, United States) 19(10):22, 4–8
10. De Faria DLA, De Souza MA (1999) Raman spectra of human skin and nail excited in the visible region. J Raman Spectrosc 30(3):169–171
11. Gniadecka M, Nielsen OF, Christensen DH, Wulf HC (1998) Structure of water, proteins, and lipids in intact human skin, hair, and nail. J Invest Dermatol 110(4): 393–398
12. Williams AC, Edwards HGM, Barry BW (1994) Raman spectra of human keratotic biopolymers: skin, callus, hair and nail. J Raman Spectrosc 25(1): 95–98
13. Barry BW, Edwards HGM, Williams AC (1992) Fourier transform Raman and infrared vibrational study of human skin: assignment of spectral bands. J Raman Spectrosc 23(11):641–645
14. Guillard E, Tfayli A, Manfait M, Baillet-Guffroy A (2011) Thermal dependence of Raman descriptors of ceramides. Part II: effect of chains lengths and head group structures. Anal Bioanal Chem 399(3): 1201–1213
15. Kessner D, Brezesinski G, Funari SS, Dobner B, Neubert RHH (2010) Impact of the long chain ω-acylceramides on the stratum corneum lipid nanostructure. Part 1: Thermotropic phase behaviour of CER[EOS] and CER[EOP] studied using X-ray powder diffraction and FT-Raman spectroscopy. Chem Phys Lipids 163(1):42–50
16. Wartewig S, Neubert RHH (2007) Properties of ceramides and their impact on the stratum corneum structure. Skin Pharmacol Physiol 20(5): 220–229
17. Williams AC, Edwards HGM, Lawson EE, Barry BW (2006) Molecular interactions between the penetration

enhancer 1,8-cineole and human skin. J Raman Spectrosc 37(1–3):361–366
18. Vavrova K, Zbytovska J, Hrabalek A (2005) Amphiphilic transdermal permeation enhancers: structure-activity relationships. Curr Med Chem 12(19):2273–2291
19. Edwards HGM, Williams AC, Barry BW (1996) Human skin. A Fourier transform Raman spectroscopic study of the Iceman. Spectrosc Eur 8(1):10, 2, 4–6, 8
20. Darvin ME, Fluhr JW, Caspers P, van der Pool A, Richter H, Patzelt A et al (2009) In vivo distribution of carotenoids in different anatomical locations of human skin: comparative assessment with two different Raman spectroscopy methods. Exp Dermatol 18(12):1060–1063
21. Lademann J, Caspers PJ, van der Pol A, Richter H, Patzelt A, Zastrow L et al (2009) In vivo Raman spectroscopy detects increased epidermal antioxidative potential with topically applied carotenoids. Laser Phys Lett 6(1):76–79
22. Hata TR, Scholz TA, Ermakov IV, McClane RW, Khachik F, Gellermann W et al (2000) Non-invasive Raman spectroscopic detection of carotenoids in human skin. J Invest Dermatol 115(3):441–448
23. Egawa M, Nomura J, Iwaki H (2010) The evaluation of the amount of cis- and trans-urocanic acid in the stratum corneum by Raman spectroscopy. Photochem Photobiol Sci 9(5):730–733
24. Egawa M, Hirao T, Takahashi M (2007) In vivo estimation of stratum corneum thickness from water concentration profiles obtained with Raman spectroscopy. Acta dermato-venereologica 87(1):4–8
25. Crowther JM, Blenkiron P, Matts PJ, Sieg A, Inventors (2007) The Procter & Gamble Company, USA. assignee. Confocal Raman spectroscopy for dermatological studies. Application: WO WO patent 2006-IB53044 2007026325. 20060831
26. Bielfeldt S, Schoder V, Ely U, van der Pol A, de Sterke J, Wilhelm K-P (2009) Assessment of human stratum corneum thickness and its barrier properties by in-vivo confocal Raman spectroscopy. IFSCC Mag 12(1):9–15
27. Zhang Shuliang L, Meyers CL, Subramanyan K, Hancewicz TM (2005) Near infrared imaging for measuring and visualizing skin hydration. A comparison with visual assessment and electrical methods. J Biomed Opt 10(3):031107
28. Stamatas GN, de Sterke J, Hauser M, von Stetten O, van der Pol A (2008) Lipid uptake and skin occlusion following topical application of oils on adult and infant skin. J Dermatol Sci 50(2):135–142
29. Matts PJ, Rawlings AV (2006) The dry skin cycle. Cosmetic science and technology series. In: Draelos ZD and Thanan LA (eds) Cosmetic formulation of skin care products, 30. pp 79–114
30. Chrit L, Bastien P, Biatry B, Simonnet JT, Potter A, Minondo AM et al (2007) In vitro and in vivo confocal Raman study of human skin hydration: assessment of a new moisturizing agent, pMPC. Biopolymers 85(4):359–369
31. Egawa M, Tagami H (2008) Comparison of the depth profiles of water and water-binding substances in the stratum corneum determined in vivo by Raman spectroscopy between the cheek and volar forearm skin: effects of age, seasonal changes and artificial forced hydration. Br J Dermatol 158(2):251–260
32. Zhang Q, Andrew Chan KL, Zhang G, Gillece T, Senak L, Moore David J et al (2011) Raman microspectroscopic and dynamic vapor sorption characterization of hydration in collagen and dermal tissue. Biopolymers 95(9):607–615
33. Edwards HGM, Williams AC, Barry BW (1995) Potential applications of FT-Raman spectroscopy for dermatological diagnostics. J Mol Struct 347:379–387
34. Osada M, Gniadecka M, Wulf HC (2004) Near-infrared Fourier transform Raman spectroscopic analysis of proteins, water and lipids in intact normal stratum corneum and psoriasis scales. Exp Dermatol 13(6):391–395
35. Bernard G, Auger M, Soucy J, Pouliot R (2007) Physical characterization of the stratum corneum of an in vitro psoriatic skin model by ATR-FTIR and Raman spectroscopies. Biochim Biophys Acta (General Subjects) 1770(9):1317–1323
36. Wohlrab J, Vollmann A, Wartewig S, Marsch WC, Neubert R (2001) Noninvasive characterization of human stratum corneum of undiseased skin of patients with atopic dermatitis and psoriasis as studied by Fourier transform Raman spectroscopy. Biopolymers 62(3):141–146
37. O'Regan GM, Kemperman PMJH, Sandilands A, Chen H, Campbell LE, Kroboth K et al (2010) Raman profiles of the stratum corneum define 3 filaggrin genotype-determined atopic dermatitis endophenotypes. J Allergy Clin Immunol 126(3):574–580, e/1
38. Moncada B, Sahagun-Sanchez LK, Torres-Alvares B, Castanedo-Cazares JP, Martinez-Ramirez JD, Gonzalez FJ (2009) Molecular structure and concentration of melanin in the stratum corneum of patients with melasma. Photodermatol Photoimmunol Photomed 25(3):159–160
39. Azrad E, Cagnano E, Halevy S, Rosenwaks S, Bar I (2005) Bullous pemphigoid detection by micro-Raman spectroscopy and cluster analysis: structure alterations of proteins. J Raman Spectrosc 36(11): 1034–1039
40. Gniadecka M, Nielsen OF, Wulf HC (2003) Water content and structure in malignant and benign skin tumours. J Mol Struct 661–662:405–410
41. Kezic S (2008) Methods for measuring in-vivo percutaneous absorption in humans. Hum Exp Toxicol 27(4):289–295
42. Broding Horst C, van der Pol A, de Sterke J, Monse C, Fartasch M, Bruning T (2011) In vivo monitoring of epidermal absorption of hazardous substances by confocal Raman micro-spectroscopy. J Dtsch Dermatol Ges 9(8):618–627
43. Baena JR, Lendl B (2004) Raman spectroscopy in chemical bioanalysis. Curr Opin Chem Biol 8(5):534–539
44. Saar BG, Contreras-Rojas LR, Xie XS, Guy RH (2011) Imaging drug delivery to skin with stimulated Raman scattering microscopy. Mol Pharm 8(3): 969–975

20 Imaging Drug Delivery to Skin with Coherent Raman Scattering Microscopy

Natalie A. Belsey, Luis Rodrigo Contreras-Rojas, and Richard H. Guy

Abbreviations

CARS	Coherent anti-Stokes Raman scattering
CN	Condenser
DMSO	Dimethyl sulfoxide
F	Filter
FTIR	Fourier transform infrared spectroscopy
HPLC	High-performance liquid chromatoraphy
LIA	Lock-in amplifier
M	Modulator
OL	Objective lens
PD	Photodiode
RA	Retinoic acid
SRG	Stimulated Raman gain
SRL	Stimulated Raman loss
SRS	Stimulated Raman scattering

20.1 Introduction

Successful treatment of dermatological disease relies on efficient drug delivery through the skin to the site of action. To optimise a formulation, the extent of drug uptake into and through the skin must be monitored. There are currently a number of different methods for investigating drug permeation through the skin [1], but all approaches suffer from various drawbacks. For example, a common technique is tape stripping, in which layers of the stratum corneum are consecutively removed using adhesive tape, for subsequent chemical analysis (e.g. by HPLC). Tapes are weighed pre and post application to the skin to determine the amount of stratum corneum removed and to calculate the concentration of drug in each layer and thus generate the drug disposition profile. This laborious technique is destructive/invasive, an undesirable trait, which also blights many other methods such as microdialysis, suction blister formation and punch biopsy. In addition, these techniques offer no insight into the penetration pathway of the drug.

Imaging techniques such as confocal microscopy [2] are useful to visualise the skin and its deeper layers to reveal the permeation route taken by the drug, for example, via the intercellular lipid matrix between corneocytes or through the hair follicles. However, the major limitation of this technique is that it is restricted to fluorescent probes. Labelling a drug with a fluorescent marker would greatly affect its physicochemical properties and therefore its permeation kinetics [3], rendering the experiment of little use.

Spectroscopic techniques such as FTIR have been used to detect chemicals in the skin [4, 5], without the requirement for chemical labelling; however, this technique is limited to the outer skin surface. Confocal Raman microspectroscopy [6–10] allows depth sectioning, but the

N.A. Belsey, DPhil • L.R. Contreras-Rojas
R.H. Guy, PhD (✉)
Department of Pharmacy and Pharmacology, University of Bath, Claverton Down Bath BA2 7AY, UK
e-mail: n.belsey@bath.ac.uk; lrcr20@bath.ac.uk; r.h.guy@bath.ac.uk

E. Berardesca et al. (eds.), *Non Invasive Diagnostic Techniques in Clinical Dermatology*,
DOI 10.1007/978-3-642-32109-2_20, © Springer Berlin Heidelberg 2014

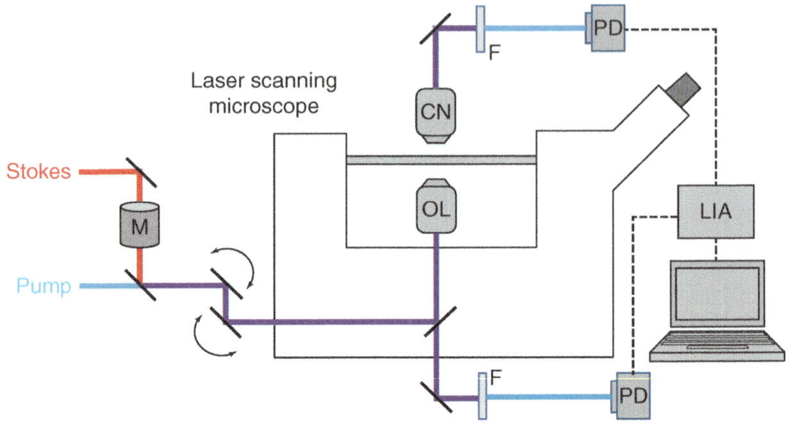

Fig. 20.1 Diagram of a laser scanning microscope set-up for SRS, showing both forward and epi-detection. *M* modulator, *OL* objective lens, *CN* condenser, *PD* photodiode, *LIA* lock-in amplifier, *F* filter

output lacks information about permeation pathways that would be possible with a visual technique, and the comparatively long pixel dwell times compromise the data integrity.

The shortcomings associated with the techniques discussed above are addressed by coherent Raman scattering microscopy [11–13].This approach provides an elegant solution by enabling optical sectioning using confocal microscopy, while utilising two lasers to probe Raman interactions rather than a single source to excite fluorescence. Mechanistic insight into the drug pathway can be gained by acquiring 3D images to reveal real-time disposition of the components of a formulation.

20.2 Principles of Coherent Raman Scattering

Stimulated Raman scattering (SRS) and coherent anti-Stokes Raman scattering (CARS) harness the Raman effect [14, 15], based on molecular polarisability, in which a photon is inelastically scattered by the sample, surrendering part of its energy and emerging with a lower frequency related to a vibrational frequency of the molecule. It can be thought of as arising from a 'virtual' transition to an excited state, before returning to a different state.

The amount of energy needed to excite a molecular vibration depends on the masses of the atoms involved in the vibration and the type of bond(s) between them and is influenced by the chemical microenvironment. Molecules with N atoms have {3N – 6}independent vibrational modes, many of which may be excited by a Raman scattering event [16]. Consequently, a Raman spectrum is highly specific and carries detailed molecular information about the sample.

For coherent Raman scattering microscopy, two incident laser beams are aligned with a frequency difference which matches the molecular vibrational frequency of the substance under investigation. These two aligned lasers are guided into the confocal microscope as illustrated in Fig. 20.1.

The Jablonski energy diagram in Fig. 20.2a demonstrates the CARS and SRS processes. For CARS, a new signal is generated at a different (the anti-Stokes) frequency. Because the signal is generated at a different wavelength to the incident laser beams, detection is relatively straightforward and can be achieved by spectral separation and a photomultiplier tube.

For SRS, the transitions result in a change in intensity between the pump and the Stokes beams, illustrated in Fig. 20.2b. The vibrational excited state of the functional group of the molecule is populated when the energy difference between pump and Stokes matches the vibrational energy of the transition. The resulting excitation requires energy, and due to the laws of the conservation of energy, the incident lasers provide the energy for the transition to the excited energy state. To achieve this, a photon from the pump beam is annihilated, while a new photon is created in the Stokes beam. Accordingly, the intensity of the pump beam decreases on interaction with the sample (stimulated Raman loss, SRL), while the Stokes beam gains energy (stimulated Raman

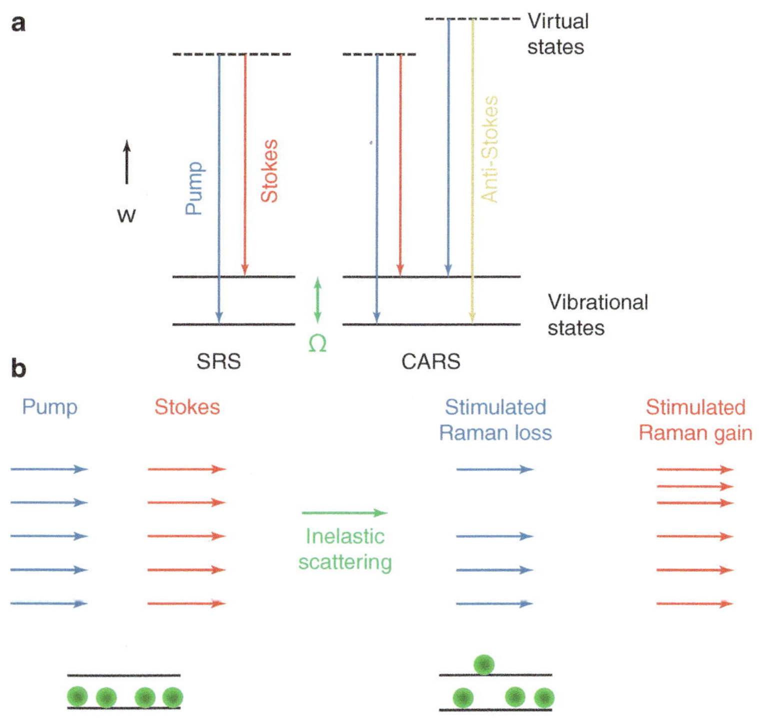

Fig. 20.2 Panel (**a**): Jablonski energy diagrams for SRS and CARS. Panel (**b**): Effect of stimulated Raman scattering on the pump and Stokes beams

gain, SRG). These two processes are collectively known as stimulated Raman scattering.

The intensity changes of the pump and Stokes beams are measured after interacting with the sample. The pump beam is filtered and then detected by a photodiode, and the stimulated Raman loss of the beam is measured using a lock-in amplifier to generate each pixel of the image. Three-dimensional images are obtained by raster scanning across the sample in conjunction with depth sectioning, achieved using a conventional confocal laser scanning microscope.

It is possible to simultaneously detect SRS and CARS if desired, since both processes occur in the sample under suitable excitation conditions. If detection of both SRS and CARS is required in the same direction, this can be achieved using a beam splitter.

20.3 Advantages of SRS over CARS

For quantitative monitoring of drug permeation through the skin, SRS is the more suitable technique. The signal output from SRS is identical to the spontaneous Raman spectrum, and the linear concentration dependence allows more straightforward quantitative analysis. In contrast, CARS generates complex vibrational spectra, for which image interpretation is much more challenging, and the concentration dependence is quadratic. CARS sensitivity is also limited by its nonresonant background, whereas SRS does not have a nonresonant background, and offers excellent vibrational contrast with chemical selectivity. Staining is not required, and there are no complications arising from photobleaching.

20.4 Applications

Stimulated Raman scattering offers unique advantages over conventional methods to investigate drug permeation through the skin. These features may be highlighted by a number of published examples.

Ketoprofen in a vehicle of deuterated propylene glycol was tracked as it permeated murine ear skin (*ex vivo*) by Saar et al. [13]. The drug and the vehicle penetration into the skin were

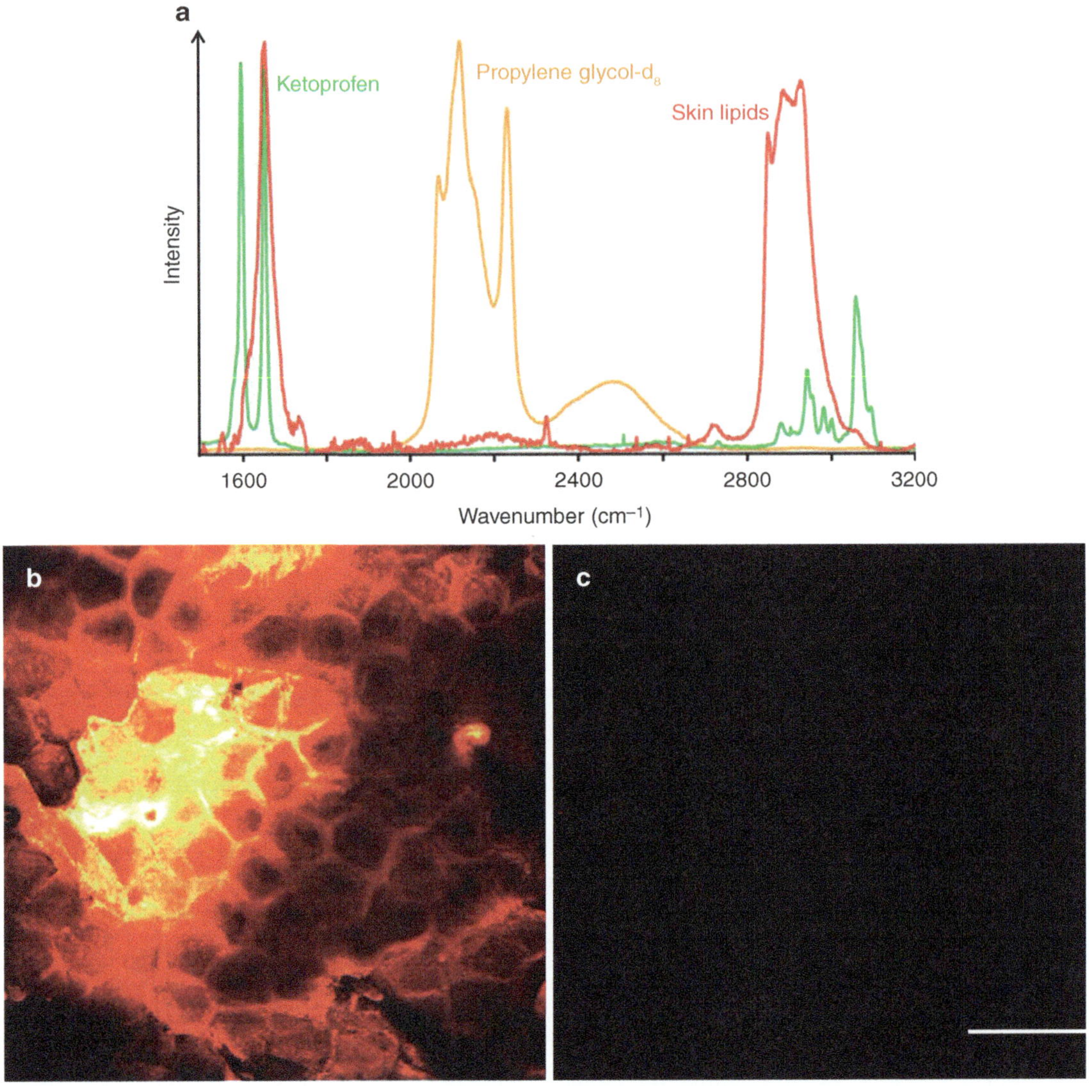

Fig. 20.3 Panel (**a**): Spontaneous Raman spectra of mouse skin lipids, ketoprofen and propylene glycol-d_8. Panels (**b**, **c**): SRS images of mouse ear stratum corneum, by tuning to the skin lipid CH_2 stretching frequency (2,845 cm^{-1}) and off-resonance (1,748 cm^{-1}), respectively. Scale bar = 50 μm

independently monitored by SRS by tuning the laser to the relevant wavelength corresponding to the peak maxima of the spontaneous Raman spectra (Fig. 20.3, panel a) to image each component (1,599 and 2,120 cm^{-1}, respectively), in addition to the skin itself, by tuning to the CH_2 stretching frequency of the skin lipids (2,845 cm^{-1}).

Figure 20.3b shows SRS contrast at 2,845 cm^{-1} (CH stretching) for skin lipids in mouse ear stratum corneum. The skin architecture is visible revealing the hexagonal-shaped corneocytes, framed by the CH_2 signal from the intercellular lipids. In addition, the off-resonance image recorded at 1,748 cm^{-1} (Fig. 20.3c) demonstrates the strong contrast achieved.

Figure 20.4a shows a time-course experiment, during which the lasers were tuned between 1,599 and 2,120 cm^{-1} to obtain contrast for ketoprofen and propylene glycol-d_8, respectively. Images were recorded at increasing depths (as labelled). Over the duration of the experiment, the signal

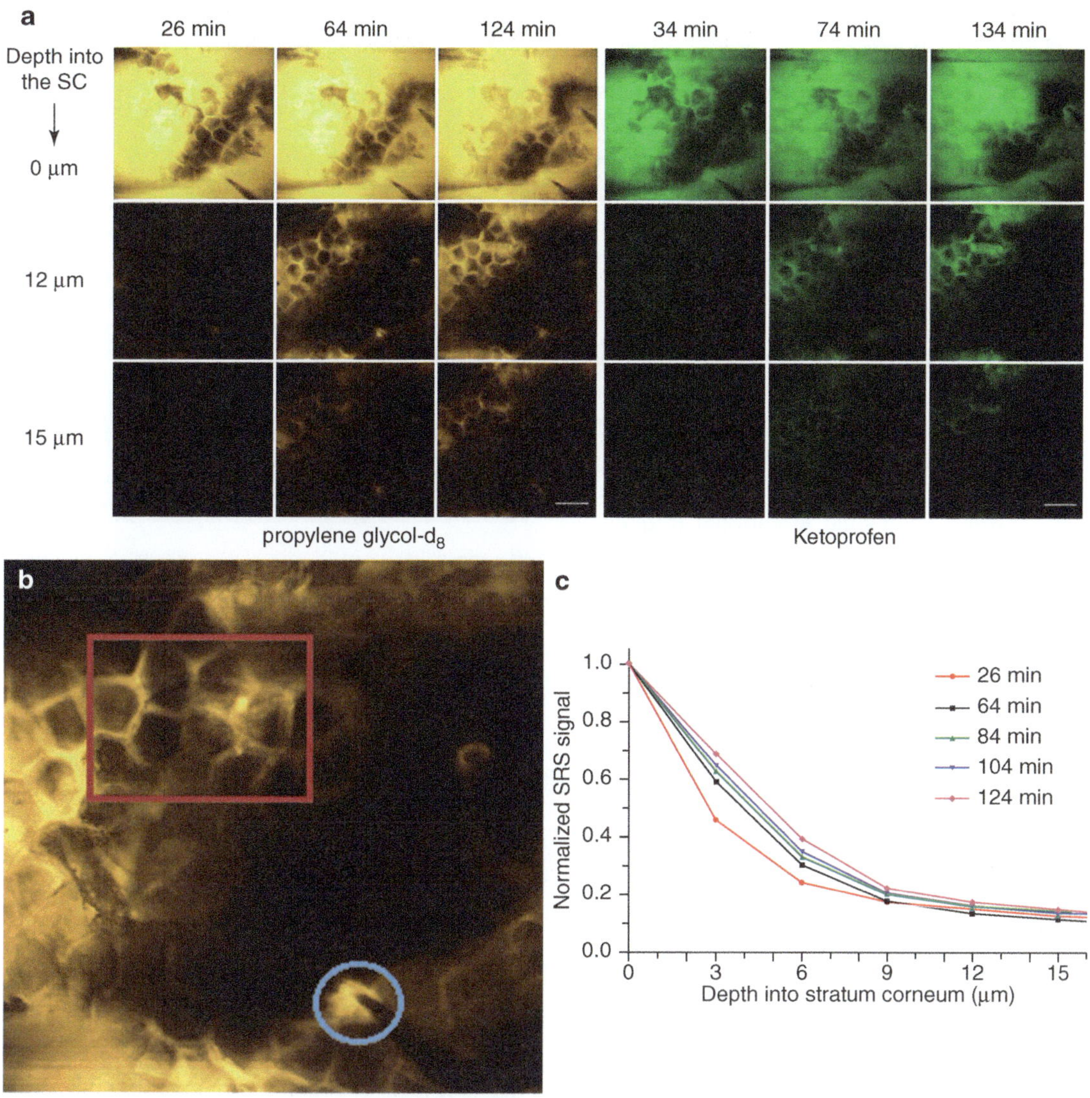

Fig. 20.4 Panel (**a**): SRS images of deuterated propylene glycol and ketoprofen penetrating the stratum corneum of an excised mouse ear. Scale bars represent 50 μm. Panel (**b**): SRS image showing contrast of propylene glycol-d_8 in the stratum corneum. Preferential pathways of penetration analysed by assessing the area outlined in *red* (intercellular, around the corneocytes) versus that in *blue* (transfollicular – the hair shaft is clearly visible). Panel (**c**): Depth profile analysis as a function of time for propylene glycol-d_8 for the region indicated by the red square in Fig. 20.4b

arising from both the ketoprofen and propylene glycol increases, as the chemicals penetrate the skin. The relative amounts of drug and excipient at each depth were determined by the relative ratio of total integrated signal, using image processing software. The profiles presented in Fig. 20.4b were normalised to one at the surface, since the concentration of the excess formulation applied on the surface does not significantly change over the duration of the experiment. It was found that propylene glycol penetrated the skin more rapidly than ketoprofen.

Figure 20.4 shows how it is possible to acquire kinetic information, by generating a chemical disposition profile in addition to mechanistic information from the images, for example, in this case, the drug penetration via a hair follicle versus the intercellular lipids of the stratum corneum. The signal from the specific regions of interest was integrated for comparison at the same depth as a function of time. It was found that the PG penetration via the hair shaft was rapid and reached steady state before the first measurement, whereas the signal from the corneocytes steadily increased over time.

A study of the permeation of retinoic acid (RA) in dimethyl sulfoxide (DMSO) also demonstrated the chemical specificity and high-resolution 3D imaging enabled by SRS microscopy [11]. DMSO, RA and skin lipids have distinct, individual vibrational frequencies of 670, 1,570 and 2,845 cm^{-1}, respectively. Mapping the drug and solvent demonstrated that the hydrophobic RA permeated exclusively through the intercellular lipids, whereas the hydrophilic DMSO is less soluble in the lipid structures and instead partitioned much more avidly into the protein phase.

An additional benefit of SRS microscopy is the visible information obtained which is not accessible when using other methods, for example, the so-called metamorphosis of topically administered drug formulations. Historically, it has been suggested that one reason for poor bioavailability is the isolation of drugs in the skin as the excipient penetrates the skin faster than the drug, leading to crystallisation. Although some indirect evidence has been reported, SRS offers direct, visible proof. Figure 20.5 shows an image of ibuprofen crystals which have formed on/in the stratum corneum of mouse skin, 25 min post application of the drug at approximately 90 % saturation in propylene glycol [13].

Although deuteration of chemicals is not an absolute requirement for this technique, it can provide significant benefit in the event that there are not a set of wavelengths at which contrast for the skin, drug and vehicle can be exclusively obtained.

The absorption of formulation by the skin can result in some swelling of the tissue, which can cause complications in experiments of long duration. However, this can be managed by recording the CH_2 stretching signal from skin lipids to track any three-dimensional movement; that is, because physiological structures in the skin can be visualised, they can be used as 'landmarks'. Thus, diffusion of chemicals can be separated from tissue movements. Laser power and pixel dwell time should be minimised to prevent damage to sensitive biological samples. In particular, tissues containing large amounts of pigment such as melanin may be more challenging or even unsuitable for imaging, since the pigments absorb more energy from the laser than non-pigmented tissue, leading to localised burning.

Fig. 20.5 Crystallisation of ibuprofen-d_3 in propylene glycol on mouse ear skin, recorded 25 min after application. SRS contrast was obtained at 2,120 cm^{-1}. Scale bar = 50 μm

SRS has also been employed to image physiological structures within the skin such as sebaceous glands [17], so there are wide-ranging possibilities for investigating drug targeting to specific appendages. In addition to the in vitro studies described above, Saar et al. have more recently reported SRS imaging performed in vivo in the forearm of a living human [12].

Coherent Raman scattering microscopy offers both visual information in addition to chemical specificity and is uniquely able to reveal information, which would not be attainable from techniques such as tape stripping.

References

1. Herkenne C, Alberti I, Naik A, Kalia YN, Mathy FX, Preat V et al (2008) In vivo methods for the assessment of topical drug bioavailability. Pharm Res 25(1):87–103
2. Pawley J (ed) (2006) Handbook of biological confocal microscopy, 3rd edn. Springer, New York
3. Potts RO, Guy RH (1992) Predicting skin permeability. Pharm Res 9(5):663–669
4. Dias M, Naik A, Guy RH, Hadgraft J, Lane ME (2008) In vivo infrared spectroscopy studies of alkanol effects on human skin. Eur J Pharm Biopharm 69(3):1171–1175
5. Hanh BD, Neubert RHH, Wartewig S, Christ A, Hentzsch C (2000) Drug penetration as studied by non invasive methods: Fourier transform infrared-attenuated total reflection, Fourier transform infrared, and ultraviolet photoacoustic spectroscopy. J Pharm Sci 89(9):1106–1113
6. Foerster M, Bolzinger M-A, Ach D, Montagnac G, Briancon S (2011) Ingredients tracking of cosmetic formulations in the skin: a confocal Raman microscopy investigation. Pharm Res 28(4):858–872
7. Caspers PJ, Lucassen GW, Carter EA, Bruining HA, Puppels GJ (2001) In vivo confocal Raman microspectroscopy of the skin: non invasive determination of molecular concentration profiles. J Invest Dermatol 116(3):434–442
8. Pudney PDA, Melot M, Caspers PJ, van der Pol A, Puppels GJ (2007) An in vivo confocal Raman study of the delivery of trans-retinol to the skin. Appl Spectrosc 61(8):804–811
9. Forster M, Bolzinger M-A, Montagnac G, Briancon S (2011) Confocal Raman microspectroscopy of the skin. Eur J Dermatol 21(6):851–863
10. Bonnist EYM, Gorce JP, Mackay C, Pendlington RU, Pudney PDA (2011) Measuring the penetration of a skin sensitizer and its delivery vehicles simultaneously with confocal Raman spectroscopy. Skin Pharmacol Physiol 24(5):274–283
11. Freudiger CW, Min W, Saar BG, Lu S, Holtom GR, He C et al (2008) Label-free biomedical imaging with high sensitivity by stimulated Raman scattering microscopy. Science 322(5909):1857–1861
12. Saar BG, Freudiger CW, Reichman J, Stanley CM, Holtom GR, Xie XS (2010) Video-rate molecular imaging in vivo with stimulated Raman scattering. Science 330(6009):1368–1370
13. Saar BG, Contreras-Rojas LR, Xie XS, Guy RH (2011) Imaging drug delivery to skin with stimulated Raman scattering microscopy. Mol Pharm 8(3):969–975
14. Raman CV, Krishnan KS (1928) A new type of secondary radiation. Nature 121(3048):501–502
15. Raman CV, Krishnan KS (1928) The optical analogue of the compton effect. Nature 121(3053):711
16. Atkins P (1998) Physical chemistry, 6th edn. Oxford University Press, Oxford, UK
17. Min W, Freudiger CW, Lu S, Xie XS (2011) Coherent nonlinear optical imaging: beyond fluorescence microscopy. Annu Rev Phys Chem 62:507–30

Assessment of Mechanical Properties of Skin by Shearwave Propagation and Acoustic Dispersion

21

Eduardo Ruvolo Jr. and Nikiforos Kollias

Abbreviations

ASTM	American Society for Testing and Material Standards
DMAE	2-(Dimethylamino)ethanol
FWHM	Full width at half maximum
IRHD	International rubber hardness degree
RRT	Resonance running time

21.1 Introduction

The role of skin as a multipurpose barrier (biochemical, biological, immunological, physical, mechanical, thermal, etc.) has been an active field of research of investigative dermatology for a long time [1]. From the mechanical point of view, it provides a viscoelastic wrapping that holds together the internal tissues and organs. Due to this viscoelastic nature, the skin can transmit mechanical tension, as well as play a role of mechanical insulation acting as a shock absorber to external impulses. The skin is always under varying amounts of tension and allows flexing and locomotion by adjusting and redistributing the applied forces. This adjustment and redistribution of forces are the result of its ability to deform both elastically (reversibly) and plastically (irreversibly) in the direction of the applied force.

The mechanical strength of the skin is primarily attributed to the dermal matrix, which consists predominantly of collagen and elastin. Collagen molecules form fibrils, fibers, and bundles, which are arranged in a basket-weave pattern. This allows the dermis to deform due to pressure and at the same time to minimize tears. The collagen bundles vary in size as we progress from the upper (papillary) to the deeper (reticular) layers of the dermis and are normally under tension that ranges from 0 to 20 N/m depending on the body site, direction, and posture [2, 3]. Ever since the nineteenth century, surgeons have recognized the fact that this tension load is directional by observing that circular punctures in skin produce elliptical holes [4]. These lines, termed "Langer's lines," represent the directions of the skin's maximum tension and have been identified over the whole body [5, 6]. They are used as the preferred direction in which surgical incisions are made so that the tension across the wound is minimal. Furthermore, it is known that dermal fibroblasts generate tension and change their orientation along the tensile direction [7]. The mechanical properties of the skin are thus expected to follow the anisotropy in tension defined by the Langer's lines.

Reduction of skin elasticity with aging is attributed to skin thinning due to loss of elastin and collagen in the dermal matrix [8, 9], as well as to loss of subcutaneous tissue mass (fat layers and muscle mass) and stiffening of collagen with increased cross-linking. Age is also expected to

E. Ruvolo Jr. (✉) • N. Kollias
Methods and Models, CPPW, Johnson and Johnson Consumer Companies and Personal Products Worldwide, 199 Grandview Rd, Skillman, NJ 08558, USA
e-mail: eruvolo@its.jnj.com

E. Berardesca et al. (eds.), *Non Invasive Diagnostic Techniques in Clinical Dermatology*, DOI 10.1007/978-3-642-32109-2_21, © Springer Berlin Heidelberg 2014

affect the directionality of the mechanical properties of the skin. During development, from infancy to adolescence, skin is expected to respond to tension isotropically to accommodate for growth, while in adulthood this isotropic behavior is expected to regress due to site-specific habituation to tension.

A number of techniques have been developed to study the mechanical properties of the skin and their dependence on aging [10, 11]. They are primarily based on the concept that assessing the applied force necessary to pull or push the skin gives an estimate of the elastic and plastic properties of the tissue, and they can be classified in the following categories: (a) instruments that generate a light vacuum on the skin surface and determine the height to which the skin may be pulled under constant suction and then the rate at which the skin returns to its original shape, (b) instruments that use two concentric cylinders that are placed in contact with the skin and measure the angular displacement under torque and the rate at which the skin returns to equilibrium once the torque is removed, and (c) instruments that assess the firmness/elasticity of the skin by determining the speed of a small mass as it bounces after striking the skin with a predetermined velocity [12]. All of the above types of instrumentation measure parameters that are affected by the mechanical properties of the skin surface as well as the deeper structures of the dermis and even the subcutaneous tissues. Therefore, the information we can get from them is not always limited to the skin. These methods do not offer any information on mechanical anisotropy.

Another type of instrument (Reviscometer® RVM 600) allows for the evaluation of the mechanical properties of a material by measuring of the propagation time of a shear wave between two sensors placed on the material surface. The distance between the sensors and the frequency of the shear wave determines the penetration depth of sampling. The Reviscometer® has been designed to sample up to a few tenth of microns in depth making it suitable for measurements of the mechanical parameters of the epidermis and the papillary layers of the dermis. The applied pulse frequency is in the audible range (4.5 kHz). The velocity of sound in a material depends on the density of the material and tension that it is under. Mechanical vibrations propagate faster at the direction of higher tension, and as with a guitar string, the higher the tension, the higher the frequency of oscillation after plucking. As the preferred disposition of the collagen fibers corresponds to the skin's cleavage lines (Langer's lines), the speed of propagation of elastic disturbances on the skin will depend strongly on orientation.

Previous investigations have shown a weak dependence of the shear wave propagation time on the skin with aging [3, 13–17]. A common denominator of these studies was that the measurements were performed at a limited number of angular positions of the probe, typically at four positions at increments of 45°. Therefore, directional features with angular width less than 45° were ignored or averaged out.

The administration of topical all-trans retinoic acid to photoaged skin reduces fine wrinkling [18]. This is associated with histological changes in both epidermis and dermis. Epidermal changes include thickening and compaction of the stratum corneum and increased proliferation of keratinocytes. In a topical treatment study with tretinoin at 0.05 % by Berardesca [19], the skin elasticity increased significantly after 4 months when compared with placebo-treated sites.

Grando [20] demonstrated that human epidermal keratinocytes express acetylcholine-gated ion channels with functional and structural characteristics similar to those of nicotinic acetylcholine receptors (nAChRs) previously believed to be expressed uniquely by neurons of the sympathetic ganglia and brain. Such nAChRs were involved in modulation of keratinocyte motility, retraction of the cytoplasmatic projections, and intercellular adhesion.

A number of compounds (agents) have been identified to have action on keratinocytes and have been described as a cellular volume contraction in vivo and in culture [21]. Among them 2-(dimethylamino)ethanol (DMAE), that is a biochemical precursor to the neurotransmitter acetylcholine, has such effect on keratinocytes. Mechanistically, this action has been shown to be due to modulation of motility and intercellular adhesion of keratinocytes.

In this chapter, we demonstrate through a number of studies the sensitivity and utility of the Reviscometer to provide unique information on the directional dependence of elastic parameters of the superficial layers of the skin and its ability to assess changes in keratinocyte morphology noninvasively. Thus we attempt to characterize the tensile forces acting on the skin surface by measuring the anisotropy of the viscoelastic properties of the skin at different ages, correlating the changes in the anisotropy with structural features of the skin and measuring morphological changes in the viable epidermis keratinocytes using reflectance confocal microscopy. To this end we established a new data acquisition methodology for the Reviscometer® with high angular resolution, and we defined a novel mechanical parameter that relates to the anisotropy and coherence of skin viscoelasticity.

21.2 The Instrument Hardware

The Reviscometer Model RVM 600 is an instrument developed by Courage Khazaka, Cologne, Germany, that measures the propagation of superficial shear waves in skin. In 1980, Prof. Hagen Tronnier [22] developed a method to measure resonance frequencies in skin. Later, in 1999, Vexler [17] described a viscoelastic skin analyzer that resembles more the commercial instrument.

A noninvasive way to assess the mechanical properties of viscoelastic materials such as the skin is by investigating the propagation of a mechanical disturbance along their surface. The speed of propagation of a pulse or of a periodic (wave) mechanical disturbance depends on the tension in the material in the direction of propagation and on the density of the material [23]. The amplitude of such a stimulus should be small so that no permanent changes are produced in the material, and the coupling between the device producing the perturbation and the medium under study, e.g., the skin, should be robust [17, 24–26].

The probe that comes in contact with the skin includes two transducers spaced by approximately 2 mm and mounted on two independent supports. One transducer generates a small amplitude acoustic pulse, <0.2 mm, in the audible range 4.5 kHz, and the second acts as the receiver. The time it takes for the acoustic pulse to travel from the transmitter to the receiver is defined as the resonance running time (RRT). The values of the RRT may be used to calculate the velocity of propagation of the acoustic pulse. If we consider the one-dimensional longitudinal model for the sonic propagation from Dahlgren [13], Young's modulus of elasticity (E) of the material is directly related to the square of the sonic transmission velocity, c:

$$E = dc^2 \cdot 10^6 \tag{21.1}$$

where E is in N/m^2, d is the density of the skin in g/cm^3, and c in m/s. Therefore, the modulus of elasticity may be determined once the velocity of propagation has been measured.

To ensure good contact (no slipping) between the transducers and the skin, a constant pressure is applied; the pressure that the probe exerts on the skin is regulated by a spring mechanism: a given amount of pressure is necessary for the measurement sequence to start. The probe is held perpendicular to the skin surface by a hollow cylindrical holder that is attached to the surface of the skin with double-sided adhesive tape. The holder is marked along its periphery at angular intervals of 45° by the manufacturer. In all the work presented in this chapter, we used a modified probe holder assembly by placing a millimeter scale on the probe and another on the holder. We carried out measurements by rotating the probe within the holder so that the lines of the two scales would align with each other. This corresponds to making measurements every 3° for a total interval of 100–180°. Custom acquisition software was developed in LabVIEW 6.0 (National Instruments Corporation, Austin, TX).

To demonstrate the need of acquiring data at increased angular resolution, we recorded the Reviscometer readings from the same skin site (upper inner arm) of a volunteer of 47 years at

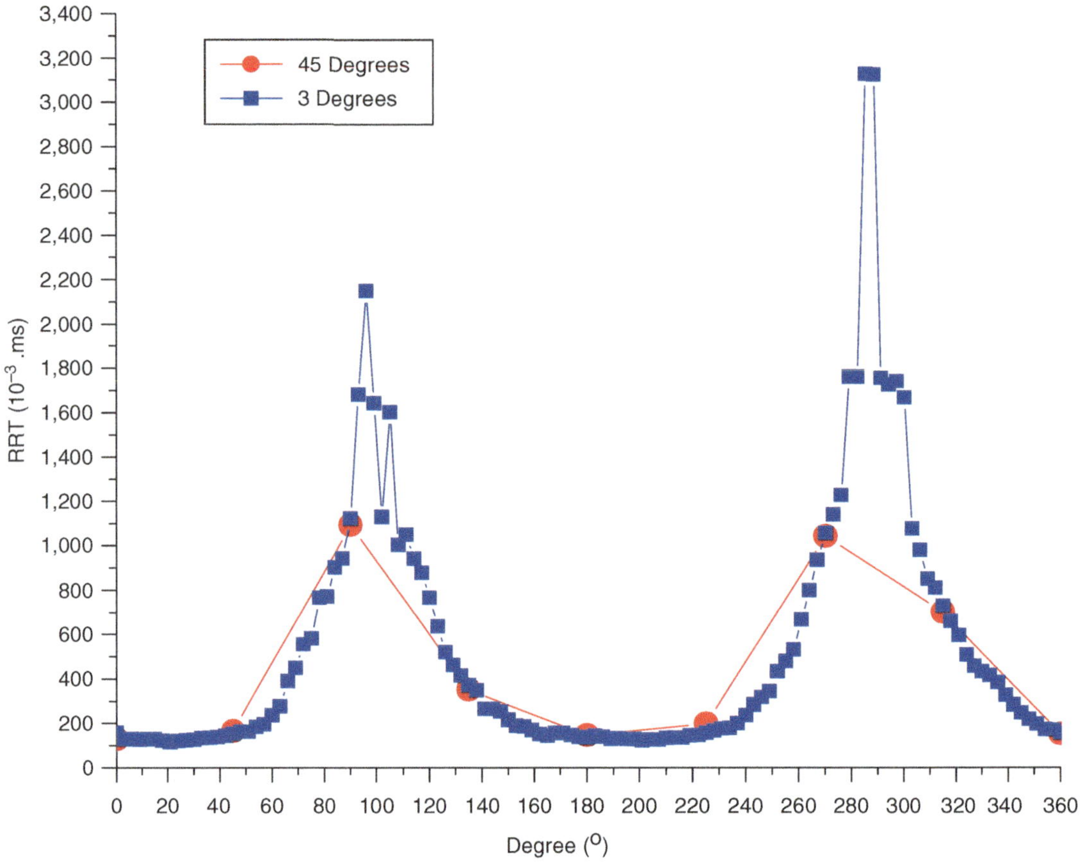

Fig. 21.1 Typical data obtained with a Reviscometer® from the upper inner arm of a subject – the data were obtained with acquisition in increments of 3 (*blue squares*) and 45° (*red circles*)

intervals of 45° and of 3° (Fig. 21.1). It is evident that the features with width less than 45° may be missed. Following the Nyquist sampling criterion, using 3° intervals we could accurately record features of at least 6° angular dispersion width. Typically for older individuals having narrower maxima, the angular dispersion width can be in the order of 10°. Taking measurements with high angular resolution increases the chances of recording the skin's radial anisotropy accurately. Increased angular resolution of viscoelasticity readings provides information that would be otherwise lost.

The dependence of the measured RRT values obtained from human skin on the angle of measurement may be described by a Gaussian curve, Fig. 21.2. Two parameters may be determined from the Gaussian fit function: the full width at half maximum (FWHM) and the amplitude of the curve which corresponds to the angular anisotropy of the RRT. The anisotropy is defined as the ratio between the maximum and minimum RRT of the measurements on a given skin site:

$$A = \frac{RRT_{max}}{RRT_{min}} \tag{21.2}$$

The definitions for minimum and maximum RRT values (RRT_{min} and RRT_{max} correspondingly) as well as the full width at half maximum (FWHM) of the fitted curve are demonstrated in Fig. 21.2; these correspond to two new mechanical parameters that may be determined in the skin based on measurements with the Reviscometer®, and they are anisotropy and angular dispersion width (FWHM).

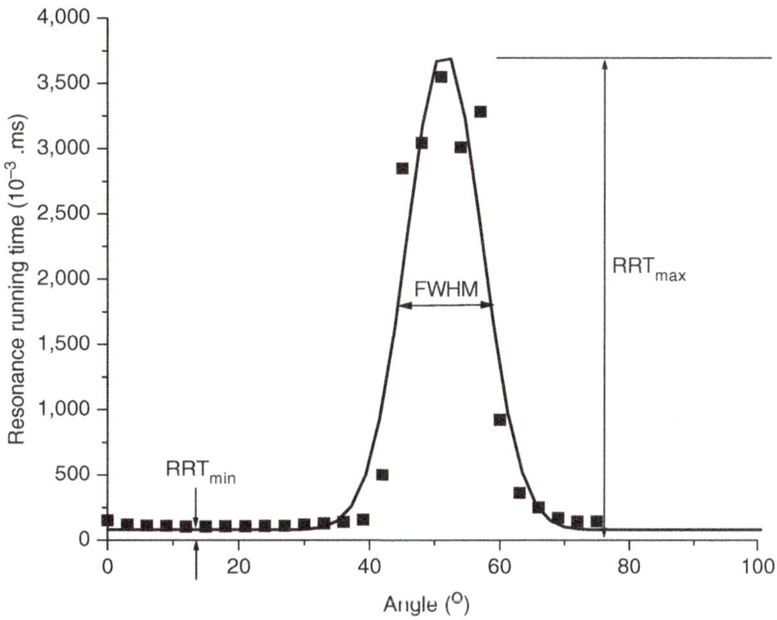

Fig. 21.2 RRT data obtained from the upper inner arm of a subject obtained in increments of 3° over a range of angles of 90° were fitted with a Gaussian curve. The relevant parameters obtained from such a fit are shown. These are RRT_{min}, RRT_{max}, and FWHM of the curve

The Reviscometer® does not work equally well on all skin sites. Skin sites with high density of vellus hairs present difficulties in the interface between the transducers and the skin. Equally difficult has been to obtain good data from skin sites where the skin is tight such as the calves and the forehead. The instrument gives very reliable and high-quality data from sites where the skin is particularly "soft" to the touch such as the upper inner arm, the neck, the abdomen, breast, thighs, and in particular the inner aspect and the sides of the torso. It is interesting to note that the sites where good measurements may be made with the Reviscometer® are also skin sites where "stretch" marks are likely to appear.

21.3 Instrument Calibration

Typical RRT reading from anatomical sites like the dorsal and ventral forearm ranges from 100 to 450×10^{-3} ms. The manufacturer supplies with the instrument an elastomer standard with a RRT$=230\pm30\times10^{-3}$ ms. The RRT value for the standard is suitable for a great majority of the values obtained in different parts of the body.

When different Reviscometer® probes were used to measure skin viscoelasticity of the same skin site, we observed the readings to vary between probes, possibly due to differences in the distance between the transducers and differences in their sensitivities. Therefore, the probes were calibrated against a wider range of viscoelastic standards. To this end we used the standard thermoplastic elastomers Durometer [27] test block kits for Shore scale A (Rex Gauge Company Inc., Buffalo Grove, IL) and Shore scale 00 (Corporate Consulting Services Instruments Inc., Akron, Ohio). These materials were chosen because of their wide response range of RRT readings (200–$1,100\times10^{-3}$ ms). Measurements were repeated ten times in each sample, and the data were averaged. The elastomers were isotropic in terms of RRT values. All measurements were taken at ambient conditions at a temperature of 18 °C and a relative humidity of 40–50 %. Table 21.1 shows the Shore type, the International Rubber Hardness Degree (IRHD) [28], and the density (g/cm^3) of the standard test blocks used in this study. Shore hardness is a measure of the resistance of material to indentation by a spring-loaded indenter. The hardness testing of plastic material is most commonly measured by the Shore (Durometer) test or Rockwell hardness test, the higher the number, the greater is the resistance to indentation. Using Eq. 21.1 we estimated the speed of sound in each elastomer as shown in Table 21.1. Figure 21.3 shows the RRT values versus the speed of sound (m/s) on the standards for two different probes. It can be observed from Fig. 21.3 that the probes have similar responses for elastomers with low RRTs but differ at high RRTs.

The standard elastomer blocks can be used to correlate the readings from one probe to the other by plotting elastomer measurements as illustrated in Fig. 21.4. Testing probes can be described by a linear fit ($r^2=0.92$). As an example, Fig. 21.5 illustrates RRT readings, as a function of the angle, taken on the upper inner arm of a 41-year-old subject with two different probes. Taking the readings from the probe with lower sensitivity, Probe 02120709, and applying the linear correction ($Y=220.1+1.94*\text{xscale}(X)$), we can see by the Fig. 21.5 that the readings are much closer to the probe with higher dynamic range.

Table 21.1 Physical parameters of the Durometer test blocks used for calibration

Shore type	IRDH	Young's modulus (MPa)	Density (g/cm^3)	Speed of sound (m/s)
00	31	0.133	0.91	384
00	46.9	0.303	0.92	574
00	54.5	0.412	0.93	669
00	63.1	0.414	0.95	660
00	77.3	0.916	0.98	965
A	29	1.024	1.06	983
A	42.8	1.801	1.22	1,215
A	50.2	2.367	1.31	1,345

21.4 Viscoelastic Properties of the Skin at Different Ages

It is well known that the human skin changes with age particularly in becoming more "loose" and less "elastic." This has been shown on the back of the hand as a "pinch" test where we observe the speed with which the skin returns to its original state after being pinched and in lateral compression between two fingers of the skin – in young subjects the skin surface produces a single "bulge" where in old subjects the skin produces a multitude of wrinkle patterns. It has been concluded that the elastic properties of the skin must change for such pronounced changes to be observed, yet the changes documented in measurements with instruments that study the deformation of the surface of the skin when exposed to a negative pressure have been shown to be of the order of 33 % over 50-year

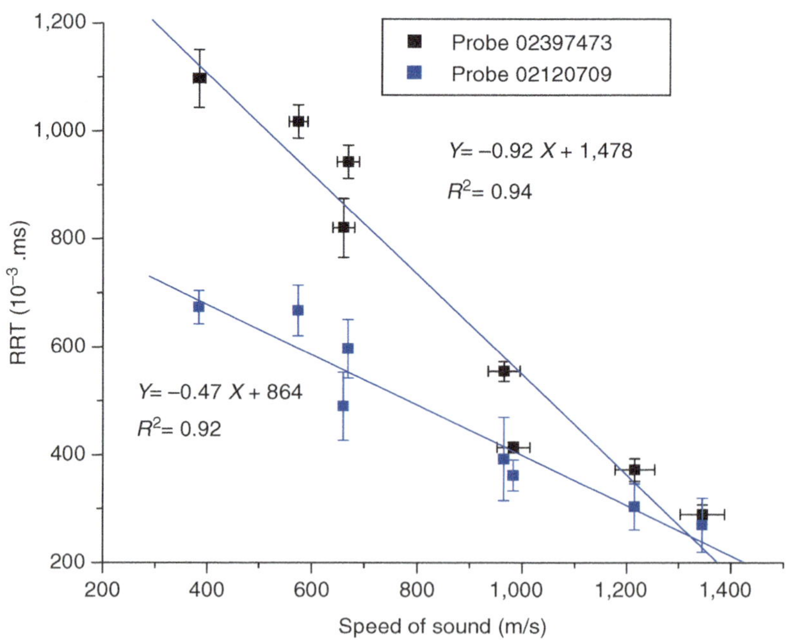

Fig. 21.3 The Reviscometer® data for two probes as supplied by the manufacturer. Figure illustrates the viscoelastic properties of different elastomers as a function of the speed of sound (m/s)

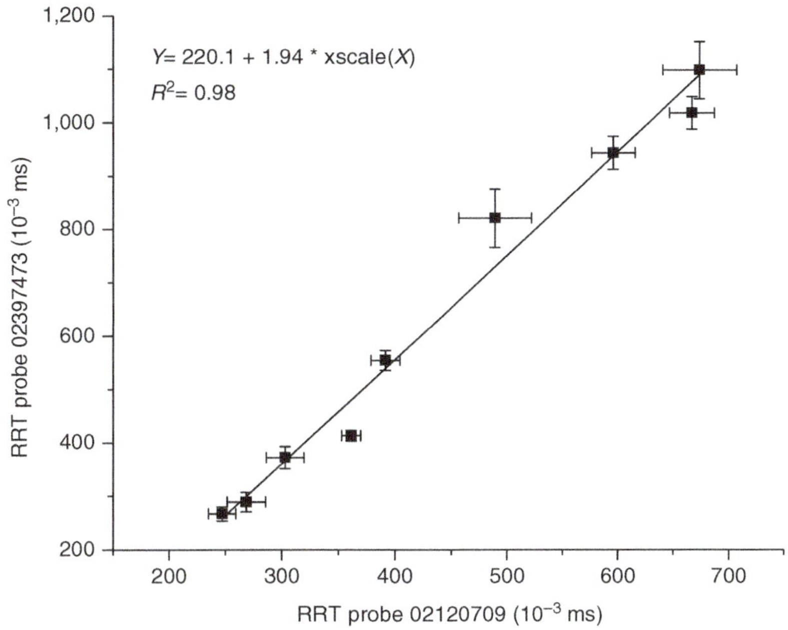

Fig. 21.4 The correlation of RRT probe values for elastomers with different hardness

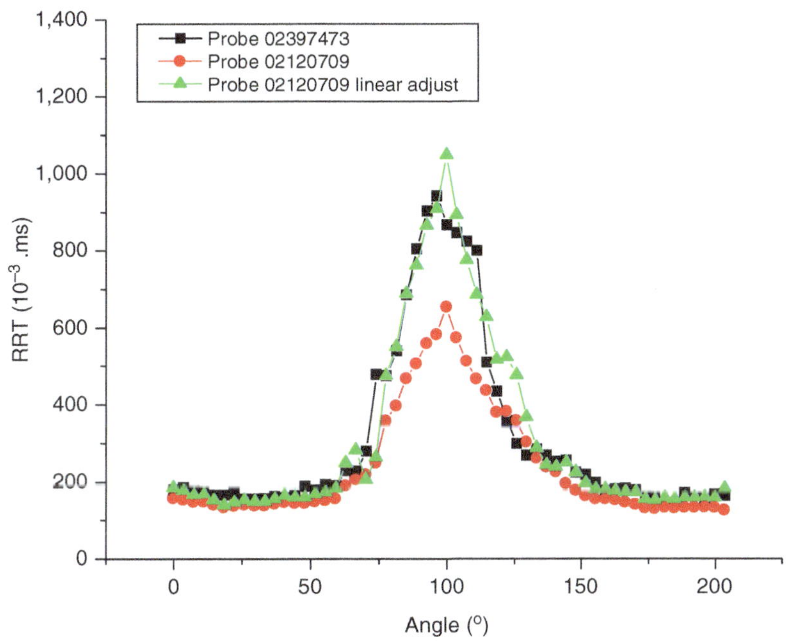

Fig. 21.5 Illustrates measurement at the upper inner arm performed with two different probes. A liner correction ($Y=220.1+1.94*X$) was applied to the probe 02120709

age interval (from 10 to 60 years). This number does not provide adequate documentation of the observed changes with age. As the Reviscometer® assesses the superficial layers of the skin, it was felt that monitoring age-related changes in the mechanical properties of the superficial skin would provide a more accurate representation of the observed changes with age.

The study protocols were approved by an independent ethics committee, and subjects were in general good health and gave informed consent. RRT measurements were taken from 239 volunteers with skin types, from very light Caucasian (types I and II) to African Americans (type VI). The volunteers were divided into 5 age groups: 0–2 years (1.8 ± 1.1, mean ± standard deviation),

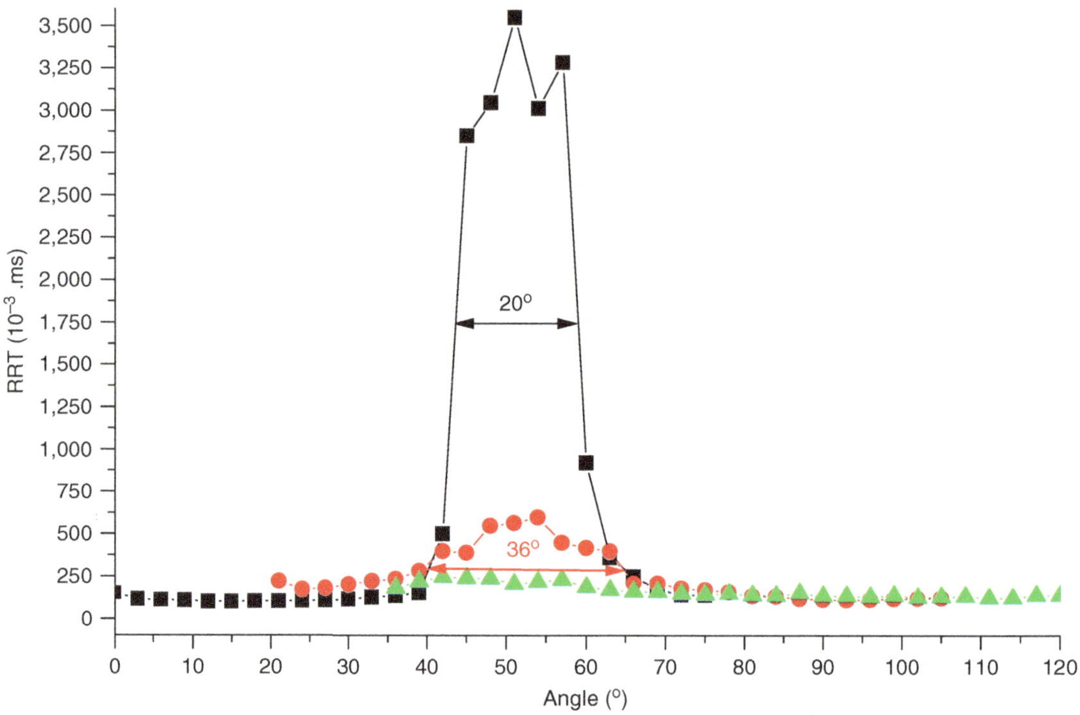

Fig. 21.6 Skin viscoelastic properties show strong age dependence. Representative resonance running time (RRT) profiles recorded from the upper inner arm for three age groups: 0–4 years (*green triangles*), 24–40 years (*red circles*), and 65–75 years (*black squares*)

14–20 years (17 ± 4.2), 24–40 years (32.5 ± 10.6), 55–60 years (57.5 ± 3.5), and 65–75 years (70 ± 7). Measurements with a Reviscometer® were taken on the upper inner arm at 15 cm from the elbow from 0° to 100° in 3° increments, where the initial orientation of the probe was chosen as the direction that gave the lowest RRT reading. Reviscometer® readings were also taken on the dorsal and ventral forearm of 117 of the volunteers.

The mean value and standard deviation were calculated for each age group and the treated sites. JMP® statistical software (SAS Institute Corporation, Cary, NC) was used to perform analysis of variance using a standard least square analysis. Statistical significance was considered for $p<0.05$.

In order to ensure the relation between visual appearance and mechanical properties in vivo, macro imaging was also performed on the upper inner arm and dorsal forearm of the 239 subjects. A video-microscope image system (HiScope®, Model KH-2400, Hirox Inc, Tokyo, Japan) was used to acquire close-up images (2.8 × 2.1 mm image size) of tested area using a 100× magnification lens (Model MX-100Z, Hirox Inc, Tokyo, Japan). This image system allows examination of the skin surface texture or of subsurface features depending on the illumination at a grazing angle. In this study grazing angle illumination was used to enhance the contrast of the skin texture.

Typical RRT profiles recorded from the upper inner arm for three different age groups are shown in Fig. 21.6. We observed that while the RRT_{min} values remained fairly constant for the age groups tested, the RRT_{max} values increased with age. Thus the anisotropy ratio given by Eq. 21.2 also increased with age (Fig. 21.7). Furthermore, the width of the RRT profiles becomes narrower, i.e., the angular dispersion width decreases, with age (Fig. 21.7). These changes are statistically significant ($p<0.01$) between the age groups. Since RRT anisotropy increases while angular

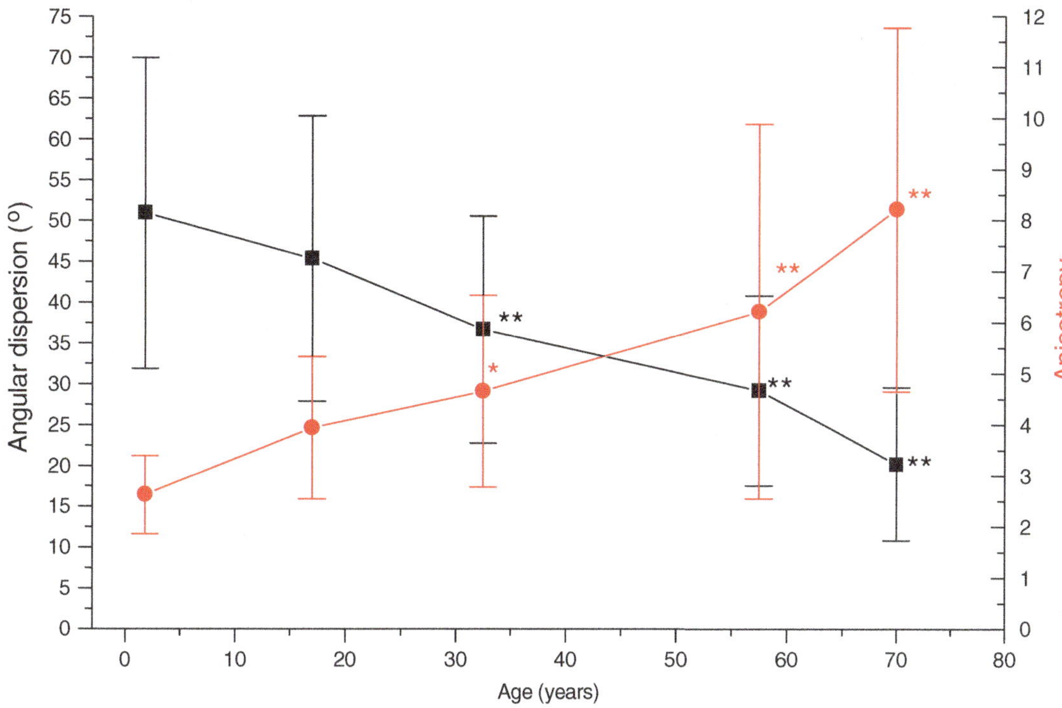

Fig. 21.7 The anisotropy ratio (*red circles*) increases, while the angular dispersion width (*black squares*) decreases with age. Measurements were taken on the upper inner arm $^{*}p<0.01$, $^{**}p<0.001$ when compared with the youngest age group

dispersion width decreases with age, we can define the ratio of these two as a single viscoelastic parameter that documents the age-dependent changes in the skin and that provides a wider dynamic range than either (Fig. 21.8). The relationship of this parameter to age is best represented by an exponential, and the data may be fitted with the equation:

$$AR = \frac{A}{ASW} = 0.041 \bullet e^{\frac{age}{26}} \qquad (21.3)$$

Where *AR* is Anisotropy Rate, *A* is Anisotropy, and *ASW* Angular Dispersion Width.

The anisotropy ratio values for each of the older age groups are statistically different compared those for the youngest group.

The mechanical and physical properties of the skin have been extensively studied [2, 11, 12]. These investigations describe the skin as being a viscoelastic medium and interpret skin with homogeneous density and isotropic elastic moduli. The mechanical properties of the skin stem from contributions of its structural components such as collagen, elastin, keratin, and the purely viscous interstitial fluids [13]. According to Eq. 21.1 the speed of shear wave propagation depends on the density of the medium in the direction of the sound propagation. Results presented in this chapter show that the shear wave propagation on the surface of the skin is anisotropic, and thus the tissue density that relates to the skin mechanical properties is also anisotropic.

The skin at the upper inner arm sites undergoes considerable changes with age in its structure and therefore its mechanical properties. It is evident in Fig. 21.6 that the RRT readings of young skin are more isotropic when compared with the other two age groups. For the 24–40-years group, there is a distinct increase in the anisotropy and for the older group a very well-defined RRT maximum.

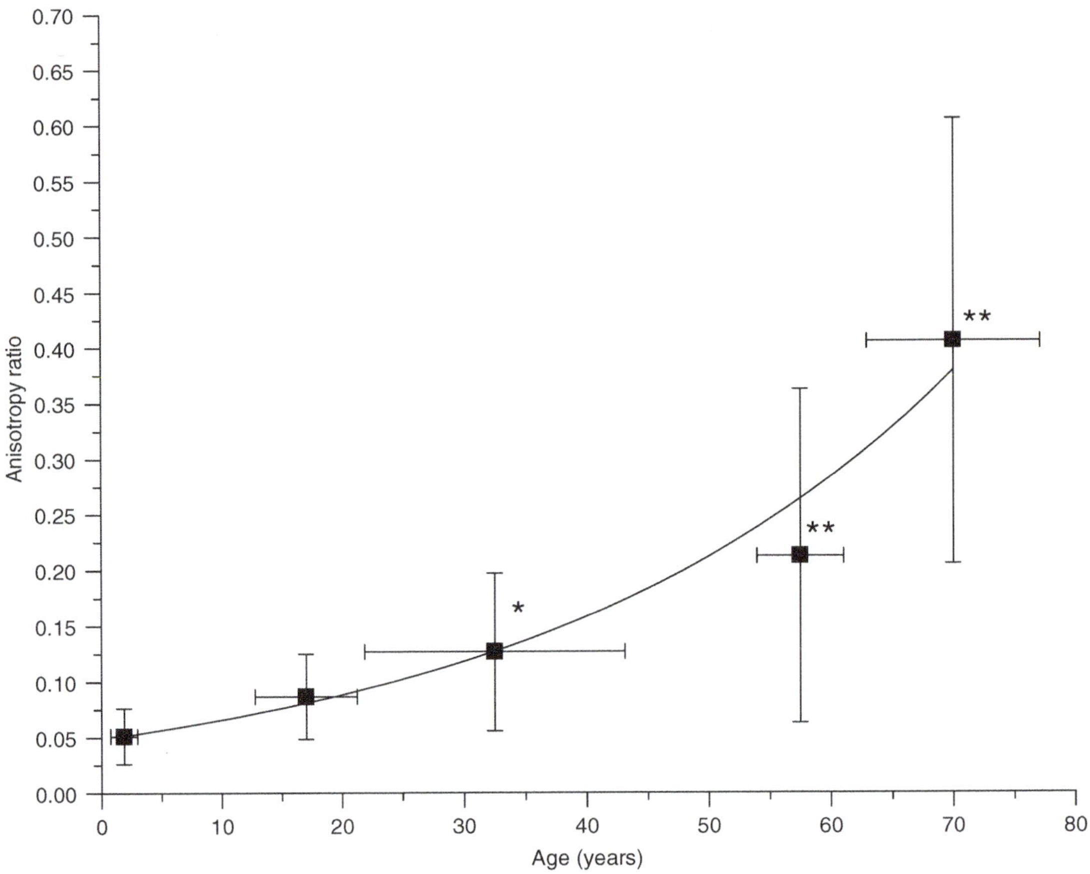

Fig. 21.8 The ratio of anisotropy to angular dispersion width is a new parameter that may be used to assess skin viscoelasticity and shows a stronger dependence on age than either of the parameters in the ratio. $*p<0.03$, $**p<0.001$ compared with the youngest age group

This may also be observed in the skin microrelief imaged by the video micro imager. Skin microrelief evolves patterns on the ventral forearm that are characteristic for different age groups. The in vivo microscopy images of Fig. 21.9 are representative for the age groups: 0–4, 14–24, 24–40, and 65–75 years. It is evident that the triangular and hexagonal patterns of the younger ages are replaced progressively with linear directional patterns presumably reflecting flattening of dermal papillae and stronger collagen fiber alignment in the papillary dermis. The change of these patterns is accompanied by decrease in viscoelastic anisotropy and increase in angular dispersion width that indicates a strong directional dependence. The anisotropy values for the four subjects imaged are 2.1 (16 month, Fig. 21.9a), 2.4 (14 years, Fig. 21.9b), 4.2 (39 years, Fig. 21.9c), and 15.6 (70 years, Fig. 21.9d), and the angular dispersion width values are correspondingly 108°, 84°, 40°, and 35°. The viscoelastic °anisotropy relates to cutaneous structural feature changes as a function of age.

We can get a better idea of the co-localization of the directionality of skin microrelief and the angular distribution of the RRT values in Fig. 21.10. A video-microscope image of the upper inner arm of a 47-year-old subject was imaged using 100x magnification. The transmitter and receiver tips of the Reviscometer transducers are drawn, to scale, and the RRT readings from the same site are overplotted in polar coordinates. Very importantly the maximum of the RRT values co-localizes with the skin surface microrelief (Fig. 21.10). Skin microrelief on the upper inner arm develops with age (Fig. 21.9) in

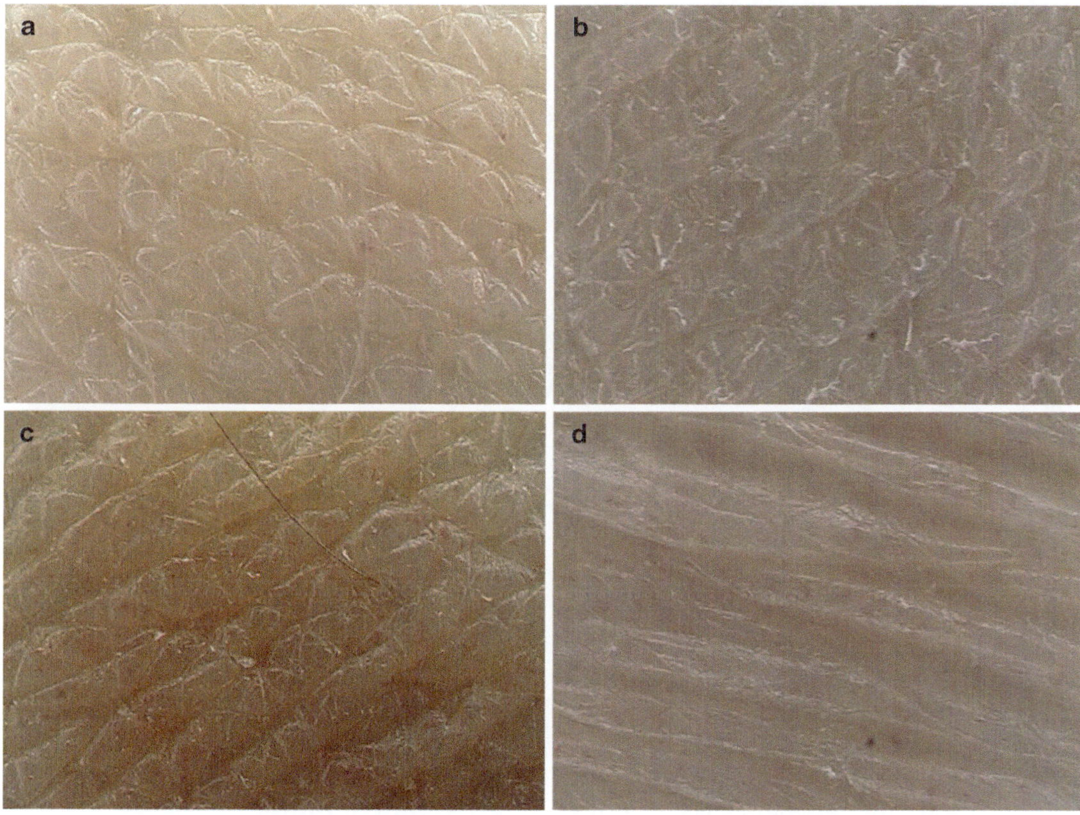

Fig. 21.9 Illustrates the cutaneous structural feature and viscoelastic changes at different ages. In vivo microscopy images from the upper inner arm representative for the age groups: (**a**) 0–4 years (anisotropy = 2.1; FWHM = 108°), (**b**) 14–24 years (anisotropy = 2.4; FWHM = 84°), (**c**) 24–40 years (anisotropy = 4.2; FWHM = 40°), and (**d**) 65–75 years (anisotropy = 15.6; FWHM = 35°)

a fashion that parallels the development of RRT anisotropy. Tensions are progressively developed with age in preferential orientations producing changes in the distribution and shape of the dermatoglyphics. In young skin the relief lines look more plump and rounded on the top and form starlike patterns. As the individuals get older, the glyphic structures become flatter and anisotropic, i.e., not symmetric in all directions, but elongated in the direction of permanent tension.

A video-microscope image (Hirox, Japan 100× lens) from the upper inner arm of a 70-year-old subject is illustrated in Fig. 21.11. This figure shows in scale the relation between the orientation of the probe (transducer tips) and the skin microrelief lines. When the transducer tips are oriented perpendicular to the relief lines as illustrated in Fig. 21.11a, the RRT value is at minimum, typically around 250×10^{-3} ms, corresponding to

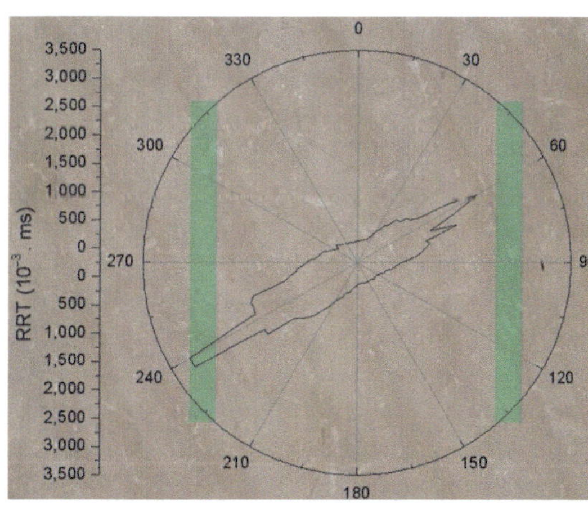

Fig. 21.10 The viscoelastic anisotropy of the skin follows the direction of microrelief lines with the line connecting the RRT maxima in a direction perpendicular to the microrelief (Langer's line). As an example a video-microscope image (100×) of the skin is shown with a plot of the RRT values superimposed plotted as a function of the angle, i.e., in polar coordinates

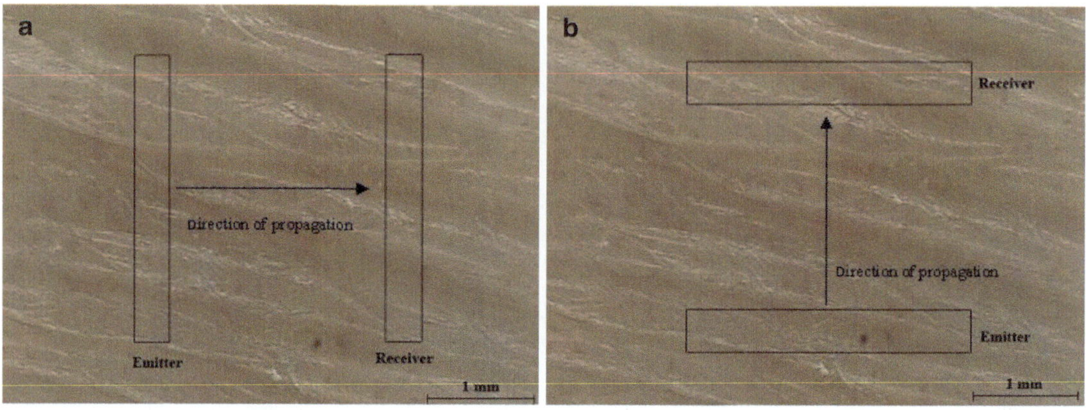

Fig. 21.11 An example of video-microscope image (HiScope, Hirox with 100× lens) of the upper inner arm from a 70-year-old subject. The figure shows in scale the Reviscometer® transducer tips in relation to skin dermatoglyphics. (**a**) The signal propagating in the direction of Langer's lines. (**b**) The signal propagating perpendicular to the direction of Langer's lines

1,800 m/s for the speed of the sound (based on the probe calibration). The tension of the skin is at maximum, and the shear waves propagate faster along this direction (along the relief lines). Conversely, a maximum of the RRT value is recorded when the transducer tips are placed parallel to the relief lines (Fig. 21.11b). An RRT value of $1,100 \times 10^{-3}$ ms corresponds to a speed of sound of 390 m/s, a value close to that of the speed of sound in the air (340 m/s). This is the direction of the lowest skin tension.

21.5 The Viscoelastic Properties of Different Anatomical Sites

Measurements of the anisotropy ratio at different body sites (neck, upper inner arm, ventral forearm, and dorsal forearm) showed that the age dependence is not the same over the whole body (Fig. 21.12). The rate of change of viscoelastic properties depends on the anatomical site. The anisotropy increases with age and most notably at anatomical sites where the skin is considered to be "tender" or "soft," such as the upper inner arm and the neck. Interestingly, the most dramatic changes happen on body sites where skin is considered to be "looser" or "softer" like the neck and the upper inner arm. The viscoelastic properties of the skin at the ventral and dorsal side of the forearm are not statistically different and do not show any significant change within the age range tested here.

21.6 Modulation of Viscoelastic Properties of the Skin by Changes in Keratinocyte Morphology

From the experiments conducted and from the layered structure of the skin, we would expect the propagation of shear waves along the superficial layers of the skin to take place within the epidermis and probably a little in the papillary dermis. To test this hypothesis, we used an agent that is known to affect keratinocyte morphology and in particular the size of keratinocytes in a confluent assembly [20] as in the viable epidermis. The agent chosen was DMAE (2-(dimethylamino)ethanol) which has been shown to produce a "shrinking" in viable keratinocytes in vitro. Assuming that a similar effect takes place in vivo, it would result in an increase in density and a corresponding change in the RRT of the shear wave. In the experiments that follow, we show the effect of application of various concentrations of DMAE on the viscoelastic properties of the skin and on the morphology of epidermal keratinocytes in vivo.

Viscoelastic measurements, confocal laser scanning microscopy (CLSM) reflectance images, and skin surface roughness measurements were

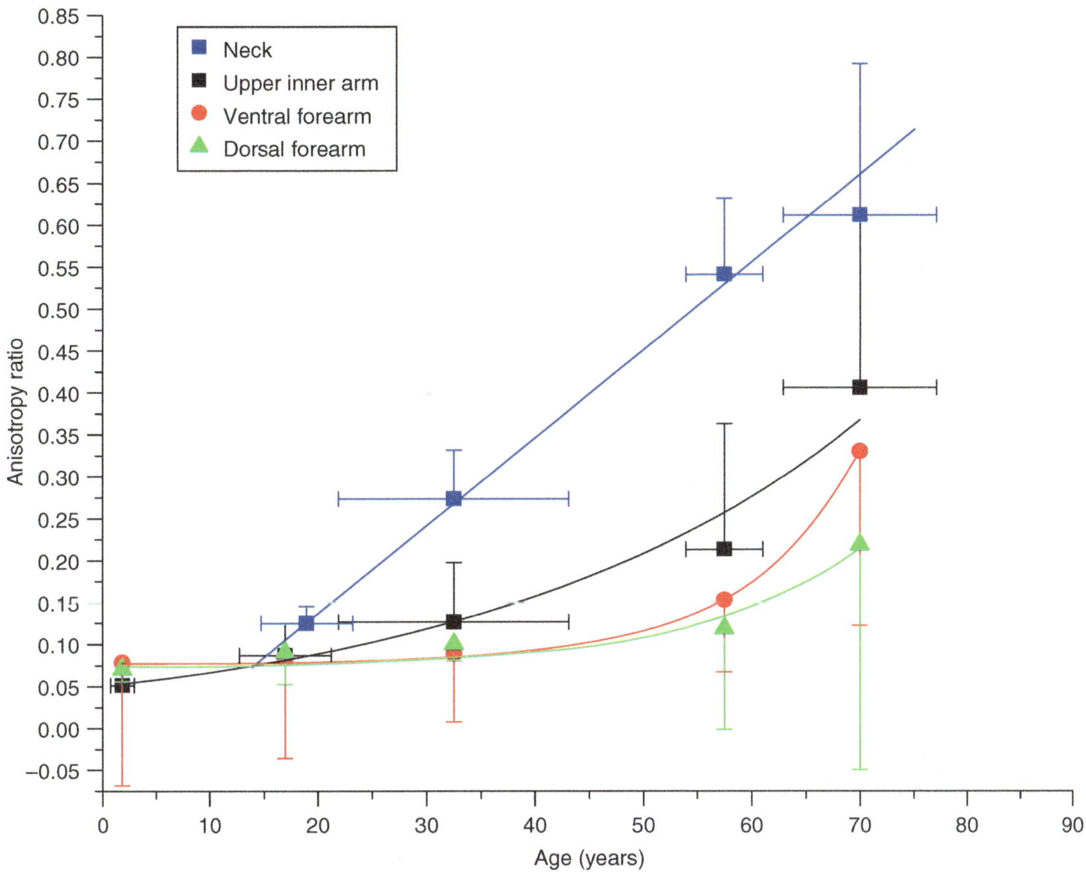

Fig. 21.12 The anisotropy ratio increases with age showing greater rate of increase for skin sites that are "softer." The anisotropy ratio is shown as a function of age for a number of anatomical sites: neck (*blue squares*), upper inner arm (*black squares*), ventral forearm (*red circles*), and dorsal forearm (*green triangles*)

taken in subjects treated with a gel formulation containing different concentrations of 2-(dimethylamino)ethanol (DMAE) (BASF, Florham Park, NJ). Gel-based formulations were prepared with 0, 0.5, 1, 2, and 3 % of DMAE. The protocol for these studies was approved by an ethics committee, and the subjects were in general good health and gave signed informed consent.

21.7 Viscoelastic Measurement of Treated Sites with DMAE: Dose Response

The Reviscometer® readings from the upper inner arm were evaluated before and after application of gel formulations containing different concentrations of DMAE. Figure 21.13a shows results obtained from a 45-year-old subject; RRT measurements were made immediately before (for untreated skin) and 35 min after the product application as a function of the probe angle (degree). Measurements were made from different skin sites each treated with a different concentration of DMAE-containing gel; the concentrations used were 0 % (placebo), 2 %, and 3 %. The effect of treatment with time is shown in Fig. 21.13b. This figure demonstrates the effect of the application of 3 % DMAE on the RRT measurements with time following application.

Relationship between the skin viscoelastic anisotropy and the concentration of DMAE was evaluated. Viscoelastic measurements were performed in 8 volunteers with skin types I–IV according to the Fitzpatrick classification, between the ages of 45 and 60 years old. Readings

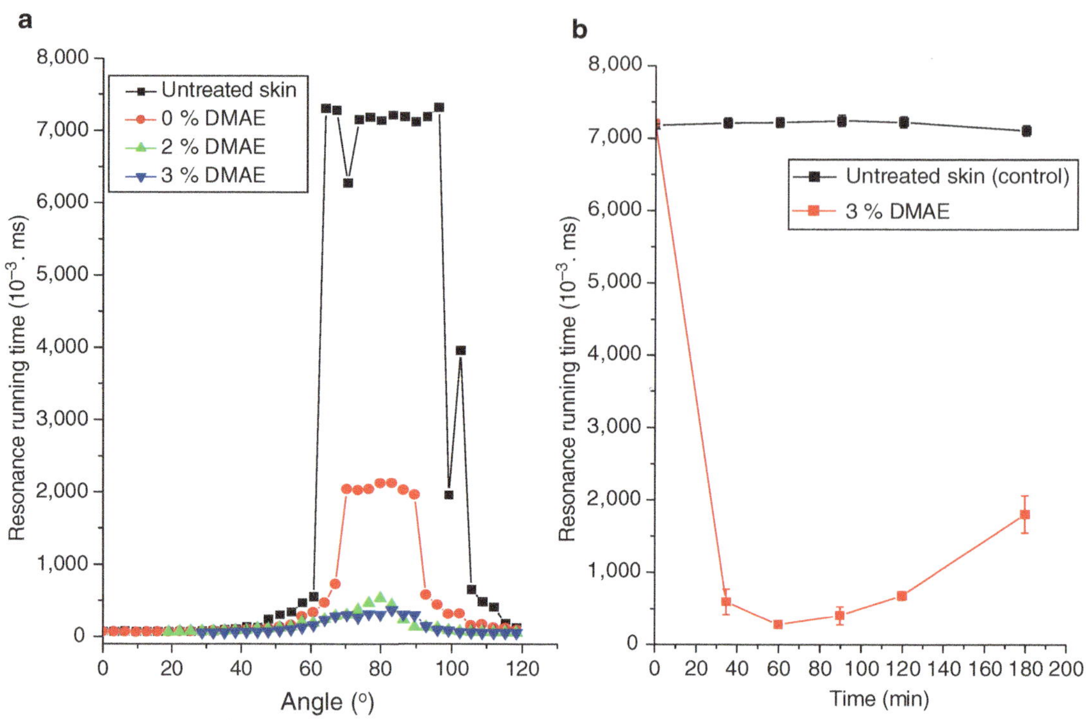

Fig. 21.13 (**a**) Resonance running time in 10^{-3} ms as a function of the probe angle (degree) measured on the upper inner arm for different concentrations of DMAE applied on skin. (**b**) Maximum value of RRT measured with Reviscometer® for sites treated with 3 % DMAE and a control site (untreated). Readings were followed for 180 min after product application. Error bars represent the standard deviation

were taken on the upper inner arm, 15 cm from the elbow with the modified RRT probe [29] where measurements were performed every 3° for a total interval of 100° obtaining the angular profile of the resonance running time (RRT) in the skin. The efficacy of the applied product was evaluated by calculating the ratio of the anisotropy using Eq. 21.2 before and after product application. The values for the anisotropy ratio and standard deviation are illustrated in Fig. 21.14

The assessment of the viscoelastic properties of the skin using the Reviscometer RVM 600® demonstrated a significant decrease in RRT values (Fig. 21.13a) for different concentration of DMAE. Uhoda et al. [31] have also studied the cutaneous tensile effects of DMAE in skin using the Reviscometer. Lower RRT values imply that the speed of sound, generated by the probe transmitter, propagates faster in the skin due to changes in local density of the skin and promoting small changes in the dermatoglyphic shape resulting in an increase of skin firmness. What is interesting to notice from the Fig. 21.13a is the fact that even a placebo gel product can significantly decrease the viscoelasticity of the skin surface (almost three-fold), a similar effect is observed after application of moisturizers. The stratum corneum (SC) is plasticized with the application of the moisturizers, the water content increases in the upper layers of the skin, and the speed of sound propagates faster when more water is added to the SC. A linear increase in skin firmness was observed with the increase of active concentration applied on skin, Fig. 21.14. A 20-fold decrease in skin anisotropy was observed with the use of the described active showing a large dynamic range for this instrument.

21.8 Morphological Changes in Keratinocytes Treated with DMAE Gel

Measurements of the changes in keratinocyte morphology were conducted using a VivaScope® 1000 Confocal Reflectance Microscope (Lucid, Rochester, NY) with a water immersion objective

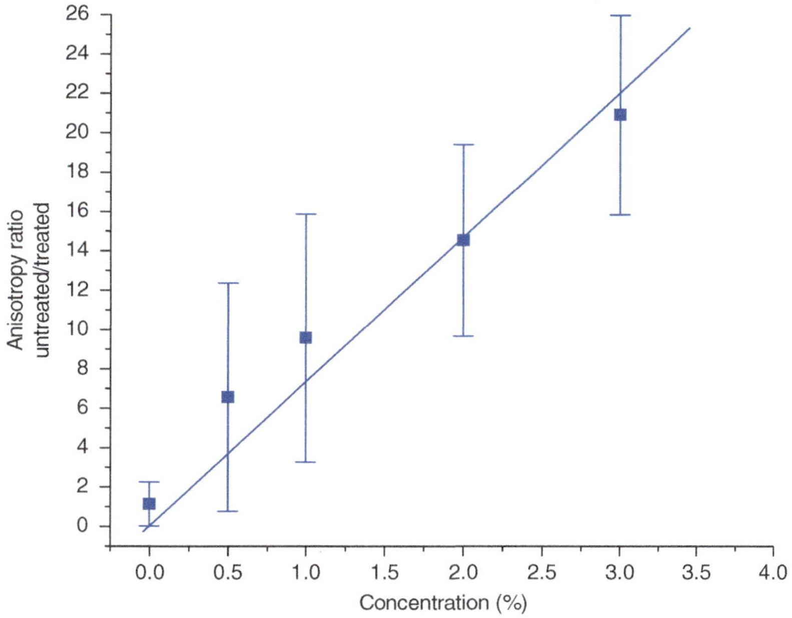

Fig. 21.14 DMAE dose response measured with the Reviscometer® RVM 600. Error bars represent the standard deviation (mean for 5 readings)

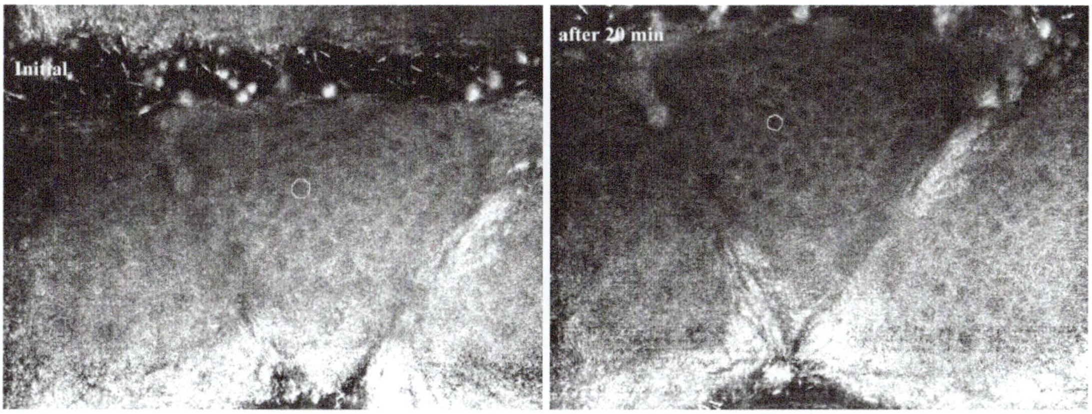

Fig. 21.15 Confocal images from the upper inner arm at 21 μm from the top of the stratum corneum. Skin was imaged initially and after 20 min of 3 % DMAE gel application

of NA 1.2–0.7 (60×). Sites on the upper inner arm were imaged immediately after sample application, 3 % DMAE gel or placebo at 2 μl/cm^2, every minute for 25 min in 8 subjects as described previously.

The confocal microscope was programmed to acquire 40 frames starting from the top of the stratum corneum and measuring into the epidermis, at incremental steps of 3 μm, allowing observation of changes in cell size/morphology within a region of 40 μm depth. As described by Rajadhyaksha [30], the size of cells varies (in diameter) with skin depth from 25 to 30 μm in the granular cell layer to 15–25 μm in the spinous cell layer. In this study cells were measured in the depth range of 20–27 μm below the top of the stratum corneum, primarily granular cells or upper spinous cells. The exact same cells were compared, at the same depth, immediately and 25 min after product application. Stacks were carefully selected to avoid any cell size differences due to the depth effect. The granular layer cells were found to undergo detectable changes; it was difficult to detect any changes in the cells of the deeper layers of the epidermis.

Figure 21.15 shows the confocal images of sections in the epidermis at the seventh step

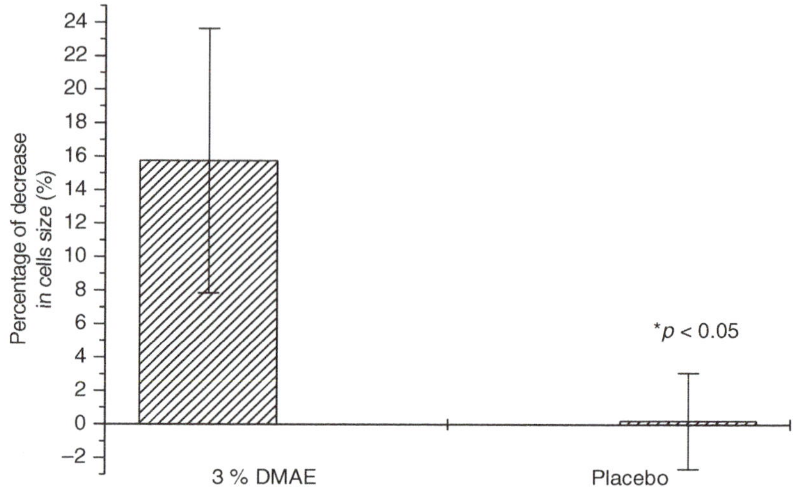

Fig. 21.16 The percent changes in the diameter of granulocytes with treatment. Twenty minutes after treatment, the average reduction in granular cell size was observed to be 15.5 % with 3 % of DMAE and almost no changes in the placebo-treated sites. Error bars represent the standard deviation of the mean

(21 μm from the stratum corneum). The white circle shows the same granular cell boundaries immediately after product application image (initial) and after 20 min.

Images were selected at approximately 20–27 μm from the surface of the skin (top of the stratum corneum), and cell sizes were measured using a linear histogram stretching routine written in IDL® (ITT Visual Information Solutions, Boulder, CO), and cell area was quantified using Image J 1.38 software (NIH, USA). The same image process analysis was applied to all acquired images.

The percentage change in cell size for sites treated with placebo and gel containing 3 % DMAE is shown in Fig. 21.16. Note that the error bar (standard deviation) is large due to variation in response among volunteers. The placebo-treated areas show very small changes in granular cell sizes.

The use of in vivo reflectance confocal microscopy was used to quantify keratinocyte cell shrinkage in the granular layer (Fig. 21.15). As the granular cells contract, there is a reshaping and rearrangement of dermatoglyphics that changes the appearance and viscoelastic properties of the skin. In average we observed a 15.5 % decrease in granular cell area, and no changes were observed in the placebo sample (Fig. 21.16). It is intriguing that despite a 15–17 % change in the cell sizes in the viable epidermis, the stratum corneum does not show an increase in roughness.

21.9 Discussion

The mechanical properties of the skin are an important factor in the determination of the “quality” of the skin. They are as important as the visual appearance of the skin. Changes in the visual appearance include pigmented lesions and minor inflammatory lesions that render the skin nonuniform. Changes in the mechanical properties involve large-area skin sites and are believed to be expressions of aging. One of the dominant characteristics of skin aging is loss of elasticity. Although the changes in the mechanical properties of the skin over several decades of life are substantial, objective measurements have failed to capture their magnitude thus far. Moreover, the mechanical properties of the skin are not isotropic (uniform in all directions), and there is a need to assess this angular anisotropy. During development, from infancy to adolescence, the skin is expected to respond to tension isotropically to accommodate for growth, while in adulthood this isotropic behavior regresses due to site-specific habituation to tension.

In this chapter, we presented a methodology of documenting the angular anisotropy of skin elasticity with high sensitivity and dynamic range using the Reviscometer® RVM 600 and its use to document changes in keratinocyte morphology elicited by topical application of DMAE with significant changes in the mechanical properties of the skin.

The method is based on determining the directional dependence of the speed of an acoustic shear wave on the skin surface at intervals of 3°. The angular distribution of the resonance running time can be described by the anisotropy and the width of the angular dispersion. We found that with increasing age, the anisotropy increases while the angular dispersion width decreases. A new mechanical parameter can be defined using the ratio of the resonance running time anisotropy by the angular dispersion. This new parameter provides a wide dynamic range that reflects more accurately the age-dependent changes of skin mechanical properties than the methods proposed so far. Using this modified Reviscometer® method, we characterized the directional characteristics of cutaneous viscoelastic properties, and we provided evidence that the directionality relates to the direction of the microrelief lines on the skin surface. These lines develop as skin ages and correspond to the tension lines described by the Langer's lines [4].

Changes in viable epidermis keratinocyte morphology and stratum corneum moisturization may be assessed and documented by these parameters. The ratio of these values provides a sensitive parameter for the assessment of the directional behavior of the skin mechanical properties. This parameter provides a large effective dynamic range capable of demonstrating close to an order of magnitude difference in skin viscoelasticity from infants up to 75 years of age and documenting efficacy of topical treatments.

The skin is inhomogeneous, and all probes produce measures of averages over the area sampled by the probe. Local inhomogeneities which may be thought of as early events in the development of any skin lesion would be missed because their effect would be minimized because of averaging.

The Reviscometer® may be capable of providing more information than the suction probe instruments because it actually provides two measures, one depending on the tension and density of the skin and the other depending on the angular anisotropy. In order to be able to capitalize on an instrument such as the Reviscometer®, we are forced to work on skin sites that show marked changes in anisotropy with age. This may be considered a shortcoming of this instrument although we have not found this to be a limitation in testing materials that affects the mechanical properties of the skin. It is unfortunate that the instrument has limited use on traditionally tested skin sites such as the ventral and dorsal forearm.

One parameter that has not been considered is the effect of the coupling of the superficial skin layers to the underlying layers. The appearance of fine lines on the skin surface upon lateral compression is also indicative of the coupling between these layers. A strong coupling would cause a loss of signal strength with distance and would render an instrument such as the Reviscometer® of limited use. On the other hand, the sites of maximum sensitivity are also skin sites that undergo maximal extension due to body motions, i.e., where the elasticity of the skin is most necessary and changes with age may result in redundancies. We have provided experimental data that show the Reviscometer® to be a very useful and sensitive instrument in the assessment of the mechanical properties of the skin and in particular of skin on sites where the skin undergoes large changes in size and shape because of motion.

References

1. Elias PM (2005) Stratum corneum defensive functions: an integrated view. J Invest Dermatol 125:183–200
2. Lanir Y (1986) Skin mechanics. In: Skalak R, Chien S (eds) Handbook of bioengineering. McGraw Hill, New York, pp 11.11–11.24
3. Nizet JL, Pierard-Franchimont C, Pierard GE (2001) Influence of body posture and gravitational forces on shear wave propagation in the skin. Dermatology 202:177–180
4. Langer K (1861) Zur anatomie und Physiologie der Haut: I. Uber die Splatbarkeit der Cutis. Sitzungber Akad Wiss Wien 44:19–46
5. Cox H (1942) The cleavage lines of the skin. Br J Surg 29:234–240
6. Gibson T, Stark H, Evans J (1969) Directional variation in extensibility of human skin in vivo. J Biomech 2:201–204
7. Takakuda K, Miyairi H (1996) Tensile behaviour of fibroblasts cultured in collagen gel. Biomaterials 17:1393–1397

8. Jansen L, Rottier PB (1957) Elasticity of human skin related to age. Dermatologica 115:106–111
9. Lavker RM, Zheng PS, Dong G (1987) Aged skin: a study by light, transmission electron, and scanning electron microscopy. J Invest Dermatol 88:44s–51s
10. Elsner P (1995) Skin elasticity. In: Bardesca E, Elsner P, Wilhelm K, Maibach H (eds) Bioengineering of the skin: methods and instrumentation. CRC Press, Boca Raton
11. Rodrigues L (2001) EEMCO guidance to the in vivo assessment of tensile functional properties of the skin. Part 2: instrumentation and test modes. Skin Pharmacol Appl Skin Physiol 14:52–67
12. Serup J, Jemec G, Grove G (2006) Handbook of non-invasive methods and the skin. CRC Press, Boca Raton
13. Dahlgren R, Elsnau W (1986) Measurement of skin condition by sonic velocity. J Soc Cosmet Chem 35:1–9
14. Davis BR, Bahniuk E, Young JK, Barnard CM, Mansour JM (1989) Age-dependent changes in the shear wave propagation through human skin. Exp Gerontol 24:201–210
15. Hermanns-Le T, Jonlet F, Scheen A, Pierard GE (2001) Age- and body mass index-related changes in cutaneous shear wave velocity. Exp Gerontol 36:363–372
16. Pereira JM, Mansour JM, Davis BR (1990) Analysis of shear wave propagation in skin; application to an experimental procedure. J Biomech 23:745–751
17. Vexler A, Polyansky I, Gorodetsky R (1999) Evaluation of skin viscoelasticity and anisotropy by measurement of speed of shear wave propagation with viscoelasticity skin analyzer. J Invest Dermatol 113:732–739
18. Varani J, Nickoloff BJ, Dixit VM, Mitra RS, Voorhees JJ (1989) All-trans retinoic acid stimulates growth of adult human keratinocytes cultured in growth factor-deficient medium, inhibits production of thrombospondin and fibronectin, and reduces adhesion. J Invest Dermatol 93:449–454
19. Berardesca E, Gabba P, Farinelle N, Borroni G, Rabbiosi G (1990) In vivo tretinoin-induced changes in skin mechanical properties. Br J Dermatol 122(4):525–529
20. Grando SA, Horton RM, Pereira EF, Diethelm-Okita BM, George PM, Albuquerque EX (1995) A nicotinic acetylcholine receptor regulating cell adhesion and motility is expressed in human Keratinocytes. J Invest Dermatol 105:774–781
21. Ruvolo E, Southall M, Bordoloi B, Lukenbach E, Lukenbach G (2009) Compositions for the treatment of signs of aging. US issued patent US20090292027 A1
22. Tronnier H (1980) Dermatologisch-phamakologische Methoden zur Prüfung kosmeticher Präparate und Grundstoffe. Ärzliche Kosmotogie 10:361–367
23. Kinsler LE, Frey AR, Coppens AB, Sanders JV (1999) Fundamentals of acoustics, 4th edn. John Wiley-Sons, Inc., New York
24. Potts RO, Chrisman DA, Burns EM (1983) The dynamic mechanical properties of human skin in vivo. J Biomech 16:365–372
25. Dorogi PL, DeWitt GM, Stone BR, Buras EM Jr (1986) Viscoelastometry of skin in vivo using shear wave propagation. Bioeng Skin 2:59–70
26. Mridha M, Odman S, Oberg PA (1992) Mechanical pulse wave propagation in gel, normal and oedematous tissues. J Biomech 25:1213–1218
27. Biscoe B, Sebastian K (1993) Analysis of the "durometer" indentation. Rubber Chem Technol 66: 827–836
28. ASTM (2003) D2240-03 Standard test method for rubber property—durometer hardness. ASTM International, West Conshohocken
29. Ruvolo EC Jr, Stamatas GN, Kollias N (2007) Skin viscoelasticity displays site- and age-dependent angular anisotropy. Skin Pharmacol Physiol 20: 313–321
30. Rajadhyaksha M, González S, Zavislan JM, Anderson R, Webb RH (1999) In vivo confocal scanning laser microscopy of human skin II: advances in instrumentation and comparison with histology. J Invest Dermatol 113:293–303
31. Uhoda I, Faska N, Robert C, Cauwenbergh G, Pierard GE (2002) Split face study on the cutaneous tensile effect of 2-dimethylaminoethanol (deanol) gel. Skin Research and Technology 8:164–167

Part III

Superficial Skin Analysis

Cyanoacrylate Skin Surface/ Follicular Stripping

22

Gérald E. Piérard, Claudine Piérard-Franchimont, Philippe Delvenne, and Pascale Quatresooz

22.1 Introduction

The stratum corneum (SC) is a dead structure. However, it exerts a unique barrier function partly protecting the living tissues from a series of environmental threats including ultraviolet light, microorganisms and irritant/toxic xenobiotics. In addition, the SC controls any excessive loss in water, electrolytes and macromolecules from the skin. In addition, the SC acts as a unique sophisticated biosensor that signals the underlying epidermis to respond to various external stimuli. Despite minimal metabolic activity, the SC corresponds to a highly specialised structure resulting from the continuous corneocyte renewal ideally keeping a steady state in the SC structure and thickness. However, corneocytes are structurally and biochemically heterogeneous.

Over a vast part of the body, the SC is typically composed of 12–16 layers of flattened corneocytes. These cells are about 1 μm thick and have a mean area reaching approximately 1,000 μm^2. Of note, the corneocyte surface area is influenced by age, anatomical location and any conditions including chemical irritation and UV insults altering the epidermal renewal. In particular, the average corneocyte size apparently increases with age. This feature is assumed to be related to a prolonged transit time of corneocytes through the SC.

Normal SC binds water and keeps its surface soft and smooth. Some of its molecular components bind water and/or prevent water evaporating from the skin surface. These compounds include the so-called natural moisturising factor (NMF), consisting of a mixture of water-soluble small molecules such as amino acids, lactate and urea, the intercellular lipids, sebum and specific protein components of corneocytes. Abnormalities in these components produce a harsh and hard SC that leads to the development of fine cracking and fissuring.

In some instances, the SC homeostasis is altered. Indeed, the SC is the repository of many biologic events that previously influenced the underlying metabolically active keratinocytes. The SC structure is further altered by diverse and repeat external insults. The genetic background, the nutritional status, some physical agents, as well as drugs, cosmetics, toiletries and other chemical xenobiotics represent additional modulators of the SC structure. Knowledge about the fine SC structure is crucial in many respects in the field of dermocosmetic science, particularly when dealing with age-related xerosis and effects of surfactants, emollients and squamolytic agents.

G.E. Piérard, MD, PhD (✉)
C. Piérard-Franchimont • P. Delvenne • P. Quatresooz
Laboratory of Skin Bioengineering and Imaging,
Department of Dermatopathology,
University Hospital of Liège, CHU,
Sart Tilman, Liège B-4000, Belgium
e-mail: gerald.pierard@ulg.ac.be

E. Berardesca et al. (eds.), *Non Invasive Diagnostic Techniques in Clinical Dermatology*,
DOI 10.1007/978-3-642-32109-2_22, © Springer Berlin Heidelberg 2014

22.2 Cyanoacrylate Skin Surface and Follicular Stripping

Cyanoacrylate skin surface stripping (CSSS) is a time-honoured method [1]. The method came into existence when high bond glues became available. After its initial description, it was soon applied for diagnostic purposes [2]. Sampling on polyethylene strips was a decisive improvement in its subsequent development [3]. Indeed, such plastic sheet is preferable to glass slide for two main reasons. It is indeed easier to get a close modelling of curved body areas. In addition, the adhesion of the SC is such that corneocytes are not lost during the laboratory procedures.

The CSSS method consists of depositing a drop of cyanoacrylate liquid adhesive onto a supple transparent sheet of terephthalate polyethylene, 175 μm thick, cut to the size of a conventional coverslip (1.5×6 cm). The sampling material is presently commercially available as a kit (S-Biokit, C+K electronic, Cologne, Germany). The material is pressed firmly onto the target site of the skin. After 15–30 s, a sheet of SC is easily harvested. Because the adhesion relies on a chemical reaction, the depth of harvested SC is determined by the depth of penetration of the adhesive before polymerisation. The cleavage level is exclusively located inside the SC [4, 5].

CSSS can be performed on any part of the body, with two main provisos. On the one hand, sampling from a hairy area is painful because of pulling out hairs, and the CSSS quality is inadequate owing to the erratic contact with the SC. It is therefore advisable to shave these areas before any CSSS harvesting. On the other hand, the natural intercorneocyte cohesion on the palms and soles is usually stronger than the cyanoacrylate bond, thus impairing the collection of a uniform sheet of corneocytes. However, a CSSS sampling on these sites is possible in certain physiopathologic conditions in which the SC texture is compromised. Of note, oozing and eroded lesions cannot be assessed by CSSS.

When vellus hairs are present on the examined site, they are captured with the CSSS. In addition, CSSS collects follicular casts corresponding to the horny material present at the opening of the pilosebaceous follicles at the skin surface. This sampling method has been specifically called follicular biopsy [6]. It is therefore possible to assess the density of the follicles per unit of surface area and to observe the presence of follicular hyperkeratosis (ketosis), as well as comedones, *Trichostasis spinulosa*, intrafollicular bacteria and mites [4–16]. The so-called skin pores corresponding to follicular or sudoral openings at the skin surface are possibly explored using CSSS [17].

22.3 Global Aspect of Normal Skin on CSSS

CSSS of normal skin reveals a regular network of high-peaked crests corresponding to the skin surface microdepressions composed of the primary, secondary and tertiary order lines [1–5, 18]. Their patterns are typical for specific parts of the body. The primary lines of the skin surface correspond to grooves in the latticework papillary relief at the dermoepidermal junction [19–21]. In young individuals, intersections of primary and secondary lines delimit regularly shaped polyhedral plateaus. On stretching of the skin surface a realignment of these lines occurs. With aging, this network progressively loses its configuration, aligning itself preferentially along the skin tension lines and ending by disappearing in the shallow wrinkles [4]. It is therefore possible to indirectly assess the texture of the superficial dermis on CSSS. As a result, dermal aging, corticosteroid-induced atrophy, sclerosis, striae distensae, scars and many other changes in the connective tissue are conveniently observed noninvasively using CSSS. Such morphologic assessment of the skin microrelief is possibly quantified by computerised image analysis using any profilometry method [4, 5].

Cytologic characteristics of corneocytes are hardly visible on CSSS unless histologic dyes are used [4, 5, 22]. A number of stains are suitable. The most useful and simplest one is a mixture of toluidine blue and basic fuchsin in 30 % ethanol [4]. Each corneocyte contains a water-insoluble protein complex made predominantly of a highly

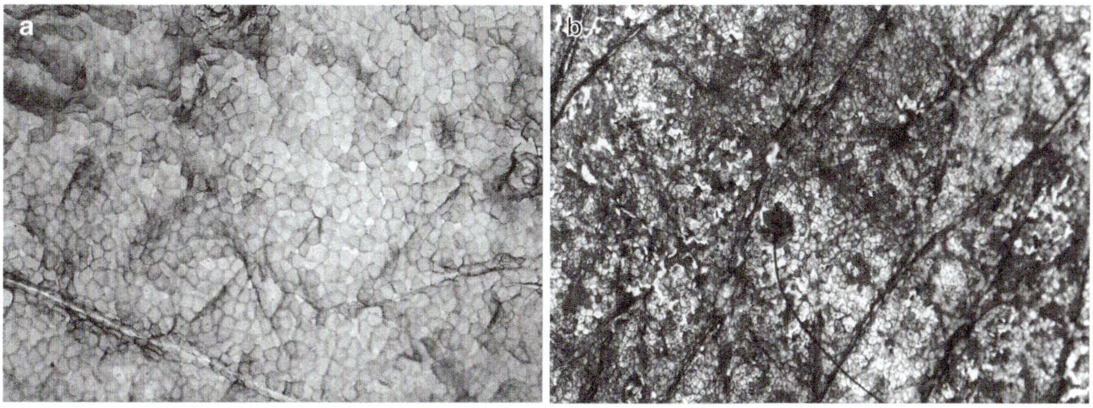

Fig. 22.1 Corneosurfametry bioassay. (**a**) Normal control. The limits of some corneocytes are visible. (**b**) Harsh surfactant. The fragile immature corneocytes are heavily stained in an irregular pattern

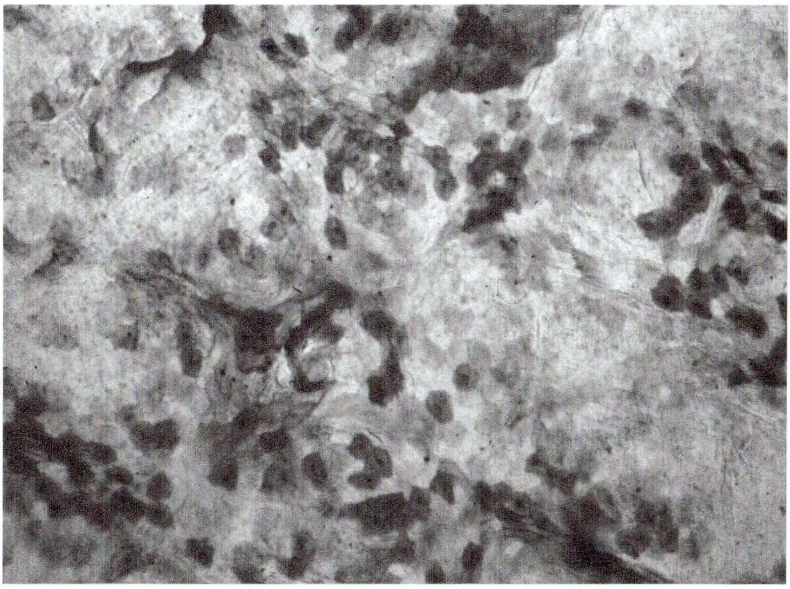

Fig. 22.2 Parakeratotic cells dispersed in the stratum corneum

organised keratin microfibrillar matrix. Such a structure is encapsulated in a protein and lipid-enriched shell. The cornified cell envelope shows differences in maturation among corneocytes. Basically, two distinct types of cornified cell envelopes are distinguished, namely, the fragile immature envelopes and the rigid mature ones [23, 24]. The former cells are recognised on CSSS used in the corneosurfametry bioassay (Fig. 22.1).

Normal skin exhibits a regular cohesive pattern of adjacent anucleated corneocytes. Their boundaries are clearly stained by a thin polyhedral rim. Parakeratotic cells are usually rare and they are not clustered on healthy skin (Fig. 22.2). They are recognised by the presence of a nucleus central to the polyhedral cell.

Saprophytic microorganisms are normally confined to the skin surface and the appendages. Thus, they are encased in the cyanoacrylate bond during sampling and they are not accessible to the staining procedure. As a result, the surface microflora is not seen on CSSS. By contrast, microorganisms present in the follicular casts can be collected distinctly from the skin surface microflora by scraping out the horny spiky structures appending to the CSSS. Viability of the intrafollicular bacteria is possibly assessed using flow cytometry [25, 26].

Sensitive skin is a condition of reduced cutaneous tolerance to environmental factors (cold, heat, wind, wool topical products, etc.). Clinical manifestations consist mainly of subjective symptoms and sensory irritation including discomfort, itching, stinging and burning sensations. No specific signs are discernable on regular CSSS.

22.4 Diagnostic CSSS in Inflammatory Dermatoses

The SC exhibits both biologic efficiency and aesthetic qualities, but it is rarely considered as a structure causing serious disablement when diseased. However, some 50 % of the clinical workload of dermatologists deals with disorders causing scaling and/or thickening of the SC. Psoriasis, the ichthyotic disorders, the various eczematous diseases and the xerosis panel are such disturbances. These relatively common conditions are characterised by abnormal epidermal maturation and scaling. Despite their frequency in most populations, and the problems they cause, there has been relatively little research on the clinical consequences of the abnormal cornification. CSSS help exploring these disorders. Obviously, the diagnostic indications for CSSS are limited to disorders characterised by SC involvement. The most common conditions that can be diagnosed by CSSS are summarised in Table 22.1 [2–5, 27–33].

Definite diagnoses are reached in superficial infectious and parasitic skin diseases [2–5, 27, 29, 31]. Morphologic examination, possibly combined with fungal cultures, can be carried out to identify these types of diseases. By essence, infectious agents that are made visible on CSSS are not those adhering to the skin surface (see above), but rather those invading the SC. Fungi, including yeasts and dermatophytes, show their typical morphology, forming clusters or network of globular or filamentous structures.

Table 22.1 Indications for surface biopsy

1. Superficial infections
Molluscum contagiosum
Bacterial diseases (impetigo, erythrasma, etc.)
Dermatophytosis
Candidosis
Pityriasis versicolor
2. Superficial parasitoses
Scabies
Demodicidosis
Oxyuriasis
3. Xeroses and erythemato-squamous, spongiotic and parakeratotic dermatitides
Xerosis, ketosis and ichthyosis
Eczema – contact dermatitis
Atopic dermatitis
Pityriasis rosea
Id reaction
Psoriasis
Seborrheic dermatitis
4. Tumours
Malignant melanoma
Melanocytic nevus
Dysplastic nevus
Seborrheic keratosis

In the group of parasitic disorders, scabies may pose a problem at the time of sampling [2, 5]. In fact, this diagnosis can be established only when the mite, its eggs or its dejecta are present in the sample. Duplicate or a series of CSSS samplings should therefore be taken from suspected scabies burrows. The first one intends to remove the roof of the burrow. The following ones have a better chance to collect the parasite. Any samples taken outside such parasitic lesion, for example, from non-specific prurigo, will be unhelpful because the diagnosis will merely suggest the presence of a spongiotic dermatitis [4, 5, 31]. Demodex mites are conveniently recognised [4, 5, 15] and highlighted in the follicular casts by the Fite stain [4].

Non-infectious erythemato-squamous disorders include spongiotic and parakeratotic dermatoses and xeroses [4, 5, 29, 31]. Spongiotic dermatoses represent superficial inflammatory reactions responsible for spongiosis, microvesiculation and serosity leakage inside the SC. Contact dermatitis, atopic dermatitis and pityriasis rosea are examples that belong to this group of diseases [5]. Parakeratotic dermatoses encompass id reactions, chronic eczema and stable psoriasis. Seborrheic dermatitis comes within this category, particularly when *Malassezia* yeasts are rare. In active psoriasis, clusters of neutrophils are found on top of parakeratotic foci.

22.5 Diagnostic CSSS in Cutaneous Neoplasms

Some epithelial neoplasms display characteristic aspects on CSSS [4, 5]. Seborrheic keratoses show spotty lenticular foci of soft hyperkeratosis. Widening of shallow furrows with hyperkeratosis is present. Samples of actinic keratosis often exhibit irregular thickness with interfollicular parakeratosis and xerosis. Actinic porokeratosis is revealed by the rim of cornoid lamellation and loss of the normal microrelief inside the lesion. Verrucous surfaces overlying melanocytic nevi and dermatofibromas are less pathognomonic, but sharp circumscription by a normal-looking surrounding skin and uniformity of the changes in the texture of the SC are usually seen in a benign neoplasm.

In CSSS taken from pigmented neoplasms, melanin can be found inside corneocytes or in atypical melanocytes. Melanin located only inside corneocytes is a feature of benign neoplasms, such as lentigines and solar lentigines. Presence of atypical melanocytes in the SC is strongly suggestive of malignant melanoma, but also, in rare instances, of a benign melanoacanthoma [4, 5, 34–36]. Thus, CSSS proves to be sensitive and specific in the distinction between malignant melanoma and benign melanocytic tumours such as common melanocytic nevi, dysplastic nevi or pigmented seborrheic keratoses [27]. For research purposes, karyometry of neoplastic melanocytes can be performed on CSSS [35, 36]. Basal cell carcinomas and squamous cell carcinomas do not exhibit specific or suggestive features on CSSS.

22.6 CSSS Analytic Measurements

22.6.1 Xerosis Grading

Some aspects of disease severity and/or improvement are conveniently assessed noninvasively on CSSS showing specific features in the SC. An example is given by xeroses which correspond to various forms of predominantly orthokeratotic hyperkeratosis [37]. Such SC structure corresponds to the so-called dry skin, although the condition recalls some aspects of the ichthyoses [4, 5, 31, 37–39]. Several grades of orthokeratotic hyperkeratosis are detected on CSSS [5]. Type 0 refers to the absence of hyperkeratosis, except for some discrete focal accumulation of corneocytes in the primary lines of the skin. Type 1a corresponds to a continuous linear hyperkeratosis of the primary lines. Type 1b is characterised by hyperkeratosis predominant at the site of adnexal openings either at hair follicles or at acrosyringia. Type 2 corresponds to focal hyperkeratosis of the skin surface plateaus representing less than 30 % of the surface of the sampling. Type 3 resembles type 2 but with an altered area over 30 % of the skin CSSS. Type 4 is defined by a homogeneous and diffuse hyperkeratosis with persistence of the primary lines. Type 5a resembles type 4 but with loss of recognisable primary lines. Type 5b corresponds to the most heterogeneous and diffuse hyperkeratosis with loss or marked remodelling of the primary line network.

22.6.2 Corneofungimetry

Some quantifications of disease severity and therapeutic activity can be performed on CSSS using computerised image analysis. Quantifications of the fungal load in dermatomycosis can be performed similarly to what described in the corneofungimetry bioassay [30–33].

22.6.3 Corneomelametry

Melanin is present in corneocytes of normal skin in phototype V and VI individuals. The dusty melanin

Fig. 22.3 Fluorescent dye persistence in a stratum corneum renewal test

load is specifically revealed using argentaffin-staining procedures. The relative darkness of these CSSS can be assessed using corneomelametry [40–42]. This method consists of measuring the reduction of light transmission through the CSSS using a photometric device designed for photomicroscopy. It is important to distinguish melanin-laden anucleated corneocytes from neoplastic dendritic melanocytes having migrated inside SC overlying a malignant melanoma.

22.6.4 Stratum Corneum Dynamics

The dynamics of SC renewal is conveniently assessed using CSSS collected about 10 days after topical application of a fluorescent or a coloured dye. The more the SC renewal is rapid, the less stain remains present on the CSSS.

Dansyl chloride is a time-honoured fluorescent compound for the SC. For years, the test relied on daily assessment of the decline in the clinical fluorescence [43]. The rate of SC renewal was determined by the duration of the fluorescence persistence. However, this clinical test proved to be difficult to interpret because it was not easy to clinically evaluate with precision the moment of fluorescent loss. This was due to the uneven fade-out of fluorescence persistence. The CSSS method is a variant performed at a fixed time after dansyl chloride application. The fluorescence pattern is quantified using image analysis [44]. The shallow skin lines represent a typical site for residual fluorescence (Fig. 22.3).

Fluorescence fading is assessed in vivo after application of topical products and interpreted as an effect on the keratinocytes proliferation [45]. However, this procedure represents a pitfall when the test product extracts dansyl chloride from the SC [46]. The correct procedure should begin with the application of the test product for a dozen of days. In a second step dansyl chloride should be applied without any further applications of the test product. Such a procedure allows to disclose any boosting effect on the epidermis without any risk for artefactual dansyl chloride extraction.

A risk of allergy and systemic resorption of dansyl chloride is possible. Hence, there is some limitation for its use, particularly in subjects involved in a series of similar tests. Dihydroxyacetone was offered as a surrogate SC marker [47].

22.6.5 Comedometry

Comedometry allows the computerised quantification of the number and size of follicular casts present on CSSS (Fig. 22.4). This method

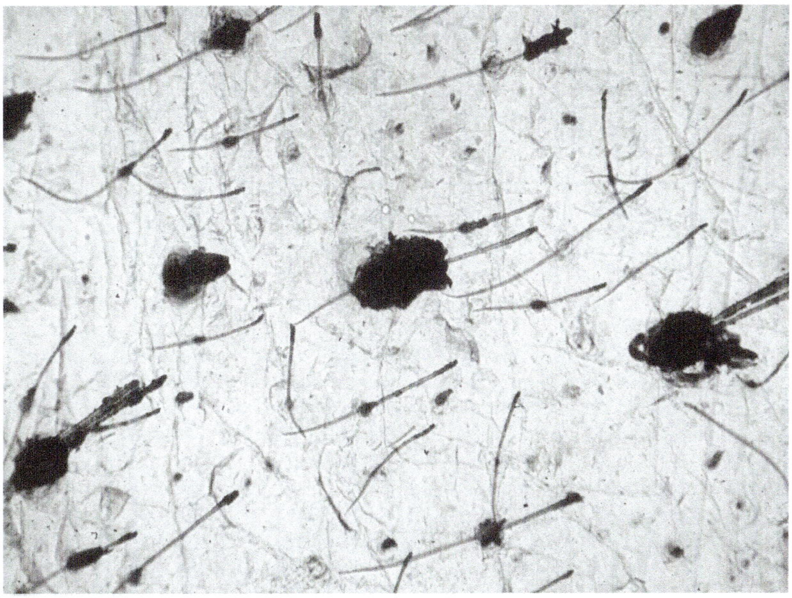

Fig. 22.4 Comedometry showing a few microcomedones (follicular casts) dispersed in an otherwise normal stratum corneum

finds application in the comedogenesis and comedolysis-related disorders and treatments [9, 10, 14, 25, 26]. Comedometry on human skin appears more relevant than animal (rabbit ear) models of comedogenesis. There are large interindividual differences in the number of horny follicular casts among subjects. When an exogenous comedogenic factor is involved, the vast majority of the follicles are similarly affected. By contrast, endogenous comedogenic factors typically affect at variable degree a minority of follicles. The sensitivity of the method is such that microcomedolysis is possibly objectivated after a few days or weeks of adequate treatment.

Lipid-sensitive foils are conveniently used to assess the sebum output at the skin surface. It is possible to combine this method with CSSS [48]. In a first step, the foil is applied to the skin for a limited period of time not exceeding 1 h. The outlines of the foil are ink marked on the SC. In a second step following removal of the foil, a CSSS is collected from the very same skin site. The ink mark is visible on this sampling. The CSSS and the foil are then exactly superposed using the ink mark as an adjusting mark. The dual samplings are examined under the microscope and submitted to image analysis considering the darker horny follicular casts and the clear transparent sebum spots. Correlations are possibly established between the follicular pore sizes, microcomedones and the follicular sebum output [48].

Conclusions

Beside conventional biopsies and cytology of exudates, imprints and scrapings, CSSS provide useful information in the field of dermatopathology and skin pharmacology. This simple and noninvasive method allows the clinician to avoid invasive biopsy within limits of well-defined indications. Less than 3 min are necessary between sampling and examination. There are evident features and subtle characteristics discernible in the structure of the SC that enable a diagnosis to be made in a variety of skin diseases. It is important to stress that no single criterion should usually be relied upon for a definitive diagnosis on CSSS. Rather a constellation of clues should be sought. Quantifications are made possible on CSSS using computer-assisted image analysis.

Acknowledgements This work was supported by a grant from the "Fonds d'Investissement de la Recherche Scientifique" of the University Hospital of Liège. No other sources of funding were used to assist in the preparation of this manuscript. The authors have no conflicts of interest that are directly relevant to the content of this review. The authors appreciate the excellent secretarial assistance of Mrs. Ida Leclercq and Marie Pugliese.

References

1. Marks R, Dawber RPR (1971) Skin surface biopsy: an improved technique for the examination of the horny layer. Br J Dermatol 84:117–123
2. Agache P, Mairey J, Boyer JP (1972) Cyanoacrylate stratum corneum stripping. Interest in skin physiology and pathology. J Med Lyon 53:1017–1022
3. Lachapelle JM, Gouverneur JC, Boulet M, Tennstedt D (1977) A modified technique (using polyester tape) of skin surface biopsy. Br J Dermatol 97:49–52
4. Piérard-Franchimont C, Piérard GE (1987) Assessment of aging and actinic damages by cyanoacrylate skin surface stripping. Am J Dermatopathol 9:500–509
5. Piérard-Franchimont C, Piérard GE (1985) Skin surface stripping in diagnosing and monitoring inflammatory, xerotic and neoplastic diseases. Pediatr Dermatol 2:180–184
6. Mills OH, Kligman AM (1983) The follicular biopsy. Dermatologica 167:57–63
7. Holmes RL, Williams M, Cunliffe WJ (1972) Pilosebaceous duct obstruction and acne. Br J Dermatol 87:327–332
8. Marks R, Dawber RPR (1972) In situ microbiology of the stratum corneum. Arch Dermatol 105:216–221
9. Mills OH, Kligman AM (1982) A human model for assaying comedolytic substances. Br J Dermatol 107:543–548
10. Mills OH, Kligman AM (1982) A human model for assessing comedogenic substances. Arch Dermatol 118:903–905
11. Groh DG, Mills OH, Kligman AM (1992) Quantitative assessment of cyanoacrylate follicular biopsies by image analysis. J Soc Cosmet Chem 43:101–112
12. Pagnoni A, Kligman AM, El Gammal S, Stoudemayer T (1994) Determination of density of follicles on various regions of the face by cyanoacrylate biopsy: correlation with sebum output. Br J Dermatol 131: 862–865
13. Piérard GE, Piérard-Franchimont C, Goffin V (1995) Digital image analysis of microcomedones. Dermatology 190:99–103
14. Letawe C, Boone M, Piérard GE (1998) Digital image analysis of the effect of topically applied linoleic acid on acne microcomedones. Clin Exp Dermatol 23:56–58
15. Forton F, Seys B, Marchal JL, Song AM (1998) Demodex folliculorum and topical treatment: acaricidal action evaluated by standardized skin surface biopsy. Br J Dermatol 138:461–466
16. Uhoda E, Piérard-Franchimont C, Piérard GE (2003) Comedolysis by a lipohydroxyacid formulation in acne prone subjects. Eur J Dermatol 13:65–68
17. Uhoda E, Piérard-Franchimont C, Petit L, Piérard GE (2005) The conundrum of skin pores in dermocosmetology. Dermatology 210:3–7
18. Hashimoto K (1974) New methods for surface ultrastructure: comparative studies of scanning electron microscopy, transmission electron microscopy and replica method. Int J Dermatol 13:357–381
19. Piérard GE, Hermanns JF, Lapière CM (1974) Stereology of the dermo-epidermal interface. Scanning electron microscopy observation of the basement membrane plasticity. Dermatologica 149:266–273
20. Piérard GE, Franchimont C, Lapière CM (1980) Aging, its expression in the skin microanatomy. Int J Cosmet Sci 2:209–214
21. Rochefort A, Makki S, Agache P (1986) Anatomical location of human skin furrows. Clin Exp Dermatol 11:445–449
22. Marks R (1972) Histochemical applications of skin surface biopsy. Br J Dermatol 86:20–26
23. Hirao T, Denda M, Takahashi M (2001) Identification of immature cornified envelopes in the barrier-impaired epidermis by characterization of their hydrophobicity and antigenicities of the components. Exp Dermatol 10:35–44
24. Harding CR, Long S, Richardson J, Rogers J, Zhang Z, Bush A, Rawlings AV (2003) The cornified cell envelope: an important marker of stratum corneum maturation in healthy and dry skin. Int J Cosmet Sci 25:157–167
25. Piérard-Franchimont C, Gaspard V, Lacante P, Rhoa M, Slachmuylders P, Piérard GE (2000) A quantitative biometrological assessment of acne and hormonal evaluation in young women using a triphasic low-dose oral contraceptive containing gestodene. Eur J Contracept Reprod Health Care 5:275–286
26. Piérard-Franchimont C, Goffin V, Arrese JE, Martalo C, Braham C, Slachmuylders P, Piérard GE (2002) Lymecycline and minocycline in inflammatory acne a randomized, double-blind study on clinical and in vivo antibacterial efficacy. Skin Pharmacol Appl Skin Physiol 15:112–119
27. Whiting DA, Bisset EA (1974) The investigation of superficial fungal infections by skin surface biopsy. Br J Dermatol 91:57–65
28. Knudsen EA (1975) The areal extent of dermatophyte infection. Br J Dermatol 92:413–416
29. Piérard GE, Piérard-Franchimont C, Dowlati A (1992) Skin surface biopsy in clinical and experimental dermatology. Rev Eur Dermatol MST 4:445–466
30. Piérard GE, Piérard-Franchimont C, Arrese EJ (1993) Comparative study of the activity and lingering effect of topical antifungals. Skin Pharmacol 6:208–214
31. Piérard-Franchimont C, Piérard GE (1995) Surface biopsies and skin diseases. Rev Med Liege 50:7–15
32. Arrese JE, Fogouang L, Piérard-Franchimont C, Piérard GE (2002) Euclidean and fractal computer-

assisted corneofungimetry. A comparison of 2% ketoconazole and 1% terbinafine topical formulations. Dermatology 204:222–227
33. Piérard GE, Piérard-Franchimont C, Quatresooz P (2010) Updating corneofungimetry: a bioassay exploring dermatomycoses and antifungal susceptibility. Mycopathologia 169:27–35
34. Piérard GE, Piérard-Franchimont C, Arrese EJ et al (1989) Cyanoacrylate skin surface strippings as an improved approach for distinguishing dysplastic nevi from malignant melanomas. J Cutan Pathol 16: 180–182
35. Piérard-Franchimont C, Piérard GE (1989) The value of morphometry and of surface biopsy in the detection of malignant melanoma. Rev Med Liege 44: 610–614
36. Piérard GE, Ezzine Sebai N, Fazaa B et al (1995) Karyometry of malignant melanoma cells present in skin strippings. Skin Res Technol 1:177–179
37. Piérard-Franchimont C, Piérard GE (1984) Xeroses: structure of harsh skin. Int J Cosmet Sci 6:47–54
38. Piérard GE (1996) EEMCO guidance for the assessment of dry skin (xerosis) and ichthyosis: evaluation by stratum corneum strippings. Skin Res Technol 2:3–11
39. Piérard-Franchimont C, Piérard GE (2002) Beyond a glimpse at seasonal dry skin. A review. Exog Dermatol 1:3–6
40. Hermanns JF, Petit L, Piérard-Franchimont C, Paquet P, Piérard GE (2002) Assessment of topical hypopigmenting agents on solar lentigines of Asian women. Dermatology 204:281–286
41. Petit L, Piérard GE (2003) Analytic quantification of skin lightening by a 2% hydroquinone-cyclodextrin formulation. J Am Acad Dermatol Venereol 17: 546–549
42. Thirion L, Piérard-Franchimont C, Piérard GE (2006) Whitening effect of a dermocosmetic formulation. A randomized double-blind controlled study on melasma. Int J Cosmet Sci 28:263–267
43. Janssen LH, Hojko-Tomoko MT, Kligman AM (1974) Improved fluorescence staining technique for estimating turnover of the human stratum corneum. Br J Dermatol 90:9–12
44. Piérard GE (1992) Microscopic evaluation of dansyl chloride test. Dermatology 185:37–40
45. Ridge BD, Batt MD, Palmer HE, Jarett A (1988) The dansyl chloride technique for stratum corneum renewal as an indicator of changes in epidermal mitotic activity following topical treatment. Br J Dermatol 118:167–174
46. Paye M, Smion A, Piérard GE (1994) Dansyl chloride labelling of stratum corneum: its rapid extraction from skin can predict skin irritation due to surfactants and cleansing products. Contact Dermatitis 30:91–96
47. Piérard GE, Piérard-Franchimont C (1993) Dihydroxyacetone test as a substitute for the dansyl chloride test. Dermatology 186:133–137
48. Piérard GE (1987) Rate and topography of follicular sebum excretion. Dermatologica 175:280–283

Corneosurfametry and Corneoxenometry

23

Gérald E. Piérard, Claudine Piérard-Franchimont, and Marie-Annick Reginster

23.1 Stratum Corneum and Chemical Xenobiotics

The interaction between surfactants and the stratum corneum (SC) has many-sided aspects. Lipid removal, protein denaturation, corneocyte swelling, impaired SC barrier function, and SC roughness induction are among the best recognized effects. Testing skin compatibility of surface-active agents usually relies on a large panel of volunteers. Such a procedure is costly and time consuming. As a result, several alternative methods were designed both in vivo and in vitro.

Skin washing with a cleanser is occasionally responsible for both sensorial and chemical irritations. Most individuals complaining with sensorial or chemical irritation find it necessary to select skin cleansers using a trial-and-error approach. However, a number of in vivo methods and in vitro tests were described for the prediction of potential surfactant irritancy. These include the so-called pH rise of bovine serum albumin, the corneocyte swelling, the collagen swelling, and the zein solubilization tests [1]. Under some experimental conditions, however, data gained by these regular tests failed to correlate with in vivo observations [2]. Such pitfalls suggested that other relevant in vitro methods would be welcome.

G.E. Piérard, MD, PhD (✉)
C. Piérard-Franchimont • M.-A. Reginster
Laboratory of Skin Bioengineering and Imaging,
Department of Dermatopathology,
University Hospital of Liège, CHU,
Sart Tilman, Liège B-4000, Belgium
e-mail: gerald.pierard@ulg.ac.be

23.2 Corneosurfametry

The interaction between the SC and various chemical xenobiotics is conveniently assessed on cyanoacrylate skin surface strippings (CSSS). Corneosurfametry (CSM) was coined after corneocyte, surfactant, and metry. It refers to effects of surfactants and wash solutions on the SC [3–8]. For this purpose, CSSS are harvested from the inner forearms of healthy volunteers. A solution of the test product or its neat formulation is sprayed over a series of CSSS which are placed in plastic trays covered by lids. After a given period of incubation at controlled temperature, the samples are thoroughly but gently rinsed in running tap water, air-dried, and stained for 3 min in a toluidine blue-basic fuchsin solution. Thereafter, the samples are copiously rinsed with water and dried prior to perform color quantification using reflectance colorimetry. Indeed, surfactants remove lipids and denaturate corneocyte proteins, thus disclosing chemical sites available for staining reactivity. A combined dotted and rimmed pattern is visible on corneocytes at the microscopic examination.

Using quantitative reflectance colorimetry (Chroma Meter CR400, Minolta, Osaka, Japan), the mean values of luminancy (L^*) and Chroma C^* are calculated from measurements made on

E. Berardesca et al. (eds.), *Non Invasive Diagnostic Techniques in Clinical Dermatology*,
DOI 10.1007/978-3-642-32109-2_23, © Springer Berlin Heidelberg 2014

three sites from each sample placed on a white reference tile. Mild surfactants exhibiting little denaturating effect on corneocytes give a combination of high L^* values and low Chroma C^* values. Typically, L^* decreases and Chroma C^* increases with the irritancy potential of the product. The difference between the L^* and Chroma C^* values of each sample represents the colorimetric index of mildness (CIM). The CSM index (CSMI) of a given test product corresponds to the difference in color between water-treated control samples and those samples exposed to the test product. It is conveniently calculated according to the following formula:

$$\mathrm{CSMI} = \Delta \mathrm{L}*2 + \Delta \mathrm{C}*20.5$$

Increasing the temperature at which CSM is run increases the SC reactivity [9]. The microwave CSM variant is a more rapid procedure [10]. In this procedure, CSSS are immersed in a flask containing the test surfactant solution. Samples are then placed in a microwave oven containing a 500-ml water load. Microwave CSM is conveniently run at 750 W for 30 s. The next steps are identical to the regular CSM procedure.

Responsive CSM is a variant method where skin has been preconditioned before CSSS sampling [11]. For instance, the method is based on repeat subclinical injuries by surfactants monitored in a controlled forearm immersion test. At completion of the preliminary in vivo preconditioning procedure, CSSS are harvested for a regular or microwave CSM bioassay using the same surfactant as in the preconditioning in vivo procedure. Preconditioning the skin by this way increases CSM sensitivity helping to discriminate among mild surfactants [11]. In this context, subjects with atopic dermatitis commonly show increased CSM reactivity [12]. Similarly, some individuals with sensorial irritation exhibit increased CSM reactivity [13].

Shielded CSM was designed for testing the so-called skin protection products [14]. Such products claiming for a barrier effect should theoretically shield against noxious agents. In shielded CSM, regular CSSS are first covered by the test skin protection product ahead from performing CSM using a reference surfactant. Comparative screenings of skin protection products are conveniently performed using shielded CSM without exposing volunteers to any potential hazards linked to in vivo testing.

Animal CSM can be performed in a way similar to human CSM [15]. The method is available for safety testing of cleansing products specifically designed for some animal species. In addition, any interspecies differences in skin reactivity to surfactants are conveniently assessed [15].

23.3 Corneoxenometry

The corneoxenometry (CXM) bioassay was named after corneocyte, xenobiotic, and metry. It was introduced as a convenient approach to explore the effect of some chemical xenobiotics other than surfactants on human SC [16, 17]. The basic procedure is similar to CSM and its variants. The main CXM indication resides in testing skin irritation while avoiding any in vivo hazards. Another indication concerns the comparative assessment of penetration enhancers commonly used in topical formulations [18]. Still another CXM indication deals with the determination of a dose-effect relationship for agents active on the SC structure and function [19].

CXM was used for testing a series of chemicals harmful to the SC [16–20]. The bioassay entails collection of CSSS from normal human skin. The harvested SC sheet, uniform in thickness, is subjected to the ex vivo action of the selected xenobiotics. Series of CSSS covered in excess by their respective chemicals are kept for 2 h at room temperature in a closed environment, for instance an oven, to prevent evaporation of the test solution. Samples are then thoroughly rinsed under running tap water, air-dried, and stained with a toluidine blue-basic fuchsin solution at pH 3.45 for 3 min. Any lipid removal and protein denaturation induce increased dye binding on corneocytes. It has been shown that harsh compounds to the skin considerably increase the intensity of staining of the CSSS [1, 4, 11, 13, 16, 17, 20, 21]. After placing the samples on a white reference tile, reflectance colorimetry (Chroma Meter CR400 Minolta) is used to derive the L^* and Chroma C^* values. Colorimetric data objectively quantify the CXM bioassay.

The colorimetric index of mildness (CIM) is calculated as previously defined [11, 16, 17, 20, 21] following CIM = L^* − Chroma C^*. The relative irritancy index (RII) is calculated following [RII = 1 − [(CIM product) (CIM water)$^{-1}$]. Obviously, RII is not a direct measure of the barrier function. However, it correlates with clinical signs of irritancy. In fact the bioassay explores the combined effects of lipid removal and disorganization and of protein denaturation as well. Hence, any rise in RII is a clue for SC damage responsible for barrier function impairment.

23.4 Penetration Enhancer Testing

One of the upmost important functions of the epidermis is the formation of a well-structured barrier between the body and the ingress of potentially noxious xenobiotics. The latter compounds correspond to environmental contaminants, chemical irritants, toxins, and still others. The barrier function is vital to keep constant the internal living tissues. Much research was performed to understand the skin barrier function of the SC. In some instances, however, chemical penetration enhancers (absorption enhancers or accelerants) represent an attractive potential overcoming the barrier efficacy and increasing drug penetration through the SC. Penetration enhancers typically induce a temporary and reversible decrease in the skin barrier properties. They act in a number of ways, some of which altering the solubility properties or disrupting the ordered nature of the epidermal lipids [22]. Other molecules alter the intercorneocyte cohesiveness.

The desirable attributes for penetration enhancers are varied [22–24]. The compound should be pharmacologically inert without any effect at receptor biologic sites. The risk for irritation, allergy, and toxicity should be minimal. The enhancer should be compatible, both chemically and physically, with drugs and vehicles in the dosage form. It should possess a rapid onset of action with a predictable duration of activity. In addition, the effects should be completely and rapidly reversible upon the product removal from the skin. Furthermore, the effects should ideally be unidirectional, allowing only the ingress of specific xenobiotics without loss of any endogenous component from the body. The penetration enhancer should be cosmetically acceptable, odorless, inexpensive, tasteless, and colorless.

Despite the wide range of purported penetration enhancers, there is no chemical combining all of the desirable attributes. Some chemical enhancers are specifically designed for this purpose such as 1-dodecylazacycloheptan-2-one (laurocapram or Azone®). Other compounds are more common constituents of topical formulations such as surfactants and solvents. The relative efficacies of enhancers towards distinct drugs have been largely explored and compared [25].

In vivo testing with penetration enhancers was performed safely by some researchers in contrast to others who reported severe cell damage in the epidermis and even skin necrosis [26]. Such hazards called for ex vivo predictive bioassays on human skin or SC [26, 27].

The two classes of penetration enhancers, namely, the solvent type and the lipid fluidizer type, are conveniently combined to reach synergistic effects [28, 29]. In complex formulations, each component possibly acts in distinct ways, precluding any determination of specific interactions. Binary and ternary mixtures were reported to be more active than single-penetration enhancers [30]. However, the ideal combination activity of the chemicals is difficult to predict unless a more precise knowledge of the involved mechanisms has been deciphered. In sum, there is a need for accurate assessments of the SC permeability alterations in order to design safe, reliable, and effective formulations [31]. CXM has shown its predictive value in this field.

23.5 Corneoxenometry and Dose–Response Effect of Chemical Penetration Enhancers

A dose–response effect was searched for ethanol and laurocapram using the CXM bioassay [19]. In the same study, other assessments were performed using a gel formulation (propylene carbonate, hydroxypropyl cellulose, butylhydroxytoluene, ethanol, glycerol) containing 10 % propylene glycol and a combination of three other enhancers, namely, N-acetyl-L-cysteine (NAC), urea, and salicylic acid. The three latter penetration enhancers were incorporated in various proportions with keeping the sum of their respective concentrations at the 20 % level.

Data from CXM appeared reproducible and sensitive enough to disclose significant differences between formulations [19]. Both the nature and concentration of penetration enhancers governed the RII values. For each test formulation, the interindividual variability was reasonably low. Linear dose-effect responses were obtained with ethanol in the range 0–100 % and laurocapram in the range 0–5 %. The 10 % propylene glycol-based gel exhibited a wide range in RII values when supplemented with NAC, urea, and salicylic acid. NAC exhibited a moderate effect on CXM. RII rose with increasing amounts of urea replacing NAC. The RII rise was stronger when using salicylic acid instead of urea. The combination of salicylic acid and urea proved to be more active than salicylic acid alone.

23.6 Corneoxenometry and Organic Solvents

The effects of organic solvents were studied in many instances [32, 33]. In particular, they were compared using CXM [16]. Series of CSSS were immersed for 1, 5, 10, 30, 60, or 120 min in vials containing deionized water or an organic solvent including chloroform, ethanol, hexane, methanol, chloroform-methanol (2:1, v/v), hexane-ethanol (2:3, v/v), and hexane-methanol (2:3, v/v). After contact with the selected solvent for one of the predetermined duration, CSSS were thoroughly rinsed under running tap water for 20 s, air-dried, and stained for 3 min with toluidine blue-basic fuchsin dyes.

The ranking from the least to the most aggressive solvent according to the mean CIM was as follows: hexane (40.7), ethanol (26.5), methanol (23.5), hexane-ethanol (23.3), chloroform (20.8), chloroform-methanol (15.5), and hexane-methanol (7.8) [16]. CIM values showed that the effect of hexane-methanol on SC was significantly higher ($p<0.01$) than those of any other solvent with the exception of chloroform-methanol. Chloroform-methanol is well known as the most potent extraction mixture for lipids in biological samples. However, it did not reach the top rank using the CXM bioassay [16]. Such a finding further illustrated the fact that organic solvents may alter other biological components which in turn affect the CXM data. No significant difference was yielded between ethanol, methanol, and hexane-ethanol, but each of them was significantly ($p<0.05$) more active than hexane. The influence of exposure time between SC and solvents showed some solvent-related differences. However, all correlation were significant ($p<0.01$) and best fitted a logarithmic relationship. It appeared that most of the changes in CIM were reached within 10 min for each solvent.

Despite interindividual inconsistencies in corneocyte alterations, significant differences were reported among solvents using the CXM bioassay [16]. The test organic solvents are recognized to extract lipids [26, 27, 34–37]. In addition, alterations in the SC other than pure lipid extraction are likely [27]. Large interindividual differences in CIM were found for any of the solvents or mixtures [16] suggesting the variability in the overall lipid extraction by these compounds [26]. The alterations induced in the human SC by solvents at the CXM bioassay were indeed reported to be more variable in extent that those induced by diluted surfactants as shown at the CSM bioassay on normal subjects [13, 16].

Conclusion

CSM and its CXM variant appear as relevant and predictive bioassays for assessing the overall effect of single and combined chemical xenobiotics. They are cheap, rapid, minimally invasive, and relevant to human skin reactivity. In addition, the reproducibility, specificity, and sensibility are quite high. These bioassays therefore represent valuable screening tests proposed as an alternative to animal testing and hazardous human testing.

The CSM and CXM bioassays allow to assess the influence of the contact time between the SC and noxious chemicals. The time range between 1 and 120 min appears relevant following available information about the kinetics of lipid extraction from human SC. The CSM and CXM data are in line with a handful of other procedures. However, they do not explore the effects of xenobiotics on both the living epidermis and the nature and intensity of inflammation that could result in irritant dermatitis.

Acknowledgements This work was supported by a grant from the "Fonds d'Investissement de la Recherche Scientifique" of the University Hospital of Liège. No other sources of funding were used to assist in the preparation of this manuscript. The authors have no conflicts of interest that are directly relevant to the content of this review. The authors appreciate the excellent secretarial assistance of Mrs. Ida Leclercq and Marie Pugliese.

References

1. Goffin V, Paye M, Piérard GE (1995) Comparison of in vitro predictive tests for irritation induced by anionic surfactants. Contact Dermatitis 33:38–41
2. Zeidler U (1986) Physico-chemical in vitro methods for determination of the skin compatibility of surfactants. J Soc Cosmet Chem (Japan) 20: 17–26
3. Piérard GE, Goffin V, Piérard-Franchimont C (1994) Corneosurfametry: a predictive assessment of the interaction of personal care cleansing products with human stratum corneum. Dermatology 189:152–156
4. Piérard GE, Goffin V, Piérard-Franchimont C (1994) Squamometry and corneosurfametry in rating interactions of cleansing products with stratum corneum. J Soc Cosmet Chem 45:269–277
5. Piérard GE, Goffin V, Hermanns-Lê T, Arrese JE, Piérard-Franchimont C (1995) Surfactant-induced dermatitis. A comparison of corneosurfametry with predictive testing on human and reconstructed skin. J Am Acad Dermatol 33:462–469
6. Piérard GE (1996) EEMCO guidance for the assessment of dry skin (xerosis) and ichthyosis: evaluation by stratum corneum strippings. Skin Res Technol 2:3–11
7. Henry F, Goffin V, Maibach HI, Piérard GE (1997) Regional differences in stratum corneum reactivity to surfactants: quantitative assessment using the corneosurfametry bioassay. Contact Dermatitis 37: 271–275
8. Xhauflaire-UHoda E, Loussouarn G, Haubrechts C, Léger DS, Piérard GE (2006) Skin capacitance imaging and corneosurfametry. A comparative assessment of the impact of surfactants on stratum corneum. Contact Dermatitis 54:249–253
9. Goffin V, Letawe C, Piérard GE (1996) Temperature dependent effect of skin-cleansing products on human stratum corneum. J Toxicol-Cutan Ocul Toxicol 15:125–130
10. Goffin V, Piérard GE (2001) Microwave corneosurfametry and the short-duration dansyl chloride extraction test for rating concentrated irritant surfactants. Dermatology 202:46–48
11. Uhoda E, Goffin V, Piérard GE (2003) Responsive corneosurfametry following in vivo preconditioning. Contact Dermatitis 49:292–296
12. Goffin V, Piérard GE (1996) Corneosurfametry and the compromised atopic stratum corneum. Arch Dermatol Res 288:489–491
13. Goffin V, Piérard-Franchimont C, Piérard GE (1996) Sensitive skin and stratum corneum reactivity to household cleaning products. Contact Dermatitis 34: 81–85
14. Xhauflaire-Uhoda E, Macarenko E, Denooz R, Charlier C, Piérard GE (2008) Skin protection creams in medical settings: successful or evil? J Occup Med Toxicol 3:15–20
15. Goffin V, Fontaine J, Piérard GE (1999) Comparative surfactant reactivity of canine and human stratum corneum. A plea for the use of the corneosurfametry bioassay. Altern Lab Anim 27:103–109
16. Goffin V, Letawe C, Piérard GE (1997) Effect of organic solvents on normal human stratum corneum. Evaluation by the corneoxenometry bioassay. Dermatology 195:321–324
17. Goffin V, Piérard GE, Henry F, Letawe C, Maibach H (1997) Sodium hypochlorite, bleaching agents and the stratum corneum. Ecotoxicol Environ Saf 37: 199–202
18. Goffin V, Henry F, Piérard-Franchimont C, Piérard GE (2000) Penetration enhancers assessed by corneoxenometry. Skin Pharmacol Appl Skin Physiol 13: 280–284
19. Xhauflaire-UHoda E, Piérard-Franchimont C, Piérard GE (2008) Effect of various concentrations of glycolic acid at the corneoxenometry and collaxenometry bioassays. J Cosmet Dermatol 7:194–198
20. Goffin V, Piérard-Franchimont C, Piérard GE (1998) Shielded corneosurfametry and corneoxenometry: novel bioassays for the assessment of skin barrier products. Dermatology 196:434–437
21. Piérard GE, Piérard-Franchimont C (1996) Drug and cosmetic evaluations with skin stripping. In: Maibach HI (ed) Dermatologic research techniques. CRC Press, Boca Raton, pp 133–149
22. Hadgraft J, Walters KA (1994) Skin penetration enhancement. J Dermatol Treat 5:43–47
23. Hadgraft J (1984) Penetration enhancers in percutaneous absorption. Pharm Int 5:252
24. Woodford R, Barry BW (1986) Penetration enhancers and the percutaneous absorption of drugs: an update. J Toxicol-Cutan Ocul Toxicol 5:165–175
25. Williams AC, Barry BW (1992) Skin absorption enhancers. Crit Rev Ther Drug Carrier Syst 9: 305–353

26. Lavrijsen AP, Higounenc IM, Weerheim A, Oestmann E, Tuinenburg EE, Boddé HE, Ponec M (1994) Validation of an in vivo extraction method for human stratum corneum ceramides. Arch Dermatol Res 286:495–503
27. Abrams K, Harvell JD, Shriner D, Wertz P, Maibach H, Maibach HI, Rehfeld SJ (1993) Effect of organic solvents on in vitro human skin water barrier function. J Invest Dermatol 101:609–613
28. Wotton PK, Mollgaard B, Hadgraft J et al (1985) A vehicle effect on topical drug delivery. Effect of azone on the cutaneous penetration of metronidazole and propylene glycol. Int J Pharmacol 24: 19–26
29. Ward AJI, Du Reau C (1991) The essential role of lipid bilayers in the determination of stratum corneum permeability. Int J Pharm 74:137–146
30. Rojas J, Falson F, Courraze G, Francis A, Puisieux F (1991) Optimization of binary and ternary solvent systems in the percutaneous absorption of morphine base. STP Pharmacol Sci 1:70–75
31. Diembeck W, Beck H, Benech-Kieffer F, Courtellemont P, Dupuis J, Lovell W, Paye M, Spengler J, Stelling W (1999) Test guidelines for in vitro assessment of dermal absorption and percutaneous penetration of cosmetic ingredients. Food Chem Toxicol 37:191–205
32. Peck KD, Ghanem AH, Higuchi WI (1994) Hindered diffusion of polar molecules through and effective pore radii estimates of intact and ethanol treated human epidermal membrane. Pharm Res 11: 1306–1314
33. Garcia N, Doucet O, Bayer J, Zastrow L, Marty JP (2000) Use of reconstructed human epidermis cultures to assess the disrupting effect of organic solvent on the barrier function of excised human skin. In Vitr Mol Toxicol 13:159–171
34. Bligh EG, Dyer WJ (1959) A rapid method of total lipid extraction and purification. Can J Biochem Physiol 37:911–917
35. Scheuplein R, Ross L (1970) Effect of surfactants and solvents on the permeability of epidermis. J Soc Cosmet Chem 21:853–873
36. Deffond D, Saint Léger D, Lévêque JL (1986) In vivo measurements of epidermal lipids in man. Bioeng Skin 2:71–85
37. Imokawa G, Akasaki S, Hattori M, Yoshizuka N (1986) Selective recovery of deranged water-holding properties by stratum corneum lipids. J Invest Dermatol 87:758–761

Sebumetry and Sebumtape

24

Bernard Gabard, André O. Barel, and Peter Clarys

24.1 Introduction

Sebum is the general term defining the lipids excreted by the sebaceous glands and spreading on the surface of the skin. These skin surface lipids (SSL) are in fact a mixture of epidermal lipids and lipids from the sebaceous glands (sebaceous lipids). The quantity and the composition of SSL are not the same on different areas of the human body. Epidermal lipids are found on the whole body and are the sole component of SSL in anatomical regions where no or only few sebaceous glands are present. High quantities of SSL are present on cutaneous areas with many sebaceous glands such as the face (forehead, nose and cheeks), the scalp and the upper parts of the trunk and of the back. Here the proportion of sebaceous lipids may be important (up to 95–97 %) and the one of epidermal lipids negligible (3–5 %).

Numerous publications dealing with the sebaceous gland and sebumetry have been already published [1–8].

B. Gabard (✉)
Iderma Scientific Consulting,
Basel, CH-4053, Switzerland
e-mail: b.gabard@iderma.ch

A.O. Barel • P. Clarys
Laboratory of Human Biometry and Biomechanics,
Free University of Brussels (VUB),
Brussel, Belgium
e-mail: andreodilon.barel@skynet.be;
pclarys@vub.ac.be

24.2 Skin Surface Lipids

24.2.1 Epidermal Lipids

Epidermal lipids originate from the keratinocytes. Quantity and the composition of the intracellular lipids in the keratinocytes vary during the maturation process. In the basal layer, keratinocytes contain relatively few lipids. During the maturation process, the quantity of lipids is greatly increased, and their composition changes towards a significant proportion of ceramides (30–35 %), less fatty acids (20–25 %) and cholesterol (20–25 %), and little triglycerides (10–15 %) [9, 10]. Free fatty acids and ceramides are typical markers of epidermal lipids. In the upper epidermal layers, keratinocyte lipids are stored in vesicles called lamellar bodies (Odland bodies) [2, 10, 11]. These vesicles migrate towards the periphery of the cells, and their content is liberated into the extracellular space by exocytosis. As a consequence, relatively large multilayer structures are formed which fill in the intercellular space in the stratum corneum. During desquamation and exfoliation of the corneocytes, these lipids are liberated at the surface of the skin [9, 10].

Epidermal lipids are a key component of the stratum corneum. They ensure the cohesion between the multiple cellular layers of the stratum corneum and are responsible for the barrier function of the skin [10]. As a rule, the transepidermal water loss of normal skin remains very low (below 10 $g/m^2 \times h$) [2, 9].

E. Berardesca et al. (eds.), *Non Invasive Diagnostic Techniques in Clinical Dermatology*,
DOI 10.1007/978-3-642-32109-2_24, © Springer Berlin Heidelberg 2014

On the skin surface, the quantity of epidermal lipids remains very low and varies with the anatomical area. On the legs and forearms, 5–20 μg/cm² have been noted [12]. Measurement of SSL on areas where no or only very few sebaceous glands are present is difficult because the measured values are near the lower measurement limit of measuring equipments (near 5 μg/cm²). Measurements reveal a great intra- and interindividual variability.

24.2.2 Sebaceous Lipids

Sebum is a mixture of lipids found on the skin surface. It is produced by the sebaceous gland, which is located in the dermis and associated with a hair follicle (pilosebaceous system). The secretion of the gland takes place in the infundibulum of the follicle [4, 5].

A short note on the physiology of the sebaceous gland is necessary to understand the basics of sebum excretion. Sebaceous glands are holocrine glands, meaning that the glandular secretion consists of cells from the gland itself. Lipid excretion takes place by disintegration of entire cells, the sebocytes. Sebocytes in the basal cellular layer proliferate, become lipid loaded during differentiation and migrate towards the opening of the gland. Through disintegration, lipids and cellular debris are discharged in the sebaceous canal and then into the infundibulum. Transit time from the basal layer via the infundibulum to the skin surface amounts to 2–3 weeks. The infundibulum may build a reservoir with great quantities of sebum [1–8, 13, 14].

The activity of the sebaceous gland is characterised by 4 phases: sebum production (secretion rate), storage of the sebum in the infundibulum, excretion on the skin surface (excretion rate) and interaction of the sebum with the skin surface. A shiny and fatty skin appearance may result from an excess of sebum interacting with the cutaneous surface [15].

The pilosebaceous follicles feature different volumes in their sebaceous gland part and according to their dimensions and the type of follicle as well: vellus follicle, terminal follicle (telogen hair) and sebaceous follicle (a large gland with minute hair) [4]. The skin density of pilosebaceous follicles depends from the anatomical location: 200–500/cm² on the scalp, 300–1,500/cm² on the face and on shoulders and upper trunk 50–100/cm² [4].

The composition of sebum differs considering "native" sebum freshly synthesised and "mature" sebum collected after excretion and spreading on the skin surface. Freshly synthesised "native" sebum does not remain within the infundibulum for a long time. Its chemical composition has not been altered by lipases from the resident microbial flora of the pilosebaceous follicle. Native human sebum does not contain any free fatty acids or ceramides, but squalene (10–14 %), waxes (23–29 %), triglycerides (20–60 %) and some cholesterol (1–5 %). Squalene is a characteristic component present in relatively stable concentration (12–15 %) [4, 12, 14]. Epidermal lipids do not contain squalene or only traces [9, 10, 12]. "Mature" sebum on the skin surface results from the action of bacterial lipases partially hydrolysing triglycerides during sebum flow to the skin surface and from a small contamination by epidermal lipids (3–5 %). Chemical oxidation reactions may also take place. All this explains the presence of cholesterol, of diglycerides and particularly of free fatty acids (5–40 %) in mature sebum.

Anatomically, sebum is abundant on the scalp, on facial skin (forehead [mainly the central part], nose, cheeks and chin), neck, shoulders and upper trunk (both sides). Other anatomical locations present only low to very low sebum levels. On the face, one speaks from the T-zone (forehead, nose, perioral areas and chin) [16, 17]. Because the amount of sebum remains generally constant for each individual on these specific skin areas, it has been called "casual level" (SCL: sebum casual level) [4, 5]. SCL amounts to 100–200 μg/cm² in normal subjects and up to >500 μg/cm² in hyperseborrheic subjects [12, 16, 17].

24.3 Regulation of Sebaceous Gland Activity

Several endocrine factors, internal factors such as gender, age and ethnic group, and external factors such as temperature, relative humidity and season all may influence sebaceous gland function and sebum secretion.

24.3.1 Endocrine Regulation

Androgens stimulate dose-dependently sebaceous function. In men the main androgen is the plasma free testosterone circulating level. In women, main stimulating components are androstenedione and dehydroepiandrosterone (DHEA), which both are enzymatically converted into testosterone by the sebocyte. After entering the cell, testosterone is converted into dihydrotestosterone (DHT) by a type I 5α-reductase. DHT is a more potent androgen than testosterone and binds to a specific nuclear receptor after entering the nucleus in form of a cytosolic macromolecular complex. Estrogens decrease dose-dependently sebum production, but their action is weaker than the enhancing effect of androgens. In women, sebaceous gland function is decreased during the estrogenic follicular phase of the menstrual cycle and increased during the premenstrual luteal phase [3, 4, 13, 14, 18–21].

24.3.2 Influence of Age

During the early months of life and just after birth, a strong rise in sebum production occurs with a maximum at 1 month after birth. Then sebum secretion progressively disappears with 6 months. From 6 months to puberty, SCL remains low. At puberty, a new surge of sebum secretion occurs till adult age, with consequent cutaneous troubles well known from teenagers: oily skin and acne [3, 4, 13, 14].

After 50 years of age, SCL and sebum secretion slowly decrease in men. In women, the menopause is characterised by a strong decrease in both parameters [22]. Although sebum production decreases with age, the dimensions of the follicles do not change or may even greatly increase. Age-related changes in SCL are sometimes weak and difficult to characterise and depend on the anatomical location [22–25].

There are some studies indicating a dependency of the decrease in sebaceous gland function with age depending of the ethnic group [26, 27].

24.3.3 Influence of Gender

According to a past publication, mean sebum secretion values are significantly higher in men aged 20–60 years, whereas this difference is temporarily reversed just before puberty, which occurs sooner in girls [19]. The difference between both sexes in sebaceous gland activity becomes more important at ages 50–70 years because the secretion strongly decreases in women although remaining almost constant in men [18, 25].

Recent publications tend to show that the differences between men and women are less important than the differences due to belonging to a particular ethnic group [27].

24.3.4 Influence of Ethnic Group

Former studies point to a decrease of sebum production in the order Black skin > Caucasian skin > Asian skin [4]. More recent publications have dealt with ethnic group influence on sebaceous function [28, 29]. SCL in Caucasian women is lower than in Japanese women [30], whereas SCL on the forehead of Afro-American women is lower than in Caucasian women, although this difference was not significant [31]. More detailed investigations of the ethnic specificity of sebaceous gland activity on the forehead and cheek in Afro-American, Latin American (Hispanic), Chinese and Caucasian women did not find differences in sebaceous secretion between the four groups. Sebaceous gland density was lower in Latin American (Hispanic) and Chinese women compared to Afro-American and Caucasian women. More recently, no significant differences were noted in SCL in African, Caribbean and Caucasian women living in Europe [32]. All these results should be considered in view of the fact that variations due to climatic conditions could override variations due to the ethnic group [26–29].

24.3.5 Circadian Variations

Sebum secretion on the forehead and cheek undergoes circadian variations [3, 4, 15, 33]. Generally, a maximum in SCL is observed towards 12–14 h and a minimum at 4–7 h [33, 34]. These results have been confirmed on the scalp [35].

24.3.6 Seasonal Variations

Environmental temperature strongly influences SCL. An increase of 1 °C in cutaneous temperature leads to a 10 % increase in sebum secretion [36]. At 40 °C environmental temperature, sebum is very fluid; at 10 °C environmental temperature it is quite viscous [4, 5]. The question arises if the increase in SCL at higher environmental temperatures is due to a stimulation of secretion or simply to a better flowing and a faster excretion due to the decreased viscosity of the sebum. In any case, sebaceous secretion is higher in the spring and summer months than in fall and winter months [3, 4, 17, 33, 36].

Sweating affects SCL. Subjects with a low level of sweating also show low SCL and reciprocally [4, 5]. Sebum interacts with sweat on the skin surface and diminishes the quantity of lipids absorbed by porous polymer foils. For this reason, measurements of SCL should be conducted under environmental conditions where no sweating occurs [7].

24.4 Quantitative Parameters for Measuring Sebaceous Gland Function

Two quantitative parameters are now used to measure sebaceous gland function: the SCL (sebum casual level; already quoted in Sect. 24.2.2) and the rate of secretion SER (sebum excretion rate). Additionally, new sebum collecting techniques allow the determination of active sebaceous gland density.

24.4.1 Sebum Casual Level SCL

SCL is the quantity of skin surface lipids expressed in micrograms per square centimetre ($\mu g/cm^2$). SCL corresponds to the saturation level measured after refatting of the skin surface treated before measurement with an organic solvent (such as alcohol or hexane) and with a nonionic detergent to remove as completely as possible SSL. Generally, refatting is considered completed 3–4 h thereafter, showing then a plateau value [5, 7].

As already indicated SCL amounts to 100–200 $\mu g/cm^2$ in normal subjects and up to >500 $\mu g/cm^2$ in hyperseborrheic subjects.

24.4.2 Sebum Excretion Rate SER

After thorough removal of the surface lipids, refatting begins rapidly. Thereafter, it gradually slows to reach the plateau value after 3–5 h. Refatting kinetics follow an exponential curve [4, 5]. SER is defined as the quantity of lipids produced in a given time on a given surface and is expressed as micrograms per square centimetre and minute ($\mu g/cm^2 \times min$).

SER amounts to 0.5–2.5 $\mu g/cm^2 \times min$ on the forehead. On the scalp and on the cheeks, SER varies between 0.1 and 0.8 $\mu g/cm^2 \times min$ [4, 5].

At times beginning after removal of skin surface lipids, SER corresponds to the release of a pool of sebum already secreted and stored in the upper part of the pilosebaceous duct. SER represents an excretion rate and not the glandular secretion rate [37]. Knowing the quantity of sebum stored in the infundibulum may be important. Because early excretion during refatting mainly originates from this reservoir, it is possible to determine this quantity by collecting sebum without interruption after thoroughly defatting of the skin surface. SER represents the spontaneous refatting rate during early times after defatting which corresponds to sebum coming from the infundibulum. During the following times, SER gradually diminishes to reach a plateau which then represents the sebaceous gland secretion because the infundibulum has been emptied. The quantity of sebum in the infundibulum is calculated by subtracting the amount secreted from the total quantity of sebum sampled before stabilisation of the SER. Such determination is only possible if a quantitative analysis of SCL is conducted [4, 5, 37].

24.4.3 Surface Density of Active Sebaceous Glands

The collection of sebum by mean of a polymer film (see below Sect.24.5.4) allows the visualisation and quantification of the density of follicular reservoirs containing sebum [4, 5, 7]. The number determined is generally lower than the total number of sebaceous glands because not all glands are active at the same time. The results are expressed in the number of active glands per square centimetre, mean surface of active glands and percentage of total active glands in reference to the total surface considered.

24.5 Measurement of Skin Surface Lipids

Several parameters may be measured concerning skin surface lipids by using one or several techniques described below:

- Quantity of skin surface lipids (SSL, SCL) or of epidermal lipids (on anatomical regions where sebaceous glands are active, mainly SCL is measured).
- Chemical composition of collected lipids.
- Sebum excretion rate (SER) respectively surface lipids refatting rate.
- Surface density of active sebaceous glands.

24.5.1 Extraction of Skin Surface Lipids by Organic Solvents

Collection of surface lipids by direct application of organic solvents on the cutaneous surface is an old method which is not frequently used anymore and is not more recommended for various reasons [3, 5]. For this purpose, solvents such as ether, methanol, ethanol, hexane and acetone or mixtures such as methanol-chloroform are introduced in a defatted glass or plastic hollow cylinder applied on the skin for a given time. The choice of the solvent is important, and some solvents may be aggressive to the skin. After a given contact time, the solvent is collected and evaporated, and lipids are weighed and may be solubilised in another suitable solvent for analysis (e.g. HPLC) or any further manipulation.

24.5.2 Collection of Skin Surface Lipids by Absorbent Paper

A paper with known constant lipid-absorbing properties is used. The type of absorbent paper used is critical. After defatting of the paper with ether, it is firmly pressed for a given time (e.g. 3 h) on the skin surface. The paper is weighed before and after impregnation, and the lipids may be extracted by an appropriate solvent for further analysis [5, 37–39]. This method is precise but slow and laborious. It has been extensively used in the past. Many factors need to be controlled during the test, e.g. constant paper quality, paper saturation, possibility of different absorption of the single SSL components and extraction of the lipids [5, 39].

24.5.3 Collection of Skin Surface Lipids on Grounded Plastic Film

A rapid, simple and accurate method for quantification of SCL was developed in the 1970s by using a series of clean ground glass plates applied under constant pressure on the skin during 30s [40]. After collection of the surface lipids, the ground glass plate becomes more translucent. The variation of light transmission across the plates was shown to be a function of the fat deposited on them. This allowed a precise determination of the SCL. The device used at that time, the Lipometre®, was never commercialised.

The same principle was applied on a grounded plastic film. The device is the Sebumeter® (Courage-Khazaka, Cologne, Germany), which uses a grounded plastic ribbon-shaped film becoming more or less translucent depending on the quantity of SSL collected after 30s application on the skin surface under constant pressure (6.6 ± 1.2 N, which is relatively high). Light transmission is measured through an optical reader before application (zero values) and after collection. Results are given in arbitrary units proportional to the quantity of lipids collected by the ribbon in μg/cm^2. An innovation is to provide the plastic ribbon in a rewinding cassette allowing up to 450 consecutive measurements. The measuring head of the cassette exposes a 64 mm^2 measuring section of the ribbon.

A calibration may be conducted by applying on the ribbon or on the skin surface known quantities of synthetic lipids such as artificial sebum or Vaseline having similar viscosity to human sebum [5]. A specific cassette with a known light transmission value is provided by the manufacturer.

Several factors must be taken in account during measurement of SCL with the Sebumeter®. The first collection is not a constant fraction of the SCL. As a rule, it is estimated that an average of about 40 % of the total skin surface lipids is absorbed with one sampling (captation factor), [1–3, 5]. This proportion depends among others from the roughness of the plastic film and from temperature. The temperature for sampling is at equilibrium between skin temperature (32 °C) and probe's temperature (room temperature, usually 22 °C). At 26 °C, sebum viscosity increases by 17 %, which may influence collection [5]. Skin roughness and application pressure also influence contact quality. It is generally admitted that under similar strictly controlled measurement conditions, measurements results may be compared.

Repeated collection on the same skin site shows that the sampled quantity decreases exponentially with the number of collected samples. This may however not be the case on hyperseborrheic surfaces where the first samplings saturate the plastic film [5].

24.5.4 Collection of Skin Surface Lipids on Absorbent Polymer Film

Still following the same line of thinking, a microporous hydrophobic polymer film becomes translucent after lipid absorption [5–8]. The white film absorbs the sebum at follicular opening of active sebaceous glands, loses the colour and becomes transparent on the spots having been in contact with fat. These spots become readily visible on a black background. Single spot surface is proportional to the quantity of secreted lipids during the contact time.

Different techniques are available. Originally, the polymer film offered was glued on the skin surface by an adhesive coating (Sebutape Adhesive Patches®, CuDerm, Dallas, Texas, USA). Polymer films without adhesive are now available such as Instant Sebutape® (CuDerm, 1.1 × 1.1 cm) and Sebufix F16® (Courage-Khazaka, Cologne, Germany; 1.7 × 1.7 cm). Application time are different, at least 30 min (generally about 1 h) for the Sebutape® (because the adhesive coating slows lipid absorption by the polymer) and 30s to 1 min for Instant Sebutape® and Sebufix®.

Evaluation of the results is best made by computerised image analysis. The parameters measured are number of spots, total surface of the black spots and mean surface of the spots. Because the exact percentage of the lipids collected on the skin surface is not known, these parameters are expressed as number of spots per cm^2, individual or total surface of spots in cm^2 or in arbitrary units.

Several factors influence the results. It has been shown that the spots gradually enlarge if the polymer film is not evaluated immediately or within short time (<1 h) after removal from the skin surface. This is due to lateral spreading of the collected sebum within the polymer film and is especially pronounced if large spots are present [5, 7]. Immediate values may be considerably lower than values measured 24 h later. Longer storage leads to spots becoming whitish and more difficult to distinguish or even undetectable from their surroundings [7]. Storage in a freezer at −30 °C will delay but not completely prevent the lateral increase and the turning whitish of the spots.

It may be added that collection of sebum by a microporous polymer film is very suitable for analysis of lipid components after extraction by appropriate solvents [41–43].

24.5.5 Correlation Between the Results Obtained by the Different Measuring Techniques

No systematic investigations of the correlation between the results obtained by the different measuring techniques have been conducted. Only a few investigations are available which show that sometimes conclusions from one measuring technique may not be valid for another one. Most results are available for plastic film (Sebumeter®) measurements and polymer tape (Sebutape®, Instant Sebutape® or Sebufix®) collection, as these two techniques are now most frequently used.

Scalp sebum measurements conducted with the grounded glass plate method were shown to be well correlated with the results obtained by Sebumeter® measurements [44].

Sebumeter® measurements of SCL were shown to be linearly and strongly correlated to measurements of sebum after absorption on a polydimethylsiloxane tape and analysis via sophisticated chromatographic techniques [41]. However, in a former study, different results were obtained during investigations of the effect of a topical treatment by Sebumeter® measurement and Sebutape® collection [45].

To explain these discrepancies, it has been hypothesised that correlation may be good for intermediate levels of sebum secretion (e.g. 150–350 mg/cm^2), but the results may diverge as secretion levels become more extreme (i.e. hypo- and hypersecretion). Furthermore, tape collection and plastic film collection correspond to sebum from different sources (which is not the case if one considers a grounded glass plate and the plastic film): Sebutape® (or Instant Sebutape® and Sebufix®) absorbs the free sebum present in the upper part of the infundibulum, whereas the plastic film from the Sebumeter® collects SSL (all the sebum present in the follicular reservoir plus any on the skin surface).

Conclusion

In conclusion, before engaging in any SCL measurements, one should bear in mind that sebaceous gland function is very variable between individuals and influenced by many physiological and external factors. Most important are endocrine regulation, variations with age, gender and most probably ethnic group belonging. Circadian variations must be taken into account. External factors influencing sebaceous gland function are most importantly temperature, relative humidity and the season of the year (climate).

Therefore, measurements of sebaceous gland function and of SCL require strict control of different measuring conditions such as temperature, relative humidity (no sweating), time (circadian rhythm) and time after skin cleansing. Recent measurement techniques using grounded plastic ribbon and/or microporous polymer films applied for a short time on the skin surface have made investigations very easy. But meaningful results are obtained only under strictly controlled experimental conditions.

If this is the case, intra-subject coefficients of variability are quite low for repeated measurements on the same individual (10–15 %) but clearly higher for interindividual comparisons (up to 40 %).

References

1. Saint-Léger D (1989) Quantification of skin surface lipids and skin flora. In: Lévêque JL (ed) Cutaneous investigation in health and disease. Marcel Dekker, New York, pp 153–182
2. Clarys P, Barel AO (1995) Quantitative evaluation of skin surface lipids. Clin Dermatol 13:307–3021
3. Piérard GE, Piérard-Franchimont C, Marks R, Paye M, Rogiers V (2000) EEMCO guidance for the in vivo assessments of skin greasiness. Skin Pharmacol Appl Skin Physiol 13:372–389
4. Agache P (2004) Sebaceous physiology. In: Agache P, Humbert P (eds) Measuring the skin. Springer, Berlin, pp 271–280
5. Agache P (2004) Sebaceous function assessment. In: Agache P, Humbert P (eds) Measuring the skin. Springer, Berlin, pp 281–290
6. Miller DL (2006) Measurement of excreted sebum using sebum-absorbent film and optical reader: the tape analyzer. In: Serup J, Jemec GBE, Grove GL (eds) Handbook of non-invasive methods and the skin, 2nd edn. Taylor & Francis, Boca Raton, pp 831–834
7. El Gammal C, El Gammal S, Pagnoni A, Kligman AL (2006) Quantification of sebum output using

sebum-absorbent tapes (Sebutapes®). In: Serup J, Jemec GBE, Grove GL (eds) Handbook of non-invasive methods and the skin, 2nd edn. Taylor & Francis, Boca Raton, pp 835–840

8. O'goshi K (2006) Optical measurement of sebum excretion using opalescent film imprint: the Sebumeter®. In: Serup J, Jemec GBE, Grove GL (eds) Handbook of non-invasive methods and the skin, 2nd edn. Taylor & Francis, Boca Raton, pp 841–846
9. Madison KC (2003) Barrier function of the skin: "La raison d'être" of the epidermis. J Invest Dermatol 121:231–241
10. Feingold K (2009) The outer frontier: the importance of lipid metabolism in the skin. J Lipid Res 50: S417–S422
11. Schmitz G, Müller G (1991) Structure and function of lamellar bodies, lipid-protein complexes involved in storage and secretion of cellular lipids. J Lipid Res 32:1539–1570
12. Green RS, Downing DT, Pochi PE, Strauss JS (1970) Anatomical variation in the amount and composition of human skin surface lipids. J Invest Dermatol 54:240–247
13. Thiboutot DM (2004) Regulation of human sebaceous glands. J Invest Dermatol 123:1–12
14. Smith KR, Thiboutot DM (2008) Sebaceous gland lipids: friend or foe? J Lipid Res 49:271–281
15. Rizer RL (1999) Oily skin: claim support strategies. In: Elsner P, Merk HE, Maibach H (eds) Cosmetic controlled efficacy studies and regulation. Springer, Berlin, pp 81–91
16. Lopez S, Le Fur I, Morizot F, Heuvin G, Guinot C, Tschachler E (2000) Transepidermal water loss, temperature and sebum levels on women's facial skin follow characteristic patterns. Skin Res Technol 6: 31–37
17. Youn SW, Na JI, Choi SY, Huh CH, Park YC (2005) Regional and seasonal variations in facial sebum secretions: a proposal for the definition of combination skin type. Skin Res Technol 11: 189–195
18. Burton JL, Cartridge M, Shuster S (1973) Variation in sebum excretion during the neutral menstrual cycle. Acta Derm Venereol 53:81–84
19. Pochi PR, Strauss JS (1974) Endocrinologic control of the development and activity of the human sebaceous gland. J Invest Dermatol 62:191–201
20. Piérard-Franchimont C, Piérard GE, Kligman AM (1991) Rhythm of sebum excretion during the menstrual cycle. Dermatologica 182:211–213
21. Muizzuddin N, Marenus KD, Schnittger SF, Sullivan M, Maes DH (2005) Effect of systemic hormonal cyclicity on skin. J Cosmet Sci 56:311–321
22. Man MQ, Xin SJ, Song SP, Cho SY, Zhang XJ, Tu CX, Feingold KR, Elias PM (2009) Variation of skin surface pH, sebum content and stratum corneum hydration with age and gender in a large chinese population. Skin Pharmacol Physiol 22:190–199
23. Marrakchi S, Maibach HI (2007) Biophysical parameters of skin: map of human face, regional, and age-related differences. Contact Dermatitis 57:28–34
24. Wilhelm KP, Cua AB, Maibach HI (1991) Skin aging: effect on transepidermal water loss, stratum corneum hydration, skin surface pH, and casual sebum content. Arch Dermatol 127:1806–1809
25. Piérard GE, Piérard-Franchimont C, Lê T, Lapière CM (1987) Patterns of follicular sebum excretion rate during lifetime. Arch Dermatol Res 279: S104–S107
26. Tur E, Brenner S (2009) Skin and gender. In: Baran R, Maibach HI (eds) Textbook of cosmetic dermatology, 3rd edn. Informa Healthcare, New York, pp 13–16
27. Couturaud V (2009) Biophysical characteristics of the skin in relation to race, sex and site. In: Barel AO, Paye M, Maibach HI (eds) Handbook of cosmetic science and technology, 3rd edn. Informa Healthcare, New York, pp 5–24
28. Wesley NO, Maibach HI (2009) Racial (ethnic) differences in skin properties: the objective data. In: Baran R, Maibach HI (eds) Textbook of cosmetic dermatology, 3rd edn. Informa Healthcare, New York, pp 523–539
29. Saggar S, Wesley NO, Moulton-Levy N, Maibach HI (2009) Ethnic differences in skin properties: the objective data. In: Barel AO, Paye M, Maibach HI (eds) Handbook of cosmetic science and technology, 3rd edn. Informa Healthcare, New York, pp 41–57
30. Aramaki J, Kawanas S, Effendy I (2002) Differences of skin irritation between Japanese and European women. Br J Dermatol 146:1052–1056
31. Grimes P, Edison BL, Green BA, Wildnauer RH (2004) Evaluation of inherent differences between African-American and White skin surfaces using subjective and objective measures. Cutis 73:392–396
32. Fotoh C, Elkyat A, Mac S, Sainthillier JM, Humbert P (2008) Cutaneous differences between Black, African and Caribbean mixed race and Caucasians women: biometrical approach of the hydrolipidic film. Skin Res Technol 14:327–335
33. Le Fur I, Reinsberg A, Lopez S, Morizot F, Mechkouri M, Tschachler E (2001) Analysis of circadian and ultradian rhythms of skin surface properties of face and forearm of healthy women. J Invest Dermatol 117:718–724
34. Piérard-Franchimont C, Henry F, Loussouarn G, Saint-Léger D, Piérard GE (2007) Chronophysiologie circadienne du cuir chevelu. Pathol Biol 6:283–287
35. Verschoore M, Poncet M, Krebs B, Ortonne JP (1993) Circadian variations in the number of actively secreting sebaceous follicles and androgen circadian rhythms. Chronobiol Int 10:349–359
36. Cunliffe WJ, Burton JL, Shuster S (1970) Effect of local temperature variations on sebum excretion. Br J Dermatol 83:650–654
37. Cunliffe WJ, Taylor JP (2006) Gravimetric technique for measuring sebum excretion rate (SER). In: Serup J, Jemec GBE, Grove GL (eds) Handbook of non-invasive methods and the skin, 2nd edn. Taylor & Francis, Boca Raton, pp 847–852
38. Strauss JS, Pochi PE (1961) The quantitative gravimetric determination of sebum production. J Invest Dermatol 36:293–298

39. Schaefer H, Kuhn-Bassius H (1970) Methodik zur quantitativen Bestimmung der menschlichen Talg Sekretion. Arch Klin Exp Dermatol 238:429–435, German
40. SaintLéger D, Berrebi C, Duboz C, Agache P (1979) The lipometre. An easy tool for rapid quantification of skin surface lipids (SSL) in man. Arch Dermatol Res 265:79–85
41. Sisalli A, Adao A, Lebel M, Le Fur I, Sandra P (2006) Sorptive tape extraction-a novel sampling method for the in vivo study of the skin. LC GC Eur 9(nr 1): 33–39
42. Robosky LC, Wade K, Woolson D (2008) Quantitative evaluation of sebum lipid components with nuclear magnetic resonance. J Lipid Res 49:686–792
43. Camera E, Ludovici M, Galante M, Sinagra JL, Picardo M (2010) Comprehensive analysis of the major lipid classes in sebum by rapid resolution high-performance liquid chromatography and electrospray spectrometry. J Lipid Res 51:3377–3388
44. Black D, Lagarde JM, Auzoux C, Gall Y (1998) An improved method for the measurement of scalp sebum. In: Elsner P, Barel AO, Berardesca E, Gabard B, Serup J (eds) Skin bioengineering techniques and applications in dermatology and cosmetology (Current Problems in Dermatology). Karger, Basel, pp 61–68
45. Piérard-Franchimont C, Martalo O, Richard A, Rougier A, Piérard GE (1999) Sebum rheology evaluated by two methods in vivo: split-face study of the effect of a cosmetic formulation. Eur J Dermatol 9:455–457

Biophysical Assessment of Skin Desquamation and Scaliness Using Tape Strips and Adhesive Discs

25

Kristien De Paepe, Yvan Vander Heyden, and Vera Rogiers

Abbreviations

a.u.	Arbitrary units
DI	Desquamation index
EW	Elbow
FA	Forearm
LL	Lower leg
NMFs	Natural moisturizing factors
SC	Stratum corneum
SRRC	Scaling/roughness/redness/cracks

K. De Paepe (✉)
Department of Toxicology, Dermato-Cosmetology and Pharmacognosy, Department of Analytical Chemistry and Pharmaceutical Technology, Center for Pharmaceutical Research (CePhar), Faculty of Medicine and Pharmacy, Vrije Universiteit Brussel (VUB), Laarbeeklaan 103, Brussels BE-1090, Belgium
e-mail: kdepaepe@vub.ac.be

Y. Vander Heyden
Department of Analytical Chemistry and Pharmaceutical Technology, Center for Pharmaceutical Research (CePhar), Faculty of Medicine and Pharmacy, Vrije Universiteit Brussel (VUB), Laarbeeklaan 103, Brussels BE-1090, Belgium

V. Rogiers
Department of Toxicology, Dermato-Cosmetology and Pharmacognosy, Center for Pharmaceutical Research (CePhar), Faculty of Medicine and Pharmacy, Vrije Universiteit Brussel (VUB), Laarbeeklaan 103, Brussels BE-1090, Belgium

25.1 Stratum Corneum and Skin Desquamation

25.1.1 Introduction

The outer part of the human skin is the epidermis, which consists of different layers that continuously renew themselves due to cell proliferation and differentiation, finally leading to the formation of the stratum corneum (SC). Eventually, flat corneocytes desquamate from the surface as single cells or small scales. In healthy skin, the total process takes approximately 1 month [13]. The SC forms an effective barrier against transepidermal water loss. Indeed, corneocytes are tightly joined by lamellar lipid bilayers – mainly consisting of ceramides, free fatty acids, and cholesterol – which are covalently bound to cell membrane proteins [27]. Corneodesmosomes are the intercellular junctions between corneocytes responsible for the cohesion and mechanical strength of the SC. Mainly due to enzymatic activation of an array of serine, cysteine, and aspartic proteases, desmosomal degradation takes place and induces the physiological desquamation process [22, 30]. An optimal water and lipid environment and pH gradient are key factors for the enzymatic activity of the serine proteases, which are initially produced as inactive proenzymes [12, 13, 27, 33]. Otherwise, an insufficient breakdown will lead to conglomerates of corneocytes, visible as large scales and squames [34]. Together with various hyperkeratotic skin pathologies, the so-called senile xerosis and winter dry skin are

E. Berardesca et al. (eds.), *Non Invasive Diagnostic Techniques in Clinical Dermatology*, DOI 10.1007/978-3-642-32109-2_25, © Springer Berlin Heidelberg 2014

characterized by an uneven, flaky horny layer with difficultly desquamating scales [14, 20]. Therefore, parameters having an effect on desquamation and the occurrence of dry scaly skin may originate as well from genetic predisposition, as from chronobiological skin aging, anatomic location, and environmental factors [4, 6, 22, 25].

Besides the desquamation process, the SC structure is also responsible for the barrier function of the skin against penetrating xenobiotics. In this context, diffusion rate studies and the determination of the reservoir function of the skin are important [20]. Different methods have been described, including skin surface biopsies by cyanoacrylate stripping, skin scraping, and – most commonly used – tape stripping [10, 11, 18, 26].

25.1.2 Methodologies to Evaluate Skin Desquamation and Scaliness

Skin scaliness can be assessed either clinically or biophysically, using the so-called skin bioengineering techniques. Direct skin measurements are preferred because clinical observations might lack objectivity and discrimination power for borderline cases [4, 29]. Biophysical techniques are applicable for the measurement of a whole range of skin parameters and can be used as well by dermatologists as researchers, particularly when properties including high precision and accuracy, easy handling, time saving, and economical prices are present.

Different approaches and methodologies to measure skin scaliness and desquamation in a quantitative way have been reviewed by Lambers and Pronk [16]. They consist of (1) assessment of the desquamation rate and SC turnover, (2) measurement of the intracorneal cohesion, and (3) quantification of scaling [1]. When collection of corneocytes is carried out by adhesive tape material, the methodology is described as a so-called squamometric technique [23]. Some in vivo methods without tape stripping also exist. They are mainly based on epiluminescence microscopic photography or light reflection [16, 24]. These methods are beyond the scope of this chapter and will not be further discussed.

25.1.3 Squamometric Measurements

A classical way to perform squamometric measurements is to sample corneocytes using adhesive-coated transparent discs, followed by quantitative color measurements after staining of the squames [7, 24]. Microscopic evaluation of the colored samples in combination with a four-point-scale scoring system minimizes overestimation of the chromametric measurements. When making use of such a scoring system, very dry skin produced a high amount of scaling, while normal skin was characterized by only a few small areas of cells or even a fine layer of single cells [7]. Significant overestimation of colorimetric values may occur due to coloring artifacts. Since scales are heterogeneously spread and stained on the adhesive disc, any slight deviation in positioning of the measuring probe may cause important changes in the colorimetric values measured. This means that removing of the excess stain is a critical point since too intense rinsing could remove adhering scales [16].

Consequently, measuring techniques without coloring would be preferable. Therefore, another option is to produce digitized, video-captured images of the collected strips and to calculate the desquamation index (DI) with an image analysis program [4, 29]. The DI is then calculated using the following equation: $6\text{DI}=2A+\sum_{n=1}^{5}T_n \times (n-1)$ with A = the percentage of area covered by scales, T_n = the percentage of scales in relation to thickness, and n = the scale thickness (marked from 1 to 5) [1, 29].

25.1.4 Stratum Corneum Tape Stripping

Tape stripping, as a technique for sampling and evaluating SC layers, is well known as being simple and useful. It is applied in a broad range of dermatological and cosmetic studies, including analysis of SC lipids, enzymes, cytokines, growth factors, and natural moisturizing factors (NMFs) (reviewed by [27, 32]). Also after penetration of topically applied products, analysis of successively removed SC layers appears to be a suitable method to assess the quality and efficacy of product formulations and to perform quantitative measurements of pharmaceuticals and their metabolites [11, 15].

For analytical purposes, results are best expressed as a ratio to the amount of sampled SC, either for each single tape strip, or as a function of SC depth [18]. Quantification of SC is mostly performed gravimetrically [21] or by protein determination [10]. Both methods, however, have their limitations. The weighing procedure is time consuming and may be prone to interference by some moisture and traces of sebum or SC lipids [10, 32]. On the other hand, the commonly used colorimetric method for protein determination, as established by Dreher and coworkers, has as drawback that the colored samples cannot be further used for other bioassays [32]. Different methods to quantify the amount of SC removed from the skin have been comprehensively reviewed by Lademann et al. [15].

25.2 Fields of Application of Tape Strips and Adhesive Discs

Besides their applicability in research for sampling of SC layers and investigating the bioavailability and bioequivalence of topical drugs and other constituents, tape strips and discs are an important tool for the visualization of dry, flaky skin, guiding consumers in their usage of specific skin moisturizing products. Making use of adhesive strips before and after product application, the efficacy of both moisturizers and exfoliants can be evaluated [2, 4]. It is also a suitable method for the classification of cleansing products for their potential irritative properties and skin compatibility [16]. Sampling tapes may also be used in a variety of tape-stripping protocols as well for skin barrier integrity measurements as for human nail studies [3, 31].

25.2.1 Key Considerations When Using Tape Strips and Discs

Various brands of tape material, applicable for skin tape stripping, exist. They differ in surface area and shape, as well as in composition and properties of the adhesive material [5, 17]. Skin application of commercialized tapes usually is a noninvasive technique (i.e., sequential in-depth stripping), meaning that it is superficial, painless, and without any health risk. Regardless the adhesive tape material used, strips – ensuring a reproducible rigidity and adhesion – can uniformly sample the upper SC layers. The advantage of clear, transparent strips is the visibility of the adhering corneocytes and scales. Furthermore, they allow staining in case of histological or colorimetric analysis [18]. An important remark is that adhesive strips should always be handled by the tabbed edge – preferably by a pair of tweezers – to avoid interference with fingerprints. They should be used on clean, dry skin surfaces, applying a constant pressure during an appropriate time period, which needs to be specified in the protocol. Removal of the strips should be done with a single continuous movement [5, 17]. As being important for all biophysical methods, a number of working conditions need to be taken into account [16]: (1) an air-conditioned room with constant temperature (20–22 °C) and humidity (40–60 %); (2) an acclimatization period for the volunteers, preferably in a stressless environment and the test zones should be kept uncovered; and (3) dust-free frames and closed boxes to protect from direct light in case of storage of tape strips [8]. Corneofix® strips and D-SQUAME® discs are two examples of adhesive tape materials that will be further discussed.

25.2.2 Product Information

Corneofix® F20 Desquamation Collector Foils from Courage + Khazaka electronic GmbH

(Köln, DE) are adhesive transparent strips with dimensions of 2.00×1.95 cm and a thickness of 0.1 mm. As collector foils, the strips can be used in combination with the Visiometer® SV600 or the Visioscan® VC98 camera, in order to quantify the number, size, and area covered with flakes, and the DI [www.courage-khazaka.de]. The Visioscan® VC98 (UVA light source) can also be applied in vivo, producing images from which the degree of scaling can be scored [14].

D-SQUAME® Skin Surface Sampling Discs are commercialized by CuDerm Corporation (Dallas, TX, USA). The diameter of these clear polymer discs is 2.2 cm (area of 3.8 cm^2). The sampling discs can be used in combination with the D-SQUAME® Skin Indicator, being an adhesive sampling device with a dark background which allows quick estimation of dry skin flakes [www.cuderm.com].

When a nondestructive method is preferred to determine the protein content of the collected SC samples, quantification can be performed by infrared (850 nm) densitometry. The SquameScan™ 850A densitometer is commercialized by Heiland electronic (Wetzlar, DE) and can be used with both D-SQUAME® discs and Corneofix® strips. Compared to a colorimetric technique, it not only is time saving but also allows subsequent analysis of the strips, e.g., for enzyme activities. As described in the product information, infrared light prevents thermal denaturation of adhering molecules and effects of ambient light. As the measured diameter is 1.5 cm, more than half of the tape strip area is being covered [www.heilandelectroic.de]. This compact infrared densitometer has been validated by comparing the SC optical absorption with the extracted SC proteins. The optical absorption of SC tape strippings showed to be linearly proportional to their protein content, making this nondestructive technique a convenient measuring tool. In addition, the method revealed to be sensitive enough to discriminate between SC samples collected from hydrated and less hydrated skin, making it also applicable for evaluating moisturizing products [32].

25.3 Practical Applicability of the Corneofix® F20 Technology

As stated in various review papers, there is no doubt that SC desquamation and hydration are considered to be relevant parameters in the evaluation of overall skin condition [6, 12, 27]. Increased water content in the upper layers of the skin undoubtedly results in a softer skin with a more pleasing appearance [4]. Manuskiatti and coworkers could find a significant correlation between SC hydration and the quantity of scaliness, estimated from the DI when D-SQUAME® discs were being used [19]. However, making use of biophysical measurements, the expected negative correlation between both the SC hydration and scaliness should be interpreted with much care [23].

Therefore, in order to understand the efficacy of skin moisturizers, a comparative quantitative evaluation of the skin DI and capacitance values, measured at different skin sites of young and aging skin, was performed [9]. Adhesive Corneofix® F20 strips – pressed on the skin (300 g/cm^2) for 10 s – were used for the collection of SC corneocytes and squames and quantitatively analyzed by a light transmission technique. Measuring devices were from Courage+Khazaka electronic GmbH (Köln, DE). The methodology used was validated earlier and checked for its applicability [8]. The specific settings of the apparatus were optimized in relation to (1) shutter time; (2) calculation area with a default of 2 (5.66×5.61 mm) – meaning the largest calculation area but still avoiding artifacts induced by borders and shadowing effects – and (3) basic gray level taking into account blank values. When applied on the ventral forearms of 15 healthy females (Fitzpatrick phototypes II–III, age range 20–35 years old), squamometric measurements showed symmetry between corresponding test sites on both forearms. Consequently, in product application studies, one skin test zone can be used for treatment, while the corresponding side may serve as control area. Sequential in-depth stripping showed significantly decreased DI values, accompanied by increased capacitance levels.

However, no statistical difference could be observed between the DI values of the first and second Corneofix® sample, making comparisons possible before and after product application on the same test zone [8].

Examples of Corneofix® strips collected from young and aging skin, respectively, are given in Fig. 25.1. Before strip application, visual grading of the test zone was performed making use of the scaling rating of the "specified symptom sum score" SRRC (scaling/roughness/redness/cracks) system [28]. The median values of the visual scoring are given in Table 25.1.

The results of the biophysical measurements are shown in Fig. 25.2. When comparing the different measuring sites in both populations, the following ranking for DI values could be found: EW > LLb > LLa > FA. Although a more scaly skin was visually assessed by the SRRC scoring system after the exposure to the acetone/ether solution (see Table 25.1), significant differences for scaliness between LLa and LLb within one age group could not be statistically confirmed using biophysical measurements. Also hydration measurements revealed no significant differences between both conditions used for the lower leg.

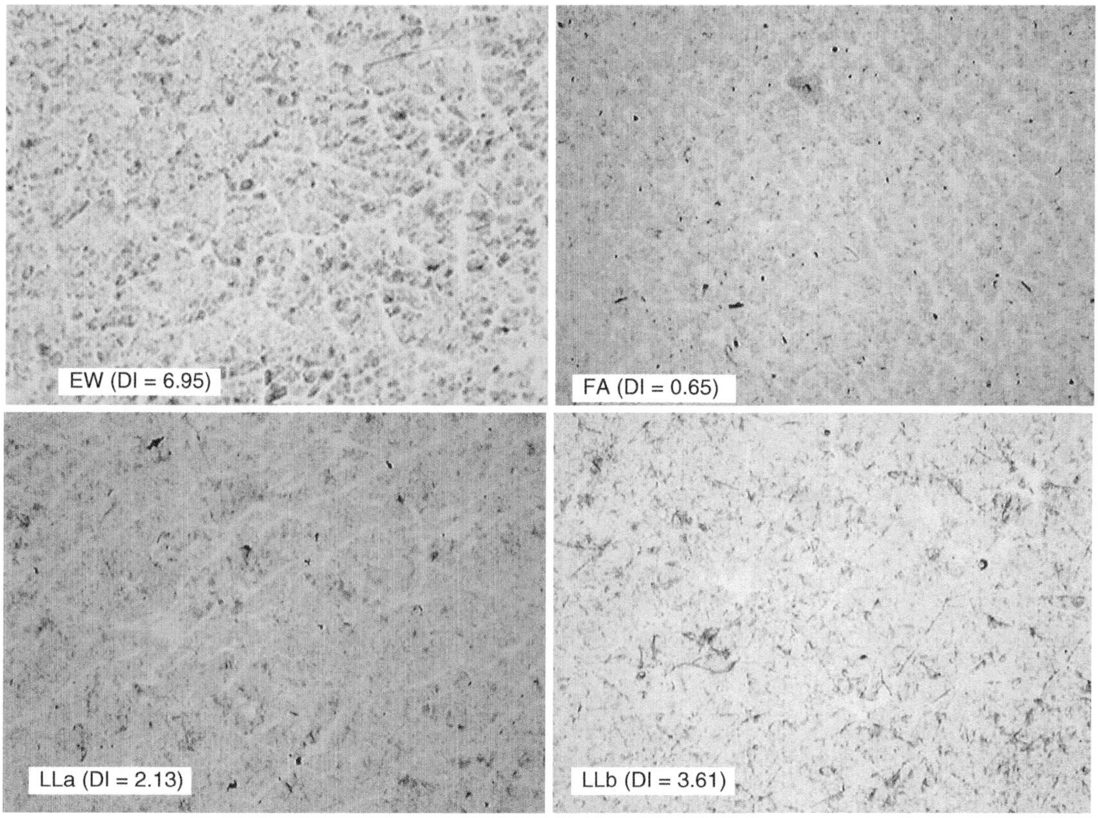

Fig. 25.1 Examples of Corneofix® strips (image taken by Visiometer® SV600) collected from young (mean age 25±4 years old) and aging skin (mean age 52±2 years old), indicating the measured DI values. Measuring sites: elbow (EW), inner forearm (FA), and lower leg (LLa). The test site on the lower leg was also measured 10 min after 1 min exposure to acetone/ether 1:1 v/v (LLb). All samples were collected during a study period of 1 month in autumn to avoid seasonal variations

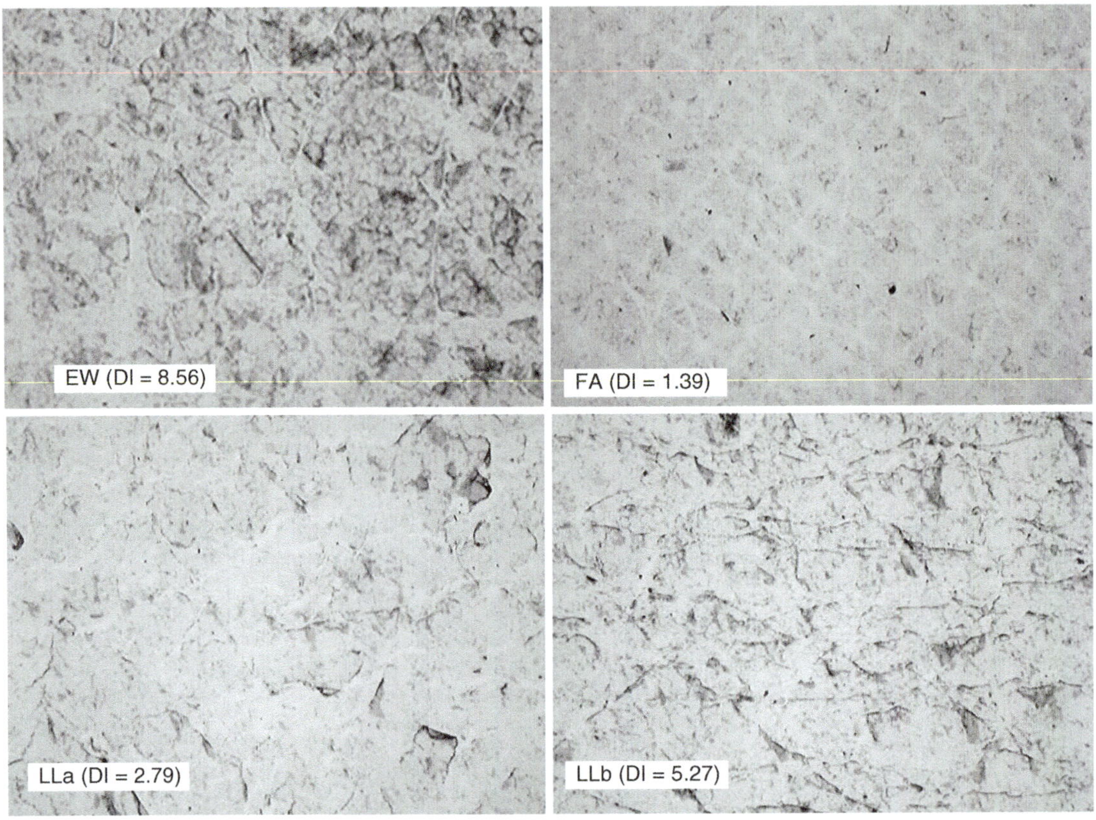

Fig. 25.1 (continued aging skin)

Table 25.1 Visual scaling scores (median values) as clinically assessed (SRRC), $n=15$ for both populations (0 = absent, 1 = slight, 2 = moderate, 3 = severe, 4 = extreme)

	Measuring sites			
Skin type	EW	FA	LLa	LLb
Young skin	3	0	1	2
Aged skin	3	1	2	3
Significance (p value)	NS	NS	0.009	0.005

Statistical significance $p<0.05$ (nonparametric Mann–Whitney test and Wilcoxon signed-rank test)
NS not significant, *SRRC* scaling/roughness/redness/cracks

When comparing the capacitance measurements between both age groups, significant differences were seen for LLa ($p=0.002$, unpaired parametric t-tests) with higher skin hydration values for the young participants. All other comparisons were not statistically different. These observations were confirmed by published literature data, emphasizing that differences in skin hydration between age groups and the occurrence of xerosis are mainly observable at the lower legs [28].

A Pearson correlation coefficient (r) was used to describe the correlation between DI and skin hydration. For some – but not all test sites – a significant negative correlation could be found for hydration and scaliness measurements. However, these results should be interpreted with care, and they only indicate that skin hydration values increase when the DI decreases. This has been explained by the effect of flaky corneocytes on the skin surface which disturb capacitance measurements of the lower SC layers [25]. In this respect, as shown in Fig. 25.3, a somewhat stronger linear regression was found for the younger group ($R^2=0.995$) compared to the aged group ($R^2=0.858$) [9].

Based on this limited set of results, the Corneofix® technique revealed to be a useful method to define skin desquamation. It can be an extra tool for claim support, providing objective

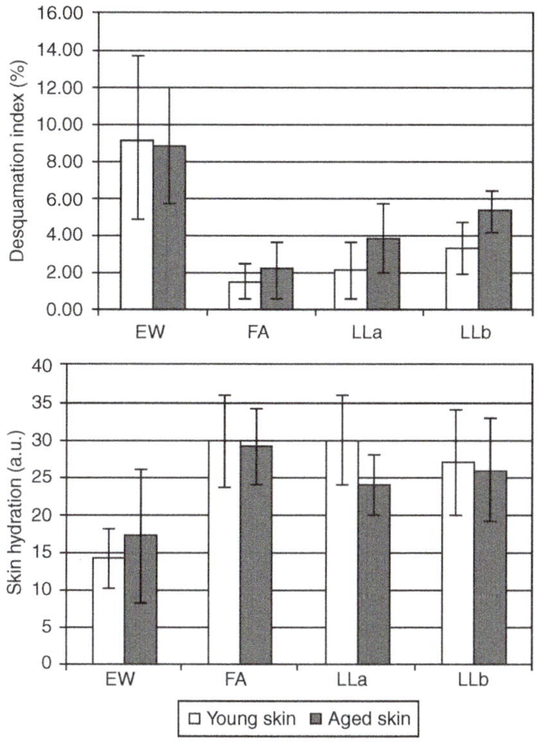

Fig. 25.2 Mean (± St. Dev.) DI values expressed in % (Corneofix®F20/Visiometer®SV600) and hydration measurements expressed in arbitrary units (a.u.) (Corneometer® CM825) of two different age groups (21–35 and 49–55 years old) with 30 healthy female volunteers (Fitzpatrick phototypes II–III). Biophysical measurements were carried out in a climate-controlled room with a constant temperature of 20.2±0.4 °C and relative humidity of 45.9±1.5 %

evidence for the efficacy of moisturizing products applied under dry to very xerotic skin conditions.

Acknowledgments The authors would like to thank Goedele Alaerts and Hayate Hajioui for their valuable practical assistance during skin measurements. Part of the described research work was supported by a grant of IWT-Vlaanderen – Flemish Institute for the Promotion of Scientific-Technological Research in Industry.

References

1. Agache P (2004) Metrology of the stratum corneum. In: Agache P, Humbert P (eds) Measuring the skin. Springer, Berlin/Heidelberg, pp 101–111
2. Bashir SJ, Dreher F, Chew AL, Zhai H, Levin C, Stern R et al (2005) Cutaneous bioassay of salicylic acid as a keratolytic. Int J Pharm 292:187–194
3. Berthaud F, Boncheva M (2010) Correlation between the properties of the lipid matrix and the degrees of integrity and cohesion in healthy human stratum corneum. Exp Dermatol 20:255–262
4. Black D, Boyer J, Lagarde JM (2006) Image analysis of skin scaling using D-Squame® samplers: comparison with clinical scoring and use for assessing moisturizer efficacy. Int J Cosmet Sci 28:35–44
5. Breternitz M, Flach M, Prässler J, Elsner P, Fluhr JW (2007) Acute barrier disruption by adhesive tapes is influenced by pressure, time and anatomical location: integrity and cohesion assessed by sequential tape stripping; a randomized, controlled study. Br J Dermatol 156:231–240

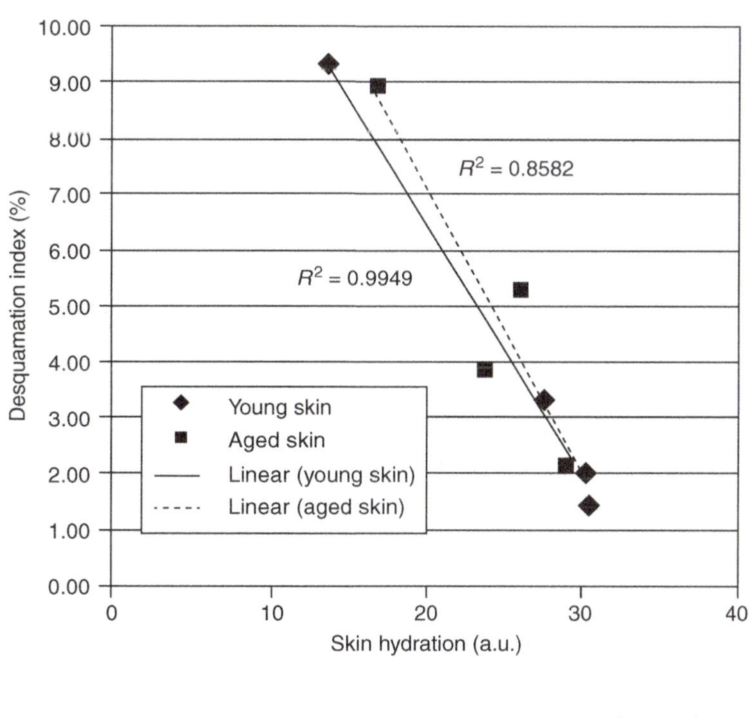

Fig. 25.3 Regression analysis for DI (%) and skin hydration (a.u.)

6. Byrne AJ (2010) Bioengineering and subjective approaches to the clinical evaluation of dry skin. Int J Cosmet Sci 32:410–421
7. De Paepe K, Janssens K, Hachem JP, Roseeuw D, Rogiers V (2001) Squamometry as a screening method for the evaluation of hydration products. Skin Res Technol 7:184–192
8. De Paepe K, Rogiers V (2005) Corneofix F20® a new technology to define skin desquamation. In: Conference proceedings ISBS world congress, Philadelphia, 28 Sept–1 Oct 2005
9. De Paepe K, Rogiers V (2007) Comparative quantification of stratum corneum scaling and hydration state in young and aging skin. In: Proceedings of the IFSCC conference, Amsterdam, 24–26 Sept 2007
10. Dreher F, Modjtahedi BS, Modjtahedi SP, Maibach HI (2005) Quantification of stratum corneum removal by adhesive tape stripping by total protein assay in 96-well microplates. Skin Res Technol 11:97–101
11. Escobar-Chávez JJ, Merino-Sanjuán V, López-Cervantes M, Urban-Morlan Z, Piñón-Segundo E, Quintanar-Guerrero D et al (2008) The tape-stripping technique as a method for drug quantification in skin. J Pharm Pharm Sci 11:104–130
12. Harding CR, Watkinson A, Rawlings AV, Scott IR (2000) Dry skin, moisturization and corneodesmolysis. Int J Cosmet Sci 22:21–52
13. Houben E, De Paepe K, Rogiers V (2007) A keratinocyte's course of life. Skin Pharmacol Physiol 20:122–132
14. Kim JH, Kim BY, Choi JW, Kim SO, Lee HS, Park KC et al (2012) The objective evaluation of the severity of psoriatic scales with desquamation collecting tapes and image analysis. Skin Res Technol 18:143–150
15. Lademann J, Jacobi U, Surber C, Weigmann HJ, Fluhr J (2009) The tape stripping procedure evaluation of some critical parameters. Eur J Pharm Biopharm 72: 317–323
16. Lambers H, Pronk H (2002) Biophysical methods for stratum corneum characterization. In: Förster T (ed) Cosmetic lipids and the skin barrier. Marcel Dekker, New York/Basel, pp 185–225
17. Löffler H, Dreher F, Maibach HI (2004) Stratum corneum adhesive tape stripping: influence of anatomical site, application pressure, duration and removal. Br J Dermatol 151:746–752
18. Maibach HI (2005) Stratum corneum adhesive tape stripping. In: Marks R, Matts P, Leévêque JL (eds) Stratum corneum: the vital structure. Stratum Corneum Group, Cardiff, pp 75–77
19. Manuskiatti W, Schwindt DA, Maibach HI (1998) Influence of age, anatomic site and race on skin roughness and scaliness. Dermatology 196:401–407
20. Marks R (2004) The stratum corneum barrier: the final frontier. J Nutr 134:2017S–2021S
21. Marttin E, Neelissen-Subnel MT, De Haan FH, Boddé H (1996) A critical comparison of methods to quantify stratum removed by tape stripping. Skin Pharmacol 9:69–77
22. Milstone LM (2004) Epidermal desquamation. J Dermatol Sci 36:131–140
23. Piérard GE (1996) EEMCO guidance for the assessment of dry skin (xerosis) and ichthyosis: evaluation by stratum corneum strippings. Skin Res Technol 2:3–11
24. Piérard GE, Piérard-Franchimont C, Saint-Léger D, Kligman AM (1992) Squamometry: the assessment of xerosis by colorimetry of D-Squame adhesive discs. J Soc Cosmet Chem 47:297–305
25. Piérard-Franchimont C, Piérard GE (2002) Beyond a glimpse at seasonal dry skin: a review. Exog Dermatol 1:3–6
26. Piérard-Franchimont C, Piérard GE (2004) Cyanoacrylate skin surface stripping for visualizing stratum corneum structures and dynamics. In: Agache P, Humbert P (eds) Measuring the skin. Springer, Berlin/Heidelberg, pp 74–79
27. Rawlings AV (2003) Trends in stratum corneum research and the management of dry skin conditions. Int J Cosmet Sci 25:63–95
28. Serup J (1995) EEMCO guidance for the assessment of dry skin (xerosis) and ichthyosis: clinical scoring systems. Skin Res Technol 1:109–114
29. Schatz H, Kligman AM, Manning S, Stoudemayer T (1993) Quantification of dry (xerotic) skin by image analysis of scales removed by adhesive discs (D-squames). J Soc Cosmet Chem 44:53–63
30. Simon M, Bernard D, Caubet C, Guerrin M, Egelrud T, Schmidt R et al (2002) Corneodesmosomal proteins are proteolysed in vitro by both SCTE and SCCE – two proteases which are thought to be involved in desquamation. In: Marks R, Lévêque JL, Voegeli R (eds) The essential stratum corneum. Martin Dunitz, London, pp 57–61
31. Tudela E, Lamberbourg A, Cordoba Diaz M, Zhai H, Maibach HI (2008) Tape stripping on a human nail: quantification of removal. Skin Res Technol 14:472–477
32. Voegeli R, Heiland J, Doppler S, Rawlings AV, Schreier T (2007) Efficient and simple quantification of stratum corneum proteins on tape strippings by infrared densitometry. Skin Res Technol 13:242–251
33. Watkinson A, Harding C, Moore A, Coan P (2001) Water modulation of stratum corneum chymotryptic enzyme activity and desquamation. Arch Dermatol Res 293:470–476
34. Watkinson A, Smith C, Coan P, Wiedow O (2000) The role of pro-SCCE and SCCE in desquamation. IFSCC Magazine 3:45–49

Stripping Techniques: Tape Stripping

26

Razvigor Darlenski, Joachim W. Fluhr, and Jürgen Lademann

Abbreviations

SC	Stratum corneum
TEWL	Transepidermal water loss
TS	Tape stripping

Core Messages

- Tape stripping (TS) is a simple and efficient method for the assessment of SC physiology and the quality and efficacy of topically applied formulation on the skin.
- The major uses of TS include the study of the penetration and the storage of topically applied exogenous substances the physiology of the SC (integrity and cohesion), epidermal wound healing, and the excretion of endogenous substances.
- The sequential removal of parts of the SC is used as a model of standardized acute barrier disruption and barrier repair kinetics.
- Different variables influence the TS procedure and should be considered when performing the technique.

R. Darlenski
Department of Dermatology,
Tokuda Hospital, Sofia, Bulgaria

J.W. Fluhr, MD (✉) • J. Lademann
Department of Dermatology, Venerology and Allergology, Center of Experimental and Applied Cutaneous Physiology,
Charité – Universitätsmedizin Berlin,
Charitéplatz 1, Berlin 10117, Germany
e-mail: joachim.fluhr@charite.de

26.1 Introduction

According to the definition found in the Merriam-Webster Dictionary, the verb "to strip" means "to remove covering, or surface matter from" [1]. In dermatological research the removal of the skin's outermost layers has been used for decades [2]. In 1939, Wolf first described the simple technique of (tape stripping) TS to get more information about the morphological details of stratum corneum (SC) removed by strips [3]. The subsequent removal of the SC using adhesive tapes has emerged as a useful technique in many studies. This minimally invasive procedure can be used in both in vivo and in vitro studies, in humans and animals, e.g., mice, rats, guinea pigs, and pigs [4, 5]. It is useful in the study of:

- The penetration and the storage of topically applied exogenous substances.
- The physiology of the SC.
- The sequential removal of parts of the SC as a model of standardized acute barrier disruption and barrier repair kinetics.
- Epidermal wound healing.
- The excretion of endogenous substances.

The SC is the outermost layer of the epidermis with a thickness of approximately 10–20 μm and consists of akaryotic corneocytes and intermediate bilamellar lipid layers. Approximately 90 % of the barrier function resides in the SC. The SC became "alive" over the last decades, and its understanding evolved from the simple two-compartment system ("brick-and-mortar" model) to a system with a regulated metabolic activity,

E. Berardesca et al. (eds.), *Non Invasive Diagnostic Techniques in Clinical Dermatology*,
DOI 10.1007/978-3-642-32109-2_26, © Springer Berlin Heidelberg 2014

its link to deeper part of the skin and ultimately as a biosensor for external factors to regulate, e.g., proteolytic activity, DNA synthesis, and lipid synthesis [6]. Previously considered as immunologically inert, the residential cells of the epidermis (keratinocytes and corneocytes) can secrete preformed cytokines (i.e., interleukin-1 alpha) upon skin barrier disruption [7]. SC and its main components, i.e., the corneocytes, the intercorneal bilamellar lipids and the cornified envelope (CE), are considered as the rate-controlling structures for the transcutaneous xenobiotic delivery [8, 9]. The dynamic evolution of the concept for SC structure and functions has been at least partly studied by the TS method.

26.2 Tape Stripping Technique

26.2.1 The Adhesive Tape

Different types of adhesive tapes can be used to perform effective TS procedure. There are several commercially available such as D-SQUAME® (CuDerm Corp., Dallas, TX, U.S.A.) and Corneofix® (Courage + Khazaka GmbH, Cologne, Germany). Furthermore, adhesive tapes used in household conditions such as Tesafilm®, Scotch tape®, and others can be used especially when the main focus is barrier disruption. In addition, self-prepared tapes with variable or fixed length are also applicable. The skin tolerability to the selected tape should be proven in advance, e.g., no allergy to the adhesive could compromise the study. A uniform composition and distribution of the adhesive layer on the tape strip should be guaranteed. The properties, e.g., transparence and flexibility, of the adhesive tape are determined by the study investigator in accordance with the defined requirements of the study.

26.2.2 Test Site

The test site should be hair trimmed in advance in order to avoid the uneven adhesion of the tape to the skin surface. No shaving is advised as this could interfere with the SC by removing part of it. Scar tumor lesions on skin folds should be avoided. If investigating skin physiology, no application of skin care products and soaps to the test site is advisory for a week or at least 24 h before the study enrollment, the so-called washout phase [10]. Marking the test site by skin marker is necessary to assure the application of the adhesive tape at the exact same site. Sometimes the mark on the skin must be renewed after removal of several strips.

The effect of the anatomical site was investigated by measuring transepidermal water loss (TEWL) after TS and the assessment of the protein amount with the strips [11]: in regions with a thinner SC, such as the cheek, the water permeability barrier of the skin was vulnerable to sequential tape stripping, i.e., the inverse of TEWL values (1/TEWL) displayed linear correlation to the removed thickness of SC. TS at the cheek and back leads to a faster increase of TEWL values than on the forearm and upper arm. After 22 strips, more protein was removed on the cheek compared to the forearm and upper arm but less compared to the back.

26.2.3 Environmental Conditions

The effect of different environment-related variables such as room temperature and air relative humidity has not been challenged. However we believe that the general requirements for setting up a skin physiology study should be followed [12]. From a practical point of view, sweating resulting from high room temperature could hinder the dense contact of the adhesive tape to the skin. Therefore the procedure should be performed in a climatized room with temperature 18–21 °C and relative humidity of 40–60 %. Acclimatization time of the study volunteers – 20–30 min prior to the first strip – is advisory.

26.2.4 Application of the Tape

The pressure used for adhering the tape on the skin surface is crucial for the procedure [10]. Different methods exist: (1) a constant weight, (2) a spatula, (3) thumb covered with a glove by pressing the tape gently in a circular movement and then again in a gentle sideways movement across the tape, and (4) by means of stamps which contain a spring providing constant pressure. In addition a roller which reduces the influence of skin furrows could be used together with the standardized pressure method. On an adjacent site the skin should be stretched with three fingers before and during the application of the tape which was pressed onto the skin. Another protocol suggests the application of a roller for pressing the tape onto the skin. During the roller movement the skin is stretched so that furrows and wrinkles disappear. The film to be applied comes in contact with the flat surface, thus enabling a homogeneous removal of the corneocyte layers. Using different methods makes the direct comparison between laboratories difficult. The pressure methods should be described in detail in the study protocol and the material and method section.

26.2.5 Removal of the Tape

After pressing the tapes onto the skin, they are removed in one swift movement. Forceps should be used. A constant velocity should always be applied. A slowing down or stopping of the procedure could lead to an increase in the SC amount adhered on the tape strip, whereas an increase in speed could result in a decreased amount of corneocytes [11].

After removal of the tapes, they are ready to be further investigated by different means: weighing, spectroscopy, and microscopy. The first tape strips contain almost a complete cell layer of corneocytes and remnants of exogenous compounds, e.g., cosmetic products. With increasing the tape stripping number, the corneocytes attached to the tape become less. In pharmacokinetic studies, the first tape strip is often discarded because it represents unabsorbed drug on the skin surface [13]. However, this amount is required to calculate the concentration of substance recovered within the SC when applying a definite dosage. In Fig. 26.1 the tape stripping procedure is presented, with a roll being used to press the tape unto the skin surface. Figure 26.2 shows a microscopic image of the removed tape

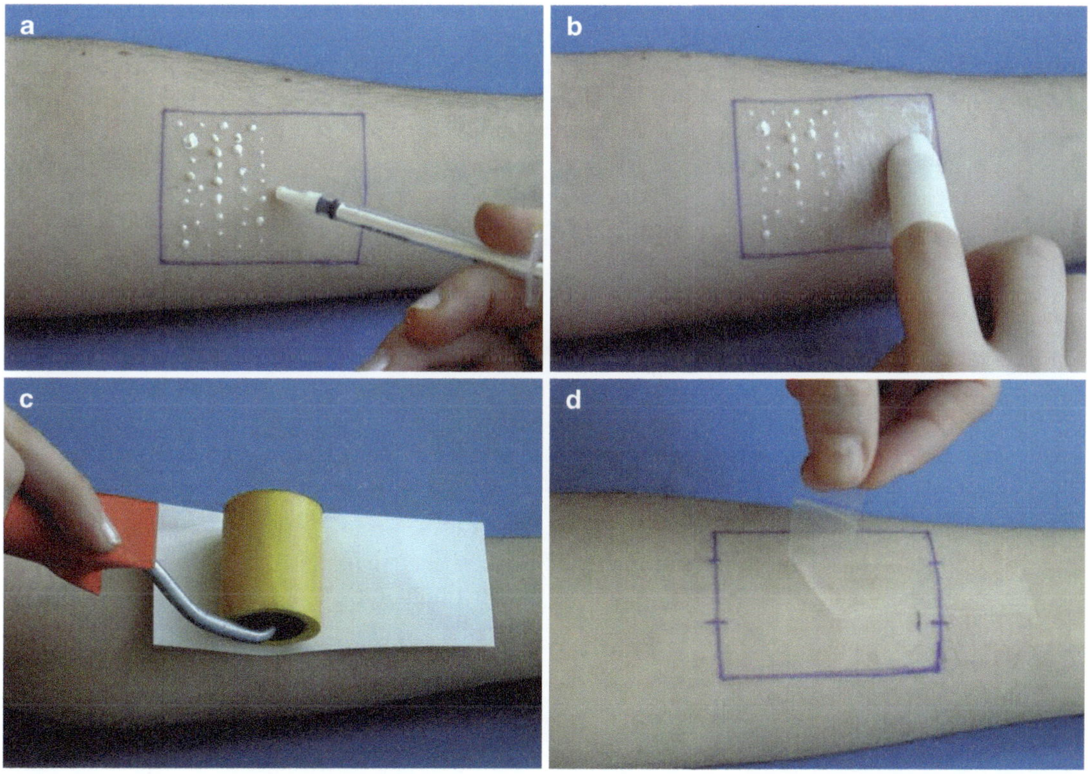

Fig. 26.1 The tape stripping procedure comprises the following steps: homogeneous application of the formulation (**a**), homogeneous distribution of the formulation (**b**), pressing the tape strip unto the skin after an application time (**c**), and removing it with one quick move (**d**)

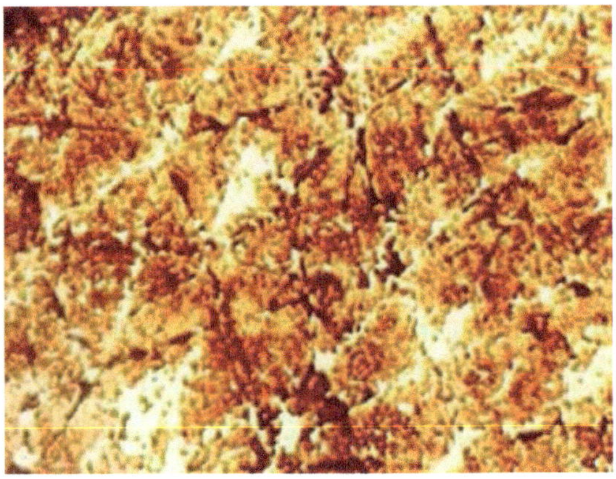

Fig. 26.2 Microscopic image of a tape strip showing that the tape is covered with corneocytes

strip, which contained amounts of corneocytes and parts of the topically applied substances, which had penetrated into this specific cell layer.

26.3 Variables Influencing the Procedure

A variety of factors can potentially influence the TS procedure. The effects of anatomical site, the climate conditions, the material of the tape, and the applied pressure have been discussed above. Other factors normally influencing SC properties such as age, gender, circadian rhythms, and season have not been directly studied as variables affecting TS.

26.3.1 Skin Microrelief

In general, the skin has an uneven surface with the prominent *area cutanea* and the concave *sulci interpapillares* (furrows). This situation is presented in Fig. 26.3 showing the skin surface structure imaged by noninvasive optical coherence tomography. This can cause the removal of SC parts from different cell layers with one tape strip. This disadvantage can be compensated by using a pressing roller and an adhesive tape with high flexibility. In addition, a swelling of the tissue during the procedure of complete SC removal was observed resulting in a smoothing of the skin during in vivo experiments [14].

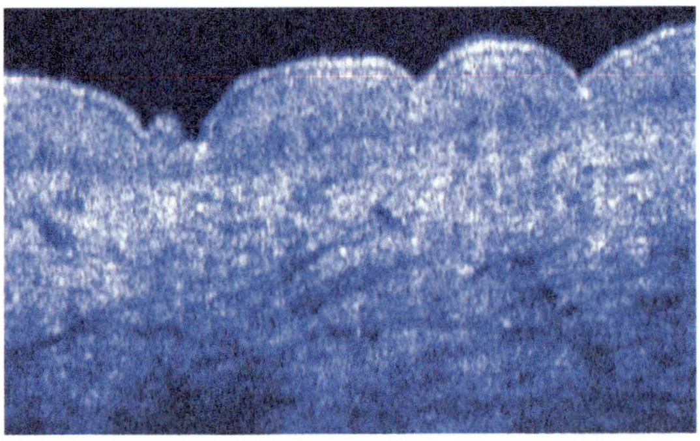

Fig. 26.3 Skin surface structure analyzed by noninvasive optical coherence tomography

26.3.2 Effect of Topically Applied Formulation

Different amounts of SC can be removed with the first tape strips in studies comparing different topically applied vehicles [15]. The properties of the topically applied formulations have a strong influence on the amount of the removed SC. For instance, the application of an alcohol solution enhances the adhesion of the SC to the strips, while application of an oily formulation leads to a reduction of adhesion [16]. More strips are required to remove a comparable amount of the SC. Oily formulations are distributed homogeneously on the corneocytes, while in the case of an alcohol-based formulation, the solution is mainly located in the furrows and wrinkles.

26.3.3 Occlusion

Evidence exists that occlusion influences the TS procedure. Both the amount of protein removed and the TEWL values after TS at occluded test site (24 h – Finn Chamber® with a filter paper disk soaked with distilled water) were significantly higher than at the non-occluded corresponding skin region [10]. Occlusion results in retention of water and causes degradation of intercellular proteins, mainly the corneodesmosomes [17]. After occlusion, it is easier to remove the tapes, and an increased amount of protein is removed per tape.

26.4 Examples of Practical Tape Stripping Applications

26.4.1 SC Integrity and Cohesion

The integrity and cohesion of the SC can be investigated by adhesive tape methods. The number of tape strips required to induce a predefined degree of barrier disruption (e.g., to raise TEWL values to a certain level) is used to characterize the SC integrity [10]. On the other hand, the SC cohesion is evaluated by the amount of SC removed by sequential tape stripping. An increase of the removed mass indicates a decrease of the cohesion of SC, and in the opposite, lower amount of SC removed with increasing tape number is due to the stronger cohesion between the cells in deeper SC layers. Generally the tape stripping removes cell layers maximally down to the glistening layer corresponding to approximately 90 % of the SC.

26.4.2 Distribution of Topically Applied Substances

Both superficial distribution of topically applied substances on the skin surface and depth concentration profiles can be assessed by TS. A good correlation was found between the amount present in the SC 30 min after application of the drug (as estimated by tape stripping technique) and the urinary excretion over the next 4 days [16]. It was concluded that the method is useful in estimation of systemic exposure risk, although no precise data on the absorption rate of the substance can be established. A correlation was set between the levels of the drug estimated in the tape strips and the pharmacodynamic effect of topically applied drugs on the skin [18, 19]. Skin blanching effect is used to compare the potency of topically applied glucocorticosteroids. For topical betamethasone dipropionate 0.05 % the increasing amount of drug in the tape-stripped SC correlated with an increased skin blanching score [19]. Despite the promising results tape stripping has certain limitations in its use in bioavailability assays. Further efforts regarding the standardization are needed to validate the tape harvesting as a method for the assessment of topical drug pharmacokinetics [20].

26.4.3 Ex Vivo Determination of the Protection Efficacy of Sunscreens

The effect that the concentration and distribution of sunscreen formulation in the SC can be transferred to tape strips was used for the determination of a universal sun protection factor of sunscreens [21]. Cell layers of the SC are removed after application and penetration of sunscreens. The tape strips are measured in a special UV–Vis spectrometer. A correlation of the homogeneity of distribution with the in vivo sun protection factor was confirmed [21].

Conclusion

TS is a generally safe and relatively noninvasive approach with multiple applications in dermatological research. Unification and standardization of the methodology need to be implemented. Results should be interpreted carefully, and the effect of different instrument-, subject-, and environment-related variables must be considered.

References

1. Merriam Webster's dictionary (2012) Frederick C. Mish (Ed), http://www.merriam-webster.com/
2. Pinkus H (1951) Examination of the epidermis by the strip method of removing horny layers. I. Observations on thickness of the horny layer, and on mitotic activity after stripping. J Invest Dermatol 16(6):383–386
3. Wolf J (1939) Stratum desquamans der menschlichen epidermis. Z Mikrosk Anat Forsch 46:170–202
4. Loffler H, Dreher F, Maibach HI (2004) Stratum corneum adhesive tape stripping: influence of anatomical site, application pressure, duration and removal. Br J Dermatol 151(4):746–752
5. Dickel H, Goulioumis A, Gambichler T et al (2010) Standardized tape stripping: a practical and reproducible protocol to uniformly reduce the stratum corneum. Skin Pharmacol Physiol 23(5):259–265

6. Harding CR (2004) The stratum corneum: structure and function in health and disease. Dermatol Ther 17(Suppl 1):6–15
7. Wood LC, Elias PM, Calhoun C, Tsai JC, Grunfeld C, Feingold KR (1996) Barrier disruption stimulates interleukin-1 alpha expression and release from a preformed pool in murine epidermis. J Invest Dermatol 106(3):397–403
8. Elias PM, Choi EH (2005) Interactions among stratum corneum defensive functions. Exp Dermatol 14(10):719–726
9. Herkenne C, Alberti I, Naik A et al (2008) In vivo methods for the assessment of topical drug bioavailability. Pharm Res 25(1):87–103
10. Breternitz M, Flach M, Prassler J, Elsner P, Fluhr JW (2007) Acute barrier disruption by adhesive tapes is influenced by pressure, time and anatomical location: integrity and cohesion assessed by sequential tape stripping. A randomized, controlled study. Br J Dermatol 156(2):231–240
11. Ghadially R, Brown BE, Sequeira-Martin SM, Feingold KR, Elias PM (1995) The aged epidermal permeability barrier. Structural, functional, and lipid biochemical abnormalities in humans and a senescent murine model. J Clin Invest 95(5):2281–2290
12. Fluhr JW (2011) Practical aspects of cosmetic testing. How to set up a scientific study in skin physiology. Springer, Berlin/Heidelberg
13. Sheth NV, McKeough MB, Spruance SL (1987) Measurement of the stratum corneum drug reservoir to predict the therapeutic efficacy of topical iododeoxyuridine for herpes simplex virus infection. J Invest Dermatol 89(6):598–602
14. Kandárová H, Richter H, Liebsch M, Lademann J (2007) Stratum corneum architecture of reconstructed human skin models monitored by fluorescent confocal laser scanning microscopy laser. Laser Phys Lett 4:308–311
15. Jacobi U, Meykadeh N, Sterry W, Lademann J (2003) Effect of the vehicle on the amount of stratum corneum removed by tape stripping. J Dtsch Dermatol Ges 1(11):884–889
16. Lademann J, Jacobi U, Surber C, Weigmann HJ, Fluhr JW (2009) The tape stripping procedure–evaluation of some critical parameters. Eur J Pharm Biopharm 72(2):317–323
17. Haftek M, Teillon MH, Schmitt D (1998) Stratum corneum, corneodesmosomes and ex vivo percutaneous penetration. Microsc Res Tech 43(3): 242–249
18. Caron D, Queille-Roussel C, Shah VP, Schaefer H (1990) Correlation between the drug penetration and the blanching effect of topically applied hydrocortisone creams in human beings. J Am Acad Dermatol 23(3 Pt 1):458–462
19. Pershing LK, Silver BS, Krueger GG, Shah VP, Skelley JP (1992) Feasibility of measuring the bioavailability of topical betamethasone dipropionate in commercial formulations using drug content in skin and a skin blanching bioassay. Pharm Res 9(1): 45–51
20. Darlenski R, Sassning S, Tsankov N, Fluhr JW (2009) Non-invasive in vivo methods for investigation of the skin barrier physical properties. Eur J Pharm Biopharm 72(2):295–303
21. Teichmann A, Pissavini M, Ferrero L et al (2006) Investigation of the homogeneity of the distribution of sunscreen formulations on the human skin: characterization and comparison of two different methods. J Biomed Opt 11(6):064005

27 Stratum Corneum Quantification

Frank Dreher

27.1 Tape-Stripping Technique

27.1.1 Experimental Procedure

SC tape stripping is carried out by pressing an adhesive tape onto the skin surface and removing it by tearing off [1, 2]. Ideally, the tape is placed onto a previously defined skin surface area (e.g., marked by lines, mask) by applying a constant pressure (e.g., 100 g cm^{-2}) using a weight or spring system over a constant and reproducible time period (e.g., 3–10 s; depending on adhesive tape). Afterwards, the tape is removed with a single continuous motion with forceps. The application and removal procedure may be repeated up to more than 100 times at the same site. Commonly used tapes for skin tape stripping are stationary tapes (e.g., Tesa® Film No. 57330-00000-02, Tesa, Germany), medical tapes (e.g., Transpore®, 3 M Co., USA), or for such a purpose specially designed tapes (e.g., D-SQUAME®, CuDerm Inc., USA; Corneofix®, Courage & Khazaka GmbH, Germany). Tapes differ in shape, size, composition, and adhesive properties. Following tape stripping, the solute contained in the SC can be extracted and measured using appropriate analytical methods such as liquid chromatography.

F. Dreher, MD
Neocutis, Inc., 3053 Fillmore Street #140,
San Francisco, CA 94123, USA
e-mail: frank.dreher@neocutis.com

27.1.2 Parameters Influencing SC Removal

Numerous parameters have been described to influence the amount of SC removed by a single tape strip. Differences in adhesive properties between tape brands will result in significantly different amounts of SC removed per surface unit [3]. Pressure [4], time course between application and removal [5], as well as velocity of tape removal process [4] also influences SC amount removed. Additionally, SC removal can depend on intrinsic skin properties related to age, gender, ethnicity, skin region, and skin condition (e.g., moist vs. dry, oily vs. nonoily, healthy vs. pathologic). Moreover, the SC amount removed by tape stripping varies according to the depth. In general, the first strips remove the largest amounts of SC because they remove the more loosely packed squamous cells. Decreasing amounts of SC are removed with increasing strips since the cohesiveness between corneocytes increases with depth [6]. In addition, application of topical products prior to tape stripping further influences the amount of SC removed by sequential tape stripping [7]. For instance, the vehicle components may alter both adhesive properties of the tape, at least for the very first strips, as well as cohesiveness between corneocytes.

27.1.3 Quantification of SC Removal

As described before, the amount of SC removed by tape stripping is variable and depends on how

E. Berardesca et al. (eds.), *Non Invasive Diagnostic Techniques in Clinical Dermatology*,
DOI 10.1007/978-3-642-32109-2_27, © Springer Berlin Heidelberg 2014

the tape-stripping procedure is performed as well as on SC characteristics and conditions. As a consequence, the amount of SC removed by tape stripping is generally not proportional to the number of strips removed. The quantification of the amount SC removed is therefore pivotal for the majority of studies involving tape stripping. The SC removal can be determined using various methods including weighing, protein quantification, and optical methods.

27.1.3.1 Weighing

For a long time, weighing was the method of choice to measure the amount of SC removed on a tape strip [8]. Thereby, tapes are weighted before and after stripping and the amount of SC is given by weight difference. High precision balances are required since a rather low amount of SC is removed per square centimeter of tape. However, weighing is time consuming and is biased by water absorption or desorption during the weighing procedure [8]. Furthermore, after topical product application, the weighing of SC is only reliable to some extent since the tape strips may also contain applied vehicle and active. Therefore, newer methods to help accurately quantify SC removal were introduced in the last years.

27.1.3.2 Protein Quantification

As an alternative to weighing, a simple colorimetric method based on commercially available protein assays was first described by Dreher and Maibach in 1998 [9]. Briefly, the total protein assay (according to Lowry or Bradford) was carried out after immersing the SC containing tapes in a one molar sodium hydroxide solution to extract the soluble SC protein fraction (SC is mainly composed of corneocytes filled with keratins) and subsequently neutralizing the solution with one molar hydrochloric acid. The neutralization was realized since the protein assays are not compatible with acidic conditions. The extracted SC proteins, or their hydrolysates, remain in solution after neutralization. This quantification method allows determining as little as a few micrograms of SC adhering to a single tape strip in an accurate and reproducible manner. Furthermore, with the exception of protein containing products and a few chemicals interfering with the protein assay, the absorption of active and product excipients into the SC after topical application does not interfere with this method. Importantly, the water content of the SC adhering to the tape strip has no influence on the colorimetric assay. Besides the D-SQUAME® tape, for which this assay has originally been developed, other tapes can be used and are compatible. Other than performing the protein extraction with each tape strip, the extraction can also be performed with pooled strips such as with five strips, what allows to significantly shortening the time of analysis. Alternatively, the protein assay can also be realized in 96-well microplates [10].

When performing bioavailability studies, the above method is particularly suited for hydrophilic chemicals, which are chemically stable under alkaline conditions. Hydrophilic chemicals can be easily extracted from the SC adhering to tape strips and can therefore be analyzed using appropriate analytic methods such as high-pressure liquid chromatography in parallel with the SC quantification. However, this method may be less valuable for hydrophobic chemicals and for chemicals not stable under the SC extraction conditions. In that case, optical methods should be used to quantify the SC amount removed, followed by the extraction of the solute from the tape strip using appropriate solvents.

27.1.3.3 Optical Methods

One year later, Lademann and Weigmann presented a method based on the measure of UV/VIS absorbance to determine the SC amount adhering on tape strips [11]. Unlike the method through protein quantification, this technique does not require any treatment of the tapes. As a consequence, the entire tape strip remains intact and can be used for subsequent analysis including solute extraction. In this method, SC determination is performed at 430 nm using a modified, double-beam UV/VIS spectrophotometer with a 1 × 1 cm light beam. The reference beam chamber contains an unused tape. The absorbance at this wavelength originates from light reflection, scattering, and diffraction by corneocyte aggregates on the tape and was reported to correlate to the SC weight removed by tape stripping. However, this method has some disadvantages including the interference with chemicals absorbing in the wavelength range of corneocyte absorbance at 430 nm. Additionally, potential changes in reflection, scattering, and diffraction properties at 430 nm of the SC due to uptake of water or oils after topical product application may further affect the outcome of the measure.

More recently, another optical and nondestructive method was presented by Voegeli and coworkers [12]. This method measures SC absorption with a diode-emitting light with a peak wavelength of 850 nm using a purpose-developed instrument (SquameScan™ 850A, Heiland Electronic GmbH, Germany). This method prevents thermal denaturation of biomolecules, reduces light scattering of the tapes, reduces ambient light, and measures no molecular absorption of SC compounds. Absorption data was shown to correlate with protein content on the tape and seemed to be less interference prone than pseudo-absorbance measurements in the visible light range at 430 nm, what emphasizes the validity of this newer near-infrared light-based method. In the meanwhile, this method was applied also for tape stripping on human [13] and porcine ear skin ex vivo [14, 15] as well as to determine the distribution of a drug within the stratum corneum [16].

27.2 Summary and Conclusion

The application of tape-stripping technique is well established in dermatopharmacological research, and the technique is appreciated as the most useful method to remove SC allowing investigation of its structure, properties, and functions. However, despite apparent simplicity, the tape-stripping technique entails several technical problems and care has to be taken to avoid misleading conclusions when interpreting data. For instance, results given as a function of tape strip number or pooled tape strips have to be interpreted with care, since the amount of SC removed by tape stripping is generally variable and depends on numerous factors related to tape-stripping procedure and SC properties. Therefore, SC removal by tape stripping should be quantified with accurate and reliable methods. As compared to weighing, protein quantification and optical methods represent more accurate methods. Consequently, those newer methods are valuable tools in the assessment of cutaneous bioavailability and the determination of bioequivalence of dermatology drugs.

References

1. Pinkus H (1951) Examination of the epidermis by the strip method of removing horny layers. J Invest Dermatol 16:383–386
2. Surber C, Schwarb FP, Smith EW (1999) Tape-stripping technique. In: Bronaugh RL, Maibach H (eds) Percutaneous absorption: drugs – cosmetics – mechanisms – methodology, vol 97, 3rd edn, Drugs and the pharmaceutical sciences. Dekker, New York, pp 395–409
3. Tsai JC, Weiner ND, Flynn GL, Ferry J (1991) Properties of adhesive tapes used for stratum corneum stripping. Int J Pharm 72:227–231
4. Löffler H, Dreher F, Maibach HI (2004) Stratum corneum adhesive tape stripping: influence of anatomical site, duration and removal. Br J Dermatol 151: 746–752
5. Tokumura F, Ohyama K, Fujisawa H, Suzuki M, Nukatsuka H (1999) Time-dependent changes in dermal peeling force of adhesive tapes. Skin Res Technol 5:33–36
6. King CS, Barton SP, Nicholls S, Marks R (1979) The change in properties of the stratum corneum as a function of depth. Br J Dermatol 100:165–172
7. Tsai JC, Cappel MJ, Weiner ND, Flynn GL, Ferry J (1991) Solvent effects on the harvesting of stratum corneum from hairless mouse skin through adhesive tape stripping in vitro. Int J Pharm 68: 127–133
8. Marttin E, Neelissen-Subnel MTA, De Haan FHN, Boddé HE (1996) A critical comparison of methods to quantify stratum corneum removed by tape stripping. Skin Pharmacol 9:69–77
9. Dreher F, Arens A, Hostynek JJ, Mudumba S, Ademola J, Maibach HI (1998) Colorimetric method for quantifying human *stratum corneum* removed by adhesive-tape-stripping. Acta Dermatol Venerol (Stockh) 78:186–189
10. Dreher F, Modjtahedi BS, Modjtahedi SP, Maibach HI (2005) Quantification of stratum corneum removal by adhesive tape stripping by total protein assay in 96-well microplates. Skin Res Technol 11: 97–101

11. Weigmann HJ, Lademann J, Meffert H, Schaefer H, Sterry W (1999) Determination of the horny layer profile by tape stripping in combination with optical spectroscopy in the visible range as a prerequisite to quantify percutaneous absorption. Skin Pharmacol Appl Skin Physiol 12:34–45
12. Voegeli R, Heiland J, Doppler S, Rawlings AV, Schreier T (2007) Efficient and simple quantification of stratum corneum proteins on tape strippings by infrared densitometry. Skin Res Technol 13:242–251
13. Hahn T, Hansen S, Neumann D, Kostka KH, Lehr CM, Muys L, Schaefer UF (2010) Infrared densitometry: a fast and non-destructive method for exact stratum corneum depth calculation for in vitro tape-stripping. Skin Pharmacol Physiol 23:183–192
14. Klang V, Schwarz JC, Hartl A, Valenta C (2011) Facilitating in vitro tape stripping: application of infrared densitometry for quantification of porcine stratum corneum proteins. Skin Pharmacol Physiol 24:256–268
15. Klang V, Schwarz JC, Lenobel B, Nadj M, Auböck J, Wolzt M, Valenta C (2012) In vitro vs. in vivo tape stripping: validation of the porcine ear model and penetration assessment of novel sucrose stearate emulsions. Eur J Pharm Biopharm 80:604–614
16. Schwarz JC, Klang V, Hoppel M, Wolzt M, Valenta C (2012) Corneocyte quantification by NIR densitometry and UV/Vis spectroscopy for human and porcine skin and the role of skin cleaning procedures. Skin Pharmacol Physiol 25:142–149

Utilisation of Infrared Densitometry in Stratum Corneum Research

28

Rainer Voegeli and Anthony V. Rawlings

Abbreviations and Synonyms

AD	Atopic dermatitis
BCA	Bicinchoninic acid
BGG	Bovine gamma globulin
BSA	Bovine serum albumin
DNP	Dinitrophenyl
IR	Infrared
KLK	Kallikrein
LEKTI	Lympho-epithelial Kazal-type serine protease inhibitor
LLOQ	Lower limit of quantification
NMF	Natural moisturising factor
PCA	2-Pyrrolidone-5-carboxylic acid
SC	Sratum corneum
TEWL	Transepidermal water loss
UCA	Urocanic acid
uPA	Urokinase

R. Voegeli (✉)
R&D Personal Care, DSM Nutritional Products Ltd., P.O. Box 2676, Bldg. 203.4/86, CH-4002 Basel, Switzerland
e-mail: rainer.voegeli@dsm.com

A.V. Rawlings
AVR Consulting Ltd., 26 Shavington Way, Kingsmead, Northwich, Cheshire, CW9 8FH, UK
e-mail: tonyrawlings@aol.com

28.1 Methods for Quantification of SC on Tape Strippings

For most investigations the determination of the amount of SC mass adhering to each tape stripping is required as all SC analytes are reported as a ratio to the amount of SC collected to account for the amount of SC isolated. This not only allows the comparison between samples, i.e. amount of analyte per amount of SC, but also allows concentration profiling of endogenous and exogenous molecules with depth of the SC. Changes in SC levels are known for SC corneodesmosomal proteins [1, 2], lipids [1], NMF compounds [3] and enzymes [4–6] for instance. However, a nondestructive determination of SC mass is a prerequisite to enable the measurement of other analytes on the same tape.

28.1.1 SC Protein Estimation by Gravimetric Analysis

Weighing of tape strippings for SC estimation is time consuming and laborious because the tapes have to be weighed with a sensitive microbalance before and after stripping under constant humidity conditions [7]. The average amount of SC removed reaches only approx. 10–20 μg/cm^2 tape in the topmost cell layers. However, the gravimetric method is complicated by the presence of other analytes such as

E. Berardesca et al. (eds.), *Non Invasive Diagnostic Techniques in Clinical Dermatology*,
DOI 10.1007/978-3-642-32109-2_28,

sebum lipids, sweat components and any components of topically applied creams or lotions. As a result more direct methods of protein determination have been used.

28.1.2 SC Protein Estimation by Colorimetric Analysis

Many workers have used colorimetric methods to determine SC protein levels from tape strippings [1, 2, 4–6] but Dreher et al. [7, 8] performed extensive studies on the subject reporting a well-defined linear relation between the amount of SC weighed for each tape stripping and the quantity of SC determined by Bio-Rad DC protein microassay after extraction. Correlation coefficients (R^2) higher than 0.95 were obtained. Consequently many scientists use protein determinations to quantify the amount of SC removed.

28.1.3 SC Protein Estimation by Pseudo-absorption at 430 nm

Weigmann et al. proposed a method for the evaluation of the amount of SC on tape strippings based on determination of the pseudo-absorption of corneocytes [9]. The pseudo-absorption is influenced by absorption, scattering and reflection properties of corneocytes. It can be measured in the complete UV/VIS spectral range, but the signal decreases with increasing wavelengths. It was proposed to measure the pseudo-absorption at a wavelength of 430 nm (Fig. 28.1), which is outside the UV absorption of most molecules and where sufficiently high intensities of the pseudo-absorption can be obtained. The spectroscopic measurements were performed with a special spectrophotometer, with a measuring area of 1.5 cm² enabling a reliable integration of the inhomogeneous distribution of the corneocytes on the removed tape strippings. The

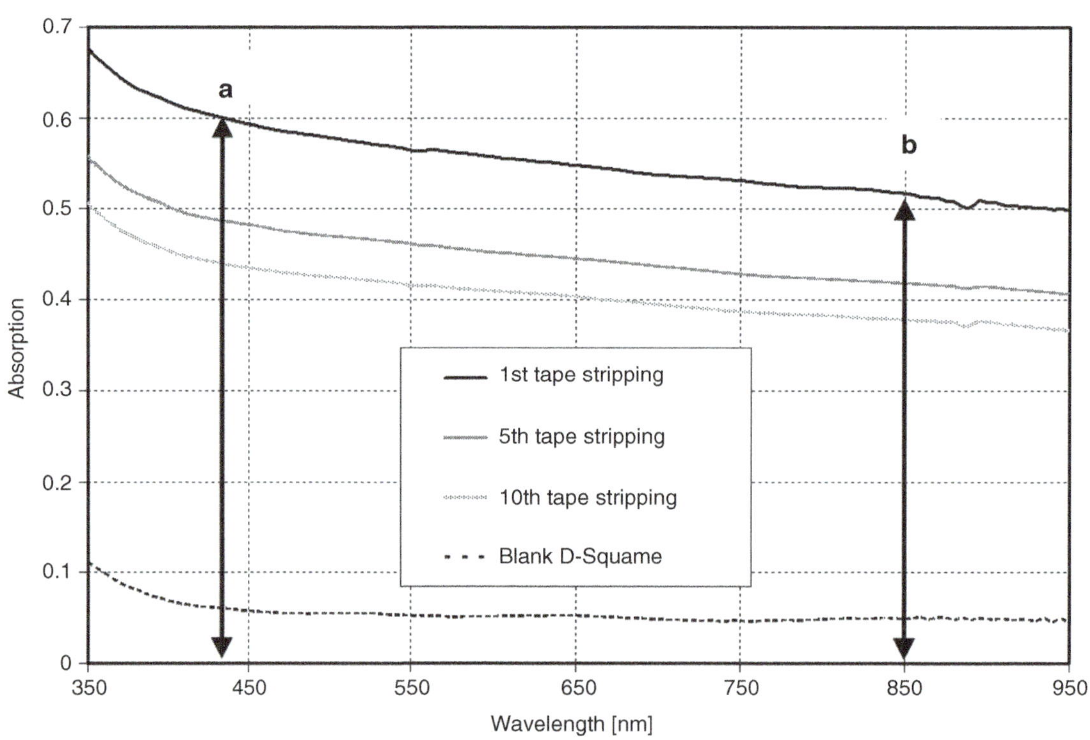

Fig. 28.1 UV/VIS/NIR spectra of a blank D-SQUAME® tape, a first, a fifth and a tenth tape stripping; (*a*) wavelength of the so-called pseudo-absorption at 430 nm and (*b*) peak wavelength of SquameScan 850A at 850 nm (Modified from [16])

pseudo-absorption at 430 nm correlates with the weight of SC ($R^2 = 0.99$), the protein absorption at 278 nm ($R^2 = 0.92$) and the absorption at 652 nm of trypan blue-stained proteins ($R^2 = 0.99$) [9–14].

Lademann et al. also developed an optical device measuring pseudo-absorption at 430 nm of corneocytes on tape strippings while using an automated slide projector to routinely assess samples. The lamp of the slide projector was substituted by a halogen lamp and an interference filter ($\lambda = 430$ nm). A photodiode was positioned behind the lens system. The mechanical handling system of the slide projector was used to change between samples automatically. When compared with UV/VIS spectrophotometric measurements described above, a correlation coefficient of $R^2 = 0.97$ was demonstrated [12], and with protein quantified with the Bio-Rad Protein Assay Kit, R was calculated as 0.86 [15].

28.1.4 SC Protein Estimation by IR Densitometry

28.1.4.1 Historical Background

The primary driving force for developing a novel, cheap, easy to use, high-throughput and nondestructive technique for the measurement of SC on tape strippings was our interest to investigate serine protease profiles in the SC as devices and techniques described in the literature did not completely meet our aims (Table 28.1).

Table 28.1 Overview of common methods used for the determination of SC amount on tape strippings

Method	Sources of errors
Gravimetry	
Differential weighing	Nonspecific weight increase due to sweat, lipids, formulation excipients
Optical methods	
Protein absorption at 278 nm	Nonspecific absorption due to other substances at 278 nm
Pseudo-absorption at 430 nm	Nonspecific pseudo-absorption due to sweat, lipids, formulation excipients, absorption due to coloured substances
Trypan blue staining, absorption at 652 nm	Low sensitivity due to background colouring of tapes; low resolution of small differences in degree of coverage of the tape
Protein quantification	
Colorimetric quantification of NaOH extractable protein by a Lowry, Bradford or BCA assay	Destructive, therefore determination of depth and concentration of penetrating drug on different skin sites; nonspecific reaction with a wide range of substances, e.g. penetrating drugs

Modified from [18]

Pseudo-absorption: decrease of transmitted light due to reflection, diffraction and scattering

The idea using densitometry for SC quantification was born at the Gala Dinner of the Stratum Corneum Conference IV, in June 2004 in Paris together with David Miller (CuDerm) and Gary Grove (cyberDERM) after some glasses of enjoyable red wine. Following the conference we found, thanks to David, the small company Heiland electronic (Wetzlar, Germany) which designed and produced a first prototype of an IR densitometer according to our specifications within few months. A good correlation between the signal of the device and the protein mass resulting from a very first preliminary trial in July 2005 convinced us that we were on the right track. After a few small adaptations, the IR densitometer, named SquameScan 850A, was ready, and we presented on its use at the World Congress on Noninvasive Studies of the Skin in Wilmington, DE, USA, in September 2005.

28.1.4.2 Description and Specification

SquameScan 850A (Fig. 28.2) was developed for standard D-SQUAME® disks (CuDerm Corporation, Dallas, USA) with a diameter of 22 mm. SquameScan 850A has a compact design (weight: 1 kg, length × width × height: 200 × 100 × 100 mm). The instrument is equipped with a diode-emitting light with a peak wavelength of 850 nm which minimises any potential thermal denaturation of biomolecules. It also prevents the influence of ambient light absorption on the measurement result, reduces light scattering of the tapes and measures no molecular absorption of SC compounds. A blank D-SQUAME® has a constant absorption level between 450 and 850 nm (Fig. 28.1), and the curve progressions

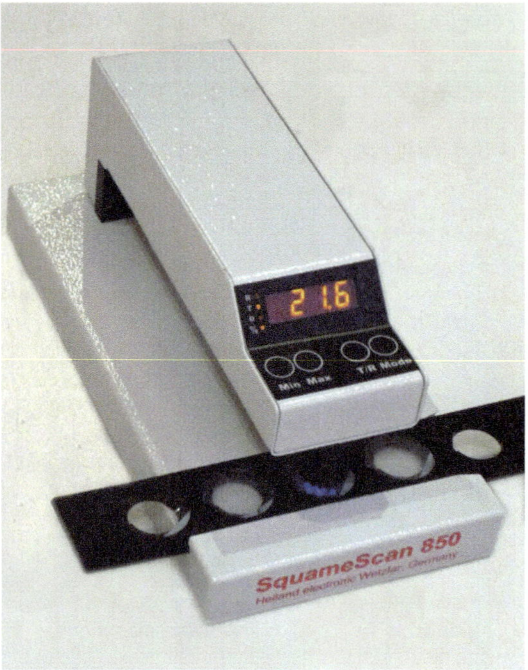

Fig. 28.2 The compact infrared densitometer SquameScan 850A with an inserted sample applicator [16]

of the tape strippings are parallel and rather flat around 850 nm, proving the wavelength range around 850 nm useful for the intended application. The measurement range of the SquameScan 850A is 0–40 % absorption with a resolution of 0.1 %. The diameter of the circular slit is 1.5 cm, corresponding to 1.8 cm^2, which covers 45 % of the area of a standard D-SQUAME® disk. The disks are placed in an easily and user-friendly way adhesive side up, into a sample applicator. A constant absorption value is reached 3 min after tape stripping. The sample applicator takes up to 10 tapes allowing a fast determination. Moreover the sample applicator avoids contamination and mixing up of the tape strippings during measurement and secures their alignment. Most recently sample applicators for Corneofix® tapes (Courage & Khazaka electronic GmbH, Cologne, Germany) have been developed [16].

The mean variation of absorption due to possible inhomogeneous distribution of corneocytes on the tape strippings was 0.0040 ± 0.0016 which corresponded to a mean percentage variation of 1.99 ± 0.66 %. This was determined by sampling the volar forearm and measuring the initial IR absorption followed by a second measurement at a 90° angle. SquameScan 850A is calibrated by setting the absorption of an empty sample holder to 0 % absorption. The absorption of an ambient light filter provided for calibration by the manufacturer should then give a reading of 33.8 %. The absorption of an empty tape is then subtracted from the signal from the SC stripping protein measurement to correct for background noise.

28.1.4.3 Correlation of IR Absorption as a Function of Colorimetrically Determined Protein Content

The protein content per tape stripping was determined with the Micro BCA Protein Assay Kit (Pierce Biotechnology Inc., Rockford, IL, USA). Initially we used bovine gamma globulin (BGG) as standard protein [16] because according to Fluhr et al. [17] it correlates best with human SC. However, recently we've compared BGG, bovine serum albumin (BSA) and human plantar SC extracts in the Micro BCA Protein Assay Kit. Before the protein assay the human plantar SC samples were dissolved for 1 h at 37 °C in 1 N sodium hydroxide. Interestingly the absorption of human plantar SC protein was 35 % smaller than the one of BGG and BSA (Fig. 28.3), which is similar to the data of Hahn et al. [18]. Obviously plantar SC protein shows a reduced protein reactivity in the Micro BCA Protein Assay.

To evaluate the correlation between IR absorption and SC protein in different individuals, we enrolled 12 healthy Caucasian subjects (aged 27–50 years, Fitzpatrick skin phototypes II–III, six females and six males) and took 20 sequential tape strippings from the volar forearm. The correlation of IR absorption and protein curves was similar irrespective of tape stripping number. The overall correlation ($n = 238$) between IR absorption and protein content of forearm measurements was $R^2 = 0.852$ [16]. Although the data distribution in the different subject groups varied, the regression was always quite similar and independent of gender, age, skin hydration rate, skin pH and varying skin areas. The correlations

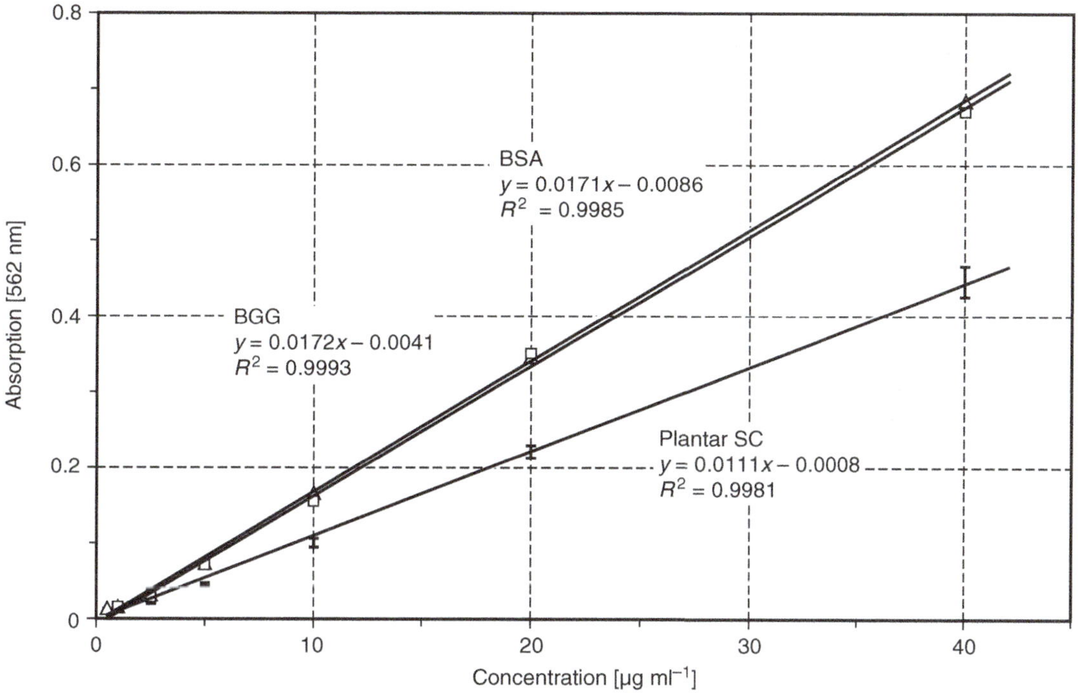

Fig. 28.3 Calibration curves of human plantar SC extracts of five donors compared to BGG and BSA solution at 550 nm using the Micro BCA Protein Assay

reached were statistically significant. The linear regression for IR absorption as a function of protein content was $y = 0.623x + 2.703$ for BGG as the standard protein (Fig. 28.4) [16].

28.1.4.4 Lower Limit of Quantification

The lower limit of quantification (LLOQ) was introduced by Hahn et al. [18]. The LLOQ can be calculated from the fivefold background noise of an empty tape. They demonstrated that only negligible amounts of SC were left on the skin's surface when reaching the LLOQ of SquameScan 850A. As the cumulative IR absorption of all tape strippings is related to the total SC thickness, the depth into the SC can be easily calculated from the sum of IR absorption (c.f. following paragraph, Eqs. 28.1, 28.2, 28.3, 28.4, and 28.5) [19].

It was shown that IR densitometry is well suited for determining the endpoint of almost complete SC removal as it allows fast in-process monitoring of the results. Tape stripping does not completely remove all layers of the SC, it only removes the SC to the SC typically at the glistening layer. Consequently, this method allows for adjusting the total number of tape strippings that are performed for a single skin site during the experiment. Thereby, the SC depth profiles can be determined more accurately [18].

28.1.4.5 SC Thickness and Depth Estimation and Calculations Using IR Densitometry

Boncheva et al. [20] estimated the depth within the SC reached by sequential tape stripping of the same skin site. They correlated the cumulative amount of protein progressively removed by tape stripping and the SC thickness estimated from the confocal Raman water profiles collected after removing each tape. The linear correlation ($R = 0.79$) between these values indicates that the removal of 100 μg/cm² protein corresponds to removal of 1.9 ± 0.2 μm from SC (Fig. 28.5).

Melero et al. [19] summarised SC thickness and depth calculations using IR densitometry as follows:

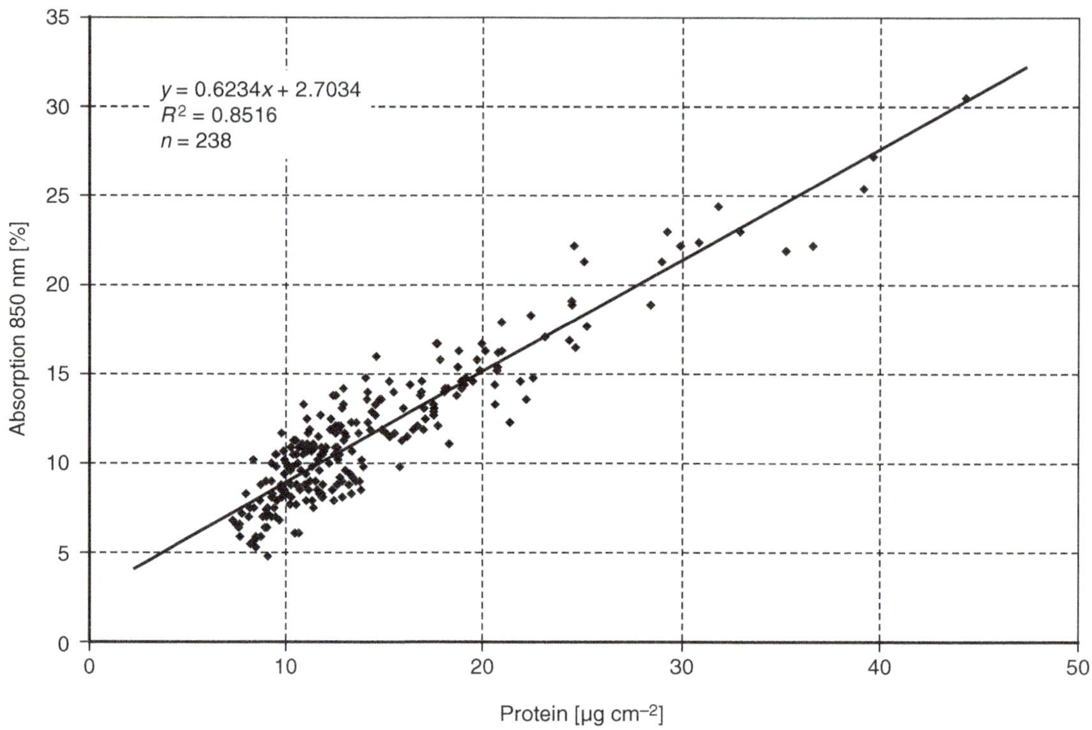

Fig. 28.4 Linear regression analysis of D-SQUAMES® obtained by 20 sequential tape strippings from the volar forearm of 12 subjects based on BGG calibration curve [16]

Corrected absorption value for a single tape stripping:

$$x_i = x_i^* - x_0 \tag{28.1}$$

i: tape stripping number.
x_i: IR absorption of a tape stripping and corrected for absorption of an empty tape.
x_i^*: IR absorption measured, non-corrected value.
x_0: IR absorption of an empty tape measured.

IR absorption value for pooled strippings:

$$x_{pool(n)} = \sum_{i=h}^{j} x_i \tag{28.2}$$

i: tape stripping number.
h: first stripping belonging to the nth pool.
j: last stripping belonging to the nth pool.
$x_{pool(n)}$: sum of IR absorption of pool n containing tape strippings $h-j$.
x_i: IR absorption of ith tape stripping measured and corrected for empty tape.

Relative thickness of SC removed per pool:

$$d^*_{rel(n)} = \sum_{i=h}^{J} xi / Xn_{max} \text{ with } Xn_{max} = \sum_{i=1}^{n_{max}} x_i \tag{28.3}$$

h: first stripping belonging to the nth pool.
j: last stripping belonging to the nth pool.
$d^*_{rel(n)}$: relative SC thickness removed per pool n.
n_{max}: total number of tape strippings.
x_i: IR absorption of ith tape stripping measured and corrected for an empty tape.
X_{max}: IR absorption of total number of tapes n_{max}.

Absolute SC thickness of the nth pool:

$$d^*_n = d_{total} \bullet d^*_{rel(n)} \tag{28.4}$$

d^*_n: absolute SC thickness of the nth pool.
d_{total}: total SC thickness.
$d^*_{rel(n)}$: relative SC thickness removed per pool n.

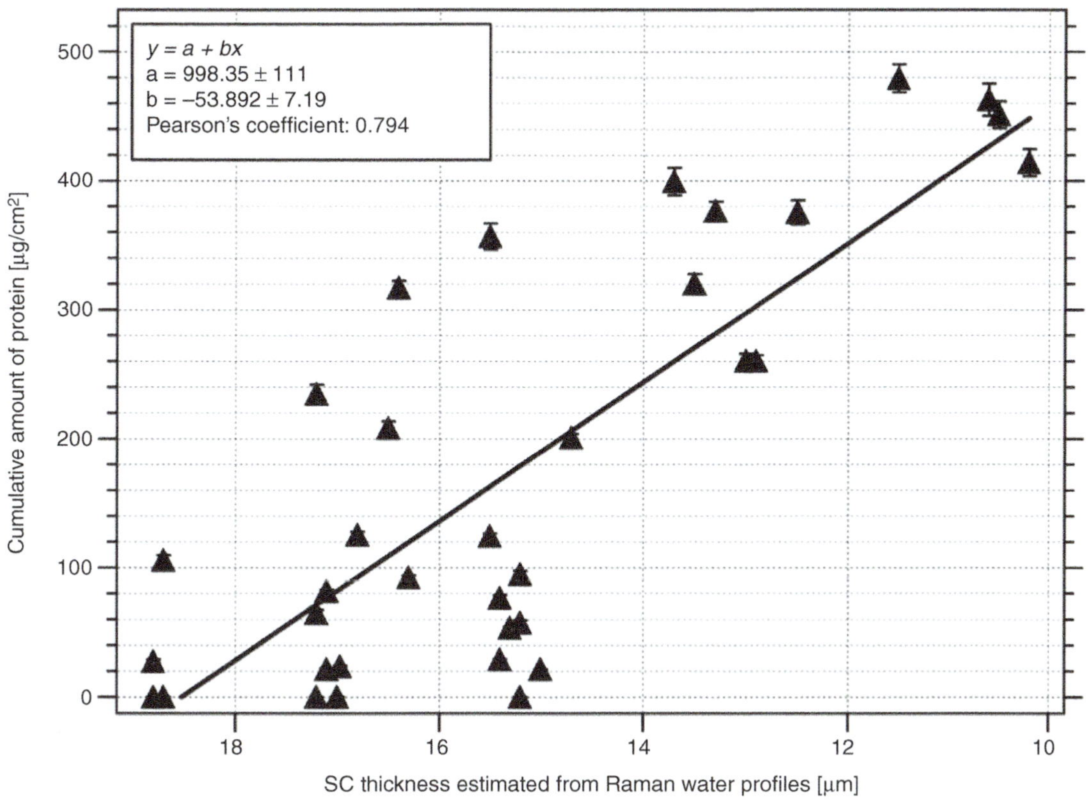

Fig. 28.5 Estimation of the depth within SC reached by sequential tape stripping of the same site of five subjects. Correlation between the cumulative amount of protein progressively removed by tape stripping and the SC thickness estimated from confocal Raman water profiles collected after removing each tape. The line is the best fit to the experimental points (Modified from [20])

SC depth after n_n pools of tape strippings:

$$d^*n_n = \sum_{n=1}^{n_n} d^*n \quad (28.5)$$

$d^*_{n_n}$: SC depth after n_n pools of tape strippings.
d^*_n: thickness of the nth pool.

28.2 Recent Applications Using IR Densitometry

Tape stripping is a minimally invasive technology that can broadly be used in dermatological research. In fact the process of tape stripping is a special type of superficial SC 'biopsy', which is very easily accessible. A colleague once mentioned: 'tape stripping is like the work of archaeologists, instead of digging into the dirt you dig into the SC'. However, in contrast to archaeologists who are unravelling history, skin biologists can use tape stripping to investigate not only the past but also the present time, and far better they sometimes open a window to the future.

IR densitometry enables a fast and convenient quantification not only of SC protein removed but also for SC depth determination, and thus depth profiling of endogenous and exogenous molecules can be followed as well as the endpoint determination for 'complete' SC removal. In the following we briefly review very recent papers using this technology.

28.2.1 Studies on Skin Penetration

As a result of the European REACH Programme (Regulation (EC) No. 1907/2006), in vitro skin penetration studies are of special interest not only for drug administration authorities but also for evaluation of toxicological substances. In this context Hahn et al. [18] challenged SquameScan 850A for the first time for in vitro applications with human abdominal and breast skin obtained from plastic surgery and correlated their measurements with optical microscopy observations. IR densitometry provided accurate depth results both for freshly excised skin and for skin stored frozen for up to 3 months. Their results showed that IR densitometry is also well suited for determination of the endpoint for complete SC removal. This is important as it is not advisable to work with a fixed total number of tape strippings. The method allows the adjustment of the total number of tape strippings that is needed for removing the SC quantitatively of an individual skin sample. IR densitometry improves the quality of skin concentration-depth profiles.

Later Klang et al. [21] established calibration curves of SC proteins removed by tape strippings from the porcine ear skin. The IR absorption of SC protein on tape strippings was correlated with the protein content determined with the Micro BCA Protein Assay Kit (Pierce Biotechnology Inc.) after extraction of the tapes. The obtained linear regressions ($n=240$) confirm that IR densitometry is suitable for the quantification not only of human but also of porcine SC proteins. Correlation coefficients (R^2) were 0.812 for Corneofix® tapes (Courage & Khazaka) and 0.732 for D-SQUAMES® (CuDerm). The pattern of protein removal observed with porcine SC differs from that of human SC, which necessitates specific evaluation of porcine SC samples and a working protocol that takes this into account. A comparison of the calibration curves showed that the slope of the presented porcine correlations was less steep compared with the human curves. While human corneocytes are removed as a consistent layer of finely distributed cell aggregates, the porcine corneocytes are removed in large clusters or patches. In pig skin, the cells are organised in polygonal clusters or columns that comprise the whole of the epidermis and are separated by intercluster regions or ‘canyons’ [22]. High protein density as primarily observed for the first tape strippings removed will therefore lead to comparatively inhomogeneous protein coverage. The following adhesive films showed more homogeneous protein coverage. Tapes with lower adhesive power are therefore more suitable for IR densitometric analysis of porcine SC protein. It was shown that SquameScan 850A leads to satisfyingly accurate results for porcine ear skin despite significant protein clustering. The pseudo-absorption measured by IR densitometry was found to be suitable for quantification of corneocyte aggregates on individual tapes. The presented data will facilitate future analysis of porcine SC proteins during in vitro tape stripping.

Klang et al. [23] also performed comparative tape stripping experiments in which the skin penetration of curcumin and fluorescein sodium from conventional microemulsions and hydrogels was investigated. The trends observed for the skin penetration into porcine ear skin were highly representative for the in vivo situation on human skin, confirming that the porcine ear is an excellent in vitro model for tape stripping experiments. Moreover, the validity of the IR densitometric approach for the quantification of both human and porcine SC proteins was further consolidated from these studies.

Moreover, Melero et al. [24] investigated the release kinetics skin diffusion of nicotine from five different nortriptyline patches marketed for smoking cessation on heat-separated human epidermis using Franz diffusion cells. The thickness of SC removed has been calculated using IR densitometry.

Based on their in vivo penetration studies of topically applied sunscreens (Parsol® 1789), microparticles and corticosteroids (clobetasol propionate), Lademann et al. concluded that IR densitometry is well suited for the determination of the amount of corneocytes on tape strippings and that the most promising methods for the determination of the amount of corneocytes on the removed tape strippings are the analysis of the IR absorption at 850 nm and of the pseudo-absorption at 430 nm [25].

28.2.2 Determination of NMF Levels and Relationship with Filaggrin Mutations and Atopic Dermatitis

Carriers of loss-of-function mutations in the filaggrin gene have reduced levels of natural moisturising factor (NMF) in the SC. The concentration of NMF components which are formed from filaggrin proteolysis in the SC have been proposed to be a useful as a biomarker of the filaggrin genotype. Kezic et al. [26] investigated the levels of the NMF components, 2-pyrrolidone-5-carboxylic acid (PCA) and urocanic acid (UCA) for subjects with the two most common filaggrin mutations (R501X and 2282del4). Tape-stripped SC was quantified with SquameScan 850A, and PCA and UCA were measured by HPLC after extraction of the tapes. The most significant difference between the filaggrin genotypes was found for PCA. The mean values of PCA were 0.18, 0.50 and 1.64 mmol · g^{-1} SC protein in homozygous, heterozygous and wild-type genotypes, respectively. Thus, the concentration of PCA in the SC collected by tape stripping was shown to be a feasible biomarker of the filaggrin genotype.

As filaggrin gene loss-of-function mutations have been shown to represent the strongest so far known genetic risk factor for atopic dermatitis, Angelova-Fischer et al. [27] investigated SC integrity, SC cohesion and barrier recovery after controlled mechanical and irritant-induced barrier abrogation of nonlesional and lesional skin of AD patients harbouring the European R501X, 2282del4, 3702delG, R2447X or S3247X FLG variants. SC integrity was assessed by measurement of transepidermal water loss (TEWL) and SC cohesion by quantification of removed protein following sequential tape stripping by IR densitometry. Tape stripping revealed distinct genotype-related impairment of both SC integrity and SC cohesion. The results provide evidence for discernible filaggrin-related barrier integrity phenotypes in atopic eczema.

28.2.3 Determination of SC Protein Oxidation and Effect of UV Exposure

Protein carbonyl groups in the SC may be used as a biomarker for photo-stress of the skin. Date et al. [28] determined protein carbonyls by a sensitive optical technique based on surface plasmon resonance. After SC collection by tape stripping and quantification by IR densitometry, the protein carbonyls were reacted with 2,4-dinitrobenzenesulfonic acid dehydrate, and the quantity of dinitrophenylated (DNP) protein carbonyls was determined using an anti-dinitrophenyl (anti-DNP) antibody. The mass of DNP-protein carbonyl was measured using surface plasmon resonance. A significant difference was observed in the 'protein carbonyl/total protein' ratio between the sun-protected area of the mid-ventral arm and the sun-exposed area of the upper cheek. The carbonylation ratio is suggested as a useful index of photo-stress of the skin.

28.2.4 Depth Profiling of Serine Protease Activities in the SC on Different Body Sites

We studied healthy Caucasian subjects in the winter months of the year and compared SC cohesion, SC hydration and TEWL with depth profiles of enzymatic activities of desquamatory serine proteases (trypsin-like and chymotrypsin-like kallikreins) and inflammatory serine proteases (plasmin-, urokinase- (uPA) and SC tryptase-like proteases) on SC extracts of ventral forearm and facial tape strippings[6]. Although there was no difference in SC hydration, SC cohesion, as determined via IR densitometry (Fig. 28.6) [16], and TEWL together with protease activities were increased on the cheek (Table 28.2a, b). While desquamatory enzymes have 2.5–4 times greater activities, the inflammatory enzymes showed 7–8 times more activity. Except for the chymotrypsin-like kallikreins, all tested protease activities positively correlated with TEWL and negatively with SC hydration [29]. On the forearm all

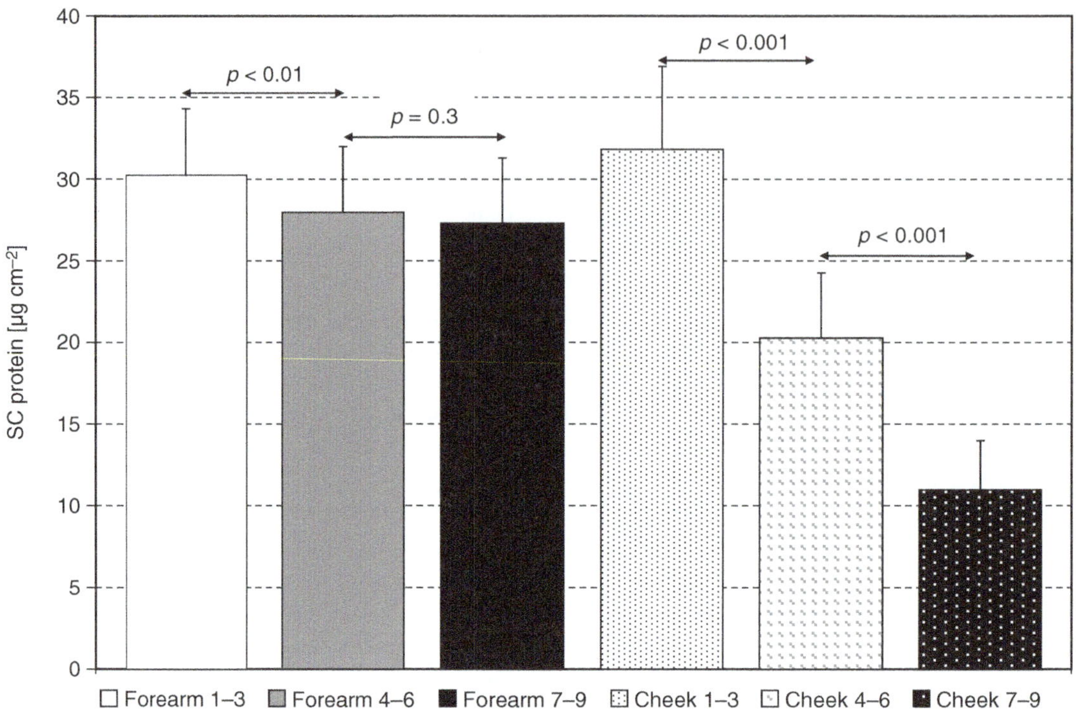

Fig. 28.6 Cohesion of healthy forearm and cheek SC. Protein content (mean per tape ± SEM, $n = 14$ subjects) on tape stripping pools (tapes 1–3, 4–6, 7–9) of the forearm and of the cheek (Modified from [6])

serine proteases showed a distinct gradient with highest activity in the outermost layers of the SC (Fig. 28.7), whereas on the cheek area the activity gradients were less distinct except for chymotrypsin-like kallikrein activity which, like the samples from the forearm, showed increased activation towards the surface layers of the SC. There was a greater trypsin-like kallikrein than a chymotrypsin-like kallikrein activity on both body sites.

Many authors [30, 31] have described that processing of kallikreins occur within early SC development. Conversely, using a sensitive approach, we have established that both trypsin- and chymotrypsin-like kallikreins show increased activities towards the surface layers of the SC on samples from the forearm skin. Similarly, Chang-Yi et al. have found higher chymotrypsin- and trypsin-like protease activities in the outer part of the SC [32]. Presumably, the processing of these enzymes is occurring much later than assumed by others. Similar activation processes can be observed for chymotrypsin-like kallikreins on samples from the face, but on this body site, trypsin-like kallikreins seem to be activated consistently throughout the inner and outer SC. Differences in the control mechanisms of activation of these enzymes may occur on these two different skin areas.

Although the subjects of our study had clinically normally looking facial skin, TEWL values were clearly above the values of the forearm (29 vs. 14 $g \cdot m^{-2} \cdot h^{-1}$) indicating a barrier disturbance (Table 28.3a). We propose that the elevated serine protease activities on the face are due to a subclinical micro-inflammatory or pre-inflammatory condition induced, e.g. by environmental influences. Thus, serine proteases can be considered as key markers for underlying and sometimes non-observable skin abnormalities.

Reduced SC thickness, corneocyte size [33, 34] and maturity together with elevated SC cohesion and TEWL indicate that facial SC is composed predominantly of stratum compactum which is probably due to increased

Table 28.2 (a) Biometric data of the subjects and (b) SC serine protease activities on 9 consecutive tape stripping extracts of a healthy forearm and facial skin

(a)			
Biometric data	**Forearm**	**Cheek**	
Skin surface pH	4.75±0.08	4.93±0.04	
SC hydration [DPM]	104±4	110±3	
TEWL [$g \cdot m^{-2} h^{-1}$]	13.5±0.9	28.6±2.8*	
SC cohesion [μg protein cm^{-2}], 9 tape strippings	28.5±4.0	21.0±4.0*	
(b)	**Protease activity [μU/mg SC protein]**		
Serine protease	**Forearm (F)**	**Cheek (C)**	**Factor C/F**
Trypsin-like kallikreins	3.20±0.47	12.05±1.57*	3.8
Chymotrypsin-like kallikreins	1.45±0.35	3.77±0.48*	2.6
SC tryptase-like	1.62±0.58	13.44±1.50*	8.3
Plasmin-like	1.39±0.29	9.91±1.25*	7.1
Urokinase-like	0.92±0.16	6.99±1.31*	7.6

Modified from [6]
Mean±SEM; $n=14$ per group
*p <0.01 for cheek vs. forearm

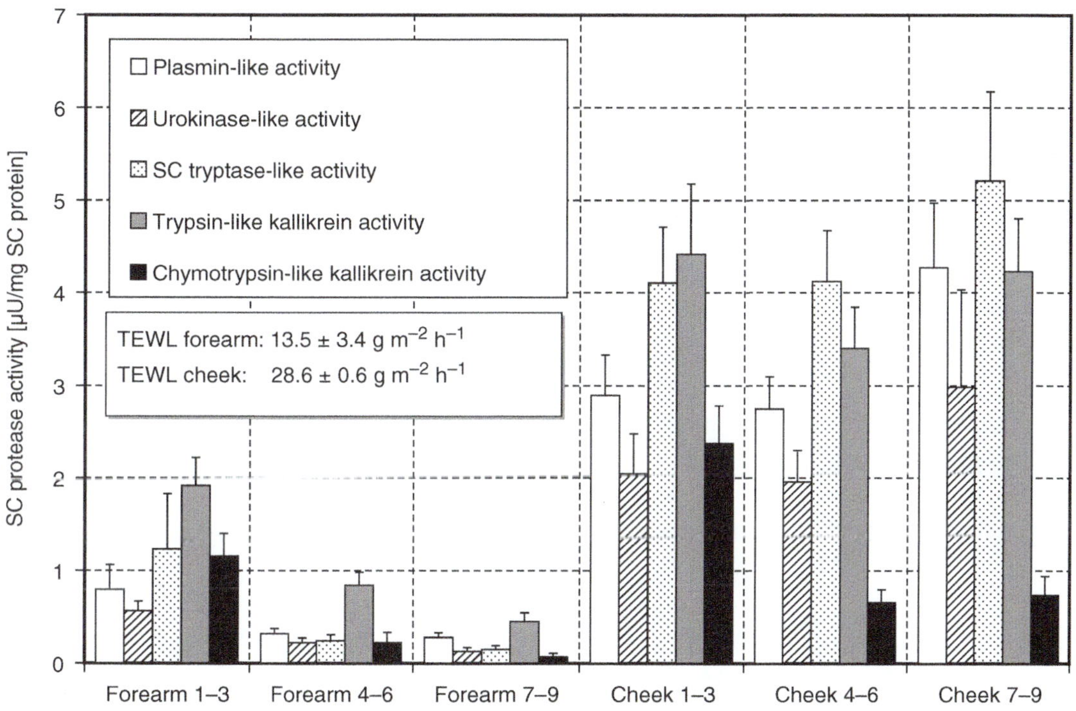

Fig. 28.7 Depth profiling of serine protease activity (mean±SEM, $n=14$ subjects) on tape stripping pools (tapes 1–3, 4–6, 7–9) of healthy forearm and cheek SC. The TEWL of the cheek is twice as much the one of the forearm, indicating an impaired barrier (Modified from [6])

proteolytic activity which leads to premature corneodesmosomal degradation. Facial skin is one of the most exposed skin sites of the body and is submitted to numerous environmental attacks for long periods each day as it is not protected by clothing. Paradoxically, facial skin is particularly sensitive because it possesses a thin epidermis and a thin SC [35].

Table 28.3 (a) Biometric data of the subjects; (b) SC serine protease mass and (c) activities on 15 consecutive tape stripping extracts of healthy (H), nonlesional (NL) and lesional (L) atopic dermatitis skin

(a)					
Biometric data	**Healthy**	**Nonlesional**	**Lesional**		
Local SCORAD	0.0±0.0	0.2±0.4	7.0±1.2		
Irritation score	0.0±0.0	0.0±0.0	5.8±1.8		
Mexameter data	203±60	162±30	323±73*		
Skin surface pH	6.1±0.7	6.4±0.7	6.0±0.4		
SC hydration [AU]	29.6±3.2	26.9±5.0	13.4±1.5*		
TEWL [$g \cdot m^{-2} h^{-1}$] before tape stripping	8.2±1.9	12.5±3.4	41.8±9.5*		
TEWL [$g \cdot m^{-2} h^{-1}$] after 20 tape strippings	34.9±5.5	27.3±6.9	81.3±7.1		
SC thickness [μm]	17 ±1.13	20 ±2.84	10 ±1.80*		
SC cohesion [μg protein cm^{-2}], 20 tape strippings	20.8±1.4	17.8±2.6	7.2±1.7		
(b)	**Protease mass [ng protease/mg SC protein]**				
Serine protease	**Healthy**	**Nonlesional**	**Lesional**	**Factor L/H**	**Factor L/NL**
KLK5 (trypsin-like)	234±69	151±34	190±75	0.8	1.3
KLK7 (chymotrypsin-like)	21.9±5.2	18.3±3.9	62.0±26.0	2.8	3.4
KLK11 (trypsin-like)	7.18±0.80	7.20±1.07	29.0±9.7**	4.0	4.0
KLK14 (trypsin- and chymotrypsin-like)	0.26±0.11	0.28±0.12	0.21±0.09	0.7	0.8
Plasmin	0.54±0.14	1.09±0.41	45.0±10.7**	83	41
Urokinase	0.036±0.007	0.040±0.010	0.052±0.018	1.4	1.3
(c)	**Protease activity [μU/mg SC protein]**				
Serine protease	**Healthy**	**Nonlesional**	**Lesional**	**Factor L/H**	**Factor L/NL**
Trypsin-like kallikreins	1.40±0.27	1.26±0.25	7.29±1.48**	5.2	5.8
Chymotrypsin-like kallikreins	0.84±0.18	0.42±0.09	1.62±0.28*	1.9	3.8
SC tryptase-like	0.71±0.29	0.87±0.32	39.4±10.8*	55	45
Plasmin-like	0.53±0.11	1.22±0.34	36.4±10.4*	69	30
Urokinase-like	0.40±0.13	0.46±0.12	3.25±0.75*	8.1	7.1
Leukocyte elastase-like	0.01±0.01	0.16±0.16	3.13±1.45*	n.a.	n.a.

Modified from [37]
Mean±SEM; $n=6$ per group
*p <0.05; **p <0.01 for L vs. H

28.2.5 Serine Protease Profiling in Acute Eczematous Atopic Skin

Atopic dermatitis (AD) is a chronic inflammatory disease associated with changes in SC structure and function. Paradoxically, the barrier may be less functional both due to a thickening and a thinning of the SC. The breakdown of epidermal barrier function in AD is associated with changes in corneocyte size and maturation, desquamation, lipid profiles together with their packing states and some protease activities. In this latter respect, changes in the expression and activities of serine proteases in the skin have been reported in epidermal samples taken from subjects with AD [36–40].

We have compared physiological changes in acute eczematous lesional skin of AD patients with nonlesional AD skin and skin of healthy subjects on ventral forearms and determined protease mass levels (KLK5, KLK7, KLK11, KLK14, plasmin and uPA) and enzyme activities on 15 consecutive tape strippings together with traditional bioinstrumental approaches and IR densitometry [37, 40]. While TEWL levels were elevated before and after tape stripping in lesional skin compared with nonlesional and healthy skin, the amount of SC removed by sequential tape stripping was decreased in lesional skin, indicating increased intracorneocyte cohesion. By correlating reciprocal TEWL values and the amount of SC removed [41], we estimated a significantly thinner SC in lesional atopic skin compared with both nonlesional and healthy skin (Table 28.3a).

We observed that there were little if no differences in the mass levels of KLK5, KLK14 and uPA between healthy, nonlesional and lesional skin. However, the absolute KLK5 mass levels extracted from the tape strippings were large, whereas KLK14 and uPA levels were very low and close or even at detection limit. Moreover, there was a trend of increasing KLK7 levels (3×), and the levels of KLK11 (4×) and plasmin (83×) were significantly elevated in lesional skin (Table 28.3b). Although mass levels of serine proteases tended to be higher in the deeper SC layers for lesional skin, there was no change in the mass levels of any enzyme measured with increasing depth from either nonlesional or healthy skin (data not shown). Unfortunately we did not succeed to establish useful antibody constructs for leukocyte elastase, tryptase and KLK8. KLK8 has been of interest as it has been suggested to be involved in pathogenesis of inflammatory skin diseases [42].

In contrast to the measurements on the levels of protease mass, the enzyme activities of all the tested SC serine proteases were elevated in lesional skin (Table 28.3c). The order compared to healthy skin was plasmin (69×), SC tryptase-like enzyme (55×), uPA (8×), trypsin-like kallikreins (5×) and chymotrypsin-like kallikreins (2×). Leukocyte elastase showed considerable activity in lesional skin indicating an infiltration of neutrophils, whereas in the SC samples from nonlesional and healthy skin, no activity was found. Analysing these enzymes with depth showed increased extractable protease activity towards the surface of the SC in nonlesional and healthy skin, while conversely, the activities of inflammatory proteases were elevated in the deeper layers of lesional SC, whereas the desquamatory proteases showed no obvious gradient (data not shown).

The mass levels of KLK14 in lesional skin were very low and not elevated compared with nonlesional or healthy skin. These results would suggest that the increases in SC chymotrypsin-like kallikrein activity we observed were probably due to the increased mass levels of KLK7 (Table 28.3b, c) [31, 43]. As we did not measure the mass levels of all SC trypsin-like kallikreins and did not observe increased KLK5 mass levels in lesional skin, other trypsin-like kallikrein activities are probably contributing to the increased SC trypsin-like kallikrein activity. KLK11 would appear to be a candidate but not KLK14.

For the first time we report greatly increased mass levels and activity of plasmin in SC extracts from atopic patients with acute eczema. The correlation between mass levels and activity was statistically highly significant ($R^2=0.822$, $p=10^{-5}$). uPA, the major plasminogen activator, was found to be present at very low concentration within the studied SC layers although uPA-like activity levels were increased in lesional skin. This fact raises the question: is plasminogen activated at the basal layer only by uPA/uPAR or are there other plasminogen activators than uPA within the SC such as KLK5 [44]? Not much is known about the interaction of desquamatory kallikreins and proteases of the thrombostasis axis. However, Yoon et al. [45] demonstrated the potential for important regulatory interactions between these two major protease families. It is hypothesised that activation of proKLK7 and/or proKLK11 by plasmin in AD and vice versa may occur and that these KLKs may play a role in the inflammatory cascade in AD.

As there exists a discrepancy between protease mass levels and the corresponding enzyme activities, the following may explain the reasons why [1]: that antibodies do not differentiate between inactive zymogen, active protease and inactive protease/inhibitor complex and [2] that enzyme substrates may exhibit a certain unspecificity.

The reason for the increased protease mass levels and activities in AD is not known. The study results imply that there is insufficient amount of LEKTIs (lympho-epithelial Kazal-type serine protease inhibitors), or any other inhibitor, to inhibit the elevated protease activity in chronic lesions [46]. Increased expression may occur due to kallikrein gene polymorphisms such as those reported for KLK7 by Vasilopoulous et al. [47]. Also barrier perturbation can lead to increased inflammatory protease levels, e.g. uPA after tape stripping the SC of a mouse model of AD [48]. We have shown SC protease activities are also increased in barrier-compromised conditions even in non-diseased skin such as the face [6, 29]. However, we did not observe any impaired barrier function or increased serine protease activity in nonlesional skin sites, implying that the impairment of barrier function is not intrinsic in subjects with AD as has been reported by Kikuchi et al. [49].

The question remains, do these elevated protease levels impair SC or epidermal functioning? Statistically significantly increased irritation, increased TEWL and decreased hydration levels were found in lesional skin. It is highly likely that the elevated levels of serine proteases in patients with active lesions were contributing to their apparently thinner SC. The increased cohesion in the remaining SC probably indicates that the desquamation is occurring above the stratum compactum and not at the stratum compactum/ stratum granulosum interface. The thinning of the SC in atopic eczema was established using IR densitometry for the first time.

Conclusion

IR densitometry is an easy to use and nondestructive technique for the convenient measurement of the optical absorption of SC tape strippings which has been shown to be linearly proportional to their protein content. Thus the corresponding SC densitometric-protein or SC weight content calibration curves can be used for a fast indirect evaluation of tape strippings. As this is a nondestructive method, the unmodified tapes can then be used for further investigations. Absorption measurement of tape strippings in the IR range (850 nm) seems to be less interference prone than pseudo-absorbance measurements in the visible light range at 430 nm. It is also easier to perform than gravimetrical tape stripping evaluation. The method is of tremendous value to researchers using tape stripping in both in vivo and also in vitro applications. Although SquameScan 850 is an indirect measure for SC proteins leading to semi-quantitative data, its advantages are in its speed of analysis and its nondestructive nature. However, even though there are indications that the data distribution in different subject groups is quite similar, the variation of data in more heterogeneous study populations in context of age, body area, race and different skin conditions and diseases remains to be investigated.

References

1. Rawlings AV, Watkinson A, Rogers J, Mayo A, Hope J, Scott IR (1994) Abnormalities in stratum corneum structure, lipid composition, and desmosome degradation in soap-induced winter xerosis. J Soc Cosmet Chem 45:203–220
2. Simon M, Bernard D, Minondo A-M, Camus C, Fiat F, Corcuff P et al (2001) Persistence of both peripheral and non-peripheral corneodesmosomes in the upper stratum corneum of winter xerosis skin versus only peripheral in normal skin. J Invest Dermatol 116(1):23–30
3. Rawlings AV (2003) Trends in stratum corneum research and the management of dry skin conditions. Int J Cosmet Sci 25(1–2):63–95
4. Harding CR, Watkinson A, Rawlings AV, Scott IR (2000) Dry skin, moisturization and corneodesmolysis. Int J Cosmet Sci 22(1):21–52
5. Declercq L, Muizzuddin N, Hellemans L, Van Overloop L, Sparacio R, Marenus K et al (2002) Adaptation response in human skin barrier to a hot and dry environment. J Invest Dermatol 119(3):716
6. Voegeli R, Rawlings AV, Doppler S, Heiland J, Schreier T (2007) Profiling of serine protease activities in human stratum corneum and detection of a stratum corneum tryptase-like enzyme. Int J Cosmet Sci 29(3):191–200
7. Dreher F, Arens A, Hostynek JJ, Mudumba S, Ademola J, Maibach HI (1998) Colorimetric method for quantifying human stratum corneum removed by adhesive-tape stripping. Acta Derm Venereol 78(3):186–189
8. Dreher F, Modjtahedi BS, Modjtahedi SP, Maibach HI (2005) Quantification of stratum corneum removal by

adhesive tape stripping by total protein assay in 96-well microplates. Skin Res Technol 11(2):97–101
9. Weigmann HJ, Lademann J, Schanzer S, Lindemann U, von Pelchrzim R, Schaefer H et al (2001) Correlation of the local distribution of topically applied substances inside the stratum corneum determined by tape-stripping to differences in bioavailability. Skin Pharmacol Appl Skin Physiol 14(Suppl 1): 98–102
10. Weigmann H, Lademann J, Meffert H, Schaefer H, Sterry W (1999) Determination of the horny layer profile by tape stripping in combination with optical spectroscopy in the visible range as a prerequisite to quantify percutaneous absorption. Skin Pharmacol Appl Skin Physiol 12(1–2):34–45
11. Jacobi U, Kaiser M, Richter H, Audring H, Sterry W, Lademann J (2005) The number of stratum corneum cell layers correlates with the pseudo-absorption of the corneocytes. Skin Pharmacol Physiol 18(4):175–179
12. Lademann J, Ilgevicius A, Zurbau O, Liess HD, Schanzer S, Weigmann HJ et al (2006) Penetration studies of topically applied substances: optical determination of the amount of stratum corneum removed by tape stripping. J Biomed Opt 11(5):054026
13. Lindemann U, Weigmann HJ, Schaefer H, Sterry W, Lademann J (2003) Evaluation of the pseudo-absorption method to quantify human stratum corneum removed by tape stripping using protein absorption. Skin Pharmacol Appl Skin Physiol 16(4):228–236
14. Alikhan A, Maibach H (2010) Biology of stratum corneum: tape stripping and protein quantification. In: Farage MA, Miller KW, Maibach H (eds) Textbook of aging skin. Springer, Berlin
15. Bornkessel A, Flach M, Arens-Corell M, Elsner P, Fluhr JW (2005) Functional assessment of a washing emulsion for sensitive skin: mild impairment of stratum corneum hydration, pH, barrier function, lipid content, integrity and cohesion in a controlled washing test. Skin Res Technol 11(1):53–60
16. Voegeli R, Heiland J, Doppler S, Rawlings AV, Schreier T (2007) Efficient and simple quantification of stratum corneum proteins on tape strippings by infrared densitometry. Skin Res Technol 13(3):242–251
17. Fluhr JW, Kao J, Jain M, Ahn SK, Feingold KR, Elias PM (2001) Generation of free fatty acids from phospholipids regulates stratum corneum acidification and integrity. J Invest Dermatol 117(1):44–51
18. Hahn T, Hansen S, Neumann D, Kostka KH, Lehr CM, Muys L et al (2010) Infrared densitometry: a fast and non-destructive method for exact stratum corneum depth calculation for in vitro tape-stripping. Skin Pharmacol Physiol 23(4):183–192
19. Melero A, Hahn T, Schaefer UF, Schneider M (2011) In vitro human skin segmentation and drug concentration-skin depth profiles. Methods Mol Biol 763:33–50
20. Boncheva M, de Sterke J, Caspers PJ, Puppels GJ (2009) Depth profiling of stratum corneum hydration in vivo: a comparison between conductance and confocal Raman spectroscopic measurements. Exp Dermatol 18(10):870–876
21. Klang V, Schwarz JC, Hartl A, Valenta C (2011) Facilitating in vitro tape stripping: application of infrared densitometry for quantification of porcine stratum corneum proteins. Skin Pharmacol Physiol 24(5):256–268
22. Carrer DC, Vermehren C, Bagatolli LA (2008) Pig skin structure and transdermal delivery of liposomes: a two photon microscopy study. J Control Release 132(1):12–20
23. Klang V, Schwarz JC, Lenobel B, Nadj M, Aubock J, Wolzt M et al (2012) In vitro vs. in vivo tape stripping: Validation of the porcine ear model and penetration assessment of novel sucrose stearate emulsions. Eur J Pharm Biopharm 80(3):604–614
24. Melero A, Garrigues TM, Alos M, Kostka KH, Lehr CM, Schaefer UF (2009) Nortriptyline for smoking cessation: release and human skin diffusion from patches. Int J Pharm 378(1–2):101–107
25. Lademann J, Weigmann H-J, Schanzer S, Meinke M, Sterry W, Patzelt A (2010) Analysis of the penetration of topically applied substances into the human skin. SÖFW 136(11):2–8
26. Kezic S, Kammeyer A, Calkoen F, Fluhr JW, Bos JD (2009) Natural moisturizing factor components in the stratum corneum as biomarkers of filaggrin genotype: evaluation of minimally invasive methods. Br J Dermatol 161(5):1098–1104
27. Angelova-Fischer I, Mannheimer AC, Hinder A, Ruether A, Franke A, Neubert RH et al (2011) Distinct barrier integrity phenotypes in filaggrin-related atopic eczema following sequential tape stripping and lipid profiling. Exp Dermatol 20(4):351–356
28. Date A, Shimakura T, Sasaki M, Yamaguchi M (2012) An analytical technique for measuring protein carbonyl in the stratum corneum using surface Plasmon resonance. Int J Cosmet Sci 34(1):81–85
29. Voegeli R, Rawlings AV, Doppler S, Schreier T (2008) Increased basal transepidermal water loss leads to elevation of some but not all stratum corneum serine proteases. Int J Cosmet Sci 30(6):435–442
30. Ekholm IE, Brattsand M, Egelrud T (2000) Stratum corneum tryptic enzyme in normal epidermis: a missing link in the desquamation process? J Invest Dermatol 114(1):56–63
31. Brattsand M, Stefansson K, Lundh C, Haasum Y, Egelrud T (2005) A proteolytic cascade of kallikreins in the stratum corneum. J Invest Dermatol 124(1):198–203
32. Chang-Yi C, Takahashi M, Tezuka T (1997) 30-kDa trypsin-like proteases in the plantar stratum corneum. J Dermatol 24(8):504–509
33. Hadgraft J, Lane ME (2009) Transepidermal water loss and skin site: a hypothesis. Int J Pharm 373(1–2):1–3
34. Machado M, Hadgraft J, Lane ME (2010) Assessment of the variation of skin barrier function with anatomic site, age, gender and ethnicity. Int J Cosmet Sci 32:397–409

35. Proksch E (2008) Protection against dryness of facial skin: a rational approach. Skin Pharmacol Physiol 22(1):3–7
36. Komatsu N, Saijoh K, Kuk C, Liu AC, Khan S, Shirasaki F et al (2007) Human tissue kallikrein expression in the stratum corneum and serum of atopic dermatitis patients. Exp Dermatol 16(6):513–519
37. Voegeli R, Rawlings AV, Breternitz M, Doppler S, Schreier T, Fluhr JW (2009) Increased stratum corneum serine protease activity in acute eczematous atopic skin. Br J Dermatol 161:70–77
38. Redoules D, Tarrouxa R, Assalita MF, Perié JJ (1999) Characterisation and assay of five enzymatic activities in the stratum corneum using tape-strippings. Skin Pharmacol Appl Skin Physiol 12(4):182–192
39. Tarroux R, Assalit MF, Licu D, Périé JJ, Redoulès D (2002) Variability of enzyme markers during clinical regression of atopic dermatitis. Skin Pharmacol Appl Skin Physiol 15:55–62
40. Voegeli R, Doppler S, Joller P, Breternitz M, Fluhr JW, Rawlings AV (2011) Increased mass levels of certain serine proteases in the stratum corneum in acute eczematous atopic skin. Int J Cosmet Sci 33(6):560–565
41. Kalia YN, Albertia I, Naika A, Guya RH (2001) Assessment of topical bioavailability in vivo: the importance of stratum corneum thickness. Skin Pharmacol Appl Skin Physiol 14(Suppl 1):82–86
42. Kishibe M, Bando Y, Terayama R, Namikawa K, Takahashi H, Hashimoto Y et al (2007) Kallikrein 8 is involved in skin desquamation in cooperation with other kallikreins. J Biol Chem 282(8):5834–5841
43. Stefansson K, Brattsand M, Ny A, Glas B, Egelrud T (2006) Kallikrein-related peptidase 14 may be a major contributor to trypsin-like proteolytic activity in human stratum corneum. Biol Chem 387(6):761–768
44. Debela M, Beaufort N, Magdolen V, Schechter NM, Craik CS, Schmitt M et al (2008) Structures and specificity of the human kallikrein-related peptidases KLK 4, 5, 6, and 7. Biol Chem 389(6):623
45. Yoon H, Blaber SI, Evans DM, Trim J, Juliano MA, Scarisbrick I et al (2008) Activation profiles of human kallikrein-related peptidases by proteases of the thrombostasis axis. Protein Sci 17:1998–2007
46. Roedl D, Traidl-Hoffmann C, Ring J, Behrendt H, Braun-Falco M (2009) Serine protease inhibitor lymphoepithelial kazal type-related inhibitor tends to be decreased in atopic dermatitis. J Eur Acad Dermatol Venereol 23(11):1263–1266
47. Vasilopoulos Y, Cork MJ, Murphy R, Williams HC, Robinson DA, Duff GW et al (2004) Genetic association between an AACC insertion in the 3'UTR of the stratum corneum chymotryptic enzyme gene and atopic dermatitis. J Invest Dermatol 123(1):62–66
48. Descargues P, Deraison C, Prost C, Fraitag S, Mazereeuw-Hautier J, D'Alessio M et al (2006) Corneodesmosomal cadherins are preferential targets of stratum corneum trypsin- and chymotrypsin-like hyperactivity in netherton syndrome. J Invest Dermatol 126(7):1622–1632
49. Kikuchi K, Kobayashi H, O'goshi K-I, Tagami H (2006) Impairment of skin barrier function is not inherent in atopic dermatitis patients: a prospective study conducted in newborns. Pediatr Dermatol 23(2):109–113

Part IV

Skin Mechanics

Cutometer®

29

Hristo Dobrev

29.1 Introduction

The human skin possesses a complex structure and various functions which ensure the entity between the organism and the environment. Mechanical properties of the skin are of major importance for its protective function. They vary in accordance with age, sex and body sites, in some physiological and pathological skin conditions, and change due to different external and therapeutic influences. Considerable progress in the quantification of the skin mechanical functions had been achieved for the past 20 years through the introduction of modern non-invasive bioengineering methods and devices which provide the researchers with objective, quantitative, sensitive and reproducible measurements in vivo.

The Cutometer® (Courage + Khazaka Electronic GmbH, Cologne, Germany) is a well-recognized commercial device for measurement of the biomechanical properties of the skin. In this chapter, based on our own experience and review of the literature, we aimed to present the practical application of the Cutometer® method.

H. Dobrev, MD, PhD
Clinic of Dermatology and Venereology, University Hospital "St. George", 15A V. Aprilov St., 4002, Plovdiv, Bulgaria

Department of Dermatology and Venereology, Medical University, Plovdiv, Bulgaria
e-mail: hristo_dobrev@hotmail.com

29.2 Measuring Equipment and Principle

29.2.1 Equipment

The Cutometer® is designed as separate or combined device.

1. The Cutometer® SEM 575 (Fig. 29.1), which is the successor of the Cutometer® SEM 474, is an independent device that includes two basic parts [7, 9, 40, 70].
   - Main unit, a metal housing containing vacuum pump with pressure sensor and microelectronics. The load of the vacuum (negative air pressure), the rate of its increase or decrease, the duration of suction (on time) and relaxation (off time) and the number of measuring cycles during one measurement can be defined through the Cutometer® software. The resolution of applied pressure is equal to 1 mbar.
   - Measuring probe, a handheld probe containing suction head, optical measuring system and microelectronics. The suction head is centred in the probe and has a standard circular aperture of 2-mm diameter (test area of about 3 mm^2). Optional probes with apertures of 4, 6 and 8 mm are available on request. The measuring system inside the probe head is noncontact and consists of a light source (light-emitting diode) and a light recipient, as well as two opposing glass prisms, which project the

E. Berardesca et al. (eds.), *Non Invasive Diagnostic Techniques in Clinical Dermatology*,
DOI 10.1007/978-3-642-32109-2_29, © Springer Berlin Heidelberg 2014

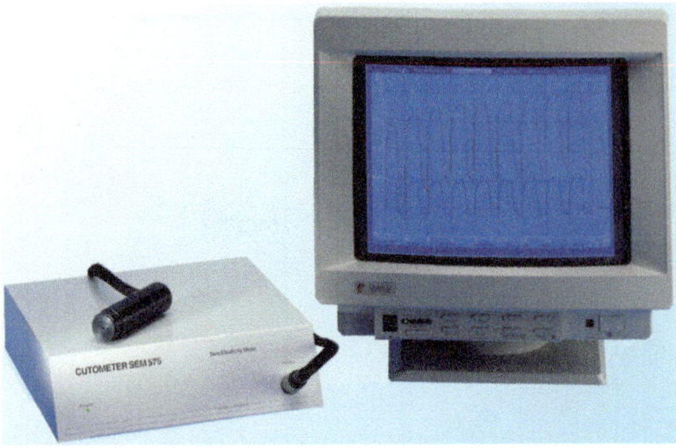

Fig. 29.1 Cutometer® SEM 575 (With permission from Courage and Khazaka GmbH, Cologne, Germany)

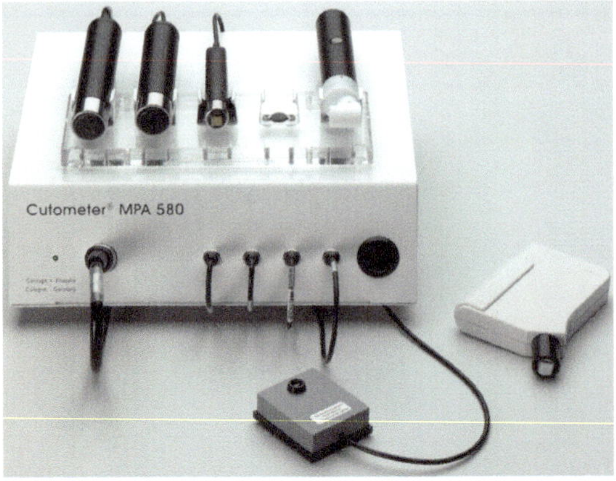

Fig. 29.2 Cutometer® MPA 580 (With permission from Courage and Khazaka GmbH, Cologne, Germany)

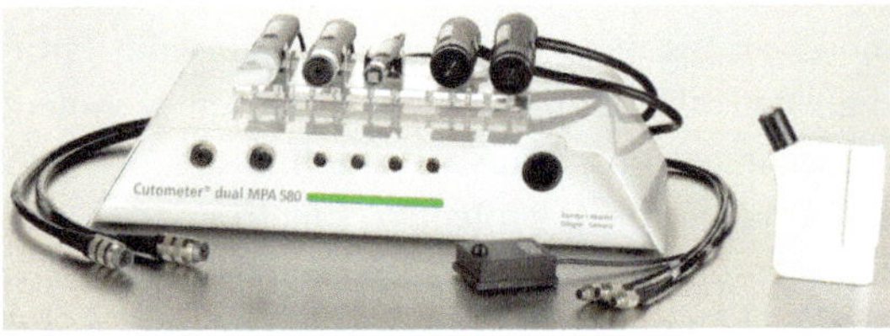

Fig. 29.3 Cutometer® Dual MPA 580 (With permission from Courage and Khazaka GmbH, Cologne, Germany)

light from emitter to recipient. The changes of the infrared light beam intensity during the measurement are converted into millimetres (from 0 to 3.0 mm) and calculated with a resolution equal to 2 μm. The measuring probe is connected to the main unit through an air tube and an electric cable. During the measurement, it is held perpendicularly to the skin surface under constant pressure ensured (provided) by an elastic spring. The calibration data are stored in the memory of each probe which allows fast and easy exchange of probes with different apertures.

2. The Cutometer® MPA 580 (Fig. 29.2), that is currently available, combines the vacuum box for the Cutometer® with the built-in Sebumeter SM 815 and modular Multi Probe Adapter System which makes possible to connect up to 4 additional measuring probes as well as an ambient condition sensor. The Cutometer® Dual MPA 580 (Fig. 29.3) is the last generation device that allows the connection of two Cutometer® probes (with different diameters) at the same time.

29.2.2 Principle

The measuring principle of the Cutometer® is based on a suction method that consists of the measurement of vertical deformation of the skin surface after application of vacuum (Fig. 29.4) [7, 9, 40, 70].

A defined negative air pressure is applied perpendicular to the skin through the opening of the probe for a selected time period. The evaluated skin surface is sucked into the aperture of the probe, and the resulted vertical deformation is measured by the optical measuring system inside the probe. The changes of light intensity are proportionally related to the penetration depth of the skin and are displayed on the monitor as curves in a coordinate system (extension/time or pressure/extension).

29.3 Software

The Cutometer® is designed for operation with an IBM-compatible PC via USB port. The last version of the Windows software allows storage of various data regarding the volunteer, date and time of experiment, skin area, external temperature and relative humidity, type of probe used and mode of measuring technique. The obtained

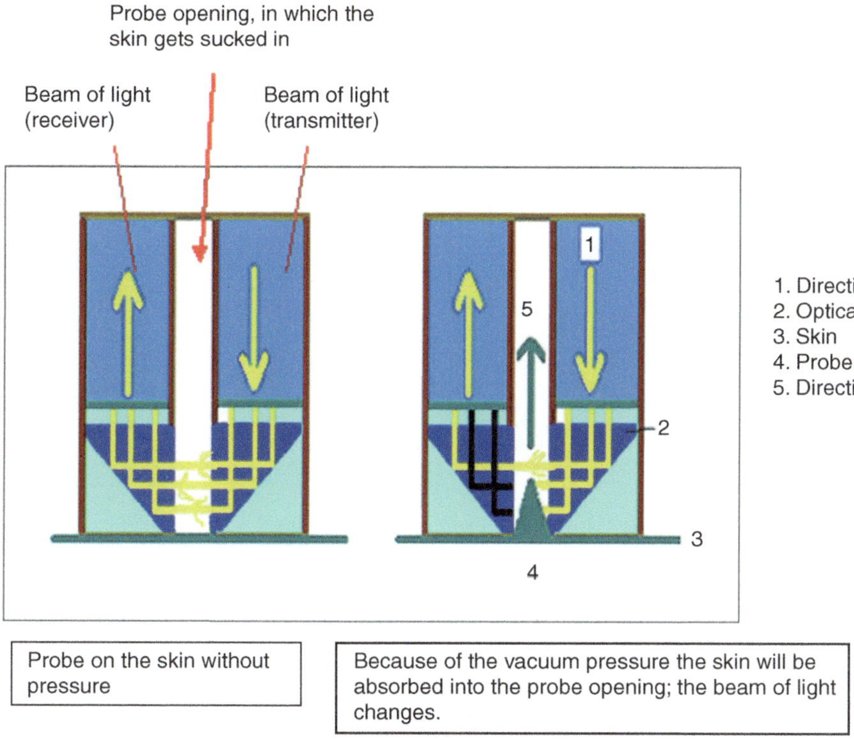

Fig. 29.4 Schematic view of the measuring system within the Cutometer® probe [56] (With permission from Courage and Khazaka GmbH, Cologne, Germany)

results are automatically calculated and displayed as curves and values [56].

The Cutometer® measuring cycle consists of suction phase and relaxation phase. It can be applied once or several times. The following parameters of the measurement can be exactly defined by the software:

- Pressure. The load of the air-negative pressure (vacuum) can be chosen between 20 and 500 mbar.
- Rate. The rate of increase or decrease of the air-negative pressure can be selected between 10 and 100 mbar/s.
- On-Time. The time when the air-negative pressure is applied (suction interval) can be chosen between 0.1 and 60 s.
- Off-Time. The time when the air-negative pressure is applied no longer (relaxation interval) can be chosen between 0.1 and 60 s.
- Repetition. The number of measuring cycles (suctions) included in one measurement can be varying between 1 and 99.
- Pre-Time. This function allows setting a short interval between pressing the start key and the beginning of the measurement.
- Preconditioning Time. This function allows (in the strain–time mode) to pretension the skin by applying a preliminary suction during a short time (0.1 s) before the real measurement is carried out.

The most used settings include air-negative pressure between 400 and 450 mbar, on-time and off-time intervals between 2 and 5 s and repetitions between 1 and 10.

29.3.1 Measuring Techniques

There are two measuring techniques available [7, 9, 56]:

1. In the strain–time mode, the deformation of the skin (in millimetres) is showed as a function of time (in seconds). This mode is mostly used in research studies.

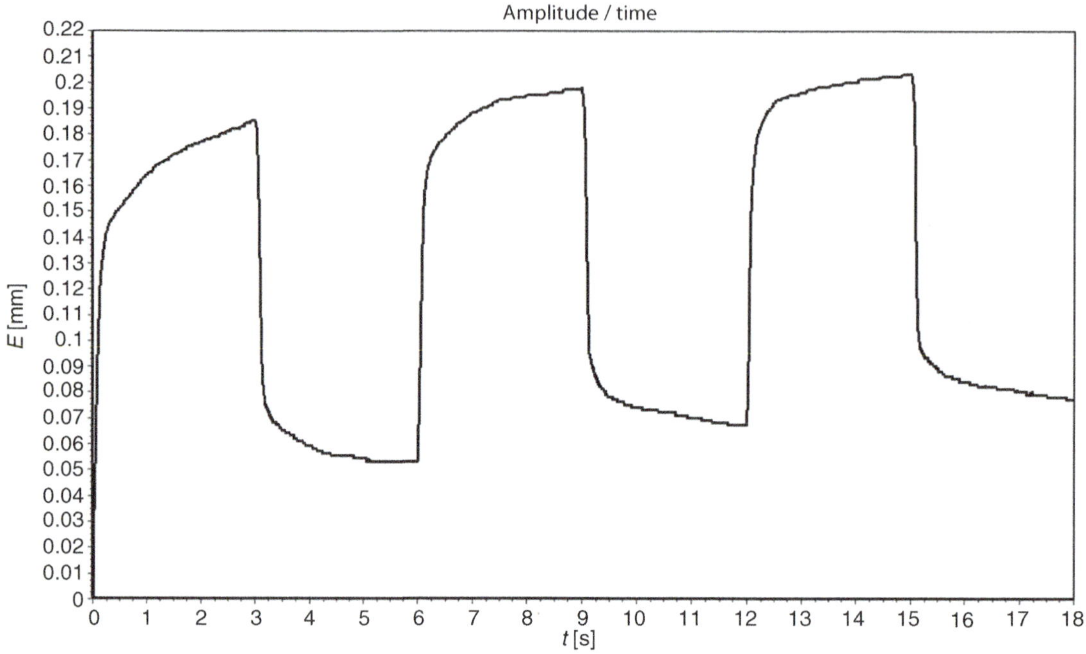

Fig. 29.5 An example of a curve obtained in mode 1 (*E*/*T* mode; extension/time)

2. In the stress–strain mode, the deformation of the skin (in millimetres) is showed as a function of the vacuum (in millibar).

29.3.2 Measuring Modes

There are four measuring modes available based on different combination of the measurement parameters [7, 9, 56].

29.3.2.1 Mode 1: Measurement with Constant Negative Pressure (–)

The measurement cycle consists of suction and relaxation phases. During the first phase, the skin is drawn into the probe with constant negative pressure set under "Pressure" within the interval set under "On-Time". In the second phase, the negative pressure is switched off, and the relaxation of the skin is determined within the interval set under "Off-Time". With the command "Repetition", the number of measuring cycles can be chosen. The skin deformation is displayed as a function of time (Fig. 29.5). The measurement mode 1 is most important and is predominantly used for research studies in the field of dermatology and cosmetics.

29.3.2.2 Mode 2: Measurement with Linear Increase and Linear Decrease in Negative Pressure (/\)

The measurement cycle consists of three phases. At the start, the negative pressure is zero. During the first phase, the skin is drawn into the probe with linearly increasing negative pressure set under commands "Pressure" and "Rate" in the menu "Parameter". This phase is succeeded by linearly decreasing negative pressure. In the last phase, the skin properties are evaluated when no negative pressure is applied for an interval set under "Off-Time". With the command "Repetition" the number of measuring cycles can be chosen. The skin deformation can be displayed both as a function of time (Fig. 29.6a) and as a function of negative pressure (Fig. 29.6b).

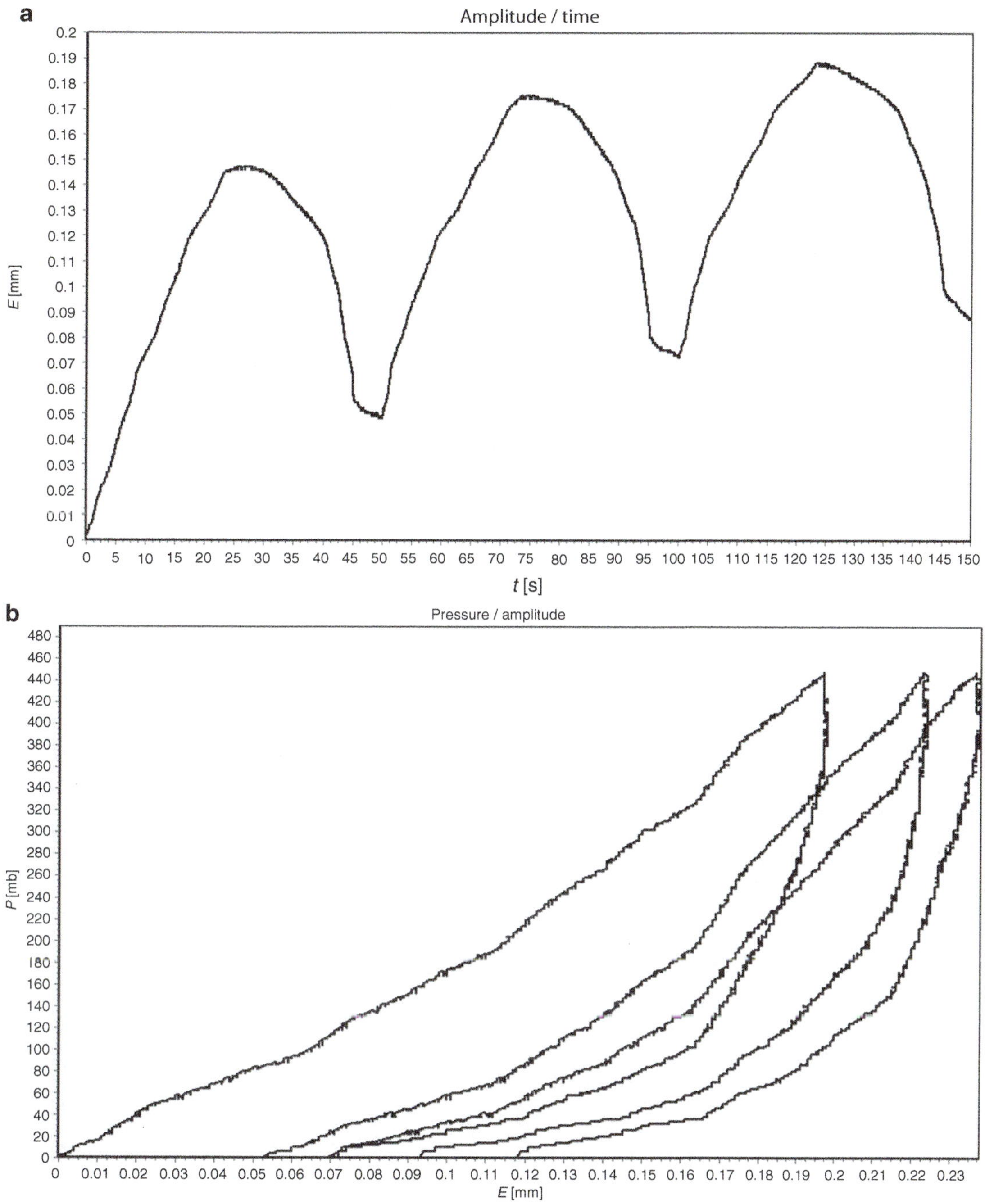

Fig. 29.6 (**a**) An example of a curve obtained in mode 2 (*E*/*T* mode; extension/time). (**b**) An example of a curve obtained in mode 2 (*P*/*E* mode; pressure/extension)

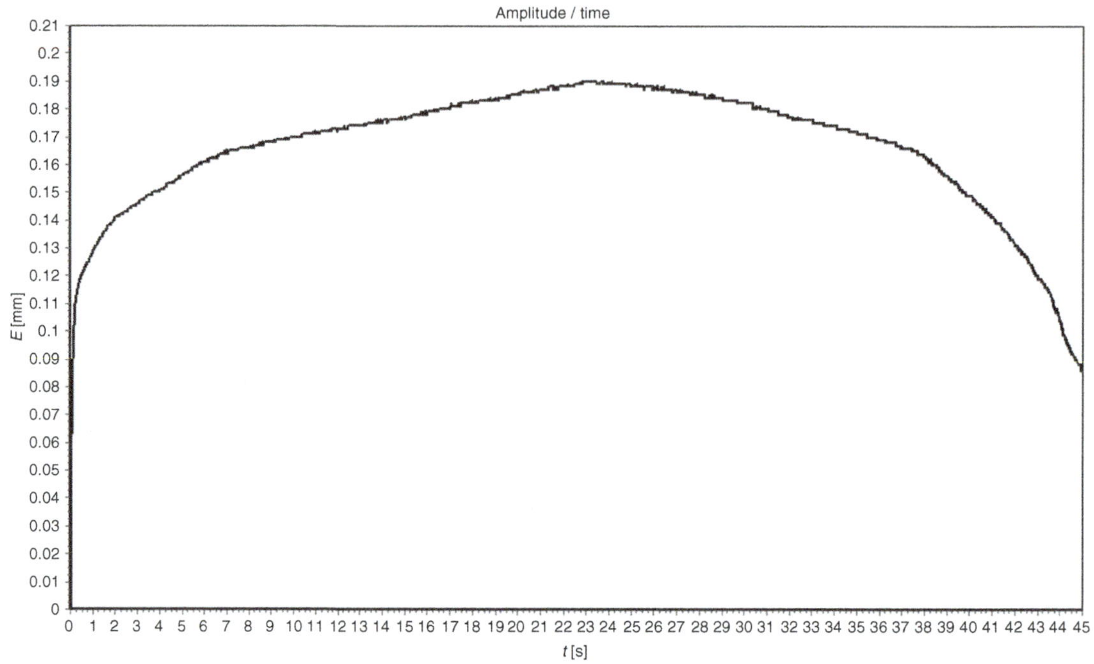

Fig. 29.7 An example of a curve obtained in mode 3 (*E*/*T* mode; extension/time)

29.3.2.3 Mode 3: Measurement with First Constant and Then Linear Decrease in Negative Pressure (–\)

The measurement consists of one measuring cycle which combines the first phase of mode 1 and the second phase of mode 2. The "On-Time" can be selected, whereas the "Off-Time" results from the selected "Pressure" and "Rate". A relaxation time (here "Off-Time") cannot be set as repetitions due to the combination of full pressure and slight release are impossible. The skin deformation is displayed as a function of time (Fig. 29.7).

29.3.2.4 Mode 4: Measurement with a Linear Increase in Negative Pressure and Then a Sudden Cessation of the Negative Pressure (/)

The measurement consists of one measuring cycle which combines the first phase of mode 2 and the second phase of mode 1. The "On-Time" results from the selected "Pressure" and "Rate", whereas the "Off-Time" can be set. A repetition is impossible due to the combination of slightly increasing pressure and full release. The skin deformation is displayed as a function of time (Fig. 29.8).

In the literature measuring modes 2, 3 and 4 have no significance. There are few studies using measuring mode 2. According to Dobrev [18], the application of measuring mode 2 does not provide any advantage and only burdens the vacuum pump of the device.

29.3.3 Skin Mechanical Parameters

The new version of Cutometer® MPA 580Q software (v.1.3.6.16) allows calculating three groups of skin mechanical parameters designated as "R-parameters", "F-parameters" and "Q-parameters". In addition, there is a possibility for advanced users to edit the calculation formula for "R-parameters" according to their needs [56].

In this section, we describe in details the mechanical parameters derived from skin deformation curves obtained using measurement mode 1. Most detailed information concerning

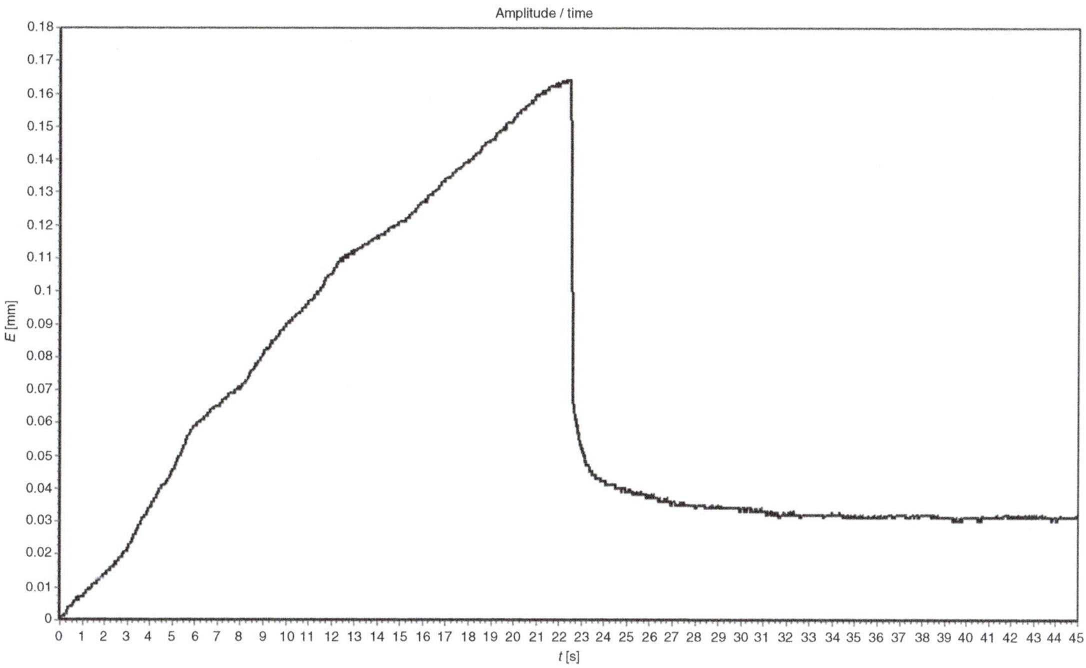

Fig. 29.8 An example of a curve obtained in mode 4 (*E*/*T* mode; extension/time)

the mechanical parameters determined by measurement modes 2, 3 and 4 can be obtained from the manufacturer's Information and Operating Instructions for the Cutometer® [56].

29.3.3.1 Mode 1 (Single Strain–Time Curve)

R-Parameters

The skin deformation curve obtained with Cutometer® includes two main parts generated during the suction phase and relaxation phase, respectively (Fig. 29.9). Each of them is composed of rapid deformation representing an elastic section, followed by a viscoelastic and finally a viscous section. These parts are designated as follows: immediate deformation (Ue), delayed deformation (Uv), immediate retraction (Ur) and delayed retraction (Ua–Ur). Values of Ue and Ur are taken at fixed intervals of time, respectively, 0.1 s after application of a suction and 0.1 s after removal of negative pressure [5, 7, 9, 56].

Based on these parts, the Cutometer® software calculates automatically the following parameters:

R0 = Uf, the final deformation (skin distensibility or skin extensibility).

R1 = Uf − Ua, the residual deformation at the end of 1st measuring cycle (resilient distension).

R2 = Ua/Uf, the ratio of total retraction to total deformation, which is called gross elasticity of the skin, including viscous deformation = gross elasticity, including viscous deformation) (overall elasticity).

R5 = Ur/Ue, the net elasticity without viscous deformation.

R6 = Uv/Ue, the ratio between delayed and immediate deformation, which indicates the relative contributions of the viscoelastic plus viscous and the elastic distension to the total deformation (viscoelastic ratio, the ratio of viscoelastic to elastic distension).

R7 = Ur/Uf, the ratio of immediate retraction to the total deformation, which is called biological elasticity (the ratio of immediate retraction to total distension).

R8 = Ua, the final retraction after removal of the vacuum (total recovery of the skin).

F-Parameters

The software calculates two surfaces ("areas") designated as F-parameters: (Fig. 29.8)

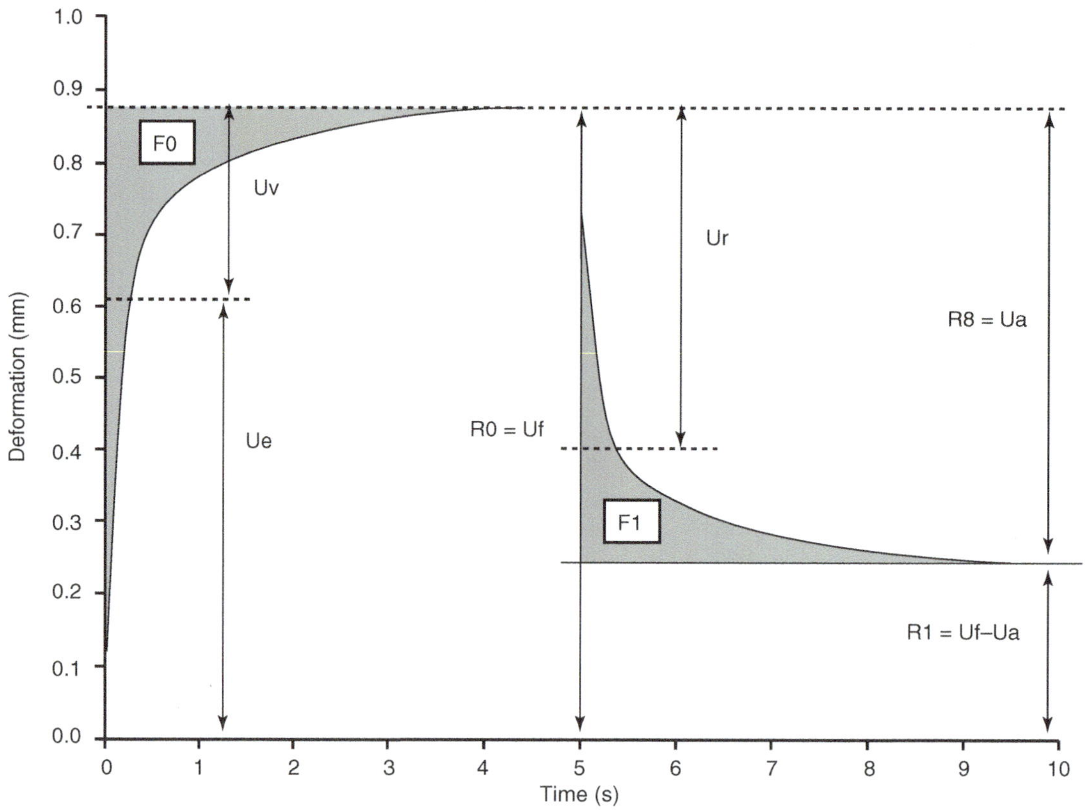

Fig. 29.9 Skin deformation curve obtained with CutoMPA 580. Aperture, 2 mm; suction time, 5 s, relaxation time, 5 s, repetition, 1

F0 = the surface between the real curve and the value corresponding to the maximal deformation Uf when going from start of suction to cease of suction.

F1 = the surface between the real recovery curve and the value corresponding to the maximal recovery going from cease of suction to cease of measurement.

Q-Parameters

The Q-parameters are developed by the scientist Di Qu et al. [56, 83] and have recently been added in the calculation formula. They could only be obtained in mode 1 for an equal suction and relaxation time. To receive these parameters, two horizontal lines have to be spread at the graph – the first one spreads through the highest point (R0), and the second one spreads through the inflexion point that is the point in time at which the recovery curve deviates from its initial linearity. At this point, the as QE and QR are divided (Fig. 29.10).

The Q-parameters include:

Q0 = the maximum recovery area, i.e. the area under the highest point (R0).

QE = the elastic recovery area of the skin.

QV = the viscous recovery area of the skin.

The parameters calculated by the Cutometer® software include:

Q0 = Q0; the maximum recovery area, i.e. the area under the highest point (R0).

Q1 = QE/Q0; the elastic recovery of the skin.

Q2 = Q2 = QV/Q0; the viscous recovery of the skin.

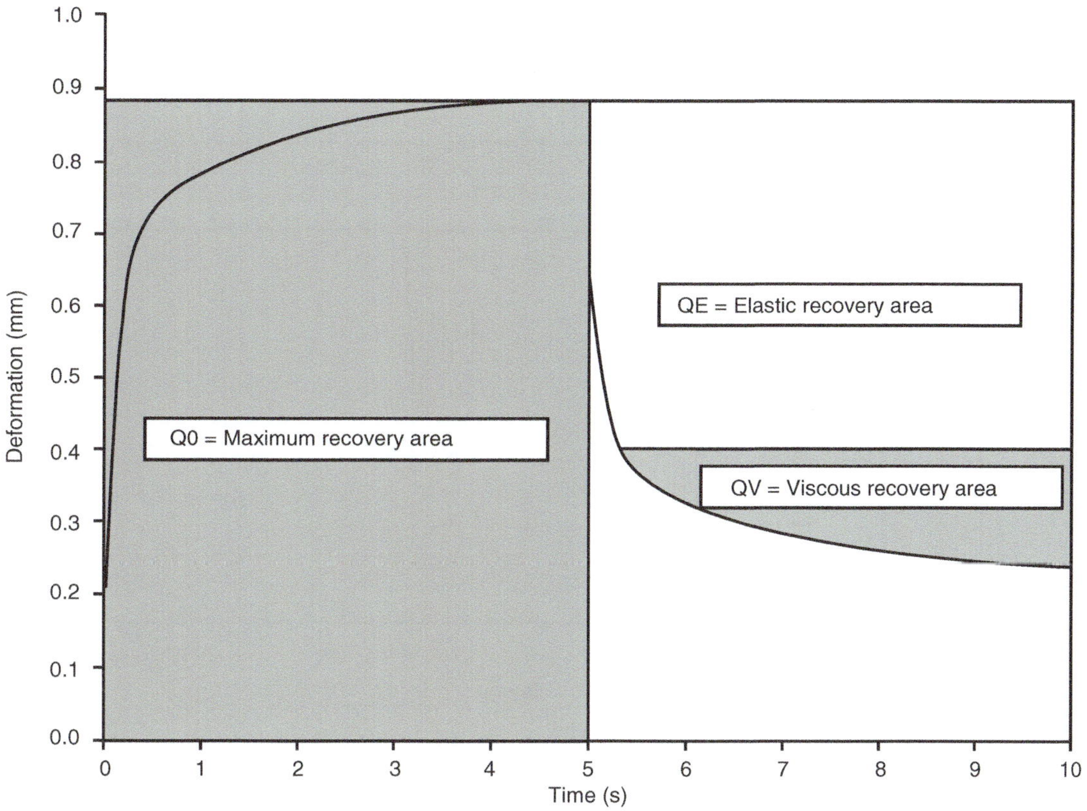

Fig. 29.10 Skin deformation curve obtained with CutoMPA 580. Aperture, 2 mm; suction time, 5 s, relaxation time, 5 s, repetition, 1

Q3 = (QE + QV)/Q0; the total viscoelastic recovery of the skin, i.e. overall skin elasticity.

29.3.3.2 Mode 1 (Repetitive Strain–Time Curve)

When repetitive suctions are applied, the subsequent curves are similar to the first one. However, they are progressively shifted vertically upward as a consequence of the slow return of the skin to the original state (Fig. 29.11) [56].

R-Parameters

The following additional parameters are calculated by the software:

R3 = last maximum amplitude (last maximal deformation).

R4 = last minimum amplitude (last residual deformation).

R9 = R3 − R0, the difference in maximal skin deformation between the last and the first suction called hysteresis (H).

F-Parameters

For curves taken in mode 1 with a minimum of 10 repetitions, the software calculates three additional surface parameters ("areas"). For this purpose, the curves are wrapped with an "envelope" function. Above and below the curves, the envelope curves appear as logarithmical average of maximum and minimum amplitudes. The F-parameters are designated as follows (Fig. 29.12):

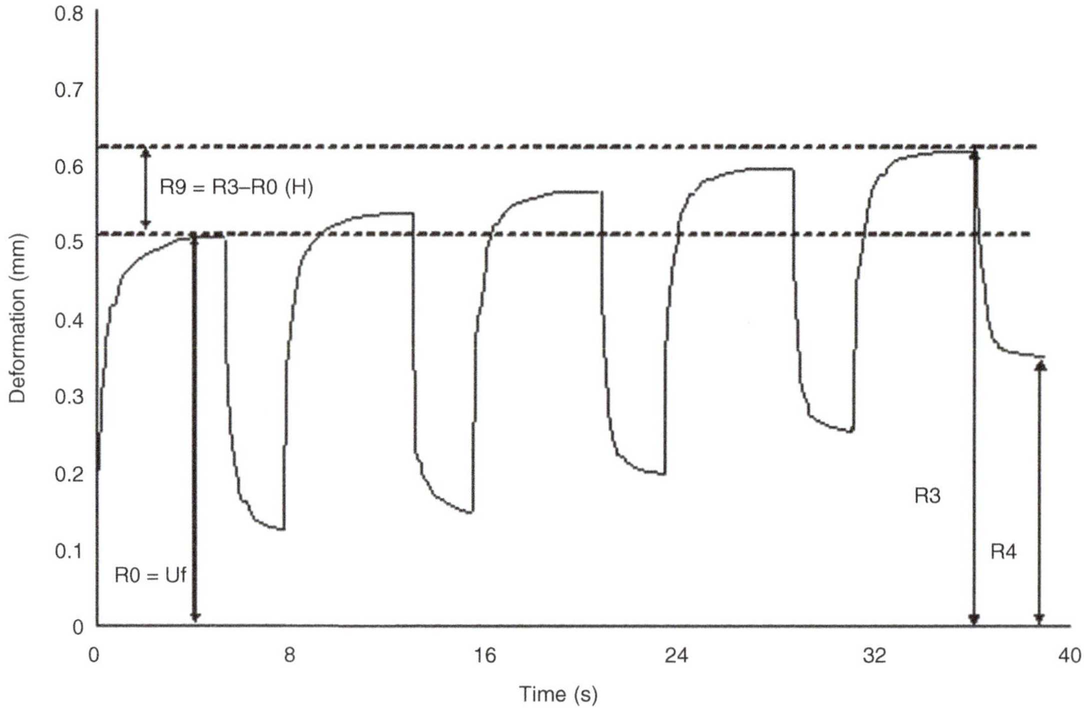

Fig. 29.11 Skin deformation curve obtained with CutoMPA 580. Aperture, 2 mm; suction time, 4 s, relaxation time, 2 s, repetition, 5

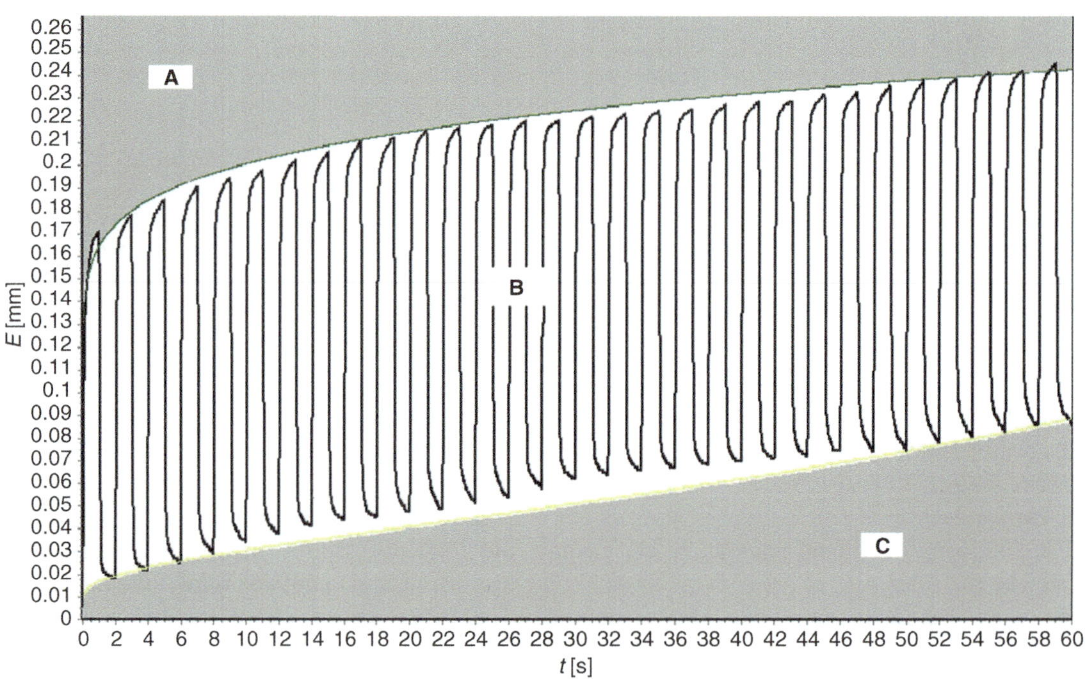

Fig. 29.12 Skin deformation curve obtained with CutoMPA 580. Aperture, 2 mm; suction time, 1 s, relaxation time, 1 s, repetition, 30 (F2 = area A; F3 = area B; F4 = area B + area C)

F2 = area above the upper envelope curve (the surface between the real curve and the value corresponding to the maximal deformation R3 after 10 cycles when going from start of suction to cease of the 10 cycles).

F3 = area within the envelope curves (the surface between the repetitive curves).

F4 = area below the upper envelope curve (the complete area, limited by the upper envelope curve).

29.4 Use of Cutometer®

29.4.1 Factors Influencing Measurements

29.4.1.1 Probe Aperture

At a constant vacuum, the degree of skin deformation (i.e. absolute parameters) directly correlates with the aperture diameter of the measuring probe and inversely correlates with the skin thickness. When using identical vacuum, the values of the absolute parameters Ue, Uv, Uf, Ur and R measured with 2-mm diameter probe are lower compared to values measured with 8-mm diameter probe. The differences in the values of relative parameters Ua/Uf, Ur/Ue, Ur/Uf and Uv/Ue measured with both probes are minimal [5, 26, 74].

The small aperture (2-mm diameter) measuring probe determines the mechanical properties of the epidermis and partly of the papillary dermis. It is applicable at any anatomical region and is most appropriate for measurement of healthy skin and studying the changes after the application of topical products.

The medium aperture (4- and 6-mm diameter) measuring probes determine the mechanical properties of the outer skin layers.

The large aperture (8-mm diameter) measuring probe determines the mechanical properties of the whole skin (derma and hypoderma). It is appropriate for evaluation of skin diseases with predominantly changes in the dermis (systemic sclerosis, scleredema of Buschke, psoriasis, keloids and erysipelas). The use of 8-mm diameter probe may be difficult for measurements at anatomic regions with thin and flabby skin (i.e. medial surfaces of the arm in elderly individuals), skin over bones or convex areas (i.e., forehead, temporal region, chest, dorsum of the hand and phalanx) as well as at the presence of residual lipid film on the skin surface shortly after the application of topical products [26, 30].

We consider that the simultaneous use of at least two probes with different apertures, i.e. both 2-mm and 8-mm probes, gives more complex information about the mechanical properties of the skin [26, 30].

29.4.1.2 Pressure (Vacuum)

For a constant opening of the probe, the absolute mechanical parameters are directly correlated with the intensity of vacuum applied, whereas the relative parameters are less independent of load for most anatomical regions [5]. Cua et al. [12] have found that Ur/Uf and Uv/Ue tended to increase with increasing loads. According to Wickett [99], the application of 200 mbar of vacuum leads to more sensitivity to moisturizing effects compared to 500 mbar of negative pressure.

29.4.1.3 Test Site

Volar forearms are considered the most appropriate test site generating reproducible measurement results using probes with different apertures. Significant differences in the skin mechanical parameters between both forearms have not been established [26, 30].

29.4.1.4 Measurement Scheme

The application of one and the same measurement scheme for study of one and the same skin condition makes the obtained results comparable. According to us, the most appropriate measurement scheme comprises the single application of constant vacuum of 400 mbar for 5 s followed by a 5-s relaxation time. This scheme produces enough meaningful results, does not overburden the device and shortens the duration of measurements [26, 30].

29.4.1.5 Preconditioning of the Skin

Barel et al. [5] have measured higher values of relative elastic parameters Ur/Ue and Ur/Uf using the pretension mode of the Cutometer®, indicating that preconditioned skin recovers more of its elastic deformation. They also consider that under pretension, the values of the skin deformation parameters are more reproducible and accurate. Dobrev [26] did not find any significant changes in the skin mechanical parameters using pretension of the skin, except for a nonsignificant tendency toward higher values.

29.4.1.6 Environmental Factors

It is recommended to perform the measurements under the same controlled room conditions. Temperature of 20–24 °C and relative humidity of 40–50 % are preferable. The tested person needs at least 15–20 min to acclimatize [56, 99].

29.4.2 Results Interpretation

The mechanical parameters determined by Cutometer® reflect the condition and the changes in skin structure and composition. They provide meaningful information about its major properties such as [5, 7, 30, 32, 70, 90]:

- Skin distension (stiffness), i.e. the ability of the skin to undergo distension or the skin resistance to change of shape under the influence of stress.
- Skin elasticity, i.e. the ability of the skin to recover the original shape after deformation.
- Skin viscoelasticity, i.e. the time-dependent deformation with a “creep” phenomenon and nonlinear stress–strain properties with “hysteresis”. The creep is characterized as a slowly increasing deformation of the skin in function of the time when a constant stress is applied. The hysteresis is related to the observation that after interrupting the stress the skin does not immediately return to its initial position and remains slightly deformed. In this way, the stress–strain curve obtained during suction time will not be superposed by the curve obtained during relaxation time.

The measurement mode 1 is most used in research studies. That is why the results obtained using mode 1 are explained in details in this section.

29.4.2.1 R-Parameters

According to their calculation, the R-parameters are divided into two groups [7, 56]:

- Absolute parameters: Ue, Uv, Uf, Ur, Ua and R1.
- Relative parameters: Ua/Uf, Ur/Uf, Ur/Ue and Uv/Ue.

The absolute parameters are measured in millimetres, while the relative parameters are presented with a number, which represents a ratio between the values of two absolute parameters – the maximal value is 1 (100 %).

It is considered that the absolute parameters are dependent on the skin thickness, which varies with age, sex and body region. That is why for comparison studies, they should firstly be standardized for skin thickness determined by ultrasound. Because this is not always possible, the ratios of absolute parameters, i.e. the relative parameters, should be compared. It is accepted that they do not depend on skin thickness and can be compared between subjects, anatomical regions and time points [5, 9].

Nevertheless, we suggest all measured skin mechanical parameters to be considered simultaneously. The reason is that relative parameters are composed of two parts – numerator and denominator, and one and the same value could be found as a result of an increase of a numerator or a decrease of a denominator and vice versa [26, 30].

According to their biological informativeness, R-parameters are divided into three groups [7, 26, 30]:

1. Elastic parameters:
    - Absolute parameters – Ue, Uf and Ur.
    - Relative parameters – Ua/Uf, Ur/Ue and Ur/Uf.
2. Viscoelastic parameters:
    - Absolute parameters – Uv and H.
    - Relative parameters – Uv/Ue.
3. Mixed parameters:
    - R1 and R4

The final skin distension Uf consists of two components – immediate distension Ue (elastic part) and delayed distension Uv (viscoelastic part).

Ue is related to the stretching of collagen and elastic fibres and reflects the skin thickness and rigidity.

Ue decreases during the skin ageing. At sun-protected areas, this is due to the decrease in elastic properties of collagen bundles as a result of fragmenting and increased number of intermolecular binds, whereas at sun-exposed areas this is due to the thickening of the skin as a result of advanced elastosis. Ue is also decreased in disorders characterized by skin thickening and induration. The enlarged volume of the skin as a result of dermal oedema (oedematous phase of scleroderma, psoriasis, erysipelas) or deposition of collagen bundles and glycosaminoglycans in the dermis (indurative phase of scleroderma, scleredema of Buschke, keloids) restricts the skin possibilities for deformation after application of vacuum [17–19, 45].

Ue is increased in some inherited diseases of connective tissue such as Ehlers–Danlos syndrome, which is due to alterations in collagen tissue and thinning of the skin [27, 32, 50]. The application of moisturizers and emollients induces an increase in Ue which is due to the softening of corneal layer and improvement of the plasticity of epidermal layer [25, 65].

Delayed distension (Uv) is attributed to the movement of the interstitial fluid throughout the fibrous network in the dermis.

Uv increases after application of moisturizing agents [6, 25], which is related to the epidermal hydration and improved plasticity of corneal layer. Uv increases in the presence of inflammatory dermal oedema (erysipelas, lymphoedema) [4, 18]. It is also increased in elderly skin because of the decrease of the viscosity due to the decrease in proteoglycans content during skin ageing [22, 26].

Uv could be decreased in scleredema of Buschke and keloids because of the increase of interstitial fluid viscosity as a result of accumulation of proteoglycans [17, 20, 45]. Lower Uv is also observed in the indurative phase of scleroderma and Ehlers–Danlos syndrome. In the first case, this is due to the increased stiffness of the skin, whereas in the second case, this is due to the thinning of the skin [19, 27, 50].

The alterations in Ue and Uv could be unidirectional (epidermal hydration, keloids) [20, 25] or heterogeneous (chronological and photoaging, UV-light irradiation, systemic sclerosis, scleredema of Buschke, erysipelas and lymphoedema, psoriasis, Ehlers–Danlos) [4, 18, 19, 24, 25, 27, 30, 50]. The changes in both elastic and viscoelastic part could influence the value of final skin distension (Fig. 29.13).

Uf could be increased at the expense of the increase in:

- Both parts Ue and Uv (epidermal hydration) [25].
- The elastic part Ue (Ehlers–Danlos syndrome) [27].
- The viscoelastic part Uv (dermal oedema in erysipelas) [18].

Uf could be decreased at the expense of the decrease in:

- Both parts Ue and Uv (keloids) [20].
- The elastic part Ue (photoaging, scleroderma, scleredema of Buschke, psoriasis) [17, 19, 22, 24, 26].

Uf could be preserved despite the changes in its parts:

- Decrease in Ue is accompanied with increase in Uv (chronological ageing, UV-light irradiation) [22, 26, 29].

Viscoelastic to elastic ratio (Uv/Ue) represents the distribution between elastic and viscoelastic parts of skin deformation. The increase of Uv/Ue indicates the prevalence of viscoelastic over elastic part of skin deformation. This can mainly be due to the increase of Uv, for example, in erysipelas [18], or decrease of Ue, in oedematous phase of scleroderma [19], scleredema of Buschke [17, 45] and lymphoedema of the lover limbs [4], for example. During skin ageing, Uv/Ue progressively increases because of the simultaneously increase of Uv and decrease of Ue [21, 22]. Identical changes are observed after UV-light irradiation [29].

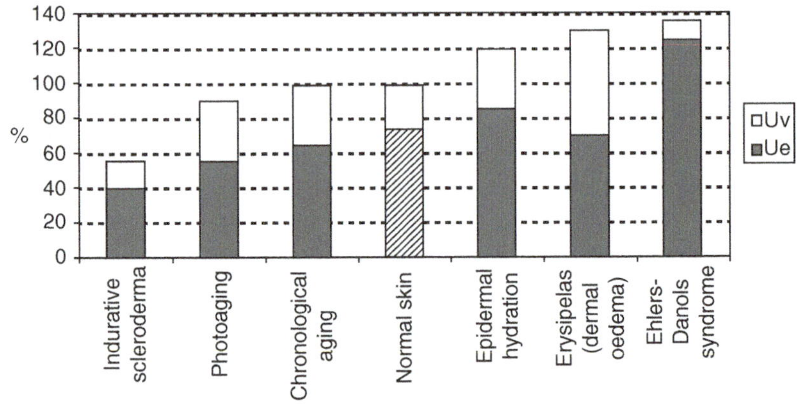

Fig. 29.13 Changes in immediate distension Ue and delayed distension Uv of the skin

A direct correlation has been established between Uv and Uv/Ue [26, 30].

Immediate retraction (Ur), gross elasticity (Ua/Uf), net elasticity (Ur/Ue) and biological elasticity (Ur/Uf) are related to the function of elastic fibres and represent the skin ability to restore its initial position after deformation. A direct correlation has been established between them [26, 30].

The elastic parameters are decreased in elderly individuals due to chronological ageing and photoaging of the skin [21, 22], after UV-light irradiation [29] and in diseases, which are characterized by increased thickness and indurations of the epidermis, dermis or whole skin such as psoriasis [24], erysipelas [18] and colloids [20].

The elastic parameters increase in varying degree after the application of moisturizers and emollients, which is due to the effects on the mechanical properties of the corneal layer rather than the effects on the elastic fibres and in inherited diseases of the connective tissue such as Ehlers–Danlos syndrome [27, 50].

In oedematous phase of scleroderma and in scleredema of Buschke, the measurements with 8-mm diameter probe explore a relative increase in the elastic parameters regardless of skin thickening. This phenomenon can be explained by the "lubricating" action of the dermal oedema and decreased friction between the fibres [17, 19].

Mechanical parameters Ua/Uf and Ur/Uf better characterize the elastic properties of the skin than Ur/Ue, because they include the viscous part of skin deformation, too [30].

Hysteresis (H) is a viscoelastic parameter. When a few consecutive suctions are applied (i.e. 3–10 suctions, 3 s/3 s), H reflects the water content of the skin. Using 8-mm diameter probe, higher values of H have been found on psoriasis and erysipelas plaques which is due to the inflammatory dermal oedema [18, 24]. H and Uv were decreased on the irradiated skin in patients undergoing telegamma therapy for breast cancer [23].

When a lot of repeated suctions are applied (i.e. 30 suctions, 1 s/1 s) at one and the same anatomic region, H characterizes skin fatigue. The age-related decline in skin elasticity results in marked fatigue of adult skin rather than of young skin [33].

Residual deformation *R* is a mixed parameter because it reflects both elastic and viscous properties of the skin. Its interpretation is somewhat difficult and it has been little reported in the literature. In healthy individuals R increases with age as well as directly correlates with the parameters Uv, Uv/Ue, R8 and H [26]. In some diseases, which are characterized by thickening of the skin such as psoriasis, scleredema of Buschke and keloids, using 8-mm diameter probe *R* is decreased, while in others such as erysipelas *R* is increased [17, 20, 24].

In general, an inverse correlation exists between the changes in elastic and viscoelastic parameters, while between the single parameters in each group, the relationship is direct [30].

We suggest the following Cutometer® R-parameters, which characterize the main mechanical properties of the skin, to be always analysed: Ue and Uf (distensibility), Ua/Uf and Ur/Uf (elasticity) and Uv and Uv/Ue (viscoelasticity) [26, 30, 32].

29.4.2.2 F-Parameters (Area-Parameters)

The surface parameters F0 and F1 reflect the viscous part of skin deformation. A completely elastic material will show the complete area (total area and F1 are the same). However, these parameters are not well-known in the scientific literature [56].

The surface parameters F2, F3 and F4 are also not familiar to scientists. We studied the age-related changes in skin fatigue applying multiple suctions at one and the same anatomic region and found that adult skin is characterized by significantly higher values of F2 and lower F3 compared to young skin [33]. F4 is considered a firmness parameter [56].

29.4.2.3 Q-Parameters

The surface Q-parameters reflect the elastic and viscous recovery of the skin. It is considered that Q0 (maximum recovery area) will go down with more firmness of the skin. Qu et al. [83] found that overall skin elasticity (Q3) and elastic recovery (Q1) decreased significantly, whereas the viscous recovery (Q2) did not show significant change with age. There was a marked decrease in Q3 and Q1, whereas Q2 was higher in the sun-exposed skin.

29.5 Practical Applications

The Cutometer® is widely used for study in the mechanical properties of healthy skin, their changes under the influence of various internal and external factors, for clinical diagnosis and monitoring, efficacy testing and claim support for medical and cosmetic topical products.

29.5.1 Study of Healthy Skin

29.5.1.1 Influence of Sex

Generally, no significant sex-dependent differences in skin mechanical parameters have been reported [5, 12, 21]. However, the menopause is associated with more expressed increase in distensibility and viscosity and decrease in elasticity of the female skin. The application of hormone replacement therapy is able to significantly reduce the climacteric-associated loss of skin elasticity [51, 75, 78, 82, 93]. Using the new introduced Q-parameters, Qu et al. [83] found that female subjects exhibited greater elastic recovery and lower viscous recovery than male subjects.

29.5.1.2 Influence of Age

Aged skin is characterized by significantly lower elastic and higher viscoelastic parameters. At all anatomic regions, the decrease in skin elasticity and the increase in skin viscoelasticity significantly correlate with the age [5, 9, 12, 21, 51, 58, 87, 94].

29.5.1.3 Influence of Body Region

The regional differences in skin mechanical properties determined by Cutometer® are mainly due to the differences in skin thickness and sun exposure. The absolute parameters are more influenced than the relative mechanical parameters [5, 12, 26, 41, 48, 53, 94, 100].

29.5.1.4 Influence of External Factors

Chronic sun exposure and UV-light irradiation produce a decrease in skin extensibility and elasticity and an increase in skin viscoelasticity. These alterations have been reported at facial and dorsal vs. volar forearm skin and are accompanied with increased skin thickness [22, 26, 94]. Similar changes in skin mechanical properties have been observed on irradiated skin in patients undergoing telegamma therapy for breast cancer [23].

Cutometer® has been used for investigation of the skin mechanical properties in astronauts before and after a long-term mission in the International Space Station [95].

29.5.2 Study of Diseased Skin

Mechanical properties of the skin are altered in many dermatological diseases. The Cutometer® allows quantifying these alterations in details, and the changes of mechanical parameters determined are valuable for diagnosis, assessment of severity, monitoring of progression and evaluation of treatment in skin diseases characterized by thickening or thinning and induration or softness of the skin (Fig. 29.14).

The Cutometer® has been used to study mechanical properties of the skin affected by systemic sclerosis [19, 42, 54, 69], Raynaud's phenomenon [36, 80], localized scleroderma [2, 14], scleredema of Buschke [17, 45], eosinophilic fasciitis [28, 86], psoriasis vulgaris [24], erysipelas [18] and lymphoedema of the lower legs [4], keloids [20] and hypertrophic scars [39, 44, 66, 67], eczema [37], striae distensae [52, 77], Ehlers–Danlos syndrome [27, 43, 50], diabetes mellitus [47, 68, 80, 96], acromegaly [10], gravitational syndrome [81], type 1 neurofibromatosis [64], spinal cord injury [73] and adult groin hernias [72].

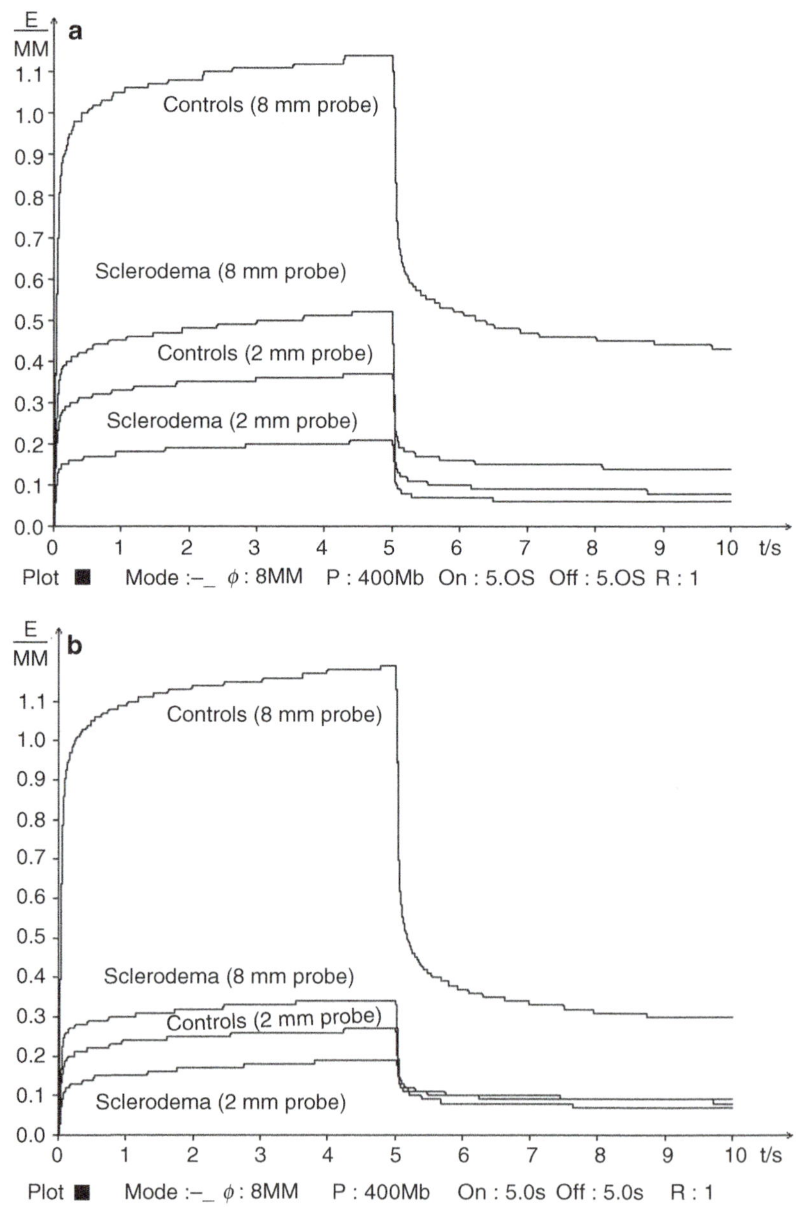

Fig. 29.14 Examples of skin deformation curves obtained with Cutometer® in some skin diseases: (**a**) Systemic sclerosis (measuring mode 1; skin deformation mode extension/time). (**b**) Scleredema of Buschke (measuring mode 1; skin deformation mode extension/time). (**c**) Keloids (measuring mode 1; skin deformation mode extension/time). (**d**) Erysipelas of the lower leg (measuring mode 1; skin deformation mode extension/time). (**e**) Erysipelas of the lower leg (measuring mode 2; skin deformation mode extension/time). (**f**) Erysipelas of the lower leg (measuring mode 2; skin deformation mode pressure/extension). (**g**) Psoriasis (measuring mode 1; skin deformation mode extension/time). (**h**) Skin fatigue (measuring mode 1; skin deformation mode extension/time)

Fig. 29.14 (continued)

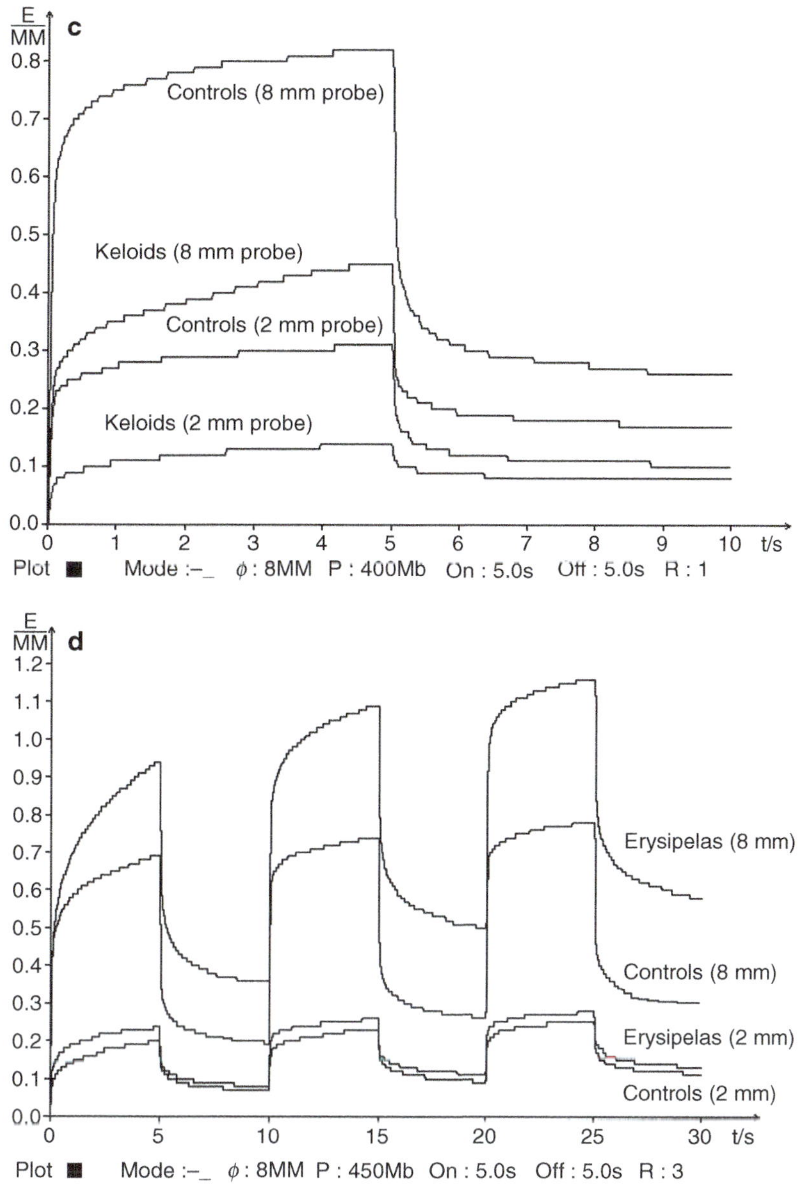

Fig. 29.14 (continued)

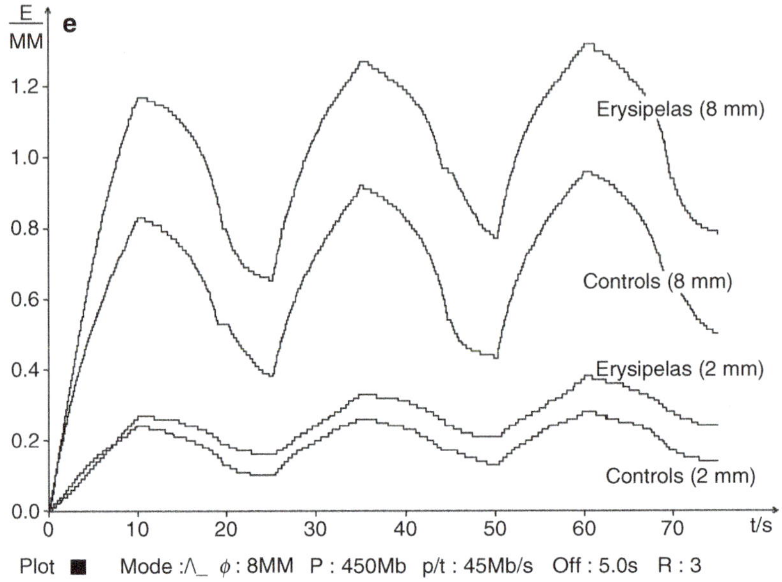

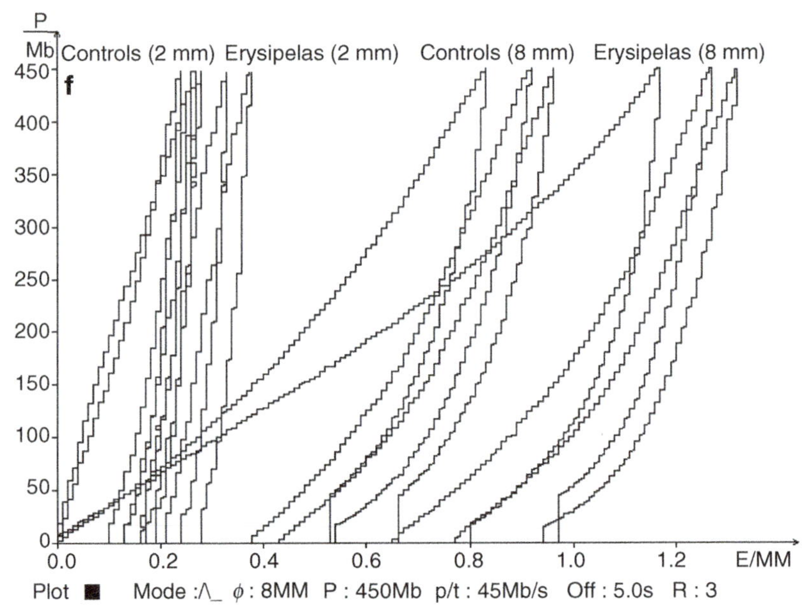

Fig. 29.14 (continued)

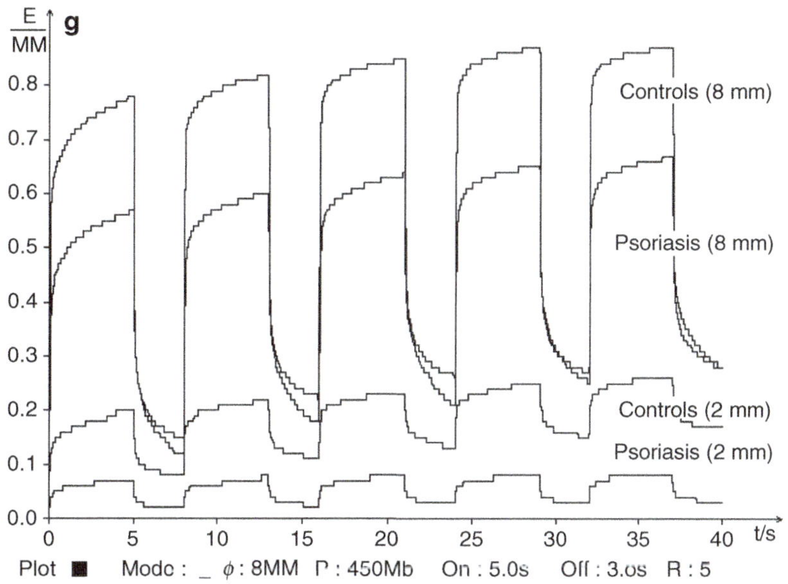

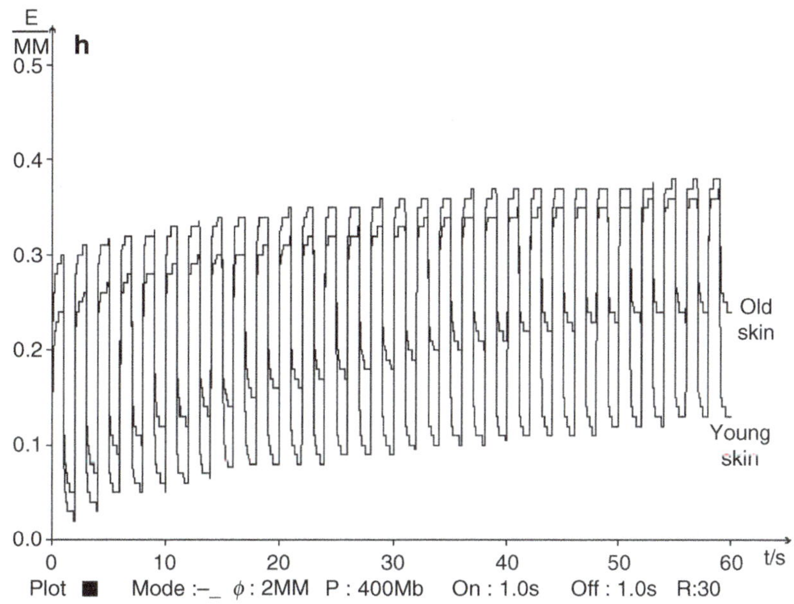

Results observed suggest that:

- The Cutometer® is more sensitive than the human perceptions and could detect minimal and initial changes in skin mechanical properties. The measurements could identify patients with secondary Raynaud's phenomenon at risk of developing subsequently systemic sclerosis [19, 36].
- The measurement values correlate well with clinical scoring systems and could be used for evaluation of the degree of skin involvement [17, 19, 42].
- Since the improvement in skin condition is accompanied by changes in skin mechanical parameters toward values of normal healthy skin, their measurements could be used for monitoring disease progress and treatment response, also [17, 18, 20, 28, 31, 45].
- Moreover, the Cutometer® measurements are able to differentiate the oedematous from the indurative phase of scleroderma as well as indurative phase of scleroderma from SB as well as the firm no pitting oedema in SB from the soft pitting oedema in erysipelas [17–19].

29.5.3 Study of Product Efficacy

The Cutometer® measurements could be useful for product efficacy and claim support studies. They have been used for evaluation of the effects of cosmetic products such as moisturizers and emollients [3, 6, 25, 65], anti-ageing creams containing different active ingredients [8, 34, 46, 62, 89], photoprotective creams [29], plant extracts [1, 35, 55, 88], chemical peelings [76], nutritional supplementation [60, 63], dietary bee pollen supplementation [92], intradermal [71] and oral [79] administration of growth factor and mucopolysaccharide polysulphate [98].

Measurements with Cutometer® were useful for exploring the mechanisms for improving skin mechanical properties after short-term and long-term application of cosmetic products. The single application of emulsions improves the plasticity of epidermal corneal layer by increasing its hydration (urea and other humectants; predominantly raised viscoelastic parameters) or by decreasing the intracorneal cohesion (alpha hydroxy acids; predominantly raised elastic parameters). Multiple applications of moisturizing cream containing plant extracts and oils improve the plasticity of the skin by increasing its water content (both raised elastic and viscoelastic parameters), while the cream containing pentapeptides increases skin firmness by inducing the accumulation of newly synthesized collagen (raised elastic and reduced viscoelastic parameters) [25, 34, 35, 38].

Cutometer® measurements have been used for assessment of the activity of different topical corticosteroids as ointments and creams. However, the alteration in skin mechanical properties observed was related to the effects of vehicles rather than to the effects of active substances [16].

Distante et al. [15] have conducted an interesting study using Cutometer® and other objective measurements. They aimed to evaluate if cosmetic product's packaging and strongly claimed efficacy attributes can influence the objectively measured efficacy. The results obtained suggest that the packaging characteristics cannot be a key factor for improving the biophysical skin properties related to anti-ageing and restoring effects.

29.5.4 Study of Treatment Efficacy

The Cutometer® measurements have been used for objective and quantitative evaluation of disease progress and treatment efficacy in many skin diseases such as localized scleroderma treated with phototherapy [14], scleredema of Buschke [17, 45], eosinophilic fasciitis treated with corticosteroids [28], psoriasis treated with dithranol ointment [24] and with topical corticosteroids and hydrocolloid occlusive dressings [31], erysipelas treated with regular treatment [18], eczema treated with corticosteroids [37], keloids treated with intralesional triamcinolone acetonide [59] and cryosurgery [20], anti-keloidal products [101], haemodialysis [13], burn wounds treated with composite and split-thickness skin grafts [44, 91, 97], topical dressings for wounds [84], CO2 therapy and liposuction for adipose tissue accumulation [11], skin resurfacing with pulsed carbon dioxide laser [57], skin rejuvenation treatment using hyaluronic acid-based gel of non-animal origin [85], liposuction [49] and cellulite treatment with a TriPollar radiofrequency device [61].

Conclusions

The Cutometer® is an easy to handle non-invasive suction device for evaluating the mechanical properties of the skin. It provides the users with objective, quantitative, reproducible and meaningful data on the elastic and viscoelastic properties of healthy and diseased human skin as well as their changes under the influence of various external factors, therapeutical and cosmetic products. The Cutometer® is now recognized as a standard tool in dermatological and cosmetic research.

References

1. Akhtar N, Zaman SU, Khan BA, Amir MN, Ebrahimzadeh MA (2011) Calendula extract: effects on mechanical parameters of human skin. Acta Pol Pharm 68(5):693–701
2. Andres C, Kollmar A, Mempel M, Hein R, Ring J, Eberlein B (2010) Successful ultraviolet A1 phototherapy in the treatment of localized scleroderma: a retrospective and prospective. Br J Dermatol 162(2):445–447, Epub 2009 Aug 8
3. Auriol F, Vaillant L, Machet L, Diridollou S, Lorette G (1993) Effects of short-time hydration on skin extensibility. Acta Derm Venereol 73(5):344–347
4. Auriol F, Vaillant L, Pelucio-Lopes C, Machet L, Diridollou S, Berson M, Lorette G (1994) Study of cutaneous extensibility in lymphoedema of the lower limbs. Br J Dermatol 131(2):265–269
5. Barel AO, Lambrecht R, Clarys P (1998) Mechanical function of the skin: state of the art. In: Elsner P, Barel AO, Berardesca E, Gabard B, Serup J (eds) Skin bioengineering. Techniques and applications in dermatology and cosmetology, vol 26. Karger, Basel, pp 69–83, Curr Probl Dermatol
6. Barel AO (2002) Product testing: moisturizers. In: Elsner P, Berardesca E, Wilhelm K-P, Maibach HI (eds) Skin biomechanics. CRC Press, Boca Raton, pp 241–256
7. Barel AO, Courage W, Clarys P (2006) Suction chamber method for measurement of skin mechanics: the new digital version of the cutometer. In: Serup J, Jemec GBE, Grove G (eds) Handbook of non-invasive methods and the skin. CRC Taylor & Francis, Boca Raton, pp 583–591
8. Berardesca E, Gabba P, Farinelli N, Borroni G, Rabbiosi G (1990) In vivo tretinoin-induced changes in skin mechanical properties. Br J Dermatol 122(4):525–529
9. Berndt U, Elsner P (2002) Hardware and measuring principle: the cutometer. In: Elsner P, Berardesca E, Wilhelm K-P, Maibach HI (eds) Bioengineering and the skin. Skin biomechanics. CRC Press, Boca Raton, pp 91–97
10. Braham C, Betea D, Piйrard-Franchimont C, Beckers A, Piйrard GE (2002) Skin tensile properties in patients treated for acromegaly. Dermatology 204(4):325–329
11. Brandi C, D'Aniello C, Grimaldi L, Caiazzo E, Stanghellini E (2004) Carbon dioxide therapy: effects on skin irregularity and its use as a complement to liposuction. Aesthetic Plast Surg 28(4):222–225
12. Cua AB, Wilhelm KP, Maibach HI (1990) Elastic properties of human skin: relation to age, sex, and anatomical region. Arch Dermatol Res 282(5):283–288
13. Deleixhe-Mauhin F, Piйrard-Franchimont C, Rorive G, Piйrard GE (1994) Influence of chronic haemodialysis on the mechanical properties of skin. Clin Exp Dermatol 19(2):130–133
14. de Rie MA, Enomoto DN, de Vries HJ, Bos JD (2003) Evaluation of medium-dose UVA1 phototherapy in localized scleroderma with the cutometer and fast Fourier transform method. Dermatology 207(3):298–301
15. Distante F, Pagani V, Bonfigli A, Rigano L, Fluhr J (2007) Objective evaluation of the placebo effect in cosmetic treatments. A randomized controlled study. Int J Cosmet Sci 29:64
16. Dobrev H (1996) In vivo noninvasive study of the mechanical properties of the human skin after single application of topical corticosteroids. Folia Med (Plovdiv) 38(2):11–17
17. Dobrev H (1998) In vivo study of skin mechanical properties in scleredema of Buschke. Acta Derm Venereol 78(2):103–106
18. Dobrev H (1998) Use of cutometer to assess dermal oedema in erysipelas of the lower legs. Skin Res Technol 4(3):155–159
19. Dobrev HP (1999) In vivo study of skin mechanical properties in patients with systemic sclerosis. J Am Acad Dermatol 40(3):436–442
20. Dobrev H (1999) Non-invasive monitoring of the mechanical properties of keloids during cryosurgery. Acta Derm Venereol 79(6):487–488
21. Dobrev H (1999) Age related changes in the mechanical properties of human skin. Dermatol Venereol (Bulgaria) 38(2):21–25
22. Dobrev H (2000) Photoaging and skin elasticity. Research Reports of the Union of Scientists in Bulgaria – Plovdiv. Annual, Series B. Nat Sci Humanit 1:117–120

23. Dobrev H (2000) Influence of telegammatherapy on the skin physiology. In: 7the national congress of dermatology and venereology, Sofia, 11–13 May 2000. pp 28 (Abstract)
24. Dobrev H (2000) In vivo study of skin mechanical properties in psoriasis vulgaris. Acta Derm Venereol 80(4):263–266
25. Dobrev H (2000) Use of cutometer to assess epidermal hydration. Skin Res Technol 6(4):239–244
26. Dobrev H (2000) Value of non-invasive bioengineering investigation of the skin mechanical properties in vivo. Doctoral thesis, Plovdiv
27. Dobrev H (2001) Syndrome Ehlers-Danlos – mechanical properties of the skin. In: 10th annual Sofia dermatological days, vol 40(2). Sofia, 2–3 Nov 2001, p 27. Dermatol Venereol (Bulgaria) (Summary)
28. Dobrev H. (2001) Fasciitis eosinophilica – mechanical properties of the skin. In: 10th annual Sofia dermatological days, vol 40(2), Sofia, 2–3 Nov 2001, p 28. Dermatol Venereol (Bulgaria) (Summary)
29. Dobrev H (2001) Evaluation of the inhibitory activity of topical indomethacin, betamethasone valerate and emollients on UVL-induced inflammation by means of non-invasive measurements of the skin elasticity. Photodermatol Photoimmunol Photomed 17(4):184–188
30. Dobrev HP (2002) A study of human skin mechanical properties by means of cutometer. Folia Med (Plovdiv) 44(3):5–10
31. Dobrev H (2002) Treatment of psoriasis vulgaris with hydrocolloid occlusive dressings in combination with betamethasone dipropionate 0.05% cream. Scientific Researches of the Union of Scientists – Plovdiv. Series D Medicine. Pharm Stomatol 1: 103–106
32. Dobrev H (2002) Mechanical properties in other dermatological diseases. In: Elsner P, Berardesca E, Wilhelm K-P, Maibach HI (eds) Skin biomechanics. CRC Press, Boca Raton, pp 215–228
33. Dobrev H (2005) Application of cutometer area parameters for the study of human skin fatigue. Skin Res Technol 11(2):120–122
34. Dobrev H (2005) The effects of topically applied Matrixyl, natural grape seed and avocado oils on skin surface, hydration and elasticity. III. Spring symposium of the European Academy of Dermatology & Venereology (EADV), Sofia, 19–22 May 2005. Book of Abstracts: p 73
35. Dobrev H (2005) Evaluation of the efficacy of a Rooibos extract containing anti-wrinkle cream. III. Spring symposium of the European Academy of Dermatology & Venereology (EADV), Sofia, 19–22 May 2005. Book of Abstracts: p 84
36. Dobrev H (2007) In vivo study of skin mechanical properties in Raynaud's phenomenon. Skin Res Technol 13(1):91–94
37. Dobrev H. (2008) Evaluation of therapy effect in patients with eczema by measuring of water content and mechanical properties of lesional skin. In: VIII national congress of the Bulgarian Dermatological Society with international participation, Albena, 2–5 Oct 2008 (Poster 25)
38. Dobrev H (2009) How do cosmetics improve the skin mechanical properties? In: 18th congress of the European Academy of Dermatology & Venereology (EADV), Berlin, 7–11 Oct 2009.. J Eur Acad Dermatol Venerol (Suppl.)
39. Draaijers LJ, Botman YA, Tempelman FR, Kreis RW, Middelkoop E, van Zuijlen PP (2004) Skin elasticity meter or subjective evaluation in scars: a reliability assessment. Burns 30(2):109–114
40. Elsner P (1995) Skin elasticity. In: Berardesca E, Elsner P, Wilhelm K-P, Maibach HI (eds) Bioengineering of the skin: methods and instrumentation. CRC Press, Boca Raton, pp 53–64
41. Elsner P, Wilhelm D, Maibach HI (1990) Mechanical properties of human forearm and vulvar skin. Br J Dermatol 122(5):607–614
42. Enomoto DNH, Mekkes JR, Bossuyt PMM, Hoekzema R, Bos JD (1996) Quantification of cutaneous sclerosis with a skin elasticity meter in patients with generalized scleroderma. J Am Acad Dermatol 35:381–387
43. Flagothier C, Goffin V, Hermanns-Lκ T, Piñrard GE, Quatresooz P (2007) A four-generation Ehlers-Danlos syndrome with vascular dissections. Skin ultrastructure and biomechanical properties. J Med Eng Technol 31(3):175–180
44. Fong SS, Hung LK, Cheng JC (1997) The cutometer and ultrasonography in the assessment of postburn hypertrophic scar–a preliminary study. Burns 23(Suppl 1):S12–S18
45. Grudeva-Popova J, Dobrev H (2000) Biomechanical measurement of skin distensibility in scleredema of Buschke associated with multiple myeloma. Clin Exp Dermatol 25(3):247–249
46. Heinrich U, Garbe B, Tronnier H (2007) In vivo assessment of Ectoin: a randomized, vehicle-controlled clinical trial. Skin Pharmacol Physiol 20(4):211–218, Epub 2007 May 23
47. Hashmi F, Malone-Lee J, Hounsell E (2006) Plantar skin in type II diabetes: an investigation of protein glycation and biomechanical properties of plantar epidermis. Eur J Dermatol 16(1):23–32
48. Hashmi F, Malone-Lee J (2007) Measurement of skin elasticity on the foot. Skin Res Technol 13(3):252–258
49. Henry F, Van Look R, Goffin V, Fissette J, Pierard GE (1996) Mechanical properties of skin and liposuction. Dermatol Surg 22(6):566–568
50. Henry F, Goffin V, Piñrard-Franchimont C, Piñrard GE (1996) Mechanical properties of skin in Ehlers-Danlos syndrome, types I, II, and III. Pediatr Dermatol 13(6):464–467
51. Henry F, Piñrard-Franchimont C, Cauwenbergh G, Piñrard GE (1997) Age-related changes in facial skin contours and rheology. J Am Geriatr Soc 45(2):220–222
52. Henry F, Piñrard-Franchimont C, Pans A, Piñrard GE (1997) Striae distensae of pregnancy. An in vivo

biomechanical evaluation. Int J Dermatol 36(7): 506–508

53. Ishikawa T, Ishikawa O, Miyachi Y (1995) Measurement of skin elastic properties with a new suction device (I): relationship to age, sex and the degree of obesity in normal individuals. J Dermatol 22(10):713–717
54. Ishikawa T, Tamura T (1996) Measurement of skin elastic properties with a new suction device (II): systemic sclerosis. J Dermatol 23(3):165–168
55. Kapoor S, Saraf S (2010) Assessment of viscoelasticity and hydration effect of herbal moisturizers using bioengineering techniques. Pharmacogn Mag 6(24):298–304
56. Khazaka D (2010) Information and operating instructions for the cutometer® MPA 580 and the software cutometer® MPA Q. Courage + Khazaka Electronic GmbH, Cologne
57. Koch RJ, Cheng ET (1999) Quantification of skin elasticity changes associated with pulsed carbon dioxide laser skin resurfacing. Arch Facial Plast Surg 1(4):272–275
58. Krueger N, Luebberding S, Oltmer M, Streker M, Kerscher M (2011) Age-related changes in skin mechanical properties: a quantitative evaluation of 120 female subjects. Skin Res Technol 17(2):141–148
59. Krusche T, Worret WI (1995) Mechanical properties of keloids in vivo during treatment with intralesional triamcinolone acetonide. Arch Dermatol Res 287(3–4):289–293
60. Manosroi A, Chutoprapat R, Abe M, Manosroi W, Manosroi J (2012) Anti-aging efficacy of topical formulations containing niosomes entrapped with rice bran bioactive compounds. Pharm Biol 50(2): 208–224
61. Manuskiatti W, Wachirakaphan C, Lektrakul N, Varothai S (2009) Circumference reduction and cellulite treatment with a TriPollar radiofrequency device: a pilot study. J Eur Acad Dermatol Venereol 23(7):820–827, Epub 2009 Apr 8
62. Marini A, Grether-Beck S, Jaenicke T, Weber M, Burki C, Formann P, Brenden H, Schцnlau F, Krutmann J (2012) Pycnogenol® effects on skin elasticity and hydration coincide with increased gene expressions of collagen type I and hyaluronic acid synthase in women. Skin Pharmacol Physiol 25(2):86–92, Epub 2012 Jan 21
63. McCall-Perez F, Stephens TJ, Herndon JH Jr (2011) Efficacy and tolerability of a facial serum for fine lines, wrinkles, and photodamaged skin. J Clin Aesthet Dermatol 4(7):51–54
64. Mimoun N, Razzouq N, Wolkenstein P, Moreno JC, Marty JP, Lantieri L, Astier A, Paul M (2006) Evaluation of skin viscoelasticity in type 1 neurofibromatosis patients. Skin Pharmacol Physiol 19(1):22–27, Epub 2005 Oct 20
65. Murray BC, Wickett RR (1996) Sensitivity of cutometer data to stratum corneum hydration level. A preliminary study. Skin Res Technol 2:167–172
66. Nedelec B, Correa JA, Rachelska G, Armour A, LaSalle L (2008) Quantitative measurement of hypertrophic scar: intrarater reliability, sensitivity, and specificity. J Burn Care Res 29(3):489–500
67. Nedelec B, Correa JA, Rachelska G, Armour A, LaSalle L (2008) Quantitative measurement of hypertrophic scar: interrater reliability and concurrent validity. J Burn Care Res 29(3):501–511
68. Nikkels-Tassoudji N, Henry F, Letawe C, Pierard-Franchimont C, Lefebvre P, Pierard GE (1996) Mechanical properties of the diabetic waxy skin. Dermatology 192(1):19–22
69. Nikkels-Tassoudji N, Henry F, Piйrard-Franchimont C, Piйrard GE (1996) Computerized evaluation of skin stiffening in scleroderma. Eur J Clin Invest 26(6):457–460
70. O'Goshi K (2006) Suction chamber method for measurement of skin mechanics: the cutometer. In: Serup J, Jemec GBE, Grove G (eds) Handbook of non-invasive methods and the skin. CRC Taylor & Francis, Boca Raton, pp 579–582
71. Ono I (2011) A study on the alterations in skin viscoelasticity before and after an intradermal administration of growth factor. J Cutan Aesthet Surg 4(2): 98–104
72. Pans A, Pierard GE, Albert A, Desaive C (1997) Adult groin hernias: new insight into their biomechanical characteristics. Eur J Clin Invest 27(10): 863–868
73. Park JW, Seo CH, Han SH, Lee YG (2011) Sympathetic influence on biomechanical skin properties after spinal cord injury. Spinal Cord 49(2):236–243, Epub 2010 Sep 21
74. Piйrard GE, Nikkels-Tassoudji N, Piйrard-Franchimont C (1995) Influence of the test area on the mechanical properties of skin. Dermatology 191(1):9–15
75. Piйrard GE, Letawe C, Dowlati A, Piйrard-Franchimont C (1995) Effect of hormone replacement therapy for menopause on the mechanical properties of skin. J Am Geriatr Soc 43(6):662–665
76. Piйrard GE, Henry F, Piйrard-Franchimont C (1996) Comparative effect of short-term topical tretinoin and glycolic acid on mechanical properties of photodamaged facial skin in HRT-treated menopausal women. Maturitas 23(3):273–277
77. Piйrard GE, Nizet JL, Adant JP, Camacho MA, Pans A, Fissette J (1999) Tensile properties of relaxed excised skin exhibiting striae distensae. J Med Eng Technol 23(2):69–72
78. Piйrard GE, Vanderplaetsen S, Piйrard-Franchimont C (2001) Comparative effect of hormone replacement therapy on bone mass density and skin tensile properties. Maturitas 40(3):221–227
79. Piйrard-Franchimont C, Henry F, Crielaard JM, Piйrard GE (1996) Mechanical properties of skin in recombinant human growth factor abusers among adult bodybuilders. Dermatology 192(4):389–392
80. Piйrard-Franchimont C, Nikkels-Tassoudji N, Lefиbvre P, Piйrard GE (1998) Subclinical skin

stiffening in adults suffering from type 1 diabetes mellitus. A comparison with Raynaud's syndrome. J Med Eng Technol 22(5):206–210
81. Piñrard-Franchimont C, Letawe C, Fumal I, Van Cromphaut I, Piñrard GE (1998) Gravitational syndrome and tensile properties of skin in the elderly. Dermatology 197(4):317–320
82. Piñrard-Franchimont C, Cornil F, Dehavay J, Deleixhe-Mauhin F, Letot B, Piñrard GE (1999) Climacteric skin ageing of the face–a prospective longitudinal comparative trial on the effect of oral hormone replacement therapy. Maturitas 32(2): 87–93
83. Qu D, Masotti CJ, Seehra GP (2007) Effect of age and gender on the viscoelastic properties of skin. J Soc Cosmet Chem 58(2):197–198
84. Rennekampff HO, Rabbels J, Reinhard V, Becker ST, Schaller HE (2006) Comparing the Vancouver scar scale with the cutometer in the assessment of donor site wounds treated with various dressings in a randomized trial. J Burn Care Res 27(3):345–351
85. Reuther T, Bayrhammer J, Kerscher M (2010) Effects of a three-session skin rejuvenation treatment using stabilized hyaluronic acid-based gel of non-animal origin on skin elasticity: a pilot study. Arch Dermatol Res 302(1):37–45
86. Romano C, Rubegni P, De Aloe G, Stanghellini E, D'Ascenzo G, Andreassi L, Fimiani M (2003) Extracorporeal photochemotherapy in the treatment of eosinophilic fasciitis. J Eur Acad Dermatol Venereol 17(1):10–13
87. Ryu HS, Joo YH, Kim SO, Park KC, Youn SW (2008) Influence of age and regional differences on skin elasticity as measured by the cutometer. Skin Res Technol 14(3):354–358
88. Saraf S, Jeswani G, Deep Kaur C, Saraf S (2011) Development of novel herbal cosmetic cream with curcuma longa extract loaded transfersomes for antiwrinkle effect. Afr J Pharm Pharmacol 5(8): 1054–1062
89. Segger D, Schuŋlau F (2004) Supplementation with Evelle improves skin smoothness and elasticity in a double-blind, placebo-controlled study with 62 women. J Dermatolog Treat 15(4):222–226
90. Serup J (2002) Mechanical properties of human skin: elasticity parameters and their relevance. In: Elsner P, Berardesca E, Wilhelm K-P, Maibach HI (eds) Bioengineering and the skin. Skin biomechanics. CRC Press, Boca Raton, pp 41–47
91. Sнn P, Stupka I, Brychta P (2010) Evaluation and comparison of composite and split-thickness skin grafts using cutometer mpa 580. Ann Burns Fire Disaster 23(4):208–213
92. Soyun C, Chong-Won C, Chong-Hyun W, Dong HL, Kwang HC, Chung JH (2006) Dietary bee pollen supplementation does not improve signs of photoaging in human skin in vivo. Korean J Invest Dermatol 13(4):120–124
93. Sumino H, Ichikawa S, Abe M, Endo Y, Ishikawa O, Kurabayashi M (2004) Effects of aging, menopause, and hormone replacement therapy on forearm skin elasticity in women. J Am Geriatr Soc 52(6): 945–949
94. Takema Y, Yorimoto Y, Kawai M, Imokawa G (1994) Age-related changes in the elastic properties and thickness of human facial skin. Br J Dermatol 131(5):641–648
95. Tronnier H, Wiebusch M, Heinrich U (2008) Change in skin physiological parameters in space–report on and results of the first study on man. Skin Pharmacol Physiol 21(5):283–292, Epub 2008 Jul 28
96. Yoon HS, Baik SH, Oh CH (2002) Quantitative measurement of desquamation and skin elasticity in diabetic patients. Skin Res Technol 8(4):250–254
97. van Zuijlen PP, van Trier AJ, Vloemans JF, Groenevelt F, Kreis RW, Middelkoop E (2000) Graft survival and effectiveness of dermal substitution in burns and reconstructive surgery in a one-stage grafting model. Plast Reconstr Surg 106(3):615–623
98. Wanitphakdeedecha R, Eimpunth S, Manuskiatti W (2011) The effects of mucopolysaccharide polysulphate on hydration and elasticity of human skin. Dermatol Res Pract 2011:807906, Epub 2011 Jun 30
99. Wickett RR (2002) Standardization of skin biomechanical measurements. In: Elsner P, Berardesca E, Wilhelm K-P, Maibach HI (eds) Bioengineering and the skin. Skin biomechanics. CRC Press, Boca Raton, pp 179–185
100. Wilhelm K-P, Maibach HI (2002) Mapping mechanical properties of human skin. In: Elsner P, Berardesca E, Wilhelm K-P, Maibach HI (eds) Skin biomechanics. CRC Press, Boca Raton, pp 187–197
101. Worret W-I (2002) Antikeloidal products. In: Elsner P, Berardesca E, Wilhelm K-P, Maibach HI (eds) Bioengineering and the skin. Skin biomechanics. CRC Press, Boca Raton, pp 257–268

30 Ballistometer

Gregor B.E. Jemec

30.1 Introduction

The skin has multiple functions. Among these, it is an elastic envelope which provides physical and chemical protection for the body. In order to do so, the mechanical properties of the skin are of great importance in order to allow the necessary flexibility and resilience of the skin. The mechanism by which the skin adapts itself to movement and continues to provide a functional barrier has been comparatively little studied in view of the importance of this function. Most studies have been structural, leaving considerable scope for more functional studies, and measurement of the actual mechanical properties of the skin, the 'skin mechanics', may therefore provide much relevant information.

Numerous methods have been developed for the study of skin mechanics based on the principles of materials testing, i.e. the application of force and the assessment of subsequent distortion or deformation. This approach mirrors the clinical approach where physicians press, squeeze or pinch the skin to gain an impression of the mechanical properties. This approach can be more or less structured as exemplified in, e.g. the Rodnan score used in scleroderma, in which skin stiffness is assessed clinically in different predetermined sites and scores added for a combined overall score [1]. An engineering approach however suggests that the process may be standardised further and precision increased substantially.

G.B.E. Jemec, MD, DMSc
Department of Dermatology, Roskilde Hospital,
Health Sciences Faculty, University of Copenhagen,
Køgevej 7-13, DK-4000 Roskilde, Denmark
e-mail: gbj@regionsjaelland.dk

30.2 The Ballistometer

In the ballistometer, force is applied by dropping an object onto the surface of the skin with a known force and measuring the recoil of the object as it bounces off the surface. A similar approach was widely used in metallurgy and has been used in ophthalmology [2, 3]. Numerous sources of variation exist in this approach such as the force used, the angle of the force, the hardness of the material and the shape of the object dropped onto the surface. The ballistometer measures the rebound energy, and in practical terms, the height of the 'bounce' is measured and plotted against time providing data on the absorption and dissipation of the force in the material.

If an inflexible body weighing P is dropped onto a fixed surface from the height H, it rebounds to a height H'. The energy of the falling object is

$$E = PH$$

The energy returned from the surface is correspondingly

$$E' = PH'$$

And the ratio between the applied energy and the returned energy, interpreted as a sign of the mechanical properties of the material tested, e.g. skin is

E. Berardesca et al. (eds.), *Non Invasive Diagnostic Techniques in Clinical Dermatology*,
DOI 10.1007/978-3-642-32109-2_30, © Springer Berlin Heidelberg 2014

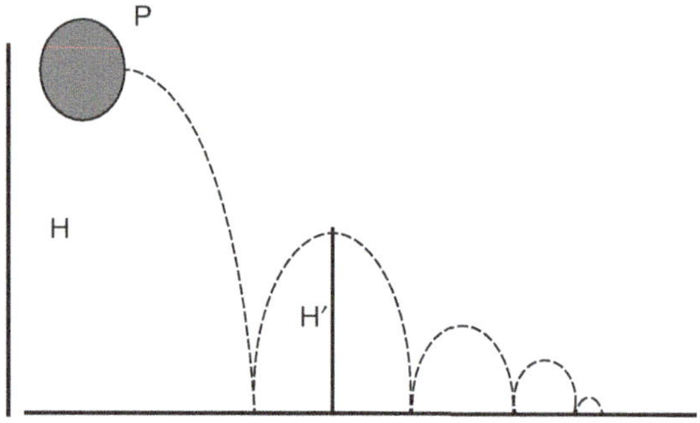

Fig. 30.1 Measurements used for calculating *e*

$$e = E'/E = H'/H\%$$

30.3 Studies

The first paper on ballistometry of the skin was published by Tosti et al. in 1977 using a custom-built ballistometer to explore the potential of the method in healthy volunteers and patients. The authors measured 12 skin areas in 46 healthy volunteers aged 8–80 years. An unspecified number of skin lesions were also tested.

Not all data were reported in the manuscript, but the authors state that no significant difference was seen in left–right comparative measurements of symmetrical areas.

A good correlation was found between age and decreasing e ($r = -0.819$ and -0.875 on the forehead and dorsum of the hand, respectively). The values of e decreased more rapidly in UV-damaged skin, while increased water content of the skin as in, e.g. oedema increased *e*.

In 1991 Fthenakis et al. [4] described their experience with their custom-built ballistometer. They evaluated the variability in skin mechanics due to age and body sites and attempted to modify the mechanical properties by treatment with a vitamin A palmitate containing cream. These authors were also able to identify age-related changes and suggest that the mechanical properties of the skin may be changed by vitamin A palmitate but not a simple moisturizer.

No new studies of ballistometry were published until Jemec et al. [5] published a study comparing measurements of the mechanical properties of the skin using a commercial ballistometer (Torsional ballistometer, Dia-Stron, UK) and a commercial suction cup device (Dermaflex, Cortex Technologies, Denmark). The suction cup method measured distensibility, elasticity, and hysteresis (creep). Ballistometry measured indentation (peak penetration depth of the probe tip beneath the skin surface level), alpha (rate of energy damping), the coefficient of restitution (CoR, bounce height relative to the start height) and area (the area between the bounce profile and the skin zero, i.e. the sum of the area under the curve described by the probe vs. the resting level of the surface of the skin).

Three sites (palm, the ventral and dorsal forearm) were measured in 17 healthy volunteers and the data compared. A moderate degree of correlation was found between the methods (Spearman test, rs = 0.315–0.540), while internal correlation between different measures obtained with one method was higher (Spearman test, rs = 0.375–0.967). Age-related changes in the palm were identified using ballistometry, which however failed to identify age-related changes in the other regions studied. Conversely, the suction cup device was able to identify age-related changes on the forearm but not in the palm.

The suction cup method parameters correlated significantly with the ballistometry parameters, except for the hysteresis. The coefficient of variation of both methods (CV = 0.02–0.35) was within the range obtained with other non-invasive methods. It was concluded that the methods describe different characteristics of the tissue, with ballistometry describing predominantly stiffness, while the suction cup method was considered predominantly to describe elasticity.

In 2007 Caisey et al. among other parameters investigated age-related variations, the mechanical properties of the lips using the commercially available ballistometer (Dia-Stron, UK); however, only the alpha value was considered. The authors found that mechanical damping of the tissue increases with age and appears not to be modified by menopause and/or HRT.

Conclusion

In contrast to suction cup techniques, ballistometry has attracted very scant attention in studies of the skin. The potential advantage of the method is its practical simplicity, as no special attachments to the skin are necessary and the speed of data acquisition is high. A commercial device is available (http://www.diastron.com/default.html).

The method however also has some weaknesses. One weakness is that it is rarely used, and the literature therefore correspondingly sparse. The other main weakness is the interpretation of data. In skin the value obtained using this measure is therefore a composite value reflecting distortion of elastic components, formation of heat, compression and viscous displacement. The exact interpretation of the data is therefore complex [6].

Although the simplicity and intuitive approach of the method are attractive aspects, the complexity of the data interpretation and the very limited dermatological literature on ballistometry of the skin makes the method less attractive. For the interested researcher, it is necessary to provide a body of methodological data to integrate the method into the existing literature which is mainly based on other methods, as well as provide a very stringent experimental setting before attempting to interpret the data.

References

1. Silman A, Harrison M, Brennan P (1995) Is it possible to reduce observer variability in skin score assessment of scleroderma? The ad hoc International Group on the Assessment of Disease Outcome in Scleroderma. J Rheumatol 22:1277–1280
2. Tosti A, Campagno G, Fazzini ML, Vilardita S (1977) A ballistometer for the study of the plastoelastic properties of skin. J Invest Dermatol 69:315–317
3. Hargens CW (2006) Ballistometry. In: Serup J, Jemec GBE, Grove GL (eds) Handbook of non-invasive methods and the skin, 2nd edn. CRC Press, Boca Raton, pp 627–632
4. Fthenakis CG, Maes DH, Smith WP (1991) In vivo assessment of skin elasticity using ballistometry. J Soc Cosmet Chem 42:211–222
5. Jemec GBE, Selvaag E, Ågren M, Wulf HC (2001) Measurement of the mechanical properties of skin with ballistometer and suction cup. Skin Res Technol 7:122–126
6. Caisey L, Gubanova E, Camus C, Lapatina N, Smetnik V, Leveque JL (2008) Influence of age and hormone replacement therapy on the functional properties of the lips. Skin Res Technol 14:220–225

Part V

Water and Stratum Corneum Hydration

31 TEWL, Closed-Chamber Methods: AquaFlux and VapoMeter

Bob Imhof, Perry Xiao, and Irena Angelova-Fischer

31.1 Introduction

TEWL – TransEpidermal Water Loss – is recognised as the main indicator of skin barrier function. Of course, TEWL only gives information about the water barrier since other substances will generally have different penetration properties. The reasons why TEWL is used almost universally for skin barrier assessment are (1) because water transpiration is clinically relevant and (2) because measurements are straightforward, safe, noninvasive and economical.

Since the 1970s, the open-chamber method of measurement [1] has established itself as the main method for TEWL measurement and a de facto standard against which newer technologies are judged. However, the open-chamber method is known to suffer from a number of limitations, including vulnerability to disturbance from ambient air movements, calibration inconsistencies, angular dependence, temperature dependence and contact pressure dependence. Disturbance by external air movements can be eliminated by using a closed measurement chamber, but this design change may affect other aspects of the measurement. This chapter describes two commercial closed-chamber instruments, the AquaFlux and VapoMeter, within the context of the established open-chamber method.

31.2 Closed-Chamber TEWL Instruments

Two commercial closed-chamber TEWL instruments have recently been introduced: the AquaFlux (Biox Systems Ltd, England) and the VapoMeter (Delfin Technologies Ltd, Finland). The two instruments use different technologies, from measurement method to practical implementation, as described below.

31.2.1 AquaFlux

The AquaFlux uses the same steady-state diffusion-gradient measurement principle as the open-chamber method. However, its measurement chamber is closed in order to eliminate disturbance by external air movements. The apparent contradiction between continuous flux measurement in a closed measurement chamber is resolved by a condenser that continuously removes water vapour by converting it to ice [2, 3].

B. Imhof, BSc, PhD, FInstP, CEng (✉)
Biox Systems Ltd, Technopark Building,
90 London Road, London SE1 6LN, UK
e-mail: bob@biox.biz

P. Xiao, BEng, MSc, PhD
Faculty of ESBE, London South Bank University,
103 Borough Road, London SE1 0AA, UK

Biox Systems Ltd, Technopark Building,
90 London Road, London SE1 6LN, UK
e-mail: xiaop@lsbu.ac.uk

I. Angelova-Fischer, MD, PhD
Department of Dermatology, University of Lübeck,
Ratzeburger Allee 160, Lübeck 23538, Germany
e-mail: irena.angelova-fischer@uk-sh.de

E. Berardesca et al. (eds.), *Non Invasive Diagnostic Techniques in Clinical Dermatology*,
DOI 10.1007/978-3-642-32109-2_31, © Springer Berlin Heidelberg 2014

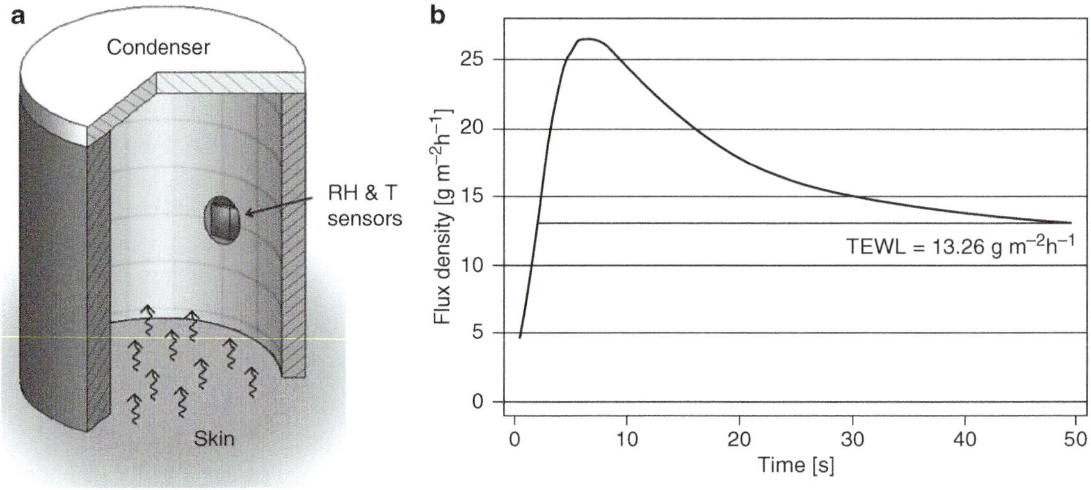

Fig. 31.1 Schematic illustration of (**a**) the measurement chamber and (**b**) a typical flux time-series signal of the condenser-chamber AquaFlux

The AquaFlux measurement chamber, illustrated schematically in Fig. 31.1a, is in the form of a cylinder about the size of a thimble. Its lower end acts as a measurement orifice that is placed into contact with the skin. Its upper end is closed with a condenser that is maintained below the freezing temperature of water by means of an electronic Peltier cooler.

The condenser controls the humidity in the chamber independently of ambient conditions. It acts as a vapour sink by forming ice on its surface, thus creating a zone of low humidity in its immediate vicinity. By contrast, the skin is a vapour source, creating a zone of higher humidity in its immediate vicinity. This humidity difference causes water vapour to migrate from source to sink by passive diffusion, leading to a linear distribution of humidity parallel to the chamber axis under steady conditions. The water vapour flux is calculated from measurements of this humidity gradient and Fick's first law of diffusion.

The humidity gradient is calculated from two humidity values at two spatially separated points. One value is calculated from the readings of relative humidity (RH) and temperature (T) of a sensor mounted in the chamber wall. The other value comes from the condenser, where the humidity can be calculated from its temperature without needing a second RH sensor.

A flux time-series curve recorded during a typical volar forearm TEWL measurement is shown in Fig. 31.1b. The peak at short times is caused partly by ambient humidity entering the measurement chamber before it is sealed against the skin and partly by any unbound skin surface water (skin surface water loss, SSWL). The measurement is ended automatically by the software when the user-set target precision is reached. The software also calculates the area under the peak, above the TEWL level, which is equal to the quantity of excess moisture captured, 2.24 μg in this case.

The AquaFlux is a benchtop, mains-powered instrument consisting of a handheld probe, a base unit and a USB connection to a computer. All measurement, signal evaluation, display and storage functions are performed by the computer.

31.2.2 VapoMeter

The VapoMeter uses the *unventilated-chamber* method of measurement [4–7] illustrated in Fig. 31.2.

The measurement chamber, illustrated in Fig. 31.2a, is in the form of a cylinder about the size of a thimble. One end of the chamber is closed; the other end has a measurement orifice that can be placed in contact with the skin. The

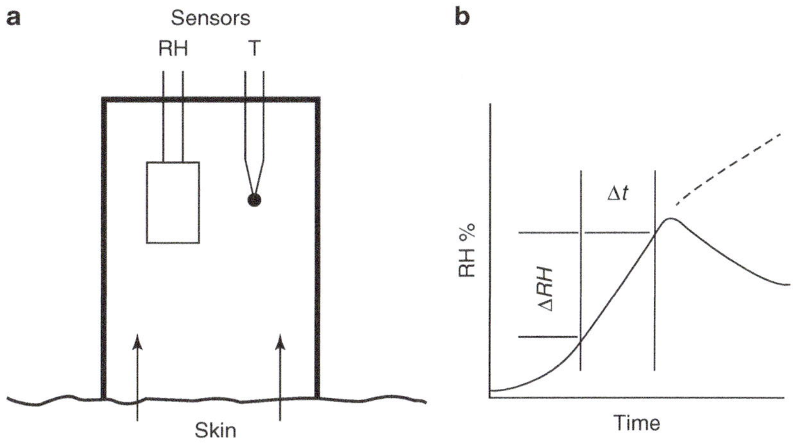

Fig. 31.2 Schematic illustrations of (**a**) the measurement chamber and (**b**) a typical signal of the unventilated-chamber VapoMeter

chamber is equipped with sensors for RH and T. Water vapour from the skin surface collects in the chamber from which it cannot escape. This causes the humidity to rise with time as illustrated in Fig. 31.2b. TEWL is calculated from the rate of increase of humidity, indicated in Fig. 31.2b by the intervals ΔRH and Δt. The measurement is ended automatically by the firmware, after which the chamber needs to be lifted from the skin to allow the accumulated water vapour to escape (solid curve), otherwise the humidity would continue to rise towards saturation level (dashed curve).

The VapoMeter is a self-contained, battery-powered instrument. It is equipped with a push button, a small LCD display and a beeper. All measurement, display and signal evaluation functions are performed by the firmware of its embedded microprocessor, with a Wi-Fi connection used optionally for data transfer to a computer.

31.3 Operating Characteristics

The two instruments use different measurement methods and their operating characteristics are consequently quite different, as described below.

31.3.1 Measurement Procedures

The AquaFlux works with a computer, which continually calculates water vapour flux density from sensor readings sampled approximately twice per second. After switch-on, the instrument needs typically 15–30 min for the zero-flux baseline to settle before it is ready for use. Thereafter, a measurement is initiated by moving the probe to the site of interest and pressing the probe button. The software controls the measurement, looking for a steady flux level according to user-set criteria. When these are met, the measurement is stored automatically, and the instrument is immediately ready for the next measurement (site hopping).

The VapoMeter manufacturer recommends that the instrument be acclimatised to ambient conditions of RH and T before starting measurements. When ready, the following sequence constitutes a measurement:

1. Hold the instrument well away from any water vapour sources, including skin.
2. Press the probe button to initiate a 3-s countdown indicated by three short beeps.
3. At the subsequent long beep, make leak-tight contact with the skin. Do not apply too much pressure.
4. Hold the instrument steady during measurements.
5. At the next long beep, the measurement is complete. Remove the instrument from the skin surface and record the reading on the LCD display, if the Wi-Fi connection is inactive.

After the measurement, the accumulated water vapour needs to be allowed to escape before the next measurement can be started.

31.3.2 Water Vapour Accumulation

Water vapour accumulation is a problem with all closed-chamber methods. With the AquaFlux, the accumulated water vapour is stored as ice on the condenser. The condenser also controls the humidity in the measurement chamber, thus eliminating post-measurement recovery delays. Ice accumulation causes a gradual reduction of sensitivity, by ~1 % for every 2.5 mg of accumulated ice. To put this into context, it would take ~500 typical volar forearm TEWL measurements at 10 $gm^{-2}\ h^{-1}$ to accumulate this quantity of ice. Therefore, ice removal is unnecessary during a normal working day. The procedure for ice removal takes about 20 s, with a further ~10 min required for the readings to stabilise.

With the VapoMeter, the humidity in the chamber increases during a measurement, and the accumulated water vapour needs to be allowed to escape before the next measurement can be started. This process is controlled by the instrument and can take between 20 and 90 s, depending on prior use and other factors.

31.3.3 Recorded Data

Both AquaFlux and VapoMeter display and record TEWL, ambient T and ambient *RH*. In addition, the AquaFlux displays and records SSWL and the measured flux density time-series curves. These curves are important for validating TEWL measurements, because they show exception conditions that may have interfered with the correct calculation of TEWL. Examples of exception conditions include sweat gland activity and imperfect sealing of the measurement head against the skin.

31.3.4 Mobility

Closed-chamber technology makes it possible for TEWL measurements to be used more flexibly, on-site in the workplace, clinics and other less than ideal environments. Mobility may therefore be an important instrument attribute. Of the two instruments featured in this chapter, the VapoMeter is the more mobile, being independent of mains power, not requiring a computer connection and needing only a relatively short period of acclimatisation to ambient conditions before being ready for measurement. The AquaFlux is mains powered, needs to be connected to a computer and requires typically 15–30 min of pre-measurement settling time. Mobility in this case can be achieved by the use of a small trolley for the instrument and laptop, together with an uninterruptible power supply to keep the system going as it is moved from room to room.

31.4 Instrumental Performance

The quality of measurements in different applications depends partly on the instrument manufacturer and partly on the user. The manufacturer is responsible for optimising the performance of the instrument, calibration, software and associated ancillary apparatus and for providing clear instructions on how to use these to best effect. The user is responsible for ensuring that the equipment is correctly used and maintained and that the skin sites are correctly prepared and acclimatised. Several performance assessments involving closed-chamber TEWL instruments have recently been published [8–12]. In this section we will highlight the most pertinent findings.

31.4.1 Method Validity

The validity of a new method of TEWL measurement is usually assessed by comparing equivalent readings with those from an open-chamber instrument. This is because the open-chamber method is seen as a de facto standard, because of both its long history of use and its perceived minimally invasive nature. The most basic method of comparison is correlation. However, correlation is a poor test of validity because there may be hidden factors that affect either one or both methods being correlated.

One such hidden factor is the finite measurement range of TEWL instruments. For TEWL instruments using the diffusion-gradient measurement principle (open chamber and condenser

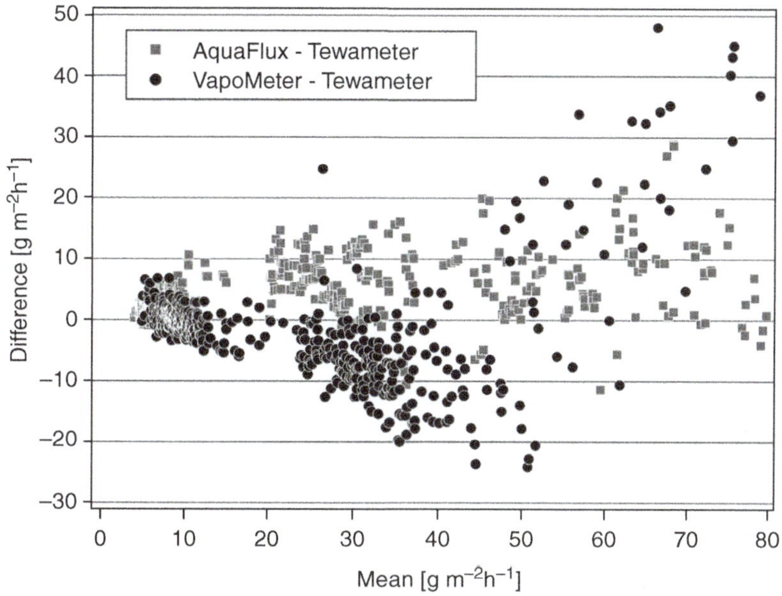

Fig. 31.3 Tukey mean-difference comparisons of in vivo AquaFlux and VapoMeter TEWL measurements (The AquaFlux data are from [13]; the VapoMeter data from [11])

chamber), the maximum flux density that can be measured is limited by the RH of the air immediately adjacent to the skin surface. A diffusion gradient is established because the skin at one end of such a measurement chamber is a source of water vapour, whereas the atmosphere or the condenser at the opposite end is a sink. Therefore, a finite water vapour flux causes the humidity immediately adjacent to the skin surface to increase, whereas the humidity at the opposite end is little affected. The measurement range limit is reached as the air immediately adjacent to the skin surface approaches saturation at 100 % *RH*. For unventilated-chamber instruments, the considerations are different. The non-steady-state nature of this method makes it difficult to predict its measurement range. However, it is generally acknowledged that such instruments respond to higher flux levels than diffusion-gradient instruments. But irrespective of measurement method, measurement range is not constant, but depends on temperature (*RH* decreases as T increases) and the diameter of the orifice in contact with the skin (smaller orifice diameters can be used to increase measurement range).

The comparisons are presented in Fig. 31.3, using the nonparametric Tukey mean-difference method [14], where the means of equivalent pairs of readings with the two instruments being compared are plotted on the x-axis and the differences between them on the y-axis. This is a more sensitive method of comparison than the standard correlation plot, because it shows deviations (differences) between equivalent readings against the assumed best estimate (the mean) of the *true value* of the quantity being measured.

The mean flux density shown in Fig. 31.3 is restricted to the range 0–80 gm^{-2} h^{-1} to remain comfortably within the measurement range of all three instruments used in the tests. Systematic bias is evident between the open-chamber Tewameter TM300 (Courage + Khazaka electronic GmbH, Germany) readings and those of the two closed-chamber instruments. For the condenser-chamber AquaFlux, there is a flux-independent offset of ~4.0 gm^{-2} h^{-1}, which indicates a calibration difference. For the unventilated-chamber VapoMeter, there are flux-dependent trends, with increasingly negative differences in the range 0–40 gm^{-2} h^{-1}, followed by a reversed trend above ~50 gm^{-2} h^{-1}.

31.4.2 Repeatability

The repeatability attribute of a TEWL instrument gives a measure of the extent to which small differences in readings are meaningful in terms of skin properties. Ideal TEWL instrument would

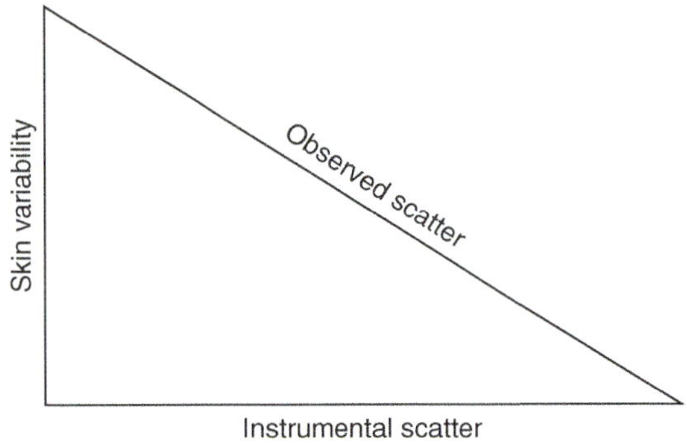

Fig. 31.4 Illustration of how instrumental scatter and skin variability combine in Gaussian statistics to produce an observed scatter that is larger than either one alone

give identical readings when repeatedly measuring the same flux. Real TEWL instruments give different readings (scatter) because of random noise, sensor hysteresis, interference and other factors.

The effect of instrumental repeatability on in vivo TEWL measurements is difficult to assess, because skin is heterogeneous and variable, depending on individual, site, time and preconditioning. The observed scatter of TEWL readings is therefore a superposition of instrumental scatter and skin variability. These are independent sources of variability, which combine as variances in Gaussian statistics (i.e. the *squares* of the standard deviations or CVs are added) as illustrated in Fig. 31.4.

According to Fig. 31.4, if the instrumental scatter is small compared with skin variability, then the observed scatter mainly reflects the variability property of the skin. Conversely, if the instrumental scatter is large compared with skin variability, then the observed scatter mainly reflects instrumental repeatability. It is therefore important to know the repeatability properties of the instrument when interpreting skin measurements.

Instrumental repeatability can be measured in isolation, if a steady flux source is used in place of in vivo skin. Imhof et al. [15] measured the repeatability characteristics of an AquaFlux and a VapoMeter using a novel inverted wet-cup flux source. The main finding, from 200 repeat measurements for each instrument, is that the scatter of AquaFlux readings (CV = 0.93 %) is approximately ten times smaller than that of the VapoMeter (CV = 10.3 %). These values are compatible with the manufacturers' published specifications.

Significant differences in the repeatability of in vivo TEWL measurements are also evident in the data presented in Fig. 31.3. Of course, the scatter in these readings cannot be attributed solely to the instruments under test since they include (1) the scatter in the reference Tewameter readings and (2) skin variability. Nevertheless, the scatter of AquaFlux-Tewameter differences (standard deviation = 4.6 $gm^{-2}\ h^{-1}$) is half that of the equivalent VapoMeter-Tewameter differences (standard deviation = 9.2 $gm^{-2}\ h^{-1}$).

31.4.3 Measurement Time

Measurement time is an important attribute, especially in large studies or studies involving young children. With both instruments, the measurement is started manually by the probe button and stopped automatically in compliance with preprogrammed termination criteria. Their different measurement methods (AquaFlux, steady-state diffusion; VapoMeter, time rate of increase of humidity) also affect measurement time.

For the AquaFlux, measurement time depends on the required measurement precision, skin and ambient conditions. The user-set measurement precision affects measurement time because high precision requires more time for the readings to settle. Skin and ambient conditions affect measurement time because they affect the height and duration of the transient flux peak. A typical example is illustrated in Fig. 31.1b, where the measurement time was 49 s for a target precision of <0.075 $gm^{-2}\ h^{-1}$ standard deviation over a 5-s moving average of the most recent flux readings.

For the VapoMeter, measurement time is 7–16 s, as determined by a proprietary firmware algorithm without user-adjustable parameters. It must be borne in mind, however, that a VapoMeter measurement is of lower precision than an AquaFlux measurement and that it does not provide any means, such as the flux time-series curves of the AquaFlux, to ascertain whether it is affected by sweat gland activity, skin surface water or other measurement errors. With the VapoMeter, the only means of verifying a measurement is to repeat it.

Also important when considering measurement time is the waiting time before the next measurement can be started. With the AquaFlux the next measurement can be started without delay. With the VapoMeter there is a firmware-controlled 20–90-s post-measurement recovery delay before the next measurement can be started.

31.5 TEWL Guidelines for Closed-Chamber Instruments

The established TEWL guidelines [16, 17] are specific to open-chamber instruments, with many of their recommendations irrelevant for closed-chamber instruments. More up-to-date guidance is provided by [18] which is, however, specific to measurements in nonclinical settings. Below is a summary of how these guidelines relate to the instruments described in this chapter:

1. Acclimatisation
   Ambient temperature should be regulated between 20 and 22 ±1 °C to avoid sweat gland activity. RH should be close to 40 % and not exceed 60 %. Skin acclimatisation under these conditions is recommended for at least 15–30 min prior to measurement. These recommendations are generally valid.
2. Instrument preparation
   The AquaFlux needs typically 15–30 min of stabilisation time prior to use. The instrument can be left running through the night, if an early start is needed. The VapoMeter needs little preparation prior to use. The main concern is to ensure that its temperature remains stable, since it has been shown that its readings change by ~1.3 ± 1.1 $gm^{-2}\ h^{-1}$ per °C [8].
3. Holding the probe
   Heat transfer from the hand can give significant measurement errors with some instruments. This is insignificant with the AquaFlux (temperature sensitivity ~0.05 ± 0.06 $gm^{-2}\ h^{-1}$ per °C), but the VapoMeter has a ~26 times higher temperature sensitivity (see above), and it is recommended that this probe always be handled with an insulating glove or other indirect means [18].
4. Contact pressure
   The contact pressure applied between the measurement head and skin has been found to influence measurements made using open-chamber instruments [19], but this has more to do with measurement head design than skin barrier property change. In contrast, closed-chamber instruments have been shown to have minimal sensitivity to contact pressure [3, 8]. Adequate constant pressure should be applied to ensure (1) a leak-free contact between the probe and skin and (2) that the probe does not slide over the skin.
5. Surface orientation
   Open-chamber measurements need to be restricted to horizontal surfaces because of interference from natural convection air movements. For closed-chamber instruments, external air movements do not interfere with measurements. It is therefore tempting to assume that such instruments are unaffected by probe angle. However, natural convection can occur inside measurement chambers, depending on temperature distribution, geometry and orientation. Angular dependence may therefore be an issue with closed-chamber instruments also.

   For the VapoMeter, Cohen et al. found that measurements vary with orientation [20] although unpublished data from the manufacturer (private communication) show no statistically significant orientation dependence. For the AquaFlux, the probe can be used with all surface orientations with little effect on sensitivity, providing the probe is held correctly, with the sensors remaining above the chamber axis [3].

31.6 Summary and Conclusion

Closed-chamber instruments bring a new dimension to skin barrier measurement because they overcome the main limitation of open-chamber instruments, namely, their vulnerability to disturbance from external air movements. This makes it possible for TEWL measurements to migrate away from the well-controlled laboratory environment, into the workplace or clinic. This and other aspects of their design liberate them from many of the restrictions and precautions recommended in the now outdated TEWL guidelines [16, 17].

The two closed-chamber instruments described in this chapter differ in measurement principle, concept and design. The AquaFlux, which uses the condenser-chamber method of measurement, is a benchtop, mains-powered instrument. The VapoMeter, which uses the unventilated-chamber method of measurement, is a self-contained, battery-powered instrument. Their performance characteristics are also quite different, with accuracy, sensitivity and repeatability the main features of the AquaFlux; speed and mobility the main features of the VapoMeter.

Acknowledgments We thank Markus Steiner of Aberdeen University, Scotland, for providing the raw Tewameter-VapoMeter comparison data used in Fig. 31.3 and Jouni Nuutinen of Delfin Technologies Ltd, Finland, for providing unpublished VapoMeter angular response data.

References

1. Nilsson GE (1977) Measurement of water exchange through skin. Med Biol Comput 15:209–218
2. Imhof RE, Berg EP, Chilcott RP, Ciortea LI, Pascut FC (2002) New instrument for measuring water vapour flux density from arbitrary surfaces. IFSCC Mag 5(4):297–301
3. Imhof RE, De Jesus MEP, Xiao P, Ciortea LI, Berg EP (2009) Closed-chamber transepidermal water loss measurement: microclimate, calibration and performance. Int J Cosmet Sci 31:97–118
4. Wallihan EF (1964) Modification and use of an electric hygrometer for estimating relative stomatal apertures. Plant Physiol 39:86–90
5. Miller DL, Brown AM, Artz EJ (1981) Indirect measures of transepidermal water loss. In: Marks R, Payne PA (eds) Bioengineering and the skin. MTP Press, Lancaster, pp 161–171
6. Tagami H, Kobayashi H, Kikuchi K (2002) A portable device using a closed chamber system for measuring transepidermal water loss: comparison with the conventional method. Skin Res Technol 8:7–12
7. Nuutinen J, Alanen E, Autio P, Lahtinen M, Harvima I, Lahtinen T (2003) A closed unventilated chamber for the measurement of transepidermal water loss. Skin Res Technol 9:85–89
8. De Paepe K, Houben E, Adam R, Wiesemann F, Rogiers V (2005) Validation of the VapoMeter, a closed unventilated chamber system to assess transepidermal water loss vs. the open chamber Tewameter. Skin Res Technol 11:61–69
9. Farahmand S, Tien L, Hui X, Maibach HI (2009) Measuring transepidermal water loss: a comparative in vivo study of condenser-chamber, unventilated-chamber and open-chamber systems. Skin Res Technol 15(4):392–398
10. Elkeeb R, Hui X, Chan H, Tian L, Maibach HI (2010) Correlation of transepidermal water loss with skin barrier properties in vitro: comparison of three evaporimeters. Skin Res Technol 16(1):9–15
11. Steiner M, Aikman-Green S, Prescott GJ, Dick FD (2011) Side-by-side comparison of an open-chamber (TM 300) and a closed-chamber (Vapometer (TM)) transepidermal water loss meter. Skin Res Technol 17:366–372
12. Shah JH, Zhai H, Maibach HI (2005) Comparative evaporimetry in man. Skin Res Technol 11:205–208
13. Angelova-Fischer I, Fischer TW, Zillikens D (2009) Die Kondensator-Kammer-Methode zur nicht-invasiven Beurteilung von irritativen Hautschäden und deren Regeneration: eine Pilotstudie. Derm Beruf Umwelt 57(3):125
14. Bland JM, Altman DG (1986) Statistical methods for assessing agreement between two methods of clinical measurement. Lancet 1:307–310
15. Imhof RE, Xiao P, Berg EP, Ciortea LI (2005) Rapid measurement of TEWL with a condenser-chamber instrument. In: 15th international meeting of the ISBS and 2nd ioint international meeting of ISBS and ISSI, Philadelphia, 2005, pp 1–7. Available from: http://www.biox.biz/Library/Conference/ConfContribDetails10.php
16. Pinnagoda J, Tupker RA, Agner J, Serup J (1990) Guidelines for transepidermal water loss (TEWL) measurement. A report from the Standardization Group of the European Society of Contact Dermatitis. Contact Dermatitis 22:164–178
17. Rogiers V (2001) EEMCO guidance for the assessment of transepidermal water loss in cosmetic sciences. Skin Pharmacol Appl Skin Physiol 14:117–128
18. Du Plessis J, Stefaniak A, Eloff F, John S, Agner T, Chou T-C et al (2013) International guidelines for the in vivo assessment of skin properties in non-clinical settings: part 2. Transepidermal water loss and skin hydration. Skin Res Technol 19(3):265–278
19. Barel AO, Clarys P (1995) Comparison of methods for measurement of transepidermal water loss. In: Serup J, Jemec GBE (eds) Handbook of non-invasive methods and the skin. CRC Press, Boca Raton, pp 179–184
20. Cohen JC, Hartman DG, Garofalo MJ, Basehoar A, Raynor B, Ashbrenner E et al (2009) Comparison of closed chamber and open chamber evaporimetry. Skin Res Technol 15:51–54

Transepidermal Water Loss (TEWL) 32

Joachim W. Fluhr and Razvigor Darlenski

Abbreviations

SC	Stratum corneum
TEWL	Transepidermal water loss

Core Messages

- Skin comprises multiple protective functions including which permeability barrier.
- Transepidermal water loss (TEWL) is a reliable in vivo parameter for the assessment of epidermal permeability barrier function.
- Different in vivo methods for the assessment of TEWL exist: closed ventilated, and open chamber, the latter with the widest popularity in clinical and experimental research.
- Practical guidelines should be followed when measuring TEWL.
- A number of instrument-, subject-, and environment-related variables influence TEWL measurement.

J.W. Fluhr, MD (✉)
Department of Dermatology, Charité – Universitätsmedizin Berlin, Charitéplatz 1, 10117 Berlin, Germany
e-mail: joachim.fluhr@charite.de

R. Darlenski
Department of Dermatology and Venereology, Tokuda Hospital, Sofia, Bulgaria

32.1 Introduction

The skin separates the inner part of our body against the potentially harmful environment. The skin barrier protects the human body against many external stressors, namely, physical stress (e.g., mechanical, thermal, UV radiation), chemical stress (e.g., tensides, prolonged water exposure, solvents), and environmental conditions [1]. Furthermore the skin as a barrier prevents the organism from loss of essential components such as ions, water, and serum proteins. The skin as a barrier also reflects internal processes, diseases, disease activity, and some of the lifestyle, manifested in intrinsic and extrinsic aging. The skin has also sociocultural functions and plays an important role in communication and self-expression.

Beyond its structural components SC is interesting to both clinicians and researchers for its dynamic properties. This encompasses the passage of water and electrolytes from the viable epidermis at one hand and the active and passive transport of exogenous substances (xenobiotics, drug molecules, irritants, allergens, and cosmetic ingredients) on the other hand.

E. Berardesca et al. (eds.), *Non Invasive Diagnostic Techniques in Clinical Dermatology*,
DOI 10.1007/978-3-642-32109-2_32, © Springer Berlin Heidelberg 2014

The aspiration for quantification and qualification dates since the ancient societies. The Vitruvian Man of Leonardo da Vinci is an example of the pioneer efforts to standardize certain measurement units using different body part ratios. Nowadays a demand for standardizing and unifying evaluation procedures in biomedical research is widely accepted. Different noninvasive methods for the in vivo investigation of the skin barrier properties have been developed in the past decades.

Transepidermal water loss (TEWL) is the most prominent in vivo parameter for the evaluation of epidermal permeability barrier function. It is comprised by the insensible perspiration which is based on the diffusion of body water through the SC. Minimizing thermal sweating, hence, is crucial for quantifying TEWL under basal condition. A low TEWL is a characteristic feature of a healthy skin protective state. TEWL measurements can be used to assess the inside-out barrier but also to indirectly predicting the influence of topically applied substances and pharmaceutical compounds at the skin surface [2].

32.2 Measuring Principle and Devices

Measurement of TEWL is typically used to assess objectively barrier function [3, 4]. Thus, the use of this parameter is reasonable in objective disease evaluation scores, such as in atopic dermatitis [5, 6]. There are different methods for TEWL measurement, the unventilated-chamber (closed) method, the ventilated-chamber method, and the open-chamber method. The unventilated chamber potentially occludes the skin and thus is incapable of performing continuous measurement. A certain drawback of the ventilated-chamber method using dry or moistened carrier gas is the interference with the microenvironment over the skin surface. Open-chamber method does not interfere with the microclimate and does not occlude the skin; therefore it is reliable and useful tool for both single and continuous measurements of the evaporative loss from the skin surface. This chapter is focused mainly on the open-chamber measurement.

There are two commercially available open-chamber devices that are widely used in research and cosmetic science: Tewameter® (Courage + Khazaka electronic GmbH, Cologne, Germany) and DermaLab® (Cortex Technology ApS, Denmark).

The measuring principle behind the open-chamber devices is Fick's diffusion law revealing the mass per cm^2 being transported in a defined period of time and calculated by the formula:

$$\frac{dm}{dt} = -D \cdot A \cdot \frac{dp}{dx}$$

where

A = surface in m^2.

m = water transported (in g).

t = time (h).

D = diffusion constant (=0.0877 g/m h mmHg).

p = vapor pressure of the atmosphere (mmHg).

x = distance from skin surface to point of measurement (m).

A schematic overview of an open-chamber device is presented in Fig. 32.1. Density gradient is measured indirectly by two pairs of sensors (temperature and relative humidity) and is analyzed by a microprocessor. In some of the commercially available devices (e.g., Tewameter), the sensors can be preheated in order to achieve steady state values in a minimum of time. After computer calculation the TEWL is displayed in g/m^2/h.

Calibration of the measuring device is critical for performing adequate measurements [7]. It should be performed in accordance with the manufacturer's instruction. In addition some of the manufacturers request a regular accuracy check and calibration to be performed in certain intervals. Adherence to these recommendations is strongly advised.

Due to the discrepancies in the basic principle of the measurement devices, the results obtained from different devices cannot be directly compared. A comparison between three closed-chamber instruments and four open-chamber devices was performed in different models (human skin in vivo, hairless mice in vivo, ex vivo mouse skin model using gravimetric assay) and across a wide range of perturbations (mild, moderate, severe) [8]. TEWL measurements, with both open and closed systems, correlated significantly

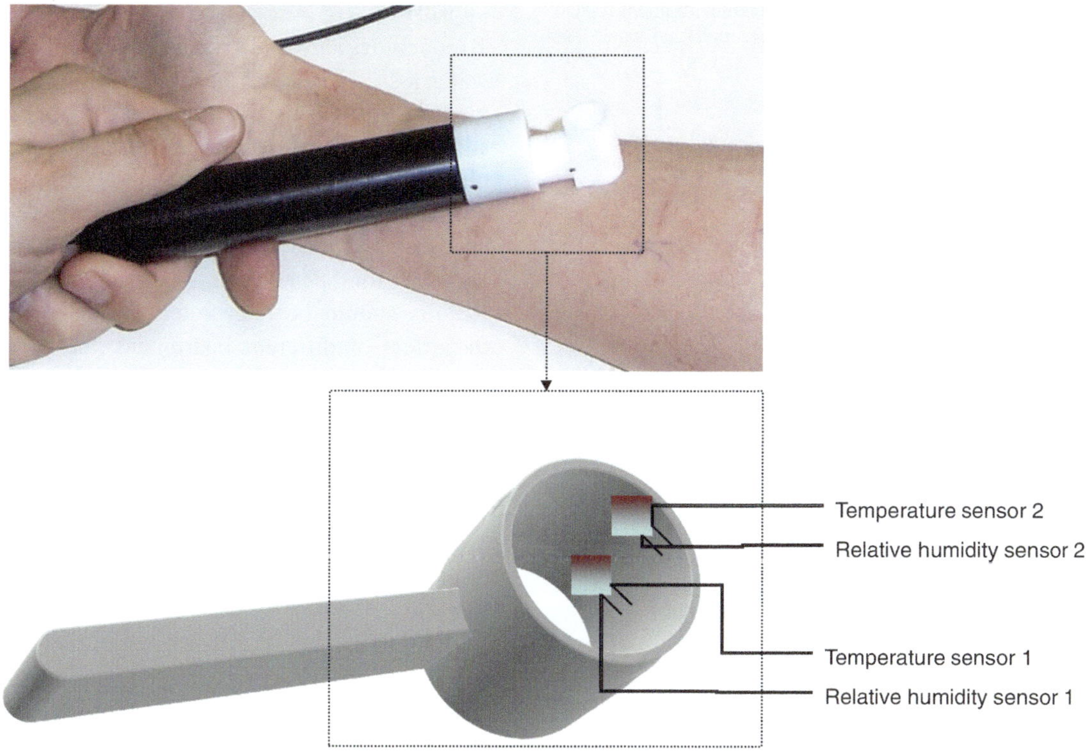

Fig. 32.1 TEWL measurement and a schematic overview of the open-chamber device measuring probe

with absolute rates of water loss, assessed gravimetrically. A high Pearson correlation coefficient was detectable for data from all instruments vs. gravimetrically assessed TEWL. Nonetheless, all methods could be influenced by the microclimatic changes near the skin surface. Hence, the measurements must be performed in climatized rooms with controlled air temperature and relative humidity without direct airflow into the test field.

32.3 TEWL Measurement

TEWL is a very sensitive parameter and could be influenced by numerous variables. Its assessment should be performed in accordance to a study protocol. A comprehensive guide on the practical use of noninvasive biophysical methods in skin physiology and cosmetic studies has recently been published [9]. Nevertheless herein some practical aspects in performing the TEWL measurement include:

- Perform measurements always (when possible) in a climatized room with temperature 18–21 °C and relative humidity of 40–60 %.
- Allow adequate acclimatization time of the study volunteers – 20–30 min prior to the first measurement.
- Keep in mind the circadian rhythm: perform measurements at the same daytime and season; avoid measurements during summer.
- Avoid direct airflow and direct light (could result in temperature increase) at the test site.
- Consider inter- and intra-individual variability in TEWL when calculating the size of the of the study population.
- Allow an equilibration time of about 20 s until a steady state is reached due to the heating of the sensors (shorter for devices with the preheating function).
- Allow for the vapor remnants in the probe to dry after each measurement (seconds), when performing repeated measurements.

Table 32.1 Influence of individual- and environment-related variables on the measurement of transepidermal water loss (TEWL)

Variable		Influence on TEWL measurement
Individual related	Age	+
	Gender	–
	Race/ethnicity	+/–
	Anatomical site	+
	Skin temperature	+
	Sweating	+
	Intake of vasoactive substances (drugs, caffeine, nicotine)	+
Environment related	Air convection	+
	Ambient temperature	+
	Humidity	+
	Direct light	+
	Season	+
	Circadian rhythms	+

Used symbols: "+" influencing, "–" no influence, "+/–" controversial data

- Perform at least two consecutive measurements of neighboring areas of each test site but avoiding the measure from the exactly same site due to possible occlusion.
- Place the probe perpendicular to the skin surface applying a minimal constant pressure.
- Do not moisten and damp the probe.
- The interval from the last product (cosmetic, topical drug) application should be at least 12 h; otherwise the occlusion effect of the product itself or its remnants is measured instead of its effects on the epidermal barrier properties.
- The interval from the last skin cleansing should be at least 2–4 h.

32.4 Variables Influencing the Measurement

Beyond instrument- and environment-related variables, a number of individual-originating factors influence the measurement and should be taken into account, e.g., anatomical site, sweating, and skin surface temperature [3, 4]. Table 32.1 summarizes the possible factors influencing TEWL measurement.

Conclusion

To conclude, TEWL also known as "insensible water loss" is universally recognized to be a measure of skin barrier function at baseline and after experimentally induced barrier perturbations and following the application of topical substances on the skin surface. Open-chamber methods are accepted as standard approach for the in vivo assessment of TEWL. Results should be interpreted critically and the effect of different instrument-, subject-, and environment-related variables must be considered.

References

1. Elias PM, Choi EH (2005) Interactions among stratum corneum defensive functions. Exp Dermatol 14(10): 719–726
2. Darlenski R, Sassning S, Tsankov N, Fluhr JW (2009) Non-invasive in vivo methods for investigation of the skin barrier physical properties. Eur J Pharm Biopharm 72(2):295–303
3. Pinnagoda J, Tupker RA, Agner T, Serup J (1990) Guidelines for transepidermal water loss (TEWL) measurement. A report from the Standardization Group of the European Society of Contact Dermatitis. Contact Dermatitis 22(3):164–178
4. Rogiers V (2001) EEMCO guidance for the assessment of transepidermal water loss in cosmetic sciences. Skin Pharmacol Appl Skin Physiol 14(2): 117–128
5. Angelova-Fischer I, Bauer A, Hipler UC et al (2005) The objective severity assessment of atopic dermatitis (OSAAD) score: validity, reliability and sensitivity in adult patients with atopic dermatitis. Br J Dermatol 153(4):767–773
6. Sugarman JL, Fluhr JW, Fowler AJ, Bruckner T, Diepgen TL, Williams ML (2003) The objective severity assessment of atopic dermatitis score: an objective measure using permeability barrier function and stratum corneum hydration with computer-assisted estimates for extent of disease. Arch Dermatol 139(11):1417–1422
7. Miteva M, Richter S, Elsner P, Fluhr JW (2006) Approaches for optimizing the calibration standard of Tewameter TM 300. Exp Dermatol 15(11): 904–912
8. Fluhr JW, Feingold KR, Elias PM (2006) Transepidermal water loss reflects permeability barrier status: validation in human and rodent in vivo and ex vivo models. Exp Dermatol 15(7): 483–492
9. Fluhr JW (2011) Practical aspects of cosmetic testing. How to set up a scientific study in skin physiology. Springer, Berlin/Heidelberg

Skin Capacitance

33

André O. Barel and Peter Clarys

33.1 General Introduction

The presence of an adequate amount of water in the stratum corneum is important for the following properties of the skin: general appearance of a soft, smooth, well-moisturized skin in contrast to a rough and dry skin and a flexible skin in contrast to a brittle and scaly skin and of an intact barrier function [1–27]. There is no universally accepted theory for explaining the situation of dry skin. Some consider dry skin related to disorders of corneocyte adhesion and desquamation (rough and scaly surface), modifications in the composition of certain epidermal lipids, or disorders of the water-retaining properties of the horny layer [28, 29]. Few data confirm that a situation of dry skin is linked solely to a diminution of the water content of the horny layer. However, the positive physiological effects of topical hydrating products in relieve of the condition of dry skin is universally confirmed [14–26]. As a consequence, given the fact that the presence of an adequate amount of water is an essential prerequisite for the maintenance of the normal structure and properties of the stratum corneum, the in vivo determination of the hydration degree of the horny layer is an important factor in the characterization of the normal horny layer but also in the assessment of the efficacy of various topical hydrating products. It has been assumed for a long time that the electrical properties of the upper part of the skin are related to the water content of the horny layer [3–26]. The total impedance of the skin Z, when the skin is submitted to an alternating current of frequency F, depends on the contribution of the resistance R and the capacitance C, placed in parallel responding to the following relation: $Z = (R^2 + 1/2\Pi FC^2)^{1/2}$, [2, 12, 18, 23, 29, 30]. By using an adequate design of the measuring electrodes, frequency of the current and design of the oscillating electronic circuit, the apparatus measures either (1) the conductance contribution (reciprocal of resistance) of the impedance, (2) the capacitance contribution of the impedance, or (3) the total impedance (combination of both parameters) [18, 29]. A certain number of instruments are commercially available using different techniques, widely used in dermato-cosmetic research and development, and were described in different handbooks and publications [3–5, 9, 10, 15, 17–19, 22–27, 30, 31].

This chapter is mainly concerned with the description of the measurements of the epidermal capacitance using the capacitance method as developed by Courage-Khazaka for the Corneometer [19]. Different aspects of the use of this capacitance instrument will be described in this chapter. An in vitro design was used for calibration, consisting of a determination of measurement depth, an estimation of the effect of dielectric constant of various solvents, the effect of ions, and

A.O. Barel, PhD (✉) • P. Clarys, PhD
Laboratory of Human Biometry and Biomechanics, Faculty of Physical Education and Physiotherapy, Free University of Brussels (VUB), Pleinlaan 2, B-1050, Brussels, Belgium
e-mail: anbarel@vub.ac.be

E. Berardesca et al. (eds.), *Non Invasive Diagnostic Techniques in Clinical Dermatology*,
DOI 10.1007/978-3-642-32109-2_33, © Springer Berlin Heidelberg 2014

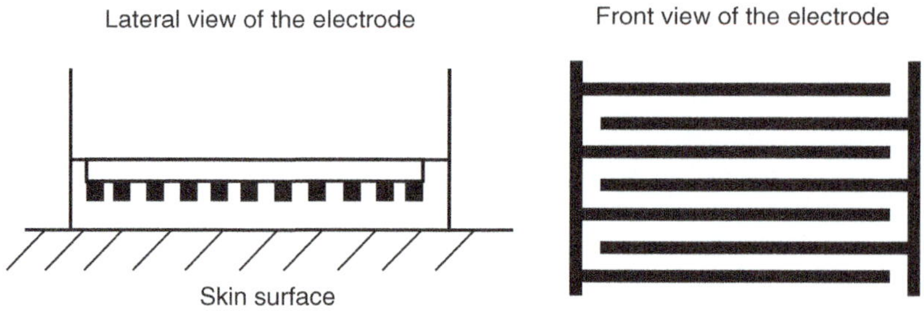

Fig. 33.1 Schematic representation of the measuring probe of the capacitance apparatus (Corneometer CM 825). Front and lateral views of the interdigital electrodes. The electrodes are covered with a vitrified material and consequently not in direct contact with the skin surface (Taken from Barel and Clarys [30], with permission)

desorption hydration measurements. An in vivo design was used to evaluate the accuracy, intraindividual repeatability, range of measurements, and influence of probe application pressure.

33.2 Principle of Electrical Measurement

33.2.1 Measuring Probe and Capacitance Hydration Units

By using an adequate design of the measuring electrodes, frequency of the current, and design of the oscillating electronic circuit, the Corneometer apparatus measures primarily the capacitance contribution of the skin in contact with the measuring electrodes [12, 18, 24, 30].

The measuring probe consists of an interdigital grid of gold-covered electrodes. The active part of the electrode covers a surface of 7×7 mm (Fig. 33.1). The electrodes are 50 μm wide with an interdigital spacing of 75 μm. The electrode is covered by a low dielectric-vitrified material with a thickness of 20 μm. The total probe surface (surface area 0.95 cm^2) is applied on the skin with a constant pressure using a spring system. The measuring probe of the instrument, which used an analog signal in the past, has been updated to digital technology, resulting in a better stability and less interference [32]. When an alternating current is applied at the surface of the skin, the form and depth of the electric field present in the skin depends on the geometry of the electrodes and the dielectric material covering the electrodes (constant capacitance C_0) and of the capacitance of the skin in contact with the electrode (variable capacitance C). As a consequence, the whole system (electrode and the upper parts of the epidermis) works as a variable capacitor. The total capacitance is influenced by changes in the dielectric constant of the biomaterial in contact with the electrode. The total capacitance of this system varies as a function of the degree of hydration of the skin (mostly in the horny layer). A resonating system in the instrument measures the frequency shift of the oscillating system, which results from the changes in the total capacitance of the skin surface. In the case of the analog and digital probe of the Corneometer, the frequency shifts from 0.95 MHz for a hydrated medium to 1.15 MHz for a dry medium [18–30]. The variable and total capacitance of the skin surface is converted into arbitrary units (a.u.) of skin hydration. The Corneometer provides, as all other electrical methods, an indirect measure for hydration of the stratum corneum and no absolute values of water (percent nor weight of water per volume) are given. The a.u. do not correspond to electrical capacitance units (Farad). The mathematical relation between a.u. of capacitance hydration and the amount of water is complex and certainly not linear. The theoretical range of the instrument varies from 0 to 120 a.u. In the latest version, the hydration probe is connected to a central multiprobe unit (MPA), which operates on a PC using adapted Windows software developed by Courage-Khazaka [33]. To warrant a constant pressure of the probe on the skin surface, a spring system is incorporated. According to the manufacturer, the pressure of the old analog probe is in the range of

1.1–1.8 N, whereas the new digital probe operates at a lower pressure (about 1 N or less) [33].

33.3 In Vitro Measurements

33.3.1 In Vitro Calibration of the Corneometer

The Corneometer is factory calibrated using an in vitro method [34]. The principle of this in vitro calibration system developed for the Corneometer by Courage-Khazaka is based on the use of a Courage-Khazaka reference filter pad soaked with a calibration solution (saturated NaCl solution in water). With this impregnated filter pad, one obtains a maximal hydration value of 120±5 a.u. When covering the surface of the impregnated pad with one layer of a polyurethane foil of 15 μm thickness, one should obtain a hydration value of about 20±5 a.u. A calibration certificate is furnished by the manufacturer with, respectively, the high reference values (120±5 a.u.) and low reference values (20±5 a.u.) [34].

33.3.2 In Vitro Effect of Ions on the Hydration Measurements

The effect of ions on the capacitance hydration measurements was analyzed on the Courage-Khazaka calibration filter pad [34] saturated with various aqueous solutions with different ion concentration, ranging from ultrapure distilled water to a saturated NaCl solution (Table 33.1) [35, 36]. The experimental capacitance hydration values (mean value of 119.8±0.2 a.u.) agree very well with the proposed nominal maximum value given by the manufacturer. The influence of electrolytes on capacitance hydration measurements is minimal: only 1.8 % increase in capacitance value was observed when comparing ultrapure distilled water (no salts) with a saturated NaCl solution. As claimed by the manufacturer, the capacitance hydration measurements are not influenced by the presence of electrolytes, which constitutes one of advantages of the capacitance method, compared to other measurements methods. Presence of salts in hydration products which are applied to the skin have only minimal influence on the measurements [37].

Table 33.1 In vitro capacitance hydration measurements on the Courage-Khazaka filter pad saturated with different aqueous solutions

Solution	Corneometer (a.u.)
Ultrapure distilled water	117.7±0.4
Ordinary distilled water	118.8±0.4
Tap water	118.2±1.9
0.15 M NaCl	119.1±0.5
C-K calibration solution	119.8±0.2

Adapted from Clarys et al. [36]
Mean values and SD

Table 33.2 In vitro capacitance hydration measurements (analog and digital probe) in function of the number of plastic sheets covering the Courage-Khazaka calibration filter pad saturated with distilled water

Number of layers	Analog probe (a.u.)	Digital probe (a.u.)
0	117±2	117.5±1.2
1	20±1	19.4±1.8
2	12±1	9.3±0.5
3	9±1	5.4±0.8
4	8±1	4.6±0.7
5	7±1	4.0±0.5

Adapted from Clarys et al. [36]
Mean value and SD of 20 measurements

33.3.3 In Vitro Determination of Depth of Hydration

The manufacturer of the Corneometer claims a measuring depth of 10–20 μm corresponding to the first layers of the stratum corneum [37]. In order to determine the depth of hydration measurements, measurements were performed on the Courage-Khazaka calibration filter pad saturated with distilled water. Consecutively, the filter pad was covered with an increasing number of sheets of polyurethane plastic foil of 15 μm thickness each.

Hydration values were obtained with, respectively, 0, 1, 2, 3, 4, and 5 sheets of plastic foil (see results in Table 33.2), [35, 36].

The experimental capacitance hydration values obtained with the analog (20±1 a.u) and digital

Table 33.3 In vitro capacitance hydration measurements (digital probe) on Courage-Khazaka calibration filter pads saturated with liquids of different dielectric constants

Solvent	Dielectric constant	Corneometer (a.u.)
Mineral oil	2–3	11±2
2-Ethyl-1-hexanol	10.3	20[a]
Cyclohexanol	15.0	33.1±0.5
1-Butanol	17.8	58±3
2-Propanol	18.9	41[a]
1-Propanol	21.8	63±6
Ethanol	24.9	76±6
Ethanol 90 %	30.4	60[a]
Methanol	32.6	85±3
Ethanol 50 %	52.5	99±5
Ethanol 50 %	52.5	83[a]
Ethanol 66 %	61.7	102±5
Water	80.2	118±2
Correlation r (p)		0.922 ($p<0.01$)

Adapted from Clarys et al. [36]
Mean values and SD
[a]Adapted from Bielfeldt et al. [39]

probe (19.4±1.8 a.u.) when covering the saturated filter pad with one foil of 15 μm thickness agree very well with the proposed nominal value given by the manufacturer (20±5 a.u.) and consequently can be used as an in vitro calibration of the instrument. As shown in Table 33.2, a reduction of about 92 % in hydration values (analog probe) and 95 % (digital, probe) were obtained when covering the calibration pad with three layers of foil (45 μm thickness). These results indicate that the penetration depth of the electrical field with both probes of the Corneometer is of the order of magnitude of 45 μm, much higher compared to the claims of the manufacturer (10–20 μm) [37].

33.3.4 In Vitro Capacitance Hydration Measurements on a Filter Pad Saturated with Organic Liquids of Different Dielectric Constant

The capacitance hydration values (digital probe) for mineral oil, various alcohols, and water/ethanol solutions are shown in Table 33.3. A significant correlation was found ($r=0.922$) between the dielectric constant and capacitance values. These capacitance hydration data obtained with the digital probe confirm the reported relation between capacitance hydration obtained by the older analog probe and the dielectric constant of the liquid saturating the filter pad [35, 36, 38, 39]. Clearly these in vitro measurements on a filter pad saturated with solvents of different dielectric constant contribute to validate the eventual use of an organic solvent of known dielectric constant as a possible external standard for calibration of the method [35, 36].

33.3.5 In Vitro Desorption Measurements on Cellulose Filter or Courage-Khazaka Calibration Filter Pad

After impregnation of the porous filter pads with a known amount of distilled water, the desorption of the filter due to evaporation of the water was followed by capacitance hydration measurements (digital probe). Simultaneously, the remaining weight of the water content in the filter was determined using a precision balance. Figure 33.2 shows the desorption curves of the filter pad impregnated with a certain amount of water as followed by capacitance hydration and weight in function of time [36]. A significant high correlation was obtained between the amount of water present in the filter and capacitance hydration values ($r=0.89$, $P<0.001$). However, Fig. 33.2 indicates that at higher water content, the capacitance curve is not following the curve of the amount of water as indicated by weight. Similar water desorption curves were observed with the capacitance measurements as followed with the older analog probe [35] and with the conductance measurements [17, 27]. Again, these in vitro desorption curves confirm that capacitance measurements are related to the amount of water present in the filter pads.

33.4 In Vivo Measurements

33.4.1 Range of Hydration Measurement of the Capacitance Instrument

When considering different skin sites in healthy volunteers, a complete range of hydration values going from very dry (foot sole, hand palm), dry

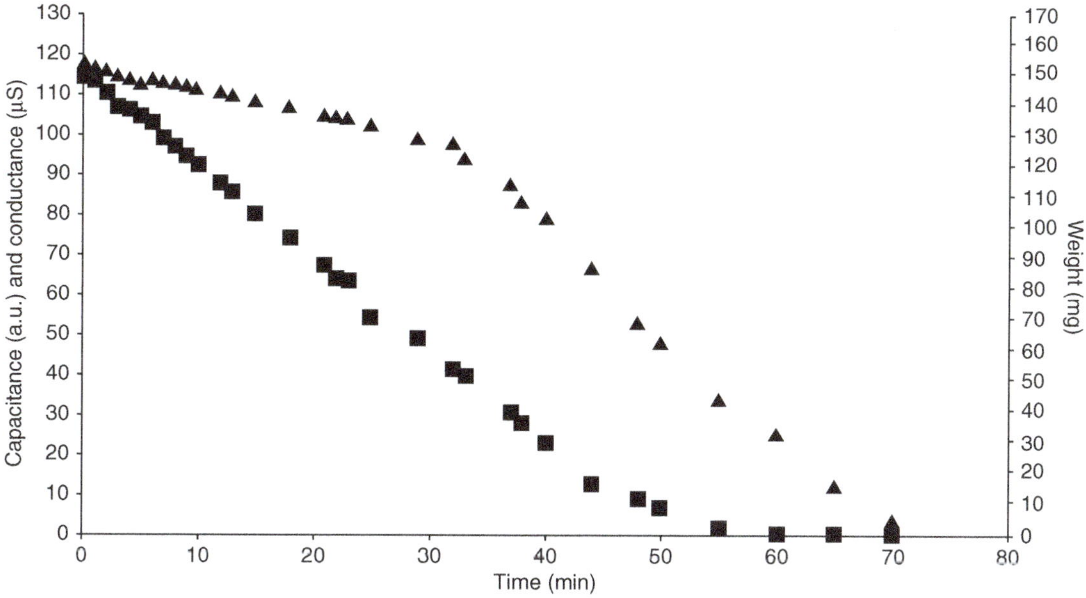

Fig. 33.2 Water desorption curves performed on cellulose filter disk after initial application of 150 µl water on the filter. Capacitance values (a.u.), *square symbols* and weight of the adsorbed water (mg), *triangle symbols* in function of time (Adapted from Clarys et al. [36])

(knee front, forearm dorsal, volar upper arm), hydrated (volar forearm, leg, calf, cheek, and abdomen), to very hydrated (forehead, forearm volar) was envisaged. For each chosen skin site, a test area (2×2 cm) was delimitated. It was assumed that the chosen test area presents more or less a homogenous hydration status. In the selected areas, in order to evaluate the intraindividual variation, 20 consecutive capacitance hydration measurements were carried out. Particular care was taken to avoid consecutive measurements on the same skin spot in order to avoid an occlusive effect provoked by the probe surface. The observed hydration values (see Table 33.4) range from 4.4 to 90.8 a.u., with the digital probe and from 8.3 a.u. (very dry) to 88.8 a.u. as measured with the analog probe [36]. A high significant correlation ($r=0.99$, $P<0.001$) was observed between the analog sensor and the digital sensor technology of the Corneometer. This is completely in line with the results observed in a previous multicenter study in Germany (DGK Skin Task Group) [40]. This group proposed a new definition of, respectively, very dry, dry, and normal state of hydration of the skin, based on capacitance measurements using the Corneometer CM 825 apparatus. Three categories of moisture-related skin types were proposed: very dry skin corresponds to hydration values below 30 a.u., dry skin corresponds to values between 30 and 40 a.u., and normal skin corresponds to values above 40 a.u. [40].

33.4.2 Intraperson Coefficient of Variation

The mean value of the intraperson coefficient of variation as analyzed in function of the different skin regions, corresponding to a wide range of hydration values, demonstrates a negative relation with the hydration values (see Fig. 33.3a, b) [36]. The correlations were, respectively, $r=-0.66$ for the analog probe and $r=-0.63$ for the digital probe. The strongly reduced variability noticed for the capacitance hydration measurements at the well-hydrated skin areas may be an indication for the decreasing sensitivity of the instrument at the higher end of the measurement scale. The mean value of the coefficient of variation over the total range of hydration values is 8.7±2.3 % for the analog probe and 7.2±3.1 % for the digital probe. The intraperson coefficients of variation for the capacitance method are smaller than those observed for the conductance method (C.V.=10.0±3.3 % for the Skicon-200EX) [36].

Table 33.4 In vivo intraindividual measurements of hydration on different body sites in 20 healthy volunteers corresponding to a wide range of levels of hydration of the stratum corneum with corresponding coefficient of variation

Capacitance	CV	Capacitance	CV
Digital a.u.	%	Analog a.u.	%
4.4±0.4	9.1	8.3 ±0.9	11.3
4.6±0.7	14.2	14.3 ±4.0	9.9
14.9±1.6	10.7	19.0 ±2.2	11.6
20.8±2.8	13.5	24.8 ±2.4	9.7
23.3±1.8	7.7	27.5±3.3	12
24.9±2.4	9.6	27.7 ±2.5	9.0
25.6±2.2	8.6	28.7 ±2.7	9.4
25.7±2.9	11.3	29.0 ±2.8	9.7
28.0±1.7	6.1	31.8 ±3.7	11.6
30.2±1.6	5.2	33.9 ±2.1	6.2
30.8±1.8	5.8	36.8 ±4.6	12.5
33.3±1.5	4.5	33.9 ±2.1	6.2
37.3±1.3	3.5	36.8 ±4.6	12.5
37.7±2.7	7.2	41.7 ±3.1	7.4
46.4±3.4	7.3	41.9 ±2.4	5.6
51.3±2.3	4.4	44.6 ±3.6	8.1
53.8±5.4	10.0	49.7 ±5.5	11.0
57.3±4.5	7.8	52.3 ±4.6	8.8
58.8±5.0	8.5	57.0 ±5.0	8.8
61.1±2.6	4.3	62.4±4.2	6.8
62.9±1.7	2.7	65.4 ±5.2	8.0
64.8±1.8	2.8	68.2 ±5.0	7.3
71.5±3.5	4.9	71.3 ±5.6	7.8
87.7±5.2	5.9	83.9 ±5.2	6.2
90.8±3.5	3.8	88.8±2.1	2.4

Adapted from Clarys et al. [36]

Intraperson coefficient of variation in %, mean values, and SD of 20 measurements

Correlation between digital and analog Corneometer: $r=0.98$, $p<0.05$

33.5 Influence of External Factors

33.5.1 Influence of Probe Application Pressure on In Vivo Capacitance Measurements

The measuring probe involves the application of the electrode at a well-defined pressure on the skin in order to ensure a constant contact between the probe and the skin surface [6, 39, 42–46]. Practically, the manufacturer tries to ensure this constant pressure by means of a spring system. For the Corneometer, the realization of the capacitance measurement is indicated by an acoustic signal triggered by an electromagnetic contact in the probe during displacement of the spring. The nominal values of the force of the probe are between 1.0 and 1.8 N, which corresponds to the full displacement of the spring [33]. When considering the validation of the capacitance measuring method, one must take into account the possible influence of the differences in the spring force of the measuring probe on the hydration values obtained. Inter-instrumental differences were observed between the analog probe (1.1–1.8 N) and the digital probe (about 1.0 N or less) [39, 41–46]. Furthermore, it is possible to trigger the hydration measurement at, respectively, lower pressure (incomplete compression of the spring) and at higher pressure (complete compression of the spring) for both type of probes (analog probe; low applied pressure $=1.16\pm0.07$ N and high applied pressure $=1.80\pm0.07$ N and digital probe; low applied pressure $=0.52\pm0.04$ N and high applied pressure $=0.78\pm0.06$ N). Consequently it is possible to carry out capacitance hydration measurements, respectively, at a low and a much higher force of application of the probe on the skin surface [42].

Comparative capacitance measurements (analog and digital probe) were carried out with, respectively, half-way (low pressure) and full displacement (high pressure) of the spring in vivo. A complete range of hydration values going from very dry to well-hydrated skin was investigated (Table 33.5).

Significantly higher hydration capacitance measurements were observed when applying the digital probe with a higher pressure on the skin surface. The percent increase of hydration was very high (40 %) for a very dry/dry skin, moderately higher (20 %) for a normal skin, and low (5–8 %) for a hydrated skin. Similar results were observed with the analog probe [42].

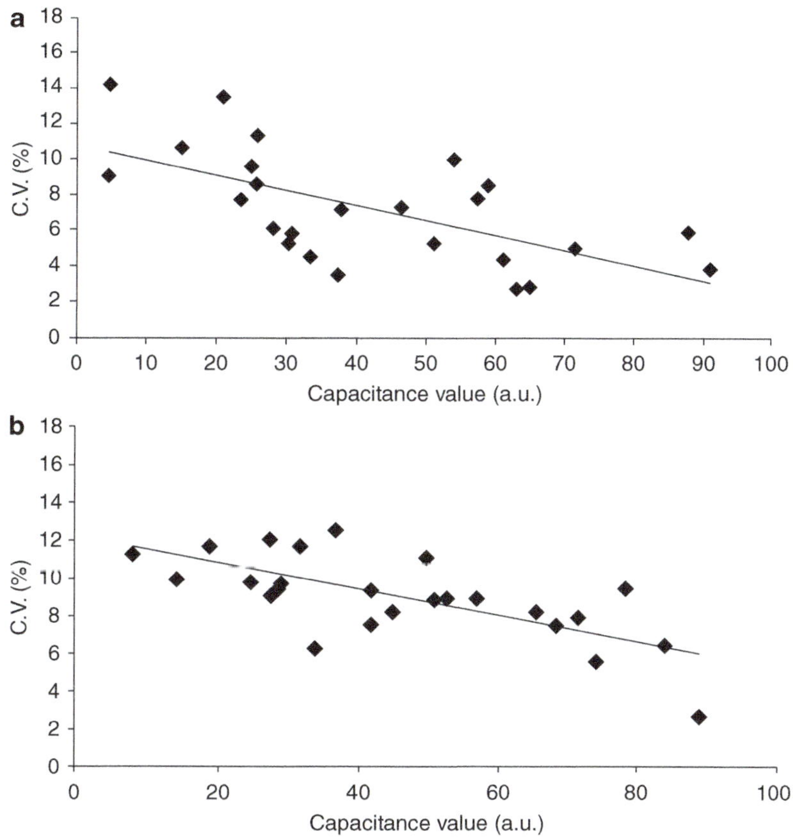

Fig. 33.3 (**a**, **b**) The coefficient of intraperson variability (CV in %) of capacitance hydration measurements on different skin sites. (**a**) Analog probe and (**b**) digital probe (Adapted from Clarys et al. [36])

When considering in vivo experiments with both probes, the pressure effect is considerable for measurements at dry skin surfaces, and the pressure effect decreases as the hydration status of the skin surface increases. Similar results were observed when examining five different Corneometer probes with different spring forces [39]. Several hypotheses can be forwarded to explain this hydration status – dependent pressure effect. At a low pressure, the probe surface may not be completely in contact with the irregular rough surface of the dry skin, resulting in insufficient capacitive material. At a higher pressure, complete contact may be obtained, resulting in a better response to the applied current. Another hypothesis is that at higher pressure, deeper layers contribute in the measurement. This phenomenon may be more pronounced in dry skin, where

Table 33.5 In vivo influence of pressure application of the probe on capacitance values (digital probe) on 10 different skin sites in 20 volunteers

Low pressure (a.u.)	High pressure (a.u.)	Percent increase
8.8±1.8	12.6±1.8*	43
16.3±2.1	23.2±2.8*	42
22.4±3.4	31.4±3.5*	40
23.4±1.9	30.6±1.8*	31
27.5±4.5	37.5±5.5*	35
32.7±2.1	40.5±1.4*	24
57.4±4.6	65.0±2.6*	13
57.7±4.8	63.7±2.6*	10
65.4±7.7	69.9±6.9	7
73.7±3.0	75.8±3.4*	3

Adapted from Clarys et al. [42]
Mean values and SD. Percent increase when comparing a low- and a high-pressure application on the skin
*t-test ($p < 0.05$); when comparing a low and a high pressure of application on the skin surface

the water gradient between the uppermost layers and the deeper layers is considerable. High pressure may compress the uppermost dry layers and collect information from deeper, more hydrated layers. Such an effect may be weaker or minimal in a well-hydrated skin.

These results demonstrate the importance of the physical contact between the probe and the skin surface and the application of a constant pressure of the probe on the skin surface, suggesting clearly, the possible influence of a manipulation effect [45]. In the process of validation of capacitance hydration measurements, a preliminary determination of the pressure of the application of the probe to be used with total compression of the spring using a balance is strongly recommended before undertaking a study with different instruments or when comparing the results of different laboratories.

33.5.2 Influence of Environmental Temperature, Relative Humidity and Season

Several publications described the environment-related variables, which could influence electrical hydration measurements. Hence, guidance methods for standardization were suggested [18, 30, 43–45].

The hydration of the skin using the analog probe was examined under constant temperature (20 ± 2 °C) in function of an increase in relative humidity (from 37 to 87 %). A linear relation ($r = 0.98$) was observed between hydration, as measured by the capacitance method and the external relative humidity [30]. As a consequence, capacitance hydration measurements must be performed in an experimental room where the relative humidity is kept more or less constant (ideally 50 ± 5 % RH). Similarly, a steady increase of capacitance was observed in function of temperature above 22 °C, which corresponds to a higher hydration of the horny layer and the beginning of invisible perspiration [44]. The temperature of the experimental room must be kept constant and preferably below 22–23 °C (ideally 20–21 °C).

It is well known that during the winter, a dry, rough, sometimes scaly skin may be present in older people or other sensitive persons at the lower legs and other anatomic skin sites (winter xerosis). This clinical picture is ideal for evaluating a moisturizing effect. In contrast the skin is optimally hydrated in summer due to the generally high external humidity. This situation is not suited to demonstrate the favorable effect of moisturizing products. Seasonal variations of temperature and humidity have to be taken in account when doing long time hydration studies. Due to possible large climatic variations in the summer period, testing should be avoided in the months of June, July, and August in the north-western countries [18, 30, 43, 45].

Conclusions

Different methods of in vitro calibration systems of capacitance hydration measurements on cellulose and filter pads were proposed by three research groups [35, 38, 39] and by the manufacturer [34] and can be used nowadays for a preliminary calibration. Using this in vitro method, significant correlations were established between capacitance hydration values and the quantity of water adsorbed on the filter discs or pad and the quantity of water adsorbed on the filters and with the dielectric constant of the liquids saturating these filters. Finally the detection of the depth of measuring of the capacitance method was determined.

The new digital version of the Corneometer shows a broad range of hydration readings (from 4.4 to 90.8 a.u.) with low intraperson variability. The mean value of coefficient of variability over the whole range of hydration values is 7.2 ± 3.1 %. Compared to the older analog system, the new digital hydration probe shows a lower pressure of application on the skin surface: generally less than 1.0 N. The probe is very sensitive for evaluating dry/very dry skin conditions. However, similar to the older analog probe, the new device is less sensitive at very high levels of hydration. As recommended by many authors, well-standardized experimental conditions concerning season, ambient conditions, physical contact of the probe with the skin, etc. must be taken in account in order to obtain valid, accurate, and reproducible experimental data [43–45].

Acknowledgments The authors thank Mr. Gabriel Khazaka and Mrs. Diana Khazaka, Courage-Khazaka Electronic Köln, for technical information concerning the Corneometer CM 825 and MPA.

References

1. Blank IH (1952) Factors which influence the water content of the S.C. J Invest Dermatol 18:433–440
2. Schwan HP (1957) Dielectric properties of living tissues. Adv Biol Med Phys V:147–163
3. Tagami H, Oki M, Iwatsuki K, Kanamura Y, Yamada M, Ichijo B (1980) Evaluation of skin surface hydration in vivo by electrical measurement. J Invest Dermatol 75:500–507
4. Tagami H, Kanamaru Y, Inoue K et al (1982) Water sorption–desorption test of the skin in vivo for functional assessment of the stratum corneum. J Invest Dermatol 78:425–428
5. Lévêque JL, de Rigal J (1983) Impedance methods for studying skin moisturization. J Soc Cosmet Chem 34:419–428
6. Dikstein S, Bercovici PG (1985) Measurement of skin surface capacitance at 16 Hz and at other frequencies. Bioeng Skin 1:357
7. Moseley H, English JS, Coghill GM, Mackie RM (1985) Assessment and use of a new skin hygrometer. Bioeng Skin 1:177–192
8. Salter DC (1987) Instrumental methods for assessing skin moisturization. Cosmet Toilet 102:103–109
9. Lévêque JL, Grove G, de Rigal J, Corcuff P, Kligman AM, Saint Léger D (1987) Biophysical characterization of dry facial skin. J Soc Cosmet Chem 82: 171–177
10. Tagami H (1988) Impedance measurements. In: Lévêque JL (ed) Cutaneous investigation in health and disease. Marcel Dekker, New York, pp 79–111
11. Blichman CW, Serup J (1988) Assessment of skin moisture. Acta Dermatol Venereol (Stockholm) 68:284–290
12. Barel AO, Clarys P, Wessels B, de Romsée A (1991) Non invasive electrical measurement for evaluating the water content of the horny layer: comparison between the capacitance and the conductance measurements. In: Scott RC, Guy RH, Hadgraft J, Bodde HE (eds) Prediction of percutaneous penetration – methods, measurements, modelling. IBC Technical Services, London, pp 238–247
13. Van Neste D (1991) Comparative study of normal and rough human skin hydration in vivo: evaluation with four different instruments. J Dermatol Sci 2:119–124
14. Gabard B (1994) Testing the efficacy of moisturizers. In: Elsner P, Berardesca E, Maibach HI (eds) Bioengineering of the skin: water and the stratum corneum. CRC Press, Boca Raton, pp 147–170
15. Salter DC (1994) Further hardware and measurement approaches for studying water in the stratum corneum. In: Elsner P, Berardesca E, Maibach HI (eds) Bioengineering of the skin: water and the stratum corneum. CRC Press, Boca Raton, pp 205–215
16. Loden M (1995) Biophysical methods of providing objective documentation of the effects of moisturizing creams. Skin Res Technol 1:101–108
17. Tagami H (1995) Measurements of electrical conductance and impedance. In: Serup J (ed) Handbook of non-invasive methods and the skin. CRC Press, London, pp 159–164
18. Barel AO, Clarys P (1995) Measurement of epidermal capacitance. In: Serup J (ed) Handbook of non-invasive methods and the skin. CRC Press, London, pp 165–170
19. Courage W (1995) Hardware and measuring principle. In: Elsner P, Berardesca E, Maibach HI (eds) Bioengineering and the skin: water and stratum corneum. CRC Press, Boca Raton, pp 171–175
20. Moss J (1996) The effect of 3 moisturizers on skin surface hydration. Electrical conductance (Skicon 200), capacitance (Corneometer CM420) and transepidermal water loss (TEWL). Skin Res Technol 2:32
21. Piérard GE (1997) Charactérisation des peaux sèches, la biométrologie complète la clinique. Cosmétologie 14:48–51
22. Clarys P, Barel AO, Gabard B (1999) Non-invasive electrical measurements for the evaluation of the hydration state of the skin: comparison between three commercial instruments: the Corneometer, the Skicon and the Nova DPM. Skin Res Technol 5:14–20
23. Barel AO, Clarys P, Gabard B (1999) In vivo evaluation of the hydration state of the skin: measurements and methods for claim support. In: Elsner P, Merk HF, Maibach HI (eds) Cosmetics: controlled efficacy studies and regulation. Springer, Berlin, pp 57–80
24. Bernengo JC, de Rigal J (2000) Techniques physiques de mesure de l'hydratation du stratum corneum in vivo. In: Agache P (ed) Physiologie de la peau et explorations fonctionnelles cutanées. Editions Médicales Internationales, Cachan, pp 117–162
25. Lee CM, Maibach HI (2002) Bioengineering analysis of water hydration: an overview. Exog Dermatol 1: 269–275
26. Barel AO (2002) Product testing: moisturizers. In: Elsner P, Berardesca E, Wilhelm KP, Maibach HI (eds) Bioengineering of the skin: skin biomechanics. CRC Press, Boca Raton, p 241
27. Tagami H (2006) Epidermal hydration: measurement of high frequency electrical conductance. In: Serup J, Jemec GB, Grove GL (eds) Handbook of non-invasive methods and the skin, 2nd edn. CRC Press, Boca Raton, pp 329–336

28. Clarys P, Barel AO (1995) Quantitative evaluation of skin surface lipids. Clin Dermatol 13:307–3021
29. Clarys P, Barel AO, Gabard B (2011) Fonction sébacée et métrologie de l'excrétion sébacée. In: Revuz J (ed) Cosmétologie et Dermatologie Esthétique. Elsevier Masson, Paris, pp 1–8
30. Barel AO, Clarys P (2006) Measurement of epidermal capacitance. In: Serup J, Jemec GB, Grove GL (eds) Handbook of non-invasive methods and the skin, 2nd edn. CRC Press, Boca Raton, pp 337–344
31. Gabard B, Clarys P, Barel AO (2006) Comparison of commercial electrical measurement instruments for assessing the hydration state of the stratum corneum. In: Serup J, Jemec GB, Grove GL (eds) Handbook of non-invasive methods and the skin, 2nd edn. CRC Press, Boca Raton, pp 351–358
32. Khazaka G (2008) Improvement of digital sensor technology of the Corneometer CM 825 probe. Available from http://www.courage-khazaka.de/index.php/en/all-downloads/downloads-en/file/6-brochcm825sm815e
33. Khazaka D (2008) Nominal pressure of application of the analogic and digital hydration measuring probe of the Corneometer CM 825 and MPA. Available from http://www.courage-khazaka.de/index.php/en/all-downloads/downloads-en/file/28-brochmpae
34. Khazaka G (2008) Technical information of the Courage-Khazaka calibration filter pad and the in vitro calibration of the Corneometer CM 825 (analogic and digital probes). Available from http://www.courage-khazaka.de/index.php/en/faq-en/faq-scientific-devices/60-general-questions-on-the-scientific-measurements#faqsc1
35. Barel AB, Clarys P (1997) In vitro calibration of the capacitance method (Corneometer CM 825)- and conductance method (Skicon 200) for the evaluation of the hydration state of the skin. Skin Res Technol 3:107–133
36. Clarys P, Clijsen R, Taeymans J, Barel AO (2012) Hydration measurements of the stratum corneum: comparison between the capacitance method (digital version of the Corneometer 8525) and the impedance method (Skicon 200EX). Skin Res Technol 18(3):316–323
37. Courage-Khazaka technical information of the Corneometer Cm 825 (2008) Available from http://www.courage-khazaka.de/index.php/en/faq-en/faq-scientific-devices/61-corneometer
38. Fluhr JW, Gloor M, Lazzerinin SL, Kleesz ZP, Grieshaber R, Berardesca E (1999) Comparative study of five instruments measuring stratum corneum hydration (Corneometer CM 820 and CM 825, Skicon 200, Nova DPM 9003 and Dermalab). Part I. In vitro. Part II. In vivo. Skin Res Technol 5:156–178
39. Bielfeldt S, Brandt M, Gerstenkorn A, Wilhelm KP (2010) Capacitance measurement of skin moisture: sophisticated calibration of instruments. Available from http://www.courage-khazaka.de/index.php/de/alle-downloads/downloads-de/file/221-litcm-multicenter
40. Heinrich U, Koop U, Leveneu-Duchemin MC, Osterrieder K, Bielfeldt S, Charcaut C et al (2003) Members of the DGK task Force « Skin Hydration », multicenter comparison of skin hydration in terms of physical, physiological and product dependant parameters by the capacitive method (Corneometer CM 825). J Cosmet Sci 25:31–53
41. André T, De Wan M, Lefèvre P, Thonnard JL (2008) Moisture Evaluator: a direct measure of fingertip skin hydration during object manipulation. Skin Res Technol 14:385–389
42. Clarys P, Clijsen R, Barel AO (2011) Influence of probe application on in vitro and in vivo capacitance (Corneometer CM 825) and conductance (Skicon 200 EX) measurements. Skin Res Technol 17:445–450
43. Wilhelm KP (1989) Possible pitfalls in hydration measurements. In: Elsner P, Barel AO, Berardesca E, Gabard B, Serup J (eds) Skin bioengineering: techniques and applications in dermatology and cosmetology. Karger, Basel, pp 223–234
44. Rogiers V, Derde MP, Verleye G, Roseeuw D (1990) Standardized conditions needed for skin surface hydration measurements. Cosmet Toilet 105:73–82
45. Berardesca E, European Group for Efficacy Measurements on Cosmetics and Other Topical Products (EEMCO) (1997) EEMCO guidance for the assessment of stratum corneum hydration: electrical methods. Skin Res Technol 3:126–132
46. O' Goshi KI, Serup J (2005) Inter-instrumental variation of skin capacitance measured with the Corneometer. Skin Res Technol 11:107–109

Part VI

Erythema and Blood Flow

34 The Use of Visual Grading Scales in Evaluating Skin Irritation and Sensitization: A Historical Perspective

Miranda A. Farage, Howard Maibach, Klaus E. Andersen, Jean-Marie Lachapelle, Petra Kern, Cindy Ryan, Jeanne Ely, and Amita Kanti

34.1 A History

The four cardinal signs of an inflammatory response are redness, heat, swelling, and pain and are said to have been described as early as 178 AD by Celsus [1]. Skin irritation and sensitization are types of inflammatory processes that exhibit these signs. The degree of redness, heat, swelling, and pain are proportional to the extent of the reaction. For decades patch tests with visual assessment of these signs have been the cornerstone of evaluating skin irritant and sensitization reactions. The history of patch testing has been reviewed by Devos and van der Valk [2] and, more recently, Lachapelle [3]. There are reports of patch tests being used for dermatologic evaluations as far back as 1847 when Städeler (Germany) described a method called the blotting paper strip, which he had devised to examine the effects triggered by *Anacardium occidentale*, or the cashew plant. Jadassohn (1896) (Austria) applied a mercurial preparation topically and reproduced dermatitis in sensitized subjects. Bloch (1911) (Basel) described a patch testing technique for allergens using visual assessment. In the USA, Sulzberger (1931) published on the utility of patch testing in the diagnosis of hypersensitivity [4]. In 1932, Graves described the use of patch testing in determining the etiology of some types of dermatosis reactions, specifically in identifying contact sensitization reactions from plants [5].

Unfortunately, these early publications did not provide details on the visual assessment of the

M.A. Farage, PhD (✉)
Clinical Sciences Innovation Center, The Procter & Gamble Company, Winton Hill Business Center, 6110 Center Hill Avenue, Cincinnati, OH 45224, USA
e-mail: farage.m@pg.com

H. Maibach, MD
Dermatology Department, University of California, San Francisco, CA, USA

K.E. Andersen, DMSc
Department of Dermatology and Allergy Centre, Odense University Hospital, Odense, Denmark

J.-M. Lachapelle, MD, PhD
Department of Dermatology, Catholic University of Louvain, Brussels, Belgium

P. Kern, PhD
The Procter and Gamble Company, Brussels, Belgium

C. Ryan, BS
Global Product Surveillance, The Procter & Gamble Company, Mason Business Center, 8700 Mason-Montgomery Rd, Mason 45040, USA

J. Ely, BSMD
Global Product Surveillance, The Procter & Gamble Company, Winton Hill Business Center, 6110 Center Hill Avenue, Cincinnati, OH 45224, USA

A. Kanti, PhD
Global Product Surveillance, The Procter & Gamble Company, Sharon Woods Business Center, 11530 Reed Hartman Highway, Cincinnati, OH 45224, USA

Adapted from Farage MA et al (2011) Historical perspective on the use of visual grading scale in evaluating skin irritation and sensitization. Contact Dermatitis 65:65–75. With kind permission from John Wiley & Sons publisher

E. Berardesca et al. (eds.), *Non Invasive Diagnostic Techniques in Clinical Dermatology*, DOI 10.1007/978-3-642-32109-2_34, © Springer Berlin Heidelberg 2014

Table 34.1 Early standardized visual assessment scales

Description	Numerical equivalent
(a) Human diagnostic patch test, 1939 [8]	
Spotty erythema	(+)
Few or more infiltrated papules	+, +(+), ++
… with minimal vesicles	++(+)
Typical oozing eczema in the test area	+++
Pronounced erythema, strong infiltration and closely spaced vesicles	+++(+)
…with bullae	++++
(b) Human diagnostic patch test, 1944 [9]	
Erythema on the area of application	1+
Erythema and edema	2+
Erythema, edema, papules, and a few vesicles	3+
Erythema, edema, many vesicles, and, in some cases, ulceration	4+
(c) Primary irritation, animals and humans, 1944 [10]	
Erythema and eschar formation	
Very slight erythema (barely perceptible)	1
Well-defined erythema	2
Moderate to severe erythema	3
Severe erythema (beet redness) to slight eschar formation (injuries in depth)	4
Edema formation	
Very slight edema (barely perceptible)	1
Slight edema (edges of area well defined by definite raising)	2
Moderate edema (area raised ~1 mm)	3
Severe edema (raised >1 mm and extending beyond area of exposure)	4

reactions, but generally described positive reactions as definite surface changes on the skin, or reactions that included redness, itching, and vesiculation or pustules. Some early authors acknowledged that there were varying degrees of response, and specified that a distinction should be made between a minimal reaction (mere redness) and a maximal reaction (infiltration, non-purulent, and/or purulent bulla). In the 1930s, German dermatologists standardized diagnostic patch testing to a certain degree describing both patch test technique and scoring of test reactions using grading scales [6, 7]. A standard series of patch tests for contact sensitization, with a standardized grading scale, was published by Bonnevie (1939) in Copenhagen (Table 34.1a) [8].

Early patch test appliances used gauze squares or blotting paper to hold the test material. These were applied to a strip of adhesive tape. A layer of oiled silk in between the gauze and the tape provided some occlusion. Improvements in patching appliances were described by Guild in 1939 [11] and Rokstad in 1940 [12]. These authors described the use of a glass square and a small plastic chamber, respectively, to hold the test substances in place. These early appliances developed into a number of standardized patch testing systems, such as the Finn chamber, the Webril patch, the Hill Top patch, and the Van der Bend chamber. The Finn chamber has been widely used by the International Contact Dermatitis Research Group, the North American Contact Dermatitis Group, and many national contact dermatitis research groups in routine diagnostic testing of patients with suspected contact allergies [13, 14].

In 1944, Schwartz and Peck described in greater detail a method of diagnostic patch testing for human subjects in order to identify the potential sources of dermatitis reactions [9]. They described the use of gauze squares saturated with the test substance and applied to the skin for 24 h. The gauze was covered with a layer of cellophane and adhesive tape. A four-point numerical scale was specified for scoring, with higher numbers indicating more extensive reactions (Table 34.1b). Interestingly, they also proposed the use of two diagnostic patch tests applied 10–14 days apart as a means of predicting if a material was a contact sensitizer, using the same four-point scale. This was an early reference to the use of patch testing in prophetic tests for contact sensitization. Protocols for such investigations were subsequently developed.

In using patch testing for contact allergy, authors often stress the importance of avoiding irritant reactions that can be difficult to distinguish from allergic reactions. However, patch testing and other protocols involving dermal exposure have been developed precisely for the purpose of evaluating irritation. In the same year as the Schwartz and Peck publication, in response to the introduction of many new compounds and topical preparations to the marketplace, scientists at the Food and Drug Administration (FDA) recognized a need to bring some standardization to tests conducted to study local toxicity to the skin and mucous membranes [10]. These FDA scientists described several protocols to evaluate the toxicity (acute and repeated dose) and primary irritation of dermally applied materials. They included a standard visual assessment scale for primary irritant reactions on the skin of animals and humans whereby numerical values were assigned to correspond with descriptions of the physiological reactions of redness and edema (Table 34.1c). Such a numerical scale was proposed to facilitate comparisons of different compounds and enable arithmetic interpretations.

The use of diagnostic patch testing in humans became very widespread in the 1950s and 1960s. In 1970, members of the International Contact Dermatitis Research Group (ICDRG) published a common grading scale to facilitate accurate communication of patch test results (Table 34.2a) [15]. This scale became widely used for allergic contact dermatitis.

With passage of the Consumer Product Safety Act and the creation of the Consumer Product Safety Commission (CPSC) in 1972, the regulation of consumer product safety was placed at the federal level, with a common set of uniform standards [17]. The CPSC requested that the Committee on Toxicology of the National Academy of Sciences (NAS) develop methods for evaluating dermal irritation in animals and humans. These were published in a procedures manual in 1977 [16] and used for many years by regulatory agencies, private companies, and research organizations. The test for skin irritation in animals used the visual assessment scale shown in Table 34.2b. This scale was very similar to that described by Draize et al. [10] (Table 34.1c) except for the addition of scores of "0" for no erythema and no edema.

The protocol for evaluating skin irritation in animals originally published in the 1977 NAS procedures manual, and the accompanying visual assessment scale, remained virtually unchanged for decades. Such protocols were specified by worldwide regulatory agencies as part of the risk assessment process [18–20]. In addition, the American Society for Testing and Materials (ASTM) published a version of the protocol [21]. Results of skin irritation testing in animals provided the essential data for determining if materials or products posed a hazard with regard to skin irritation and, therefore, required cautionary or warning labeling [22, 23]. This protocol is still

Table 34.2 Revised standardized visual assessment scales recommended by regulatory agencies and professional societies (1970s)

Description	Numerical equivalent
(a) Human diagnostic patch test recommended by ICDRG (1970) [15]	
Not tested	NT
Doubtful reaction	? +
Weak (non-vesicular) reaction	+
Strong (edematous or vesicular) reaction	++
Extreme reaction	+++
(b) Primary irritation in animals recommended by NAS (1977) [16]	
Erythema and eschar formation	
No erythema	0
Very slight erythema (barely perceptible)	1
Well-defined erythema	2
Moderate to severe erythema	3
Severe erythema (beet redness) to slight eschar formation (injuries in depth)	4
Edema formation	
No edema	0
Very slight edema (barely perceptible)	1
Slight edema (edges of area well defined by definite raising)	2
Moderate edema (area raised ~1 mm)	3
Severe edema (raised >1 mm and extending beyond area of exposure)	4
(c) Primary irritation in humans recommended by NAS (1977) [16]	
No response	0
Questionable or faint, indistinct erythema	1/2 or ı/
Well-defined erythema	1
Erythema with slight to moderate edema	2
Vesicles (small blisters) or papules (small circumscribed elevations)	3
Bullous (large blister), spreading, or other severe reaction	4

used by many, although since the 1990s, there has been a shift towards alternative methods that use fewer or no animals [24].

A protocol for evaluating irritant reactions in humans was also specified in the NAS procedures manual in 1977 [16]. This protocol differed from the recommendation of Draize et al., in that it specified a 4-h exposure time rather than 24 h and suggested an alternative visual assessment scale (Table 34.2c) based on one developed by Marzulli and Maibach in 1975 [25]. This integrated scoring scale for erythema and edema accommodated the wider variety of visible responses seen in humans, i.e., papules, vesicles, and spreading reactions.

The NAS standards published in the 1977 procedures manual did not specify particular protocols for evaluating contact sensitization in either animals or humans; however, several protocols were mentioned as being acceptable. For evaluations in animals, these included methods utilizing intradermal injection [10], topical application [26, 27], or a combination of injection and topical application [28]. Evaluations in humans included protocols utilizing topical application, with [29] or without [10, 30] prior irritation of the patch test site. All of these methods used visual assessment to evaluate the results, and many specified a scale to score the extent of erythema and a means for describing the degree of edema, induration, vesiculation, and necrosis.

As in the case of test protocols for skin irritation, standard protocols for evaluating contact sensitization using visual assessment scales were published worldwide by regulatory agencies [31, 32] and voluntary testing organizations [33]. Over the years, protocol variations using visual assessment scales have developed to evaluate contact sensitization in animals [34, 35] and humans [36, 37]. As with evaluations of irritation, there has been a shift in recent years towards alternative methods that use fewer animals. Methods such as the local lymph node assay are being incorporated into the overall approach for evaluating sensitization [38, 39].

For decades, visual assessment of skin reactions has been used to evaluate the safety of chemicals and preparations that contact the skin and to meet regulatory requirements for product labeling. However, product manufacturers also have a need to evaluate other aspects of dermal safety that may not be adequately addressed using the protocols discussed above. As a result, numerous testing methods have been developed to provide information on skin effects and product mildness. These protocols vary widely in exposure methods and are often tailored for specific types of products and designed to simulate or exaggerate actual exposure conditions that consumers may experience in the real world. Some examples are provided in Table 34.3. This list is not intended to be a comprehensive list of all test methods currently available or in use, nor is it an endorsement of these methods. Rather, it is meant to convey the broad and diverse nature of test protocols that have evolved to evaluate unique circumstances of exposure. While the means of exposures in these tests vary, they share a common reliance on a visual scoring system to determine skin effects.

Table 34.3 Examples of test protocols that use visual scoring

Method	Reference
Forearm wash test	[40, 41]
Repeat application tests	[42]
Repeat open application tests	[43]
Cumulative irritation tests	[43]
Chamber scarification test	[44]
The Behind-the-Knee test	[45]
Fabric wear test	[46, 47]
Extended duration home use test	[46, 47]
Soap chamber test	[48]
Chamber-scarification test	[49]

34.2 Reliability

Visual assessment of skin reactions has been the cornerstone of skin testing for decades. Reliable testing facilities make every effort to train the skin graders thoroughly and to control the grading conditions to reduce the potential for variation. However, it remains a subjective evaluation, and the question is sometimes raised around the reproducibility of the evaluation. Investigators have conducted studies demonstrating the reliability of visual scoring, as reviewed below.

Falk et al. [50] investigated agreement in visual scoring using both an irritant (sodium lauryl sulfate, or SLS) and induced photoreactions. Volunteer test subjects were patched with a concentration series of SLS (0.5–6 %) and increasing exposure to ultraviolet B (UVB) irradiation (42–126 mJ/cm^2). Test sites were evaluated by both the subjects themselves (on a 2-point scale, i.e., "positive" or "negative") and the investigators (using 5-point scales, with any score of > "0" being considered a "positive"). In the irritant patch series, self-readings correlated well with investigator determinations, with absolute agreement in 76.9 % of cases. Disagreements occurred when the subject reacted to a broad range of concentrations of SLS. In these instances, the test subjects tended to underreport the reactions. In the UVB series, absolute agreement between investigator and self-reading was found in 85 % of cases. Weighted kappa (κ) for agreement between observations was 0.76 for SLS reactions and 0.83 for UVB reactions.

Ivens et al. [51] evaluated the agreement in visual grading among 5 dermatologists using photographs of 55 patch tests performed using different allergens. The observers were asked to score the patch tests using both the ICDRG standard 5-point scale and a simplified 3-point scale ("positive," "negative," and "irritant"). Using the ICDRG scale, results showed fair agreement when pair-wise comparisons were conducted among the five dermatologists, with κ values mostly in the range of 0.42–0.68 (one outlier produced a κ value of 0.37). Agreement was much higher using the 3-point scale with kappa values for pairwise comparisons in the range of 0.61–0.80.

Uter et al. [52] also used digital photographic images to evaluate the agreement on visual scoring. These investigators used 20 different digital images of patch test reactions to contact sensitizers or irritants. These were examined by 122 participants from several countries. Participants included consultant dermatologists, researchers, practicing dermatologists, and professors in dermatology. Each participant rated each image on a 5-point scale for contact sensitization or as an irritant reaction. On average, 63.5 % of all ratings were in exact agreement. This percentage increased to 87.8 % when the ratings were considered on the basis of nonpositive (no evident reaction) and positive (any skin reaction, including faint erythema).

A systematic study of the agreement on scoring clinical erythema was conducted by Lock-Andersen and Wulf [53]. Twenty-one healthy, Caucasian volunteer subjects received a series of graded UV exposures (5 increments and no exposure) symmetrically on both buttocks, resulting in a total of 42 phototests for clinical scoring. Twenty to twenty-four hours after UV irradiation, eight physicians working in a dermatology department where phototesting is a routine procedure visually scored the erythema reactions on a 5-point ordinal scale. Scoring was done blinded with the identity of the subjects concealed. Each site was scored twice by the same physician to evaluate intra-observer agreement. Scores from different physicians allowed evaluation of interobserver variation. Observers were asked to determine the irradiation dose that produced barely perceptible erythema (+) and erythema with a well-defined border (≥1). Variability for determining barely perceptible erythema was less than the variability for determining well-defined erythema. For both inter- and intra-observer variability, the agreement at the lowest dose was high (83.5 and 90.5 %, respectively), with κ values consistent with "good agreement" (0.68 and 0.80, respectively). Less agreement was observed as the doses of UV increased. At the highest two dose increments (i.e., the fourth and fifth), agreement was "moderate." Interobserver agreement was 57.1 % (κ=0.45) and 52.9 % (κ=0.40), respectively, and intra-observer agreement was 66.7 % for both doses (κ=0.53 and 0.57, respectively).

Griffiths et al. [54] conducted an intralaboratory comparison of results from patch testing of volunteer human subjects using eight substances and 20 % SLS as a positive irritant control. Patches of test substances were applied progressively for up to 4 h to avoid unacceptably high levels of irritation. All laboratories scored reactions using the same 4-point scale; however, any degree of erythema (i.e., a reaction of "1+" or greater) was an indication of a positive reaction for irritation. Laboratories were located in the UK, Germany, the USA, and China, and studies were conducted over a 2-year period (1994–

1996). Specific protocols varied in the total overall duration of each patch application and the site of reapplication (i.e., the same site or a different site). In spite of variations in the protocols, there was agreement in approximately 80 % of the evaluations between the various laboratories with regard to the classification of materials as irritants for the purposes of labeling.

All of the above investigations lead to an overall conclusion that visual scales are reproducible with regard to classifying reactions as "negative" or "positive," meaning it is relatively easy to evaluate a reaction for the absence or presence of erythema. Evaluating the intensity of the erythema appears to be somewhat less reproducible. This would suggest that simplified grading scales may be the best approach when comparing results from different investigators or different studies. However, Basketter et al. [55] conducted a controlled evaluation of the reproducibility of patch test scoring in the testing of irritants using an expanded grading scale. Graders were trained using a photographic index of possible skin reactions and side-by-side blinded assessment with experienced graders. Human volunteers were patched (24 and 48 h) with test materials, including an irritant control (0.3 % sodium lauryl sulfate) and a negative control (water). A total of five observers were used for the studies. In each study, pairs of observers conducted scoring, each working independently. Observers were blinded with regard to the specific test or control sample patched at the individual test sites. Skin reactions were scored on a 10-point scoring scale where "0" was no reaction and "9" was an intense, strong erythema that may spread beyond the exposure site. In this scoring scale, the bottom half of the 10-point scale (scores of "0" to "5") focused on minimal or slight reactions. The top half (scores of "6" to "9") was comparable to the more traditional scoring scale in the description of reactions, such as that shown in Table 34.2c. Thus, the scoring scale was designed to differentiate between mild to medium reactions. Paired scoring was conducted for a total of 1,445 individual assessment points over the course of 3 experiments. In just over half (50.31 %), the two scorers gave the identical grade (Table 34.4). In 1,275 cases (88.24 %), scores were within 1 numerical grade of agreement, and in 1,411 cases (97.64 %), scores were within 2 numerical grades. In only 34 of the cases (2.35 %) was the discrepancy 3 or more numerical grades out of the 10-point scale. This study suggests that relatively subtle degrees of erythema could be reliably discriminated by observers trained to recognize reactions of various intensity using photographic images.

Table 34.4 Comparison of visual scoring in human patch testing: Interobserver agreement on a 0–9 scale [55]

Difference between observers[a]	Frequency over studies	
	Number	Percentage
0	727	50.31
1	548	37.92
2	136	9.41
3	16	1.11
4	10	0.69
5	3	0.21
6	3	0.21
7	2	0.14
0 or 1	1,275	88.24
0, 1, or 2	1,411	97.64
≥3	34	2.35

[a]The difference represents the degree of disagreement on the numerical scale of 0–9 between two observers scoring independently

34.3 Visual Scoring Versus Bioengineering Methods

Visual scoring can be standardized through appropriate training and experience. However, this method of data collection still involves an element of human judgement. A common question is whether a purely objective measure would increase the robustness of the protocols and enable the differentiation of more subtle skin reactions. Although bioengineering methods have come into more common use for the evaluation of skin reactions and disease states, there is little evidence that the use of such bioengineering methods improves the sensitivity and overall quality of the test result. Further, a potential limitation of bioengineering methods is that each measure evaluates only one aspect of the overall reaction. For example, increases in color intensity (redness) can be measured using spectroradiometers or chromometers. Increases in blood flow can be measured using laser Doppler flowmeters (LDF). Barrier disruption can be measured using evaporimeters. Increases in skin thickness from edema and vasodilatation can be measured using ultrasound scanners. However, only visual assessment can evaluate all aspects of the reaction simultaneously. A number of investigators have conducted comparative studies between visual scoring and the results obtained from various bioengineering methods. Some of these comparison studies are reviewed below.

In a study conducted by Willis et al. [56], 10 subjects were patch tested with a series of seven irritants and two controls. Reactions were scored visually on a 5-point scale and evaluated by LDF. The mean LDF readings correlated well with visual scores producing a Spearman rank correlation coefficient (r) 0.91 ($p<0.001$). In addition, the means from LDF measures that corresponded to an incremental increase in the visual scores were significantly different from each other, i.e., the mean LDF for the reactions that were scored "0.5" on the visual scale was significantly different from the mean for sites that were scored "0" and for sites that were scored "1" on the visual scale. However, when LDF measures for individual sites were considered rather than the means, there were some notable outliers that were not consistent with the visual scores, indicating LDF measures may not be always be reliable for individual assessments. Also, a minimum time to measure blood flow at 10 patch sites was 45 min, compared to 5 min for a visual assessment. In addition, the subject was required to lie in a relaxed position in quiet surroundings in order to obtain steady LDF readings. The authors concluded that LDF was not suitable for routine patch testing.

Seidenari and Belletti [57] compared both colorimetric and echographic procedures to visual scoring. These investigators patched 120 nickel-sensitive subjects with 40 mg of 5 % nickel sulfate and evaluated the reactions at baseline (i.e., before patching) and at 72-h post-patching using visual, echographic, and colorimetric assessments. They found that echographic evaluations showed significant increases in value for reactions scored visually as "+," "++," and "+++." Colorimetric values increased with higher visual scores. However, the intense reactions (+++) did not differ from the more moderate reactions (++).

Agner and Serup [58] applied increasing concentrations of SLS (0.12–1 %) via patch tests on the volar surface of forearm of 12 volunteer subjects. Reactions were evaluated at 24 and 48 h on a visual scale and by several bioengineering measures (transepidermal water loss, or TEWL, colorimetry, LDF, and skin thickness using an ultrasound scanner). As expected, the mean of the visual scores showed a linear dose response to increasing concentrations of SLS (statistical analysis was not reported). All bioengineering measuring methods showed a significant difference between normal skin and 0.12 % SLS and a significant linear dose response.

Lahti et al. [59] conducted a comparative study of visual assessment (4-point scale), LDF, and three reflectance erythema meters for measuring erythema reactions. Skin reactions in 15 subjects were caused by UVB irradiation or two well-known substances known to elicit non-immunological immediate contact reactions in the skin (benzoic acid, or BA, and methyl nicotinate, or MN). The correlation between visual assessment and the measured values for the BA, MN, and UVB reactions was good with all the instruments, with the Spearman's rank order correlation coefficient varying between 0.54 and 0.79 ($p<0.001$).

The use of LDF was evaluated by Wahlberg [60]. Eleven common solvents were applied undiluted to the volar surface of the forearm of one individual using two different techniques: a 5-min application inside a 20-mm-diameter glass ring or application of 0.1 ml which was allowed to spread on the surface and evaporate for 5 min. The results showed a similar rank order of potency to visual scoring. The author concluded that LDF may provide additional information in the case of marginal irritant effects; however, it cannot replace the traditional visual scoring.

In a study by Babulak et al. [61], the irritation caused by the six different treatments with soap and detergent bar solutions applied to human volar forearms was assessed by three measures: visual scoring, Minolta Chroma Meter measurements of redness, and TEWL. All three methods of assessment were in agreement and showed that two treatments, B and D, were significantly ($p \leq 0.01$) less irritating than the other four treatments (A, C, E, and F). The correlation coefficients between visual scoring and both instrumental measures were significant: $r=0.97$ for the chroma meter measurement ($p \leq 0.003$) and $r=0.86$ for TEWL ($p \leq 0.03$).

Zuang et al. [62] undertook an investigation with the goal of determining which of three noninvasive methods (LDF, evaporimeter, and colorimeter) was the best predictor of the visual scores. Fifty-one patients (Caucasian males and females), sensitized to various chemicals, participated in the study. A standard patch test series containing different test allergens was adapted to each patient's medical history and applied to the skin of the patients' backs. In total, 133 patch test reactions were evaluated on the 51 patients in the study. All three instrumental methods were useful in discriminating between any positive reaction and a negative one. However, with regard to the severity of reactions based on the visual scale of 1+ to 4+, each of the three instrumental methods misclassified skin reactions on known sensitized subjects in 36–38 % of cases. The percentage of misclassifications was reduced to 21 % when all three methods were used concurrently.

Fluhr et al. [63] evaluated 8 bioengineering methods in the assessment of irritant reactions. The goal was to determine which of these methods could detect significant skin changes induced by different types of irritation over time. Irritation reactions were induced using 5 methods: 2 % SLS (6-h patch), tape stripping (tape applied 15 times), 0.05 % tretinoin (2-h patch), 0.5 % dithranol (2-h patch), and ultraviolet exposure (6-s exposure to 30 Wm^{-2} UVB and 95 Wm^{-2} UVA). Measurements were conducted prior to irritation induction and at 24, 48, and 72 h. The results showed that visual assessment was more discriminating in detecting skin changes over time than colorimetric measurements (red-green and black-white measures), capacitance measures, and measures of TEWL. Visual assessment performed similarly to Mexameter hemoglobin measures and LDF. The only method that was more discriminating over time than visual assessment was LDF.

Magnusson and Koskinen [64] evaluated the effects of topical capsaicin treatment using several methods to determine if some were more sensitive than others in detecting skin irritation. The methods included impedance, visual assessment with measurement of the flare, panelists' descriptions of cutaneous sensations, LDF measures of cutaneous blood flow changes, and impedance monitor measurement of changes in the skin. Visual assessment, descriptions of sensations, and LDF measurements produced similar results and showed a maximum effect at approximately 30 min, with resolution after 3 h. Electrical impedance did not show any significant differences when the values at baseline were compared to those after capsaicin application.

Ollmar et al. [65] compared bioengineering methods using irritation induced by 24-h patch tests with SLS at concentrations ranging from 0.1 to 5.0 %. Skin sites were evaluated prior to treatment (baseline) and on days 1, 2, 7, and 14 post-treatment. They found close agreement between visual scoring, electrical impedance, and TEWL. Electrical capacitance and LDF produced results inconsistent with the other measures, and results using these methods showed fewer significant changes from baseline values.

Fullerton et al. [66] compared laser Doppler perfusion imaging in scoring the responses of skin sites treated with three irritant materials at increasing concentrations (SLS, retinoic acid, and nonanoic acid). Visual scoring was conducted using a 0–4 grading scale. Laser Doppler perfusion imaging was conducted after a 15-min acclimation of subjects in a temperature- and humidity-controlled environment. A dose response was observed using both the visual scoring and the LDF, indicating similar discrimination for these two assessment methods.

Wigger-Alberti et al. [67] evaluated the irritancy of 3 metal working fluids using visual scoring, TEWL, and chromometry. Two exposure methods were used to produce irritation: a 24-h patch test and a cumulative irritation test (6-h exposure each day for 13 days except on the 2-day weekend). All methods of evaluation, i.e., visual scoring, TEWL, and chromometry, were in agreement and produced the identical rank order of irritation of the 3 test materials in both exposure models.

Held et al. [68] compared visual scoring with the DermaSpectrometer in detecting contact sensitization in allergic subjects. The DermaSpectrometer measures an erythema index based on blood flow. The study was designed to determine which scoring method better differentiated between degrees of reactivity: visual scoring or measured erythema index. Patients with formaldehyde allergy were patched with a concentration series (0–10,000 ppm formaldehyde), and reactions evaluated. A correlation coefficient was calculated for the concentration of formaldehyde and either the visual score or the erythema index. The resulting visual scores showed a higher correlation coefficient ($r=0.60$) with the applied concentrations than the spectrometer measure of erythema index ($r=0.35$), demonstrating that the visual scoring was better able to discriminate between degrees of reactivity.

Andersen and Straberg [69] evaluated the utility of measuring perfusion (via LDF) and swelling (via skin fold thickness) in allergic contact dermatitis reactions in guinea pigs. Contact allergy to chlorocresol was induced in guinea pigs using a maximization method, and animals were challenged with two concentrations of the substance (1 and 0.1 %). Visual scores that were not indicative of sensitization were "0" and "1" (no and slight reactions, respectively). Scores indicative of sensitization were "2" and "3" (moderate and intense reactions, respectively). Measures of perfusion discriminated between animals that were not sensitized (i.e., scores of "0" or "1") versus animals that were sensitized (i.e., scores of "2" or "3"). Measures of swelling discriminated between the severity of the challenge response (i.e., between scores of "2" versus scores of "3"). Both measures discriminated between the challenge concentrations of 1 and 0.1 % in sensitized animals. These authors concluded that quantitative methods were useful supplemental measures in evaluating interobserver and interlaboratory differences.

34.4 Comparing the Benefits of Visual Scoring Versus Bioengineering Methods

Clearly, a key benefit to bioengineering methods over visual scoring is the fact that bioengineering methods are purely objective, with no element of human judgement. They can directly measure the specific physiological changes that may occur during skin reactions. Quantitative measures of barrier function, hydration, changes in blood flow, etc., can provide insight into biochemical changes and mechanisms of action that occur over the time course of the reaction. In addition, most bioengineering methods provide results on a linear, continuous scale, making statistics easier to apply. Visual assessment cannot directly measure these specific physiological changes nor are the visual scales continuous or necessarily linear.

Conversely, there are some clear disadvantages to the use of bioengineering methods. Some of these parameters, such as blood flow, skin thickness, and skin pigmentation, can vary between individuals or between different regions on the body. For bioengineering measures that rely on optics, such as LDF, care must be taken to avoid applying samples to test sites that may have uneven pigmentation, such as blemishes or freckles [66]. Scaling from dryness, a common condition with some irritants, may also produce optical interference [66]. Sample application on the back may not be appropriate since movement from normal respiration may create artifacts [66]. Andersen and Straberg observed effects on measures of perfusion and swelling due to tape and shaving irritation at the test sites [69]. This is only a partial list of some of the controls and considerations necessary when using bioengineering measures. A more complete discussion can be found in Breternitz et al. [70].

The use of bioengineering methods adds considerable complexity and cost to the testing protocols. For example, many of these measures can be done effectively only after the subject has acclimatized in a temperature- and humidity-controlled environment. Not only does this requirement add steps to the study, it adds to the time commitment required of volunteers considering participation in the study. Further, the time required to make the measurements is almost invariably longer than that required for a trained, visual grader to make an assessment. It may be necessary to control ambient light and temperature using some bioengineering measures. As a result, the grader must develop an additional set of skills to control and monitor these variables.

Current test protocols for skin reactions have been optimized over decades to provide results that can be assessed via visual scoring. These protocols continue to serve us well in the practical evaluation of irritation or sensitization reactions. There is little evidence that bioengineering measures improve the sensitivity or overall quality of current testing methods, as illustrated by the preceding discussion of comparative studies. However, the rapidly evolving bioengineering techniques will likely prove more useful in providing information that will deepen our understanding of the processes that occur in inflammation and skin reactions. Physiological changes that occur early in the process of irritation, such as changes in blood flow, moisture content, and pH, would be expected to occur before any reaction is visible. In fact, such changes undoubtably occur in reactions that never present with visible changes, i.e., subclinical reactions. An ability to measure these physiological changes may allow us to detect subtle skin effects [71]. We have recently reported that the use of cross polarized light to allow visualization of subsurface erythema increased the ability to distinguish between subtle skin effects caused by two very similar products following application of the materials in a Behind-the-Knee study [72]. In order to fully exploit this new technology, it is likely that specific protocols need to be developed with this objective in mind.

The usefulness of any scoring method is directly related to the purpose of the experiment. When evaluating the presence or absence of irritation or sensitization as the result of exposure to a material, bioengineering methods tend to provide information that is not necessary without providing a better measure of these endpoints. After all, there is no need to use a sledgehammer to drive a tack or a caliper to measure a kilometer. Basketter et al. [55] pointed out that when a rapid visual assessment is all that is necessary to meet the purpose of the experiment, such an assessment can be done reliably in the hands of a sufficiently trained grader. When benefits and cost are weighed in the balance, the visual assessment scales popularized by Draize and others remain an effective, practical method of evaluating skin irritation and contact sensitization reactions.

Acknowledgments The authors are grateful to Drs. F. Ajayi, K.W. Miller, M. Simone, R. Bartolo, and M. Lafranconi for the technical review of this manuscript and to Terresa L. Nusair, Ph.D., of the Health and Environmental Safety Alliance (Cincinnati, Ohio) for technical input.

References

1. Movat HZ (1979) The acute inflammatory process. In: Movat HZ (ed) Inflammation, immunity and hypersensitivity, 2nd edn. Harper & Row, Hagerstown, pp 1–162
2. Devos SA, van der Valk PG (2002) Epicutaneous patch testing. Eur J Dermatol 12(5):506–513
3. Lachapelle JM (2010) Giant steps in patch testing: a historical memoir. SmartPractice, Phoenix
4. Sulzberger MB, Wise F (1931) The contact or patch test in dermatology: its uses, advantages and limitations. Arch Derm Syphilol 23(3):519–531
5. Graves JM (1932) The patch test—its use in dermatology – report of cases. Cal West Med 36(3):157–160
6. Mayer RL (1930) Das Gewerbeekzem. Schriften aus dem Gesamtgebiet der Gewerbehygiene. Verlag von Julius Springer, Berlin, Neue Folge, Heft 30:1–89
7. Blumenthal F, Jaffé K (1933) Ekzem und Idiosynkrasie. Verlag von S. Karger, Berlin, pp 1–188
8. Bonnevie P (1939) Aetiologie und Patholgenise de Ekzemkrankheiten. Klinische Studien über due Ursachen der Ekzeme unter besonderer Berücksichtigung des Diagnosticshen Werted der Eksamproben. Busch, Copenhagen/Barth, Letapzig
9. Schwartz L, Peck SM (1944) The patch test in contact dermatitis. Public Health Rep 59:546–557
10. Draize JH, Woodard G, Calvery HO (1944) Methods for the study of irritation and toxicity of substances applied topically to the skin and mucous membranes. J Pharmacol Exp Ther 82(3):377–390

11. Guild TB (1939) Window patch test. Arch Derm Syphilol 39:807
12. Rokstad I (1940) Patch test, new modification of (the chamber method). Arch Derm Syphilol 41(4): 649–653
13. Cronin E (1980) Technique of patch testing. In: Cronin E (ed) Contact dermatitis. Churchill Livingstone, Edinburgh, pp 1–19
14. Fisher AA (1986) The role of patch testing. In: Fisher AA (ed) Contact dermatitis, 3rd edn. Lea & Febiger, Philadelphia, pp 9–29
15. Wilkinson DS, Fregert S, Magnusson B, Bandmann HJ, Calnan CD, Cronin E, Hjorth N, Maibach HJ, Malalten KE, Meneghini CL, Pirilä V (1970) Terminology of contact dermatitis. Acta Derm Venereol 50:287–292
16. National Academy of Sciences (1977) Dermal and eye toxicity tests. In: Principles and procedures for evaluating the toxicity of household substances. National Academy of Sciences, National Research Council, Washington, DC, pp 23–59
17. Consumer Product Safety Commission CPSC History. http://www.usrecallnews.com/2008/05/us-consumer-product-safety-commission-cpsc.html. Accessed 24 Mar 2010
18. U.S. Environmental Protection Agency, OPPTS Harmonized Test Guidelines Series 870 – Health Effects Test Guidelines. OPPTS 870.2500 Acute Dermal Irritation. EPA712-C-98-196 (1998) http://www.epa.gov/oppts/pubs/frs/publications/Test_Guidelines/series870.htm. Accessed 6 Apr 2010
19. OECD Guideline for the Testing of Chemicals 404 (2002) Acute toxicity: dermal irritation/corrosion. http://ecb.jrc.ec.europa.eu/testing-methods/annex5/. Accessed 2 Apr 2010
20. Title 16: Commercial Practices PART 1500—Hazardous Substances and Articles; Administration and Enforcement Regulations § 1500.41 Method of testing primary irritant substances. http://www.access.gpo.gov/nara/cfr/waisidx_00/16cfr1500_00.html. Accessed 2 Apr 2010
21. American Society for Testing and Materials (ASTM)-ASTM F 719 – 81 (Reapproved 2007). Standard Practice for Testing Biomaterials in Rabbits for Primary Skin Irritation. http://enterprise.astm.org
22. U.S. Consumer Product Safety Commission, Office of Compliance Requirements under the Federal Hazardous Substances Act: Labeling and Banning Requirements for Chemicals and Other Hazardous Substances. 15 U.S.C. § 1261 and 16 C.F.R. Part 1500 (2002) http://www.cpsc.gov/businfo/regsumfhsa.pdf. Accessed 2 Apr 2010
23. Guidance to Regulation (EC) No 1272/2008 on Classification, Labeling and Packaging of substances and mixtures (2009) 3.2.2.1.2.5 Testing methods: In vivo data. p 222. http://ecb.jrc.ec.europa.eu/documents/Classification-Labelling/CLP_Guidance_to_Regulation.pdf. Accessed 2 Apr 2010
24. OECD Guidelines for the Testing of Chemicals. Tests No. 430, 431 and 435. http://titania.sourceoecd.org/vl=2183769/cl=13/nw=1/rpsv/cw/vhosts/oecdjournals/1607310x/v1n4/contp1-1.htm. Accessed 26 May 2010
25. Marzulli FN, Maibach HI (1975) The rabbit as a model for evaluating skin irritants: a comparison of results obtained on animals and man using repeated skin exposures. Food Cosmet Toxicol 13(5): 533–540
26. Buehler EV (1965) Delayed contact hypersensitivity in the guinea pig. Arch Dermatol 91(2):171–175
27. Ritz HL, Buehler EV (1980) Planning, conduct, and interpretation of guinea pig sensitization patch tests. In: Drill VA, Maibach HI (eds) Current concepts in cutaneous toxicity. Academic, New York, pp 25–40
28. Maguire HC Jr (1973) The bioassay of contact allergens in the guinea pig. J Soc Cosmet Chem 24:151–162
29. Kligman AM (1966) The identification of contact allergens by human assay. III. The maximization test: a procedure for screening and rating contact sensitizers. J Invest Dermatol 47:393–409. doi:10.1038/jid.1966.67
30. Shelanski MV, Gabriel L (1961) Cutaneous toxicity evaluation of Air Force development materials – IV. ASD Technical Report 61–77. McGregor & Werner, Inc, Dayton. Contract No. AF 33(616)-6962, Project No. 7165, Task No. 71836
31. U.S. Environmental Protection Agency, OPPTS Harmonized Test Guidelines Series 870 – Health Effects Test Guidelines (2003) OPPTS 870.2600. Skin Sensitization. EPA 712–C–03–197. http://www.epa.gov/oppts/pubs/frs/publications/Test_Guidelines/series870.htm. Accessed 6 Apr 2010
32. OECD Guideline for the Testing of Chemicals 406 (1992) Skin sensitization. http://ecb.jrc.ec.europa.eu/testing-methods/annex5/. Accessed 2 Apr 2010
33. American Society for Testing and Materials (ASTM)-ASTM. Active Standard: D 6355 – 07. Standard Test Method for Human Repeat Insult Patch Testing of Medical Gloves. http://enterprise.astm.org
34. Klecak G (1987) Identification of contact allergens: predictive tests in animals. In: Marzulli FN, Maibach HI (eds) Dermatotoxicology, 3rd edn. Hemisphere Publishing, New York, pp 227–276
35. Andersen KE, Maibach HI (1985) Guinea pig sensitization assays: an overview. In: Andersen KE, Maibach HI (eds) Contact allergy: predictive tests in guinea pigs, vol 14, Current problems in dermatology. Karger, Basel, pp 263–290
36. Marzulli FN, Maibach HI (1987) Contact allergy: predictive testing in humans. In: Marzulli FN, Maibach HI (eds) Dermatotoxicology, 3rd edn. Hemisphere Publishing, New York, pp 319–340, Alternative ref Marzulli 1976
37. Hjorth N (1987) Diagnostic patch testing. In: Marzulli FN, Maibach HI (eds) Dermatotoxicology, 3rd edn. Hemisphere Publishing, New York, pp 307–317
38. McNamee PM, Api AM, Basketter DA, Gerberick GF, Gilpin DA, Hall BM, Jowsey I, Robinson MK (2008) A review of critical factors in the conduct and

interpretation of the human repeat insult patch test. Regul Toxicol Pharmacol 54:24–34
39. OECD Guideline for the Testing of Chemicals 429 (2010) Skin sensitization: local lymph node assay. http://ecb.jrc.ec.europa.eu/testing-methods/annex5/. Accessed 29 Oct 2010
40. Farage MA (2000) Development of a modified forearm controlled application test method for evaluating the skin mildness of disposable wipe products. J Cosmet Sci 51:153–167
41. Lukacovic MF, Dunlap FE, Michaels SE, Visscher MO, Watson DD (1988) Forearm wash test to evaluate the mildness of cleansing products. J Soc Cosmet Chem 3(9):355–366
42. Finkelstein P, Laden K, Meichowski W (1963) New methods for evaluating cosmetic irritancy. J Invest Dermatol 40:11–14
43. Hannuksela M, Heikki S (1986) The repeated open application test (ROAT). Contact Dermatitis 14: 221–227
44. Andersen KE (1996) Reproducibility of the chamber scarification test. Contact Dermatitis 43:181–184
45. Farage MA (2006) The Behind-the-Knee test: an efficient model for evaluating mechanical and chemical irritation. Skin Res Technol 12(2):73–82
46. Bannan EA, Griffith JF, Nusair TL, Sauers LJ (1992) Skin testing of laundered fabrics in the dermal safety assessment of enzyme-containing detergents. J Toxicol Cutan Ocul Toxicol 11(4):327–339
47. Rodriguez C, Calvin G, Lally C, Lachapelle JM (1994) Skin effects associated with wearing fabrics washed in commercial laundry detergents. J Toxicol Cutan Ocul Toxicol 13(1):39–45
48. Frosch PJ, Kligman AM (1979) The soap chamber test; a new method for assessing the irritancy of soaps. J Am Acad Dermatol 1:35–41
49. Frosch PJ, Kligman AM (1976) The chamber-scarification test for irritancy. Contact Dermatitis 2(6):314–324
50. Falk M, Anderson C (2010) Reliability of self-assessed reading of skin tests: a possible approach in research and clinical practice? Dermatol Online J 16(2):4
51. Ivens U, Serup J, O'goshi K (2007) Allergy patch test reading from photographic images: disagreement on ICDRG grading but agreement on simplified tripartite reading. Skin Res Technol 13(1):110–113
52. Uter W, Becker D, Schnuch A, Gefeller O, Frosch PJ (2007) The validity of rating patch test reactions based on digital images. Contact Dermatitis 57:337–342
53. Lock-Andersen J, Wulf HC (1996) Threshold level for measurement of UV sensitivity: reproducibility of phototest. Photodermatol Photoimmunol Photomed 12:154–161
54. Griffiths HA, Wilhelm K-P, Robinson MK, Wang XM, McFadden J, York M, Basketter DA (1997) Interlaboratory evaluation of a human patch test for the identification of skin irritation potential/hazard. Food Chem Toxicol 35:255–260 (Hard copy)
55. Basketter D, Reynolds F, Rowson M, Talbot C, Whittle E (1997) Visual assessment of human skin irritation: a sensitive and reproducible tool. Contact Dermatitis 37(5):218–220
56. Willis CM, Stephens CJ, Wilkinson JD (1988) Assessment of erythema in irritant contact dermatitis. Comparison between visual scoring and laser Doppler flowmetry. Contact Dermatitis 18(3): 138–142
57. Seidenari S, Belletti B (1998) The quantification of patch test responses: a comparison between echographic and colorimetric methods. Acta Derm Venereol 78(5):364–366
58. Agner T, Serup J (1990) Sodium lauryl sulphate for irritant patch testing—a dose–response study using bioengineering methods for determination of skin irritation. J Invest Dermatol 95(5):543–547
59. Lahti A, Kopola H, Harila A, Myllylä R, Hannuksela M (1993) Assessment of skin erythema by eye, laser Doppler flowmeter, spectroradiometer, two-channel erythema meter and Minolta chroma meter. Arch Dermatol Res 285(5):278–282
60. Wahlberg JE (1984) Erythema-inducing effects of solvents following epicutaneous administration to man – studied by laser Doppler flowmetry. Scand J Work Environ Health 10(3):159–162
61. Babulak SW, Rhein LD, Scala DD, Simion FA, Grove GL (1986) Quantification of erythema in a soap chamber test using the Minolta chroma (reflectance) meter: comparison of instrumental results with visual assessments. J Soc Cosmet Chem 37:475–479
62. Zuang V, Archer G, Rona C, Vignini M, Mosca M, Berardesca E (2000) Predicting visual assessment of allergic patch test reactions by non-invasive measurements. Skin Pharmacol Appl Skin Physiol 13(1): 39–51
63. Fluhr JW, Kuss O, Diepgen T, Lazzerini S, Pelosi A, Gloor M, Berardesca E (2001) Testing for irritation with a multifactorial approach: comparison of eight non-invasive measuring techniques on five different irritation types. Br J Dermatol 145(5):696–703
64. Magnusson BM, Koskinen LD (1996) Effects of topical application of capsaicin to human skin: a comparison of effects evaluated by visual assessment, sensation registration, skin blood flow and cutaneous impedance measurements. Acta Derm Venereol 76:129–132
65. Ollmar S, Nyren M, Nicander I, Emtestam L (1994) Electrical impedance compared with other non-invasive bioengineering techniques and visual scoring for detection of irritation in human skin. Br J Dermatol 130:29–36
66. Fullerton A, Rode B, Serup J (2002) Skin irritation typing and grading based on laser-Doppler perfusion imaging. Skin Res Technol 8:23–31
67. Wigger-Alberti W, Hinnen U, Elsner P (1997) Predictive testing of metalworking fluids. A comparison of 2 cumulative human irritation models and correlation with epidemiological data. Contact Dermatitis 36:14–20

68. Held E, Lorentzen H, Agner T, Menné T (1998) Comparison between visual score and erythema index (DermaSpectrometer) in evaluation of allergic patch tests. Skin Res Technol 4(4):188–191
69. Andersen KE, Straberg B (1985) Quantification of contact allergy in guinea pigs by measuring changes in skin blood flow and skin fold thickness. Acta Derm Venereol 65:37–42
70. Breternitz M, Fluhr JW, Berardesca E (2006) Technical bases of biophysical instruments used in sensitive skin testing. In: Berardesca E, Fluhr JW, Maibach HI (eds) Sensitive skin syndrome. Taylor & Francis, New York, pp 75–106
71. Farage MA (2005) Are we reaching the limits or our ability to detect skin effects with our current testing and measuring methods for consumer products? Contact Dermatitis 52(6):297–303
72. Farage MA (2008) Enhancement of visual scoring of skin irritant reactions using cross-polarized light and parallel-polarized light. Contact Dermatitis 58(3): 147–155

Surface Skin Temperature in Tests for Irritant Dermatitis

35

Miranda A. Farage, Baiyang Wang, Kenneth W. Miller, and Howard Maibach

35.1 Introduction

Testing for potential skin effects is a key part of the overall safety assessment and product claims support for many consumer products. Modern products in the tissue and paper products categories, including facial tissues, catamenial products, baby wipes, and baby and adult diapers, are inherently very mild to skin when tested even under very stringent conditions. We have been exploring various approaches to increase our abilities to differentiate between very similar, mild products in order to select the most skin compatible ingredients and materials [1]. One approach has been to increase the sensitivity of scoring reactions.

Visual assessment of skin reactions has been used for many decades and represents the cornerstone of skin irritation testing [2]. Although it is a subjective evaluation, when done by a trained individual, it is an inexpensive, reliable, and reproducible means of evaluating skin effects. However, several bioengineering methods have been developed recently that may increase our abilities to detect early, subclinical reactions.

The development of a high-precision, handheld, infrared thermographic scanner has made it feasible to conveniently measure localized changes in skin temperature such as those that may occur as a result of mild reactions to topical exposures. Measuring the temperature of the skin is noninvasive and painless and, unlike visual scoring of erythema reactions, completely objective. We investigated surface skin temperature changes as a result of mild reactions in a number of clinical study protocols developed to evaluate skin effects. The objective was to determine if changes in skin temperature were consistent with visual signs of erythema and other endpoints typically used in the protocols. In addition, we wished to determine if incorporating this endpoint into studies commonly used to screen for skin responses to consumer products would increase the ability to discriminate between similar, mild products.

35.2 Materials and Methods

35.2.1 Subjects

All protocols involving human testing were conducted in accordance with the Declaration of Helsinki [3] and were approved by the test facility's Institutional Review Board. Subjects were healthy, adult volunteers who had signed an informed consent. Participation was completely

M.A. Farage, PhD (✉) • B. Wang, MSc
K.W. Miller, PhD
Feminine Innovation Center, The Procter & Gamble Company, Winton Hill Business Center, 6110 Center Hill Road, 136, Cincinnati, OH 45224, USA
e-mail: farage.m@pg.com

H. Maibach, MD
Dermatology Department, University of California, San Francisco, CA, USA

E. Berardesca et al. (eds.), *Non Invasive Diagnostic Techniques in Clinical Dermatology*, DOI 10.1007/978-3-642-32109-2_35, © Springer Berlin Heidelberg 2014

Table 35.1 Test materials

	Exposure protocol	Test sample	Description
Experiment 1	Arm patch and BTK	Pad A	Currently marketed feminine hygiene pad[a]
		Pad B	Currently marketed feminine hygiene pad
		0.9 % NaCl	Negative irritant control
		0.1 % SLS	Positive irritant control
Experiment 2	Arm patch	Pad A	Currently marketed feminine hygiene pad
		Liner E	Currently marketed pantiliner
		Tissue D	Currently marketed bathroom tissue (unmoistened)
		0.9 % NaCl	Negative irritant control
		0.1 % SLS	Positive irritant control
Experiment 3	Arm patch	Pad C	Experimental feminine hygiene pad with petrolatum-based lotion formulation containing ZnO
		Pad G	Experimental feminine hygiene pad with petrolatum-based lotion formulation
		Lotion	Petrolatum-based lotion formulation containing ZnO (10 mg)
		Pad F	Currently marketed feminine hygiene pad
		0.9 % NaCl	Negative irritant control
		0.1 % SLS	Positive irritant control

[a]All feminine pads and pantiliner test samples were moistened with 0.3 ml of 0.9 % NaCl prior to application

voluntary. Subjects were excluded from participation if they had certain skin abnormalities or health conditions that could adversely impact the test. In some studies, subjects were identified as having self-assessed "sensitive skin" using a questionnaire previously described [4].

35.2.2 Test Materials

A variety of materials were tested in the various studies (Table 35.1). All test samples were composed of ingredients and materials that have been thoroughly evaluated for safety and skin compatibility. Test products included currently marketed feminine protection pads and pantiliners with and without lotion included on the surface and a bathroom tissue. All pads and the pantiliners were moistened with 0.3 ml of 0.9 % saline (NaCl) prior to application; however, the bathroom tissue sample was tested without moistening. In addition to the absorbent products, a lotion sample was evaluated in experiment 2. This lotion is a petrolatum-based formulation containing zinc oxide. Similar formulations have been extensively tested for skin compatibility using a number of standard protocols [5–8]. In the arm patch experiments, 0.1 % sodium lauryl sulfate (SLS) and normal saline (0.9 % NaCl) were used as positive and negative controls, respectively.

35.2.3 Evaluating Responses

For all exposure protocols, unaided visual assessment of test sites was conducted by an expert grader under a 100-W incandescent daylight blue bulb. Scoring for erythema and dryness were done using numerical scales previously described [9, 10] where "0" indicates no apparent cutaneous involvement and "4" and "6," respectively, indicate a severe reaction. The same grader was used throughout an experiment, and the grader was not aware of the treatment assignments.

Skin surface temperature was measured using a DermaTemp® infrared thermographic scanner (Exergen Corporation, Watertown, MA). Temperature determinations were conducted at the same time as the visual scoring. Subjects underwent a 15-min period of acclimation in a temperature-controlled room prior to measurement.

The statistical approaches to all analyses are given in the appropriate tables or figures. All analyses were conducted by using PC SAS® for Windows, version 9.1. Results were not adjusted for multiple comparisons.

35.2.4 Exposure Protocols

The modified arm patch has been previously described in detail [11]. In this protocol test materials were applied using an occlusive Webril® patch (Kendall LT, Chicopee MA, USA), and the relative positions of the test and control materials were randomized among the test subjects. Samples of the absorbent products were prepared from the midsection of the product and placed on the patch for application. Patches were applied lengthwise on the lateral surface between the shoulder and the elbow. Sites were marked with 0.5 % gentian violet to aid in visual grading and to ensure that patches were applied in the identical position each day for the duration of the test. Exposures consisted of 24 h per day for 4 consecutive days. Panelists were instructed to remove the patches 24 h after patch application and return to the laboratory for daily grading and reapplication of a fresh patch 30–60 min thereafter. Grading was conducted at the start of the study (baseline) and each day 30–60 min after sample removal. The post-baseline average grade (PBA) was determined for both the visual grades of erythema and dryness and the skin surface temperature readings.

The overall test design of the behind-the-knee (BTK) test protocol has been previously described [9, 10, 12]. Test materials were placed horizontally on the popliteal fossa, held in place behind the knee by an elastic knee band (Ace® knee bandage) of the appropriate size, and removed by the panelists prior to returning to the laboratory for grading and/or reapplication of test materials. Exposures consisted of 6 h/day for 5 consecutive days. Grading was conducted at the start of the study (baseline), each day 30–60 min after sample removal (afternoon or PM grading), and each morning prior to product application (morning or AM grading). The PBA was determined separately for afternoon and morning grades for both the visual grades of erythema and the skin surface temperature readings.

The bikini area shaving study was conducted to evaluate the skin compatibility of innovative razor designs in women who experience self-reported irritation after using a manual razor on the bikini area. During a 3-day acclimation period, panelists were given a specific product to use for all personal cleansing (Oil of Olay unscented bar soap for sensitive skin the Procter & Gamble Company, Cincinnati, Ohio). At the test facility on study day 1, the skin of the bikini area was graded (baseline evaluation), and panelists asked to shave the bikini area using temperature-controlled water, a specific manual razor (Venus Embrace with 5 blades, the Procter & Gamble Company, Cincinnati, Ohio, USA), and Dove sensitive soap (Unilever PLC, London, United Kingdom). Skin grading was conducted after shaving at 5 min, 24 h, and 48 h. Test sites were graded for erythema and dryness via unaided visual assessment. Measurement of skin surface temperature was conducted by evaluating three separate sites within each test area (i.e., top, middle, and bottom) in order to minimize site-specific variability. Measures from all sites were included in the average result for each time point.

35.3 Results

35.3.1 Site Variation in Baseline Skin Surface Temperatures

As mentioned in the methods section, visual scoring and skin temperature measurements were conducted prior to any treatment (i.e., at baseline).

At the arm patch sites, there was a consistent temperature variation along the arm, with the test sites nearest the shoulder having higher temperatures than those closer to the elbow (Fig. 35.1a). In the bikini area shaving study, there was a slight variation between left and right sides (Fig. 35.1b). As a result of these observations, the temperature readings at posttreatment evaluations included patch site as a variable in statistical analyses.

35.3.2 Arm Patch and BTK with Identical Products

Pads A and B were evaluated using the arm patch exposure along with negative and positive controls for irritation (0.9 % NaCl and 0.1 % SLS, respectively). Results are shown in Fig. 35.2a. Using visual scoring of erythema as an endpoint, the positive irritant control (0.1 % SLS) produced higher mean erythema scores than either test material or the negative control ($p \leq 0.05$). Neither pad sample showed evidence of irritation that differed significantly from the negative irritant control. Using the skin surface temperature as an endpoint, there were no significant differences between the samples including the positive irritant control.

The same two pad samples were tested in the BTK exposure protocol (Fig. 35.2b). At the afternoon grading, the erythema scores were significantly different when Pad A was compared to Pad B at both the afternoon grading, i.e., 30–60 min after sample removal, and at the morning grading, i.e., after approximately 18 h of recovery. The skin temperature readings did not differ at the afternoon scoring, but were significantly different at the morning scoring time point.

35.3.3 Comparison of "Normal"- to "Sensitive"-Skin Populations in Arm Patch Exposures

In two experiments, subjects were divided into two groups: one group with self-assessed "normal" skin and one with self-assessed "sensitive" skin. In the first experiment, a series of samples of absorbent products, i.e., a feminine protection pad, a bathroom tissue, and a pantiliner, were applied using the standard arm patch exposure (Fig. 35.3). Reactions were scored for erythema and dryness and for skin temperature. In both the self-assessed "normal"- (Fig. 35.3a) and "sensitive"-skin (Fig. 35.3b) groups, the positive control material (0.1 % SLS) produced higher mean erythema and dryness scores than the three test materials and the negative control ($p < 0.0001$). There was no significant difference in skin surface temperature. When the responses of the "normal"- and "sensitive"-skin groups were compared, there was no significant difference in any of the three endpoints (erythema, dryness, or skin temperature).

A similar experiment was conducted using two lotioned feminine protection pads, a non-lotioned pad, and sample lotions (Fig. 35.4). In "normal"-skin subjects (Fig. 35.4a), SLS produced higher mean erythema scores than all other test materials except Pad G ($p \leq 0.05$) and higher dryness scores than all other test samples ($p \leq 0.05$). There were no significant differences in mean temperature in the "normal"-skin subjects. In the "sensitive"-skin subjects (Fig. 35.4b), both the positive and negative irritant controls had lower mean temperatures than the test samples ($p \leq 0.05$). Once again, there were no significant differences between the "normal"- and "sensitive"-skin groups when mean erythema, dryness, and skin surface temperature were compared.

35.3.4 Result of Bikini Area Shaving Study

In the shaving study, erythema, dryness, and skin surface temperature were evaluated over time. As mentioned in the methods section, measurements were conducted on both the left and right bikini area. For each scoring time the endpoints (erythema, dryness, and skin surface temperature) were evaluated in two ways: separately for the left and right sides and with right and left sides pooled. Overall, the degree of erythema did not differ significantly over time (Fig. 35.5). However, when the right and left

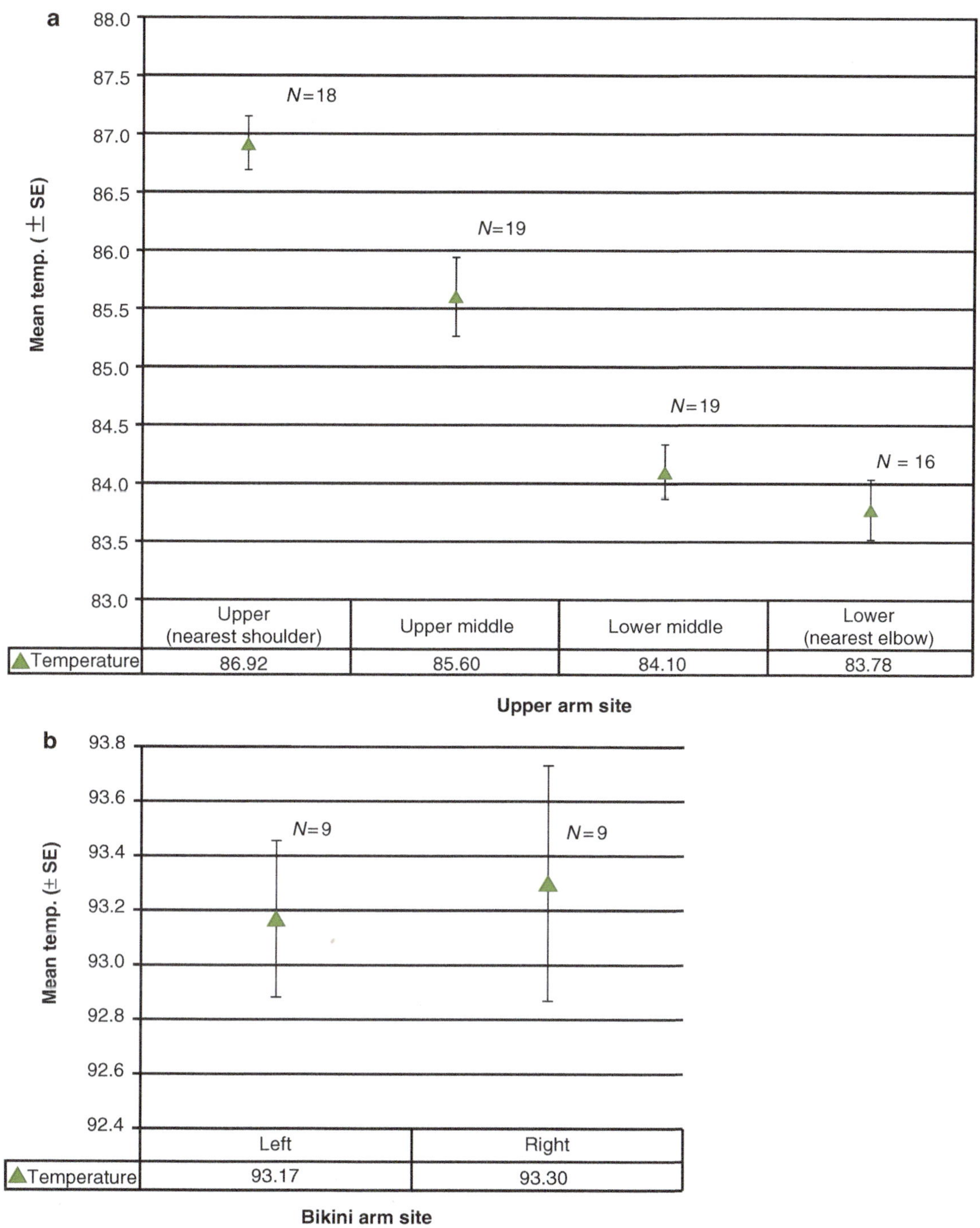

Fig. 35.1 Site variation in baseline skin surface temperatures. Prior to the first sample application (baseline), surface skin temperatures were recorded at all intended patch sites. Subjects underwent a 15-min period of acclimation in a temperature-controlled room prior to the temperature reading. The mean baseline measurement for each anatomic site (±SE) is shown in degrees Fahrenheit. (**a**) Variation in arm sites. (**b**) Variation in bikini area, right vs. left sites

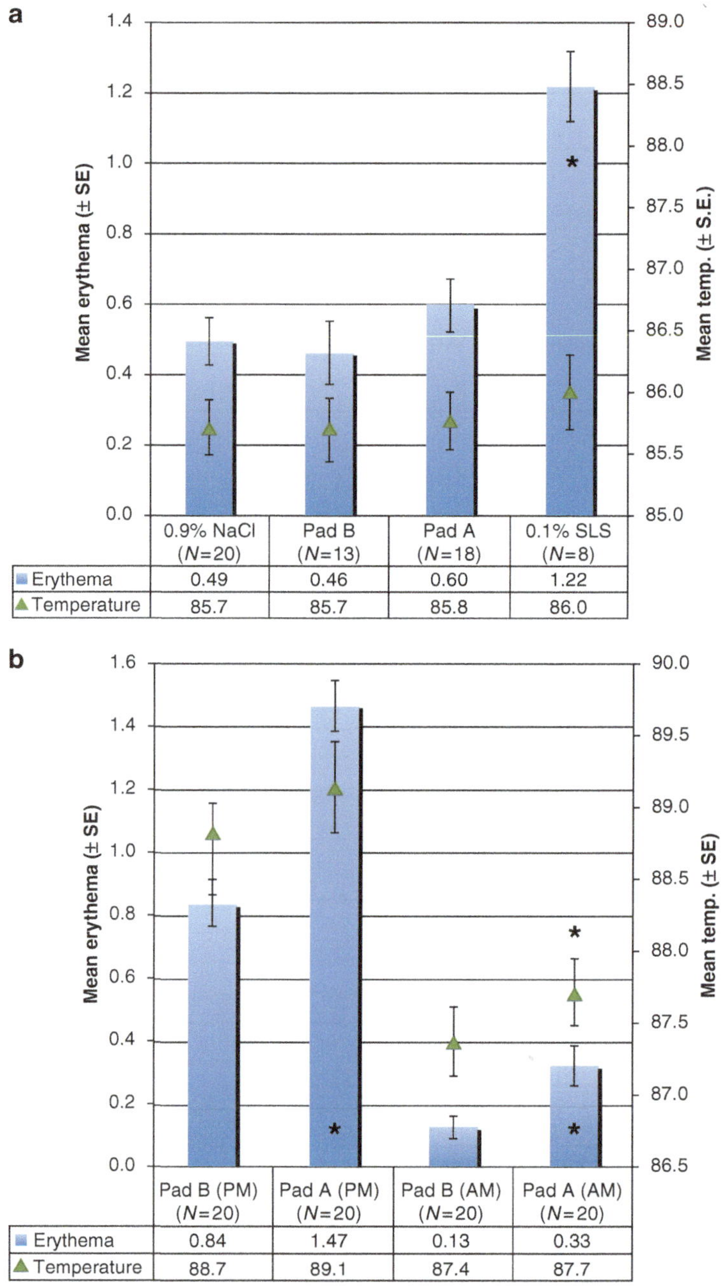

Fig. 35.2 Arm patch and BTK with identical products. In experiment 1, subjects were exposed to samples using both the arm patch and the BTK protocols, as described in the methods section. Erythema was scored visually on a scale of "0" (no apparent cutaneous involvement) to "4" (severe reaction). Skin surface temperatures were measured after acclimation, as described above. In the arm patch protocol, test sites were evaluated daily, and all daily scores were combined to determine the post-baseline average (PBA). In the BTK protocol test sites were evaluated twice daily: after sample removal (PM) and after approximately 18 h of recovery (AM). The PBA was determined separately for PM and AM evaluations. Treatment comparisons for erythema were performed using analysis of variance (ANOVA) (arm patch and BTK PM) or a stratified Cochran–Mantel–Haenszel (CMH) test (BTK AM). Treatment comparisons for skin surface temperature were performed using analysis of covariance (ANCOVA) with baseline and patch site included as variables. Significances were not adjusted for multiple comparisons. The number of test sites scored to study completion is provided in the graph. (**a**) Arm patch: * indicates the PBA for erythema for 0.1 % SLS was significantly different from Pad A, Pad B, and 0.9 % saline ($p \leq 0.05$). (**b**) BTK: * indicates the PBA for Pad A erythema (PM and AM) and skin surface temperature (AM) were significantly different from Pad B ($p \leq 0.05$)

a

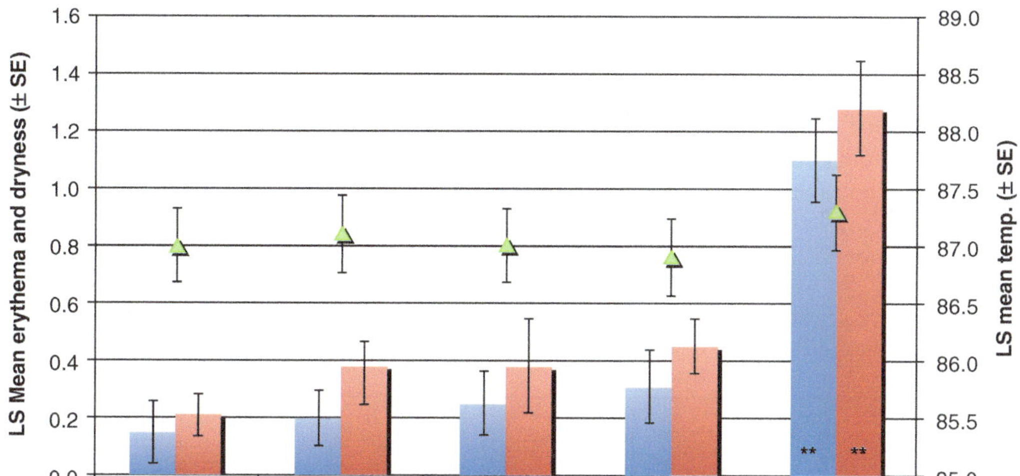

	0.9 % NaCl	Pad A	Tissue D	Liner E	0.1 % SLS
Erythema	0.15	0.20	0.25	0.31	1.10
Dryness	0.21	0.38	0.38	0.45	1.28
Temperature	87.0	87.1	87.0	86.9	87.3

b

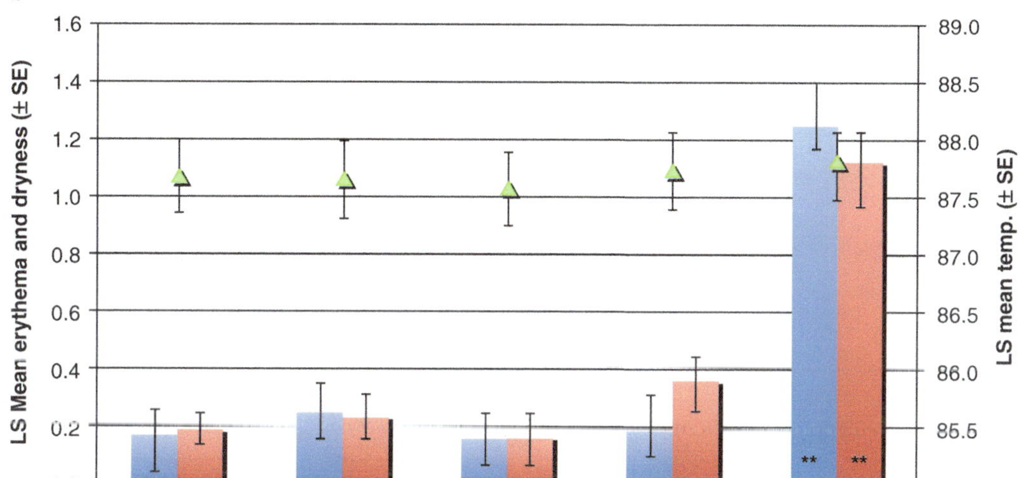

	0.9 % NaCl	Pad A	Tissue D	Liner E	0.1 % SLS
Erythema	0.16	0.23	0.15	0.18	1.29
Dryness	0.17	0.22	0.15	0.35	1.11
Temperature	87.7	87.7	87.6	87.7	87.7

Fig. 35.3 Comparison of "normal"- to "sensitive"-skin populations in arm patch exposures: absorbent products. In experiment 2, subjects were divided into two groups based on responses to a questionnaire (self-assessed "normal" and "sensitive" skin) and exposed to samples of absorbent products, i.e., a feminine protection pad, a bathroom tissue, and a pantiliner, using the arm patch protocol, as described in the methods section. Erythema and skin surface temperatures were evaluated as described in previous figure legends. In addition, test sites were evaluated for dryness on a scale of "0" (no apparent cutaneous involvement) to "6" (severe reaction). Comparisons of the Least Squares Mean (LS Mean) of the PBA for all endpoints were performed using ANCOVA with baseline scores and patch site included as variables. Significances were not adjusted for multiple comparisons. There were no significant differences between "normal-skin" and "sensitive-skin" subjects. ** indicates the PBA for erythema and dryness for 0.1 % SLS was significantly different from all other samples ($p<0.0001$). (**a**) Self-declared "normal-skin" subjects ($N=25$). (**b**) Self-declared "sensitive-skin" subjects ($N=35$)

sides were compared separately, the erythema on the left side was lower at 48 h compared to all other time points on the left side ($p<0.05$). The skin surface temperature was higher ($p<0.05$) at the pre-shave measurement than all subsequent time points (5 min, 24 h, and 48 h). There were no significant changes in dryness over time, nor were there any differences between anatomic sites (left side versus right side) for dryness or skin surface temperature evaluations (data not shown).

35.4 Discussion

In the exposure protocols evaluated in this series of experiments, changes in skin surface temperature did not reflect the changes in visual signs of erythema. In the modified arm patch studies, the positive irritant control (0.1 % SLS) produced erythematous reactions higher than reactions produced by the negative irritant control (0.9 % saline), or the mild products (Figs. 35.2a, 35.3 and 35.4). When dryness was evaluated

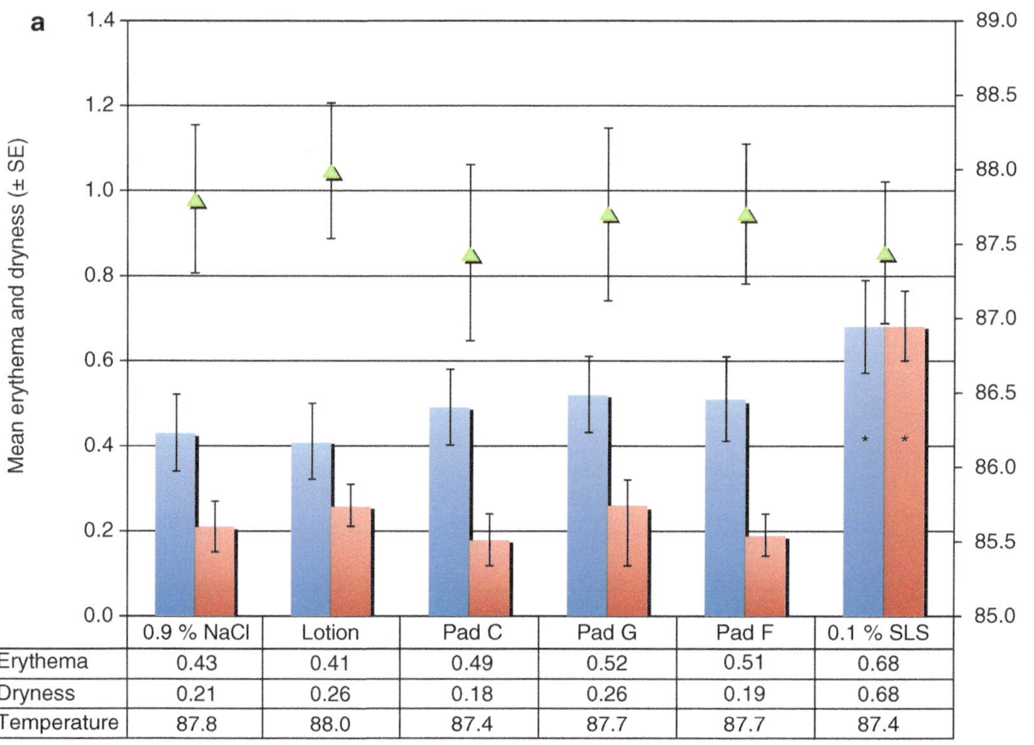

	0.9 % NaCl	Lotion	Pad C	Pad G	Pad F	0.1 % SLS
Erythema	0.43	0.41	0.49	0.52	0.51	0.68
Dryness	0.21	0.26	0.18	0.26	0.19	0.68
Temperature	87.8	88.0	87.4	87.7	87.7	87.4

Fig. 35.4 Comparison of "normal"- to "sensitive"-skin populations in arm patch exposures: lotioned feminine protection pads. In experiment 3, subjects were divided into self-assessed "normal" and "sensitive" skin and exposed to two lotioned feminine protection pads, a non-lotioned pad, and a lotion sample using the arm patch in a manner identical to that described in the previous figure legend. Comparisons of the PBA for all endpoints (erythema, dryness, and skin surface temperatures) were performed using ANCOVA with baseline and patch site included as variables. Significances were not adjusted for multiple comparisons. There were no significant differences between "normal-skin" and "sensitive-skin" subjects. (**a**) Self-declared "normal-skin" subjects ($N=17$). *indicates the PBA for erythema for 0.1 % SLS was significantly different from 0.9 % saline, lotion, Pad C, and Pad F ($p \leq 0.05$), and the PBA for dryness for 0.1 % SLS was significantly different from 0.9 % saline, lotion, Pad C, Pad G, and Pad F ($p \leq 0.05$). There were no significant differences in mean temperature. (**b**) Self-declared "sensitive-skin" subjects ($N=15$).* indicates the PBA for erythema for SLS was significantly different from 0.9 % saline, lotion, Pad C, Pad G, and Pad F ($p \leq 0.05$). In addition, erythema for Pad G was significantly different from Pad F. The PBA for dryness for SLS was significantly different from 0.9 % saline, lotion, Pad C, Pad G, and Pad F ($p \leq 0.05$). Dryness for Pad F was also significantly different from Pad C and lotion. The PBA for skin surface temperature for SLS was significantly different from lotion, Pad C, and Pad F ($p \leq 0.05$).In addition, temperature for saline was significantly different from Pad C and Pad F. Pad C was also significantly different from Pad G

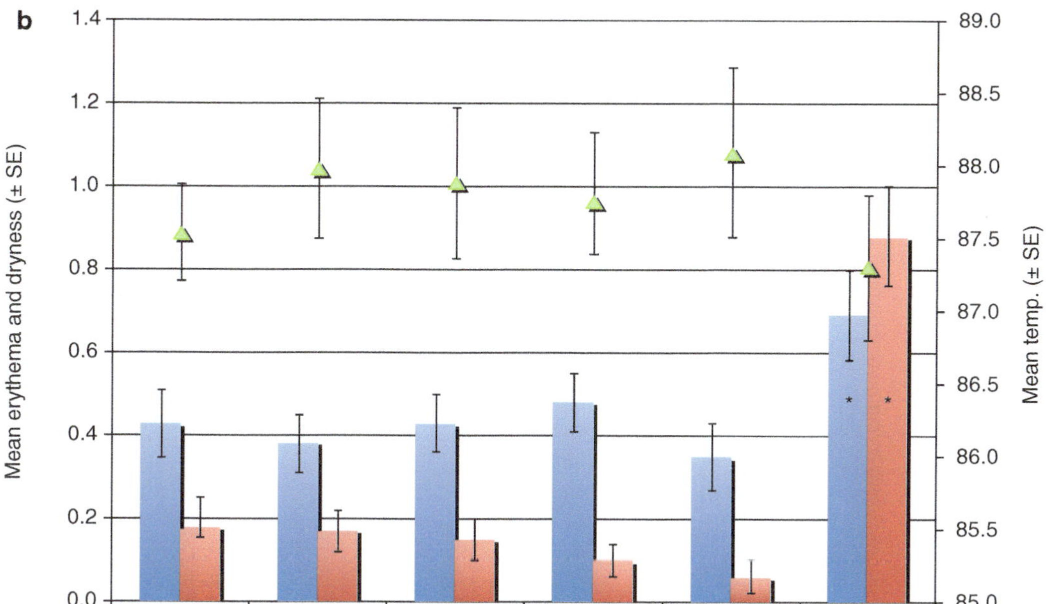

	0.9 % NaCl	Lotion	Pad C	Pad G	Pad F	0.1 % SLS
Erythema	0.43	0.38	0.43	0.48	0.35	0.69
Dryness	0.18	0.17	0.15	0.10	0.06	0.88
Temperature	87.5	88.0	87.9	87.8	88.1	87.3

Fig. 35.4 (continued)

(Figs. 35.3 and 35.4), differences were also detected for the irritant control. However, the skin surface temperatures did not vary consistently or significantly. In the BTK study (Fig. 35.2b), the two mild products produced significantly different levels of erythema at both the afternoon scoring (30–60 min after sample removal) and the morning scoring (after approximately 18 h of recovery). Skin surface temperatures differed significantly only at the morning scoring where the product that produced more erythema also produced a significant increase in mean surface temperature.

There are several possible explanations for the failure of the skin surface temperature readings to detect the low level of skin irritation that was apparent with visual scoring. Unlike the visual scores, which are always "0" at baseline, the skin temperature shows considerable variability prior to any treatment even though the conditions were controlled as much as was practical via a 15-min period of acclimation in a temperature-controlled room. In spite of these measures, the mean starting temperature varied among experiments. Although the means of baseline values (shown in Fig. 35.6 for two typical experiments) were adjusted for test site location in the statistical analysis, the adjusted means differed by as much as 2° between experiments 1 and 3, with an adjusted standard error of almost 0.5°.

The timing of the temperature measurements may not have been optimal to detect differences. Patrick et al. evaluated different irritants in a mouse ear model and found that different patterns of irritation responses were produced [14]. The peak in the increase in temperature varied from 2–3 h to 5–10 min, depending on the specific irritant used. Jacobi et al. investigated changes in skin temperature, redness, and blood flow in volunteers after application of a topical gel containing benzyl nicotinate and found a good correlation between changes in temperature and redness [15]. However, both measures were relatively short lived and were resolving by the evaluation conducted at 100 min. In our exposure

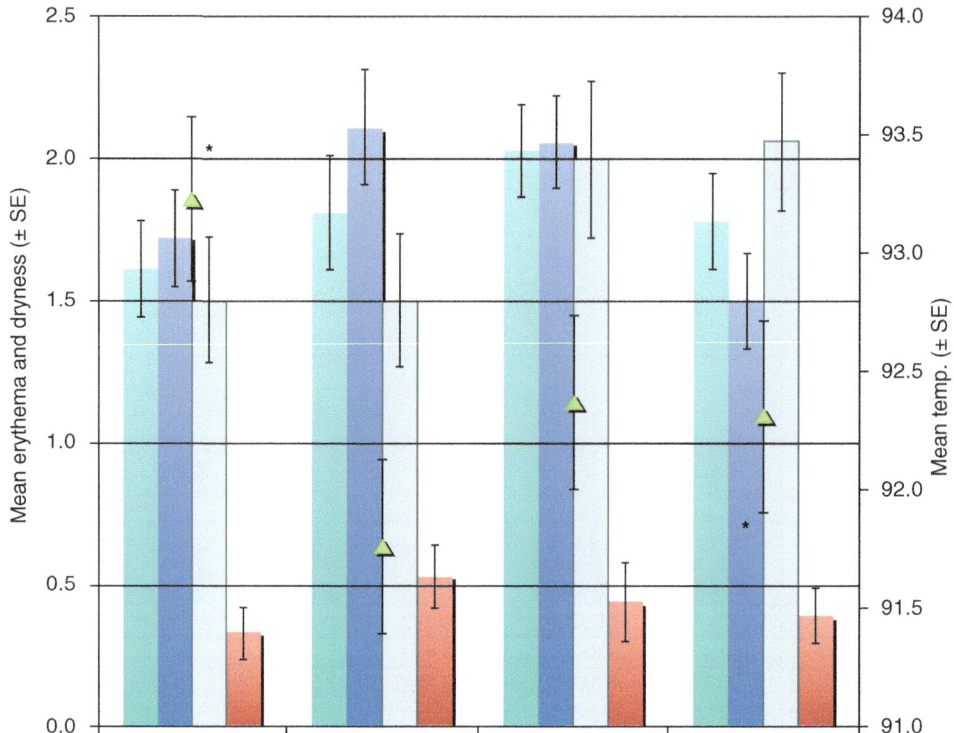

	Pre shave	5-Min post shave	24-h post shave	48-h post shave
Erythema	1.61	1.81	2.03	1.78
Erythema-Left only	1.72	2.11	2.06	1.50
Erythema-Right only	1.50	1.50	2.00	2.06
Dryness	0.33	0.53	0.44	0.39
Temperature	93.2	91.8	92.4	92.3

Fig. 35.5 Result of bikini area shaving study. Women ($N=9$) who experience self-reported irritation after using a manual razor on the bikini area were recruited for the study. After a 3-day acclimation period, subjects were asked to shave the bikini area under controlled conditions. Erythema, dryness, and skin surface temperatures of three separate sites within each test area (i.e., top, middle, and bottom) were evaluated prior to treatment and, at 5 min, 24 h and 48 h post shaving. Comparisons of averages for all subjects at all test sites were performed using a Wilcoxon signed-rank test for erythema and dryness and a generalized linear mixed model for skin surface temperature. An empirical (sandwich) variance estimator was implemented along with a small sample bias correction [13]. Significances were not adjusted for multiple comparisons. *indicates the skin surface temperature prior to shaving was significantly different ($p<0.05$) from all time points after shaving (i.e., 5 min, 24 h, and 48 h). On the left side only, erythema at 48 h was significantly different ($p<0.05$) from erythema at 5 min and 24 h

protocols it is possible that any meaningful changes in skin surface temperature may have occurred shortly after initial application of the test sample, e.g., within the first few minutes or hours, and resolved before sample removal at 24 h in the arm patch and at 6 h in the BTK.

A next step in evaluating the utility of adding skin surface temperature measurements is to alter the time course of the evaluations. However, this would necessitate a fundamental change in the current exposure protocols.

In the bikini area shaving study (Fig. 35.5), mean erythema after 48 h was lower than after 5 min or 24 h on the left side only ($p<0.05$), suggesting a certain level of handedness. This study was not designed to examine this aspect of the test, with 7 right-handed individuals and 2 left-handed individuals completing the study. In the

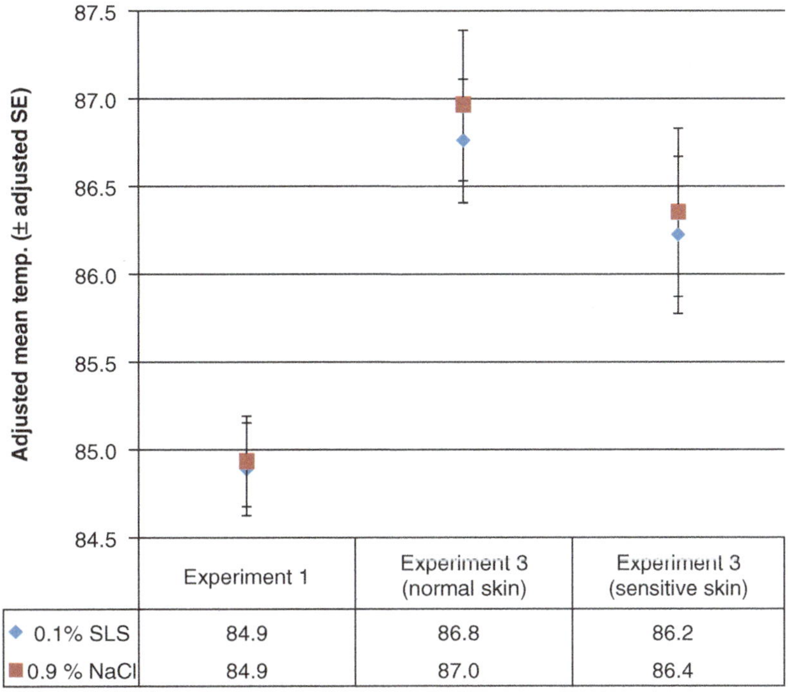

Fig. 35.6 Example of variation in baseline skin surface temperature between experiments. The mean baseline skin surface temperatures and standard errors were adjusted for test site location. Starting skin temperatures for the positive and negative irritant control sites (1 % SLS and 0.9 % NaCl) are plotted for arm patch experiments 1 and 3

future a balance of right- and left-handedness should be included as part of the protocol.

We have reported in previous studies that the BTK often detects differences between two very similar products that are not apparent using other test protocols [11, 12]. This is demonstrated once again in the results presented in Fig. 35.2. In the arm patch protocol, the positive irritant control produced higher erythema, but both product samples were equivalent to the negative irritant control, i.e., the exposure protocol did not differentiate between the two mild products based on visual scoring for erythema. In the BTK, although both products were mild to skin with relatively low mean erythema scores, it was possible to differentiate between the products. This added sensitivity in the BTK is the result of the additional component of irritation, i.e., the mechanical irritation resulting from friction.

The overall results indicate that the addition of measurements of skin surface temperature to standard exposure protocols used to evaluate potential skin responses to consumer products would not increase the ability to discriminate between similar, mild products. Modifying exposure protocols to include skin surface temperature measurements would introduce additional technical complications and expenses that would not be warranted.

Acknowledgments The authors are grateful to Ms. H. Tucker, Ms. J. Erb, Ms. M. Jones, Ms. C. Wilson, and Dr. J. Morel for their technical help and assistance and to Terresa L. Nusair, Ph.D., of the Health and Environmental Safety Alliance (Cincinnati, Ohio) for technical input.

References

1. Farage MA (2008) Need for more sensitive tools as we reach the limits of our ability to detect differences in skin effects from mild products. In: Zhai H, Wilhelm K-P, Maibach HI (eds) Marzulli and Maibach's dermatotoxicology, 7th edn. CRC Press, Boca Raton, pp 759–764
2. Farage MA, Maibach HI, Andersen KE, Lachapelle JM, Kern P, Ryan C, Ely J, Kanti A (2011) Historical perspective on the use of visual grading scales in evaluating skin irritation and sensitization. Contact Dermatitis 65(2):65–75
3. World Medical Association Declaration of Helsinki: ethical principles for medical research involving human subjects (2000) JAMA 284(23):3043–3045
4. Farage MA (2008) Perceptions of sensitive skin: changes in perceived severity and associations with environmental causes. Contact Dermatitis 59(4):226–232

5. Odio MR, O'Connor RJ, Sarbaugh S, Baldwin S (2000) Continuous topical administration of a petrolatum formulation by a novel disposable diaper. 2. Effect on skin condition. Dermatology 200(3): 238–243
6. Baldwin S, Odio MR, Haines SL, O'Connor RJ, Englehart JS, Lane AT (2001) Skin benefits from continuous topical administration of a zinc oxide/petrolatum formulation by a novel disposable diaper. J Eur Acad Dermatol Venereol 15(Suppl 1):5–11
7. Farage MA, Ebrahimpour A, Steimle B, Englehart J, Smith D (2007) Evaluation of lotion formulations on irritation using the modified forearm-controlled application test method. Skin Res Technol 13(3):268–279
8. Farage MA, Berardesca E, Maibach H (2009) Skin moisturization and frictional effects of an emollient-treated menstrual pad with a foam core. Cutan Ocul Toxicol 28(1):25–31
9. Farage MA, Gilpin D, Enane NA, Baldwin S (2001) Development of a new test for mechanical irritation: behind the knee as a test site. Skin Res Technol 7(3):193–203
10. ASTM Standard F2808, 2010 (2003) Standard test method for performing behind-the-knee (BTK) test for evaluating skin irritation response to products and materials that come into repeated or extended contact with skin. ASTM International, West Conshohocken. Available from: www.astm.org. Accessed 08 July 2011
11. Farage MA, Meyer SJ, Walter D (2004) Evaluation of modifications of the traditional patch test in assessing the chemical irritation potential of feminine hygiene products. Skin Res Technol 10(2): 73–84
12. Farage MA (2006) The Behind-the-Knee test: an efficient model for evaluating mechanical and chemical irritation. Skin Res Technol 12(2):73–82
13. Morel JG, Bokossa MC, Neerchal NK (2003) Small sample correction for the variance of GEE estimators. Biom J 4:395–409
14. Patrick E, Maibach HI, Burkhalter AE (1985) Mechanisms of chemically induced skin irritation: I. Studies of time course, dose response, and components of inflammation in the laboratory mouse. Toxicol Appl Pharmacol 81(3 Pt 1):476–490
15. Jacobi U, Kaiser M, Koscielny J, Schütz R, Meinke M, Sterry W, Lademann J (2006) Comparison of blood flow to the cutaneous temperature and redness after topical application of benzyl nicotinate. J Biomed Opt 11(1):014025

Measurements of Lymph Flow

36

Michael Weiler and J. Brandon Dixon

36.1 Lymphatic Introduction

The lymphatic vasculature is present in nearly all tissues of the body and serves three primary functions: (1) regulation of tissue fluid homeostasis through the transport of large proteins and excess interstitial fluid, (2) immune cell trafficking, and (3) lipid transport [1]. Approximately 10 % of the fluid filtered out of capillaries is transported through the lymphatic vasculature and returned to the venous circulation in the left subclavian vein. The lymphatics are forced to transport all this fluid without the benefit of a pumping organ such as the heart, and thus, the lymphatics rely upon the driving force of interstitial fluid pressure (extrinsic pump) and the rhythmic contractions of the vessels themselves (intrinsic lymphatic pump). The lymphatics begin in the interstitial spaces as draining initial lymphatics and then merge to form collecting vessels, which have an outer smooth muscle layer and are capable of contraction. Recently, it has been estimated that the lymphatic system transports as much as 8 L of fluid per day from the interstitium to the blood [2].

When the normal function of the lymphatic system deteriorates, many complications can arise. Loss of lymphatic pump function often leads to tissue fluid accumulation, fibrosis, and lipid deposition – a disease known as lymphedema [3]. Breast cancer survivors, especially those having had mastectomies and radiation are among the most at-risk populations for developing such a condition due to the lymph node and lymphatic vessel resection procedure commonly associated with cancer treatment [4]. The lymphatic system has also been implicated in several diseases including obesity [5], cancer metastasis [6], diabetes [7], and asthma [8], but the role of lymphatics in the progression of these diseases is not yet well understood.

36.2 Clinical Significance

Despite the critical roles that it performs, very little is known about the lymphatic vasculature in comparison to the blood vasculature. One of the main reasons for this knowledge gap may be the lack of in vivo imaging techniques to noninvasively visualize and obtain quantifiable information regarding lymphatic function, both in health and disease. New techniques are needed to better study lymphatic biology, elucidate the functional role of lymphatics and lymphangiogenesis in health and disease conditions, and better diagnose patients with lymphatic disease at an early stage before any resulting tissue damage is permanent. The development of new imaging modalities capable of rapid, noninvasive quantification of lymphatic pump function will play an

M. Weiler • J.B. Dixon, PhD (✉)
Wallace H. Coulter Department of Biomedical Engineering, George W. Woodruff School of Mechanical Engineering, Parker H. Petit Institute for Bioengineering and Bioscience,
Georgia Institute of Technology, GA, USA

E. Berardesca et al. (eds.), *Non Invasive Diagnostic Techniques in Clinical Dermatology*,
DOI 10.1007/978-3-642-32109-2_36, © Springer Berlin Heidelberg 2014

essential role in the next generation of lymphatic research and diagnostics.

When evaluating the various imaging modalities, it is important to consider the requirements the approach must meet for the designated lymphatic function or architecture that is being investigated. In general these can be divided into four categories: (1) spatial resolution requirements, (2) temporal resolution requirements, (3) contrast requirements, and (4) penetration depth requirements. The specification of these parameters depends on the context of the lymphatic physiology/anatomy under investigation.

In humans, lymphatic vessels range anywhere for 20 μm in diameter for the smaller capillaries to up to 5 mm for the thoracic duct and up to a few centimeters for lymph nodes [9]. The collecting lymphatics, which provide a majority of the pumping activity for promoting lymph flow, range from 100 to 600 μm. The temporal resolution needed depends on the importance of imaging this dynamic pumping function. Contractions of collecting lymphatics range from 10 to 50 contractions per minute, so frame rates of up to several frames per second are often needed to capture these dynamics [10]. Nearly all tissues are supported by a lymphatic vasculature and thus these vessels reside at various depths. There are both lymphatic capillaries and collectors in the dermal layer just a few mm below the skin as well as vessels deep in the thoracic cavity.

Most lymphatic imaging techniques rely upon the injection of a particular contrast agent into the interstitium (or directly into a vessel itself) and the subsequent drainage of this contract agent by the lymphatic vessels. When the contrast agent is of the appropriate size, or binds to something of the appropriate size (usually 3–100 nm), it is exclusively taken up into the lymphatic vessels, provided they are functioning. While essentially all lymphatic imaging techniques require contrast agents, the ideal injection should be a very small volume to minimize the effects of this added volume on interstitial fluid pressure, particularly if one is trying to capture a measurement of baseline lymphatic function. This is less important when imaging solely for architectural reasons or for sentinel lymph node mapping. Additionally, the contrast agent itself should have minimal biological effects on lymphatic function and should be rapidly cleared from the circulation. Here, we summarize imaging techniques currently used to study lymphatics in vivo, their advantages and limitations as it relates to these criteria, and suggest directions for the future.

A quick survey of the literature shows that the primary application of lymphatic imaging has been the detection of lymph nodes, rather than lymphatic vessels. This is due in part because the clinical interest in lymph node targeting as it relates to cancer far outweighs the clinical interest in lymphatic diseases. However, another reason for this disparity is that the most successful techniques used clinically for mapping nodes do not meet the spatial or temporal resolution criterion for imaging lymphatic vessel function. Thus, the title of "lymphatic" imaging in a majority of the literature (including what is discussed here) is not really focused on imaging lymphatics, but rather lymph nodes. However, most of the techniques that target lymph nodes work because there is a functional network of vessels that drains the contrast agent to the node; thus, these approaches can also indirectly be a useful tool for assessing lymphatic function and are worth mentioning here.

36.3 Lymphoscintigraphy

One of the longtime gold standards of lymphatic imaging is lymphoscintigraphy, the most common nuclear imaging method for lymphatics [11]. Lymphoscintigraphy usually serves two main purposes with regard to lymphatic imaging: sentinel lymph node (SLN) mapping for post-cancer lymph node resection surgery [12, 13] and lymphatic vessel visualization for the assessment of lymphedema progression and treatment efficacy [14]. The technique typically relies upon an intradermal or subcutaneous injection of Tc-99m sulfur colloid (11.1–111 MBq) [15], Tc-99m-labeled human serum albumin [16, 17], or Tc-99m-labeled dextran [18] given between 30 and 120 min before the imaging event. The scintigram takes a relatively long period of time to acquire, between 20 min and 2 h depending on the imaging conditions [19].

Lymphoscintigraphy can be used in the diagnosis of lymphedema by examining differences in radioactivity between affected and unaffected regions, calculating clearance rates of radionuclide injections, and by observing dermal backflow of radionuclides in lymphatic vessels. Szuba and colleagues demonstrated the ability of lymphoscintigraphy to quantitatively estimate the radioactivity ratio between the axilla of a healthy arm and an affected arm as well as identify dermal backflow patterns in lymphedema patients [20, 21]. A numerical index was also developed specifically for postmastectomy lymphedema patients in order to assess lymphedema progression based upon observations of dermal backflow and visualization of proximal lymph nodes. There was a correlation between the lymphoscintigraphic lymphedema numerical index and elevated limb volume as well as a correlation between the ratio of radioactivity and the posttreatment percentile change in edema. Taken together, these results indicate that lymphoscintigraphy can be used clinically to diagnose and assess lymphedema and that the technique has been used with moderate success in breast cancer-related lymphedema (BRCRL) patients. In addition to diagnosing lymphedema, lymphoscintigraphy has been used to evaluate new strategies for the treatment of lymphatic diseases such as autologous lymphatic transplant [22], VEGF-C treatment [23], surgical lymphatic-venous anastomoses [24], the efficacy of manual lymphatic drainage [25], and the management of chylous ascites [26] to name a few.

The most prevalent use of lymphatic imaging using lymphoscintigraphy is for SLN mapping. The sentinel lymph node is theoretically the nearest downstream lymph node from a primary tumor and, as such, collects metastatic cancer cells. Contrast agents can thus be introduced in or near the primary tumor to be taken up by the local lymphatic vessels and transported to the SLN to assess the metastatic state of the tumor [27–29]. A preoperative lymphoscintigraphy study in breast cancer patients has shown the technique to have an 87 % success rate in identifying SLNs as hot spots [30]. Recently, a new "dual mapping" procedure was performed in 60 breast cancer patients using dye injections in combination with isotopes capable of identifying SLNs in 59 patients (98.3 %) with a false-negative rate of only 1.7 % [31]. In an effort to reduce or prevent the need for lymphatic vessel resection and BCRL, Thompson et al. [32] has developed an axillary reverse mapping (ARM) technique to identify and distinguish lymphatic vessels that drain the arm instead of the breast, but a long-term study is needed to verify the technique limits the risk of BCRL without compromising nodal staining accuracy. Modi and colleagues developed a novel technique to equip lymphoscintigraphy with more quantitative capabilities [33]. After attaching a sphygmomanometer to the arm of BCRL patients and inflating to a pressure to cutoff lymphatic flow (as determined by lymphoscintigraphy), they were then able to gradually reduce this pressure until lymph flow was restored giving an estimation of lymphatic pumping. This study showed that this pumping pressure was reduced in women with BCRL providing some of the strongest evidence that the disease is strongly associated with the inability of these collecting vessels to adequately pump.

To summarize, lymphoscintigraphy is a 2-D imaging technique to visualize the lymphatic vasculature and SLNs. While clinical SLN mapping is becoming more common, evaluation of lymphatic function and the state of lymphedema remains mostly qualitative with only minimal quantitative capabilities based upon clearance rates. The major limitations of lymphoscintigraphy are poor temporal resolution resulting from long gamma camera integration times and poor spatial resolution that limits the identification of exact SLN locations. While the technique is useful in identifying lymphatic dysfunction as the general underlying cause in cases of limb swelling or chylous ascites, the resolution limitation makes it difficult to assess the exact location or functional cause of the lymphatic failure.

36.4 Positron Emission Tomography

In an effort to address the drawbacks of lymphoscintigraphy, a hybrid SPECT/CT imaging modality has been implemented in lymphatic imaging to offer higher spatial resolution (~2 mm) than traditional lymphoscintigraphy [34–37]. The technique combines positron emission tomography (PET) and X-ray CT to detect diseased lymph nodes [38–40], a process that uses the traditional intravenous injection of positron-emitting radiopharmaceuticals to specifically target molecules of interest and provides the added benefit of structural information from the X-ray CT [41]. While the technique has shown promise for detecting nodes not found by traditional lymphoscintigraphy [42], it comes with a very high cost and thus is not widely used for routine SLN mapping [43]. Also, while the 2 mm resolution limit is an improvement for mapping nodes, it is still well below the requirements needed to image the primary unit actually responsible for lymphatic flow: the lymphatic collectors.

36.5 Magnetic Resonance Imaging

MR lymphangiography is the second traditional gold standard in lymphatic imaging. The technique involves the injection (interstitial or intravenous) of various contrast agents including iron oxide particles, gadolinium-labeled diethylenetriaminepentaacetic acid (Gd-DTPA), Gd dendrimers or liposomes, and nanoparticles [44–49] and analyzes the alignment, behavior, and interaction of protons when a magnetic field is applied. MR as it relates to lymphatic imaging is most commonly used for visualization of lymphatic vessels and cancer staging of lymph nodes with the advantage of better spatial resolution than nuclear techniques. The technique's feasibility was established a decade ago and has been used with a variety of contrast techniques including injections of gadoterate meglumine, an extracellular paramagnetic agent [50], nonionic water soluble paramagnetic contrast agents gadodiamide and gadoteridol [51–53], and gadobenate dimeglumin [54] all injected into the dorsum of the foot. Recently, Ruddell et al. developed a dynamic contrast-enhanced MR imaging (DCE-MRI) technique to assess lymphatic dysfunction by measuring and dynamically mapping changes in lymph flow and drainage in mice models [55]. The technique begins with an intravenous injection of a contrast agent and then acquires serial MR images. Wash-in and wash-out curves can then be generated using specified regions of interest. In the study, they were able to show that tumor growth increased lymphatic flow through draining lymph nodes over a 3 week period.

Others have taken a different approach by exploring unenhanced MR lymphatic imaging [56–58]. Matsushima and colleagues showed in their studies that using heavily T2-weighted imaging in combination with respiratory triggering can allow for the distinction between lymphatic flow and venous flow such that they can visualize lumbar lymphatics, cisterna chili, and the thoracic duct noninvasively without contrast agents. One concern is that lymphatic flow may be distorted by other fluid flow aside from venous circulation (especially slow fluid flows), but the possibility of imaging the lymphatic vasculature without the addition of contrast agents would be very valuable since the addition of most contrast agents probably alters true lymphatic function to a certain degree.

A new technique known as lymphotropic nanoparticle-enhanced MRI (LN-MRI) may offer a solution to achieve more accurate lymph node staging [59, 60] by using a novel set of MR contrast agents possessing unique biochemical and physical properties [50, 61, 62]. MR lymphatic imaging with ultrasmall superparamagnetic iron oxide (USPIO) particles may provide adequate contrast necessary to resolve salient features of normal and cancer-positive lymph nodes [11, 19, 44, 63–67]. Harisinghani et al. used highly lymphotropic superparamagnetic nanoparticles to detect lymph node metastases with significantly improved sensitivity compared to traditional MR lymphatic imaging [68]. Likewise, Ross et al. used an intravenous injection of ferumoxtran, a lymph-node-specific contrast agent, to detect prostate cancer lymph metastases with excellent sensitivity [69]. Recently, Kimura et al. also used USPIO lymphography and differences in T2*-weighted and T1-weighted enhancement patterns to identify normal and diseased lymph nodes [70].

It is important that many of the USPIO techniques for lymph node mapping do not directly rely on lymphatics for targeting the node, but rather take advantage of the leaky vasculature present in tumors to achieve extravasation from the blood. Thus, techniques that are promising for identifying lymph nodes do not necessarily translate for imaging the lymphatic vasculature. Another challenge with these nanoparticle approaches is that there is often a trade-off between ease of uptake from the interstitium into the lymphatics and their subsequent retention in the lymph node [71].

MRI offers excellent spatial resolution, which has a lot of potential for imaging the structure and morphology of the lymphatic vasculature and lymph nodes. Perhaps the most exciting direction of MR lymphatic imaging is the potential to fully develop an unenhanced MR lymphangiography technique to visualize the lymphatic system without contrast agents. Such a breakthrough would revolutionize lymphatic imaging and greatly improve diagnostic capabilities for detecting lymphatic diseases.

36.6 Computer Tomography

X-ray computer tomography (CT) is commonly used to assess differences in tissue density between soft tissue and electron-dense bone, and contrast agents are often incorporated to help discern healthy and diseased tissue [72]. CT is commonly used as a diagnostic method for the screening and identification of tumors before surgery is performed for many cancers including breast cancer [43]. One particular subset of CT imaging, computed tomographic lymphography (CT-LG) has been used to visualize lymphatic drainage pathways and to discern enlarged or metastatic lymph nodes [11]. The current technique utilizes a class of small organic iodinated molecules that are highly water soluble (examples include iopamidol, ioxaglate, ioversol, and iohexol) to provide additional contrast for CT SLN imaging. Suga et al. employed iopamidol in CT-LG for the visualization of breast lymph vessels and the mapping of SLN locations in healthy dogs and healthy human volunteers and were able to observe the direct path of lymphatic vessels from the injection site to the SLN, detect a localization of the contrast agent in the SLN, and visualize distant nodes [73–75]. Others have built upon this technology and had similar successes using CT-LG for SLN mapping [76–78].

Despite the success of water soluble organic iodinated contrast agents in SLN detection, the molecules offer a very short imaging duration because they drain rapidly to the blood vasculature and are cleared through the renal system [11]. Additionally, the technique is limited by difficulty of injecting the necessary millimolar concentrations of contrast agents in the dermal space. In an attempt to mitigate this drawback, several new CT contrast agents are under development [19]. Rabin et al. has synthesized a polymer-coated bismuth sulfide nanoparticle as a novel CT contrast agent with a longer hemovascular half-life, but concerns remain regarding the long-term toxicity of this nanoparticle [79]. Although to date CT-LG cannot detect occult metastases, the technology has a lot of potential as a strong alternative to lymphoscintigraphy for the visualization and identification of SLNs for biopsies. The detailed anatomy of the lymphatic vasculature that CT-LG can provide may help to minimize several of the shortcomings of lymphoscintigraphy as well as increase the accuracy and sensitivity of SLN biopsy. However, the major limitation of CT-LG as compared to other imaging modalities is the inability to image dynamic lymph flow or lymphatic contractions, which is due to a lack of spatial resolution. Radiation concerns associated with CT may also limit its clinical usefulness.

It should also be noted that another X-ray-based technique similar to angiography requires the cannulation of lymphatic vessels and direct injection of contrast agents (e.g., Lipiodol, an iodinated poppy seed oil) directly into the cannulated vessels. Originally termed lymphangiography (or lymphography), it has been used postoperatively in patients with lymphatic fistulas, lymphocele, and chylothorax as a means of detecting lymphatic leakage [80]. While the term lymphangiography is now broadly used for a variety of techniques that involve the uptake of a contrast agent into a lymphatic vessel for subsequent imaging, traditional lymphographies have largely been abandoned due to the increase in availability of other techniques, the technical expertise required for vessel cannulation, and complications that have been shown to arise from the procedure [81–83].

36.7 Ultrasound

Ultrasound is the primary imaging modality used to study and diagnose secondary lymphedema resulting from filariasis, a disease common to tropical regions around the world in which parasitic nematodes enter a human host via mosquito vectors and come to reside in the lymphatic vasculature. The biggest advantages for ultrasound imaging of lymphatics are the relatively low cost and availability of equipment, the achievable penetration depth, and the lack of ionizing radiation. Ultrasound has been employed to visualize dilated lymphatic vessels in filarial patients [84, 85], and Doppler ultrasound has been used to successfully measure nematode motion in lymphatic vessels [86] and evaluate the efficacy of treatment strategies on filarial infection [87–89]. Doppler ultrasound is not regularly used in lymphatic imaging, however, because the lower number of scattering objects (e.g., cells) present in lymph flow severely limits measurements of lymph drainage patterns, while the abundance of red blood cells in the blood vasculature has allowed it to be widely used for measuring blood flow. Additionally, ultrasound poses some spatial resolution limits when compared to other techniques that make it difficult to image the smaller vessels. High-frequency ultrasound has been able to achieve spatial resolution down to 20 μm; however, a trade-off exists between spatial resolution and penetration depth with these high-frequency systems being limited to depths under 5 mm.

Despite the limited applications in lymph flow measurements, ultrasound has been used to assess the malignant state of lymph nodes based upon geometric features [44]. Since ultrasound images do not necessarily require contrast agents, the technique may serve a very unique role in mapping SLNs characterized by afferent lymph vessel blockage that would otherwise prevent contrast agents from reaching the node and produce a false-negative result. Contrast-enhanced ultrasound imaging (CEUS) [90] is much more widely used in preclinical and clinical imaging of lymphatics than ultrasound alone because the technique has the advantages of high signal-to-noise ratio, contrast sensitivity, and spatial resolution [11, 50]. The most widely used contrast agents for CEUS are submicron-sized microbubbles, which are often made with lipids, polyethylene glycol polymers, or denatured albumin [91, 92]. Microbubble contrast agents have been employed in CEUS imaging to visualize lymph nodes, lymphatic vessels, and other areas of interest in the microcirculation [45, 80, 93–97]. Like all contrasts agent approaches, the probes can rely on the drainage of the lymphatics to be delivered from the primary tumor to the node, or they can be functionalized with specific ligands for targeting the nodes and thus can be administered via the blood [98].

36.8 Optical Imaging Techniques

Optical fluorescent imaging may provide the best combination of spatial and temporal resolution for specifically visualizing lymphatic vessels and propulsive lymphatic flow out of any of the imaging modalities previously mentioned, and the technique has the added bonus of not exposing patients to ionizing radiation [44]. The basic principle is to exploit the natural function of the lymphatic system by giving an intradermal or subcutaneous injection of a fluorescent tracer and visualizing the uptake and transport of the fluorophore by the lymphatic vessels (Fig. 36.1). Fluorescent imaging allows the use of camera integration times as low as 50 ms [99], which enables excellent temporal resolution for the dynamic visualization of lymphatic uptake and transport characteristics. While dynamic pulsatile lymph flow has been recorded via intravital microscopy techniques in mesenteric and isolated lymphatic vessels [10, 100, 101], there is no other imaging modalities capable of obtaining such high-resolution lymphatic imaging noninvasively in vivo.

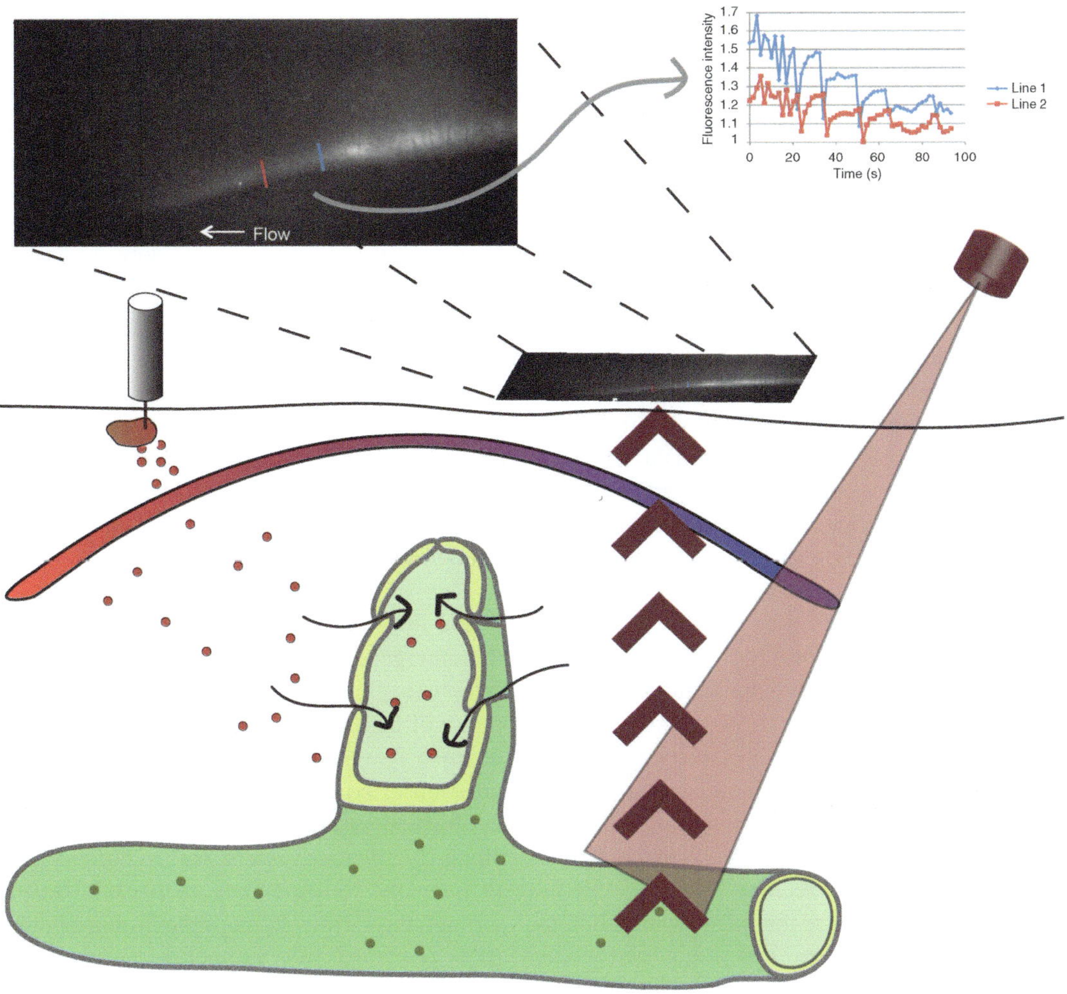

Fig. 36.1 Nearly all modes of lymphatic imaging rely on the injection of some tracer into the interstitial space of the draining lymphatic basin. If this tracer is of an optimal size (5–100 nm), it will be taken up into the lymphatics through their specialized junctions and not the blood vasculature. Some injected dyes used in lymphatic imaging (like ICG and Evan's Blue) readily bind to interstitial albumin, which then causes them to be taken up into lymphatics (as free dye is usually much smaller than the molecular weight cutoff that requires lymphatic transport). In the example shown here, NIR light excites the dye and the emission is recorded in reflection mode. Packets of fluorescence are recorded as they travel through the collecting lymphatics, and packet velocity and frequency can be quantified from the data generated

Optical fluorescent imaging of lymphatics was first reported in the form of fluorescence microlymphangiography using the fluorophore fluorescein, which is maximally excited and emits fluorescence at visible wavelengths, thus limiting its usage to imaging very superficial lymphatic vessels [102, 103]. McGreevy et al. also used a visible contrast agent, Cy5, to detect lymph nodes in swine [104]. The limited penetration depth of visible light, however, prevented the visualization of any deeper lymphatic vessels, such as collecting vessels. Quantum dots have also been employed for sentinel lymph node mapping [105], but most quantum dots are excited in the visible range, which limits the ability to image deeper collecting lymphatic vessels despite the mild improvement in tissue penetration depth over visible fluorescent dyes

[19]. Additionally, there remain serious concerns about the long-term toxicity of quantum dots due to their heavy metal content.

More recently, near-infrared (NIR) fluorescence lymphatic imaging using fluorescent dyes has gained traction. NIR imaging technologies may provide the ideal solution to functional lymphatic imaging for the purposes of disease assessment and management as they reside in the optimal range where light absorption and scattering are low in biological tissue and there is minimal autofluorescence, thus affording deeper penetration depths and excellent contrast and spatial resolution, all of which are vital for measuring lymphatic contractile properties [106]. NIR imaging with an FDA-approved fluorescent dye, indocyanine green (ICG), has been reported as an optical diagnostic dating back to the 1950s as it was originally used to assess cardiac output and hepatic function [107]. Based upon its fluorescent properties, the indications for ICG use expanded over the next 50 years to include microcirculation of skin flaps, visualization of retinal and choroidal vasculatures, pharmacokinetic analysis, object localization in tissue, tissue welding, fluorescence probing of enzymes and proteins, and tumor screening [108, 109]. ICG-based NIR imaging, a technique in which ICG is injected intradermally, excited with a laser diode, and imaged with an NIR-sensitive detector as it is taken up by the lymphatic system, has emerged as a novel method for lymphatic vasculature and sentinel lymph node mapping and for the quantitative assessment of lymphatic function in animals and humans [110–113].

ICG-based NIR imaging specifically designed for lymphatics was first reported in two separate studies by Sharma et al. [19] and Unno et al. [114] in 2007. The study by Sharma and colleagues showed, for the first time noninvasively, dynamic propulsive lymph flow in anesthetized swine. The rhythmic contractions of sequential lymphangions and the series of one-way valves force the fluid to flow in discrete "packets" along the lymphatic vessels (see graph in Fig. 36.1). The packets were recorded as traveling between 0.1 and 1 cm/s, which was orders of magnitude above the μm/s lymph velocities previously reported by fluorescence lymphangiographies in animal preparations [80, 115–117] as the previous studies were only able to image the initial capillaries and not the contractile collecting lymphatics. The study by Unno and coworkers used NIR lymphatic imaging to identify four characteristics of lymphatic flow associated with lymphedema including dermal backflow, extended fluorescence in the injection site, dilated lymph channels, and fluorescence diffusion otherwise known as leaky vessels. Sevick-Muraca et al. also showed lymphatic vessel visualization in breast cancer patients using microdose administration of ICG [118], which was significant because smaller fluorophore injections will limit the effects of the procedure itself on baseline lymphatic function.

One of the major advantages of NIR lymphatic imaging over other imaging modalities is the emergence of new quantitative measures of lymphatic pump function. Such quantifiable metrics will be enormously valuable in objectively defining lymphatic health and assessing the progression of lymphatic disease. Rasmussen et al. established discrete regions of interest (ROIs) along lymphatic vessels and recorded spikes in intensity corresponding to packets traveling through the various ROIs. If the distance between ROIs is known, metrics such as average packet velocity and packet frequency can be calculated as representative metrics of lymphatic pump function [119–121].

Unno et al. adapted a previously described technique [33] to measure lymphatic pumping pressure to NIR imaging [122]. ICG was injected into the dorsum of the foot of human volunteers, and real-time fluorescence images of lymphatic transport were recorded while a custom sphygmomanometer was used to increase the pressure around the legs until lymphatic flow was observed to have stopped. The pressure was then slowly released until lymphatic flow was restored, which was defined as the lymphatic pumping pressure, P_{pump}. Using NIR imaging of ICG transport through dermal lymphatic vessels during this procedure gives a very accurate measure of the exact time and corresponding applied pressure at which lymphatic flow is stopped.

Healthy patients were reported to have a P_{pump} of 29.3 mmHg while lymphedema patients had a significantly lower P_{pump} of 13.2 mmHg, indicating reduced lymph transport capabilities in lymphedema patients. More recently this technique was also applied to assess the decline in P_{pump} associated with aging [123], although currently the technique does not take into account changes in tissue compliance and how this might affect the transmission of the externally applied pressure to the lymphatic vessel.

ICG is the most commonly used tracer, primarily because it is already approved by the FDA. However, there are some concerns in regard to its effects on lymphatic function, as it has been observed to remain at the injection site for up to two weeks, and appears to have an effect on lymphatic function during this time. One study showed direct effects on lymphatic contraction in response to ICG in isolated vessels [124], and recently, it has been shown that ICG reduces lymphatic function and enlarges draining lymph nodes for 2 weeks in vivo in a rat tail model [125]. It appears that the concentration of injected ICG has a minimal effect on lymphatic function immediately after injection [126], but the collection of studies suggests that repeat ICG injections should be performed more than two weeks apart to avoid erroneous measurements of lymphatic function. In addition, ICG is a relatively poor fluorophore, and there are several other dyes available in the NIR that can be conjugated to albumin. Several other studies have come out investigating other probes as tracers in NIR lymphatic imaging [127–130].

NIR lymphatic imaging has the potential to revolutionize the study of lymphatic transport phenomena and the diagnosis of lymphatic disorders with the development of new imaging techniques and quantification metrics. Conceivably, NIR lymphatic imaging with ICG could be expanded to include simultaneous imaging of multiple fluorophores, perhaps including fluorophores to label lymphatic vessel walls, immune cells, and lipids to correlate lymphatic contractile function with transport capabilities. Additionally, the development of novel image processing algorithms has the potential to provide new quantitative measures to detect subtle differences in lymphatic pump function under various conditions. The emergence of such new techniques could enable the use of NIR imaging as an early stage lymphatic disease diagnostic capable of detecting changes in lymphatic function before clinical symptoms emerge and tissue damage results.

36.9 Emerging Technologies

In addition to the techniques mentioned above that have already been used to image lymphatic function (or indirectly infer the presence of lymphatics by requiring them to deliver a contrast agent to the lymph node), there are a few emerging technologies that are also worth mentioning which could soon see widespread clinical use for certain types of lymphatic imaging. Optical coherence tomography (OCT), which essentially uses an interferometer to determine the depth of scattering objects that reside in a layer of turbid media [131], has recently been used to image blood and lymphatic vessels surrounding a lymph node [132, 133]. Interestingly, while the lack of scattering objects in the lymph is usually a challenge for imaging lymphatics, the authors took advantage of this property to actually resolve the lymphatic vessel without the use of contrast agents. Another advantage of OCT is that one can achieve optical resolution similar to that in NIR imaging, with the addition of depth resolution allowing for 3-D imaging. However, one of the drawbacks of OCT is that the penetration depth is limited to about 1–2 mm. Hybrids of OCT systems are also being explored to improve the performance by delivering contrast agents that exhibit a photothermal effect when excited in the NIR. The resulting temperature change produces a phase shift that can be detected with OCT systems [134].

Another optical technique that overcomes this depth limitation is photoacoustic tomography (PAT) [135]. This technique utilizes the fact that upon absorption of light, thermoelastically induced pressure waves are formed that can be detected with ultrasound sensors. Thus, one can

Table 36.1 A comparison of the capabilities of various modalities for imaging the lymphatic vasculature

	Spatial resolution	Depth penetration	2-D/3-D	Time resolution (2-D)	Contrast agent required	Ability to measure flow	Cost
Lymphoscintigraphy	10 mm	No limit	2-D	1 image per 20 min	Yes	No	$
SPECT/CT	2 mm	No limit	2-D	1 image per 3 min	Yes	No	$$$$
MRI	Lateral: 150 μm Axial: 500 μm	No limit	3-D	1 image per minute	No but beneficial	Yes	$$$$
X-ray CT	Lateral: 1 mm Axial: 5 mm	No limit	3-D	2 fps	Yes	No	$$$
Ultrasound	Lateral: 15 μm Axial: 80 μm	3 mm and up	3-D	Up to 400 fps	Yes	Yes	$$
NIR fluorescence	1 μm	3 mm–10 cm	2-D	Up to 30 fps	Yes	Yes	$$
OCT	8 μm	1–2 mm	3-D	Up to 150 fps	No but beneficial	Yes	$$
Photoacoustic	Lateral: 5–25 μm Axial: 15–200 μm	3 cm	3-D	Up to 150 fps	Yes	Yes	$$

Values are not necessarily the theoretical limit, but the best achievements published to date for imaging lymphatics, lymph nodes, or blood vessels. It should be noted that for any given technique, the trade-offs exist between the various parameters. For example, to achieve the ideal spatial resolution, one usually has to sacrifice the frame rate and/or the penetration depth. This is particularly true for NIR fluorescence, where imaging depth significantly affects spatial resolution due to light scattering

take advantage of the contrast available from optics, while gaining the advantages of penetration depth provided by ultrasound. Given that it is easy to dermally deliver dyes that are taken up by the lymphatics, this technique could prove quite useful for imaging lymphatic function and has been used with a variety of contrast agent in animal models [136–139]. An additional advantage of PAT is that oxygenated and deoxygenated hemoglobin provide inherent contrast to blood vessels and veins allowing one to obtain high-resolution images of the vasculature while also imaging lymphatics.

Conclusion

Lymphatic diseases remain a prevalent problem throughout the world, and treatment options available for those suffering from lymphatic diseases are limited at best. While the clinical interest in lymphatic research is likely to remain low, the growing interest in the role of the lymphatics in cancer metastasis and the necessity of their existence and function for many of the nodal staging and sentinal lymph node screening approaches now in use have paved the way for technologies that could begin addressing this neglected patient population. Given the limited availability of resources for patients suffering from lymphatic diseases, particularly in developing countries, it is likely that those technologies associated with lower costs show the most promise for actually making it into the clinic. Every technology has its advantages and disadvantages (Table 36.1) and even within a given technique there are trade-offs between capabilities (e.g., resolution vs. imaging depth, resolution vs. frame rate, signal-to-noise ratio vs. frame rate, and sensitivity vs. tracer volume). Much more work is needed to fully appreciate the capabilities of these imaging approaches for studying various mechanisms of lymphatic dysfunction. Given the growing efforts in nanotechnology to provide new multifunctional probes, along with the fact that the size range of these probes makes them easy to target to draining lymphatics via simple injections, it is likely that the lymphatic tracer of the future will be functionalized to allow for a patient's lymphatics (or lymph nodes) to be imaged across a multitude of platforms, harnessing each of their unique capabilities. As we continue to grow in our ability to experimentally measure lymphatic function, it is likely that a new appreciation will arise for the role of these vessels in a variety of physiologic functions such as immune responses, lipid transport, cancer metastasis and tolerization, wound healing, and fluid balance.

References

1. Dixon JB (2010) Lymphatic lipid transport: sewer or subway? Trends Endocrinol Metabol 21: 480–487
2. Levick JR, Michel CC (2010) Microvascular fluid exchange and the revised Starling principle. Cardiovasc Res 87:198–210
3. Rockson SG (2001) Lymphedema. Am J Med 110:288–295
4. Rockson SG (2008) Diagnosis and management of lymphatic vascular disease. J Am Coll Cardiol 52: 799–806
5. Harvey N et al (2005) Lymphatic vascular defects promoted by Prox1 haploinsufficiency cause adult-onset obesity. Nat Genet 37:1072–1081
6. Alitalo K, Tammela T, Petrova T (2005) Lymphangiogenesis in development and human disease. Nature 438:946–953
7. Ji RC (2005) Characteristics of lymphatic endothelial cells in physiological and pathological conditions. Histol Histopathol 20:155–175
8. Baluk P (2005) Pathogenesis of persistent lymphatic vessel hyperplasia in chronic airway inflammation. J Clin Invest 115:247–257
9. Foldi M (2003) Foldi's textbook of lymphology: for physicians and lymphedema therapists. Elsevier GmbH, Munich, Germany
10. Dixon JB, Moore JE Jr, Cote G, Gashev AA, Zawieja DC (2006) Lymph flow, shear stress, and lymphocyte velocity in rat mesenteric prenodal lymphatics. Microcirculation 13:597–610
11. Zhang F, Niu G, Lu G, Chen X (2010) Preclinical lymphatic imaging. Mol Imaging Biol 13:599–612
12. Ege GN (1976) Internal mammary lymphoscintigraphy. The rationale, technique, interpretation and clinical application: a review based on 848 cases. Radiology 118:101–107
13. Veronesi U et al (1997) Sentinel-node biopsy to avoid axillary dissection in breast cancer with clinically negative lymph-nodes. Lancet 349:1864–1867
14. Weissleder H, Weissleder R (1988) Lymphedema: evaluation of qualitative and quantitative lymphoscintigraphy in 238 patients. Radiology 167:729–735
15. Wilting J, Becker J, Buttler K, Weich HA (2009) Lymphatics and inflammation. Curr Med Chem 16:4581–4592
16. Ristimäki A, Narko K, Enholm B, Joukov V, Alitalo K (1998) Proinflammatory cytokines regulate expression of the lymphatic endothelial mitogen vascular endothelial growth factor-C. J Biol Chem 273: 8413–8418
17. Saban MR et al (2004) Visualization of lymphatic vessels through NF-kappaB activity. Blood 104:3228–3230
18. Pepper MS, Tille J-C, Nisato R, Skobe M (2003) Lymphangiogenesis and tumor metastasis. Cell Tissue Res 314:167–177
19. Sharma R et al (2008) New horizons for imaging lymphatic function. Ann N Y Acad Sci 1131:13–36
20. Szuba A, Shin WS, Strauss HW, Rockson S (2003) The third circulation: radionuclide lymphoscintigraphy in the evaluation of lymphedema. J Nucl Med 44:43–57
21. Szuba A, Strauss W, Sirsikar SP, Rockson SG (2002) Quantitative radionuclide lymphoscintigraphy predicts outcome of manual lymphatic therapy in breast cancer-related lymphedema of the upper extremity. Nucl Med Commun 23:1171–1175
22. Weiss M, Baumeister RGH, Hahn K (2002) Post-therapeutic lymphedema: scintigraphy before and after autologous lymph vessel transplantation: 8 years of long-term follow-up. Clin Nucl Med 27:788–792
23. Szuba A et al (2002) Therapeutic lymphangiogenesis with human recombinant VEGF-C. FASEB J 16:1985–1987
24. Boccardo FM et al (2011) Surgical prevention of arm lymphedema after breast cancer treatment. Ann Surg Oncol 18:2500–2505
25. Kafejian-Haddad AP, Perez JMC, Castiglioni MLV, Miranda Júnior F, de Figueiredo LFP (2006) Lymphscintigraphic evaluation of manual lymphatic drainage for lower extremity lymphedema. Lymphology 39:41–48
26. Campisi C et al (2006) Diagnosis and management of primary chylous ascites. YMVA 43:1244–1248
27. Morton DL et al (1992) Technical details of intraoperative lymphatic mapping for early stage melanoma. Arch Surg (Chicago, Ill: 1960) 127:392–399
28. Giuliano AE, Kirgan DM, Guenther JM, Morton DL (1994) Lymphatic mapping and sentinel lymphadenectomy for breast cancer. Ann Surg 220:391–398; discussion 398–401

29. Krag D et al (1998) The sentinel node in breast cancer–a multicenter validation study. N Engl J Med 339:941–946
30. Czerniecki BJ, Bedrosian I, Faries M, Alavi A (2001) Revolutionary impact of lymphoscintigraphy and intraoperative sentinel node mapping in the clinical practice of oncology. Semin Nucl Med 31:158–164
31. Nakashima K et al (2010) Preoperative dynamic lymphoscintigraphy predicts sentinel lymph node metastasis in patients with early breast cancer. Breast Cancer 17:17–21
32. Thompson M et al (2007) Axillary reverse mapping (ARM): a new concept to identify and enhance lymphatic preservation. Ann Surg Oncol 14:1890–1895
33. Modi S et al (2007) Human lymphatic pumping measured in healthy and lymphoedematous arms by lymphatic congestion lymphoscintigraphy. J Physiol (Lond) 583:271–285
34. Ploeg IMC, Valdés Olmos RA, Kroon BBR, Nieweg OE (2008) The hybrid SPECT/CT as an additional lymphatic mapping tool in patients with breast cancer. World J Surg 32:1930–1934
35. van der Ploeg IMC, Olmos RAV, Kroon BBR, Rutgers EJT, Nieweg OE (2009) The hidden sentinel node and SPECT/CT in breast cancer patients. Eur J Nucl Med Mol Imaging 36:6–11
36. Uren RF (2009) SPECT/CT lymphoscintigraphy to locate the sentinel lymph node in patients with melanoma. Ann Surg Oncol 16:1459–1460
37. Uren RF et al (2011) SPECT/CT scans allow precise anatomical location of sentinel lymph nodes in breast cancer and redefine lymphatic drainage from the breast to the axilla. Breast. doi:10.1016/j.breast.2011.11.007
38. Schöder H et al (2006) Molecular targeting of the lymphovascular system for imaging and therapy. Cancer Metastasis Rev 25:185–201
39. O'Mahony S, Solanki C (2006) Imaging of lymphatic vessels in breast cancer–related lymphedema: intradermal versus subcutaneous injection of 99mTc-immunoglobulin. Am J
40. Delbeke D (1999) Oncological applications of FDG PET imaging: brain tumors, colorectal cancer, lymphoma and melanoma. J Nucl Med 40:591–603
41. Hoffman JM, Gambhir SS (2007) Molecular imaging: the vision and opportunity for radiology in the future. Radiology 244:39–47
42. Veenstra HJ, Vermeeren L, Olmos RAV, Nieweg OE (2012) The additional value of lymphatic mapping with routine SPECT/CT in unselected patients with clinically localized melanoma. Ann Surg Oncol 19:1018–1023
43. Nune SK, Gunda P, Majeti BK, Thallapally PK, Forrest ML (2011) Advances in lymphatic imaging and drug delivery. Adv Drug Deliv Rev 63: 876–885
44. Barrett T, Choyke PL, Kobayashi H (2006) Imaging of the lymphatic system: new horizons. Contrast Media Mol Imaging 1:230–245
45. Clément O, Luciani A (2004) Imaging the lymphatic system: possibilities and clinical applications. Eur Radiol 14:1498–1507
46. Misselwitz B (2006) MR contrast agents in lymph node imaging. Eur J Radiol 58:375–382
47. Song I, Hyeon T (2009) Inorganic nanoparticles for MRI contrast agents. Adv Mater 21:2133–2148
48. Hasebroock KM, Serkova NJ (2009) Toxicity of MRI and CT contrast agents. Expert Opin Drug Metab Toxicol 5:403–416
49. Waters EA, Wickline SA (2008) Contrast agents for MRI. Basic Res Cardiol 103:114–121
50. Ruehm SG, Schroeder T, Debatin JF (2001) Interstitial MR lymphography with gadoterate meglumine: initial experience in humans. Radiology 220:816–821
51. Lohrmann C, Foeldi E, Bartholomae J-P, Langer M (2007) Gadoteridol for MR imaging of lymphatic vessels in lymphoedematous patients: initial experience after intracutaneous injection. Br J Radiol 80: 569–573
52. Lohrmann C, Foeldi E, Speck O, Langer M (2006) High-resolution MR lymphangiography in patients with primary and secondary lymphedema. AJR Am J Roentgenol 187:556–561
53. Lohrmann C, Foeldi E, Langer M (2006) Indirect magnetic resonance lymphangiography in patients with lymphedema preliminary results in humans. Eur J Radiol 59:401–406
54. Liu N-F, Lu Q, Jiang Z-H, Wang C-G, Zhou J-G (2009) Anatomic and functional evaluation of the lymphatics and lymph nodes in diagnosis of lymphatic circulation disorders with contrast magnetic resonance lymphangiography. J Vasc Surg 49: 980–987
55. Ruddell A et al (2008) Dynamic contrast-enhanced magnetic resonance imaging of tumor-induced lymph flow. Neoplasia 10:706–713, 1 p following 713
56. Matsushima S, Ichiba N, Hayashi D, Fukuda K (2007) Nonenhanced magnetic resonance lymphoductography: visualization of lymphatic system of the trunk on 3-dimensional heavily T2-weighted image with 2-dimensional prospective acquisition and correction. J Comput Assist Tomogr 31: 299–302
57. Lu Q et al (2012) Magnetic resonance lymphography at 3T: a promising noninvasive approach to characterise inguinal lymphatic vessel leakage. Eur J Vasc Endovasc Surg 43:106–111
58. Lu Q, Xu J, Liu N (2010) Chronic lower extremity lymphedema: a comparative study of high-resolution interstitial MR lymphangiography and heavily T2-weighted MRI. Eur J Radiol 73:365–373
59. Kobayashi H et al (2005) Detection of lymph node involvement in hematologic malignancies using micromagnetic resonance lymphangiography with a gadolinium-labeled dendrimer nanoparticle. Neoplasia 7:984–991

60. Corot C, Robert P, Idée J-M, Port M (2006) Recent advances in iron oxide nanocrystal technology for medical imaging. Adv Drug Deliv Rev 58:1471–1504
61. Ruehm SG, Corot C, Debatin JF (2001) Interstitial MR lymphography with a conventional extracellular gadolinium-based agent: assessment in rabbits. Radiology 218:664–669
62. Herborn CU et al (2002) Interstitial MR lymphography with MS-325: characterization of normal and tumor-invaded lymph nodes in a rabbit model. AJR Am J Roentgenol 179:1567–1572
63. Mouli SK, Zhao LC, Omary RA, Thaxton CS (2010) Lymphotropic nanoparticle enhanced MRI for the staging of genitourinary tumors. Nat Rev Urol 7:84–93
64. Weissleder R et al (1990) Ultrasmall superparamagnetic iron oxide: an intravenous contrast agent for assessing lymph nodes with MR imaging. Radiology 175:494–498
65. Bellin MF, Beigelman C, Precetti Morel S (2000) Iron oxide-enhanced MR lymphography: initial experience. Eur J Radiol 34:257–264
66. Bellin MF et al (1998) Lymph node metastases: safety and effectiveness of MR imaging with ultrasmall superparamagnetic iron oxide particles – initial clinical experience. Radiology 207: 799–808
67. Réty F et al (2000) MR lymphography using iron oxide nanoparticles in rats: pharmacokinetics in the lymphatic system after intravenous injection. J Magn Reson Imaging 12:734–739
68. Harisinghani M, Barentsz J, Hahn P (2003) Noninvasive detection of clinically occult lymph-node metastases in prostate cancer. N Engl J Med 348:2491–2499
69. Ross RW et al (2009) Lymphotropic nanoparticle-enhanced magnetic resonance imaging (LNMRI) identifies occult lymph node metastases in prostate cancer patients prior to salvage radiation therapy. Clin Imaging 33:301–305
70. Kimura K et al (2010) High resolution MR lymphography using ultrasmall superparamagnetic iron oxide (USPIO) in the evaluation of axillary lymph nodes in patients with early stage breast cancer: preliminary results. Breast Cancer 17: 241–246
71. Kaminskas LM, Porter CJH (2011) Targeting the lymphatics using dendritic polymers (dendrimers). Adv Drug Deliv Rev 63:890–900
72. Yu SB, Watson AD (1999) Metal-based X-ray contrast media. Chem Rev 99:2353–2378
73. Suga K, Ogasawara N, Okada M, Matsunaga N (2003) Interstitial CT lymphography-guided localization of breast sentinel lymph node: preliminary results. Surgery 133:170–179
74. Suga K et al (2003) Visualization of breast lymphatic pathways with an indirect computed tomography lymphography using a nonionic monometric contrast medium iopamidol: preliminary results. Invest Radiol 38:73–84
75. Suga K et al (2004) Breast sentinel lymph node mapping at CT lymphography with iopamidol: preliminary experience. Radiology 230:543–552
76. Takahashi M et al (2008) Clinical efficacy and problems with CT lymphography in identifying the sentinel node in breast cancer. World J Surg Oncol 6:57
77. Wu H et al (2009) Preliminary study of indirect CT lymphography-guided sentinel lymph node biopsy in a tongue VX2 carcinoma model. Int J Oral Maxillofac Surg 38:1268–1272
78. Suga K et al (2005) Breast sentinel lymph node navigation with three-dimensional interstitial multidetector-row computed tomographic lymphography. Invest Radiol 40:336–342
79. Rabin O, Perez J, Grimm J, Wojtkiewicz G (2006) An X-ray computed tomography imaging agent based on long-circulating bismuth sulphide nanoparticles. Nat Mater 5:118–122
80. Kos S, Haueisen H, Lachmund U, Roeren T (2007) Lymphangiography: forgotten tool or rising star in the diagnosis and therapy of postoperative lymphatic vessel leakage. Cardiovasc Intervent Radiol 30: 968–973
81. Silvestri RC, Huseby JS, Rughani I, Thorning D, Culver BH (1980) Respiratory distress syndrome from lymphangiography contrast medium. Am Rev Respir Dis 122:543–549
82. Tiwari A, Cheng K, Button M, Myint F (2003) Differential diagnosis, investigation, and current treatment of lower limb lymphedema. Arch Surg 138:152–161
83. Vogl TJ, Bartjes M, Marzec K (1997) Contrast-enhanced lymphography. CT or MR imaging? Acta Radiol Suppl 412:47–50
84. Noroes J, Addiss D, Santos A, Mereidos Z (1996) Ultrasonigraphic evidence of abnormal lymphatic vessels in young men with adult Wuchereria bancrofti infection in the scrotal area. J Urol 156:409–412
85. Dreyer G, Addiss D, Roberts J, Norões J (2002) Progression of lymphatic vessel dilatation in the presence of living adult Wuchereria bancrofti. Trans R Soc Trop Med Hyg 96:157–161
86. Suresh S et al (1997) Ultrasonographic diagnosis of subclinical filariasis. J Ultrasound Med 16:45–49
87. Taylor MJ et al (2005) Macrofilaricidal activity after doxycycline treatment of Wuchereria bancrofti: a double-blind, randomised placebo-controlled trial. Lancet 365:2116–2121
88. Dreyer G et al (1996) Ultrasonographic assessment of the adulticidal efficacy of repeat high-dose ivermectin in bancroftian filariasis. Trop Med Int Health 1:427–432
89. Noroes J et al (1997) Assessment of the efficacy of diethylcarbamazine on adult Wuchereria bancrofti in vivo. Trans R Soc Trop Med Hyg 91:78–81

90. Gramiak R, Shah PM, Kramer DH (1969) Ultrasound cardiography: contrast studies in anatomy and function. Radiology 92:939–948
91. Voigt J-U (2009) Ultrasound molecular imaging. Methods 48:92–97
92. Dayton PA, Rychak JJ (2007) Molecular ultrasound imaging using microbubble contrast agents. Front Biosci 12:5124–5142
93. Nielsen KR et al (2009) Sentinel node detection in melanomas using contrast-enhanced ultrasound. Acta Radiol 50:412–417
94. Wisner ER et al (2003) Sentinel node detection using contrast-enhanced power Doppler ultrasound lymphography. Invest Radiol 38:358–365
95. Goldberg BB et al (2004) Sentinel lymph nodes in a swine model with melanoma: contrast-enhanced lymphatic US. Radiology 230:727–734
96. Sever AR et al (2012) Percutaneous removal of sentinel lymph nodes in a swine model using a breast lesion excision system and contrast-enhanced ultrasound. Eur Radiol 22:545–550
97. Sever A et al (2009) Preoperative localization of sentinel lymph nodes using intradermal microbubbles and contrast-enhanced ultrasonography in patients with breast cancer. Br J Surg 96:1295–1299
98. Hauff P, Reinhardt M, Briel A, Debus N, Schirner M (2004) Molecular targeting of lymph nodes with L-selectin ligand-specific US contrast agent: a feasibility study in mice and dogs. Radiology 231:667–673
99. Weiler M, Kassis T, Dixon JB (2012) Sensitivity analysis of near-infrared functional lymphatic imaging. J Biomed Opt 17:066019
100. Dixon JB, Zawieja DC, Gashev AA, Cote G (2005) Measuring microlymphatic flow using fast video microscopy. J Biomed Opt 10:064016
101. Dixon JB, Gashev AA, Zawieja DC, Moore JE Jr, Cote G (2007) Image correlation algorithm for measuring lymphocyte velocity and diameter changes in contracting microlymphatics. Ann Biomed Eng 35:387–396
102. Fischer M, Costanzo U, Hoffmann U (1997) Flow velocity of cutaneous lymphatic capillaries in patients with primary lymphedema. J Vasc 17:143–149
103. Mellor R et al (2000) Enhanced cutaneous lymphatic network in the forearms of women with postmastectomy oedema. J Vasc Res 37:501–512
104. McGreevy JM, Cannon MJ, Grissom CB (2003) Minimally invasive lymphatic mapping using fluorescently labeled vitamin B12. J Surg Res 111:38–44
105. Soltesz EG et al (2005) Intraoperative sentinel lymph node mapping of the lung using near-infrared fluorescent quantum dots. Ann Thorac Surg 79:269–277; discussion 269–277
106. Rao J, Dragulescu-Andrasi A, Yao H (2007) Fluorescence imaging in vivo: recent advances. Curr Opin Biotechnol 18:17–25
107. CHERRICK GR, STEIN SW, LEEVY CM, DAVIDSON CS (1960) Indocyanine green: observations on its physical properties, plasma decay, and hepatic extraction. J Clin Invest 39:592–600
108. Philip R, Penzkofer A, Baumler W, Szeimies RM, Abels C (1996) Absorption and fluorescence spectroscopic investigation of indocyanine green. J Photochem Photobiol A 96:137–148
109. Saxena V, Sadoqi M, Shao J (2003) Degradation kinetics of indocyanine green in aqueous solution. J Pharm Sci 92:2090–2097
110. Ohnishi S, Lomnes S, Laurence R (2005) Organic alternatives to quantum dots for intraoperative near-infrared fluorescent sentinel lymph node mapping. Mol Imaging 4:172–181
111. Sevick-Muraca E, Rasmussen J (2008) Molecular imaging with optics: primer and case for near-infrared fluorescence techniques in personalized medicine. J Biomed Opt 13:041303
112. Hutteman M et al (2010) Clinical translation of ex vivo sentinel lymph node mapping for colorectal cancer using invisible near-infrared fluorescence light. Ann Surg Oncol 18:1006–1014
113. Sampath L, Wang W, Sevick-Muraca E (2008) Near infrared fluorescent optical imaging for nodal staging. J Biomed Opt 13:041312
114. Unno N et al (2007) Preliminary experience with a novel fluorescence lymphography using indocyanine green in patients with secondary lymphedema. J Vasc Surg 45:1016
115. Swartz M, Berk D (1996) Transport in lymphatic capillaries. I. Macroscopic measurements using residence time distribution theory. Am J Physiol 270:H324–H329
116. Berk D, Swartz MA, Leu A, Jain R (1996) Transport in lymphatic capillaries 2. Microscopic velocity measurement with fluorescence photobleaching. Am J Physiol 39:H330–H337
117. Swartz MA (2001) The physiology of the lymphatic system. Adv Drug Deliv Rev 50:3–20
118. Sevick-Muraca E et al (2008) Imaging of lymph flow in breast cancer patients after microdose administration of a near-infrared fluorophore: feasibility study1. Radiology 246:734
119. Rasmussen JC et al (2010) Human lymphatic architecture and dynamic transport imaged using near-infrared fluorescence. Transl Oncol 3:362–372
120. Rasmussen J, Tan I (2009) Lymphatic imaging in humans with near-infrared fluorescence. Curr Opin Biotechnol 20:74–82
121. Rasmussen JC, Kwon S, Sevick-Muraca EM, Cormier JN (2011) The role of lymphatics in cancer as assessed by near-infrared fluorescence imaging. Ann Biomed Eng. doi:10.1007/s10439-011-0476-1
122. Unno N et al (2010) A novel method of measuring human lymphatic pumping using indocyanine green fluorescence lymphography. YMVA 52: 946–952

123. Unno N et al (2011) Influence of age and gender on human lymphatic pumping pressure in the leg. Lymphology 44:113–120
124. Gashev AA, Nagai T, Bridenbaugh EA (2010) Indocyanine green and lymphatic imaging: current problems. Lymphat Res Biol 8:127–130
125. Weiler M, Dixon JB (2013) Differential transport function of lymphatic vessels in the rat tail model and the long term effects of Indocyanine Green as assessed with near-infrared imaging. Frontiers in Physiology 4:1–10
126. Aldrich MB et al (2012) Concentration of indocyanine green does not significantly influence lymphatic function as assessed by near-infrared imaging. Lymphat Res Biol 10:20–24
127. Davies-Venn CA et al (2011) Albumin-binding domain conjugate for near-infrared fluorescence lymphatic imaging. Mol Imaging Biol. doi:10.1007/s11307-011-0499-x
128. Proulx ST et al (2010) Quantitative imaging of lymphatic function with liposomal indocyanine green. Cancer Res 70:7053–7062
129. Luo S, Zhang E, Su Y, Cheng T, Shi C (2011) A review of NIR dyes in cancer targeting and imaging. Biomaterials 32:7127–7138
130. Karlsen TV, McCormack E, Mujic M, Tenstad O, Wiig H (2012) Minimally invasive quantification of lymph flow in mice and rats by imaging depot clearance of near-infrared albumin. Am J Physiol Heart Circ Physiol 302:H391–H401
131. Huang D et al (1991) Optical coherence tomography. Science 254:1178–1181
132. Jung Y, Zhi Z, Wang RK (2010) Three-dimensional optical imaging of microvascular networks within intact lymph node in vivo. J Biomed Opt 15:050501
133. Vakoc BJ et al (2009) Three-dimensional microscopy of the tumor microenvironment in vivo using optical frequency domain imaging. Nat Med 15:1219–1223
134. Jung Y, Reif R, Zeng Y, Wang RK (2011) Three-dimensional high-resolution imaging of gold nanorods uptake in sentinel lymph nodes. Nano Lett 11:2938–2943
135. Wang LV, Hu S (2012) Photoacoustic tomography: in vivo imaging from organelles to organs. Science 335:1458–1462
136. Yao J, Maslov K, Hu S, Wang LV (2009) Evans blue dye-enhanced capillary-resolution photoacoustic microscopy in vivo. J Biomed Opt 14:054049
137. Galanzha EI, Shashkov EV, Tuchin VV, Zharov VP (2008) In vivo multispectral, multiparameter, photoacoustic lymph flow cytometry with natural cell focusing, label-free detection and multicolor nanoparticle probes. Cytometry A 73:884–894
138. Kim C, Song K, Gao F, Wang L (2010) Sentinel lymph nodes and lymphatic vessels: noninvasive dual-modality in vivo mapping by using indocyanine green in rats—volumetric spectroscopic photoacoustic imaging and planar fluorescence imaging. Radiology 255:442–450
139. Song L, Kim C, Maslov K, Shung KK, Wang LV (2009) High-speed dynamic 3D photoacoustic imaging of sentinel lymph node in a murine model using an ultrasound array. Med Phys 36:3724–3729

Part VII

Hair and Follicles

37 Digital Imaging for Measuring of Hair Growth on the Human Scalp

Rolf Hoffmann, Holger Lüdtke, Maciej Hoffman-Wecker, and Betsy J. Hughes-Formella

37.1 Introduction

Hair loss or thinning hair is a common complaint in clinical dermatology. However, patients seeking advice for hair loss are not necessarily bald and the effects of treatment attempts can be hard to measure. Digital analysis of phototrichograms is a sensitive tool to monitor hair loss and treatment response in clinical practice as well as in experimental trials. Any analysis method should be able to reliably analyze the most important biological parameters of hair growth, such as hair density ($1/cm^2$), hair diameter (μm), hair growth rate (mm/day), or anagen/telogen ratio, just to mention a few. In addition, any technique must produce results which are representative for the whole scalp.

Since the introduction of the phototrichogram as an in vivo noninvasive method to evaluate the human hair cycle, variations of the technique have been used as a basis for quantitative analysis of hair growth [1–8]. The basic principle of digital phototrichogram imaging is the same irregardless of the system used: The growth of hairs in a defined target area is measured by analysis of images following close clipping of hairs to the scalp. Repeated imaging of the same target area over time allows assessment of the efficacy of hair-promoting drugs or products. Differences between analysis systems can be found in the parameters which can be measured and, importantly, in the methods used to mark the hairs in images prior to analysis.

This chapter provides insight into the theoretical and statistical background of hair analysis and the sensitivity of the method with reference to recent clinical trial data obtained using the TrichoScan® system.

R. Hoffmann (✉)
Tricholog GmbH, In den Eschmatten 24, Freiburg 79117, Germany
e-mail: info@tricholog.de

H. Lüdtke
DatInf® GmbH, Wilhelmstr. 42, Tübingen 72074, Germany

M. Hoffman-Wecker • B.J. Hughes-Formella
Bioskin GmbH, Burchardstrasse 17, Hamburg 20095, Germany

37.2 Measurable Hair Parameters

At the very least, any hair analysis method must be able to analyze the two most important biological parameters of hair growth: hair density ($1/cm^2$) and hair diameter (μm). However, the measurements of these two parameters in isolation may not be adequate to detect treatment changes: Assuming that all miniaturized hairs in a target area just get thicker upon treatment, then the mere calculation of the hair density would not pick up such a change. Or in case a treatment induces growth of clinically insignificant vellus hairs, the calculation of just the hair density would give rise to a misleading result. We introduced the parameter "cumulative hair thickness" (mm/cm^2) with the introduction of the TrichoScan® Research software. This parameter represents the hair mass and takes into account both hair density and hair thickness. In our view

E. Berardesca et al. (eds.), *Non Invasive Diagnostic Techniques in Clinical Dermatology*, DOI 10.1007/978-3-642-32109-2_37, © Springer Berlin Heidelberg 2014

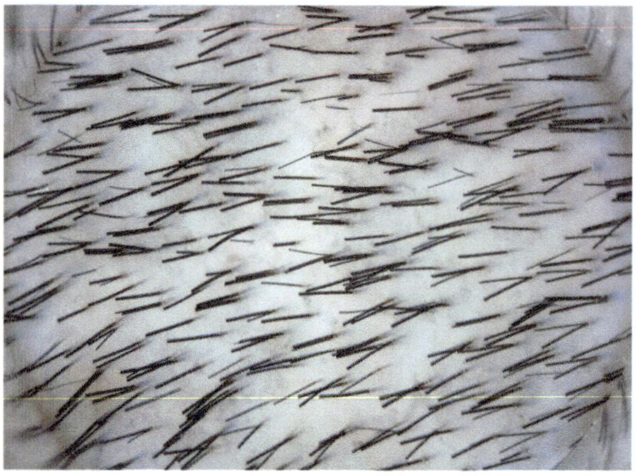

Fig. 37.1 TrichoScan® image 2 days after hair clipping. Nongrowing telogen and catagen hairs can be seen

this is the most precise parameter for hair growth analysis. Each of these parameters can be measured on a single image taken at a defined interval following hair clipping, usually 2–3 days afterwards. A baseline image immediately following clipping is not necessary.

The measurement of two other frequently reported parameters, the hair growth rate (mm/day) and anagen/telogen ratio, require that two images be taken, one on the day of hair clipping and the second 2 days later (Fig. 37.1).

Continued development of systems will allow for ease of evaluation of additional parameters and more comprehensive analysis of treatment effects. For example, the upcoming TrichoScan® Research version 4.0 will be able to calculate automatically the "thickness-adjusted trichogram" (thin hairs or vellus hairs grow slower), the number and density of follicular units, as well as the thickness variability of hair, a parameter which was introduced some years ago [9].

37.3 Manual Versus Automatic (TrichoScan®) Hair Counting

Before analysis of the hair parameters themselves, the individual hairs have to be marked on the image. If done manually this is a very time-consuming process, prone to human error. Most current systems still rely on manual marking and computer-assisted hair counts.

While there is no doubt that manual or computer-assisted hair counts, such as generated by the Canfield system, are valid and generate robust data [10], overall variability can be reduced still further by removal of the inconsistency associated with manual marking. TrichoScan® sets itself apart as the only fully automated method for the evaluation of hair growth parameters such as hair counts of total and terminal hairs, cumulative hair thickness, and growth rate [8]. Recently the validity of the TrichoScan® method was established by comparative assessment of hair growth parameters using TrichoScan® software versus manual identification of hairs prior to the final assessment of hair parameters [11]. The results using the TrichoScan® method were more reproducible with a smaller margin of operator error than with manual marking of hairs. Not surprisingly, in this investigation, considerable variability was noted for the manually marked images. Manual marking is a very tedious process and it is near to impossible to repeatedly count hundreds of hairs without variation for density, length, and thickness. The mean data variability for one evaluator who marked the same image three times ranged from 2.71 to 13.0 %, depending on the parameter. Some evaluators showed more variability than others. The correlation between different evaluators was best for total hair density and parameters related to hair length and worst for the parameters related to hair thickness. In contrast, there was no variability in repeated measures with automated marking by TrichoScan®; the software delivered completely consistent results. Keeping in mind the definition of repeatability, the maximum of the difference between two measurements on the same patient, this is a tremendous difference between manual marking of hairs and TrichoScan®. In a clinical trial setting, this would mean that a much larger sample size is required in the case of manual marking compared to fully automated evaluation.

As we were unable to find more recent published data from other systems regarding reliability, reproducibility, standard deviation, or margin of error, it is impossible to directly compare other hair growth systems to TrichoScan®.

37.4 Sensitivity

In addition to already published TrichoScan® data [12], data from a recent trial have been used, to furthermore assess the reproducibility and sensitivity of TrichoScan® in a clinical trial setting. A reference group of 57 men suffering from androgenetic alopecia were treated with 5 % topical minoxidil for 16 weeks. The data reported here were collected as comparator group data from a larger two-center double-blind Phase II study.

37.4.1 Methodology

Altogether 180 subjects were randomly assigned to three treatment groups of 60 subjects each, one of which was the minoxidil comparator group and one of which was the vehicle group. To achieve equal distribution of treatments between the investigative centers (bioskin GmbH centers in Hamburg and Berlin) block-wise randomization was applied. For the purpose here, only comparator (minoxidil) vehicle groups have been reassessed. The study was conducted double blind.

Eligible subjects were Caucasian men aged 18–49 years with Norwood/Hamilton grade III–IV androgenetic alopecia. The subjects had healthy skin on the scalp and were without disease unless it was considered irrelevant to the outcome of the study. Main exclusion criteria included previous surgical correction of scalp hair loss, hair loss due to disease or drug treatment, regular occlusion of the scalp, and treatment with another medication that might interfere with the study objective, such as finasteride within the previous 12 months, minoxidil within the previous 6 months, or treatment with other hair growth products within the previous 3 months. The study was conducted in compliance with the guidelines for good clinical practice from the International Conference on Harmonization and was approved by the review board of the Berlin Chamber of Physicians. All subjects provided written informed consent.

During the 16-week (113 days) treatment period, 1 ml of minoxidil 5 % solution (Rogaine® Extra Strength for Men, purchased from Pfizer corporation and repackaged by DPT Laboratories) or a propylene glycol/ethanol vehicle was applied between the frontoparietal and vertex areas of the scalp twice daily by the subjects at home. Two days before the beginning of treatment, the measurement area (a 1.8 cm^2, tattooed area in the center of the treatment area on the vertex of the subject's scalp) was shaved, and on day 1 (baseline), it was dyed and digital images for TrichoScan® evaluations were taken (using a Canon® PowerShot 95 camera with digital skin microscopy device). TrichoScan® images were repeated after 8 weeks (study day 57) and 16 weeks (study day 113) of treatment, and again 12 weeks following the end of treatment (study day 197). Two days before each scheduled visit, the measurement area was re-shaved. The images were analyzed at the end of the study using TrichoScan® Research by an independent observer who was blinded to the treatment received and who was also unaware of the time point in the study for each image.

The physician's global photographic assessment was conducted by assessment of standardized global scalp photographs (frontal and vertex views) taken at baseline, after 16 weeks of treatment and 12 weeks following the end of treatment. The images were taken before re-shaving the measurement areas for the TrichoScan® analyses (study days 2, 111, and 195). A 7-point rating scale was used to score change in hair growth: greatly decreased (−3), moderately decreased (−2), slightly decreased (−1), no change (0), slightly increased (+1), moderately increased (+2), and greatly increased (+3).

37.4.2 Results

Fifty-seven of the 60 subjects enrolled in the minoxidil group and 56 subjects in the vehicle group completed the study as planned. Three of the subjects in the minoxidil group and four in the vehicle group discontinued for personal reasons or reasons not related to the study products. No efficacy data were collected in these subjects. The age of the subjects included in the statistical analysis ranged from 23 to 49

Table 37.1 Mean and SD of hair growth parameters for each treatment group

Parameter	Treatment	Baseline	8 weeks	16 weeks
Total hair density [1/cm²]	Minoxidil	158 ± 48	184 ± 47	177 ± 49
	Placebo	165 ± 47	167 ± 42	164 ± 43
Terminal hair density [1/cm²]	Minoxidil	102 ± 42	113 ± 45	116 ± 42
	Placebo	111 ± 43	110 ± 39	109 ± 36
Cumulative hair thickness density [mm/cm²]	Minoxidil	8.2 ± 2.9	9.4 ± 3.0	9.3 ± 3.0
	Placebo	8.8 ± 2.9	8.9 ± 2.8	8.7 ± 2.6

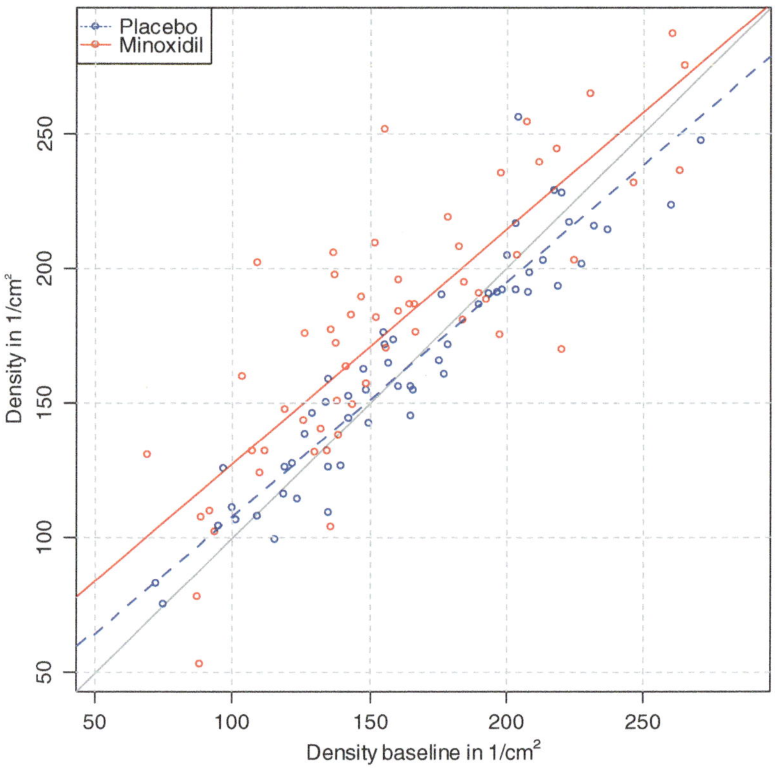

Fig. 37.2 Effect of minoxidil as analyzed with ANCOVA for hair density

(mean = 38.7, SD = 6.8) in the minoxidil group and 25 to 49 (mean = 39.7, SD = 6.4) in the vehicle group.

37.4.3 Hair Growth Parameters

Values for hair growth parameters measured using TrichoScan® are listed in Table 37.1

After treatment with 5 % minoxidil, the mean hair density increased as expected from 158 hairs/cm² at baseline to 184 hairs/cm² after 8 weeks of treatment and 177 hairs/cm² after 16 weeks of treatment ($p < 0.0001$ at both time points) (Fig. 37.2).

The corresponding values for mean terminal hair density were 102 hairs/cm² at baseline, 113 hairs/cm² after 8 weeks of treatment, and 116 hairs/cm² after 16 weeks of treatment ($p < 0.0001$ at both time points). Three months after the end of treatment (day 197) the mean hair density and total hair density had returned to slightly below the baseline values (Fig. 37.3).

Significant increases in cumulative hair thickness were also noted under treatment with minoxidil. The mean cumulative hair thickness was 8.2 mm/cm² at baseline, 9.4 mm/cm² after 8 weeks of treatment, and 9.3 mm/cm² after 16 weeks of treatment (Fig. 37.4, Table 37.1). By 12 weeks after the end of treatment, the cumulative hair

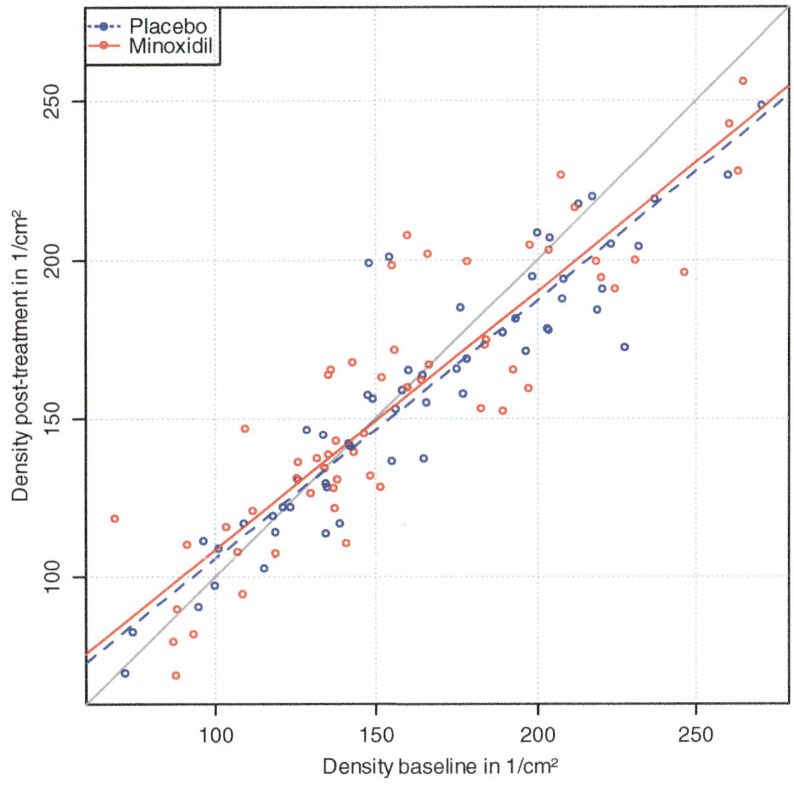

Fig. 37.3 Hair density in minoxidil group 3 months after withdrawal of treatment

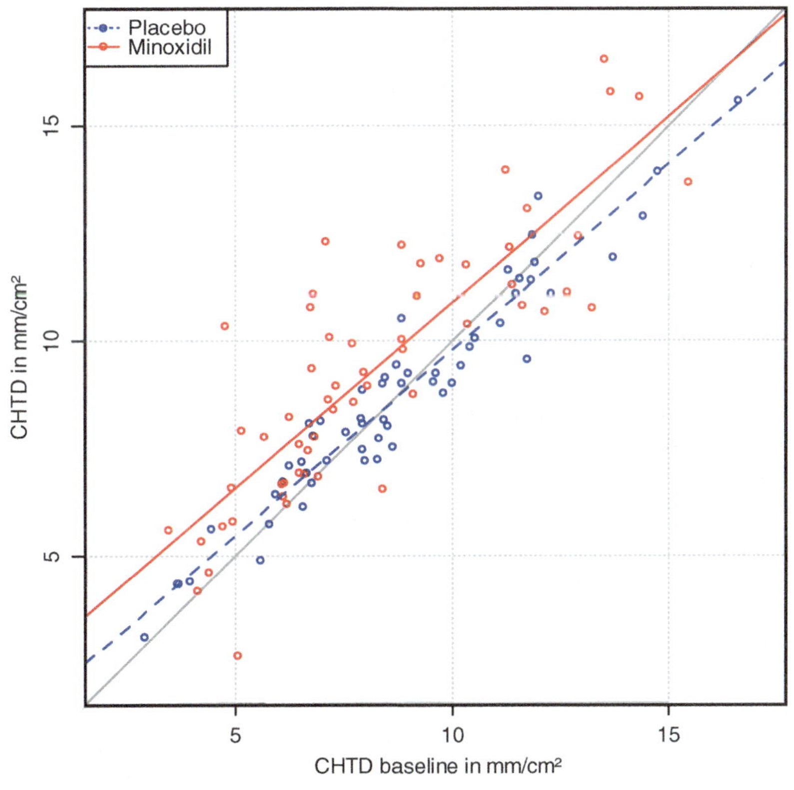

Fig. 37.4 Effect of minoxidil as analyzed with ANCOVA for cumulative hair thickness

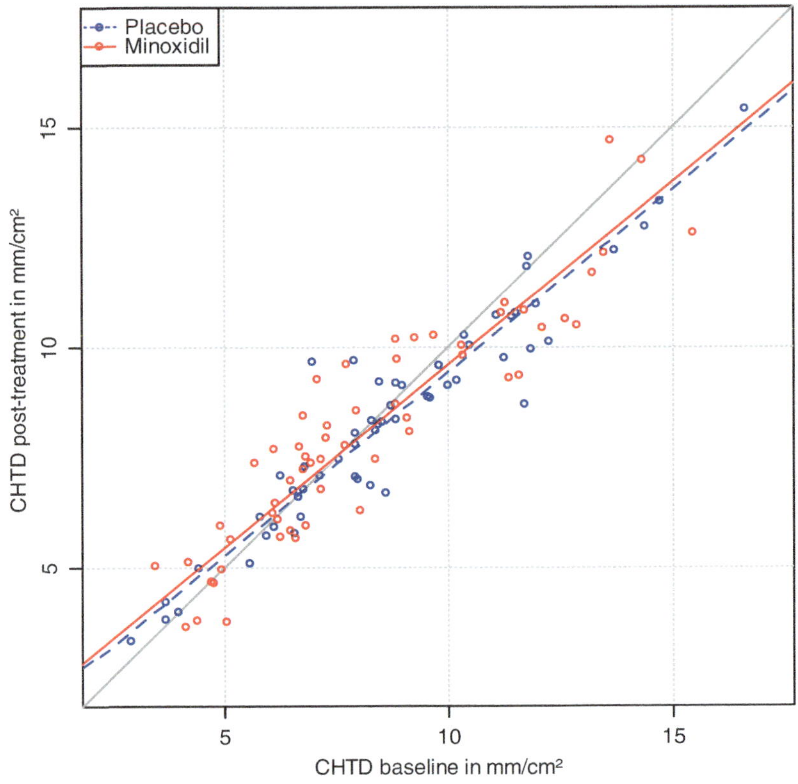

Fig. 37.5 Cumulative hair thickness in minoxidil group 3 months after withdrawal of treatment

thickness returned to baseline values (Fig. 37.5). In the subjects treated with the vehicle, mean hair density, terminal hair density, and cumulative hair thickness were similar at all assessment points over the study period (Table 37.1).

37.4.4 Correlation of Measured Hair Counts with Physician's Global Photographic Assessment

After 16 weeks of treatment with minoxidil, 29 of 56 subjects were rated as having increased hair growth based on comparison of standardized global scalp photographs by the physician. In 18 patients (32.1 %), hair growth was scored as slightly increased, in 9 (16.1 %) moderately increased, and in 2 (3.6 %) greatly increased. No change was reported in the remaining 27 (48.2 %) patients. Three months after the end of treatment, 43 of 55 (75.4 %) patients were rated as having no change in growth, 3 (5.3 %) patients having slightly increased, and 11 (19.3 %) patients having slightly decreased changes in hair growth. In the vehicle treated group, only 6 of 54 subjects were rated as having increased hair growth on the basis of the physician's photographic assessment. In 5 subjects (9.3 %), hair growth was scored as slightly increased and in 1 (1.9 %) moderately increased. No change was reported in 45 (83.3 %) of the subjects and a slight decrease in hair growth was reported in 3 (5.6 %) of the subjects. Three months after the end of treatment, 42 of 55 (76.4 %) subjects were rated as having unchanged hair growth and 13 (23.6 %) as having slightly decreased hair growth. Overall, these values are in very good agreement with the TrichoScan® results.

37.5 Minimal Size of Target Area

For hair counting the target area must be clipped. Patients already suffering from hair loss often find this objectionable and the size of the clipped area may have a significant impact on recruitability for clinical trials. The size of the target area

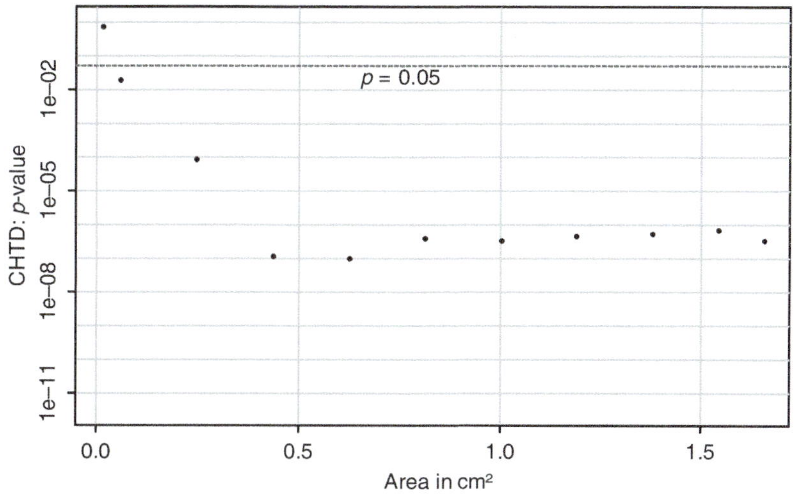

Fig. 37.6 The *p*-value as a result of the statistics of the cumulative hair thickness density (*CHTD*) is plotted against the investigated area. Note that the *p*-value is plotted logarithmically. For comparison a line is shown for $p = 0.05$

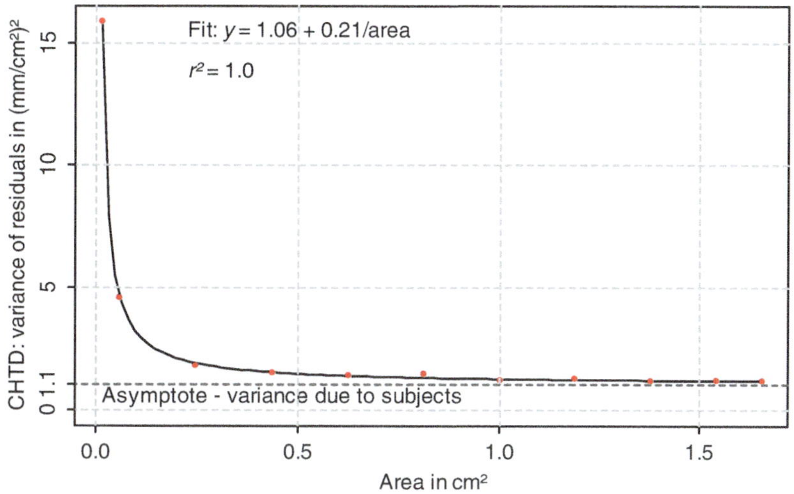

Fig. 37.7 The variance of the residuals of the statistics of the cumulative hair thickness density (*CHTD*) is plotted against the investigated area. The points are fitted with the reciprocal area (cf. formula in figure)

must bring together two contrasting needs: On the one hand, the size must be large enough to be representative for the entire scalp, and on the other, it must be small enough that the patient is willing and compliant. In our experience areas of more than 2 cm^2 are poorly tolerated by patients, making patient recruitment difficult. While it is obvious that the target area affects the sensitivity of any method, to our knowledge it is unknown what size is sufficient enough to generate robust and valid data. Therefore, we used the data set described above and reanalyzed all images of the 113 subjects (57 minoxidil group and 56 vehicle group) with different areas (0.01 cm^2, 0.06 cm^2, 0.25 cm^2, 0.43 cm^2, 0.62 cm^2, 0.81 cm^2, 1.00 cm^2, 1.19 cm^2, 1.38 cm^2, 1.54 cm^2, 1.66 cm^2) within the original image. The analyzed variable was the cumulative hair thickness density (CHTD).

For every area, the same statistical analysis, the ANCOVA described above, was performed. As the number of subjects and images is the same for all analyzed areas (i.e., the degrees of freedom are equal), the calculated p-values for the effect of minoxidil can be compared. The results are plotted in Fig. 37.6. As expected, the highest p-value can be seen for the smallest analyzed area. From about 0.5 cm^2 there is no more change of the p-value. To have a closer look at this, the variance of the residuals of the statistical model is plotted against the analyzed area (Fig. 37.7). Apparently, there is a saturation of the variance of the residuals with increasing area. The reason

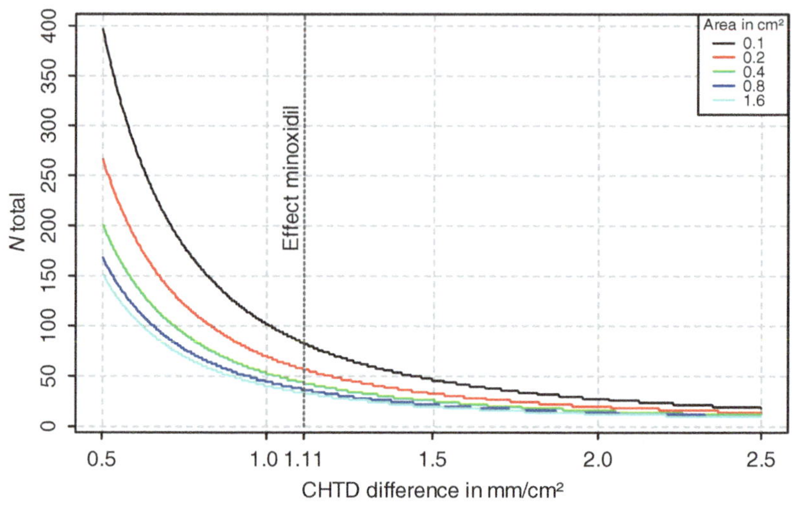

Fig. 37.8 For different areas, a sample size estimation was calculated for the cumulative hair thickness density (*CHTD*). As a result, the total number of subjects is plotted against the expected difference. The standard deviation used for the sample size calculation is calculated using the fit of the variance of the residuals in Fig. 37.7. The *vertical line* shows the effect of minoxidil, analyzed for the maximum area

is that the variation has two sources: The variation between the subjects and the variation due to the measurement error. Only the latter can be reduced by increasing the analyzed area. Of course, the variation due to subject differences cannot be diminished. Theoretically, the measurement error decreases reciprocally with the analyzed area. Therefore, the data plotted in Fig. 37.7 can be fit very convincingly with a hyperbola with an offset – the variance of the subjects. Using the variances seen in Fig. 37.7, a sample size calculation can be done. The standard deviation for each analyzed area is calculated with the help of the fitted curve. Figure 37.8 shows the total sample size plotted against the expected difference for different areas. The sample size was calculated for a two-sided, two-sample t-test with a significance level of 0.05 and a power of 0.8.

37.6 Centering of Target Area

In every trial the target area must be defined and this can be easily accomplished with a small red dot tattoo. Some investigators prefer two tattoos located on the left lower and right upper edge but outside the image, rather than a single tattoo within the image. Canfield's technology introduced the "target area hair count," which maps the target area of the same patients at different visits and via an image overlay ensuring that the tattoo is always centered and the same 1 cm² is counted. To assess whether such a technical approach is really necessary, we used the data set described above and reanalyzed all images of those 113 subjects but with three different 1 cm² areas within the original image (Fig. 37.9).

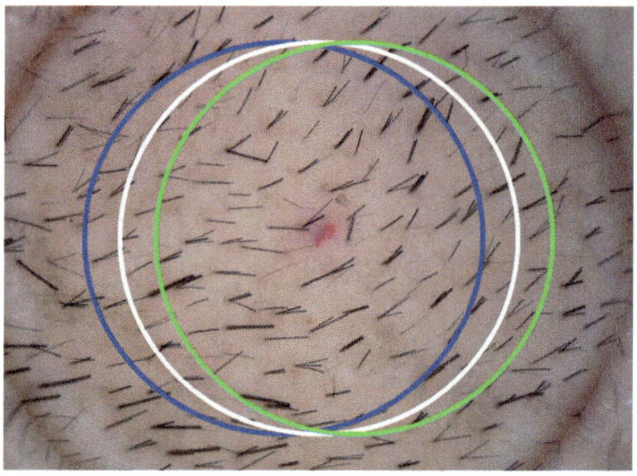

Fig. 37.9 TrichoScan® images were analyzed three more times. The *white circle* represents the situation where the tattoo is 100 % centered within a 1 cm² target area. The *blue and green circles* represent target areas where the tattoo is 1 mm left and right "off center"

Even with a shift of 2 mm, the correlation of the cumulative hair thickness density is very good (Fig. 37.10, $r = 0.991$). For the sake of simplicity, the difference between day 113 and baseline was analyzed using a two-sample t-test. The analysis was done for all nine combinations of TrichoScan® data at baseline and day 113; baseline left-shifted data with day 113 left-shifted,

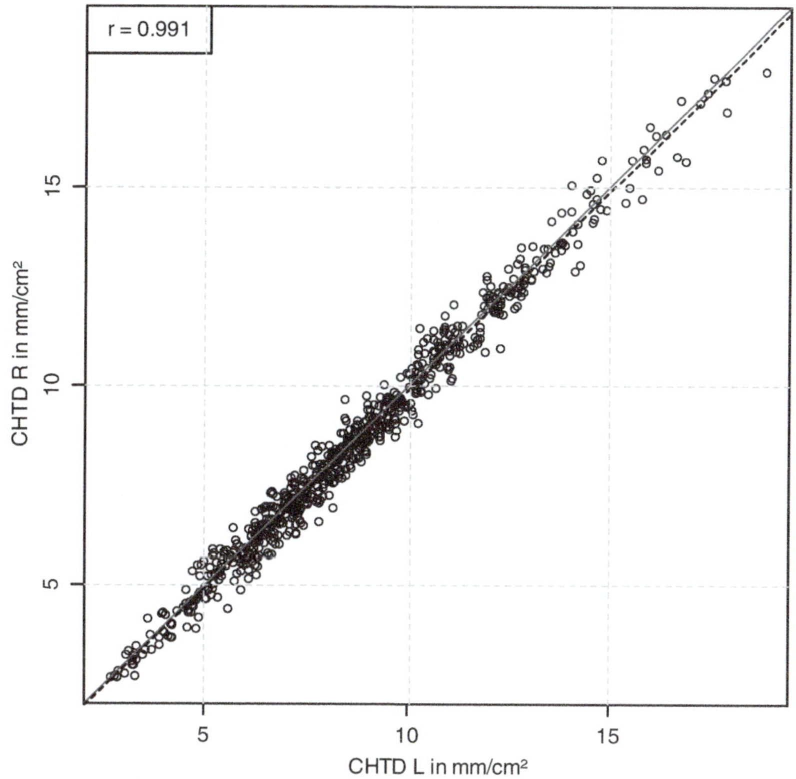

Fig. 37.10 Correlation of cumulative hair thickness density (*CHTD*) between left- and right-shifted analyzed images

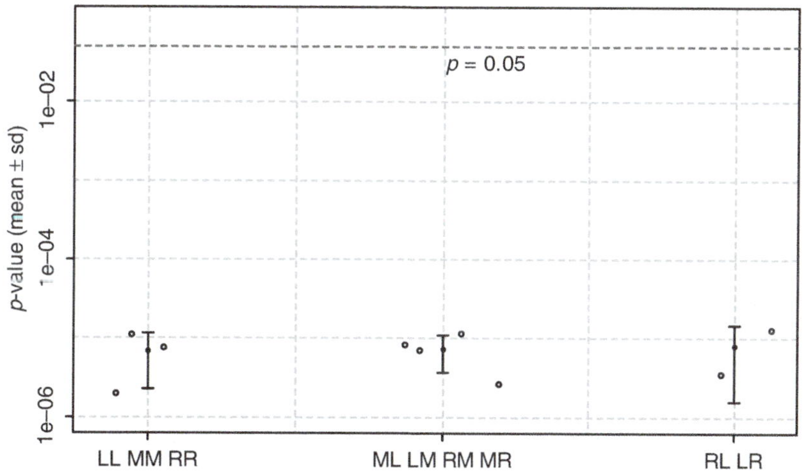

Fig. 37.11 Cumulative hair thickness density: *p*-values for effect of minoxidil of analysis using *t*-test with different shifts between baseline and measurement, mean and SD. On the left no shift (*LL*, *MM*, *RR* meaning no shift between baseline and measurement), in the middle a single shift (*ML*, *LM*, *RM*, *MR*) and at the right the maximum shift (*RL*, *LR*). Note that the *p*-value is plotted logarithmically. For comparison a line is shown for $p = 0.05$

not shifted, and right-shifted data; and so on. Actually, looking at the p-values, no difference or trend can be seen according to a shift. So, even a shift of 1 mm does not affect the result which implies that even smaller shifts caused by not exactly hitting the same area at a following measurement should not relevantly affect the results (Fig. 37.11).

37.7 Discussion

Minoxidil is reported to halt and/or slow hair loss in a majority of men with androgenetic alopecia (AGA). In a post-marketing surveillance study of 733 men with AGA, the majority (60 %) of patients reported efficacy with respect to reduction of affected scalp area, promotion of new hair growth, and improvement of hair density. Olsen [13] compared the efficacy of 2 and 5 % topical minoxidil solutions, and 5 % minoxidil ($n = 157$) was found to be significantly superior to 2 % minoxidil ($n = 158$) in terms of nonvellus hair count and scalp coverage over a 48-week treatment period. Nonvellus hairs were identified by manual marking of all pigmented hairs identified in macrophotographs of clipped hair in a 1 cm^2 target area. A concentration-dependent increase in the number of non-vellus hairs was already noted at the first assessment point following 8 weeks of treatment.

Using TrichoScan® it was possible to reproduce the increase in hair count of all hairs and terminal hairs (>40 μm diameter) following 8 weeks of treatment with 5 % minoxidil in a smaller sample size of 57 men with AGA. Further, the cumulative hair thickness, an important cosmetic parameter reflecting scalp coverage, was also shown to be significantly increased after only 8 weeks of treatment. These results were supported by the results of the physician's global photographic assessment as well as the subject assessment. The good agreement of these data with historical data for 5 % minoxidil underscores the sensitivity of the TrichoScan® methodology and its usefulness as a tool in hair growth studies. The full automation of this analysis system, which eliminates the necessity of time-consuming and more error-prone manual marking of hairs, makes it particularly attractive as a clinical research tool.

In clinical studies, one of the major factors driving sample size is variability from various sources. By lowering the variability of the of the analysis method, the study sample size can be reduced accordingly. This makes the TrichoScan® technique particularly attractive for clinical studies with treatment comparisons.

In addition, we were able to show that a visible, centered tattoo is enough to generate valid results and that a minimal target area size of 1 cm^2 is necessary to calculate robust and reliable results, although larger target area sizes are associated with lower standard deviations.

References

1. Bouhanna P (1985) The phototrichogram, a macrophotographic study of the scalp. Bioeng Skin 3:265
2. Guarrera M, Ciulla MP (1986) A quantitative evaluation of hair loss: the phototrichogramm. J Appl Cosmetol 4:61–66
3. Rushton H, James KC, Mortimer CH (1983) The unit area trichogram in the assessment of androgen-dependent alopecia. Br J Dermatol 109:429–437
4. Friedel J, Will F, Grosshans E (1989) Phototrichogram: adaptation, standardization and applications. Ann Dermatol Venereol 116:629–636
5. Van Neste DJJ, Dumrotier M, De Coster W (1989) Phototrichogram analysis: technical aspects and problems in relation with automated quantitative evaluation of hair growth by computer-assisted image analysis. In: Van Neste DJJ, Lachapelle JM, Antoine JL (eds) Trends in human hair growth and alopecia research. Kluwer, Dordrecht, pp 155–165
6. Van Neste DJ, Dumrotier M, de Brouwer B et al (1992) Scalp immersion proxigraphy (SIP): an improved imaging technique for phototrichogram analysis. J Eur Acad Dermatol Venereol 1:187–191
7. Hayashi S, Miyamoto I, Takeda K (1991) Measurement of human hair growth by optical microscopy and image analysis. Br J Dermatol 125: 123–129
8. Hoffmann R (2001) TrichoScan: combining epiluminescence microscopy with digital image analysis for the measurement of hair growth in vivo. Eur J Dermatol 11:362–368
9. de Lacharrière O, Deloche C, Misciali C et al (2001) Hair diameter diversity: a clinical sign reflecting the follicle miniaturization. Arch Dermatol 137(5): 641–646
10. Canfield D (1996) Photographic documentation of hair growth in androgenetic alopecia. Dermatol Clin 14:713–721
11. Gassmueller J, Rowold E, Frase T et al (2009) Validation of TrichoScan® technology as a fully-automated tool for evaluation of hair growth parameters. Eur J Dermatol 12:224–231
12. Gassmueller J, Hoffmann R, Webster A (2008) Treatment of androgenetic alopecia using topical fulvestrant solution in males and postmenopausal females; results from two randomized, proof-of- concept studies. Br J Dermatol 158:109–115
13. Olsen EA, Dunlap FE, Funicella T et al (2002) A randomized clinical trial of 5% topical minoxidil versus 2% topical minoxidil and placebo in the treatment of androgenetic alopecia in men. J Am Acad Dermatol 47:377–385

Tensile Properties of Hair

38

Leszek J. Wolfram

38.1 Introduction

The purpose of this chapter is to provide an overview of elastic behavior of hair focusing on its stress-strain characteristics. The measurements of tensile properties of hair proved to be particularly useful approach to elucidate the physicochemical changes in hair caused by environmental factors or brought about by cosmetic treatments. The absorption of moisture by hair strongly influences the relationship between the structural hierarchy of hair and its tensile properties becoming a dominant factor of hair elasticity.

38.2 Some Salient Aspects of Hair Structure

Emerging from the scalp, hair embodies the marvel of biomolecular engineering accomplished by its follicular origin. A smooth and shiny fiber, strong and elastic, water-repellent yet moisture absorbing, wavy, curly, or straight in shape, and spanning a gamut of melanin-derived, iridescent colors – it can truly claim the "crowning glory" epithet.

The patent morphology of hair appears quite straightforward. The outer layer of cuticle cells comprises a protective sheath around the elongated spindle-shaped cells of the cortex which forms the main shaft of the fiber and is the source of its mechanical strength. The airy cells of the medulla together with the melanosomes both situated within the cortex complement the structural profile. However, this picture, simple as it appears, is highly misleading. Except for the pigment and the medulla, both the cuticle and cortex are highly complex structures skillfully designed for functional and lasting performance of hair (Fig. 38.1).

38.3 Cuticle

The cuticle cells are in the form of flat, square sheets, 0.5 μm thick, and 50 μm in length. Their proximal portions are firmly attached to the cortex while the distal edges protrude towards the tip of the fiber. Extensive overlapping of the cells (up to 5/6 of their length) results in a thick band of cuticle layer (up to 3 μm thick) and gives the hair a ratchety appearance. These imbrications are highly functional. By interlocking with the pointing downwards cuticle cells of the inner root sheath, they contribute to the follicular anchorage of the growing hair. The imbricated surface also serves as a self-cleaning fixture. As the hairs grow and move relative to each other, the outward pointing cuticular edges facilitate removal of trapped dirt particles and desquamated cell from the scalp.

In the course of the process of hair formation, a stratified structure develops within each cuticle cells displaying layers of different textures

L.J. Wolfram
1310 Jones St., San Francisco, CA 94109, USA
e-mail: ljmaw@aol.com

E. Berardesca et al. (eds.), *Non Invasive Diagnostic Techniques in Clinical Dermatology*,
DOI 10.1007/978-3-642-32109-2_38, © Springer Berlin Heidelberg 2014

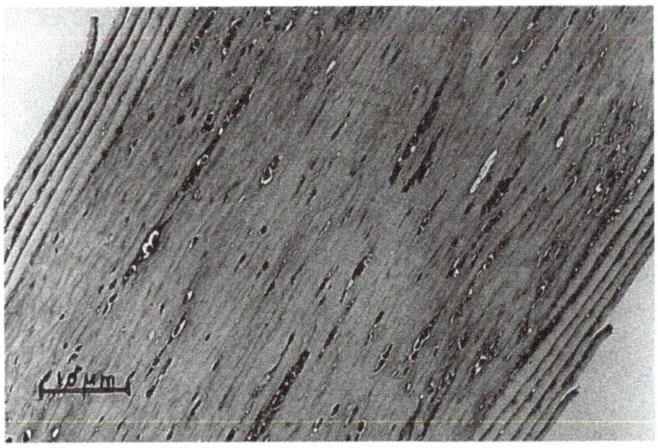

Fig. 38.1 Transmission electron micrograph of longitudinal hair section

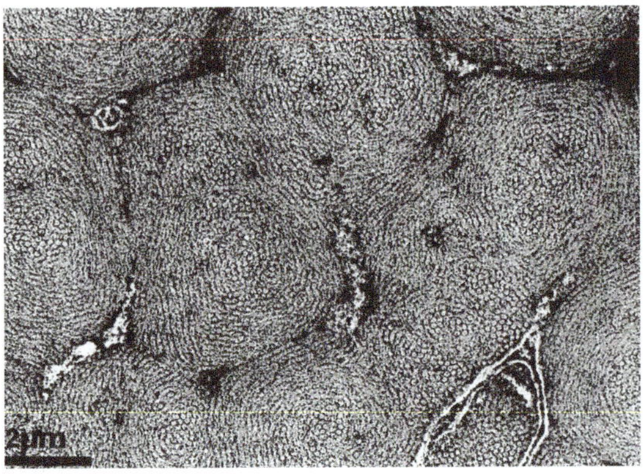

Fig. 38.2 Transmission electron micrograph of hair filament/matrix texture (Courtesy of Dr. J. Sikorski)

forming an internal mosaic of proteins, lipids, and carbohydrates. The outer surface of each cuticle is covered by a thin (15 nm) continuous membrane of the epicuticle below which a cystine-rich A layer abuts the major component of the cuticle—the exocuticle. The latter is also densely cross-linked by disulfide bonds in contrast to the adjacent, hydrophilic layer of the endocuticle. Each cuticle is surrounded and separated from its neighbors by the cell membrane complex (CMC). The description of the structural elements of the cuticle would be incomplete without mentioning the 18-methyleicosanoic acid (18-MEA). An important lipid constituent of the cuticular CMC, the 18-MEA, is covalently linked to its protein components [1]. It has been ascertained that a thin layer of 18-MEA is also grafted onto the outer surface of each cuticle [2]. The presence of such lipid film confers to the hair surface attribute of low friction with concomitant hydrophobic character.

It is noteworthy how well the cuticle sheath has been adapted to meet the environmental challenge. A water-repellant surface facilitates drying of hair while the cuticular imbrications keep the hair and the scalp clean. The densely cross-linked outer layers provide stability and a measure of protection against mechanical assaults while cushioning the effects of physical impacts with a softer stratum of endocuticle.

The layers of overlapping cells constitute a formidable diffusion barrier that must be taken into account when devising processes for chemical modification of hair. Yet, as most of the reactive treatments are carried out in an aqueous environment, it is fortuitous that a potentially easy passage can be channeled through the endocuticle whose low cross-link density combined with its hydrophilic character make it highly water swellable.

38.4 Cortex

The morphologically dominant and mechanically most important component of hair is made of elongated, spindle-like cells, approximately 100 μm long and 5 μm across at the maximum width. The cells are tightly bundled together and oriented parallel to the axis of the fiber. They are fused together by a modified CMC providing strong intercorticular adhesion. Viewed under the electron microscope, each cortical cell is packed with fine, long, and axially oriented uniform filaments (microfibrils) that consist of highly organized helical proteins responsible for the diagnostic X-ray diffraction pattern of α-keratins. The microfibrils are about 7 nm in diameter and are grouped into larger units called macrofibrils. By using a specific staining technique, the structural resolution of these fibrillar assemblies was ascertained. The results indicate that each macrofibril represents a structural composite consisting of rods of microfibrils embedded in cystine-rich matrix with the fibrils accounting for approximately 50–60 % of the cortex material (Fig. 38.2).

The filament/matrix texture is a hallmark of the strongest materials that have evolved in nature [3]. They are typically made up of two components with contrasting properties of strength and elasticity – the elastic, load-bearing filaments surrounded by a pliable matrix. Under load, the easily deformed matrix distributes the stress evenly over the filaments thus preventing propagation of cracks from local imperfections or rupture points.

38.5 Physical Properties

As with all materials the mechanical properties comprise a bulk property, the extensibility of the whole fiber being a summation of the extensibility of bonds holding the keratin together. In presence of water some cohesive bonds become labile allowing the structure to flow, whereas others stay unaffected. The nature of hair interaction with water is the dominant factor in the fiber response to both physical and chemical stimuli.

38.6 Water Sorption and Swelling

The chemical composition of keratin fiber lies at the root of their high affinity for water extending over the whole range of relative humidities (Fig. 38.3). The adsorption and desorption curves exhibit hysteresis. At any given RH (relative humidity) between 0 and 100 %, the hair contains less water when it is absorbed from the dry than when it undergoes desorption from the wet state. For most of the fibers, the absorption isotherms are fairly similar but diverge at the saturation point suggesting that under such conditions the water uptake and associated with it swelling might be controlled by restraints in its molecular structure. A good case in point is made by absorption characteristic of hair subjected to cosmetic treatments such as waving, bleaching, or relaxing which cause some disruption in the native structure of hair by attendant cleavage of the disulfide cross-links (Table 38.1).

Although water vapor penetrates hair readily, there is binding selectivity and accessibility restraints within the filament/matrix texture.

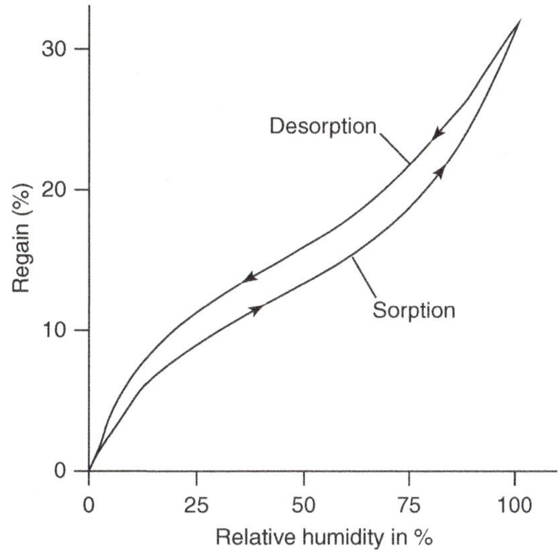

Fig. 38.3 Moisture absorption isotherms of human hair

Table 38.1 Effect of hair treatments on moisture uptake by hair (wt %)

	Moisture regain (% at)		
Treatment	65 % RH	95 % RH	Liquid water
Intact	15	32	34
Bleached	16	35	50
Waved	15	33	41
Relaxed	15	34	57

X-ray diffraction studies [4] showed conclusively that water sorbed by hair is preferentially confined to the matrix component bringing about significant swelling of this phase in comparison to the microfibrils which are relatively impenetrable by water. A consequence of this differential response is the highly anisotropic swelling behavior of hair. Thus the diametrical swelling on complete hydration (>12 %) is over one magnitude higher than the length change (1 %) of the fiber. Feughelman and Bendit [5] point out that as the biosynthesis of hair occurs in moist environment, both the microfibrils and matrix will be in mechanical equilibrium in fully hydrated hair. Upon drying, the matrix contracts resulting in only small longitudinal compression of the microfibrils.

As both the amino acid composition and fiber morphology of hairs of different ethnic origins are quite similar, one should not expect (and none have been found) any major differences in their water sorption and swelling characteristics. Recent publication by Franbourg et al. [6], however, strongly suggests that Afro-American hair equilibrates at significantly lower moisture content at corresponding relative humidities and exhibits less swelling than either Caucasian or Asian hair. As of now there is no satisfactory explanation to account for this finding except for suggestions, made by the authors, that internal lipids might be involved.

38.7 Load Extension Properties of Hair

The first comprehensive studies of the mechanical properties of α-keratin fibers were carried out by Speakman [7]. He has shown that when the fiber is stretched, the relationship between the stress and strain demonstrates three distinct regions of extension (Fig. 38.4). Up to about 2 % of strain, the load rises rapidly, and there is a linear relationship between stress and strain which is referred to as the Hookean region. As the fiber is extended further, one encounters the yield region where only a small increase in stress generates a large increase in strain (up to about 30 %). This follows an inflection in the curve; the hair becomes stiffer with increase in strain (post-yield region) and persists until it breaks. The three regions are most distinctly defined in water where the slopes of the linear portions of stress-strain curve for the Hookean, yield, and post-yield regions are in approximate ratio of 100:1:10. An increase in hydration of hair brings about progressive lowering of the Young modulus and, due to the swelling anisotropy of hair, a drastic decrease in the rigidity modulus. This dependence on moisture content is shown in Fig. 38.5 illustrating the fact that while the Young modulus decreases by factor 2.7 between 0 and 100 % relative humidity, the rigidity modulus drops by factor 15 over the same range.

These observations of differential changes in moduli with hydration with the background of the filament/matrix texture led Feughelman [8] to the postulate of a two-phase model of α-keratin fibers consisting of rigid, elastic, and water-impenetrable rods (microfibrils) parallel to the fiber axis embedded in the matrix phase which is accessible to and weaken by water.

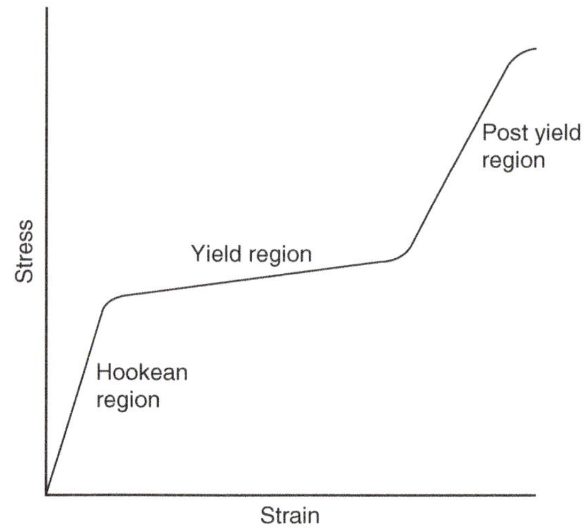

Fig. 38.4 A stress-strain curve of a keratin fiber

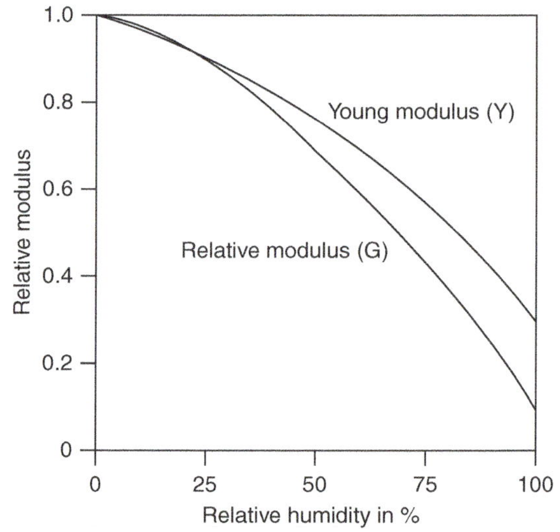

Fig. 38.5 Effect of moisture on changes in tensile and rigidity moduli of hair

Each of the regions of the stress-strain curve reflects the different character of structural bonding within the fiber. The stiffness of the Hookean region is attributed to the elastic stability of the hydrogen-bonded α-helices which, beyond the yield point, unfold reversibly into β-configuration. The covalent disulfide cross-links appear, on the other hand, to be a controlling factor in the bond network that oppose extension in the post-yield region.

Speakman's observation that the keratin fibers which are extended in water no further than the yield region have the ability to recover fully their properties when subsequently relaxed in water proved of exceptional value in assessing simply and accurately the changes in their properties associated with chemical or physical treatments. The technique involves determination of work required to stretch the hair 30 % (calibration step) followed by overnight relaxation in water, subjecting the fiber to intended treatment and restretching. The change in work (RW30) expressed as the percentage of the calibrated value provides the information of changes imparted by the treatment. As each hair has its own control, few tests suffice to generate reliable results, and there is no need to account for variation in fiber diameter. This method has been extensively employed as a test for hair damage in waving, bleaching, coloring, etc. By adjusting the pH of the testing solution, insight can be gained as to the Coulombic (electrostatic) interactions in hair, testing in solvents and their combinations with water yields information on the accessibility of hair to various reagents. Over the years, several modifications of the technique have been introduced including replacing the overnight relaxation by 1-h water exposure at 52 °C [9].

As the reactivity of the disulfide bonds in hair is significantly increased by stretching the fibers [10], a question arose whether the calibration step might sensitize the fiber to subsequent treatments. Evaluation of tensile properties of hair fibers that had been calibrated and then together with unstretched controls submitted to oxidative treatment with alkaline peroxide revealed that calibration step has no effect on the subsequent treatments [11].

It is essential to point out that the mechanical properties of hair vary with temperature and the relative humidity at which the fibers are tested. Thus, strictly adhered-to testing procedure must be followed to assure the validity and reproducibility of the results. In this respect, performing the tensile tests in water is often more convenient—the environment of 100 % is assured and the temperature control is easy to maintain. Also, due to the plasticizing effect which water exerts on keratin, the interplay of structural bonding in different region of the stress-strain curve is easier to interpret.

While the technique described above is very useful in portraying changes in hair resulting from variety of cosmetic treatments, the intermittent stress relaxation measurements permit insight into the process itself as it occurs in the fiber [12]. This is of particular value in case of reactive treatments such as waving or straightening where the knowledge of the kinetics of the process is invaluable and essential in designing potential process improvements. The coincidental advantage of such measurements is that they are carried out at very low strain levels, which mimics the conditions of product application.

38.8 Brief Survey of Tensile Testing of Hair

Simplicity and the accuracy of the measurement technique combined with the wealth of reliable information it can provide made the evaluation of tensile properties an indispensable tool in fundamental and applied hair research. The availability of modern instrumentation (Instron, Diastron) has made it easy to carry out measurements in well-controlled environment, thus expanding both the scope and reliability of collected data.

Although the calibration technique of Speakman is highly effective in exploring and characterizing, hair response to range of reactive treatments is less suitable when testing requires environment other than water and where the tensile parameters of choice are primarily breaking stress and breaking extension. The latter are relatively well comprehended by the public for whom the fiber strength is tantamount to the hair "wellness."

To provide statistical reliability, the measurements are usually carried out with large number of fibers, and a question may arise as to what extent the cuticle might be a factor in data analysis. The rationale behind the question seems reasonable bearing in mind that in case of fine fibers, the cuticle layer can account for as much as 35 % of the total mass of the hair (cuticle layer averages 3.5 μm in thickness and is diameter invariant). Also lack of any axially oriented structure within the cuticle together with the hydrophilic amorphous endocuticle would imply a weaker mechanical performance than that of cortex. Yet, all published information on this subject [13–15] clearly indicates that the tensile properties of dry

Table 38.2 Tensile properties of hair of different ethnic origin

Hair type	Breaking strength (MPa)		Breaking extension (%)	
	At 65 % RH	In water	At 65 % RH	In water
Caucasian	178–188	155–165	38–49	49–61
Asian	185–190	158–165	43–47	48–62
African	112–191	94–160	29–42	42–54

or wet hair fibers show no difference with the change of the cuticle/cortex ratio. While the tensile data for dry fibers might not be surprising as the dry modulus of cuticle is not much different from that of cortex the results for wet fibers are unexpected in view of the fact that computed modulus of wet cuticle layer is much less than that of the microfibrils [16]. It is conceivable that on hydration some ordered structure might arise in the cuticle as the latter is under significant stress from the expanding matrix of the cortex that would enhance its mechanical performance. There is some evidence of stress birefringence in swollen cuticle layers [14]

A number of comparative studies of hair of different ethnicities [17] show similarity in chemical composition that extends to the proteinaceous makeup of the filament/matrix texture, and it is not particularly surprising that tensile measurements follow suit except for the African hair which is characterized by relatively wide span of the results for both the breaking strength and breaking extension as shown in Table 38.2.

One must wonder to what extent is this due to the harvesting technique employed as well as the fact (often observed by the author) that quite a number of individual hairs show evidence of pronounced axial shifts resulting in narrow segments. The latter have always proven to be weak points at which the fibers invariably break.

The weathering of hair is a seasonal affair, and the most obvious manifestation of exposure to sunlight is hair bleaching. This photobleaching phenomenon is caused by oxidative destruction of the melanin pigment which, ironically, is both the site of light absorption and production (in presence of oxygen) of superoxide moiety involved in pigment destruction [18]. Prolonged exposure leads to significant loss of cystine and weakening of the fiber as shown by a decrease in both the Young modulus and ultimate strength. Surprisingly the loss in strength is much less than anticipated from the extent of cystine degradation. This is, apparently, due to network of crosslinks formed in the course of weathering [17, 19]

Loss of pigmentation in aging (hair graying) leads often to individual complains of yellowing, undesirable changes in texture, and perception of increased stiffness. Studies on mechanical properties of gray hair while detecting some slight increase in strength on graying at the individual level (pigmented and nonpigmented hair from the same head of hair) found no evidence at the global scale [20–22].

Heat setting of hair is a popular and convenient technique for set impartation and is readily accomplished by the use of heated rollers. A legitimate question arises as to the potential of heat damage to hair with frequent use of such appliances. Experiments were performed with two types of appliances whereby the hair tresses were set repeatedly (50 setting cycles) and their tensile breaking strength and breaking extension evaluated. No apparent changes were evident. Furthermore, the supercontraction characteristic (a sensitive test for latent damage) of both the tested fibers and controls was virtually identical suggesting acceptable safety of the setting procedures [23].

Recently the tensile measurements have been resorted to [24] in assessing the effect of daily oral intake of orthosilicic acid on hair. The study involving 48 women lasted for 9 months. Hair morphology and tensile properties were evaluated before and after treatment. There was small increase in hair thickness and corresponding change in the breaking load and elasticity but not in the breaking stress.

As an aside I would like to bring up the word "stronger." It is used often in the context of product claims, referring most often to "hair breakage." As the latter usually results from hair combing or brushing, the strength of the individual hair fibers is of little relevance. The breakage encountered in such situation is the result of the entanglements of fibers when in contact with a comb or brush and is primarily caused by interfiber friction. Lowering this friction, for example, by application of conditioner, will significantly decrease the extent of breakage yet the conditioner, clearly and assuredly, did not make the hair stronger.

References

1. Rivett DE (1984) The cell membrane complex and its influence on the properties of wool. Wool Sci Rev 3:35
2. Jones LN, Swift JA (1988) Structure of the fully formed hair shaft. In: Butler H (ed) Fundamentals of human hair science. Micelle Press, Weymouth, p 7
3. Slayter G (1962) Two phase materials. Sci Am 206:24
4. Fraser RDB, MacRae TP, Millward GR, Parry DAD, Suzuki A, Tulloch PA (1971) The molecular structure of keratins. Appl Polym Symp 18:65
5. Bendit EG, Feughelman M (1968) Keratin. In: Encyclopedia of polymer science and technology, vol 8. Wiley, New York, p 29
6. Franbourg A, LeRoy F, Escoubesh M, Leveque J-L (2007) Influence of ethnic origin of hair on water-keratin interaction. In: Berardesca E, Leveque J-L, Maibach H (eds) Ethnic skin and hair. Informa Healthcare, New York
7. Speakman JB (1927) The intercellular structure of the wool fibre. J Text Inst 18:T431
8. Feughelman M (1959) A two phase structure for keratin fiber. Text Res J 29:223
9. Feughelman M (1997) Mechanical properties and structure of alpha-keratin fibers. Nada Madjar editor. University of New South Wales Press, Sydney, p 17
10. Wolfram LJ (1965) Reactivity of disulphide bonds in strained keratin. Nature 206:304
11. Wolfram LJ, Lennhoff M (1966) The effect of chemical treatment on the tensile properties of keratin fibers. J Text Inst 57:T590
12. Garcia MN, Nadgorny EM, Wolfram LJ (1990) Letter to the editor. J Soc Cosmet Chem 41:149
13. Robbins CR, Crawford RJ (1991) Cuticle damage and tensile properties of human hair. J Soc Cosmet Chem 42:59
14. Wolfram LJ, Lindemann KO (1971) Some observations on the hair cuticle. J Soc Cosmet Chem 22:839
15. Feughelman M (1997) Mechanical properties and structure of alpha-keratin fibers. Nada Madjar editor. University of New South Wales Press, Sydney, p 6
16. Wolfram LJ, Albrecht L (1985) Torsional behavior of hair. J Cosmet Chem 30:87
17. Wolfram LJ (1981) The reactivity of human hair. A review. In: Orfanos CE, Montagna W, Stuettgen G (eds) Hair research. Springer, Berlin/Heidelberg
18. Wolfram LJ, Albrecht L (1987) Chemical-and photo-bleaching of brown and red hair. J Soc Cosmet Chem 82:179
19. Chandrasbekara MN, Ranganthaiah C (2010) Chemical and photochemical degradation of human hair. A free volume microprobe study. J Photochem Photobiol B 101:286
20. Hollfelder B, Blankenburg G, Wolfram LJ (1985) Chemical and physical properties of pigmented and non-pigmented hair. Int J Cosmet Sci 17:87
21. Kaplan PD, Polefka T, Grove G, Daly S, Jumbelic L, Harper D, Nori M, Ramaprasad R, Biancini R (2000) Grey hair: clinical investigation into changes in hair fibers with loss of pigmentation in photoprotected population. Int J Cosmet Sci 33:171
22. Gupta AB, Ghosh KG, Haldar B (1979) Comparison of the tensile properties of human scalp hair with and without pigment. Indian J Dermatol 24:55j
23. Wolfram LJ (1984) Effect of heat on hair. Letter to the editor. J Soc Cosmet Chem 35:229
24. Wickett RR, Kossmann E, Barel A, Demeester N, Clarys P, Vanden Berghe D, Calomme M (2007) Effect of oral intake of choline-stabilized orthosilicic acid on hair tensile strength and morphology in women with fine hair. Arch Dermatol Res 299:499

39 Use of Videodermatoscopy in Scalp Disorders

Franco Dinotta, Francesco Lacarrubba, Paola Di Mauro, and Giuseppe Micali

39.1 Introduction

Videodermatoscopy (VD) is a new technique (see Chap. 1) that, allowing the visualization at high magnification (×10 to ×1,000) of hair and scalp skin, has been increasingly and successfully used in the evaluation of different hair and scalp disorders (Table 39.1), either as a diagnostic device or as a useful tool to identify and visualize some pathogenetic mechanisms in investigational settings [1–8]. When used to study hair disorders, VD is sometimes defined as *trichoscopy* by some authors [4, 5, 8, 9]. The usual working magnifications range from ×20 to ×200, much greater than those provided by handheld dermoscope that allows a rapid and easy evaluation of hair but reaches only tenfold magnification, with no possibility of precise measurements or image storing [8].

39.2 Hair Loss

VD may be used in the diagnosis and follow-up of different types of hair loss (Table 39.1). In a patient presenting with this complaint, a number of variably combined VD features may be identified (Table 39.2).

F. Dinotta, MD • F. Lacarrubba • P. Di Mauro, MD
G. Micali, MD (✉)
Dermatology Clinic, University of Catania,
A.O.U. Policlinico – Vittorio Emanuele,
Via Santa Sofia, 78-95123 Catania, Italy
e-mail: cldermct@nti.it

Table 39.1 VD in hair and scalp disorders

Hair loss
Androgenetic alopecia
Alopecia areata
Telogen effluvium
Trichotillomania
Congenital triangular alopecia
Tinea capitis
Scarring alopecia
Parasitoses
Pediculosis
Scabies
Psoriasis
Hair shaft disorders

The knowledge of normal VD appearance of scalp and hair is essential. It is characterized by the presence of regularly spaced follicular openings containing 2–3 terminal hairs and 1or 2 vellus hairs inside [6] (Fig. 39.1). Normal hair shafts are uniform in shape and color [9]. About 10 % of normal hairs of the scalp are short, hypopigmented vellus hairs [9]. In dark skinned individuals, a perifollicular pigmented network (honeycomb pattern) is usually appreciated [6].

39.3 Androgenetic Alopecia

VD significantly enhances the diagnosis and the evaluation of response to therapy of androgenetic alopecia (AGA). An increased ratio of vellus hairs is characteristic for AGA [9]. The typical VD feature of AGA is represented by the so-called hair

E. Berardesca et al. (eds.), *Non Invasive Diagnostic Techniques in Clinical Dermatology*,
DOI 10.1007/978-3-642-32109-2_39, © Springer Berlin Heidelberg 2014

Table 39.2 VD features in hair loss

Yellow dots	Follicles plugged with keratin and sebum	
Black dots	Follicles containing broken hair remnants (*cadaverized hairs*)	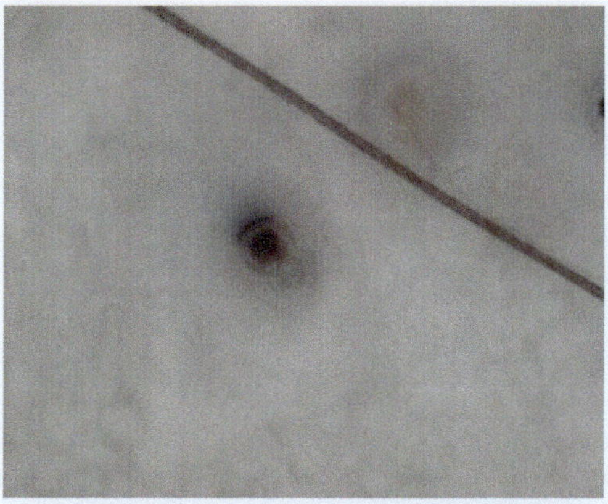
White dots	Expression of follicular fibrosis	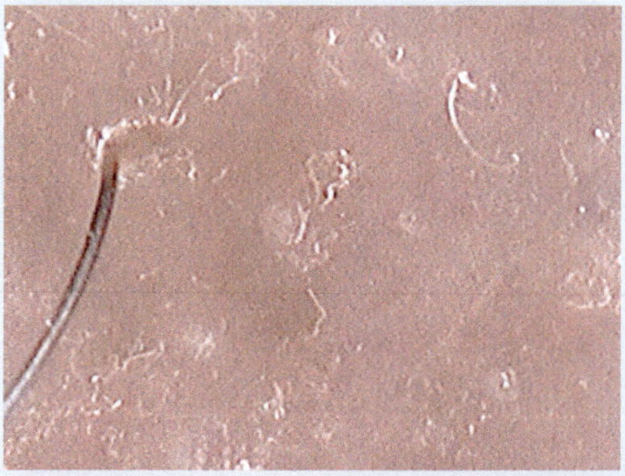
Red dots	Widened follicles plugged by keratin and surrounded by dilated vessels and extravasated erythrocytes	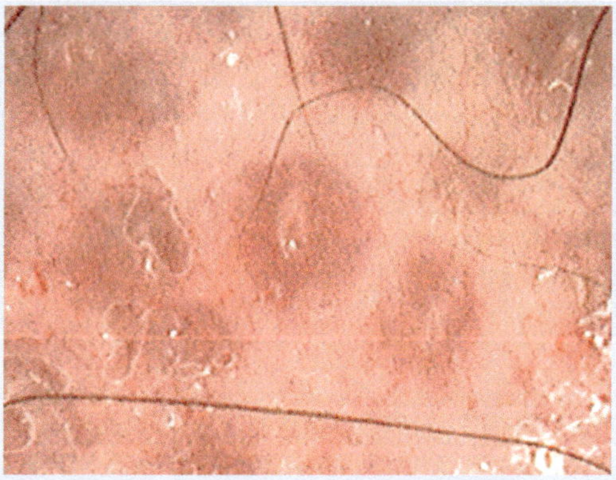

Table 39.2 (continued)

Exclamation-mark hairs	Fractured hairs whose tip is wider than the proximal portion of the shaft (*tapered hair*)	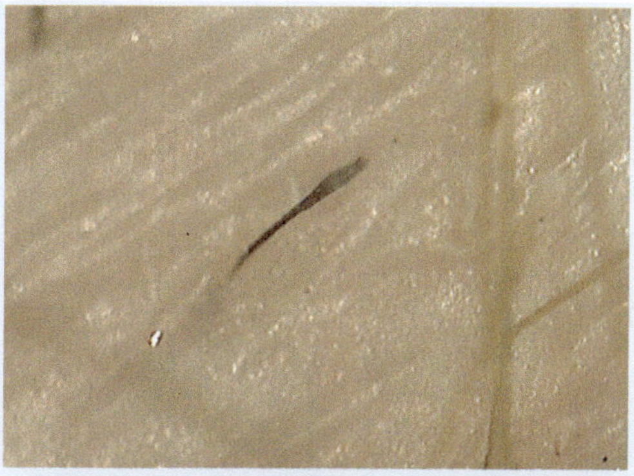
Broken hairs	Fractured hairs with uniform shaft diameter	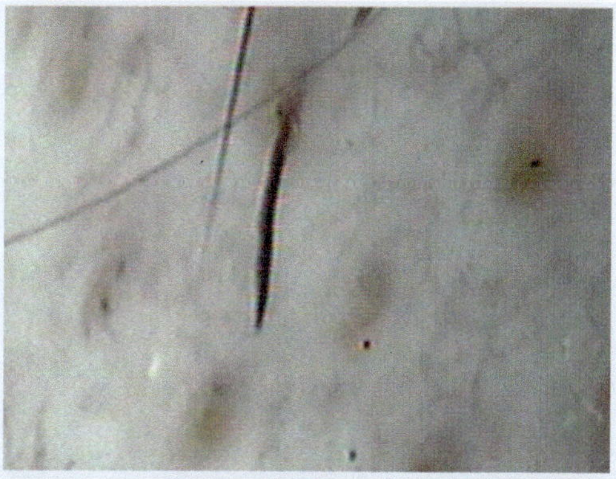
Vellus hairs	Hairs 0.03 mm or less in thickness, representing miniaturized hairs or regrowing hairs	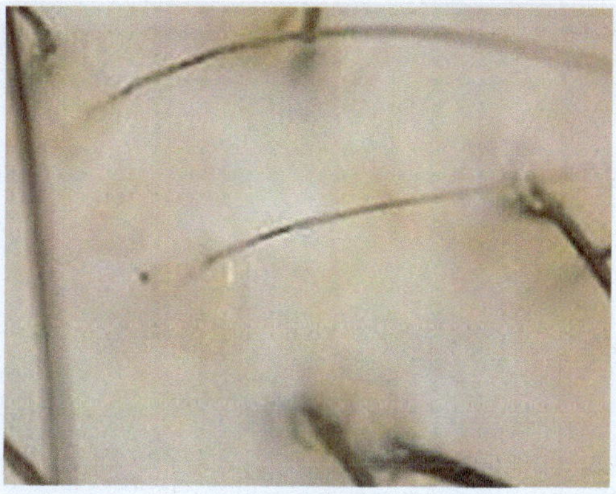

diameter diversity (HDD). It is one of the earliest signs of the disease resulting from hair miniaturization that does not equally affect all the hair follicles of the same area, resulting in the simultaneous presence of terminal, intermediate, and vellus hair (Fig. 39.2). HHD involving almost >20 % of hair is indicative of AGA. Another VD aspect in AGA is the predominance of single over multiple (from 2 to 4) hairs in the follicles compared to normal subjects [4, 6]. Additional features include brown haloes at the follicular ostium, due to a superficial perifollicular lymphocytic infiltrate in an early stage of the disease [10], and yellow dots, corresponding to the presence of intact sebaceous glands adjacent to miniaturized hair follicles [9]. Finally, a honeycomb-like pigmented network is found in scalp areas that are sun exposed with progression of baldness [3, 6].

VD is particularly useful in distinguishing female AGA from chronic telogen effluvium: some authors proposed three major criteria for female AGA, represented by more than four

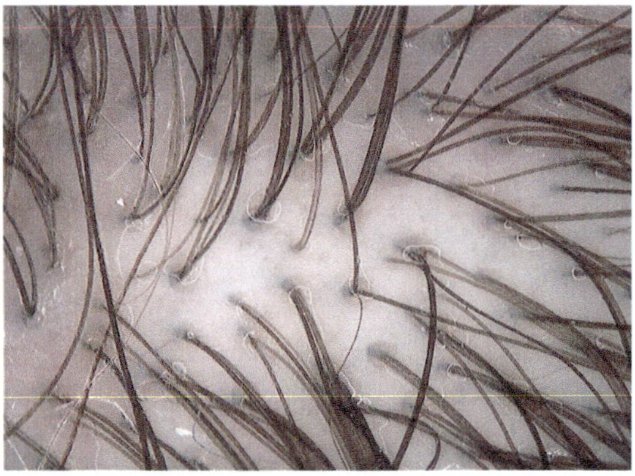

Fig. 39.1 VD appearance of normal scalp (×100)

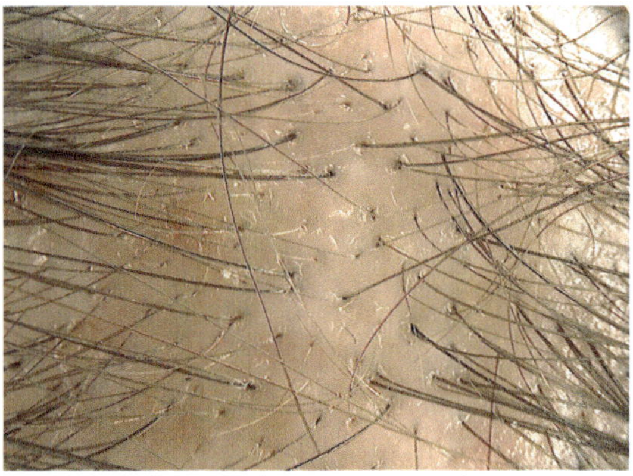

Fig. 39.2 VD of androgenetic alopecia: presence of *hair diameter diversity* (×100)

yellow dots in four images (70-fold magnification) in the frontal area, lower average hair thickness in the frontal area compared to the occiput, and more than 10 % of vellus hairs (below 0.03 mm) in the frontal area [11].

39.4 Alopecia Areata

Alopecia areata (AA) VD findings include yellow dots, dystrophic hair shafts, and hypopigmented vellus hairs (Fig. 39.3) [2, 6, 12, 13]. The evaluation of these findings is important both for diagnostic purposes as well for treatment follow-up. Yellow dots, which may be observed through the epiluminescence technique, are described as yellow to yellow-red, round, or polycyclic dots of different sizes and correspond to dilated follicular openings [2, 6, 12]. They may be observed both in acute and chronic forms of AA. Dystrophic hair shafts are indicative of anagen alterations [6, 9], and at VD observation may appear as black dots (anagen arrest causes hair shafts fracture before emergence from the scalp), broken hairs, and exclamation-mark hairs (growth of the broken shafts) and indicate strong disease activity [1, 6, 12]. Conversely, short hypopigmented vellus hairs are usually indicative of remitting disease (Fig. 39.4). In one study, yellow dots and short vellus hairs have been regarded as the most sensitive signs of the disease, while exclamation-mark hair and black dots as the most specific [12].

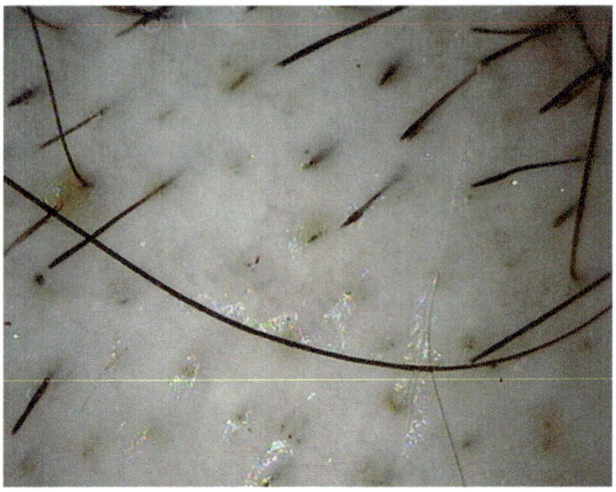

Fig. 39.3 VD of alopecia areata: presence of *yellows dots*, *black dots*, and exclamation-mark hairs, indicative of acute disease (×100)

VD may be also useful for the diagnosis of AA incognita, a form of AA characterized by acute diffuse shedding of telogen hairs in the absence of typical patches, clinically misdiagnosed as telogen effluvium or AGA [14].

39.5 Telogen Effluvium

There are no specific VD features of telogen effluvium, and the diagnosis is generally of exclusion. This frequent condition may be suspected for the presence of empty hair follicles and short, dark regrowing hairs, in the absence of features characteristic of other scalp disorders [15].

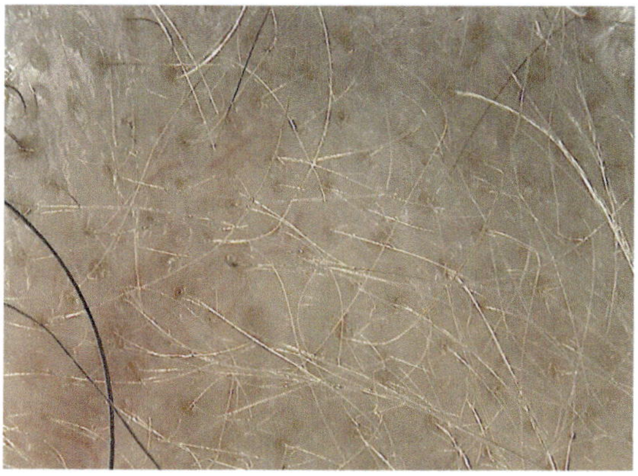

Fig. 39.4 VD of alopecia areata: presence of hypopigmented vellus hairs indicative of remitting disease (×100)

Fig. 39.5 VD of congenital triangular alopecia: vellus hairs surrounded by normal terminal hairs in the adjacent scalp (×100)

39.6 Trichotillomania

VD evaluation improves the diagnosis of trichotillomania and is a useful tool in differentiating this compulsive disorder from other form of alopecia, limiting the need for skin biopsy [2, 16]. At VD, hair appears broken at variable lengths with the extremities showing a characteristic frayed aspect derived from the habit of pulling hair out. Black dots and exclamation-mark hairs may also be observed [9, 17].

39.7 Congenital Triangular Alopecia

VD examination of the triangular-shaped alopecia, generally localized in the frontotemporal hairline, shows the presence of normal follicular openings and numerous vellus hairs surrounded by normal terminal hairs in the adjacent scalp (Fig. 39.5) [6, 18].

39.8 Tinea Capitis

The presence of specific *comma-shaped hairs* and/or *corkscrew hairs*, likely a result of subsequent cracking and bending of a hair shaft filled with the hyphae, may be observed within alopecic patches caused by tinea capitis. This finding may be seen along with broken and dystrophic hairs [9, 19, 20].

39.9 Scarring Alopecia

VD evaluation of scarring alopecia provides important diagnostic clues, although scalp biopsy is generally required to assess the final diagnosis.

In all types of scarring alopecia, VD shows absence of follicular openings at variable degree.

39.9.1 Lichen Planopilaris

VD shows whitish scales (or casts) surrounding the hair shafts up to few millimeters above scalp surface, especially at the periphery of the patch [9]. White dots, expression of follicle fibrosis, are generally observed in the alopecic areas (Fig. 39.6). Sometimes, the presence of perifollicular arborizing vessels has been reported [2, 21]. Similar findings have been described in frontal fibrosing alopecia [22]. In this variant of lichen planopilaris, characterized by frontotemporal hair recession, another characteristic finding is the abrupt interruption of the hairline, with absence of vellus hair typically observed in normal scalp (Fig. 39.7).

39.9.2 Discoid Lupus Erythematous

Several but not specific VD features have been reported, as perifollicular white scales and large yellow dots surrounded by arborizing and tortuous vessels [2, 9]. Red dots, appearing as erythematous

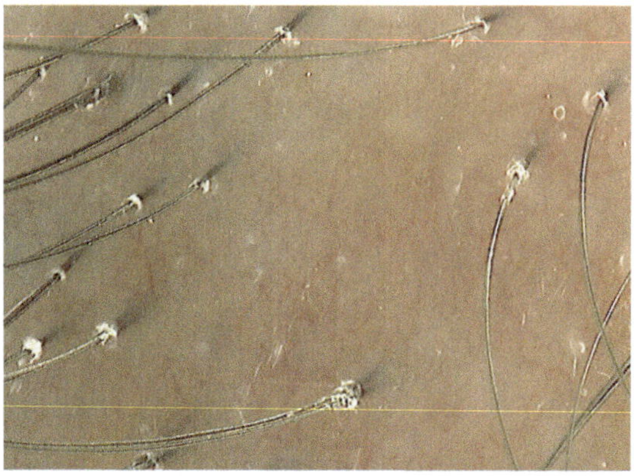

Fig. 39.6 VD of lichen planopilaris: absence of follicular openings and presence of *white dots* and *whitish scales* surrounding the residual hair shafts (×100)

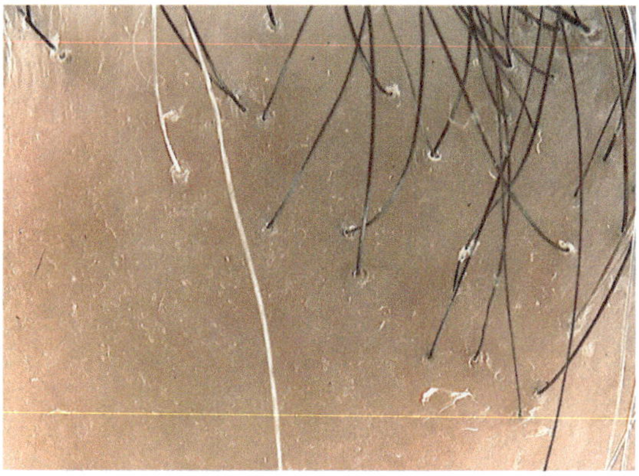

Fig. 39.7 VD of frontal fibrosing alopecia: absence of follicular openings and presence of *white dots* and *whitish scales* and abrupt interruption of the hairline (×100)

roundish structures around the follicular openings and corresponding to widened infundibula plugged by keratin and surrounded by dilated vessels and extravasated erythrocytes, seem to be related to a good prognosis (Fig. 39.8) [23].

39.9.3 Folliculitis Decalvans

In active forms, the presence of interfollicular twisted red loops, similar to those of psoriasis, has been reported. When scarring alopecia occurs, VD highlights the reduction of follicular openings and the presence of multiple hairs emerging from single and dilated follicular openings (tufted folliculitis) [2, 24, 25].

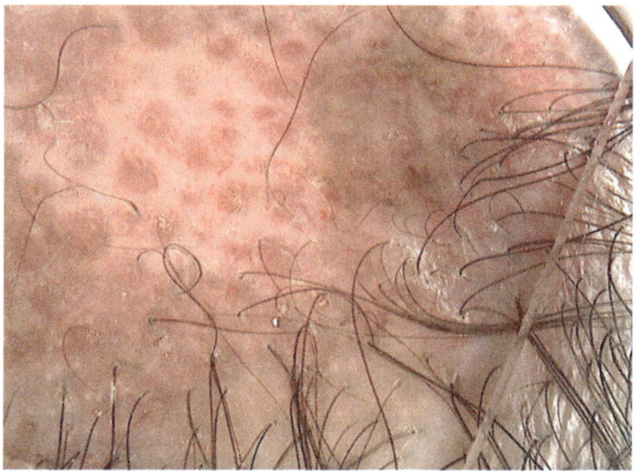

Fig. 39.8 VD of discoid lupus erythematous: presence of *red dots* (×100)

39.9.4 Cicatricial Marginal Alopecia

It is characterized by reduction in hair follicle density (loss of follicular ostia) and absence of signs of inflammation or traction [26].

39.9.5 Sarcoidosis

Although rare, scarring alopecia can occur. VD shows decreased hair density associated with perifollicular and follicular yellowish to pale-orange round spots or diffuse orange discoloration with prominent telangiectasia. Dystrophic hairs may indicate granulomatous activity [27].

39.10 Therapeutic Monitoring of Hair Loss with VD

VD represents a useful tool for monitoring treatment of hair loss [6]. In AGA, evaluation of HHD may be assessed during treatment comparing VD images of the same areas, better if previously marked. In AA, the efficacy of the treatment may be assessed by the degree of signs of disease activity (black dots and exclamation-mark hairs), presence of short hypopigmented vellus hairs, and signs of the transformation of vellus into terminal hairs (increased proximal shaft thickness and pigmentation) [13].

39.11 Parasitoses

39.11.1 Pediculosis

The diagnosis of *Pediculosis capitis* (head lice) is generally based on the clinical identification of either adult lice (*Pediculus humanus*) or viable nits through close-up examination. Louse combs and the magnifying lens represent additional diagnostic tools which increase the possibility to identify live lice [28]. VD ensures a more detailed evaluation of both mites and eggs; this is particularly useful to detect post-treatment residual eggs [29, 30]. VD unequivocally shows the presence of full, viable nits, with a rounded free ending, fixed to the hair shaft (Fig. 39.9). VD allows a rapid differentiation from empty nits, appearing as translucent structures with a flat and fissured free ending, or scales of different origin or pseudo-nits (hair casts, debris of hair spray or gel, or seborrheic scales), which appear as amorphous, whitish structures [30, 31], with remarkable effects on the therapeutic management. Furthermore, VD does not require hair pulling, so that a large scalp area can be investigated with minimal discomfort to the patient. Finally, a close and accurate VD examination may disclose the morphology and physiology of the lice themselves, together with proving the pediculocidal activity of topical agents [32]. In addition, VD examination may enhance patient compliance to therapy, showing the presence, persistence, or resolution of the infestation on a VD monitor [6].

The diagnostic efficacy of VD may be extended to *Phthiriasis pubis* infestation (pubic lice) [6]. It can be of particular importance in children, where the margin of the scalp and the eyelashes are the most common site of infestation. As observed in head lice, VD ensures a more detailed evaluation of both parasites and nits.

39.11.2 Scabies

VD is a useful, noninvasive, well-known technique for the diagnosis of scabies [7], allowing a rapid and clear in vivo detection of the diagnostic features, such as burrows and mites at magnifications ranging from ×40 to ×200 and eggs or feces at higher magnifications, up to ×600 (see Chap. 1). Scalp involvement is generally limited to newborn and immunocompromised patients, but might be responsible for some cases of persistent infestation in immunocompetent adults, representing the reservoir of the mite, as this site is generally spared from standard therapies [33]. VD can be easily used for a detailed inspection of the scalp, especially in those cases that are persistent or resistant to therapy [6].

39.11.3 Psoriasis

As extensively described in Chap. 1, VD examination of the psoriatic plaque at high magnification (>×100) shows the presence of typical tortuous and dilated *bushy* capillaries (Fig. 39.10) [34].

Fig. 39.9 VD of head lice: nits fixed to the hair shaft (×100)

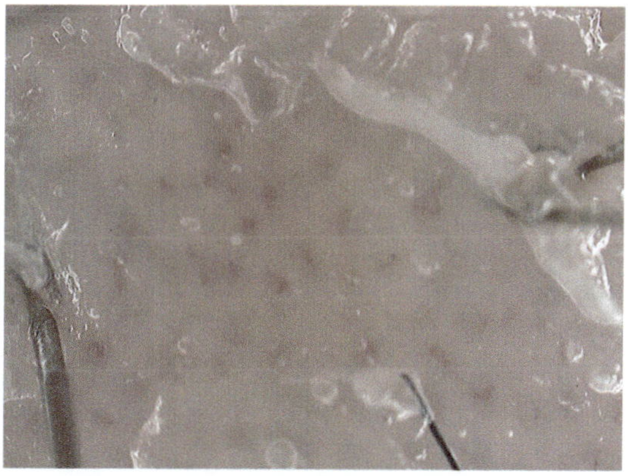

Fig. 39.10 VD of scalp psoriasis: presence of typical tortuous and dilated *bushy* capillaries (×200)

Table 39.3 VD and hair shaft disorders

Disorder	Hair shaft appearance at VD
Monilethrix	Small oval nodes of normal thickness separated by regular dystrophic constrictions, a finding described as *regularly bended ribbon sign* [38–41]
Pili torti	Regular twists of the hair shaft along the long axis [38, 41]
Pili trianguli and canaliculi (uncombable hair)	Triangular-shaped shafts with atypical longitudinal grooving [6]
Pili annulati	Alternating light and dark bands. Light bands correspond to air-filled cavities within the hair shaft [6, 9, 41]
Trichorrhexis nodosa	Diffuse white knots and a brush pattern due to hairs higher susceptibility to get fractured [6]
Trichorrhexis invaginata (bamboo hair)	Multiple ball-shaped node along hairs which causes the hairs to break off (*golf-tee hairs*) [41, 42]
Wooly hair syndrome	*Crawling snake* appearance, with short wave cycles [38]

VD may be a useful and noninvasive method for differentiating psoriasis and seborrheic dermatitis [35–37], especially when the scalp is the only affected site. In this case, the diagnosis may be sometimes difficult and may be relevant for the long-term prognosis, especially in patients with arthritis, in which deciding whether erythematous scaly plaques on the scalp are psoriasis or seborrheic dermatitis may influence the interpretation of the rheumatic symptoms. In a study VD was employed to compare capillary morphology, distribution, and density in psoriasis and seborrheic dermatitis of the scalp. Scalp psoriasis presented a homogeneous pattern with *bushy* capillaries, while scalp seborrheic dermatitis presented a multiform pattern, with mildly tortuous capillaries and only isolated *bushy* or mild dilated capillaries [36], with a conserved local micro-angio-architecture similar to healthy scalp skin. Capillary loop density was similar in both conditions and in normal scalp skin.

39.12 Hair Shaft Disorders

Hair shaft abnormalities may represent significant diagnostic findings in a number of autosomal and X-linked syndromes. VD allows an in vivo magnified observation of hair shafts without the need of pulling hair for diagnostic purposes of examination [38]. In the diagnosis of some hair shaft disorders (Table 39.3), VD gives comparable results to a microscopic examination, with the advantages of being noninvasive and quick to perform [38].

References

1. Lacarrubba F, Dall'Oglio F, Nasca MR, Micali G (2004) Videodermatoscopy enhances diagnostic capability in some forms of hair loss. Am J Clin Dermatol 5:205–208
2. Ross EK, Vincenzi C, Tosti A (2006) Videodermoscopy in the evaluation of hair and scalp disorders. J Am Acad Dermatol 55:799–806
3. Tosti A (2007) Dermoscopy of hair and scalp disorders with clinical and pathological correlations. Informa, London
4. Rudnicka L, Olzewska M, Rakowska A, Kowalska-Oledzka E, Slowinska M (2008) Trichoscopy: a new method for diagnosis of hair loss. J Drugs Dermatol 7:651–654
5. Olszewska M, Rudnicka L, Rakowska A, Kowalska-Oledzka E, Slowinska M (2008) Trichoscopy. Arch Dermatol 144:1007
6. Tosti A, Estrada BD (2010) Hair loss. In: Micali G, Lacarrubba F (eds) Dermatoscopy in clinical practice: beyond pigmented lesions. Informa Healthcare Ltd, London
7. Micali G, Lacarrubba F, Massimino D, Schwartz RA (2011) Dermatoscopy: alternative uses in daily clinical practice. J Am Acad Dermatol 64:1135–1146
8. Inui S (2011) Trichoscopy for common hair loss diseases: algorithmic method for diagnosis. J Dermatol 38:71–75
9. Rudnicka L, Olszewska M, Rakowska A, Slowinska M (2011) Trichoscopy update 2011. J Dermatol Case Rep 4:82–88
10. Deloche C, de Lacharriere O, Misciali C, Piraccini BM, Vincenzi C, Bastien P, Tardy I, Bernard BA, Tosti A (2004) Histological features of peripilar signs associated with androgenic alopecia. Arch Dermatol Res 295:422–428
11. Rakowska A, Slowinska M, Kowalska-Oledzka E, Olszewska M, Rudnicka L (2009) Dermoscopy in female androgenic alopecia: method standardization and diagnostic criteria. Int J Trichology 1:123–130
12. Inui S, Nakajima T, Nakagawa K, Itami S (2008) Clinical significance of dermoscopy in alopecia areata: analysis of 300 cases. Int J Dermatol 47:688–693

13. Mane M, Nath AK, Thappa DM (2011) Utility of dermoscopy in alopecia areata. Indian J Dermatol 56:407–411
14. Tosti A, Whiting D, Iorizzo M, Pazzaglia M, Misciali C, Vincenzi C, Micali G (2008) The role of scalp dermoscopy in the diagnosis of alopecia areata incognita. J Am Acad Dermatol 59:64–67
15. Olszewska M, Warszawik O, Rakowska A, Słowińska M, Rudnicka L (2010) Methods of hair loss evaluation in patients with endocrine disorders. Endokrynol Pol 61:406–411
16. Lee DY, Lee JH, Yang JM, Lee ES (2009) The use of dermoscopy for the diagnosis of trichotillomania. J Eur Acad Dermatol Venereol 23:731–732
17. Abraham LS, Torres FN, Azulay-Abulafia L (2010) Dermoscopic clues to distinguish trichotillomania from patchy alopecia areata. An Bras Dermatol 85:723–726
18. Iorizzo M, Pazzaglia M, Starace M, Militello G, Tosti A (2008) Videodermoscopy: a useful tool for diagnosing congenital triangular alopecia. Pediatr Dermatol 25:652–654
19. Slowinska M, Rudnicka L, Schwartz RA, Kowalska-Oledzka E, Rakowska A, Sicinska J, Lukomska M, Olszewska M, Szymanska E (2008) Comma hairs: a dermatoscopic marker for tinea capitis: a rapid diagnostic method. J Am Acad Dermatol 59:S77–S79
20. Mapelli ET, Gualandri L, Cerri A, Menni S (2012) Comma hairs in tinea capitis: a useful dermatoscopic sign for diagnosis of tinea capitis. Pediatr Dermatol 29:223–224
21. Kossard S, Zagarella S (1993) Spotted cicatricial alopecia in dark skin. A dermoscopic clue to fibrous tracts. Australas J Dermatol 34:49–51
22. Inui S, Nakajima T, Shono F, Itami S (2008) Dermoscopic findings in frontal fibrosing alopecia: report of four cases. Int J Dermatol 47:796–799
23. Tosti A et al (2009) Follicular red dots. A novel dermoscopic pattern observed in scalp discoid lupus erythematosus. Arch Dermatol 145:1406–1409
24. Braun RP, Oliviero M, Kolm I, French LE, Marghoob AA, Rabinovitz H (2009) Dermoscopy: what's new? Clin Dermatol 27:26–34
25. Lacarrubba F, Dall'Oglio F, Micali G (2011) Massive tufted hair folliculitis associated with chronic use of systemic corticosteroids. G Ital Dermatol Venereol 146:164–165
26. Tosti A, Torres F, Misciali C, Vincenzi C, Duque-Estrada B (2009) The role of dermoscopy in the diagnosis of cicatricial marginal alopecia. Br J Dermatol 161:213–215
27. Torres F, Tosti A, Misciali C, Lorenzi S (2011) Trichoscopy as a clue to the diagnosis of scalp sarcoidosis. Int J Dermatol 50:358–361
28. Micali G, Umana M, Lacarrubba F (2010) Videodermatoscopy and pediculosis. In: Micali G, Lacarrubba F (eds) Dermatoscopy in clinical practice: beyond pigmented lesions. Informa Healthcare Ltd, London
29. Bakos RM, Bakos L (2007) Dermoscopy for diagnosis of pediculosis capitis. J Am Acad Dermatol 57:727–728
30. Di Stefani A, Hofmann-Wellenhof R, Zalaudek I (2006) Dermoscopy for diagnosis and treatment monitoring of pediculosis capitis. J Am Acad Dermatol 54:909–911
31. Zalaudek I, Giacomel J, Cabo H, Di Stefani A, Ferrara G, Hofmann-Wellenhof R, Malvehy J, Puig S, Stolz W, Argenziano G (2008) Entodermoscopy: a new tool for diagnosing skin infections and infestations. Dermatology 216:14–23
32. Lacarrubba F, Nardone B, Milani M, Botta G, Micali G (2006) Head lice: ex vivo videodermatoscopy evaluation of the pediculocidal activity of two different topical products. G Ital Dermatol Venereol 141:233–235
33. Lacarrubba F, Micali G (2008) Videodermatoscopy enhances the diagnostic capability in a case of scabies of the scalp. G Ital Dermatol Venereol 143: 351–352
34. Micali G, Lacarrubba F, Musumeci ML, Massimino D, Nasca MR (2010) Cutaneous vascular patterns in psoriasis. Int J Dermatol 49:249–256
35. Rosina P, Zamperetti MR, Giovannini A, Girolomoni G (2007) Videocapillaroscopy in the differential diagnosis between psoriasis and seborrheic dermatitis of the scalp. Dermatology 214:21–24
36. Rosina P (2010) Scalp psoriasis. In: Micali G, Lacarrubba F (eds) Dermatoscopy in clinical practice: beyond pigmented lesions. Informa Healthcare Ltd, London
37. Kim GW, Jung HJ, Ko HC, Kim MB, Lee WJ, Lee SJ, Kim DW, Kim BS (2011) Dermoscopy can be useful in differentiating scalp psoriasis from seborrhoeic dermatitis. Br J Dermatol 164:652–656
38. Rakowska A, Slowinska M, Kowalska-Oledzka E, Rudnicka L (2008) Trichoscopy in genetic hair shaft abnormalities. J Dermatol Case Rep 2:14–20
39. Rakowska A, Slowinska M, Czuwara J, Olszewska M, Rudnicka L (2007) Dermoscopy as a tool for rapid diagnosis of monilethrix. J Drugs Dermatol 6: 222–224
40. Liu CI, Hsu CH (2008) Rapid diagnosis of monilethrix using dermoscopy. Br J Dermatol 159: 741–743
41. Wallace MP, de Berker DA (2010) Hair diagnoses and signs: the use of dermatoscopy. Clin Exp Dermatol 35:41–46
42. Burk C, Hu S, Lee C, Connelly EA (2008) Netherton syndrome and trichorrhexis invaginata-a novel diagnostic approach. Pediatr Dermatol 25:287–288

Part VIII

Nails

40 Nail Surface Topography and Onychochronobiology

Gérald E. Piérard and Claudine Piérard-Franchimont

40.1 Introduction

Some chronic skin disorders are responsible for concurrent nail changes. Typical examples are represented by psoriasis, lichen planus, and alopecia areata. The lifetime incidence of nail alterations in these patients potentially reaches 80 % or over. However, the nail involvement remains frequently overlooked by physicians, despite the burden to the patients caused by functional impairment of manual dexterity, pain, and psychological stress [1].

The nail plate is hard but flexible. It grows continuously and is frequently subject to a number of microtraumatisms. A healthy nail plate grossly looks smooth, but closer examination reveals it is not the case. Its specific surface microrelief has physiologic and sometimes pathologic implications. At close inspection, the regular relief is fine and only visible in children. In middle-aged individuals, the physiologic changes often remain too discrete to be readily discernable. By contrast, it is noticeable with the naked eye in a number of elderly people.

G.E. Piérard, MD, PhD (✉)
C. Piérard-Franchimont
Laboratory of Skin exhibit and Imaging,
Department of Dermatopathology,
University Hospital of Liège,
CHU SartTilman, Liège B-4000, Belgium
e-mail: gerald.pierard@ulg.ac.be

Several distinctive patterns of nail surface changes are clearly identified and distinguished by clinical inspection. However, the structure of the nail surface has attracted little attention from researchers. Descriptive reports are rarely supported by quantification of the nail plate microrelief [2, 3]. Nail alterations result from any underlying chronic dermatoses and/or from external trauma and weathering. To analyze the nail relief, it is important to have at one's disposal reliable clinical or instrumental methods identifying minimal alterations and providing qualitative and quantitative data.

40.2 Clinical Descriptive Approach

Several typical nail changes are distinguished. Their combination is used to derive a nail severity index such as in psoriasis [2]. In this condition, affected nail plates are often thick and crumble. Because the lesions are obviously visible, patients are concerned about the appearance of their nails, often causing them to avoid normal daily activities including work. In the longer term, nail involvement is possibly a signal of a more severe form of the disease [4, 5].

The nail psoriasis severity index (NAPSI) was developed as an objective and reproducible tool for estimating nail involvement [2]. NAPSI was claimed to be reproducible and easy to handle. Indeed, it represents an improvement in rating

E. Berardesca et al. (eds.), *Non Invasive Diagnostic Techniques in Clinical Dermatology*,
DOI 10.1007/978-3-642-32109-2_40, © Springer Berlin Heidelberg 2014

treatment efficacy in nail psoriasis. For this purpose, each nail is divided into four quadrants to be assessed for the presence of any psoriatic change originating from both the nail matrix (pitting, leukonychia, nail plate crumbling) and the nail bed (oil drop discoloration, onycholysis, hyperkeratosis, splinter hemorrhages). If any sign is present in all four quadrants, the nail is given a score of 4 through to a score of 0 in absence of alteration in any quadrant. Each nail is assigned both a nail matrix and a nail bed score of 0–4, which are combined to yield a total score of 0–8 for each nail. All nails are assessed with the total NAPSI score corresponding to the sum of the scores, reaching up to 80 if only the 10 finger nails are considered or up to 160 if toe nails are included. In practice, it is possible to target one single specific nail for assessing drug effects.

In medical practice, some drugs potentially alter the nail plates. For instance, retinoids commonly produce thinning and brittleness of the nail. Transverse ridging, white banding, as well as deep Beau's lines occasionally develop. Cancer chemotherapy also affects the nails. Multiple transverse white bands develop following cyclic chemotherapy. High-dose chemotherapy is responsible for Beau's lines sometimes followed by nail loss. Half-and-half nails, where the proximal part of the nail is whitish while the distal part is brownish, are not uncommon complication of cancer chemotherapy.

The typical appearance of onychomycosis presents as subungual hyperkeratosis with thickening and brownish discoloration of the nail plate. The nail plate becomes porous and crumble and it easily breaks off. In some patients, it is difficult to distinguish psoriatic onychodystrophy from fungal nail infection, but the recognition of the characteristic psoriatic onycholysis with a red-yellow halo and the so-called oil drops is helpful.

The twenty-nail dystrophy is a rather troublesome disease altering all or almost all the finger- and toe nails in children and adolescents. It combines excessive ridging of the nail plate, rough nails, and trachyonychia.

40.3 Patterns of Nail Topography

40.3.1 Longitudinal Striations

Longitudinal striations of the nail surface combine parallel grooves pulled apart by discrete projecting ridges. Such a condition is considered to be a physiologic feature when presenting as shallow grooves, usually parallel, and edged by low projecting ridges. Ridging commonly appears more prominent with aging and in association with some particular disorders.

Onychorrhexis consists of a series of narrow, longitudinal, parallel superficial striations. The nail appears as having been scratched by an awl. In some instances, dust material is ingrained into the superficial nail plate. The small linear projections extend from the proximal nail fold to the distal nail edge. Splitting of the free edge of the nail commonly ensues. In other instances, the nail ridging stops short or is interrupted at regular intervals, giving rise to a beaded appearance. In some patients a single wide longitudinal median ridge appears in cross section as a circumflex accent.

Median nail dystrophy is another uncommon condition of the thumb nails consisting of a longitudinal groove developed in the midline or just off center, starting at the cuticle and growing out of the free edge.

Any tumor located nearby the nail matrix exerts pressure and produces a single wide and deep longitudinal groove. This aspect vanishes when the tumor is removed.

40.3.2 Herringbone Nail

Oblique nail ridging, pointing centrally to meet in the midline, is an uncommon pattern particularly occurring during childhood.

40.3.3 Beau's Lines

Beau's lines correspond to transverse grooves extending between the lateral edges of the nail.

They commonly affect all nails at similar levels. The width of the transverse groove is thought to be related to the duration of the causal process downgrading the nail matrix activity for a limited period of time. Any abrupt distal limit of the groove suggests a sudden outbreak of disease. By contrast, a sloping aspect suggests a progressive onset. The proximal limit of the depression is similarly abrupt or sloped according to the rate of normalization of the nail biology.

40.3.4 Pitting and Rippling

Rosenau's depressions refer to nail pitting and rippling. Pits develop as a result of defective nail formation in punctuate areas focused in the proximal portion of the nail matrix. The nail plate surface is studded in a buckshot pattern with punctuate hollow depressions varying in number, size, depth, and shape. An arbitrary figure of five pits is considered as a physiologic condition. The depth and width of the pits relate to the extent of the matrix involvement. Their length is determined by the duration of the nail damage. Deep and irregularly shaped pits often suggest psoriasis, but they are not specific for any disease including alopecia areata and other conditions.

Pits are randomly dispersed or evenly patterned in series along one or several longitudinal lines. They are sometimes aligned in a crisscross pattern resembling the external surface of a thimble. Regular pit ting commonly leads to rippling or ridging.

40.3.5 Trachyonychia

Trachyonychia refers to a spectrum of alterations resulting in severe nail roughness as if the surface had been rubbed with sandpaper.

40.3.6 Nail Underface Ridging

The underface of the nail plate disclosed after avulsion exhibits topographic aspects unrelated to the outer surface of the same nail. Deep longitudinal striations are present and they commonly deepen with age.

40.4 Bioinstrumentation

The various metrology designs for the nail plate encompass routine methods as well as more sophisticated and experimental procedures. The examination of nail clippings is occasionally informative. The distinctive features of the nail microrelief are explored using various in vivo procedures. Clinical examination allows qualitative assessment of the gross surface topography. Low-magnification photographs under carefully controlled and repeatable conditions are conveniently used to document the nail surface microrelief. Transillumination helps disclosing some of the onychodystrophies. The method is rapid, simple, and cheap. The examination is performed in a dark environment. Digit transillumination is possible by shining a strong penlight beam upward through the pulp or alternatively by using the fiber optic from a high-power light device.

Controlled abrasion, microindentation, and sclerometry are used to investigate some physical properties of the nail surface. In particular, simulation of the worn nail by abrasion allows to assess the wear process, which is the prime target of cosmetic application of nail hardeners and nail varnishes. Sclerometry simulates by scrapping the effects of wear and tear on the nail surface. Such insults deteriorate the morphologic, mechanical, and optical properties and, thereby, the loss of the main properties of the nail.

The examination of the nail surface roughness can be performed following several noninvasive methods. High-resolution digital photography reveals some specific nail surface physical properties. The use of light filters selecting only a narrow wavelength spectrum contributes to the optical properties of the nail surface. Using polarized light reduces the glossy reflectance from the surface, thus improving visualization of fine nail dystrophies. Video microscopy is another noncontact method, helpful in revealing quantitative changes in the nail microrelief. Fringe projection is another method disclosing three-dimensional (3-D) nail surface contours.

The 3-D microscopy represents an advance in the analysis of the nail surface topography. Initially, the method aimed at controlling the quality of manufactured surfaces by assessing their roughness in a given direction. The parameters were standardized internationally. However, this method was limited to a two-dimensional (2-D) analysis of the surface structure with a signal representing roughness on a line of relief but with absence of a field image. This method was greatly improved by producing a record of a 3-D image of the surface. This achievement provides reconstructed 3-D images from an atomic scale (atomic force microscope) to macroscopic relief.

Scanning microdensitometry discriminates shadows and highlights photographic negatives of the nail surface taken under controlled procedures (light intensity, exposure time, angle of camera recording, and distance to the surface).

A technical advance was achieved with the introduction of negative replicas. Precise quantitative examinations are conveniently made in vivo on the outer portion of the nail. Alternatively, observations are performed after its avulsion or after making a negative replica. Scanning electron microscopy provides images of an unsurpassed morphologic accuracy. The examination of silicone replicas of the nail microrelief is an optional procedure. The microrelief of the nail replicas can be scrutinized using one of several methods including mechanical profilometry, autofocus laser beam profilometry, oblique illumination of the replicas, and optical measurement of the translucent replica thickness. The choice of the replica material is of importance. It has to polymerize quickly at the nail temperature, be adequately liquid to fill in all the erratic surface aspects, avoid deformation during polymerization, and avoid any artifact. The replica relief is scrutinized using a scanning tactile or optical analyzer.

The stringent use of optical profilometry or any other microtopographic assessment on nails or their replicas brings quantitative information [6]. The diversity of the typical alterations is clearly evidenced [7–10]. In addition, due to the sustained nail growth, onychochronobiology is explored using the same means at different time intervals [11, 12]. In addition, these methods are suitable for assessing the nail mechanical properties in combination with the microindentation and the sclerometry methods.

40.4.1 Static Microtopography of the Nail Surface

Quantitative assessment of the nail microtopography is usually performed on longitudinal scans [13]. Transverse scans are more difficult to interpret due to the natural nail curvature. When information has to be gained in this direction, it is recommended to examine sections no longer than 5 mm in length to minimize this pitfall. It should be kept in mind that native alterations such as shallow grooves and Beau's lines are better revealed at the proximal part of the nail (Fig. 40.1). Weathering and natural microabrasions add their effects in a cumulative way when moving toward the distal part of the nail (Figs. 40.2 and 40.3). Sources of variability such as uncontrolled nail plate wear should be discarded when appropriate. Controlled positioning of the nail is of the upmost importance for comparative purposes.

40.4.2 Dynamic Microtopography of the Nail Surface

Repeat controlled assessments overtime provide insight in onychochronobiology. By this way, the effects of therapies can be assessed. The speed of nail growth is conveniently recorded following the move of an engraved mark made at the initial examination. The linear nail growth is indeed measured following a shallow scratch in the proximal nail with a piece of hacksaw blade [14]. The nail mark is made at a defined distance (approximately

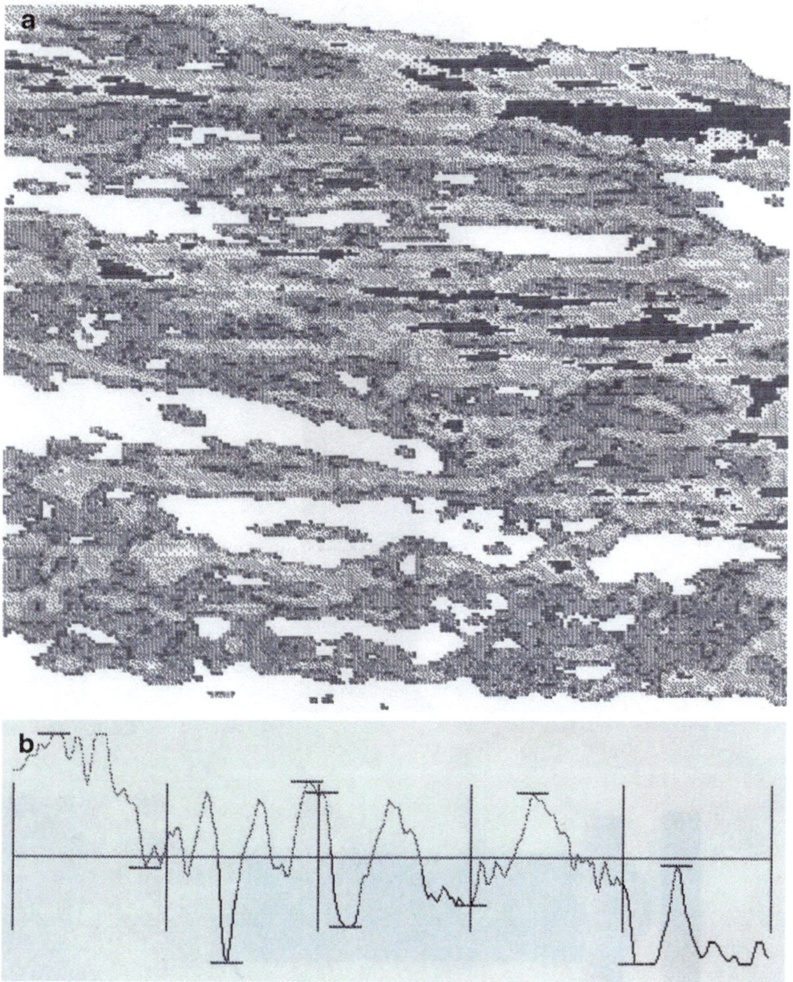

Fig. 40.1 Nail surface topography on longitudinal scans showing transverse grooves during cancer chemotherapy. (**a**) Two-dimensional field image of a nail surface recording. (**b**) Profilometric presentation

3 mm) from the proximal nail fold [15]. The subjects should refrain from pushing back their nail cuticles during the study period. After a period of approximately 1 month, the scratch has moved with the nail growth, and its distance to the proximal nail fold has increased. The average outgrowth is divided by the elapsed time to determine the rate of linear nail growth. Thus, a weekly or a daily nail growth is conveniently calculated. Measurement is made using a magnifier equipped with a linear scale reaching 0.1 mm in accuracy. Alternatively, a screw set two-pointed caliper marked the distance from the groove to the proximal nail skinfold. The distance between the points of the caliper is measured to the nearest 0.1 mm with a micrometer. Circadian variations in body temperature have been reported to influence the linear nail growth rate with a split-image range finder adapted to a trinocular microscope [14]. With such instrumentation, nail growth measurements were claimed to be possible over short intervals.

Nail growth was assessed after indenting the nail surface and measuring the volumetric change of the grooves as they reached the free edge [16]. Erosion was responsible for a 30–50 % reduction in volume [17].

Such methods allow to evaluate the rate of improvement or degradation of the nail condition.

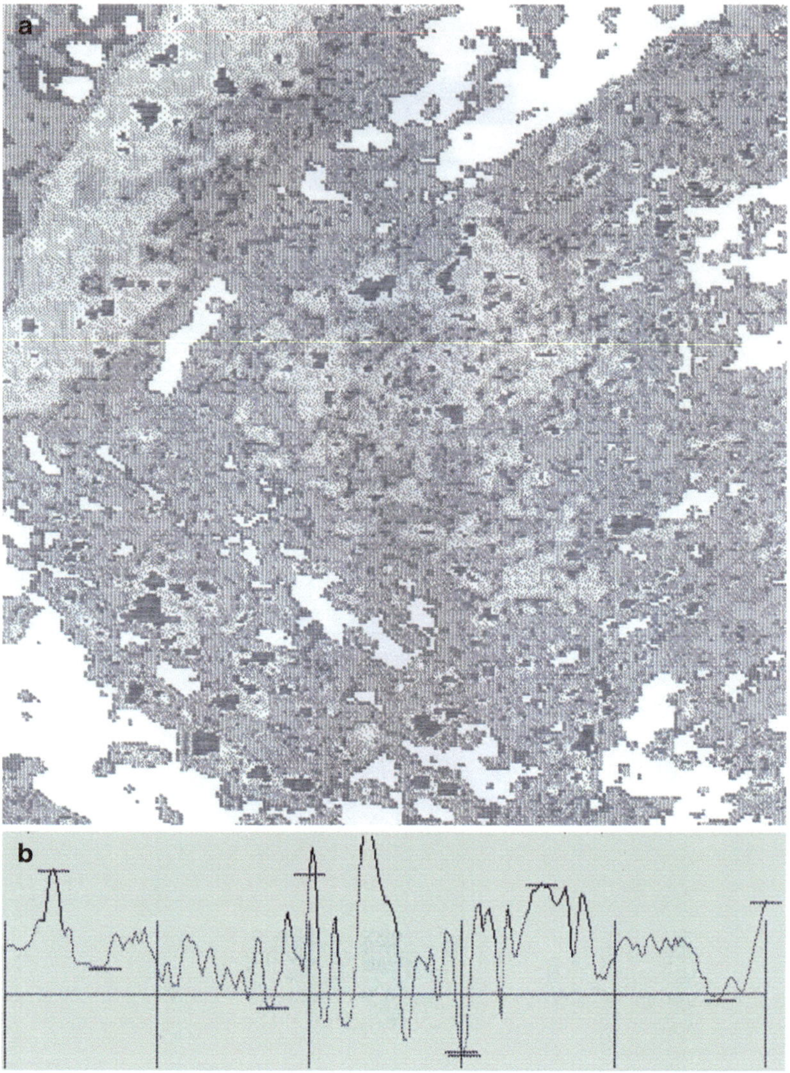

Fig. 40.2 Nail surface weathering on the proximal part of a nail in the twenty-nail dystrophy. (**a**) Two-dimensional field image of a nail surface recording. (**b**) Profilometric presentation

A relevant biological interval should be respected between successive measurements. In this consideration the speed of nail growth must be taken into consideration. Indeed, there is likely interdependence between the disclosed microtopographic changes and variations in the nail growth rate. An example is given by the Beau's lines and beaded nail [7]. Seasonal variations in the nail surface microtopography were reported to vary from insignificant to quite obvious [12].

40.4.3 Nail Microindentation

Experimental microindentation allows to assess some mechanical properties of the nail plate following a load application under controlled conditions on a small nail surface. The nail indentation is possibly measured during the test procedure. If any residual plastic deformation persists after releasing the indentation force, the imprint into the nail can be measured using profilometry.

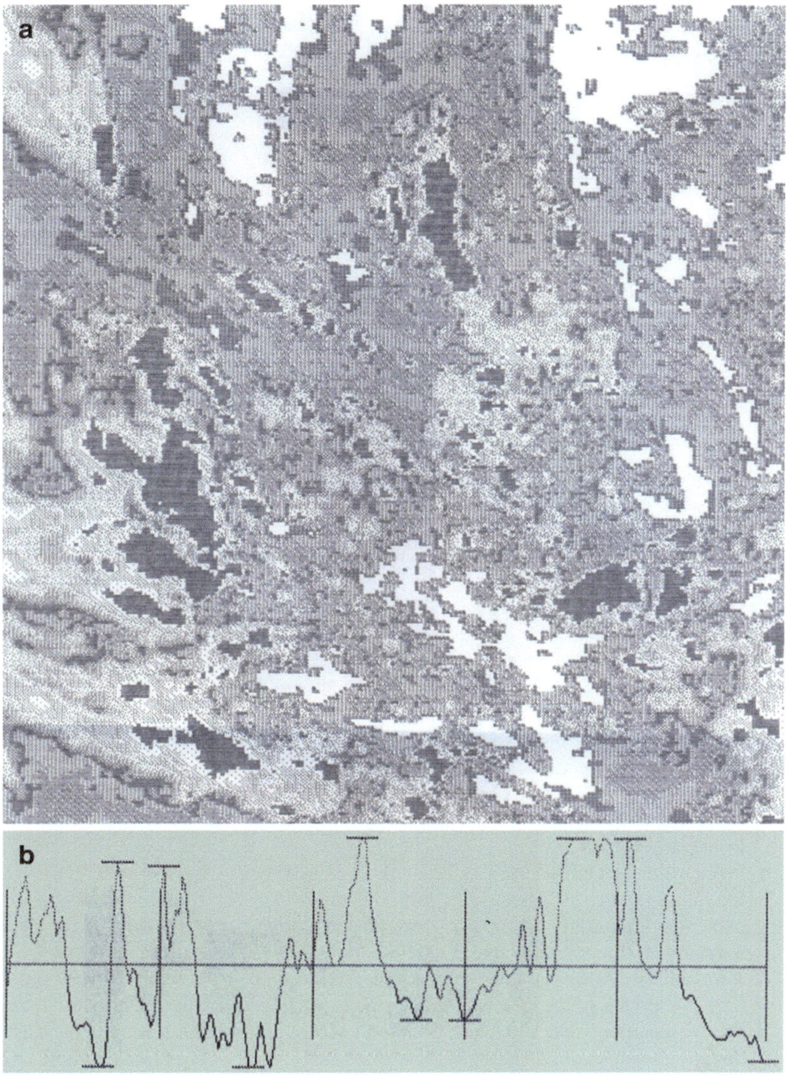

Fig. 40.3 Nail surface weathering on the distal part of a nail in the twenty-nail dystrophy. (**a**) Two-dimensional field image of a nail surface recording. (**b**) Profilometric presentation

40.4.4 Nail Sclerometry

Sclerometry deals with the dynamic assessment of the response of an object during microabrasion. In addition to the classic abrasion parameters, profilometry reveals any groove traced into the nail. The effect of nail hardening products and nail protectors can be tentatively assessed by that way. Similarly, nail softening by xenobiotics or altered health condition can be quantified.

Conclusion

Objective assessments of the nail surface topography were seldom addressed in the literature. However, the nail microrelief is variably influenced by various physiologic parameters, weathering, external trauma, and pathologic features altering the nail matrix. Some microtopographic alterations are linked to changes in the nail growth rate and in the nail hardness. Presumably some methods developed for the skin surface microtopography can be applied to the nail apparatus. These methods should highlight the features of onychochronobiology, thus providing unique information about the physiopathologic processes having altered the nail over the past weeks and months.

Acknowledgments This work was supported by a grant from the "Fonds d'Investissement de la Recherche Scientifique" of the University Hospital of Liège. No other sources of funding were used to assist in the preparation of this manuscript. The authors have no conflicts of interest that are directly relevant to the content of this review. The authors appreciate the excellent secretarial assistance of Mrs. Ida Leclercq and Marie Pugliese.

References

1. Reich K (2009) Approach to managing patients with nail psoriasis. J Eur Acad Dermatol Venereol 23:S15–S21
2. Rich P, Scher RK (2003) Nail psoriasis severity index: a useful tool for evaluation of nail psoriasis. J Am Acad Dermatol 49:206–212
3. Baran RL (2004) A nail psoriasis severity index. Br J Dermatol 150:568–569
4. Williamson L, Dalbeth N, Dockerty JL et al (2004) Extended report: nail disease in psoriatic arthritis – clinically important, potentially treatable and often overlooked. Rheumatology 43:790–794
5. Jiaravuthisan MM, Sasseville D, Vender RB et al (2007) Psoriasis of the nail: anatomy, pathology, clinical presentation, and a review of the literature on therapy. J Am Acad Dermatol 57:1–27
6. Lévêque JL (1999) EEMCO guidance for the assessment of skin topography. J Eur Acad Venereol 12:103–114
7. De Doncker P, Piérard GE (1994) Acquired nail beading in patients receiving itraconazole: an indicator of faster nail growth? A study using optical profilometry. Clin Exp Dermatol 19:404–406
8. Nikkels-Tassoudji N, Piérard-Franchimont C, De Doncker P, Piérard GE (1995) Optical profilometry of nail dystrophies. Dermatology 190:301–304
9. Piérard GE, Piérard-Franchimont C (1996) Fractal microrelief of the skin and nail. Giorn Int Dermatol Ped 8:75–79
10. Piérard-Franchimont C, Piérard GE (1998) Surface image analysis of nail alterations in juvenile pityriasis rubra pilaris. Skin Res Technol 4:34–36
11. Piérard GE, Piérard-Franchimont C (1996) Dynamics of psoriatic trachyonychia during low dose cyclosporin A treatment A pilot study on onychochronobiology using optical profilometry. Dermatology 192:116–119
12. Piérard-Franchimont C, Jebali A, Ezzine N, Letawe C, Piérard GE (1996) Seasonal variations in polymorphic nail surface changes associated with diabetes mellitus. J Eur Acad Dermatol Venereol 7:182–183
13. Piérard-Franchimont C, Piérard GE (2005) Image analysis of the nail surface. In: Serup J, Jemec G, Grove G (eds) Handbook of non-invasive methods and the skin, 2nd edn. CRC Press, Boca Raton, pp 925–928
14. Orentreich N, Markofsky J, Vogelman JH (1979) The effect of aging on the rate of linear nail growth. J Invest Dermatol 73:120–130
15. Dawber RPR (1970) Fingernail growth in normal and psoriatic subjects. Br J Dermatol 82:454
16. Heikkila H, Stubb S, Kiistala U (1996) Nail growth measurement employing nail indentation – an experimental follow-up study of nail growth in situ. Clin Exp Dermatol 21:96–99
17. De Berker D (1997) Nail growth measurement by nail indentation. Clin Exp Dermatol 22:109–111

Combined Technology for Measuring Skin Diseases with Molecular Imaging

41

Gyu Man Park, Sang Wook Son, Gun Woo Lee, Seung Han Ha, On Seok Lee, Jae Young Kim, and Chil Hwan Oh

G.M. Park, PhD
College of Life Sciences and Biotechnology, Korea University, Seoul, South Korea

S.W. Son, MD, PhD
Department of Dermatology, College of Medicine, Korea University, Seoul, South Korea

G.W. Lee, PhD
Research Institute for Skin Image, College of Medicine, Korea University, Seoul, South Korea

S.H. Ha, PhD
Center for Ultrasound Molecular Imaging and Therapeutics, Heart Vascular Institute, University of Pittsburgh Medical Center, Pittsburgh, PA, USA

O.S. Lee, PhD
Department of Radiological Science, Gimcheon University, Gimcheon, South Korea

J.Y. Kim, MS
Division of Instrument Development, Korea Basic Science Institute, Daejeon, South Korea

C.H. Oh, MD, PhD (✉)
Department of Dermatology, College of Medicine, Korea University, Seoul, South Korea

Research Institute for Skin Image, College of Medicine, Korea University, Seoul, South Korea

Department of Dermatology, Korea University Guro Hospital, Korea University College of Medicine, 97 Guro-dong gil, Guro-dong, Guro-gu, Seoul 152-703, South Korea
e-mail: choh@korea.ac.kr

Abbreviations

[^{18}F] FDG	[^{18}F]-fludeoxyglucose
[^{18}F]FHBG	[^{18}F]-fluoro-hydroxymethyl butyl guanine
CBR	Click beetle red-emitting luciferase
CCD	Charge-coupled device
CLIO	Cross-linked iron oxide
DTIC	Dacabazine
FL	Firefly luciferase
GFP	Green fluorescent protein
hFTH	Heavy-chain ferritin
HSV1-tk	Herpes simplex virus type 1 thymidine kinase
MION	Monocrystalline iron oxide
MR	Magnetic resonance
OCT	Optical coherent tomography
PET	Positron emission tomography
QDs	Quantum dots
RFP	Red fluorescent protein
SPECT	Single photon emission computed tomography
SPIO	Superparamagnetic iron oxide
SUV	Standard uptake value
TMZ	Temozolomide
USPIO	Ultra superparamagnetic iron oxide

41.1 Introduction

In vivo molecular imaging has distinct advantages compared to conventional in vitro and cell culture research. Although in vitro and cell culture studies may be the best way to define

E. Berardesca et al. (eds.), *Non Invasive Diagnostic Techniques in Clinical Dermatology*, DOI 10.1007/978-3-642-32109-2_41, © Springer Berlin Heidelberg 2014

Table 41.1 Comparison of the three molecular imaging modalities

	Instrument	Spatial resolution	Depth	Advantage	Disadvantage
Optical molecular imaging	CCD Multiphoton microscope Optical coherent tomography (OCT)	2–5 mm μm unit	1 cm ↓ 500 μm Several cm	High sensitivity Low cost	Low resolution in CCD But, high resolution in multiphoton and OCT
Radionuclear molecular imaging	SPECT PET	1–2 mm	No limit	High sensitivity Most advanced technology in the molecular imaging field	Radiologic hazard Relatively low resolution
MR molecular imaging	MRI	25–100 μm	No limit	Relatively high spatial resolution	Relatively low sensitivity

**CCD* charge-coupled device

biochemical or gene expression pathology, these studies cannot provide a dynamic visualization of the functions of particular genes or proteins within the whole-body complex. Molecular imaging offers distinct advantages in this regard. In addition, experimental animals do not need to be killed during imaging, which makes it possible to repeat the imaging in the same animal at a later time. Molecular imaging is also a useful tool for early drug discovery studies because it permits early validation and evaluation of compounds.

Molecular imaging can be categorized into three basic modalities: optical molecular imaging, radionuclear molecular imaging, and magnetic resonance (MR) molecular imaging. The various imaging instruments and technologies differ in several respects, including the depth of penetration, spatial and temporal resolution, energy expansion for image generation, and the detection threshold of the probe. Table 41.1 outlines the general characteristics of the three imaging modalities.

Optical molecular imaging consists of fluorescent and bioluminescent imaging, and it plays an important role in the molecular imaging field. In fluorescent imaging, the illuminator light excites fluorescent receptors in a living subject, and a CCD camera detects the emission light. Green fluorescent protein (GFP) and red fluorescent protein (RFP) are the most commonly used fluorescent receptors. In bioluminescent imaging, the receptor gene codes for luciferase, which is derived from fireflies. After injection of the substrate luciferin, the cells or animals containing the luciferase gene are imaged with a CCD detector. Optical molecular imaging is one of the most powerful tools to detect cellular level events at a high spatial and temporal resolution using in vivo multiphoton microscopes or optical coherent tomography (OCT). However, the penetration depth is very limited (500 μm ~3 mm). To overcome this disadvantage, an endoscopic system of multiphoton microscopes and OCT is required, and these systems could be available for every organ of the body within the near future.

Radionuclear molecular imaging has its roots in radionuclear medicine. β-ray-emitting isotopes (e.g., ^{3}H, ^{13}C) are not useful for molecular imaging because β-particles do not travel significant distances. γ-emitter isotopes (e.g.,^{99m}Tc, ^{111}In, ^{123}I, ^{131}I) could be used for molecular imaging, but they require special signal detectors, such as a gamma camera and single photon emission computed tomography (SPECT). Positron emission tomography (PET) records high-energy γ-rays that are emitted. ^{15}O, ^{13}N, ^{11}C, and ^{18}F are frequently used as positron emission isotopes. Most of these isotopes are produced in a cyclotron, but some can be produced with a generator. A labeled molecular probe or tracer can be introduced into the subject, and then PET imaging can determine the distribution and concentration of the injected molecule. However, the spatial resolution power of PET is very limited (2 mm^3). Radionuclear molecular imaging can be divided into direct and indirect methods. Direct methods use molecular probes, in which synthesis is targeted at a specific molecular marker such as a receptor, enzyme, or gene. Indirect methods use a pre-targeted molecule that is subsequently activated when specific molecular events occur. Receptor gene imaging is one of the indirect methods, and it provides a convenient tool to monitor gene expression. Herpes simplex virus type 1 thymidine kinase (HSV1-tk) is the most commonly used receptor gene in the indirect method of radionuclear molecular imaging.

MR imaging works by detecting differential contrast in soft tissue resulting from endogenous differences in water content, relaxation time, and diffusion characteristics of the tissue. To increase the intrinsic contrast generated in an MR image, superparamagnetic complexes (e.g., monocrystalline iron oxide (MION), superparamagnetic iron oxide (SPIO), ultra superparamagnetic iron oxide (USPIO), cross-linked iron oxide (CLIO)) are used to develop new contrast agents that can target specific molecular markers. Recently, ferritin has been developed as a new MR reporter gene.

A major challenge for molecular imaging is to provide the means for noninvasive detection of reporter gene expression in live animals at a high spatial and temporal resolution. Reporter genes, which lead to the generation of easily detectable markers, provide an indispensable tool for the analysis of biological function.

41.2 Optical Molecular Imaging

41.2.1 Optical Molecular Imaging Using Quantum Dots

Colloidal semiconductor quantum dots (QDs) are nanometer-sized crystals with unique photochemical and photophysical properties. The emission spectra of QDs are narrow, symmetrical, and adjustable according to their size and material composition, which allows closer spacing of different probes without substantial spectral overlap. QDs also exhibit excellent photostability and broad absorption spectra, making it possible to excite QDs of all colors at the same time with a single excitation light source [1].

Fluorescent QDs are powerful multipurpose interfaces of nanotechnology that provide long-term, multicolor imaging of cellular and molecular interactions.

41.2.1.1 QD Imaging in a Wound-Healing Model [2]

Cultured human keratinocytes were labeled with QD 655 (Quantum Dot Corporation, Hayward, CA, USA), which emits a green fluorescence. Cultured human fibroblasts were labeled with QD 565 (Fig. 41.1).

A mixture of dual-labeled keratinocytes and fibroblasts was inoculated into artificial wound sites of BALB/C nude mice. Multispectral emission images were taken with an in vivo optical imaging system that we created (Fig. 41.2). Dual labeling made it possible to distinguish between fibroblasts and keratinocytes in the same wound site (Fig. 41.3). The fluorescent signals of the QD-labeled fibroblasts diminished after 10 days (Fig. 41.4).

41.2.1.2 QD Imaging in a Zebra Fish Model [3]

The use of QDs in living organisms is just beginning to be explored, and zebra fish embryos may be a suitable vertebrate model for intravital imaging with QDs. To investigate their potential in dermatology research, we used QDs as microangiography contrast agents to visualize the cardiovascular system in zebra fish. We also investigated the pathway relationship between the cardiovascular system and the nerve network using QDs and a transgenic zebra fish line.

Quantum Dot QD605, which emits a green color under a fluorescent microscope, was used as a microangiography contrast agent. The olig2-Dsred transgenic zebra fish line, which expresses motor neurons in a red color, was used together with the QDs. Images of the QD605-injected embryos were recorded with a digital camera.

By combining the green fluorescence of QD605 and the red fluorescence of olig2-Dsred

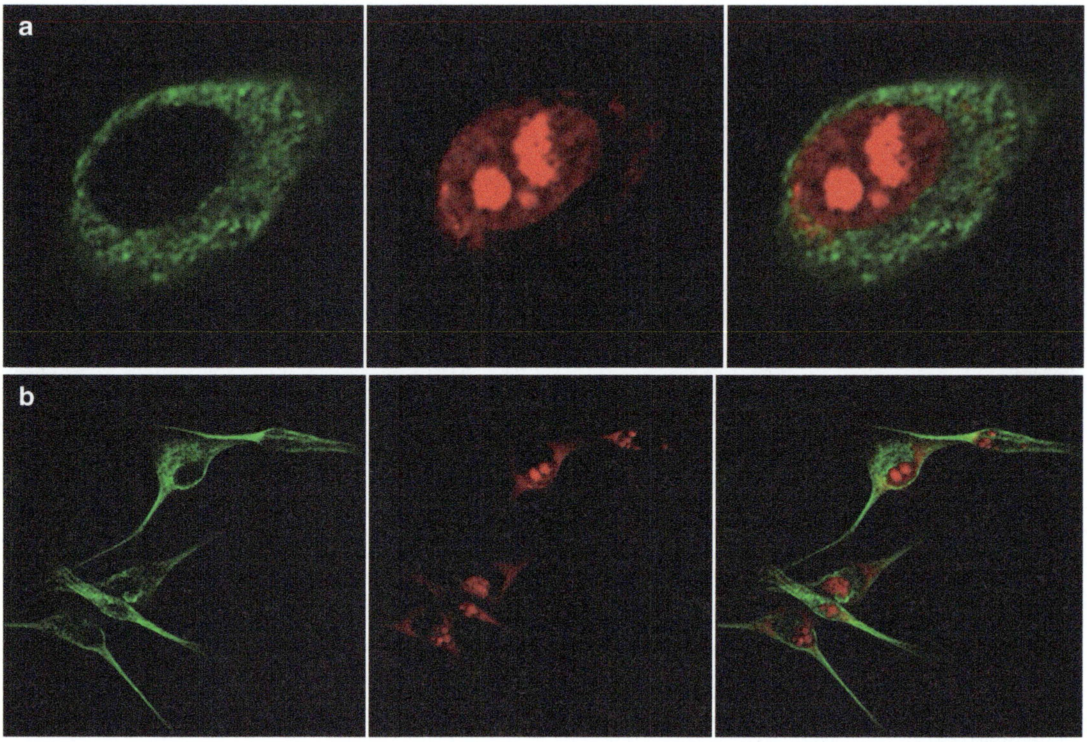

Fig. 41.1 Dual optical imaging using QDs. (**a**) Cultured keratinocytes labeled with QD 655. (**b**) Cultured fibroblasts labeled with QD 565

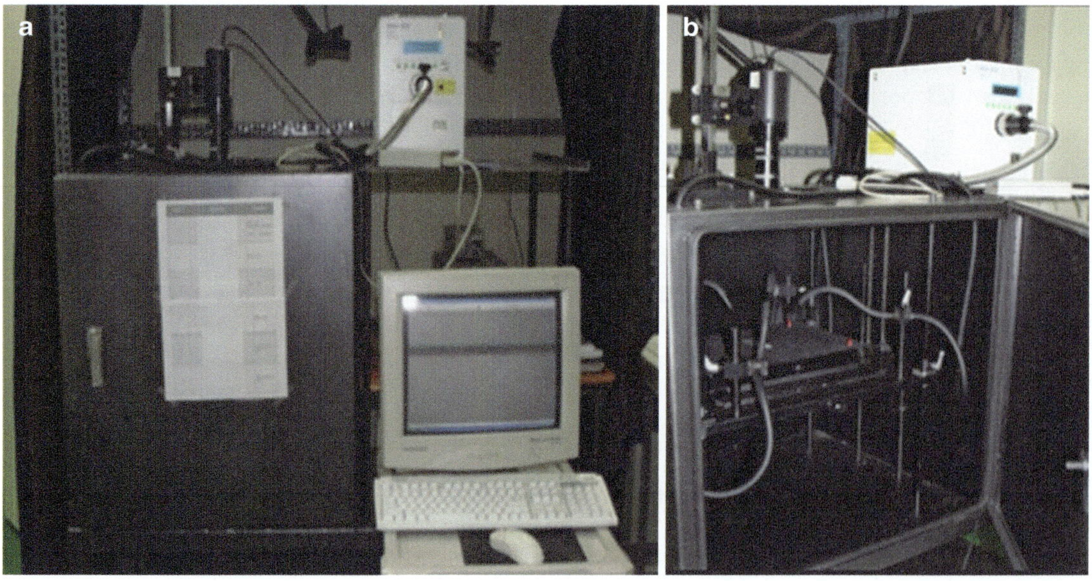

Fig. 41.2 The in vivo optical molecular imaging system that we created includes a xenon arc lamp (light source), multiple filter systems to detect specific excitation and emission spectra, and a cooled CCD detector. (**a**) Outer schematic diagram of the system and (**b**) inner schematic diagram of the system

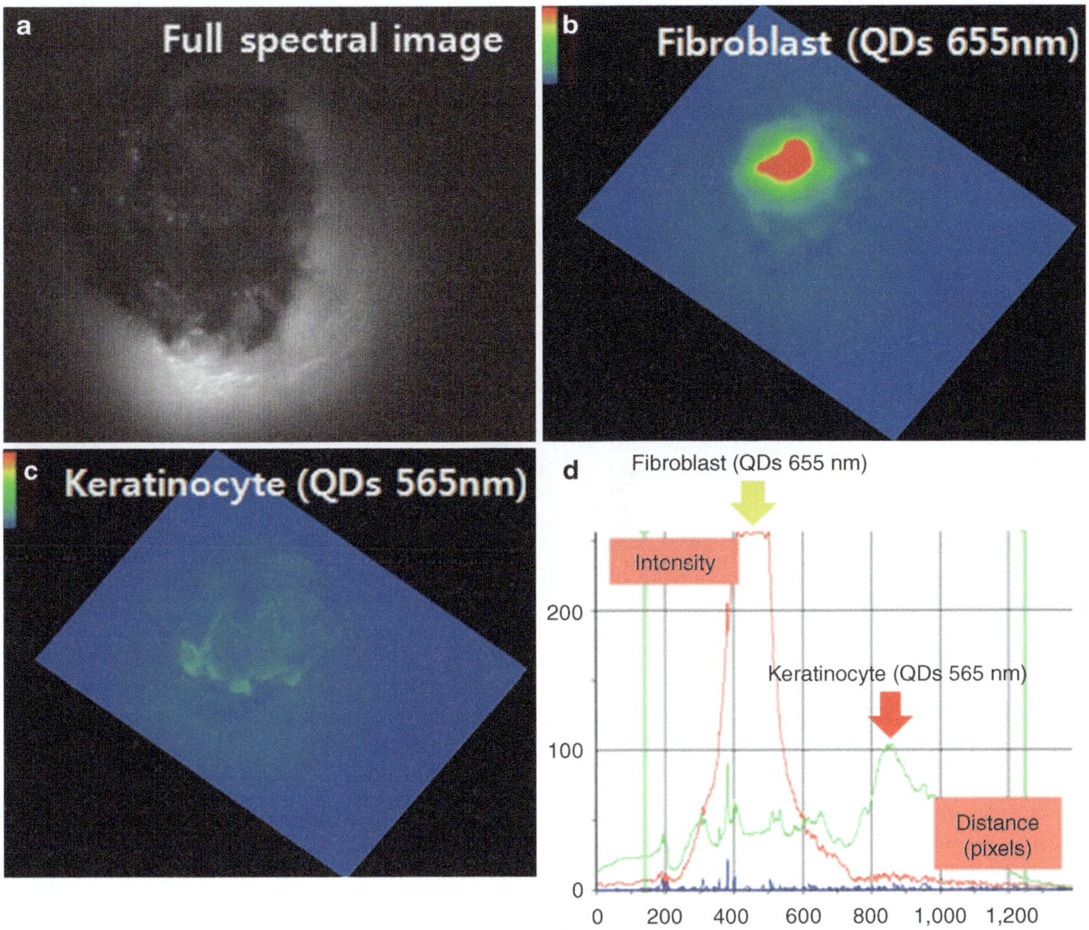

Fig. 41.3 The multiplexed image of QD-labeled fibroblast (655 nm) and keratinocyte (565 nm) cell lines. (**a**) Wound image (full spectra), (**b**) spectral image of the QD signal (655 nm) for keratinocytes, (**c**) fluorescence and QD signal (565 nm) for fibroblasts, and (**d**) intensity signal for keratinocyte (*green curve*) and fibroblast cell lines (*red curve fitting*)

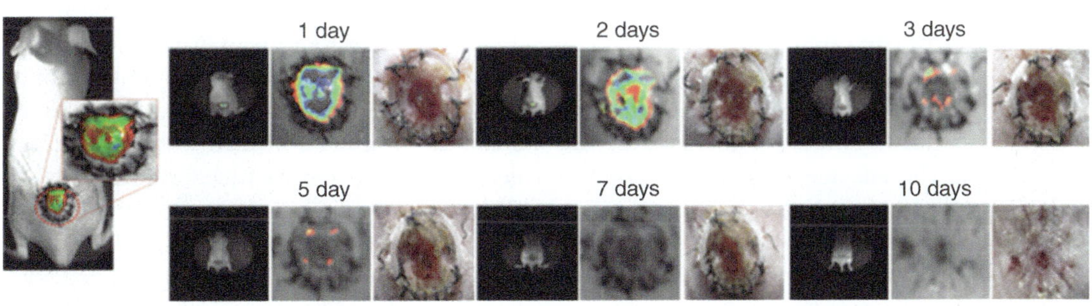

Fig. 41.4 Wound healing was monitored in the QD-labeled fibroblast for 10 days after inoculation using an in vivo optical molecular imaging system that we created

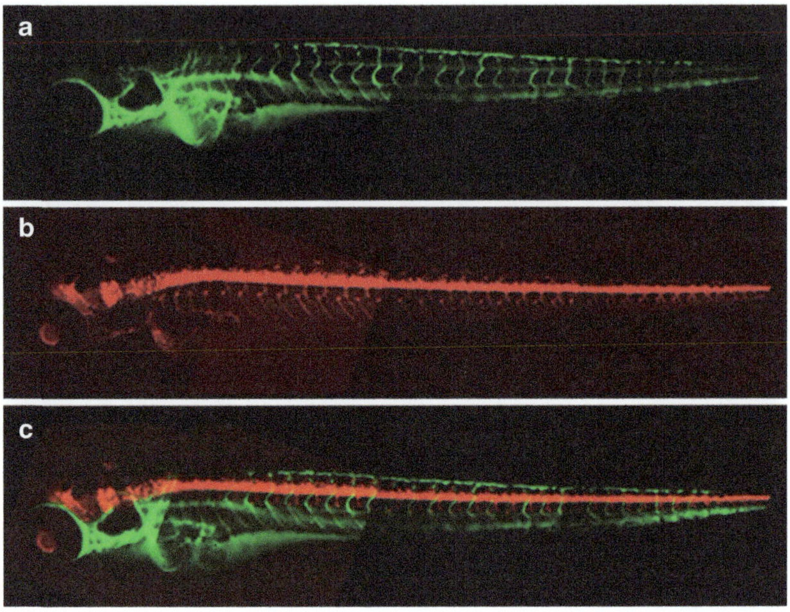

Fig. 41.5 Combining the intravital image of quantum dot microangiography and the transgenic zebra fish line. (**a**) Quantum dot microangiography. Quantum Dot QD605, which emits a *green color*, was used as a contrast agent. (**b**) Transgenic zebra fish line. The olig2-Dsred transgenic zebra fish line expresses motor neurons and their axon tracts in a *reddish color*. (**c**) By combining the green fluorescence of QD605 and the red fluorescence of olig2-Dsred transgenic zebra fish, the vascular network and the motor nervous system could be viewed together

transgenic zebra fish, we obtained detailed images of the spatial relationship between the vascular and nervous systems of zebra fish (Fig. 41.5).

Our results suggest that QDs could easily be used as bright microangiography contrast agents in living embryos. Our images of the zebra fish vascular and motor nervous systems showed a similar pattern of trajectory overall for the blood vessels and axon tracts. However, the segmented repetitive networks along the dorsoventral axis were not completely overlapped.

41.2.2 Optical Molecular Imaging Using the Luciferase Gene [4]

Luciferase is an enzyme that reacts with luciferin to produce yellow-green luminescence at 562 nm. The enzyme is frequently used as a reporter gene to measure activity of gene-regulatory sequences. The purpose of this study was to develop a model of tumor and wound healing using a luciferase system. The tumor models were generated by subcutaneous inoculation of lenti-CMV-luciferase vector-transduced melanoma cells into mice. We then evaluated the efficacy and biodistribution of luciferase-transduced fibroblast cells, which were administrated into a wound site in the wound-healing mouse model.

We used an in vivo optical imaging system to observe the xenograft tumor models after introduction of the melanoma cells; the melanoma cells expressed the luciferase gene (Fig. 41.7). We also introduced human fibroblasts to evaluate wound-healing efficacy. All cells had stable expression of lenti-CMV luciferase due to clonal selection with puromycin (Fig. 41.6).

Cultured murine melanoma B16 cells (5×10^5) in PBS were injected subcutaneously in C57BL/6 J mice. The fibroblast cell lines (5×10^5) were inoculated to the lesion directly. Some of the fibroblast cells lines were suspended in a Tissucol kit, and some were suspended in PBS.

Approximately 1 week after B16 cells injection, the mice developed a solid tumor of 5–8 mm diameter and were evaluated with optical imaging. All animals received luciferin substrates by IP injection before imaging on days 1, 5, 7, 10, 14, and 28 after inoculation (Fig. 41.7).

We demonstrated that using luciferase-transduced B16 melanoma cell inoculation in a mouse model results in long-term stable expression of luciferase up until mouse death due to cachexia. We also detected a local metastasis region after administration of luciferin substrates by IP injection (Fig. 41.8).

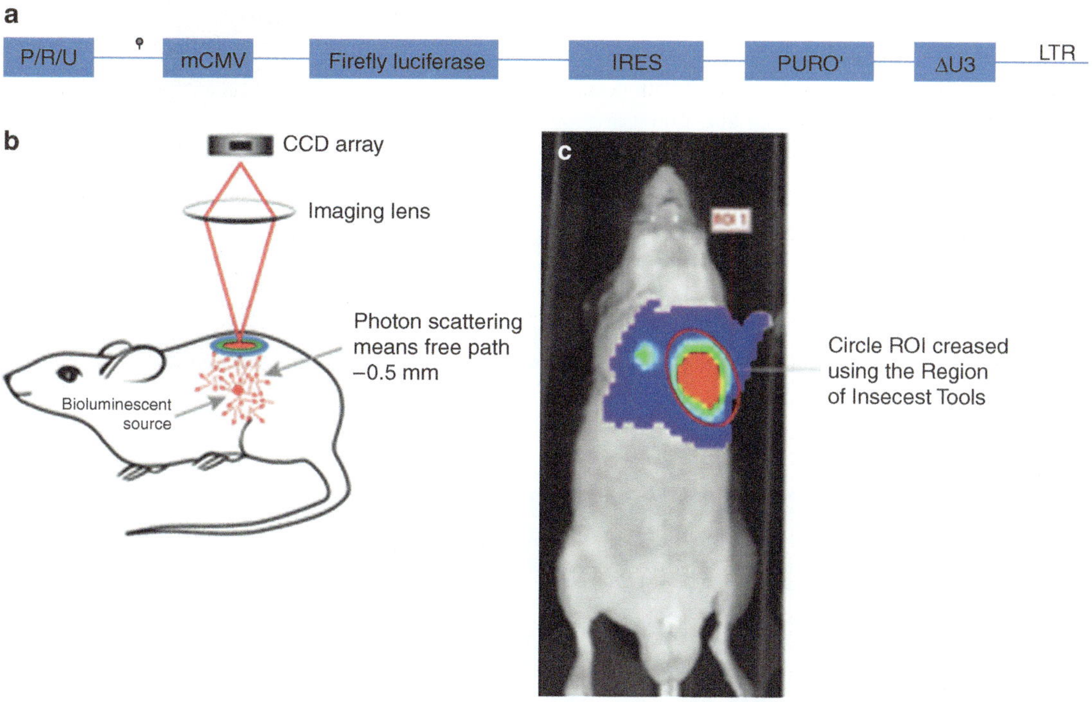

Fig. 41.6 (**a**) Structure of the lentiviral vector containing the luciferase gene. (**b**) Illustration of the photon transport of light from an internal source to the visible location on the surface of the animal. It is the surface image that is observed. (**c**) The circular region of interest (*ROI*) expressed luciferase fluorescence after luciferin was administered by IP injection

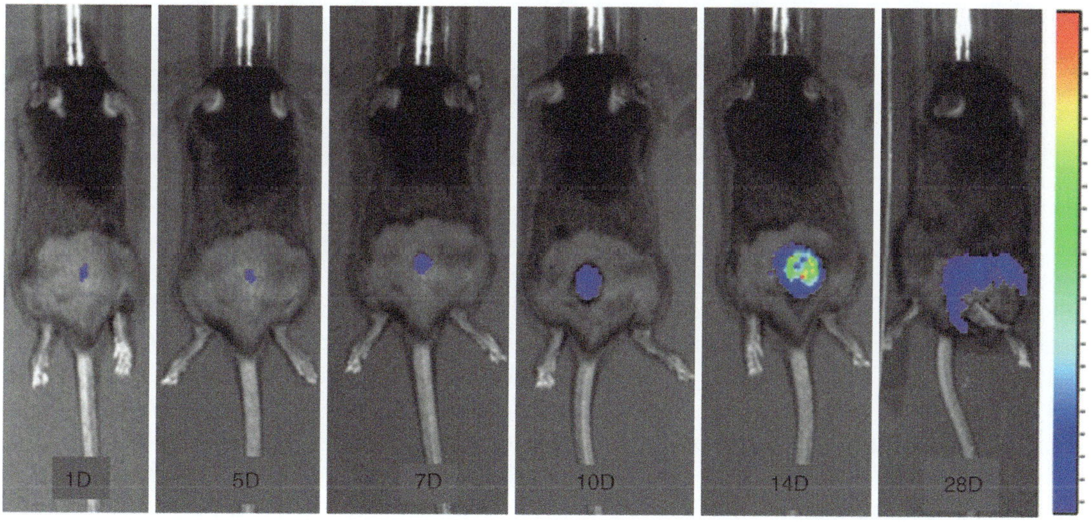

Fig. 41.7 Fluorescent images of the melanoma mouse models for 28 days after the mice were inoculated with cultured melanoma cells that expressed the luciferase gene

The lenti-CMV-firefly luciferase-transduced human fibroblast cells were inoculated into an artificial wound site on the dorsal side of a Balb/C nude mouse. The wound-healing process was monitored for 28 days (Fig. 41.9). The results suggest that luciferase-transduced fibroblasts can be used

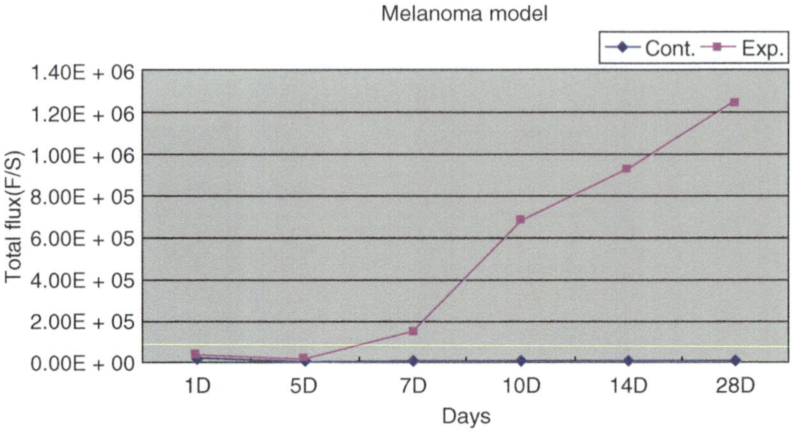

Fig. 41.8 Intensity of luminescent flux of the luciferase-expressed melanoma models until death of the mice due to cachexia

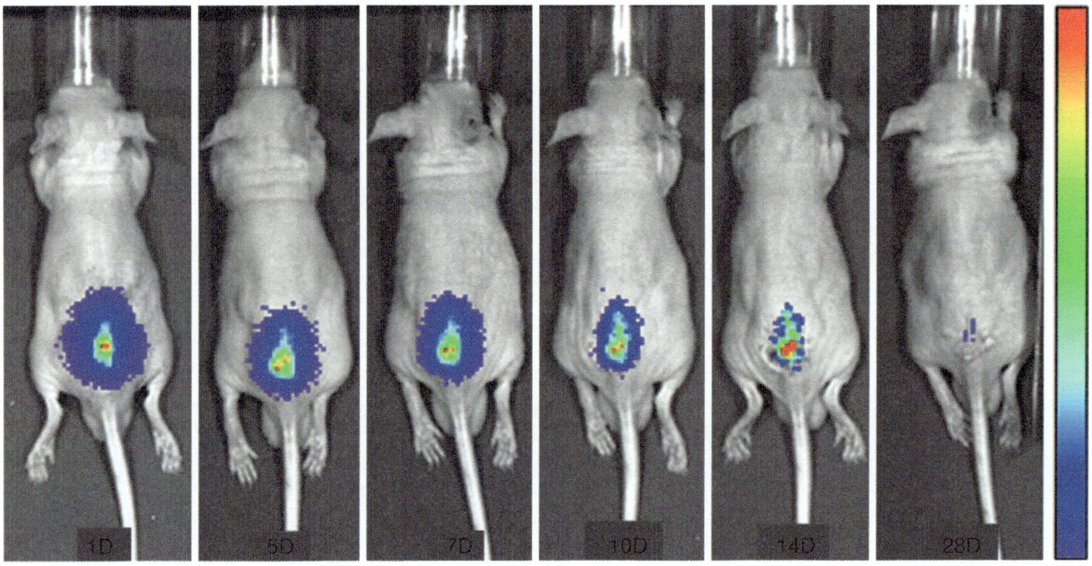

Fig. 41.9 Wound healing was monitored for 28 days after inoculation with luciferase-expressing fibroblasts

for easy, rapid real-time bioluminescence imaging at low cost. Thus, the luciferase system has many potential applications in optical imaging.

41.2.3 Detection of Apoptosis Using an Engineered Luciferase Gene

The purpose of this study was to develop an animal model of apoptosis expression in order to evaluate the efficacy of anticancer drugs for treatment of malignant melanoma using optical molecular imaging. Since caspase-3/7 activation is the main feature of apoptosis, apoptosis can be monitored in real time with a caspase-monitoring system. We used the Protease-Glo™ system (Promega Corporation, Madison, Wisconsin, USA), which contains protease activities using a genetically engineered firefly (*Photinus pyralis*) luciferase (GloSensor™-10F). The GloSensor™-10F luciferase vector consists of two domains, N-terminus and C-terminus, and these domains must be oriented in the correct direction. We made engineered luciferase, which contained a caspase-3/7 cleavage target sequence (DEVED) between reverse-oriented domains. This engineered luciferase was not active luciferase; rather, it acquired activity when two reverse-oriented domains were cleaved and then formed the exact 3-dimensional structure of luciferase. When caspase-3/7 was activated and cleaved the

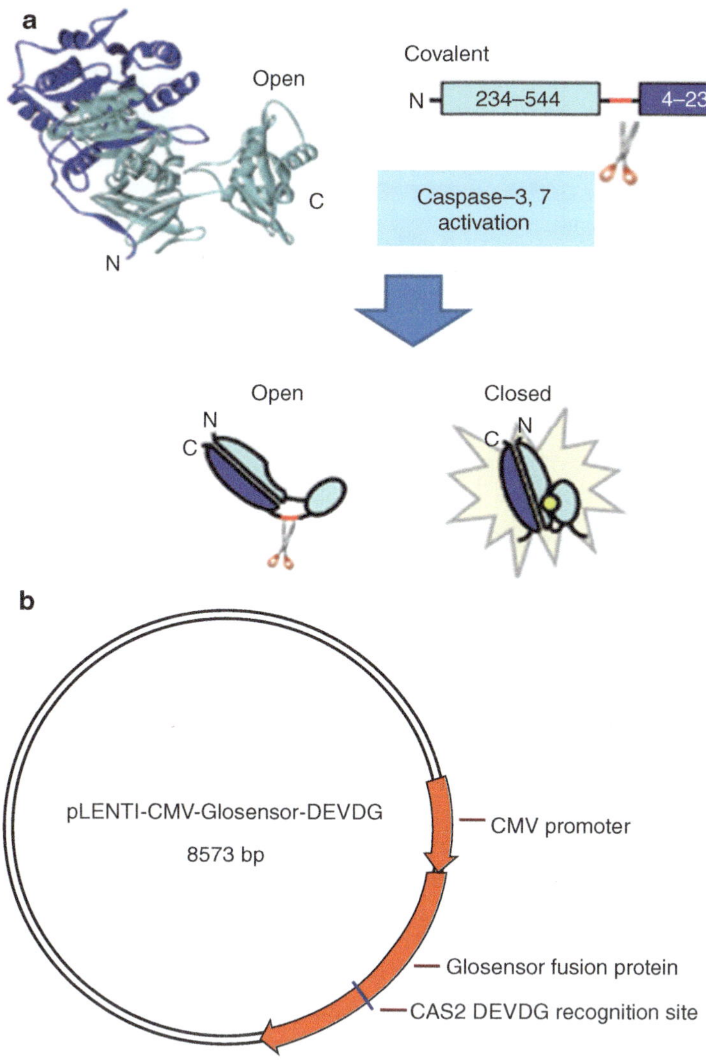

Fig. 41.10 (**a**) Measurement technique for the caspase-3/7 activity of engineered luciferase and (**b**) the vector map of pLenti-CMV-Glosensor-DEVEG

caspase-3/7 cleavage target sequence, the domains formed functionally intact luciferase (Fig. 41.10).

We introduced this engineered luciferase-encoded lentiviral vector into cultured melanoma B16 cells. When we induced apoptosis of melanoma cells with anticancer drugs, the luminescence signal intensity also increased in vitro. We administered various concentration of temozolomide (TMZ) and BCNU in vitro, and luciferase activity increased in a dose-dependent manner as a result of the anticancer drug (Fig. 41.11).

To investigate the possibility of an in vivo apoptosis-monitoring system, we subcutaneously inoculated mice with melanoma cells containing the engineered luciferase-encoded lentiviral vector. Two weeks after inoculation, a chemotherapeutic agent, TMZ or BCNU, was injected peritoneally. The TMZ- and BCNU-treated groups showed a much greater increase in fluorescence due to the induced apoptosis activity than the control group (Fig. 41.12).

41.2.4 Monitoring a Metastatic Melanoma Model and Trafficking Neural Stem Cells in Optical Molecular Imaging

To monitor metastasis and stem cell targeting properties, we used click beetle red-emitting luciferase (CBR) and 800-nm-emitting QD [5].

We monitored B16 melanoma metastasis with two different luciferase systems: firefly luciferase

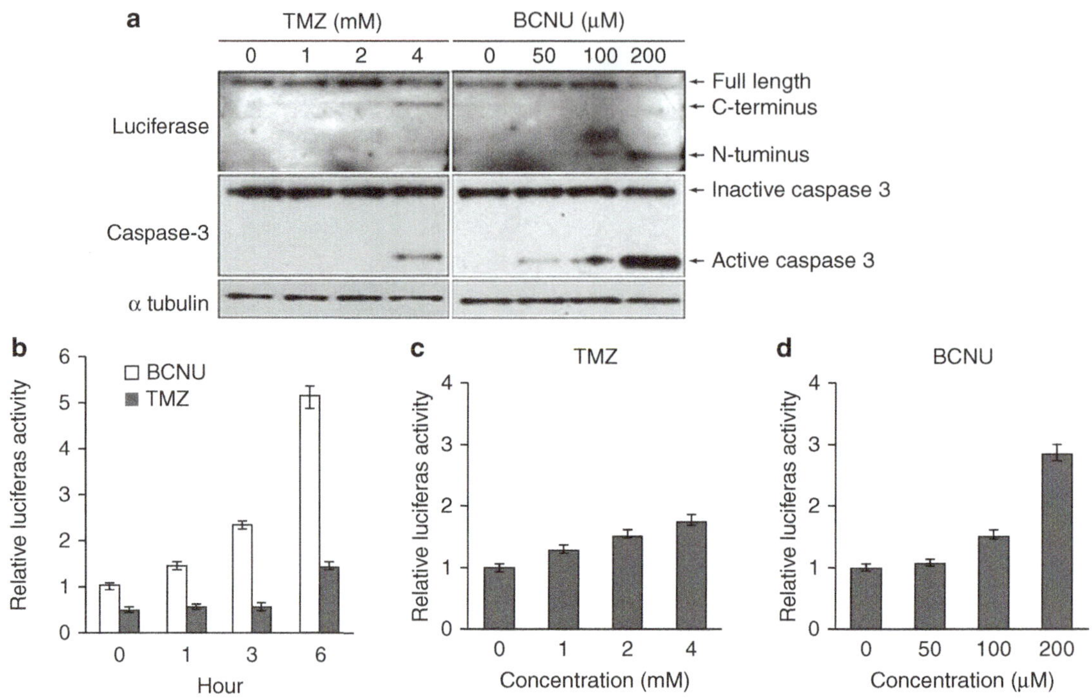

Fig. 41.11 Construction of the luciferase turn-on system for apoptosis and in vitro validation of the apoptosis system in relation to various concentrations of the anticancer drugs

(FL) and click beetle red-emitting luciferase (CBR). CBR luciferase has greater potential for in vivo optical imaging applications than FL. Because emission wavelength of CBR (800 nm) has a more longer wavelength and strong intensity than FL (562 nm), emission light from CBR penetrates more deeper in living tissue. As shown in Fig. 41.15a, CBR luciferase and FL luciferase have a similar optical density in vitro and in SC injected cancer, but CBR luciferase signals more efficiently penetrate the body after intravenous injection of CBR-labeled melanoma cells (Fig. 41.13b).

In a recent study, stem cells were moved toward a malignant tumor, and this targeting efficacy is much higher than viral or DNA targeting systems [6]. To track neural stem cell targeting, we labeled neural stem cells derived from the embryonic brain of C57BL/6 mice with 655-nm- and 800-nm-emitting QDs. B16 melanoma cells were inoculated subcutaneously in the dorsal area of nude mice. Two weeks after inoculation with the melanoma cells, QD-labeled neural stem cells were injected intravenously. Many QD-labeled neural stem cells were later located in the kidney and lungs, but other QD-labeled stem cells migrated to the site of melanoma inoculation.

41.3 Radionuclear Molecular Imaging

41.3.1 Thymidine Kinase-Based Radionuclear Molecular Imaging [7]

[^{18}F]-fludeoxyglucose positron emission tomography (FDG-PET) is widely applied for diagnostic assessment of malignancy, in tumor staging, and in therapy monitoring. Melanoma cells also show an increase in glucose transporter proteins. FDG uptake, which reflects glucose metabolic rate, is increased in cancer cells. However, FDG accumulation is not specific to tumors and can be present as a result of various conditions, including inflammation, infection, and other benign processes.

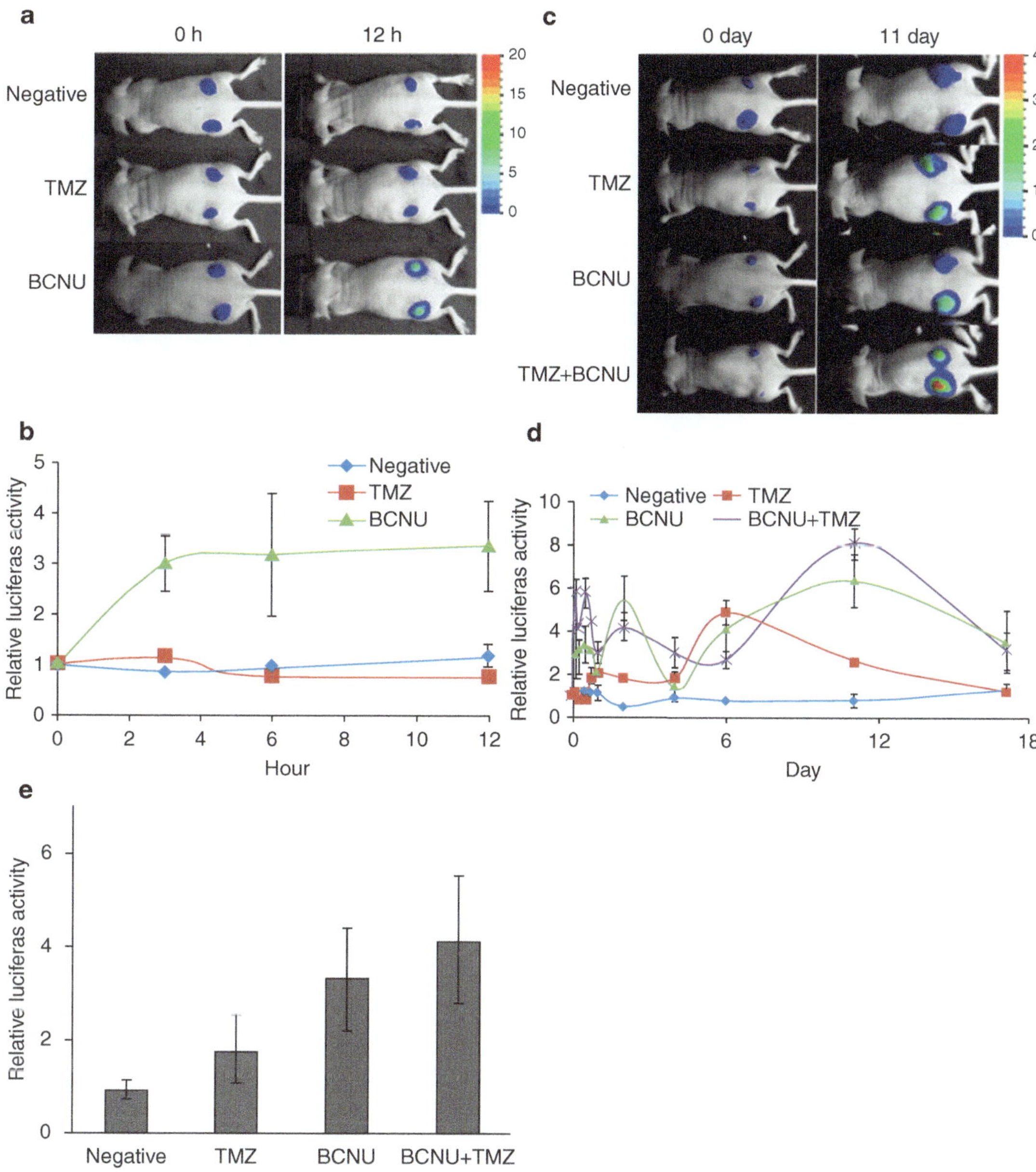

Fig. 41.12 Monitoring of in vivo luciferase activation by anticancer drugs (TMZ, BCNU) during apoptosis of the melanoma xenograft mouse models that contained engineered luciferase

A positron emission tomography (PET) reporter gene/reporter probe system has recently been developed to provide sensitive and quantitative molecular imaging. Thymidine kinase, the most commonly used receptor gene, is a marker of active tumor proliferation, and it compares favorably with the more commonly used marker, FDG.

The purpose of this study was to establish a small-animal model for radionuclear molecular imaging and to acquire basic data on the efficacy of candidate anticancer drugs for treatment of melanoma using a small-animal PET imaging analysis with [^{18}F] FHBG for herpes simplex virus 1 thymidine kinase (HSV1-tk) gene expression.

To transduce the HSV1-tk gene in a melanoma cell line, a lentiviral vector containing HSV1-tk gene was generated (Fig. 41.14).

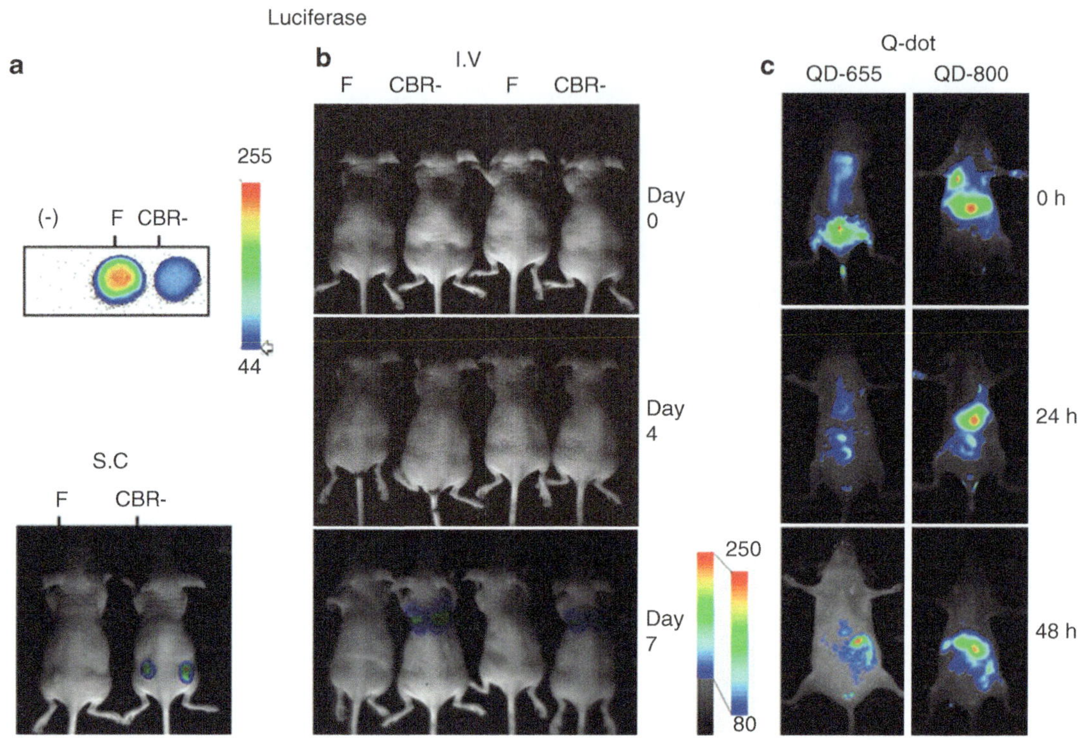

Fig. 41.13 Optical molecular images of cancer trafficking using luciferase and QDs. (**a**) Comparison of the strength of luminescence between firefly luciferase and CBR in subcutaneous xenograft melanoma models and (**b**) a metastatic melanoma model via IV injection. (**c**) QD-labeled neural stem cells migrated into the CBR-expressed melanoma xenograft site **CBR-L* click beetle red-emitting luciferase, *FL* firefly luciferase

The B16 melanoma cell line was transduced with a recombinant lentiviral vector encoding the HSV1-tk gene. B16 melanoma cells containing the HSV1-tk gene were inoculated subcutaneously into the backside of C57BL/6 J mice. Three weeks after inoculation, PET images were acquired. At 1 h after intravenous injection of 200 μCi [^{18}F] FHBG, whole-body images were obtained with a small-animal PET/CT (eXplore Vista®) scanner (Fig. 41.15).

To evaluate the efficacy of chemotherapeutics, we gave intraperitoneal injections of oregonin (a new candidate drug) and dacarbazine (DTIC) to tk gene-encoded melanoma xenograft mice. The mice injected with DTIC, oregonin, and both DTIC and oregonin showed reductions in tumor size compared with control mice. Control mice died after 25 days (Fig. 41.16).

One potential drawback during the development of new drugs is rising costs, which result from increasing preclinical evaluation expenditures. Early identification of drug failure could greatly reduce the development costs per drug.

In conclusion, our results suggest that PET imaging of a melanoma mouse model expressing the HSV1-tk gene could be used as a rapid, semi-quantitative, repetitive, noninvasive imaging tool to assess the efficacy of new anticancer candidates (oregonin) in the preclinical stage.

41.3.2 Measuring Tumor Volume Using Calipers, Stereo Image-Based Three-Dimensional Measurement, Micro-CT, and [^{18}F]FDG-PET Images in a Xenograft Melanoma Model

In tumor xenograft animal studies, calipers are commonly used to measure external tumor size.

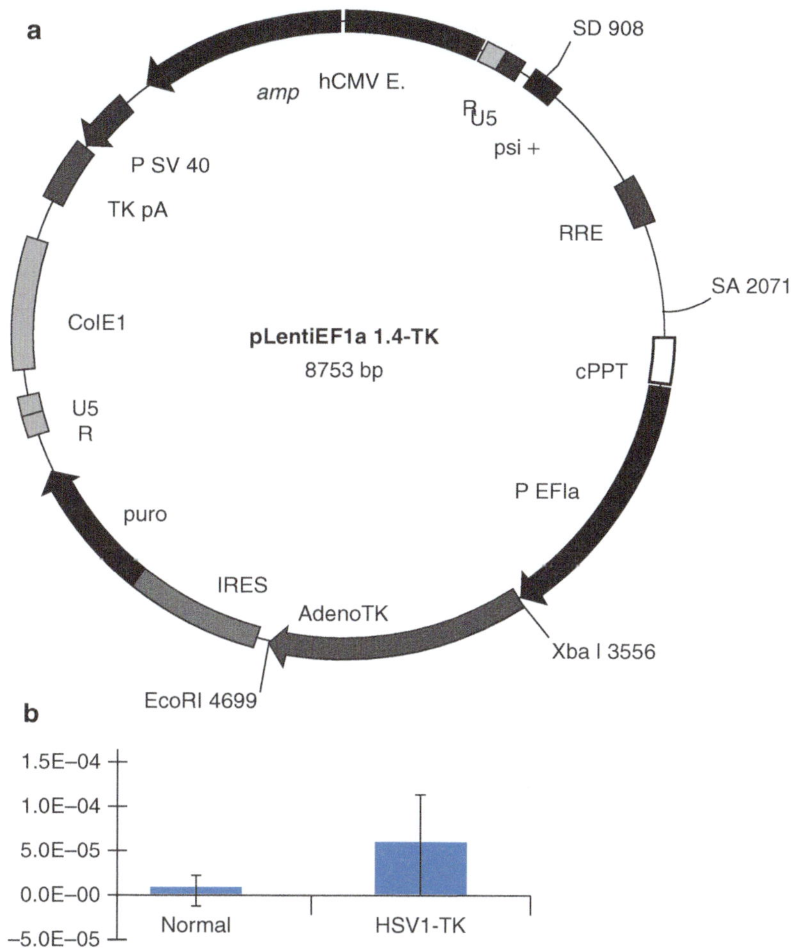

Fig. 41.14 Recombinant lentiviral vector map for transduction of HSV1-tk genes into a B16 melanoma cell line

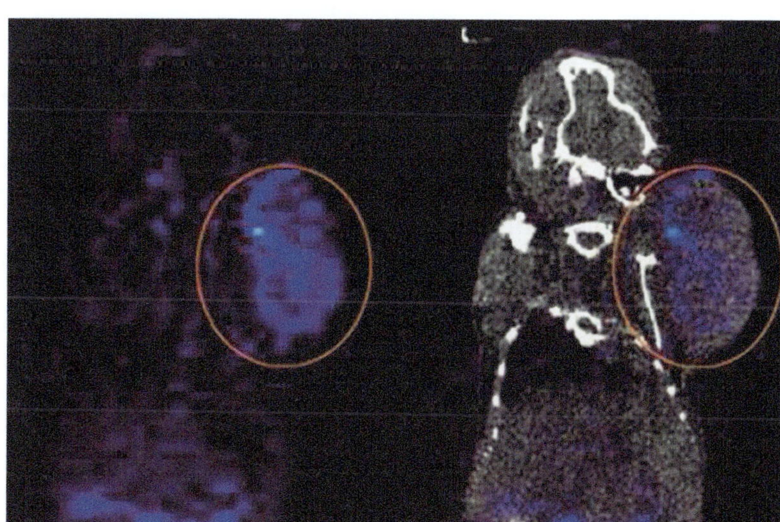

Fig. 41.15 [^{18}F] FHBG PET image (*left*, control view) and PET/CT fusion image (*right*, coronal view) in a melanoma mouse model containing the HSV1-tk gene

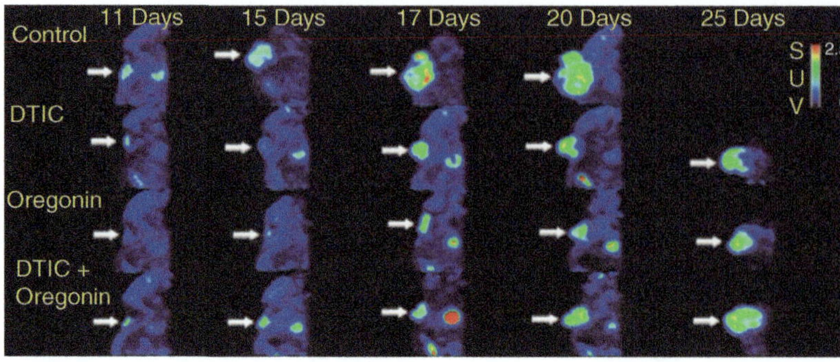

Fig. 41.16 [^{18}F] FDG micro-PET images (sagittal view, threshold$_{max}$ = 100 %, slice thickness = 0.2 mm) of subcutaneously transplanted mice under treatment with PBS, DTIC, oregonin, or DTIC + oregonin. Tumors (*arrows*) were induced by subcutaneous injection of 5×10^5 B16 melanoma cells in 50 μL PBS buffer on the backside of Balb/c nude mice on day 0. Treatment began after 11 days, and mice were treated with DTIC (50 mg/kg IP), oregonin (10 mg/kg IP), or DTIC (50 mg/kg IP) + oregonin (10 mg/kg IP)

However, many tumors are not prolate spheroids, so measurement with calipers can be inaccurate. Tumor volume and shape may also differ according to the size and type of tumor.

Xenograft tumors were generated by transplanting B16 melanoma cells into the backs of C57BL/6 J mice. We measured the volume of the xenograft tumor with calipers (Fig. 41.17a), stereo image-based three-dimensional measurement (Fig. 41.17b) [8], measurement of the standard uptake value (SUV) of [^{18}F] FDG-PET, and micro-CT (Fig. 41.17c). Stereo image-based 3-D measurement was a more reliable method than measurement with calipers. Internal changes in tumor mass and necrosis were easier to detect with micro-CT and [^{18}F] FDG-PET.

41.4 Magnetic Resonance (MR) Molecular Imaging

41.4.1 Stem Cell Trafficking in an Ischemic Brain Model Using Feridex® Labeling

Feridex® (ferumoxides) is a superparamagnetic iron oxide (SPIO) associated with dextran and approved by the US FDA as an MR contrast agent. Labeling cells with Feridex® enables monitoring of the migration, biodistribution, and behavior of stem cells in vivo with MRI. We made an animal model of brain ischemia by ligation of the midcerebral artery in mice. Neural stem cells were characterized in vitro to assess the degree of Feridex labeling that these cells achieved while still being viable (Fig. 41.18). Feridex-labeled neural stem cells were injected into the ischemic area with a stereotactic device. The Feridex-labeled neural stem cells appeared as hypointensive areas in the ischemic brain area on the iron-sensitive T_2 and T_2^*-weighted MR image (4.7T MRI, Bruker®) up to 21 days after inoculation (Fig. 41.19).

41.4.2 MR Molecular Imaging with the Ferritin Gene [9]

The classic reporter genes for MRI include β-galactosidase and the enzymes creatine and arginine kinase. However, recent studies have also supported the use of genes associated with iron homeostasis, including the transferrin receptors, tyrosinase, and ferritin [9–12].

Ferritin, an iron storage protein with ferroxidase activity was developed as a novel endogenous reporter for the detection of gene expression by magnetic resonance imaging (MRI). Elevated relaxation rates are detected by MRI when h-ferritin is expressed [13, 14].

We constructed a lentiviral vector encoding a heavy-chain ferritin gene, abbreviated as pLenti EF-1α/hFTH IRES puro (Fig. 41.20a). To generate stable cell lines that expressed heavy-chain

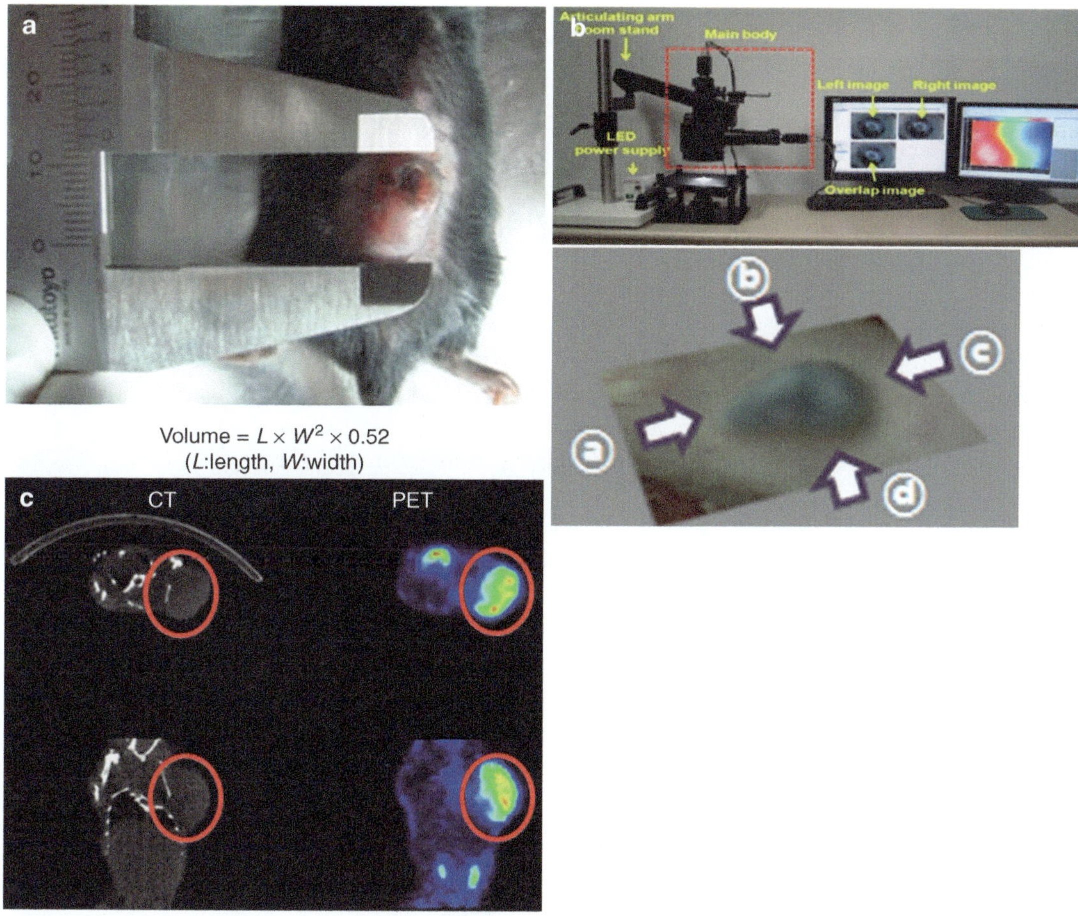

Fig. 41.17 Tumor volume measured with (**a**) calipers, (**b**) three-dimensional reconstruction of the mouse melanoma with a stereo-imaging system, (**c**) micro-CT image (*left*), and [18F] FDG micro-PET image (*right*, threshold-$_{max}$=100 %, slice thickness=0.2 mm). Tumor volume measured by [18F] FDG micro-PET imaging using the standard uptake value (SUV). The *red circle* denotes the xenograft tumor

ferritin (hFTH), B16 melanoma cells were transduced with the lentivirus supernatants in the presence of polybrene. Transduced cells were selected with puromycin in media. The ferritin gene-expressed melanoma cells were then inoculated in C57BL/6 J mice subcutaneously. Two weeks after inoculation, MR imaging of the ferritin-based tumor model was performed (Fig. 41.20b).

In the ferritin-based MR tumor model, oregonin, which is a new anticancer drug derived from plants, and DTIC were injected peritoneally. The oregonin group and the oregonin-plus-DTIC group had a greater reduction in tumor size than the control group (Fig. 41.21) [9].

41.4.3 MR Imaging in tet-Ferritin-Conditioned Double-Transgenic Mouse

A previous study reported the generation of transgenic tet-ferritin-conditioned mice that express h-ferritin under tetracycline (Tet-on) regulation [15].

We also generated TET-AcGFP-HA ferritin (tet-hfer)-conditioned transgenic mice at the Korea Research Institute of Bioscience and Biotechnology by inducing transgenic expression in the double-transgenic (dTG) offspring resulting from a crossing between AcGFP-mFTH and rtTA mice (Fig. 41.22). We confirmed the

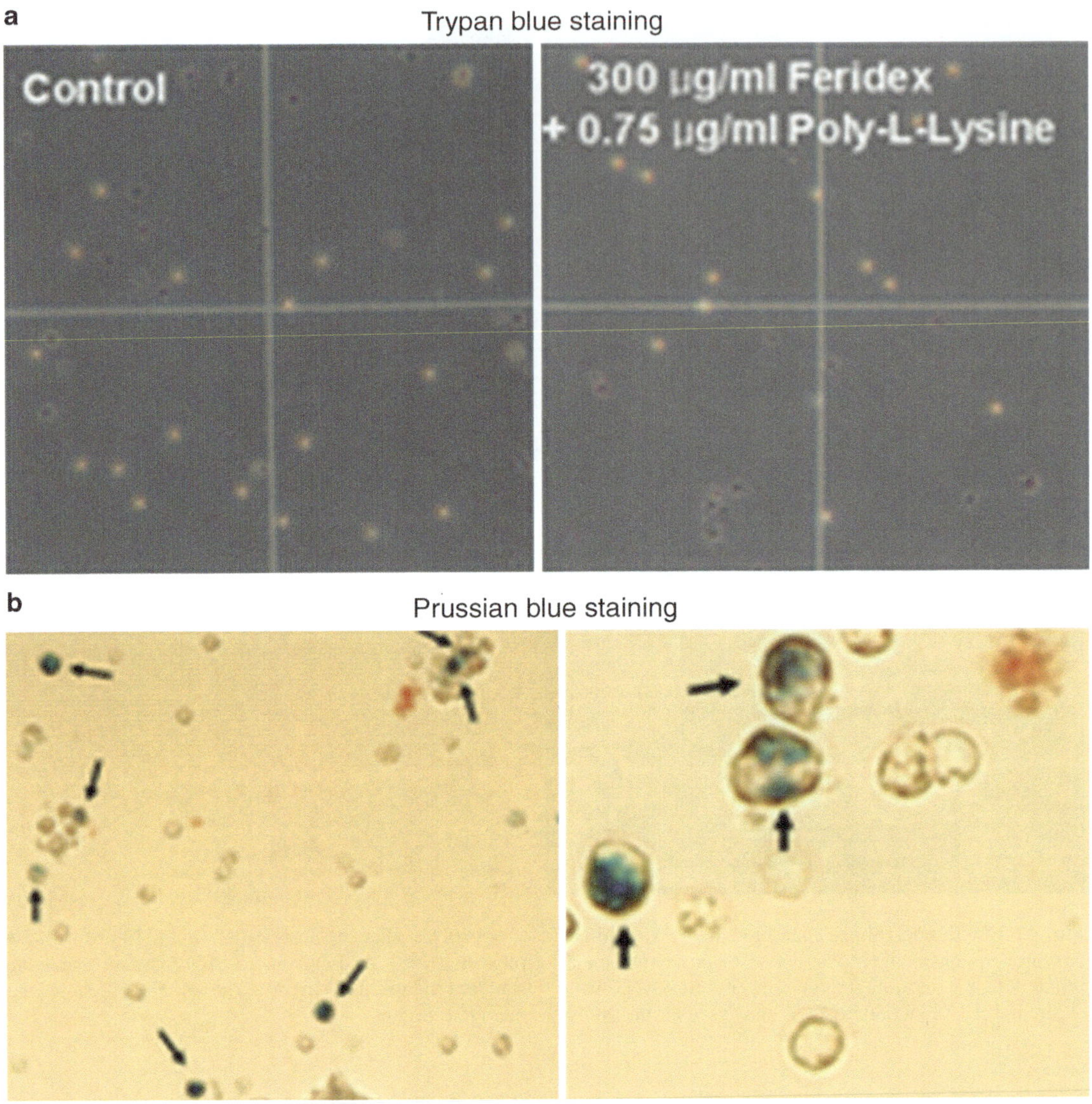

Fig. 41.18 After Feridex labeling of stem cells, (**a**) trypan blue staining was used to assess cell viability, and (**b**) Prussian blue staining was used to assess labeling efficacy

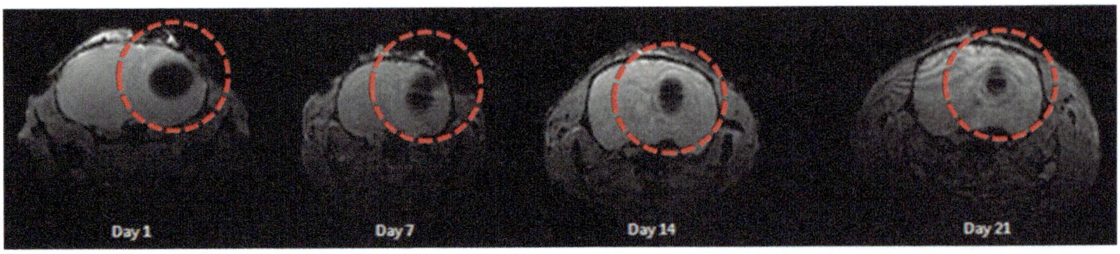

Fig. 41.19 The T_2-weighted axial images at 4.7T MR in the ischemia brain animal models up to 21 days after inoculation. The ROI was used to measure the decrease in signal intensity overtime at the injection site of the stem cells. The time course of signal decreases from days 1 to 21 after injection

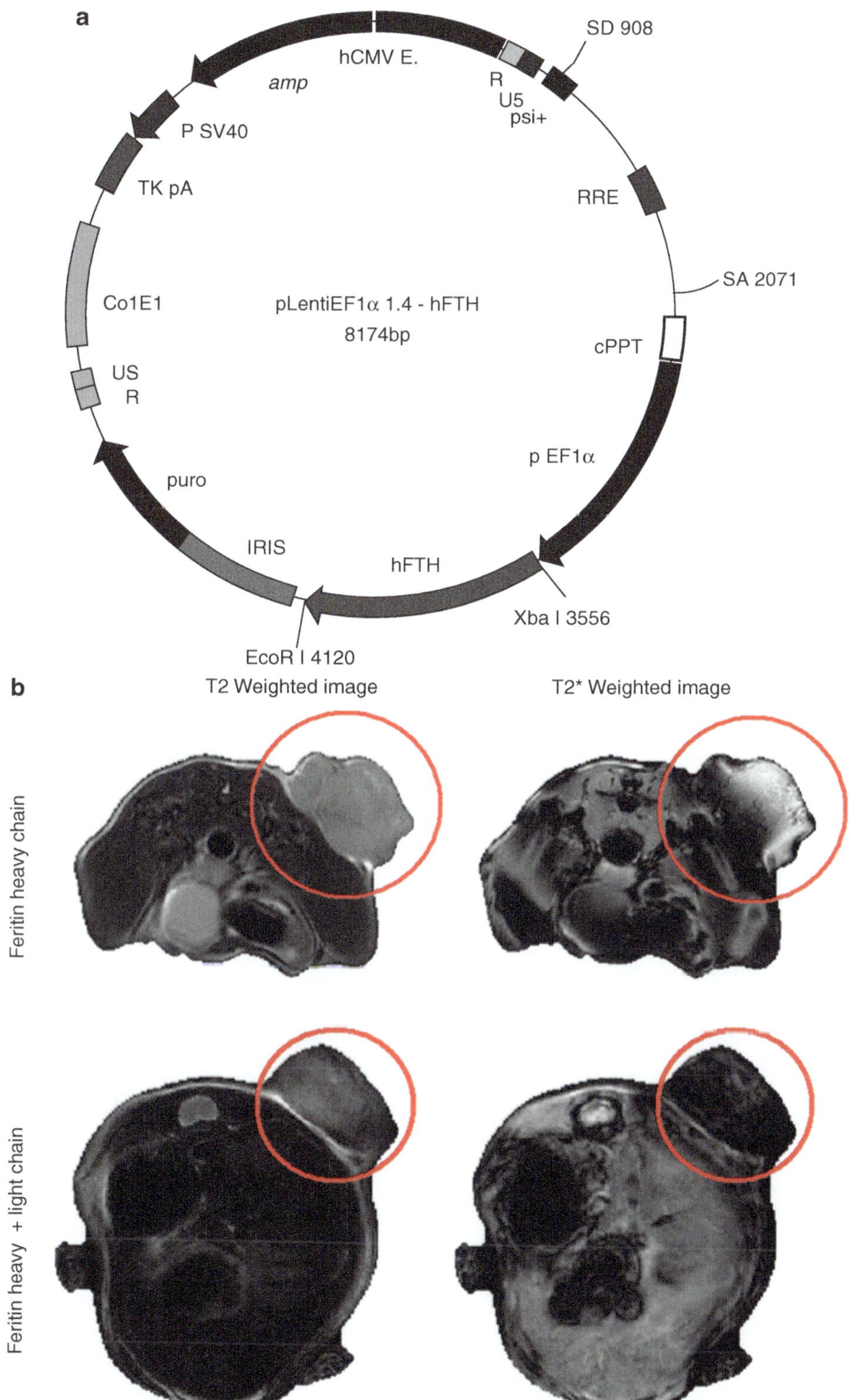

Fig. 41.20 (**a**) Lentiviral vector map of the heavy-chain ferritin gene (pLenti EF-1**α**/hFTH IRES puro). (**b**) MR images of the melanoma model containing the ferritin gene

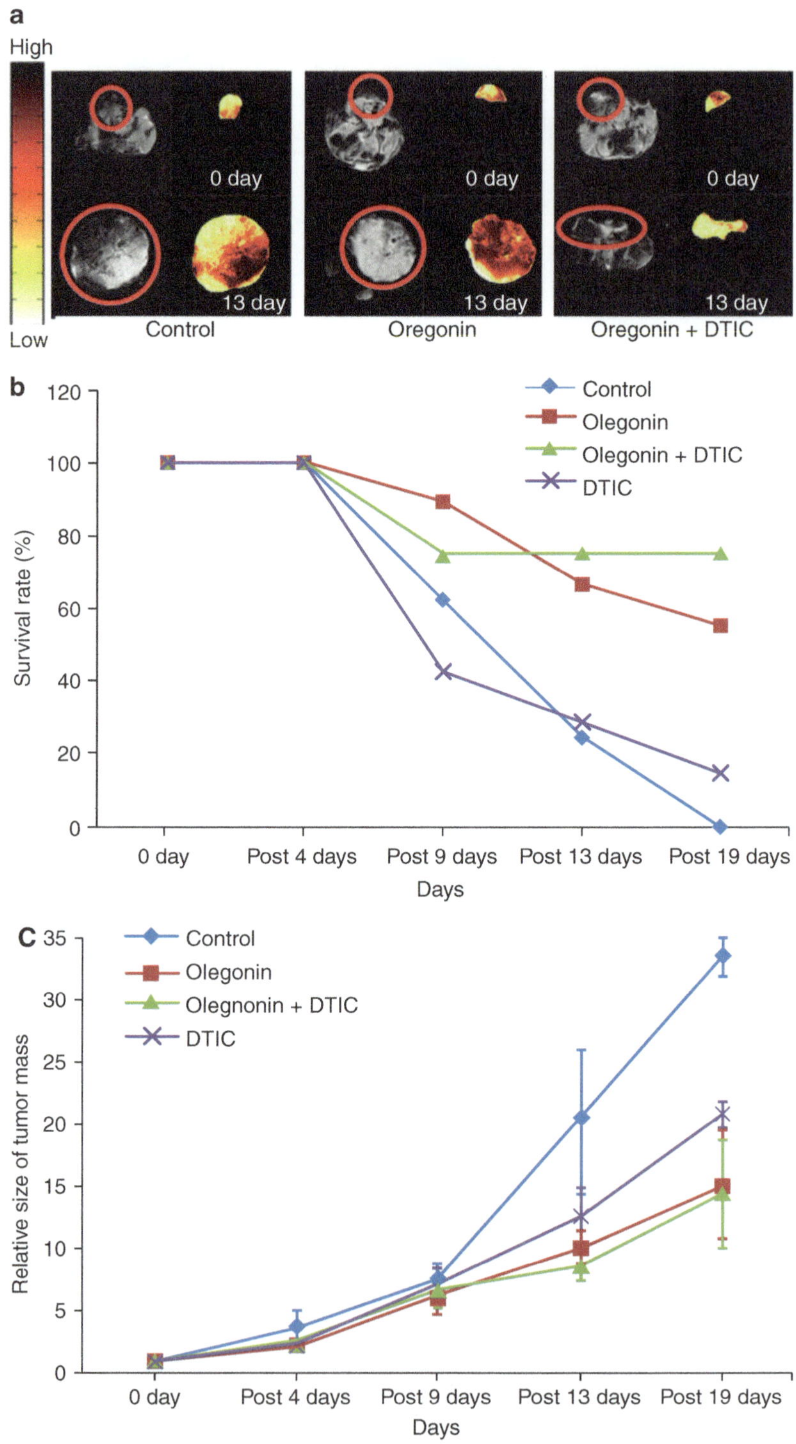

Fig. 41.21 Evaluation of anticancer drug efficacy using MR imaging. (**a**) In vivo MR imaging of B16/pLenti EF-1α/hFTH IRES puro after treatment with DTIC and oregonin. (**b**) Survival rate of the mice with melanoma after anticancer treatment. (**c**) Changes in tumor size after anticancer treatment

successful creation of tet-ferritin-conditioned mice with PCR (Fig. 41.22). Ferritin expression occurred when tetracycline was added to the drinking water (1.5 mg/ml and 5 % sucrose), and it was suppressed by tetracycline withdrawal (5 % sucrose). We imaged mice with 4.7 T (Bruker®) MRI. Tetracycline-regulated overexpression of ferritin appeared as an increased transverse relaxation rate ($1/T_2 = R_2$), especially in the liver (Fig. 41.23).

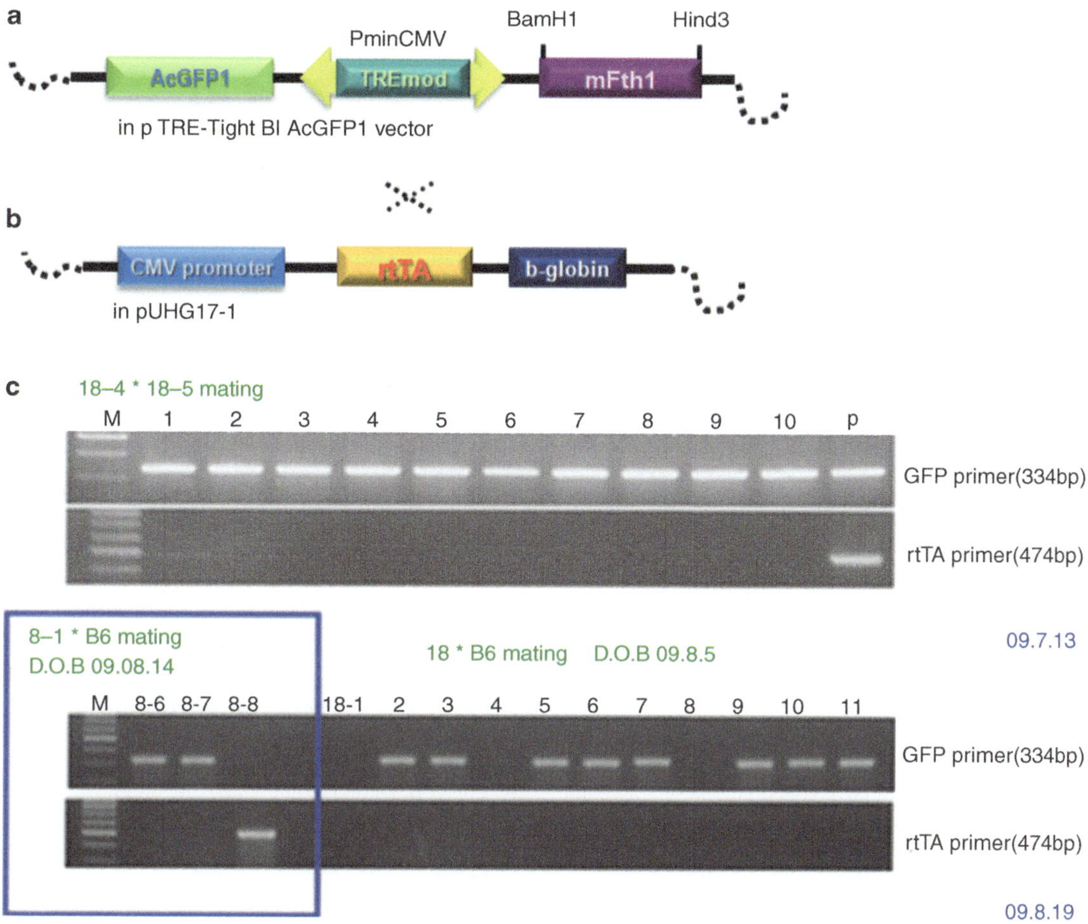

Fig. 41.22 Schematic of the Tet-AcGFP-Ha ferritin vector (**a**) and the rtTA responsive vector (**b**), mFTH, rtTA double knockout mice confirmed with PCR (**c**)

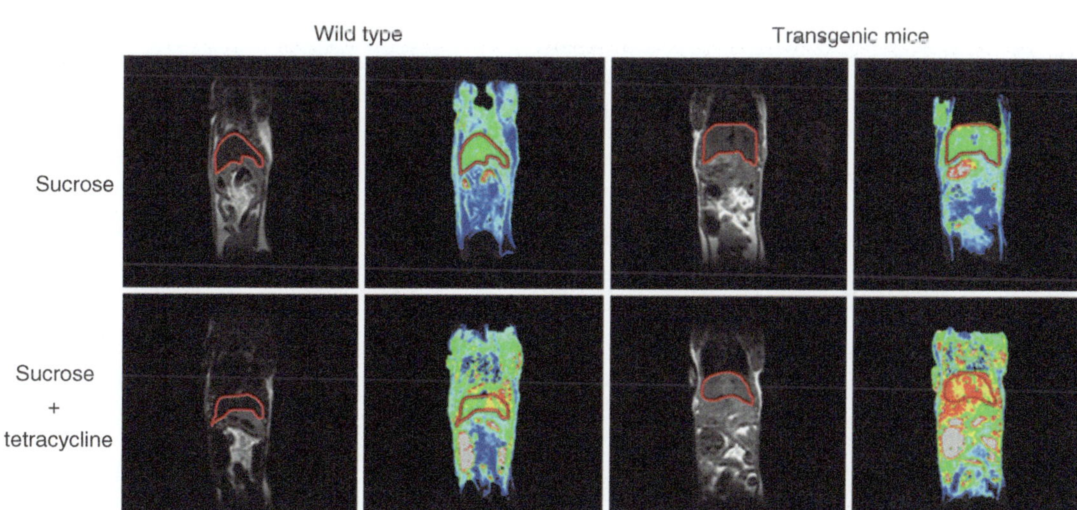

Fig. 41.23 Comparison of the MR images for wild-type and tet-ferritin transgenic mice

Acknowledgment This work was supported by the Seoul Research and Business Development Program (grant number 10574) and Basic Science Research Program through the National Research Foundation of Korea (NRF) funded by the Ministry of Education, Science, and Technology (grant number 2012R1A1A2006556).

References

1. Chan WC, Nie S (1998) Quantum dot bioconjugates for ultrasensitive nonisotopic detection. Science 281:2016–2018
2. Ha SH, Lee OS, Lee GW, Kim JY, Park GM, Moon SH, Kim JH, Son SW, Kang KH, Son YS, Oh CH (2009) In vivo optical molecular imaging for skin disease using QDs. Skin Res Technol 15:108
3. Son SW, Kim JH, Kim SH, Kim H, Chung AY, Choo JB, Oh CH, Park HC (2009) Intravital imaging in zebra fish using quantum dots. Skin Res Technol 15:157–160
4. Park GM, Kim JH, Ha SH, Lee OS, Lee GW, Kim JY, Moon SH, Son SW, Kang KH, Oh CH (2009) Noninvasive image of firefly luciferase gene expression in different animal cancer model. Skin Res Technol 15:109
5. Miloud T, Henrich C, Hämmerling GJ (2007) Quantitative comparison of click beetle and firefly luciferases for in vivo bioluminescence imaging. J Biomed Opt 12(5):054048
6. Tabatabai G, Wick W, Weller M (2011) Stem cell-mediated gene therapies for malignant gliomas: a promising targeted therapeutic approach? Discov Med 11(61):529–536
7. Kim JY, Ha SH, Lee GW, Lee OS, Park GM, Moon SH, Kim JH, Kim SE, Kang GW, Kang KH, Oh CH (2009) In vivo molecular imaging of skin cancer with thymidine kinase-transfected cell in mice using nuclear medicine imaging system. Skin Res Technol 15:98
8. Lee OS, Lee GW, Oh JS, Kim MG, Oh CH (2010) An optimized in vivo multiple-baseline stereo imaging system for skin wrinkles. Opt Commun 283(23): 4840–4845
9. Kim JH, Ha SH, Lee OS, Lee GW, Kim JY, Moon SH, Park GM, Son SW, Son YS, Moon WK, Oh CH (2009) In vivo and in vitro 4.7T MR imaging using Lenti EF-1a hFTH transfection in skin tumor model. Skin Res Technol 15:93
10. Yu J, Liu L, Kodibagkar VD, Cui W, Mason RP (2006) Synthesis and evaluation of novel enhanced gene reporter molecules: detection of beta-galactosidase activity using 19F NMR of trifluoromethylated aryl beta-D-galactopyranosides. Bioorg Med Chem 14(2):326–333
11. Walter G, Barton ER, Sweeney HL (2000) Noninvasive measurement of gene expression in skeletal muscle. Proc Natl Acad Sci U S A 97(10):5151–5155
12. Moore A, Josephson L, Bhorade RM, Basilion JP, Weissleder R (2001) Human transferrin receptor gene as a marker gene for MR imaging. Radiology 221(1):244–250
13. Weissleder R, Simonova M, Bogdanova A, Bredow S, Enochs WS, Bogdanov A Jr (1997) MR imaging and scintigraphy of gene expression through melanin induction. Radiology 204(2):425–429
14. Cohen B, Dafni H, Meir G, Harmeliny A, Neeman M (2005) Ferritin as an endogenous MRI reporter for noninvasive imaging of gene expression in C6 glioma tumors. Neoplasia 7(2):109–117
15. Cohen B, Ziv K, Plaks V, Israely T, Kalchenko V, Harmelin A, Benjamin LE, Neeman M (2007) MRI detection of transcriptional regulation of gene expression in transgenic mice. Nat Med 13(4): 498–503

Index

E. Berardesca et al. (eds.), *Non Invasive Diagnostic Techniques in Clinical Dermatology*,
DOI 10.1007/978-3-642-32109-2, © Springer Berlin Heidelberg 2014

D

X

GPSR Compliance

The European Union's (EU) General Product Safety Regulation (GPSR) is a set of rules that requires consumer products to be safe and our obligations to ensure this.

If you have any concerns about our products, you can contact us on ProductSafety@springernature.com

In case Publisher is established outside the EU, the EU authorized representative is:

Springer Nature Customer Service Center GmbH
Europaplatz 3
69115 Heidelberg, Germany

Batch number: 09225377

Printed by Printforce, the Netherlands